IRC-SET 2021

Huaqun Guo · Hongliang Ren · Victor Wang ·
Eyasu Getahun Chekole · Umayal Lakshmanan
Editors

IRC-SET 2021

Proceedings of the 7th IRC Conference on Science, Engineering and Technology, August 2021, Singapore

Springer

Editors
Huaqun Guo
International Researchers Club
Singapore, Singapore

Victor Wang
Singapore Institute of Technology
Singapore, Singapore

Umayal Lakshmanan
Agency for Science Technology and Research
Experimental Drug Development Centre
Singapore, Singapore

Hongliang Ren
Department of Biomedical Engineering
National University of Singapore
Singapore, Singapore

Eyasu Getahun Chekole
Department of Information Systems Technology and Design
Singapore University of Technology and Design
Singapore, Singapore

ISBN 978-981-16-9871-2 ISBN 978-981-16-9869-9 (eBook)
https://doi.org/10.1007/978-981-16-9869-9

This Springer imprint is published by the registered company Springer Nature Singapore Pte Ltd.
The registered company address is: 152 Beach Road, #21-01/04 Gateway East, Singapore 189721, Singapore

Organizing Committee

General Chair

Prof. Robert H. Deng, Singapore Management University, Singapore

General Co-chair

Dr. Guang Chen, Stronghold Diagnostic Lab—A*STAR, Singapore

Technical Program Chair

Dr. Umayal Lakshmanan, EDDC, A*STAR, Singapore
Dr. Eyasu Getahun Chekole, Singapore University of Technology and Design, Singapore

Publicity Chair

Dr. Noori Kim, Newcastle University in Singapore, Singapore
Dr. Aishwarya Bandla, National University of Singapore, Singapore

Publication Chair

Dr. Victor Peng Cheng Wang, Singapore Institute of Technology, Singapore

Dr. Hongliang Ren, National University of Singapore

ePoster Chair

Dr. Rong Li, Shanghai University of Sport, China
Mr. Peng Liu, A*STAR Research Entities, Singapore

Sponsorship Chair

Dr. Kasthurirengan Sampath, JOil (s) Pte. Ltd., Singapore
Dr. Yi Zhou, Singapore Institute of Technology, Singapore

Financial Chair

Dr. Yu Wang, A*STAR Research Entities, Singapore
Dr. Yongqing Zhu, Singapore University of Social Sciences, Singapore

Steering Committee

Prof. Lawrence Wai Choong Wong, National University of Singapore, Singapore
Dr. Huaqun Guo, International Researchers Club
Dr. Bhojan Anand, National University of Singapore, Singapore

International Advisory Panel

Prof. Jianying Zhou, Singapore University of Technology and Design, Singapore
Prof. Chip Hong Chang, Nanyang Technological University, Singapore
Prof. Maode Ma, Nanyang Technological University, Singapore
Prof. Bharadwaj Veeravalli, National University of Singapore, Singapore
Prof. Chau Yuen, Singapore University of Technology and Design, Singapore

TPC Members

Dr. Aishwarya Bandla, National University of Singapore, Singapore
Dr. Eyasu Getahun Chekole, Singapore University of Technology and Design, Singapore
Dr. Boon Seng Chew, Singapore Polytechnic, Singapore
Dr. Yvonne Chow, Singapore Institute of Food and Biotechnology Innovation, Singapore
Dr. Aldy Gunawan, Singapore Management University, Singapore
Dr. Noori Kim, Newcastle University in Singapore, Singapore
Dr. Umayal Lakshmanan, EDDC, A*STAR, Singapore
Dr. Rong Li, Shanghai University of Sport, China
Mr. Yong Zhi LIM, Singapore University of Technology and Design, Singapore
Mr. Peng Liu, A*STAR Research Entities, Singapore
Mr. Mark Huasong MENG, National University of Singapore, Singapore
Dr. Nithyavathy N., Kongu Engineering College, India
Prof. Foram Rajdev, Marwadi University, India
Dr. Hongliang Ren, National University of Singapore, Singapore
Dr. Kasthurirengan Sampath, JOil (s) Pte. Ltd., Singapore
Dr. Guoxian Tan, Raffles Institution, Singapore
Mr. Teik Guan Tan, Singapore University of Technology and Design, Singapore
Dr. Victor Peng Cheng Wang, Singapore Institute of Technology, Singapore
Dr. Zaiqun Yu, Institute of Chemical and Engineering Sciences, A*STAR, Singapore
Prof. Chau Yuen, Singapore University of Technology and Design, Singapore
Dr. Yi Zhou, Singapore Institute of Technology, Singapore
Dr. Yongqing Zhu, Singapore University of Social Sciences, Singapore

Preface

International Researchers Club (IRC) (www.irc.org.sg) was set up in 2001. The vision of IRC is to create a vibrant and innovative research community for Singapore and beyond, through contributions of technical specialties and occupational experiences from its members, and fostering strong networking and social interactions of expatriates and new citizens with the local community.

With the vision of IRC, it is our great pleasure to organize IRC conference on Science, Engineering and Technology (IRC-SET, www.ircset.org) for the younger talents and researchers. IRC-SET 2015 was the inaugural conference of IRC, and IRC-SET 2021 is now in its seventh run. IRC-SET conference aims to provide a platform for young and talented researchers to share their research findings, to obtain comments and feedback from experts in the field and exchange innovative ideas of leading-edge research in multi-disciplinary areas. Students from universities, junior colleges, polytechnics and secondary schools are warmly welcomed to participate in this conference to showcase and present their research projects, results and findings. Unlike other academic conferences, this conference focuses specifically on Education and Youth development and has officially been given technically sponsorship from six universities, namely National University of Singapore (NUS), Nanyang Technological University (NTU), Singapore University of Technology and Design (SUTD), Singapore Management University (SMU), Singapore Institute of Technology (SIT) and Newcastle University, Singapore. IRC-SET 2021 conference is also supported by IEEE Intelligent Transportation Systems Society (ITSS) Singapore Chapter, IEEE Broadcast Technology Society (BTS) Singapore Chapter and IEEE Singapore Section Women in Engineering (WIE) Affinity Group.

The program of IRC-SET 2021 advocates the importance of innovative technology backed by the strong foundation of science and engineering education. Exposing our young participants to the key technology enablers will encourage more interest into the fields of science, engineering and technology. To select good quality papers, the IRC-SET 2021 Call for Papers was broadcasted to universities, junior colleges, polytechnics and secondary schools. The students then submit their technical papers to the conference online system. To meet the criteria of the conference, the submitted papers (which have to follow the required template) were peer reviewed by the conference technical program committee, which comprises a few reviewers coming from the IRC researchers, professors, lecturers and teachers. Comments and recommendations based on novelty of the work, scientific, engineering and technology relevance, technical treatment plausible and clarity in writing, tables, graphs and illustrations were then provided back to the students. Through this rigorous review, the technical program committee has selected these papers to be presented in the IRC-SET 2021 conference, which are now published in this proceeding.

IRC-SET 2021 Conference was held online and consists of Opening Speeches by Prof. Tit Meng LIM (Chief Executive of Singapore Science Centre) and Prof. Robert H. Deng (Deputy Dean for Faculty and Research, Singapore Management University), Keynote Speech by Prof. Sir John O'REILLY (Chairman of A*STAR Science and Engineering Research Council) and nine parallel presentation sessions. The nine presentation sessions cover the key focus areas which include physics, chemical engineering, biomedical science, ITSS session, BTS session, WIE session, mechanical engineering, computer engineering and life sciences. In the closing ceremony, Prof. Robert H. Deng (General Chair), announced the winners of best paper awards and best presenter awards. Last but not least, Dr. Huaqun Guo (President, IRC) concluded the conference with a closing speech and also gave an introduction to the International Researchers Club.

This proceeding is dedicated to International Researchers Club and its members.

Singapore
November 2021

Huaqun Guo
Hongliang Ren
Victor Wang
Eyasu Getahun Chekole
Umayal Lakshmanan

Acknowledgments

We would like to extend our utmost gratitude to the people who have, in one way or other, inspired, aided and contributed to the successful completion of this book.

First of all, we would like to thank all members of the conference organizing committee who have contributed their time and professional knowledge to make the conference a big success.

Special thanks to all reviewers for their expertise, time, effort and timely response throughout the peer evaluation process.

We would also like to take this opportunity to thank ST Engineering Land Systems Ltd. for sponsoring the conference and supporting IRC.

Last but not least, our greatest appreciation to all IRC members for their unity.

Contents

Chapter 1
Spinal Muscular Atrophy: Evaluation of New Emerging Methods for Carrier Screening and Diagnosis

Sim Joylynn, Jinghan Xie, Tan Grace Li Xuan, Liu Chun Ping, and Lai Poh San

Abstract Spinal muscular atrophy (SMA) is the most common genetic cause of infant mortality. Homozygous deletion of *Survival Motor Neuron (SMN)* 1 causes the disease in 95% of cases, but severity varies with the copy number of *SMN2*. The nearly identical sequences of the *SMN* genes make differentiation between them difficult, and current methodologies such as multiplex ligation-dependent probe amplification (MLPA) and real-time PCR (RT-PCR) present limitations for diagnosis. In this report, we analyse SMA samples of known copy number to compare new emerging technologies (Linked-Reads, Cytoscan array and next generation sequencing) against MLPA. MLPA analysis was the only method able to report the known copy numbers of the samples, making it still the most reliable means of SMA carrier screening and diagnosis. However, the three new methodologies are still emerging technologies for routine diagnostics compared to MLPA, and with further optimisation, they could greatly improve SMA carrier screening and diagnosis.

Keywords Spinal muscular atrophy (SMA) · SMN1 · SMN2 · Linked-Reads · Cytoscan (CYT) array · Whole genome sequencing (WGS) · Multiplex ligation-dependent probe amplification (MLPA)

1.1 Introduction

Spinal muscular atrophy (SMA) is an autosomal recessive motor neuron disease with an annual incidence of about 1 in 6000 to 1 in 10,000 live births, and the carrier frequency is as high as 1 in 40 [1]. 95% of SMA patients are homozygous for

Sim J. · J. Xie (✉)
Raffles Institution, Bishan, Singapore
e-mail: 21YXIEJ710B@alumni.ri.edu.sg

Sim J.
e-mail: 21YSIMJ160E@alumni.ri.edu.sg

Tan G. L. X. · Liu C. P. · Lai P. S.
Department of Paediatrics, Yong Loo Lin School of Medicine, National University of Singapore, Queenstown, Singapore

H. Guo et al. (eds.), *IRC-SET 2021*, https://doi.org/10.1007/978-981-16-9869-9_1

SMN1 deletion. *SMN2* partially compensates for *SMN1* loss, lowering the severity of SMA. However, when two *SMN1* genes are carried in cis, this results in a silent carrier (i.e. a carrier that does not express the S*f*MA phenotype) [2, 3]. Further complicating diagnosis, *SMN1* and *SMN2* are nearly identical inversions [4], differing by only five base pairs: c.835−45G>A, c.840C>T, c.*3+100A>G, c.*3+214A>G and c.*248A>G [5]. Therefore, there is a need to (1) differentiate between *SMN1* and *SMN2*, (2) call copy number (CN) of *SMN1* and *SMN2* and (3) determine the phase of *SMN1* and *SMN2* for SMA clinical classification, prognosis, carrier identification and diagnosis [6]. Current methodologies such as multiplex ligation-dependent probe amplification (MLPA) and real-time PCR (RT-PCR) present some limitations. MLPA is unable to determine the phase of *SMN1* and *SMN2*, resulting in the inability to identify silent carriers with 2 *SMN1* on one chromosome but none on the other, and hence, false negative results in silent carriers. *SMN* genes are unevenly amplified in RT-PCR, which leads to inaccurate results. Therefore, in order to overcome these limitations, we evaluated new emerging methods—Linked-Reads, Cytoscan (CYT) array and whole genome sequencing (WGS)—against MLPA, the most commonly used method for carrier screening and diagnosis for SMA, by their ability to: 1. differentiate between *SMN1* and *SMN2*, 2. determine the CN of *SMN1* and *SMN2*, 3. locate structural variants in *SMN1* and *SMN2* and 4. phase alleles.

1.1.1 Hypothesis

We hypothesise that Linked-Reads, CYT array and WGS can overcome the limitations of current methods in determining patient or carrier status by differentiating between the two almost identical *SMN* genes, as well as calling CN in trans.

1.2 Methods

1.2.1 Sample Information

A total of six anonymised data sets from SMA patients with known *SMN1* and *SMN2* CN were provided by the National University of Singapore, Department of Paediatrics.

1.2.2 Technologies

1. Linked-Reads

DNA was sheared and put through size selection. The ChromiumTM system was then used for automated barcoded library construction. The barcoded libraries were then sequenced using Illumina Whole Exome Sequencing (WES). The data obtained was visualised on Loupe, a genome browser by 10× Chromium Platform designed for visualisation of Linked-Reads data [7]. The BAM file obtained was also visualised on Integrative Genome Browser (IGV).

2. CYT Array

Gene probes were deposited on a chip. cDNA, labelled with either green or red fluorescence, was generated from mRNA extracted. cDNA complementary base pairs with probes on the chip were analysed by fluorescence emission. The data obtained was visualised on Chromosome Analysis Suite (ChAS). Manual guides provided by Thermo Fisher Scientific Inc. were used in aiding the usage of software [8].

3. WGS

Patient DNA was sequenced through whole genome sequencing (WGS) at 40× read depth. The BAM data obtained was visualised on IGV. User guides provided by the Broad Institute (2018) were used in aiding the usage of software. Genome Reference Consortium Human Build 37 (hg19) was used as reference for WGS samples.

4. MLPA

DNA strands were denatured to separate the strands and hybridised with probes. The right probe oligo contains a stuffer sequence which is used to identify DNA pieces. The DNA was then amplified through PCR. MLPA amplicons are separated by length using capillary electrophoresis [9]. The measured fluorescence was visualised as a peak pattern and used to quantify each probe. CN was determined using probe ratio.

5. CN calling using SMN/mean read depth ratio

The c.840C>T site on exon 7 is the critical difference between *SMN1* and *SMN2*. Additionally, we were provided with the read depths of *SMN1* and *SMN2* exons 7 and 8 for each of our samples. Thus, using protocols modified from [10, 11], we determined the CN for our Linked-Reads and WGS samples by comparing the read depth of *SMN1* and *SMN2* exons 7 and 8 against each sample's overall mean read depth using the following formula.

$$\text{Copy number} = \frac{\text{Read depth of exon}}{\text{Mean read depth of sample}} \times 2$$

1.3 Results

1.3.1 Linked-Reads

Linked-Reads sequencing generates reads with an integrated barcode which traces the reads back to the original DNA molecule [12]. This allowed the reads to be mapped to the *SMN1* and *SMN2* genes, as shown from the read coverage in Fig. 1.1. The coverage of the c.840C site on *SMN1* and c.840*T* site on *SMN2* was also verified on IGV, confirming that Linked-Reads was able to differentiate between the SMN genes.

Structural variants (SVs) can be detected through calls and candidates recognised by Linked-Reads, where calls meet the higher-quality call threshold than candidates and occur in unambiguous regions of the reference genome. However, no SVs were called by Linked-Reads in our samples. A deletion of exon 7 in *SMN1* was observed in Sample *300,099* when viewing the reads in IGV, corresponding to the known *SMN1* CN of 0.

CN was also calculated by comparing the read depths of exons 7 and 8 in *SMN1* and *SMN2* against each sample's mean read depth. As seen in Table 1.1, there are discrepancies between the calculated and known CN, indicating that observed read depths underestimate the actual CN. This discrepancy could be due to difficulties in sequencing for the following reasons: 1. *SMN1* and *SMN2* genes are part of a 500 kb highly repetitive inverted duplication on chromosome 5, making it difficult to determine the organisation of this genomic region [13]; 2. high GC level of 54% in *SMN1* and *SMN2* [14], leading to a poor coverage of reads and less complete

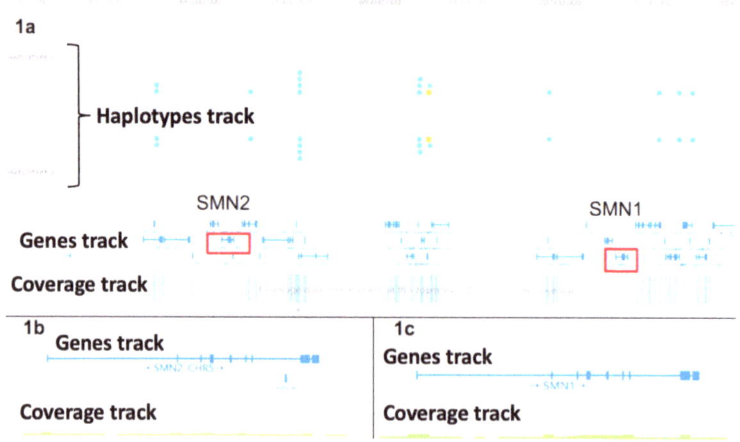

Fig. 1.1 **a** *SMN1* and *SMN2* on sample 300,097. **b**, **c** *SMN2* (left), *SMN1* (right). Green bar in the coverage track indicates read depth for the region. Genes and their exons are identified and labelled in the genes track

Table 1.1. CN call for *SMN1* and *SMN2* exons 7 and 8 of samples 300,097–99

	Calculated CN				Known CN			
Sample	SMN1		SMN2		SMN1		SMN2	
	exon 7	exon 8	exon 7	exon 8	exon 7	exon 8	exon 7	exon 8
300,097	1	1	0	0	2	2	1	1
300,098	1	2	0	1	2	2	2	2
300,099	0	1	1	1	0	0	3	2

assembly; 3. low input DNA mass of 0.4–0.5 ng, which was below the recommended range of 1–3 ng, affecting the performance of sequencing.

Linked-Reads is able to phase alleles by assembling long reads from short reads, creating a phase block by utilising continuous reliable heterozygous variants (phasing quality > 23) to connect the reads [15]. However, in this analysis, reads in *SMN1*, *SMN2* and their flanking regions in samples 300,097–30,099 were not assigned to either haplotype as there were insufficient single-nucleotide variants (SNVs) [15] present in our samples that were informative for Long Ranger to determine phase blocks.

1.3.2 CYT Array

Both *SMN* genes are labelled as *SMN1* and *SMN2* simultaneously by ChAS (Fig. 1.2), showing that ChAS is unable to differentiate between *SMN1* and *SMN2*.

Mean weighted Log2 ratio and smooth signal values were calculated by ChAS and used to determine the CN. Log2 ratio indicates gain or loss in genetic material, with a ratio of 0 indicating a CN of 2. The Log2 ratios of *SMN1* and *SMN2* in CYT34 and CYT221 are close to 0, indicating that CN for both genes in both samples is 2. Smooth signal is a smoothed calibrated estimate which can represent non-integer CN. It uses the Gaussian function to reduce noise within the array, thus allowing for a more accurate CN to be determined. The smooth signal values of *SMN1* and *SMN2* in CYT34 and CYT221 are also close to 2, corroborating the Log2 ratio calculated CN of both samples. However, there is a discrepancy between the calculated and the known CN (Table 1.2), as Cytoscan is unable to differentiate between highly

Fig. 1.2 *SMN1* and *SMN2* annotation in ChAS

Table 1.2. Mean weighted Log2 ratio, smooth signal values, calculated and known CN of samples CYT34 and CYT221

Sample/gene	Data type	Mean	Calculated CN	Known CN
CYT34/*SMN1*	Weighted Log2 ratio	– 0.03	2	2
	Smooth signal	1.969	2[a]	
CYT34/*SMN2*	Weighted Log2 ratio	– 0.051	2	1
	Smooth signal	1.942	2[a]	
CYT221/*SMN1*	Weighted Log2 ratio	0.159	2	1
	Smooth signal	2.398	2[a]	
CYT221/*SMN2*	Weighted Log2 ratio	0.191	2	3
	Smooth signal	2.402	2[a]	

[a] value rounded down to whole number

homologous regions such as *SMN1* and *SMN2* [15].

1.3.3 WGS

WGS was able to identify the *SMN1* and *SMN2* genes, as indicated by the presence of reads in these regions. Reads that align with the reference sequence are displayed in grey. However, WGS was not able to differentiate between the two genes well. Out of the five base pair differences in the *SMN2* genes, a misalignment of *SMN1* reads to *SMN2* was observed in four of these regions in sample NGS-1108 (Fig. 1.3). Critically, at the crucial c.840 site, no reads were observed in *SMN1*, but reads were mapped to C (24 reads) and T (17 reads), respectively, in *SMN2* (Fig. 1.3). This misalignment reflects WGS's inability to distinguish between homologous regions such as the *SMN* genes [16] (Table 1.3).

1.3.4 MLPA

SMN1 and *SMN2* genes were identified and differentiated by MLPA (Table 1.4) using probes specific to *SMN1* and *SMN2* exons 7 and 8. CNs were deduced from probe ratios provided by the manufacturer [9]. Sample O221 had a single copy of *SMN1* exon 7 and exon 8 and is, therefore, a carrier of SMA. Sample O34 was detected to have two copies of *SMN1*, which indicates that the patient is unaffected. However, this is only true when the *SMN1* genes occur in trans. As MLPA does not phase alleles, it cannot confirm that O34 is not a silent carrier. Sample O34 has two copies of *SMN2* exon 7, but only one copy of *SMN2* exon 8. This indicates a deletion of *SMN2* exon 8 in sample O34. Hence, the second copy of *SMN2* is not a full functional

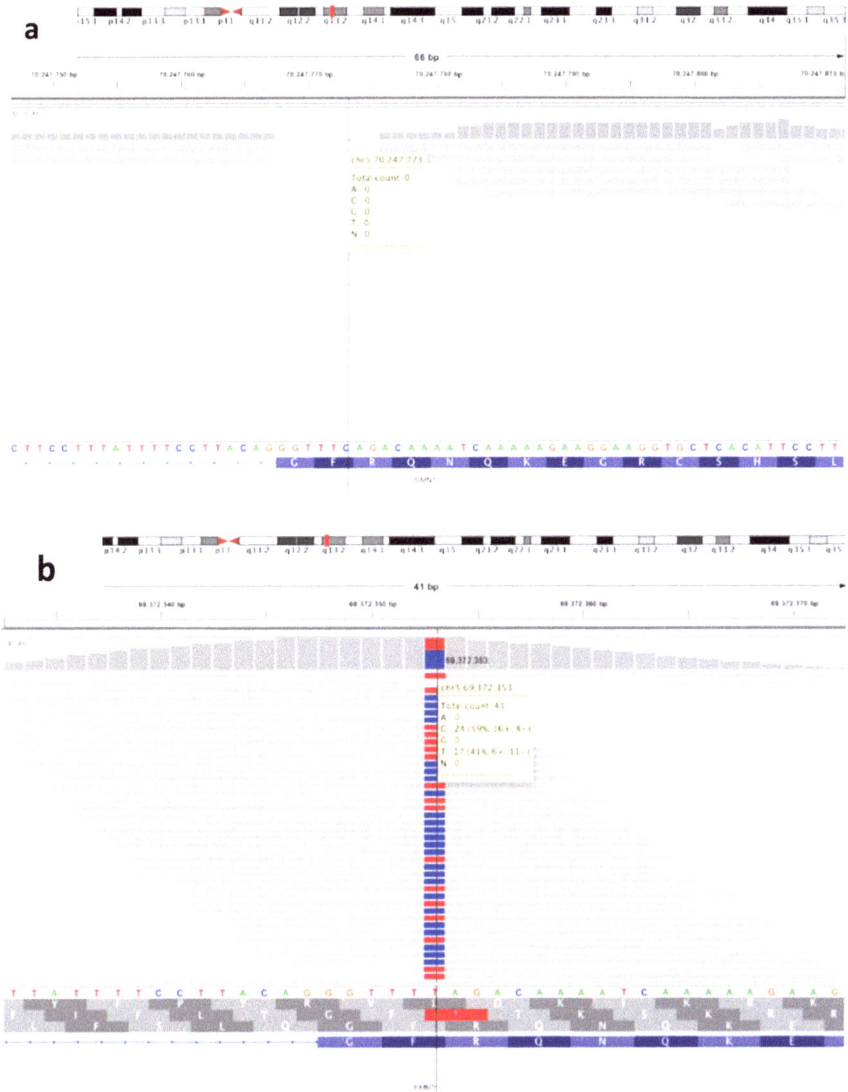

Fig. 1.3 **a** c.840C on *SMN1* of sample NGS-1108. No reads were observed. **b** c.840T on *SMN2* of sample NGS-1108. Bar is coloured in proportion to the read count of each base. Cytosine is in blue, and thymine is in red

Table 1.3. CN call for *SMN1* and *SMN2* exons 7 and 8 of sample NGS-1108

Calculated CN				Known CN			
SMN1		SMN2		SMN1		SMN2	
exon 7	exon 8	exon 7	exon 8	exon 7	exon 8	exon 7	exon 8
0	1	0	1	2	2	1	1

Table 1.4. Determined CN from probe ratio, in comparison with known CN for sample O34

	O34			O221		
Gene/exon number	Probe ratio	Calculated CN	Known CN	Probe ratio	Calculated CN	Known CN
SMN1 exon 7	0.75	2	2	0.66	1	1
SMN1 exon 8	0.89	2	2	0.61	1	1
SMN2 exon 7	0.76	2	2	2.49	5	5
SMN2 exon 8	0.44	1	1	1.97	4	4

gene. MLPA can be considered a reliable tool for determining CN, as the obtained CN was consistent with the known CN (Table 1.4).

1.3.5 Comparison of Linked-Reads, CYT Array and WGS with MLPA

These newly emerging methods—Linked-Reads, CYT array and WGS—were compared against MLPA. Linked-Reads is capable of differentiating between the highly homologous *SMN1* and *SMN2* genes, a critical factor which measures up to MLPA and sets it apart from CYT array and WGS. Linked-Reads identification of molecules is more reliable than CYT array as it uses different "identification codes" for each molecule [12], whereas CYT array uses microarray analysis, which utilises probes [17] that are similar for the highly homologous *SMN1* and *SMN2* genes. This is also a drawback of WGS as reads from *SMN1* may misalign to a highly homologous *SMN2* gene during sequence assembly due to short read length.

In this analysis, we used a simple method of estimating CN using Linked-Reads and WGS data by comparing the read depths of *SMN* exons 7 and 8 to each sample's mean read depth, due to the lack of access to sophisticated computational pipelines or software (such as those described by [18]) for determining CN. Our method gives an approximate estimation which would have to be confirmed either computationally or through wet experiments. Nevertheless, as we expect CN to be either in terms of deletion or between two to four, the exact value of the increased read depth can be rounded off. While Linked-Reads was able to detect a deletion of exon 7 in *SMN1* corresponding to sample *300,099*'s *SMN1* CN of 0, the CNs of *SMN1* and *SMN2* could not be accurately determined using Linked-Reads and WGS data, unlike in MLPA.

Due to the lack of heterozygous SNVs in close proximity within SMN regions, Linked-Reads was not able to determine the haplotype of *SMN1* and *SMN2* in our samples, which is important for identifying silent carriers of SMA. However, if there were adequate heterozygous SNVs in SMN regions, Linked-Reads would be able to determine phase blocks and resolve haplotypes [19], giving it a huge advantage over the other three methods.

1.4 Conclusion

Considering Linked-Reads' ability to differentiate between *SMN1* and *SMN2* genes, identify SNPs, and its potential ability to identify SVs and phase alleles to determine haplotypes, Linked-Reads can be viewed as a possible tool for carrier screening and diagnosis as it presents the ability to overcome limitations of MLPA and RT-PCR, which is the inability to phase *SMN1* and *SMN2* and uneven amplification of genes, respectively. Although Linked-Reads and WGS were not able to call *SMN* gene CNs accurately in this analysis, further work can be done to optimise the technology to be up to par with MLPA's ability to call CN. To overcome the limitation of low read depth, normalisation of read counts can be done to account for GC bias [19], and DNA input mass can be increased to 1 ng per library [20].

Acknowledgements We would like to thank Professor Lai Poh San, Dr Liu Chun Ping and Ms Grace Tan Li Xuan from the Department of Paediatrics for their guidance and National University Singapore (NUS) for the valuable learning opportunity.

References

1. MedlinePlus (2020). *SMN1* gene—survival of motor neuron 1, telomeric. U.S. National Library of Medicine. https://medlineplus.gov/genetics/gene/smn1/
2. Prior, T. W., Nagan, N., Sugarman, E. A., Batish, S. D., & Braastad, C. (2011). Technical standards and guidelines for spinal muscular atrophy testing. *Genetics in Medicine, 13*(7), 686–694. https://doi.org/10.1097/GIM.0b013e318220d523
3. Jones, K., & Livingston, A. (2019). *Spinal* Muscular Atrophy and the Difficult *SMN1* Gene. https://www.thermofisher.com/blog/behindthebench/spinal-muscular-atrophy-and-the-difficult-smn1-gene/

4. National Center for Biotechnology Information (NCBI) (2020, December 6). *SMN1* survival of motor neuron 1, telomeric [Homo sapiens (human)]. U.S. National Library of Medicine. https://www.ncbi.nlm.nih.gov/gene/6606
5. Monani, U. R., Lorson, C. L., Parsons, D. W., Prior, T. W., Androphy, E. J., Burghes, A. H. M., & McPherson, J. D. (1999). A single nucleotide difference that alters splicing patterns distinguishes the SMA gene *SMN1* from the copy gene *SMN2*. *Human Molecular Genetics, 8*(7), 1177–1183. https://doi.org/10.1093/hmg/8.7.1177
6. Chen, X., Sanchis-Juan, A., French, C. E., Connell, A. J., Delon, I., Kingsbury, Z., Chawla, A., Halpern, A. L., Taft, R. J., Bentley, D. R., Butchbach, M. E. R., Raymond, F. L., Eberle, M. A., & BioResource, N. I. H. R. (2020). Spinal muscular atrophy diagnosis and carrier screening from genome sequencing data. *Genetics in Medicine, 22*(5), 945–953. https://doi.org/10.1038/s41436-020-0754-0
7. Zhang, L., Zhou, X., Weng, Z., & Sidow, A. (2019). Assessment of human diploid genome assembly with 10× linked-reads data. *GigaScience, 8*(giz141). https://doi.org/10.1093/gigascience/giz141
8. Thermo Fisher Scientific Inc. (2020). Chromosome analysis suite (ChAS) v4.2. https://assets.thermofisher.com/TFS-Assets/GSD/Handbooks/ChAS_4.2_Manual.pdf
9. MRC-Holland. (n.d.). Principle of MLPA. https://www.mrcholland.com/technology/mlpa/technique
10. Ogino, S., Gao, S., Leonard, D. G. B., Paessler, M., & Wilson, R. B. (2003). Inverse correlation between *SMN1* and *SMN2* copy numbers: Evidence for gene conversion from *SMN2* to *SMN1*. *European Journal of Human Genetics, 11*(3), 275–277. https://doi.org/10.1038/sj.ejhg.5200957
11. Tattini, L., D'Aurizio, R., & Magi, A. (2015). Detection of genomic structural variants from next-generation sequencing data. *Frontiers in Bioengineering and Biotechnology, 3*, 92. https://doi.org/10.3389/fbioe.2015.00092
12. Chen, Z., Pham, L., Wu, T.-C., Mo, G., Xia, Y., Chang, P., Porter, D., Phan, T., Che, H., Tran, H., Bansal, V., Shaffer, J., Belda-Ferre, P., Humphrey, G., Knight, R., Pevzner, P., Pham, S., Wang, Y., & Lei, M. (2019). Ultra-low input single tube linked-read library method enables short-read NGS systems to generate highly accurate and economical long-range sequencing information for de novo genome assembly and haplotype phasing. BioRxiv, 852947.https://doi.org/10.1101/852947
13. Ott, A., Schnable, J. C., Yeh, C.-T., Wu, L., Liu, C., Hu, H.-C., Dalgard, C. L., Sarkar, S.
14. Berkner, S., & Lipps, G. (2007). An active nonautonomous mobile element in Sulfolobus islandicus REN1H1. *Journal of Bacteriology, 189*(5), 2145–2149. https://doi.org/10.1128/JB.01567-06
15. 10x Genomics. (n.d.). What is the difference between a phase set in the long ranger VCF and a phase block in the loupe browser? https://kb.10xgenomics.com/hc/en-us/articles/360001336291-What-is-the-difference-between-a-phase-set-in-the-Long-Ranger-VCF-and-a-phase-block-in-the-Loupe-Browser
16. Mandelker, D., Schmidt, R. J., Ankala, A., McDonald Gibson, K., Bowser, M., Sharma, H., Duffy, E., Hegde, M., Santani, A., Lebo, M., & Funke, B. (2016). Navigating highly homologous genes in a molecular diagnostic setting: A resource for clinical next-generation sequencing. *Genetics in Medicine, 18*(12), 1282–1289. https://doi.org/10.1038/gim.2016.58
17. Nature Education (2014). Microarray. https://www.nature.com/scitable/definition/microarray-202/
18. Lutgen, D., Ritter, R., Olsen, R.-A., Schielzeth, H., Gruselius, J., Ewels, P., García, J.T., Shirihai, H., Schweizer, M., Suh, A., Burri, R. (2020). Linked-read sequencing enables haplotype-resolved resequencing at population scale. https://doi.org/10.1111/1755-0998.13192
19. Chouvarine, P., Wiehlmann, L., Losada, P. M., DeLuca, D. S., & Tümmler, B. (2016). Filtration and normalization of sequencing read data in whole-metagenome shotgun samples. *PLoS ONE, 11*(10), e0165015. https://doi.org/10.1371/journal.pone.0165015

20. Ruhno, C., McGovern, V. L., Avenarius, M. R., Snyder, P. J., Prior, T. W., Nery, F. C., Muhtaseb, A., Roggenbuck, J. S., Kissel, J. T., Sansone, V. A., Siranosian, J. J., Johnstone, A. J., Nwe, P. H., Zhang, R. Z., Swoboda, K. J., & Burghes, A. (2019). Complete sequencing of the *SMN2* gene in SMA patients detects SMN gene deletion junctions and variants in *SMN2* that modify the SMA phenotype. *Human Genetics, 138*(3), 241–256. https://doi.org/10.1007/s00439-019-01983-0

Chapter 2
Traffic Strategy for Mixed Traffic in Container Terminals

Si Han Ng, Reuben Ong, and Chenhao Zhou

Abstract Over the past decade, autonomous vehicles have moved from the realms of science fiction into the real world. Container terminals have much to gain from adopting autonomous vehicles, such as increased safety and productivity. This has led to many ports around the world jumping on the bandwagon to tap their immense potential and sharpen their competitiveness. However, there remains an inevitable transitional period where both manually driven and autonomous trucks have to coexist. There is an urgent need to model mixed traffic in a container terminal environment, which consists of unsignalized intersections. Yet much of existing literature has only explored mixed traffic on highways and signalized intersections. We cannot directly apply these traffic strategies in container terminals, due to differing levels of network complexity. Here, we propose our traffic management protocol which uses a novel combination of turning policies and reserved lanes to manage mixed traffic. Turning policies limit the available turning options depending on the lane a vehicle is on, while reserved lanes segregate vehicles by type. This is a unique approach to manipulate vehicle trajectories, as efforts have traditionally been focused on autonomous vehicle control algorithms. We simulated two variants of unsignalized intersections found in Pasir Panjang Terminal, Singapore, using professional civil engineering software. Our microscopic simulation models show that our protocol offers up to a 35% reduction in travel time compared to the control, depending on intersection type, percentage of automated trucks and traffic volume. Our protocol is highly adaptable to different traffic conditions, enabling us to develop effective solutions for the dynamic conditions faced by container terminals throughout the transition period. This will help ensure Singapore's ports can fully capitalize on the benefits of automation to cope with the ever-increasing exponential demand of the global economy.

S. H. Ng (✉) · R. Ong
NUS High School of Math and Science, Singapore, Singapore
e-mail: h1610097@nushigh.edu.sg

C. Zhou
Centre of Excellence in Modelling and Simulation for Next Generation Ports, National University of Singapore (NUS), Singapore, Singapore

H. Guo et al. (eds.), *IRC-SET 2021*, https://doi.org/10.1007/978-981-16-9869-9_2

Keywords Container terminals · Mixed traffic · Autonomous vehicles · Intersection management · Unsignalized intersection · Traffic simulation · Microscopic traffic flow model

2.1 Introduction

The Port of Singapore is the world's second busiest, moving 37.2 million twenty-foot equivalent units a year as of 2019 [1]. The maritime industry is a key pillar of Singapore's economy, accounting for 7% of GDP and housing 170,000 jobs [2]. However, there loom new challenges on the horizon. While global freight demand is projected to triple by 2050, according to the International Transport Forum, not all of this growth would be spread around equally [3]. Competition is intensifying in the region, with rival ports eyeing bigger slices of the pie.

Singapore is unable to compete simply on price with other ports in the region, due to high operating costs resulting from labour scarcity and the strong Singapore dollar. Instead, Singapore needs to continually invest in technology and innovation to enhance competitiveness, notably automation. With careful planning and management, ports that adopt automation could see operating expenses fall by 25–55%, and productivity rises by 10–35% [4]. The future Tuas Mega Port is set to have large-scale deployments of automated guided vehicles (AGVs), as well as automated yard and quay cranes, remotely operated from a control centre [5]. This paper will mainly focus on autonomous trucks (AVs) which require no special infrastructure, unlike AGVs.

However, there will inevitably be a transitional phase where AVs have to coexist with traditional human-driven trucks, hence the necessity in modelling mixed traffic flow. Many attempts of modelling mixed traffic flow have been set on highways, where the technology requirements for autonomous vehicles are relatively low in comparison with local arterials [6, 7]. Various other studies have attempted to model signalized intersections. However, no previous study has been done to model mixed traffic flow in unsignalized intersections, where network performance is more dependent on the driver's awareness, experience and aggressiveness [8]. Furthermore, the characteristics of trucks, such as size and acceleration/deceleration, differ significantly from cars, which are used in much of existing research. Thus, it remains unclear whether existing strategies developed for mixed traffic flow in cities work in the port environment.

In our project, we propose and examine a protocol to control intersections within Pasir Panjang Terminal, Singapore. Our protocol assigns each lane with allowed turns, as well as implements reserved lanes for AVs. We use microscopic traffic simulation software PTV Vissim to design unsignalized intersections found in the container terminal and record the average travel time of vehicles within the network. We then use statistical tests to analyse the obtained data and determine which implementations of our protocol are effective in minimizing travel time for varying AV penetration rates and traffic volumes.

2.2 Methodology

2.2.1 Design of Protocol

Our protocol allows us to control and limit the trajectories of vehicles passing through the intersection. This enables AVs to predict the trajectory of other vehicles through the intersection based on the lane they are on, preventing unnecessary waiting if the AV knows that the oncoming vehicle from another road will not cross into their planned route. However, this restriction may result in additional delay in the network, as vehicles may have to change lanes to follow our protocol. We seek to balance these factors and find the optimal implementation.

The first component of our protocol is the turning policy, as presented in Fig. 2.1 with reference to Sharon and Stone [9]. The vehicles can only move according to the lane arrows, e.g. vehicles on a lane with a left-pointing arrow can only turn left at the intersection. The turning policies differ in terms of the number of turning options available. This has been reflected in their numbering, with a larger number indicating more turning options. A turning policy combination for an intersection consists of one for the North–South and another for the East–West Road, in that order, e.g. 1a2a.

This list is non-exhaustive and was designed with the distribution of turning ratios in mind, which we will discuss later. We also generally avoided turning policies where the turning options on one lane intersect with that of another lane on the same road, such as turning policy 4. In real-world scenarios, such a turning policy would be unsafe in a road network consisting of human drivers. It remains to be seen if it could be beneficial when AVs also ply the roads.

The second component of our protocol is an AV-only reserved lane as presented in Fig. 2.2. Vehicles can only utilize various lanes based on their type, with AVs

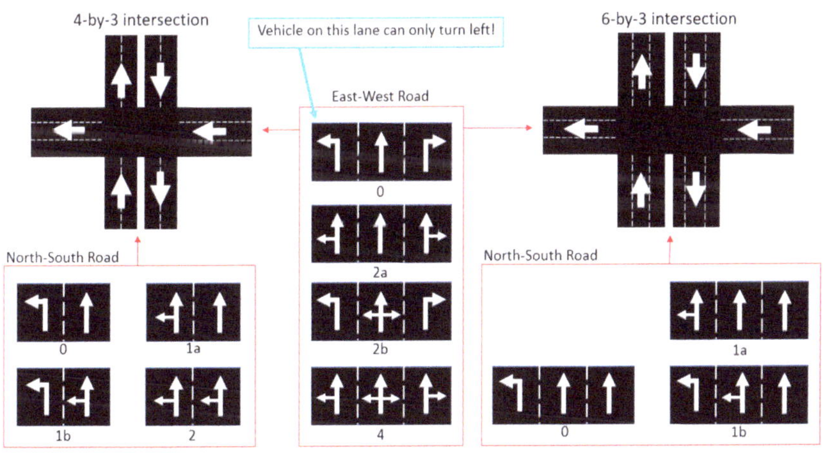

Fig. 2.1 Turning policies for North–South and East–West roads for 4-by-3 and 6-by-3 intersection

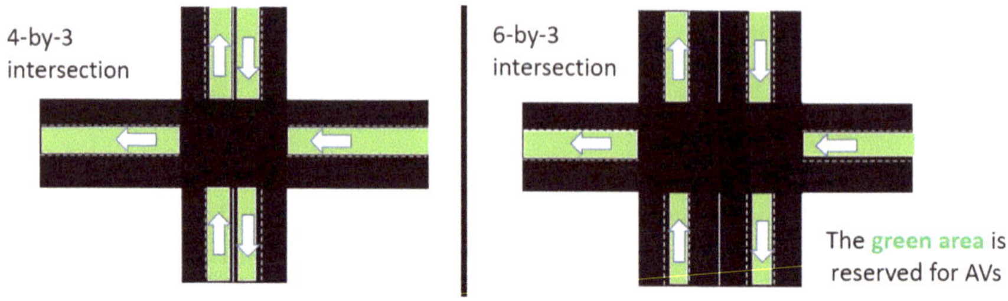

Fig. 2.2 Positioning of reserved lanes, highlighted in green, on 4-by-3 intersection (left) and 6-by-3 intersection (right)

only being able to utilize the green lanes, while human-driven trucks can use only the black lanes. The exception is within the intersection, with vehicles able to move across it freely. We contrast this with the base case where there is no restriction on which lane to use. The reserved lane is positioned to minimize the amount of lane changing required for AVs as they approach the intersection.

2.2.2 *Simulation Software and Intersection Design*

PTV Vissim software [10] is used as the simulation environment. It is one of the most popular microscopic traffic simulation software due to its user-friendly interface and adequate realism. PTV Vissim also offers the necessary features to simulate autonomous vehicles.

As presented in Figs. 2.1 and 2.2, we designed two variants of unsignalized four-way intersections located in Pasir Panjang Terminal, Singapore. For the 4-by-3 intersection, the North–South road is bidirectional with two lanes each way, while the 6-by-3 intersection has three lanes each way. Both intersections have a three-lane unidirectional road from East to West. During the design process, satellite images of the intersections were referenced.

The distribution of turning ratios for vehicles is presented in Table 2.1. The speed limit of the network is set at 30 km/h per the Maritime and Port Authority of Singapore. We prioritized traffic flow on the East–West road, as such all traffic on the North–South road must give way to that on the East–West road, regardless of vehicle type. This contrasts with the First Come First Serve policy as introduced by Dresner

Table 2.1 Turning ratios of vehicles depending on origin, regardless of vehicle type

Origin	Turning ratio		
	Left	Straight	Right
East	0.3	0.4	0.3
North	–	0.75	0.25
South	0.25	0.75	–

and Stone [11], but it is done to ensure that intersections remain a familiar and thus safer environment for human drivers. Neither AVs nor human-driven trucks have priority over the other.

2.2.3 Modelling of Vehicle Behaviour

We need to model the different driving characteristics of AVs compared to human-driven trucks. This study made use of some of the recommended and predefined settings from the CoEXist project [12] while making other modifications as deemed fit. We implemented the all-knowing driving logic for AVs, which assumes the vehicle has profound awareness and predictive capabilities, and enhanced cooperative driving behaviour with smaller headway. For the human-driven trucks, we used the predefined driving logic for heavy goods vehicles (HGVs).

Another difference between trucks would be the acceleration, deceleration and desired speed distribution. Human-driven trucks behave stochastically, due to the randomness of human driving behaviour. However, AVs behave deterministically, using predetermined profiles and moving at the desired speed without variation. As exact data is unavailable, we utilized the presets for HGVs, modifying them to fit the aforementioned requirements. For the desired speed, we set the variation as $\pm$ 5 km/h and 0 km/h for human-driven trucks and AVs, respectively. Finally, we made use of the HGV's 2D model to model both AVs and human-driven trucks, as they were good representations of their dimensions [13].

2.2.4 Evaluation Measurement

We first determined the approximate timing where the network achieves a steady state from start-up, by plotting a graph of travel time against simulation time for various simulations. This will allow us to determine when to commence data collection. Following this, we ran 30 simulation runs, with a randomly selected starting seed and increment, for each of the various implementations of our protocol. To investigate the effectiveness of our protocol, we recorded the average travel time of vehicles to traverse the intersection and conducted the relevant statistical analysis. We also simulated for different AV penetration rates, i.e. the proportion of vehicles that are AVs, of 20, 50 and 80%, as well as the traffic volume, at 600, 1000 and 1400 vehicles per hour per road.

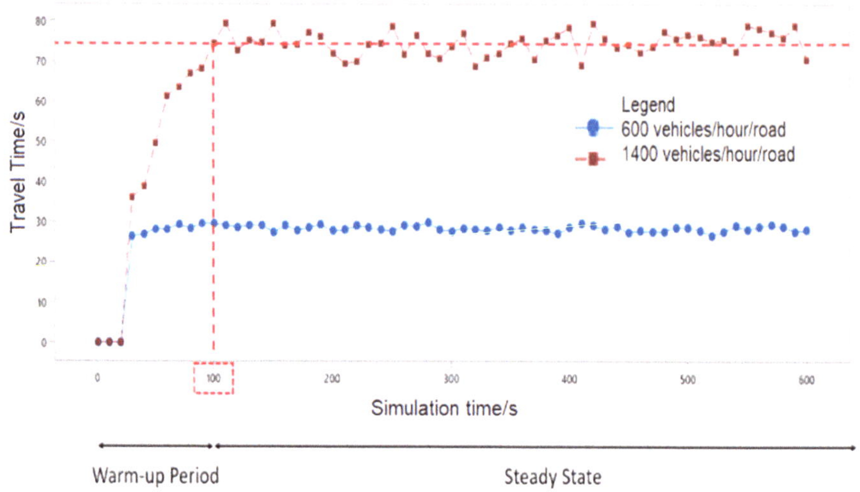

Fig. 2.3 Graph of travel time (s) against simulation time (s) for vehicles within the network. Presented are the best and worst-case scenarios in terms of time required to achieve steady state

2.3 Results

2.3.1 Determination of Steady State

We selected 56 implementations of our protocol for investigation. Using the data from various preliminary simulations, we plotted the graph of travel time against simulation time for vehicles within the network, as presented in Fig. 2.3. Depending on the volume of traffic, the graph plateaus within the range of 60–100 simulation seconds. This indicates that the network has achieved a steady state, and data collection can commence. For the sake of convenience, we set the warm-up period as 100 simulation seconds for all simulations. We do not collect any data before 100 simulation seconds are reached, as the simulation is not representative during the warm-up period.

2.3.2 Effect of AV-Only Reserved Lanes

To examine the effect of AV-only reserved lanes on travel time, we compare implementations of our protocol where the difference lies only in the presence of these reserved lanes. Selected results are presented in the boxplots of Fig. 2.4, for varying AV penetration rates and traffic volumes.

Subsequently, we need to perform statistical tests to check for significant differences in mean travel times. We conducted a one-sided two-sample t-test [14], as

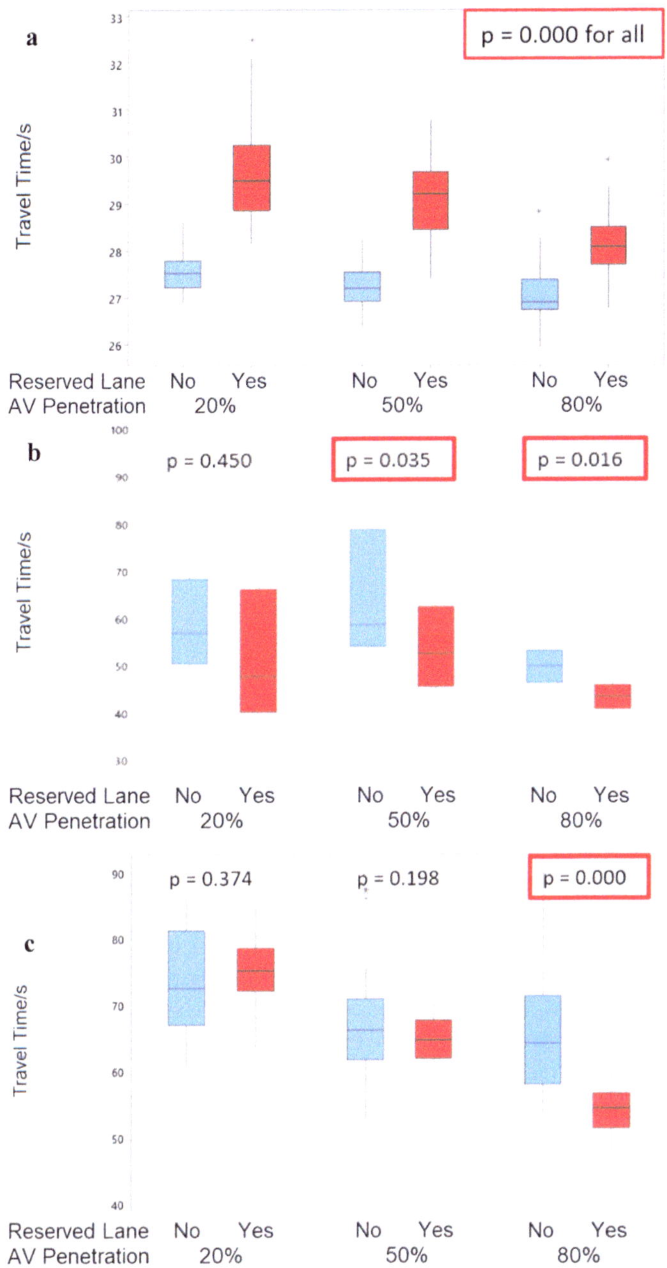

Fig. 2.4 Boxplot of travel time/s at various AV penetration rates. **a** Protocol 1a0 at 600 vehicles/h/road for 6-by-3 intersection. **b** Protocol 00 at 1000 vehicles/h/road for 4-by-3 intersection. **c** Protocol 1a0 at 1400 vehicles/h/road for 6-by-3 intersection

we wish to test the statistical significance in one and not both directions of interest. In other words, we are interested if the implementation of AV-only reserved lanes reduces or increases the mean travel time of the network. This is defined as the alternative hypothesis H1, with the null hypothesis H0 being there is no significant difference in mean travel time. For our paper, we use a significance level of 0.05. The calculated p-values can be found in Fig. 2.4 as well, rounded to 3 decimal places.

For low traffic densities of 600 vehicles/h/road, the boxplots suggest that AV-only reserved lanes increase mean travel time over networks without these lanes. We test H1: AV-only reserved lanes increase mean travel time, and obtain a p-value of 0.000, which is less than 0.05. Hence, we reject H0 in favour of H1 and conclude there is sufficient evidence to suggest this result is statistically significant. Meanwhile, for higher traffic densities of 1000 and 1400 vehicles/h/road, the boxplots suggest that AV-only reserved lanes decrease mean travel time, particularly at higher AV penetration rates. We test H1: AV-only reserved lanes decrease mean travel time. This time, we only obtain a statistically significant result (p-value less than 0.05) for higher AV penetration rates. Such a trend is observed for both 4-by-3 and 6-by-3 intersections.

2.3.3 Effect of Turning Policies

We adopt a similar approach to investigate the effect of turning policy on travel time. Within the same intersection type, AV penetration rate and traffic volume, we examine various turning policies. Selected results are presented in boxplots of Fig. 2.5.

In this case, we must use a different statistical test, as the t-test is limited to comparing two samples. We thus conduct an analysis of variance test (ANOVA) [15] to determine if there is a statistically significant difference in mean travel times between turning policies. The calculated p-values can be found in Fig. 2.5, rounded to 3 decimal places. It is important to note that ANOVA is an omnibus test and can only tell us if there are at least two significantly different groups. It cannot tell us which groups and how. Thus, we follow up ANOVA with Hsu's multiple comparisons with the best (Hsu's MCB), a post hoc test [16], to determine which turning policy produces the lowest mean travel time.

For the 4-by-3 intersection with a low traffic density of 600 vehicles/h/road, we obtain a p-value of 0.181, greater than 0.05. Hence from ANOVA, we conclude there is insufficient evidence to suggest that turning policy affects the mean travel time of the network. At higher traffic densities, however, the p-value for ANOVA is less than 0.05, and Hsu's MCB reveals that turning policy 1a2a is most effective. Meanwhile, for the 6-by-3 intersection, turning policy 1a0 generally results in the lowest mean travel time regardless of traffic density. These trends mostly hold for various AV penetration rates.

We can combine the knowledge from our findings on AV-only reserved lanes and turning policies, to obtain ideal implementations of our protocol depending on the simulation scenario. The trend is an up to 17 and 35% reduction in travel time for

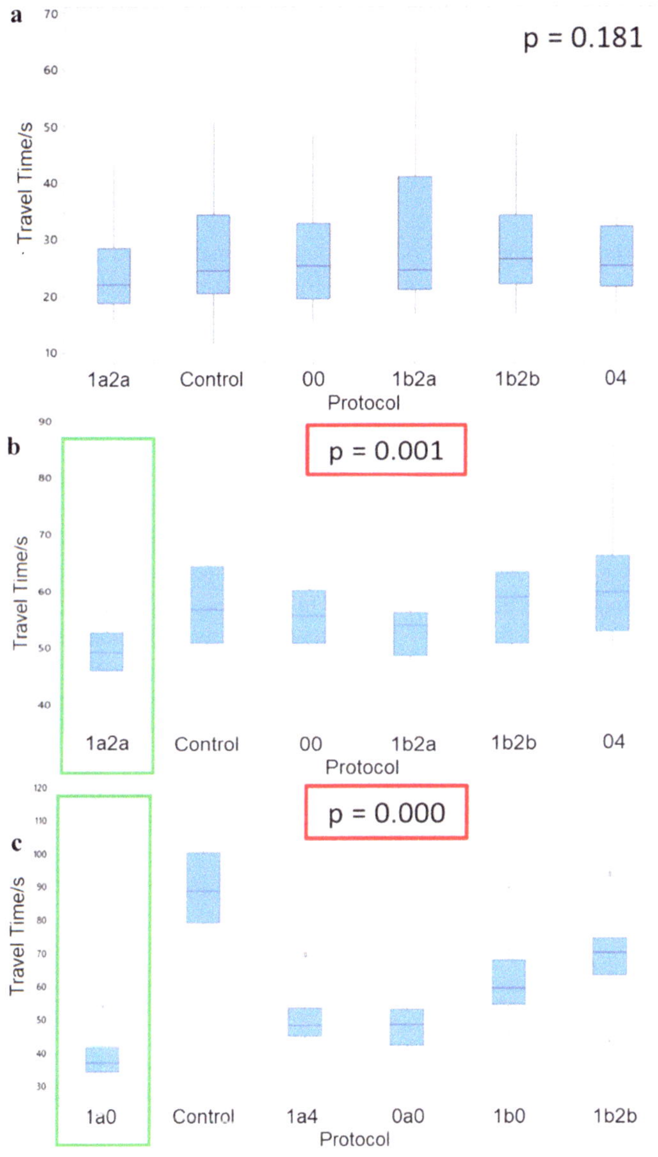

Fig. 2.5 Boxplot of travel time/s for various turning policies. **a** 80% AV penetration, 600 vehicles/h/road for 4-by-3 intersection. **b** 20% AV penetration, 1400 vehicles/h/road for 4-by-3 intersection. **c** 50% AV penetration, 1000 vehicles/h/road for 6-by-3 intersection. The *green boxes* indicate the best turning policies

the 4-by-3 and 6-by-3 intersection, respectively, in comparison with the control with our protocol not implemented.

2.4 Discussion

Broadly speaking, we can divide our simulation scenarios into four categories, as presented in Fig. 2.6. Here, we regard the 4-by-3 intersection as low complexity and the 6-by-3 intersection as high complexity.

2.4.1 *Turning Policies*

For case (III), turning policies do not have a statistically significant effect on mean travel time. During the simulations, the 4-by-3 intersection is not observed to reach capacity, allowing vehicles to freely change lanes as they approach the intersection. There is minimal obstruction to other traffic in the network, as they are spread apart. Further, the vehicles on the North–South road only need to make a maximum of one lane change to continue on their desired route. All these factors allow for lane changing behaviour without detrimental impacts on the mean travel time of the network.

However, as the simulation scenario becomes more "chaotic", as in cases (I), (II) and (IV), turning policies do have a statistically significant effect on mean travel time.

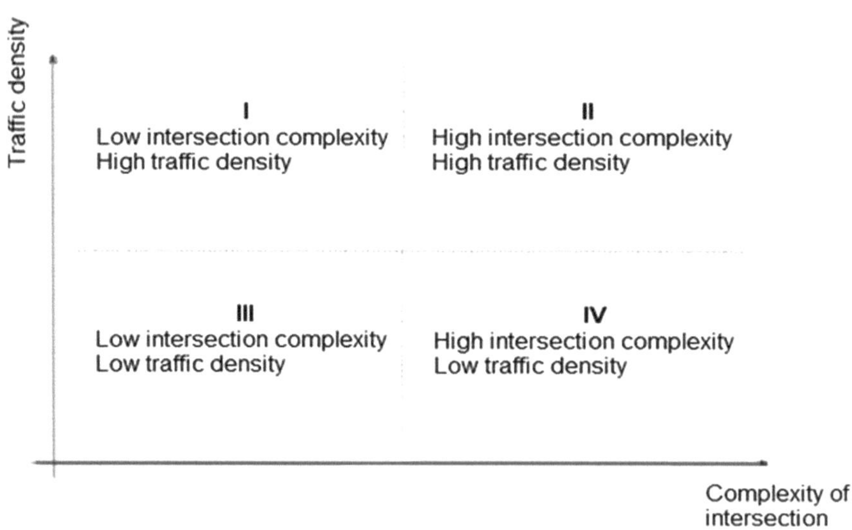

Fig. 2.6 Four categories for simulation scenarios

For cases (I) and (II), the intersection regularly reaches capacity, due to the higher traffic volume. Here, it would be ideal to minimize the conflict of vehicle trajectories, which contribute to network delay. There are two main sources of these conflicts: (1) lane changing behaviour within the road and (2) traversing the intersection itself. In (1), when the frequency of lane change increases, vehicles further down the road have to slow down or stop and wait. Meanwhile, in (2), trajectories of vehicles originating from different roads intersect, leading to vehicles on one road having to give way. These are illustrated in Fig. 2.7.

Sources (1) and (2) are inversely related, as having more turning options on a road minimizes lane changing behaviour within the road, but creates more conflicting trajectories through the intersection and vice versa. From our simulation scenarios, the effect of (2) appears more significant than (1). Hence, restrictive turning policies such as 0 and 1a would be beneficial in minimizing travel time, as it reduces the occurrence of (2) while increasing (1).

A similar concept can be applied to case (IV). Due to the low traffic density, the effect of (1) is minimal. However, as the intersection is more complex, i.e. has more lanes, there are more possible trajectories through it and thus more likelihood of conflict. Thus, restrictive turning policies also prevail for case (IV).

Another criterion for deciding turning policies is to consider the current traffic situation. Consider turning policy 1a and 1b, which have the same number of turning options, three. In our case, the majority of vehicles in the network arriving from the south are going straight, as defined in Table 2.1. Thus, in our case, turning policy 1a would offer better network performance than turning policy 1b, as the former has

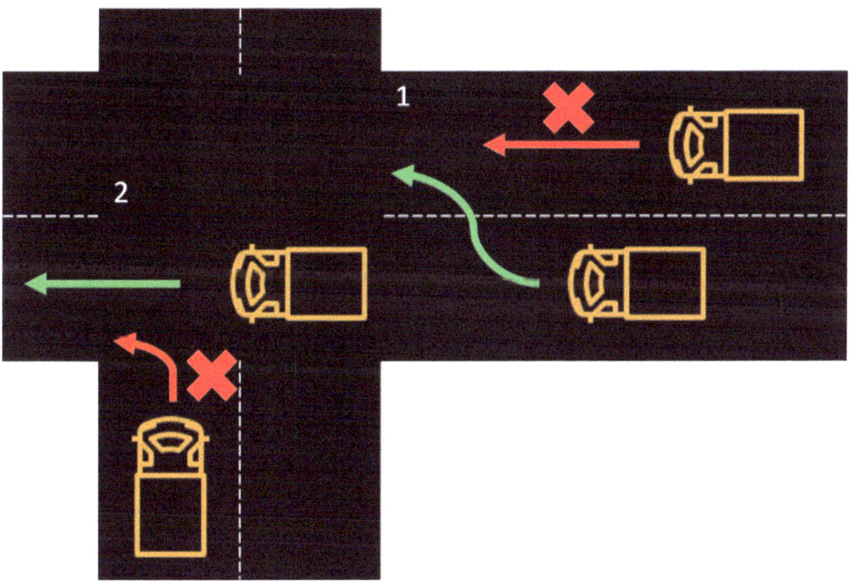

Fig. 2.7 Sources of trajectory conflicts within network

two forward lanes and one turning lane, versus the latter's one forward lane and two turning lanes. This is illustrated in Fig. 2.5b, comparing between turning policy 1a2a and 1b2a.

2.4.2 AV-Only Reserved Lanes

In cases (III) and (IV), reserved lanes severely limit the road capacity at low traffic densities. The reserved and non-reserved lanes are under-utilized at low and high AV penetration, respectively, leading to increased congestion of the network. At high traffic densities such as cases (I) and (II), network performance sees a general improvement, as the reserved lane is more utilized by AVs. As all the vehicles within the reserved lanes are AVs, they can communicate with each other. This allows for more coordinated behaviour in actions such as acceleration and lane changing. In comparison, in the absence of reserved lanes, AVs have to act more cautiously, as human-driven trucks are by nature unpredictable. As such, the mean travel time is reduced with reserved lanes.

However, we notice this trend does not always hold, as demonstrated in Fig. 2.4c. At lower AV penetration rates of 20 and 50%, there is no statistically significant decrease in mean travel time ($p > 0.05$). We hypothesize this is due to the positioning of the reserved lane in the middle of the road, restricting lane change of human-driven trucks to closer to the intersection. The fact that vehicles may need to perform two lane changes in rapid succession is likely what contributed to a drop in network performance. However, further research is necessary to determine the root cause.

2.4.3 Overall Protocol

The impact of AV-only reserved lanes on mean travel time also depends on the turning policy, particularly in the more complex 6-by-3 intersection. As demonstrated in Fig. 2.8, performing an ANOVA reveals statistically significant interaction effects between these two components of our protocol ($p < 0.05$). Reserved lanes add an additional layer of restriction to our protocol, and its impact likely depends on how restrictive the turning policy originally is. For example, at a lower AV penetration rate, pairing reserved lanes with an already restrictive policy is likely to have a more detrimental impact on network performance as compared to a more liberal policy.

Our project acts as a starting point in developing new strategies for mixed traffic in container terminals, which is an inevitable trend of the future. Our protocol does not require additional infrastructure and is thus highly attractive in a post-pandemic world where funding is limited. With this better understanding, it allows Singapore's ports to ramp up cargo handling capacity to cope with growing shipping demand due to e-commerce and continue to thrive in a dynamic global industry.

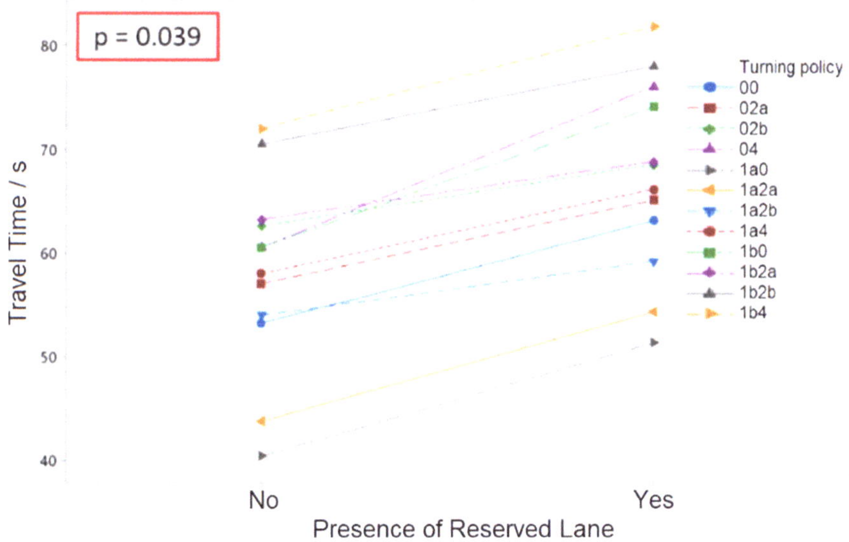

Fig. 2.8 Interaction plot for travel time/s by turning policy and reserved lanes, for 20% AV penetration, 1000 vehicles/h/road on 6-by-3 intersection

2.5 Limitations

The biggest limitation was the inability to properly calibrate the microsimulation. As our research was conducted during the COVID-19 pandemic, we were unable to physically head down to Pasir Panjang Terminal, Singapore, to observe the traffic conditions. Also, the exact vehicle profiles of AVs and human-driven trucks are not publicly available. As such, our protocol would require further verification before real-world implementation.

Another limitation was that for ANOVA, equal population variance is assumed. While ANOVA is a relatively robust procedure against violations of normality, it is still possible that the unequal population variances of different implementations of our protocol could have affected our results, depending on how large the difference is.

2.6 Conclusion and Further Studies

Using microsimulation, we showed that there is no "one-size-fits-all" solution to optimizing traffic. Rather, various conditions need to be factored in, such as the AV penetration rate and traffic density. This is where our proposed protocol shines, as it is highly adaptable. Different implementations of turning policies and AV-only reserved lanes can be used to cater to the dynamic traffic conditions within container terminals. This will ensure that mean travel time is minimized, regardless of the scenario, thereby ensuring a smooth transition to a fully automated fleet.

Our research can be extended to involve realistic testing conditions. This will allow us to better understand the impact of our protocol on container terminal road networks. Additionally, the scope can be widened to cover other types of intersection to find out if our current understanding holds true and adjusts our model accordingly to better predict mixed traffic behaviour.

Acknowledgements We would like to thank Dr. Zhou Chenhao for his expertise and our teacher mentor Mr. Li Da for his guidance during every stage of our research. Their unwavering support was invaluable for the completion of this project.

References

1. United Nations Conference on Trade and Development (2020, November). Review of maritime transport 2020. United Nations Publications. https://unctad.org/system/files/official-document/rmt2020_en.pdf
2. Facts and Trivia (2020, December 14). Maritime and port authority of Singapore. https://www.mpa.gov.sg/web/portal/home/maritime-singapore/introduction-to-maritime-singapore/facts-and-trivia
3. ITF (2019). *ITF transport outlook 2019*. OECD Publishing, Paris.https://doi.org/10.1787/transp_outlook-en-2019-en
4. Chu, F., Gailus, S., Liu, L., & Ni, L. (2020). *The future of automated ports*. McKinsey & Company. November 5, 2020. https://www.mckinsey.com/industries/travel-logistics-and-transport-infrastructure/our-insights/the-future-of-automated-ports
5. Connecting to the World: Singapore as a Hub Port (2018, July 6). Civil Service College, Singapore. https://www.csc.gov.sg/articles/connecting-to-the-world-singapore-as-a-hub-port
6. Yu, H., Tak, S., Park, M., & Yeo, H. (2019). Impact of autonomous-vehicle-only lanes in mixed traffic conditions. *Transportation Research Record: Journal of the Transportation Research Board, 2673*(9), 430–439. https://doi.org/10.1177/0361198119847475
7. Narayanan, S., Chaniotakis, M., & Antoniou, C. (2020). Factors affecting traffic flow efficiency implications of connected and autonomous vehicles: A review and policy recommendations. https://doi.org/10.1016/bs.atpp.2020.02.004
8. Wang, K., & Qin, X. (2015). Exploring driver error at intersections. *Transportation Research Record: Journal of the Transportation Research Board, 2514*(1), 1–9. https://doi.org/10.3141/2514-01
9. Sharon G., & Stone P. (2017) A protocol for mixed autonomous and human-operated vehicles at intersections. In G. Sukthankar, J. Rodriguez-Aguilar (Eds.), *Autonomous agents and multiagent systems*. AAMAS 2017. Lecture Notes in Computer Science (Vol. 10642). Cham: Springer. https://doi.org/10.1007/978-3-319-71682-4_10
10. Evanson, A. (2017). Connected autonomous vehicle (CAV) simulation using PTV Vissim. *Winter Simulation Conference (WSC), 2017*, 4420–4420. https://doi.org/10.1109/WSC.2017.8248148
11. Dresner, K., & Stone, P. (2008). A multiagent approach to autonomous intersection management. *Journal of Artificial Intelligence Research, 31*, 591–656. https://doi.org/10.1613/jair.2502
12. Sukennik, P. (2018, October). *D2.5 micro-simulation guide for automated vehicles*. In C. Fléchon (Ed.), *CoEXist*. https://www.h2020-coexist.eu/wp-content/uploads/2018/11/D2.5-Micro-simulation-guide-for-automated-vehicles.pdf

13. Hampel, C. (2019, October 2). ABB to charge driverless VDL fleet at Singapore port. Elective.Com. https://www.electrive.com/2019/10/01/abb-to-charge-autonomous-vdl-fleet-at-singapore-port/
14. Kim, T. K. (2015). T test as a parametric statistic. *Korean Journal of Anesthesiology, 68*(6), 540–546. https://doi.org/10.4097/kjae.2015.68.6.540
15. Cuevas, A., Febrero, M., & Fraiman, R. (2004). An anova test for functional data. *Computational Statistics & Data Analysis, 47*(1), 111–122. https://doi.org/10.1016/j.csda.2003.10.021
16. McHugh, M. L. (2011). Multiple comparison analysis testing in ANOVA. *Biochemia Medica*, 203–209. https://doi.org/10.11613/bm.2011.029

Chapter 3
Investigating the Effectiveness of Non-pharmaceutical Interventions on COVID-19

Wong Yin Leng Angelina, Kuai En Kai Ethan, Chieu Hai Leong, Wee Liang Chi, and Chai Kian Ming Adam

Abstract The COVID-19 pandemic has caused large regions to be locked down for extended periods, severely impacting global economies and straining healthcare systems. Such pandemics require swift and appropriate responses to slow the rate of infections and reduce stress on healthcare systems. Hence, there is an urgent need to analyse government responses to combat future epidemics more efficiently. In this project, we study non-pharmaceutical interventions (NPIs) in particular: we adapt Imperial College's COVID-19 model (Flaxman et al. in Report 13: estimating the number of infections and the impact of non-pharmaceutical interventions on COVID-19 in 11 European countries (2020), [1]; Estimating the effects of non-pharmaceutical interventions on COVID-19 in Europe (2020), [2]), by challenging their assumption that the effectiveness of NPIs are shared across all 14 European countries. We introduce two new semi-mechanistic Bayesian hierarchical models to investigate this problem from different perspectives. While this further elevates model complexity, we can now compare NPIs' effectiveness across countries. Additionally, we have fine-tuned intervention input quality by introducing more interventions and modifying the model to account for interventions only affecting a fraction of a country. Based on the analysis of government responses in specific countries, we conclude that governments should respond proactively instead of reactively to emerging pandemics and implement strict restrictions especially during the early stages. This is so that infections start slowing down, and NPIs will not need to be implemented for extended times.

Keywords Non-pharmaceutical interventions (NPIs) · Bayesian hierarchical model · COVID-19 · Reproduction number ($\boldsymbol{R}_t$)

W. Y. L. Angelina · K. E. K. Ethan (✉)
NUS High School of Math and Science, Singapore, Singapore
e-mail: h1710036@nushigh.edu.sg

C. H. Leong · W. L. Chi · C. K. M. Adam
INFO Division, DSO National Laboratories, Singapore, Singapore

H. Guo et al. (eds.), *IRC-SET 2021*, https://doi.org/10.1007/978-981-16-9869-9_3

3.1 Introduction

A. *COVID-19*

On 31 December 2019, a cluster of pneumonia cases in Wuhan, China, marked the beginning of the COVID-19 pandemic. As of 30 November 2020, there have been 63,087,137 infections, with over 1,465,368 deaths reported worldwide [3]. Several European countries, in particular, experienced large epidemics in the past few months, with daily cases increasing significantly day by day. The situation in these European countries has highlighted the severity of the COVID-19 pandemic and its impact on national health systems, the country's economy and the daily lives of its people.

The last time the world experienced a pandemic of comparable scale was the 1918 H1N1 influenza pandemic. With the help of multiple NPIs, many countries eventually managed to keep the virus under control. Measures adopted included closing schools, churches, bars and other social venues. Cities which implemented NPIs early in the epidemic were largely successful in reducing infections and mortality rates. However, an increase in cases was also observed after measures were lifted, indicating transmission resurfaced. While understanding of infectious diseases and their prevention has changed since 1918, most countries across the globe face a similar challenge today with COVID-19.

B. *Non-Pharmaceutical Interventions*

In response, multiple countries have implemented, or are implementing various NPIs to flatten the curve, maintain their healthcare systems and treat as many as possible [4]. These NPIs follow the fundamental strategy of suppressing this infectious disease, with the main aim of reducing the reproduction number (average number of secondary cases each case generates, or rate of transmission) R_t to below 1, hence reducing transmission rates and case numbers to low levels. This reduces stress on the nation's healthcare systems, resulting in lower mortality rates.

However, there remain limitations to estimating reproduction numbers for SARS-CoV-2 based on case data. Due to high proportion of undetected and asymptomatic cases, real-world infection numbers are largely unreliable. Additionally, regular changes in testing policies produce huge spikes in daily cases and, at times, negative reported cases. Hence, we studied an alternative way to estimate the course of the epidemic—using observed deaths to model infections [5, 6]. Imperial College's model does so, while accounting for NPIs implemented, resulting in a time-varying transmission rate. This is much more realistic than the constant transmission rate most models assume.

In this project, we adapt Imperial College's model [1, 2] to allow comparison of NPIs' effectiveness across countries. We introduce two new semi-mechanistic Bayesian hierarchical models to investigate this problem from different perspectives. Additionally, we have fine-tuned intervention input quality by introducing more interventions and modifying the model to account for interventions only affecting a fraction of a country.

C. *Motivation*

The main aim of our study is to gain more insights into individual countries, where the original model [1, 2] did not inform us about individual intervention effectiveness for each country. We would like to compare multiple countries and conclude which interventions were effective in which countries—and for what reasons, as well as comparing the overall effectiveness of all interventions among different countries. This would give more insight into how to better manage future pandemics, or how to improve management of COVID-19 for the near future before vaccines become widely available.

(Our analysis focuses on modelling the infection and death rate. We do not study the economic impact of the NPIs.)

3.2 Imperial College's Original Model

The summary of Imperial College's model is presented in the form of a flow chart, as shown in Fig. 3.1. In the original model, Bayesian inference is used to fit modelled deaths to observed deaths of every country. This is informed by several prior distributions, "Symptoms to Death Distribution" and "Infection to Symptoms Distribution", both of which accounts for the time lag between infection and death, as well as "Infection Fatality Ratio". The daily infections are informed by the time-varying reproduction number $R_{t,m}$ and the Serial Interval Distribution. The model calculates backwards from observed deaths to modelled deaths, which is then used to calculate modelled infections and estimate $R_{t,m}$. In Imperial College's model, $R_{t,m}$ is assumed to obey Eq. (3.1).

$$R_{t,m} = R_{0,m} \exp\left(-\sum_{k=1}^{6} \alpha_k I_{k,m,t}\right) \tag{3.1}$$

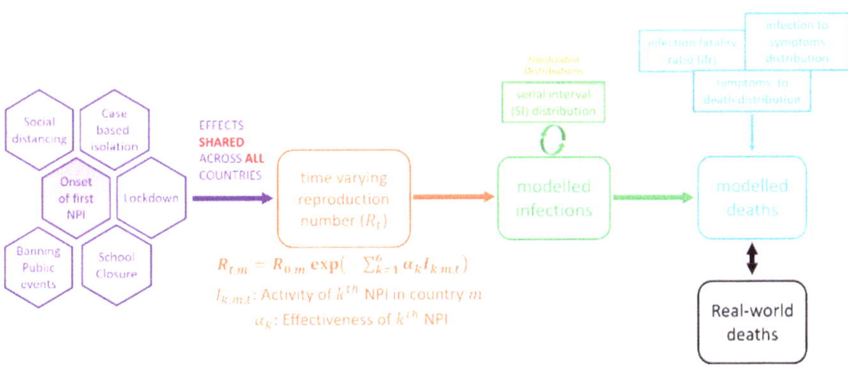

Fig. 3.1 Summary of Imperial College's model

The effectiveness of the kth NPI, α_k, is then fit to best follow the predicted real-world $R_{t,m}$.

A. *Limitations*

As Imperial College published their report in March 2020, there was little real-world infection and fatality data available. As such, they decided to treat all countries as a single data set, assuming α_k is similar across all counties. While this decreases idiosyncrasy and variance, several months later, the assumption that countries have similar NPI effectiveness is no longer true, with countries such as Belgium having much larger death rates than their European counterparts Norway.

3.3 Methodology

Our model was written with the probabilistic programming language Stan [7], and ran with Stan's Hamiltonian Monte Carlo sampler, on top of R programming language. Both models ran with six chains and 600 iterations, 300 of which is for warm-up. JavaScript was used to extract intervention data, while Python was used to process intervention data to input into the models.[1]

A. *Country and Data set*

We challenge Imperial College's main assumption by introducing two models, A and B, which treat each of the 14 European countries as distinct data sets, different from the original model's unique pooling of information from multiple countries at once.

With more real-world data, we need not worry about increased idiosyncrasy and variance.

The added advantage is increased flexibility. Due to Imperial College's key assumption, countries had to be carefully chosen to have similar circumstances, resulting in Imperial College choosing all 14 countries from Europe. However, our models are unconstrained by this and can model any country with enough reported deaths.

B. *Model A*

Model A assumes NPI's effectiveness varies with country and considers the effectiveness of the seven NPIs for **each** country, denoted by variable $\alpha_{k,m}$, representing the effectiveness of intervention k in country m. The reproduction number for country m on day t is given in Eq. (3.2), where $I_{k,m,t}$ represents active interventions on day t for each country (Fig. 3.2).

$$R_{t,m} = R_{0,m} \exp\left(-\sum_{k=1}^{7} \alpha_{k,m} \cdot I_{k,m,t}\right) \tag{3.2}$$

[1] Infection, fatality and intervention data used can be found here: https://github.com/EthanKuai/Investigating-the-Effectiveness-of-NPIs-on-COVID19/blob/main/Sources.md.

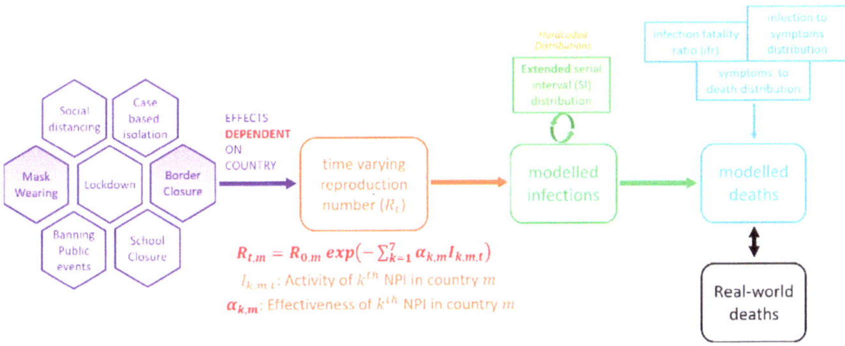

Fig. 3.2 Summary of model A

However, if a country implements multiple interventions in the same time period, they would be statistically indistinguishable from one another, increasing the variance of α of those interventions. For example, Austria and France implemented both social distancing and self-isolation from 16 March, which thus would be indistinguishable from one another. Additionally, with more variables to fit, more epochs are required for model convergence to occur.

C. *Model B*

Model A has multiple limitations involving increased variance. Hence, we counter these limitations with model B, which considers the overall impact of **all seven** NPIs for **each country**, denoted by variable α_m, representing the overall effectiveness of all interventions in country m. The reproduction number is given in Eq. (3.3). This allows us to directly compare two countries, even if they execute entirely different interventions, while controlling variance (Fig. 3.3).

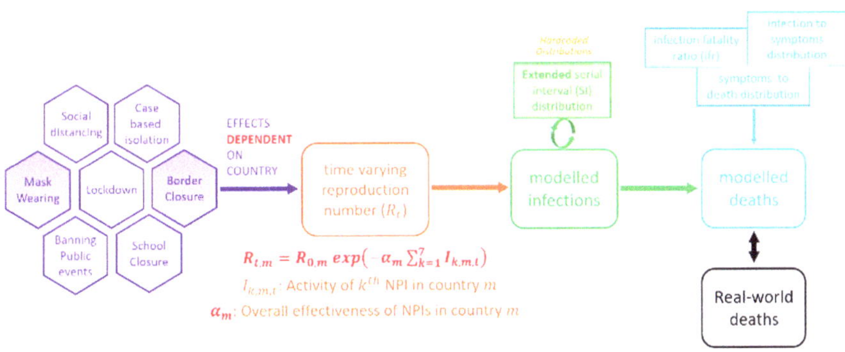

Fig. 3.3 Summary of model B

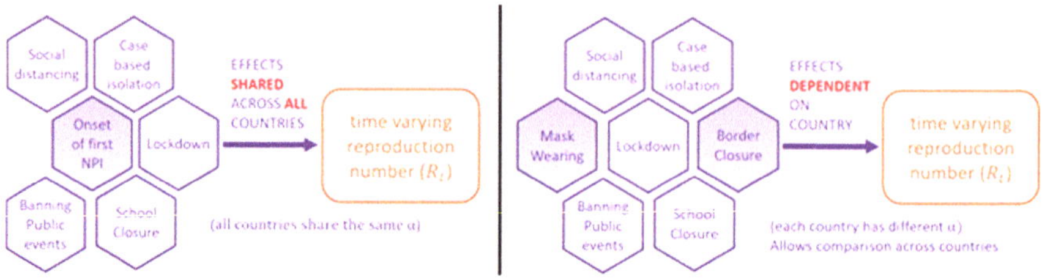

Fig. 3.4 New interventions "mask wearing" and "border closure"

$$R_{t,m} = R_{0,m} \exp\left(-\alpha_m \cdot \sum_{k=1}^{7} I_{k,m,t}\right) \tag{3.3}$$

However, Model B assumes all interventions of a country have similar effectiveness, which would increase variance of α_m and decrease convergence rate. Additionally, from the results of this model, one would be unable to conclude the effectiveness of individual interventions of a country and cannot compare between different interventions.

D. *Interventions*

Intervention data was taken until 2 December 2020. During this period, the countries have implemented more NPIs, including "Mask Wearing" and "Border Closure". We have updated the list of interventions to include them, while removing "First Intervention", which would have little effect in the long run. The comparison between the interventions considered in the original and updated model is shown in Fig. 3.4.

In the original model, all countries are treated as a single data set, thus there is multiple "First Intervention" for the model to fit its effectiveness. However, as both models A and B treat each country as distinct data sets, "First Intervention" only appears once in each data set, resulting in high variance of α of "First Intervention".

The original model assumed interventions affected the entire country and would never end once started. As many European countries only implemented interventions in highly affected regions of the country, we have collected intervention data down to individual states and towns. A Python program was used to process such data, producing a sequence of numbers between 0 and 1 (inclusive) for every intervention in every country, each representing how active a country's intervention is on a given date ($I_{k,m,t}$). 0 represents an inactive intervention, while 0.5 and 1 represent an intervention affecting half the population and the whole country, respectively. This allows interventions fed into the model to better depict the real world.

E. *Serial Interval*

Serial interval models the time between successive infections in the same chain of transmission. Serial interval data from Imperial College's model is taken from multiple studies [8, 9] based on real-world contact tracing data. However, to extend it to longer durations, we perform gamma regression, getting our new

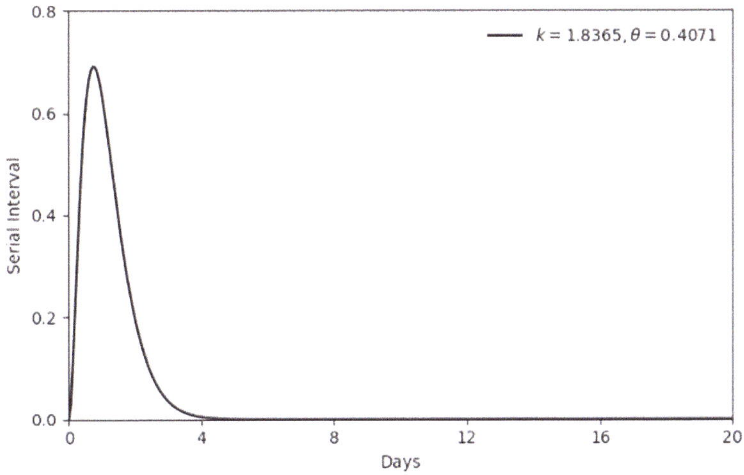

Fig. 3.5 Plot of serial interval, gamma(1.8365, 0.4071)

serial interval to be Gamma($k = 1.8365$, $\theta = 0.4071$), quantised to each day t with: $g_t = \int_{\tau=t-0.5}^{t+0.5} \frac{\tau^{k-1} e^{-\tau/\theta}}{\Gamma(k)\theta^k} d\tau$. The updated serial interval plot is shown in Fig. 3.5.

3.4 Results

A. *Norway*

From Figs. 3.6 and 3.7, Norway was observed to have the largest per cent reduction in R_t for majority of its interventions and the lowest final R_t. With the help of European news articles, we analyse the situation in Norway to understand why its interventions were significantly more effective than the other European countries.

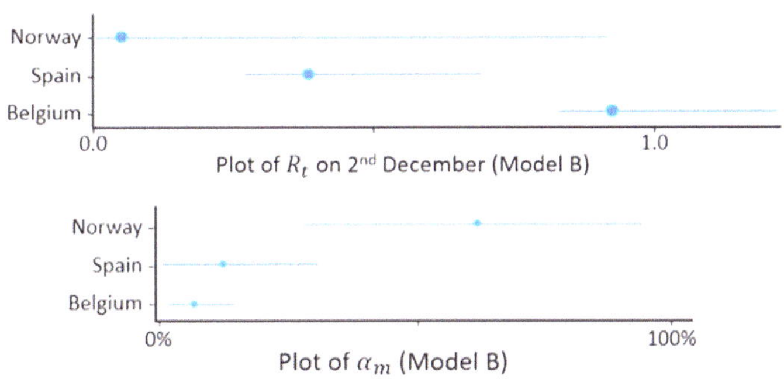

Fig. 3.6 Model B's results for Norway, Spain and Belgium

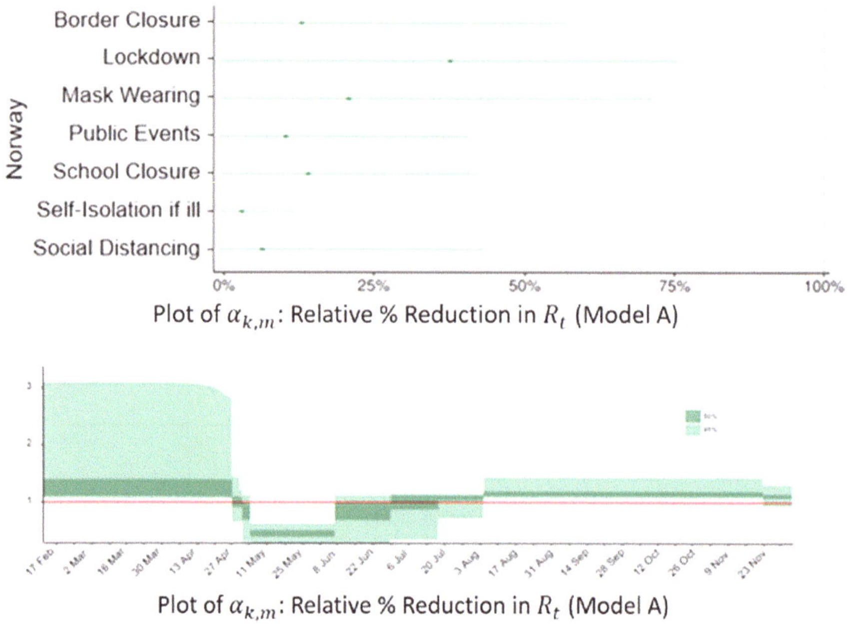

Fig. 3.7 Model A's results for Norway

Compared to other countries such as Sweden, Norway has done an impressive job of strictly enforcing interventions when the virus first emerged in the country. On 12 March, when the reproduction number in Norway was at 2.5, the Norwegian Government immediately imposed the strictest emergency measures at that time—all schools, bars and parlours were closed and those entering the country were put in a 2-week quarantine. This was a key step in combatting the virus, as Norway was able to prevent large-scale community transmission at heavily populated areas. Norway's quick and effective response to the growing pandemic allowed its reproduction number to drop from 2.5 on 12 March to 0.5 by early April.

During the resurgence period, Norway was also quick to identify the more affected regions, implementing a lockdown and enforcing mask wearing in Oslo. While less-affected regions were still able to keep up the country's economy, the lockdown in affected regions again effectively reduced transmission of the virus. As its neighbours were hit hard by the second wave of the pandemic, Norway kept its borders closed, keeping infection numbers low.

B. *Spain*

We also analyse the interventions in Spain, particularly "Border Closure" measures, which was observed to be more effective than the other interventions in Spain and result in the largest per cent reduction in R_t among all 14 countries (as observed from Fig. 3.8).

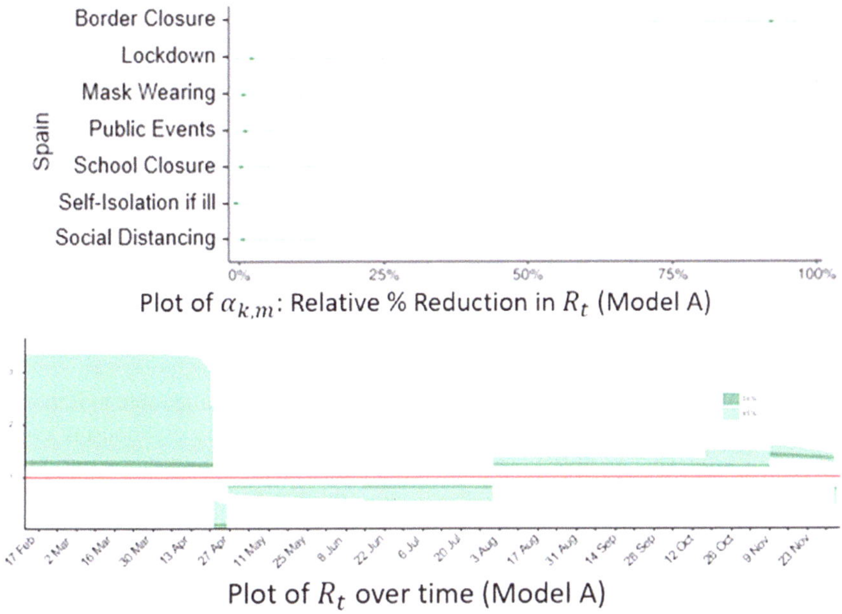

Fig. 3.8 Model A's results for Spain

COVID-19 first hit Europe in Italy, one of Spain's neighbouring countries. The Spanish government responded quickly, being the first European country to start closing its borders to Italy on 10 March, heavily reducing the number of imported cases. When it spread to other parts of Europe, Spain imposed travel restrictions on all non-residents from 16 March.

Unlike other countries, Spain kept its borders closed until infection numbers were brought down, and the virus was kept under control. This was certainly more effective than the 14-day quarantine implemented in some other countries as they eased border restrictions.

During the resurgence period, with improved testing policies, Spain was able to keep its borders open while requesting all travellers to have a negative COVID-19 test and 14-day quarantine to enter the country. As infection rates increase towards end October, Spain was also one of the few countries to implement border closures again.

These border closures were essential in ensuring no carriers of the virus entered the country, assisting the government in contact tracing efforts and for the government to be able to focus on reducing community transmission, bringing the reproduction number down significantly over time. The government responded quickly and kept borders closed for an appropriate amount of time, hence resulting in a significantly higher per cent reduction in R_t as compared to other interventions.

C. *Belgium*

From Fig. 3.6, Belgium was observed to have one of the lowest per cent reductions in R_t for majority of its interventions and the highest final R_t value. Back in April, Belgium was also already observed to have one of the highest mortality rates in the world. We study the possible reasons that led Belgium to become one of Europe's most hardly hit countries by the coronavirus.

First, Belgium is one of the smallest yet most densely populated countries in Europe. This results in a higher initial reproduction number, which was estimated to be at 3.4 before strict measures were implemented in Belgium. It also puts a strain on "Social Distancing" measures, accounting for a considerably low per cent reduction in R_t (as observed in Fig. 3.9).

Second, Belgium's location in Europe also makes it more vulnerable to imported cases. At the heart of Europe, Belgium is well connected to neighbouring countries, and Europeans are able to travel in and out of Belgium by train or car. Since "Border Closure" measures were focused more on international flights and entry-checks which were more concentrated in airports, Belgium was unable to account for several imported cases at other entry-points of the country early on during the pandemic. Hence, despite the strict travel flight restrictions implemented in Belgium, "Border Closure" measures were ineffective in reducing R_t (as observed in Fig. 3.9).

Third, Belgium's multi-layered government applied a regional approach to COVID-19 measures, throwing citizens into confusion. As some regions imposed stricter restrictions compared to neighbouring regions, it sparked outrage in its

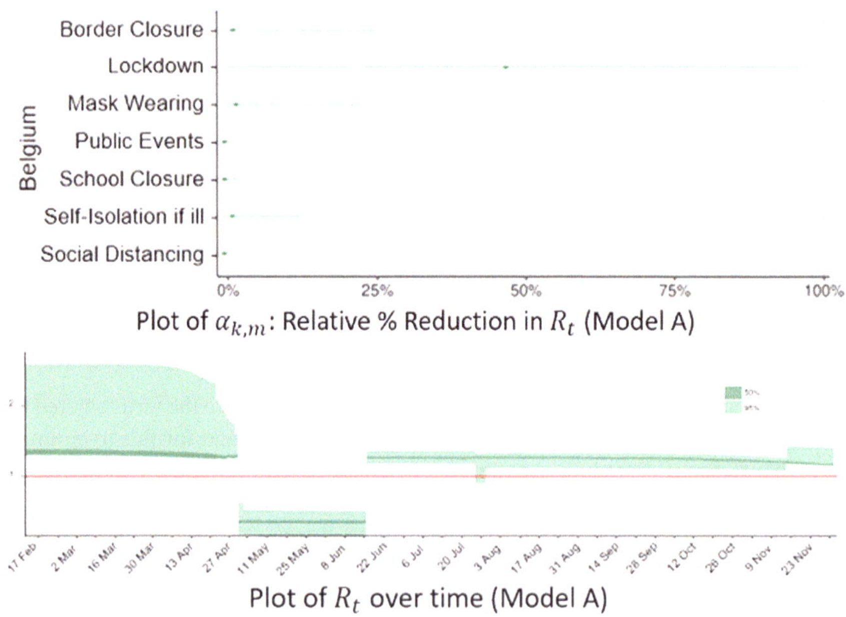

Fig. 3.9 Model A's results for Belgium

citizens, severely affecting the government's efforts in enforcing the NPIs implemented. Anti-coronavirus protest gatherings also resulted in more congested areas for large-scale community transmission, further increasing the reproduction number in Belgium.

D. *Intervention-specific*

Our results from Model A for all 14 countries imply that major NPIs such as "School Closure", "Lockdown" and "Border Closure" substantially reduced reproduction numbers.

E. *Model-specific*

3.5 Model A

As different interventions have wildly different $\alpha_{k,m}$ compared to others, this model allows us to gain additional insights into the effectiveness of different interventions by each country. For example, in the Netherlands, School Closure has an $\alpha_{k,m}$ of around 90%, which is much higher than the other interventions in the Netherlands, which are mostly $< 25\%$.

3.6 Model B

Certain countries have extremely small error bars in α_m while others have error bars which stretch close to the full range, meaning Model B has a tendency of "hit and miss". Notably, if all interventions have similar effectiveness in Model A, the α_m of Model B would have an extremely low variance. However, the same is not true for its inverse (Fig. 3.10).

3.7 Future Studies

3.7.1 Unresolved Limitations

First, the model assumes $R_{t,m}$ is constant throughout the country. There tends to be higher urban density in cities than rural areas, thus this is not true. Second, the model assumes $R_{t,m}$ varies only with a change of interventions implemented. However, movement of people throughout the country would change urban densities of different regions and consequently $R_{t,m}$. While this can be controlled to some extent by interventions such as lockdowns and border closures, it still varies significantly. Third, the model assumes α is constant throughout time. For example, people might comply to

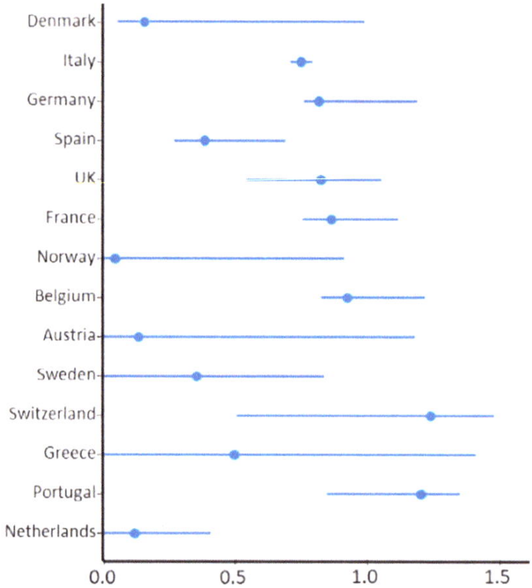

(a) Plot of R_t on 2nd December (Model B)

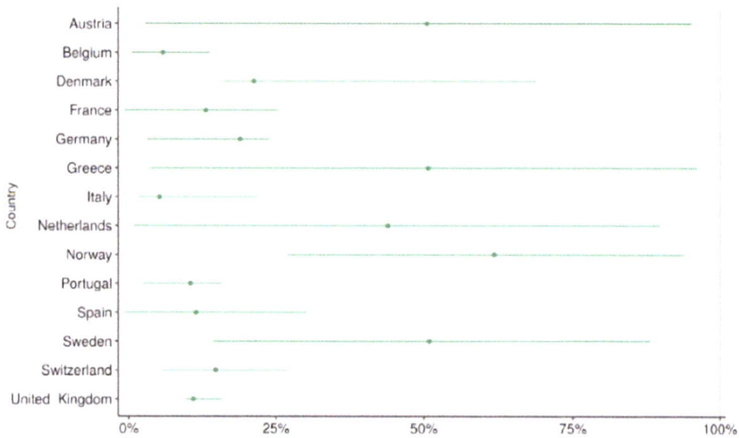

(b) Plot of α_m (Model B)

Fig. 3.10 Model B's results for all 14 European countries

lockdown rules initially, but when it is extended, some may get restless and protest, defeating the purpose of the lockdown. Additionally, if a second lockdown order is issued, our model will assume both lockdowns have the same α, which is not always the case. Fourth, prior distributions such as Serial Interval Distribution, Infection-to-onset Distribution and Onset-to-death Distribution vary throughout country and time, reducing accuracy. Fifth, as the model relies on observed deaths, countries with high cases yet few deaths like Singapore [10] cannot be modelled as it does not give the model sufficient information to predict cases.

3.7.2 Impacts on Results

First, pre-emptive interventions are penalised. If an intervention is implemented before the outbreak occurs, while it might have reduced $R_{t,m}$ significantly and lowered numbers in the outbreak, the model would register the intervention as increasing $R_{t,m}$ and thus have low-to-negative α. Second, if a country does not update interventions for an extended duration, the model is forced to believe that $R_{t,m}$ remains constant throughout. However, due to all three limitations, it is very likely a second outbreak will occur, as the governments have greater difficulty in enforcing restrictions. This is most obviously seen in Model A's Sweden data, where the model believes $R_{t,m} \leq 1$ even though the cases are on the rise.

3.7.3 Possible Improvements

(a) For limitations 1 and 2, a possible solution is to take mobility data and split countries into multiple sub-regions of similar urban density, using mobility data to simulate people movement. Another possible study is to analyse mobility data within countries to determine the effectiveness of curfews, lockdowns, and self-isolation in different countries density, using mobility data to simulate people movement, allowing for a more in-depth analysis of changes in $R_{t,m}$.

3.8 Conclusion

By modifying Imperial College's COVID-19 model, through the addition of the semi-mechanistic Bayesian hierarchical models, we present novel perspectives in the study of COVID-19 infection rates. Our model gives a clearer overview of the efficacies of NPIs based off the 14 European countries studied and provides insights into the ideal approaches governments should adopt to combat future pandemics.

After analysing selected European governments' responses, we conclude that major NPIs like "School Closure", "Lockdown" and "Border Closure" substantially reduced reproduction numbers. "Border closure" has also proven to be more effective than the "14-day quarantine", possibly because it enables governments to focus on minimising community transmission. Most importantly, governments should also respond proactively instead of reactively to emerging pandemics and implement strict restrictions especially during the early stages.

Acknowledgements The authors would like to gratefully acknowledge Dr. Chieu Hai Leong, Mr. Wee Liang Chi and Dr. Chai Kian Ming Adam from DSO National Laboratories INFO Division, for their support and guidance throughout this project. The authors would also like to thank NUS High School teacher mentors, Ms. Ye Xiajuan and Dr. Chiam Sher-Yi, for their encouragement and support during the entirety of this project. Additionally, the authors would like to acknowledge "Learning in graphical models" [11] which was referenced in much of the literature review on Bayesian modelling.

References

1. Flaxman, S., et al. (2020). Report 13: Estimating the number of infections and the impact of non-pharmaceutical interventions on COVID-19 in 11 European countries. https://www.imperial.ac.uk/media/imperial-college/medicine/mrc-gida/2020-03-30-COVID19-Report-13.pdf, https://doi.org/10.25561/77731
2. Flaxman, S., et al. (2020). Estimating the effects of non-pharmaceutical interventions on COVID-19 in Europe. https://www.nature.com/articles/s41586-020-2405-7
3. Worldometer. https://www.worldometers.info/.
4. Ferguson, N. M., et al. (2020). Report 9: Impact of non-pharmaceutical interventions (NPIs) to reduce COVID-19 mortality and healthcare demand. https://www.imperial.ac.uk/media/imperial-college/medicine/sph/ide/gida-fellowships/Imperial-College-COVID19-NPI-modelling-16-03-2020.pdf , https://doi.org/10.25561/77482
5. Fraser, C. (2007). Estimating individual and household reproduction numbers in an emerging epidemic. https://pubmed.ncbi.nlm.nih.gov/17712406/, https://doi.org/10.1371/journal.pone.0000758
6. Jombart, T., et al. (2020). Inferring the number of COVID-19 cases from recently reported deaths. https://doi.org/10.1101/2020.03.10.20033761v1
7. Stan Development Team (2019). Stan modeling language users guide and reference manual, version 2.25. https://mc-stan.org/
8. Ferguson, N. M., et al. (2009). Household transmission of 2009 pandemic influenza a (H1N1) virus in the united states. https://doi.org/10.1056/NEJMoa0905498
9. Hens, N., et al. (2020). Estimating the generation interval for coronavirus disease (COVID-19) based on symptom onset data. https://www.ncbi.nlm.nih.gov/pmc/articles/PMC7201952/, https://doi.org/10.2807/1560-7917.ES.2020.25.17.2000257
10. https://www.reuters.com/article/health-coronavirus-singapore-explainer-idUSKBN2680TF
11. Jordan, M. I. (1998). Learning in graphical models. https://doi.org/10.1007/978-94-011-5014-9#about
12. Unwin, H. J. T., et al. (2020). State-level tracking of COVID-19 in the united states. https://www.nature.com/articles/s41467-020-19652-6
13. Vollmer, M. A. C., et al. (2020). Report 20—Using mobility to estimate the transmission intensity of COVID-19 in italy: A subnational analysis with future scenarios. https://www.imperial.ac.uk/mrc-global-infectious-disease-analysis/COVID-19/report-20-italy/
14. Mellan, T. A., et al. (2020). Report 21—Estimating COVID-19 cases and reproduction number in brazil. https://www.imperial.ac.uk/mrc-global-infectious-disease-analysis/COVID-19/report-21-brazil/
15. Walker, P. G., et al. (2020). Report 12: The global impact of COVID-19 and strategies for mitigation and suppresion. https://www.imperial.ac.uk/media/imperial-college/medicine/mrc-gida/2020-03-26-COVID19-Report-12.pdf, https://doi.org/10.25561/77735
16. Lourenco, J., et al. (2020). Fundamental principles of epidemic spread highlight the immediate need for large-scale serological surveys to assess the stage of the sars-cov-2 epidemic. https://doi.org/10.1101/2020.03.24.20042291v1.full.pdf

17. Verity, R., et al. (2020). Estimates of the severity of COVID-19 disease. /https://doi.org/10.1101/2020.03.09.20033357v1.full.pdf
18. Bellman, R., & Harris, T. E. (1952). On age-dependent binary branching processes. https://www.jstor.org/stable/1969779?seq=1, https://doi.org/10.2307/1969779
19. Bellman, R., & Harris, T. E. (1948). On the theory of age-dependent stochastic branching processes. https://www.ncbi.nlm.nih.gov/pmc/articles/PMC1079176/, https://doi.org/10.1073/pnas.34.12.601

Chapter 4
Investigating the Genetic Etiology of Disease in a Patient with Aplastic Anemia

Beverly Low Ying Tong, Lee Shi Mei Charmaine, Tay Jia Ying, Grace Tan Li Xuan, Liu Chun Ping, Lai Poh San, Eng Soo Yap, and Tung Moon Ley

Abstract This study investigates the molecular basis of a male patient presenting with aplastic anemia (AA) and related symptoms of macrocytosis and mild thrombocytopenia. In addition, this patient also presents symptoms not usually found in AA patients, such as fatty liver, liver cirrhosis with portal hypertension, diffuse cerebral and cerebellar atrophy, and congenital left sensorineural hearing loss. As such, it is hypothesized that the patient has AA that is secondary to inherited bone marrow failure syndromes related to telomere biology disorders. Thus, to identify the underlying genetic cause of the disease, relative telomere length (RTL) of proband and family members was determined by qPCR, followed by identification of disease-causing variants through next-generation sequencing of the proband. The proband has a RTL of 0.11 (− 5.96 SD) which is shorter than the 1st percentile (− 2.33 SD). As RTL analysis indicates significant telomere shortening in the patient, it is likely that the patient has a telomere biology disorder. 8 variants in genes (*DKC1, ATXN3, PTPRQ, ABCB4, DIAPH3, TBP, PARP1*) associated with our patient's phenotype, 2 of which were previously reported and were also shortlisted as potential candidates. However, upon curation, these variants were found to be of uncertain significance, and the genetic cause of our patient's condition remains elusive. Nonetheless, we were able to confirm that the patient had significant shortening of telomeres.

B. L. Y. Tong (✉) · L. S. M. Charmaine · T. J. Ying
Raffles Institution, Bishan, Singapore
e-mail: beverly.low03@gmail.com

G. T. L. Xuan · L. C. Ping · L. P. San
Department of Paediatrics, National University of Singapore, Singapore, Singapore

E. S. Yap
Department of Lab Medicine, National University Hospital, Singapore, Singapore

T. M. Ley
Department of Haematology-Oncology, National University Cancer Institute, Singapore, Singapore

H. Guo et al. (eds.), *IRC-SET 2021*, https://doi.org/10.1007/978-981-16-9869-9_4

4.1 Background and purpose

4.1.1 Case Presentation

A male patient in his 30s presented main symptoms of aplastic anemia (AA), macrocytosis, and mild thrombocytopenia. The patient also presented symptoms uncharacteristic of AA, namely fatty liver, liver cirrhosis with portal hypertension, diffuse cerebral and cerebellar atrophy, and congenital left sensorineural hearing loss.

4.1.2 Aplastic Anemia (AA)

AA is a life-threatening bone marrow failure disorder [1] that can present at any age [2]. The incidence of AA ranges from 1.5 to 7 cases per million people, and the median age at diagnosis ranges from 25 to 60 years globally [3]. In most cases, AA is caused by immune-mediated destruction of hematopoietic stem cells (HSCs), resulting in pancytopenia [4]. However, the patient in this study not only presents symptoms associated with acquired AA such as macrocytosis and thrombocytopenia, but also liver cirrhosis with portal hypertension and fatty liver, which have been associated with certain inherited bone marrow failure syndromes (IBMFSs) that are telomere biology disorders (TBDs) [5, 6]. TBDs are characterized by short telomeres and may cause AA [7]. Telomere loss in HSCs can result in early apoptosis of these cells [8] and decrease their proliferative potential in vitro [9], causing them to be unable to mature and differentiate into blood cells [10]. This patient also presents with diffuse cerebral and cerebellar atrophy and congenital left sensorineural hearing loss (SHL) that are uncharacteristic of AA. Thus, it is possible that the patient's phenotype might be attributed to AA that is secondary to IBMFS [11] including dyskeratosis congenita (DC), Diamond Blackfan anemia (DBA), and Shwachman-Diamond syndrome (SDS), occurring from germline mutations inherited from parents or arising *de novo* [12].

4.1.3 Importance and Challenges of an Accurate Diagnosis

It is crucial to provide an accurate diagnosis and distinguish between acquired AA and AA resulting from IBMFS to effectively manage symptoms and improve patient outcomes [13] as clinical treatments, and approaches are different for the two diagnoses [7, 11]. Patients with congenital disorders may also present additional symptoms that must be treated. Thus, an accurate diagnosis is required to rule out other underlying disorders, primarily IBMFS, and assess for specific etiologies and associations [11]. However, it is difficult to distinguish between inherited AA and IBMFS, especially if the patient has a *de novo* mutation, or a mutation with low-disease penetration, or is not presenting any other classical congenital anomalies [11]. Besides, there are gaps in literature, where the list of genetic mutations that causes specific genetic disorders is incomplete, in addition to a large number of genes reported to be involved in a specific disease. Due to genotype heterogeneity, a specific genetic mutation can also give rise to multiple diseases. Additionally, symptoms that are not unique to a particular disease can overlap and vary in severity, making it challenging to provide an accurate diagnosis.

4.1.4 Aim

The aim of this project is to investigate the underlying cause and genetic etiology of AA in a patient presenting with multiple associated symptoms. This will be done by (1) determining the relative telomere length measurements and (2) identifying presence of any pathogenic genetic variants.

4.2 Hypothesis

Based on this patient's primary presentation of AA and taking into consideration that he presented other symptoms that are not characteristic to acquired AA, as detailed above, it is hypothesized that a telomere biology disorder may be involved. From literature review, AA can be a symptom of several IBMFSs that are TBD, including DC [14] and SDS [15].

4.3 Materials and Method

4.3.1 Literature Review

Literature review was initially conducted to examine the spectrum of clinical symptoms of IBMFS such as DC and AA. The respective causal genes, methods to identify the causes of the diseases, treatment options, and all overlapping symptoms were identified. The testing methods for IBMFS and AA such as next-generation sequencing for genetic testing, flow-FISH, and quantitative PCR for telomere length testing were also reviewed to determine the most suitable methodologies to be used as laboratory investigations to support a diagnosis.

4.3.2 Relative Telomere Length (RTL) Assay

The RTL was obtained from quantitative PCR (qPCR) data, using extracted genomic DNA from peripheral blood of patient, brother, mother, and father. Two references (NM914) and (NM917) were used as the positive controls, and a no-template control was used as the negative control. Telomere qPCRs and single-copy gene (*HBG* gene) qPCRs were performed in separate wells using primers [16, 17] as follows. Telomere (A) (5′-CGG TTT GTT TGG GTT TGG GTT TGG GTT TGG GTT TGG GTT-3′) and Telomere (B) (5′-GGC TTG CCT TAC CCT TAC CCT TAC CCT TAC CCT TAC CCT-3′); *HBG1* (5′-GCT TCT GAC ACA ACT GTG TTC ACT AGC-3′) and *HBG2* (5′-CAC CAA CTT CAT CCA CGT TCA CC-3′). A final volume of 10μl amplification reaction for both qPCRs using the QuantiNova SYBR Green PCR Kit - Qiagen 208056 contained 5 μl of 2x SYBR PCR Master Mix, 0.2 μl forward primer (10μM), 0.2 μl reverse primer (10 μM), 0.1 μl template DNA (10 ng/μl), and 4.5 μl RNase-Free water. Each sample was run in triplicates. PCR cycling conditions for telomere amplification started with 1 cycle of 95 °C denaturation for 2 min, followed by 35 cycles of 95 °C for 5 and 10 s at 58 °C for telomere amplification, and 56 °C for *HBG* amplification [16]. The PCR efficiency was > 90%, and the linear correlation coefficient values for both reactions were > 0.99.

4.3.3 RTL Determination

Telomere (T) signals and single-copy gene (S) signals were obtained from qPCR. The T/S ratio of each individual run was calculated by dividing the telomere starting quantity by the single-copy gene *HBG* starting quantity [16]. The mean of the triplicates was then calculated to give the mean T/S ratio, which represents the RTL value. Next, the standard deviation of the triplicates and the coefficient of variation (CV) were calculated to ensure the quality of data collected [16].

4.3.4 Statistical Analysis of RTL Methodology

Z-scores of patient and brother who were in their 30 s as well as father and mother who were in their 60 s were calculated using the formula $Z\text{-score} = (X - \mu)/\sigma$, where X = RTL of the patient; μ = the mean RTL of age-matched controls; and σ = the standard deviation (SD) of age-matched controls [18].

4.3.5 Next-Generation Sequencing (NGS)

As no suitable candidates were identified, the variants were re-filtered using subsequent sets of less stringent criteria. 2nd pass candidates were filtered for MAF < 0.05 instead of MAF < 0.01 before continuing with the previously described pipeline. 3rd pass candidates were filtered for MAF < 0.01 with non-coding and synonymous variants removed, however, these variants were prioritized for curation and pathogenicity prediction based on the genes' association with the patient's phenotype, as well as the results of computational predictive programs. The computational predictive programs used were SIFT, PolyPhen2-HDIV, PolyPhen2-HVAR, MutationTaster, mutation assessor, and FATHMM for exonic missense variants and NetGene2, BDGP, MutationTaster, varSEAK for intronic variants near a splice site. 4th pass candidates were filtered for non-synonymous variants occurring in AA genes with a MAF of < 0.01, but inclusive of non-coding variants, before curation and pathogenicity prediction (Fig. 4.1; Table 4.1).

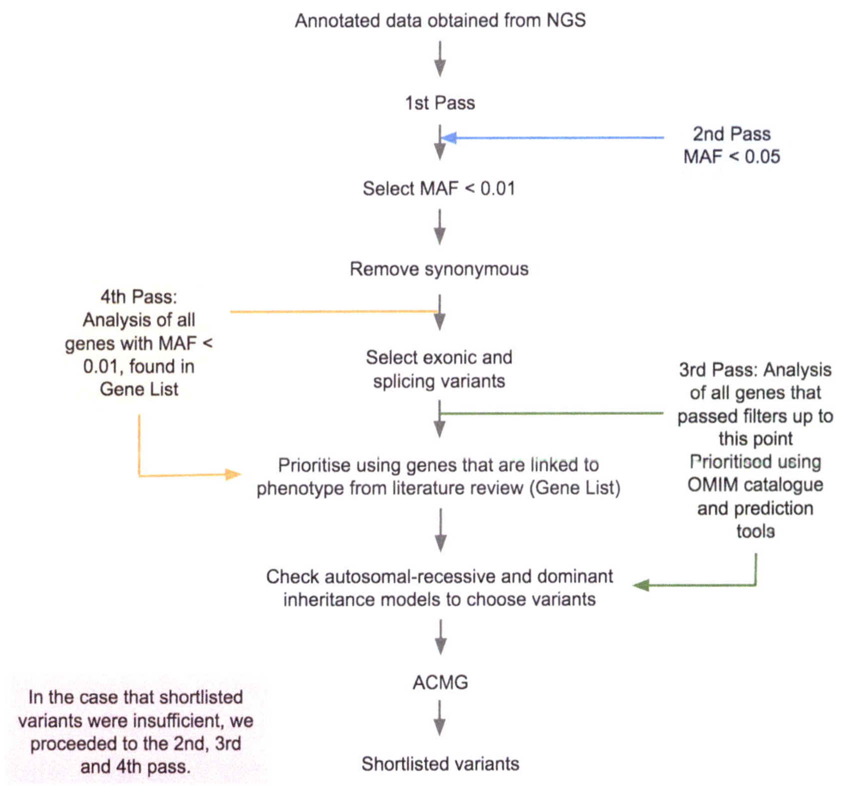

Fig. 4.1 NGS pipeline: Pathogenic variants were identified by passing annotated data obtained from NGS through the pipeline containing 4 passes

Table 4.1 RTL results showing raw RTL and Z-score calculated relative to age-matched normal controls

Sample	RTL	Z-score/SD
Patient	0.11	– 5.96
Patient's brother	0.13	– 5.56
Patient's mother	0.14	– 7.50
Patient's father	0.16	– 6.66

4.4 Results and Discussion

4.4.1 RTL Z-Score Analysis

The Z-score compared the telomere measurement in each individual with the age-matched mean and SD of the values in the normal controls, accounting for the known wide inter-individual telomere length variability. Telomere shortening and TBD were considered when the Z-score was below the 10th percentile of a normal distribution (– 1.28 SD) [19]. The patient was observed to have a RTL of 0.11 (– 5.96 SD), which is shorter than the 1st percentile (– 2.33 SD) and has been cited by literature to be 91% specific for DC [20]. However, the patient also presents with left SHL, diffuse cerebral, and cerebellar atrophy which have not been linked to DC, suggesting alternative underlying genetic mutations. Both the patient's mother and father also have RTLs shorter than the 1st percentile when compared to age-matched normal controls, suggesting the possibility of a TBD. However, only the patient and his brother presented symptoms while their parents are phenotypically normal. Telomere length decreases progressively with advancing age [21], but the patient and his brother who are in their 30 s have shorter RTL than their parents who are in their 60 s. Hence, both patient and brother might be suffering from a TBD that causes shorter RTL. The brother has milder symptoms and a longer RTL compared to the patient, indicating that the brother might have a milder form of TBD. The mother has a shorter RTL of 0.14 as compared to the father who has a RTL of 0.16. Thus, it is possible that the mother is an asymptomatic mutation carrier [7] for an X-linked TBD such as DC, while telomere shortening in the father may be due to other factors such as chronic stress, poor lifestyle choices [22], and paternal age at birth [23]. To identify potential genetic causes of the patient's phenotype, exome sequencing was conducted on the patient's genome.

4.4.2 NGS Analysis

Table 4.2 describes the 8 variants shortlisted from patient's WES data of which the *DKC1, ATXN3, ABCB4,* and *PTPRQ* variants will be discussed in detail. A homozygous *DKC1* variant NM_001363.3:c.771+3A>G was identified in the patient. This mutation has not been reported in any variant database and is absent in any

Table 4.2 8 shortlisted variants with ACMG curation

Gene	Disease and inheritance	Variant	Variant zygosity	ACMG	Prior reports/ClinVar	Deleterious predictions
DKC1	Dyskeratosis congenita, X-linked [25]	NM_001363.3:c 771+3A>G	Homo	VUS: PM2	Variant not reported	2/4
ATXN3	Machado-Joseph disease (AD) [29]	NM_004993:exon10:c.915_916insCAGCAGCAGCAGCAGCAGCAGCAGCAGCAGCAGC: p.G306fs	Compound het	VUS: PM2	Variant not reported	NA
ATXN3	Machado-Joseph disease (AD) [29]	NM_004993:exon10:c.915_916insCAGCAGCAGCAGCAGCAGCAGCAGCAGCAGCAG: p.G306delinsQQQQQQQQQQQG	Compound het	VUS: PM2	Variant not reported	NA
ABCB4	Cholestasis, progressive familial intrahepatic 3 (AR); gallbladder disease 1 (AD) [30]	NM_000443:exon13:c.A1529G:p.N510S	Het	VUS: PM2, PP3, PP5	Reported likely pathogenic; pathogenic; uncertain significance	6/7
PTPRQ	Deafness, autosomal recessive 84A (AR); deafness, autosomal dominant 73 (AD) [31]	NM_001145026:exon3:c.C206G:p.S69C	Het	VUS: PM2, PP3	Variant not reported	4/7

(continued)

Table 4.2 (continued)

Gene	Disease and inheritance	Variant	Variant zygosity	ACMG	Prior reports/ClinVar	Deleterious predictions
DIAPH3	Auditory neuropathy, autosomal dominant, 1 (AD) [32]	NM_001042517:exon16:c.C1787T:p.P596L	Het	VUS: PM2, PP3, BP6	Reported benign	4/7
TBP	{Parkinson disease, susceptibility to}; spinocerebellar ataxia 17 (AD) [33]	NM_003194:exon3:c.222_223insCAGCAGCAG:p.Q74delinsQQQQ	Het	VUS: PM2	Variant not reported	NA
PARP1	Fanconi anemia [34]	NM_001618:exon21:c.A2819G:p.K940R	Het	VUS: PM2, PP3	Variant not reported	5/7

population databases used. Since this mutation occurs 3 bases away from an intron-exon boundary, it is predicted to affect splicing by disrupting the highly conserved sequences that define the intron-exon boundary [24]. MutationTaster predicts that this variant causes a gain in splice donor site, with a score of 0.76. varSEAK predicts that this variant is deleterious with likely loss of splice site. BDGP and NetGene2 predict that this variant is not likely to cause a loss or gain of splice sites, with NetGene2 predicting that confidence of the donor splice site (direct strand) decreases from 0.83 to 0.71. However, BDGP and NetGene2 both show that this variant occurs close to the predicted splice site and might affect splicing through activating cryptic splice sites or affecting the binding of splicing enhancers and silencers [24]. Thus, this variant is likely to affect splicing and compromise the function of the *DKC1* gene product dyskerin, which is a component of the telomerase complex, possibly causing a defect in the maintenance of telomeres [25].

DKC1 is associated with X-linked dyskeratosis congenita (DC) [25], with 20–25% of DC cases being caused by a mutation in the *DKC1* gene [26]. DC is a TBD and has also been associated with liver cirrhosis and cerebellar ataxia [27]. Approximately 80–90% of patients with DC develop AA before age 30, with the severity of bone marrow failure varying [28]. While the patient does not display all the symptoms of DC such as the classical triad of nail dystrophy, skin hyperpigmentation, and oral leukoplakia [20], DC has a broad phenotypic spectrum [20], so patients may not display the full or classical phenotype [20]. Based on our ACMG curation, this variant is classified as variant of uncertain significance (VUS): PM2. However, considering that the variant occurs close to the predicted splice site, likely affecting splicing and protein function, and since the patient's phenotype corresponds with the symptoms of DC, including AA, thrombocytopenia, liver cirrhosis, and cerebellar ataxia, in addition to the short telomeres below the 1st percentile that is characteristic of DC patients [18], we consider this variant to be a strong causative candidate for our patient's condition.

Two novel compound heterozygous *ATXN3* variants NM_004993:exon10:c.915_916insCAGCAGCAGCAGCAGCAGCAGCAGCAGCAGCAGCAGCAGCA:p.G306fs and NM_004993:exon10:c.915_916insCAGCAGCAGCAGCAGCAGCAGCAGCAGCAGCAGCAG:p.G306delinsQQQQQQQQQQQG were also observed in the patient. Mutations in *ATXN3* have been associated with Machado-Joscph disease (MJD) [29, 35], also known as spinocerebellar ataxia, which includes symptoms of cerebellar atrophy. While the patient does not present the full range of symptoms of MJD including eye abnormalities and limb ataxia, MJD has wide clinical variability [29]. There is variability in age of onset and severity of disease, with age of onset varying from early teens to late-adulthood [36]. In this variant, there is an expansion of 26 CAG repeats and a CA insertion, resulting in a frameshift mutation. MJD is known to result from an expansion of a (CAG)*n* repeat in the *ATXN3* gene [37]. Additionally, a-1 frameshift with an expanded CAG repeat has been reported to have a deleterious effect on *Drosophila* and mammalian neurons since a GCA frame would be translated, producing polyalanine instead of polyglutamine [38].

The patient harbored a heterozygous *ABCB4* variant, NM_000443:exon13:c.A1529G:p.N510S. While homozygous mutations in *ABCB4*, also known as *MDR3*, have been associated with cholestasis, progressive familial intrahepatic 3 autosomal recessive [30], heterozygous mutations in *ABCB4* variants have been reported to result in less severe clinical patterns [39]. Mutations in the *ABCB4* gene present clinical symptoms including adult idiopathic cirrhosis [39] and drug-induced liver injury [40]. As such, this mutation in the *ABCB4* gene might be related to the fatty liver and liver cirrhosis with portal hypertension that ML1 presents with.

Our analysis also identified a heterozygous *PTPRQ* variant, NM_001145026:exon3:c.C206G:p.S69C, in the patient. PTPRQ has been associated with deafness, autosomal dominant 73 [31]. This variant occurs in the Fibronectin Type III (FN3) domain. PTPRQ has 17 FN3 domains which mediate binding to extracellular proteins [41]. Mutations in the FN3 domain can lead to malformations of shaft connectors and incomplete maturation of cochlear hair bundles, resulting in SHL [42]. Hence, mutations in this domain are likely to deleteriously impact function. Since mutations in the *PTPRQ* gene have been associated with the patient's phenotype of congenital left SHL, and the affected FN3 domain may result in SHL, this variant may be another potential contributing candidate for our patient's phenotype.

Although 8 variants were shortlisted for the patient, we were unable to classify any of them as clinically pathogenic under the ACMG guidelines. There are limitations of WES, where there may be damaging deep intronic pathogenic variants, copy number variations, or chromosomal relocations that occurred but were undetected. To address these limitations, whole-genome sequencing can be performed to identify variants in the entire genome. If no pathogenic variants are identified, further studies on potential underlying cause of the patient's disease can be made through investigating the phenotype and genotype of the patient's relatives using WES. This would enable us to analyze the cosegregation of variants with the disease in the family, aiding in the classification of pathogenicity of variants with uncertain significance. Furthermore, in vitro or in vivo functional studies can be carried out on shortlisted variants to help determine their pathogenicity.

Although the causal gene underlying a TBD could not be conclusively identified, the telomere measurements study points to the latter as being associated with the condition in the patient. It is noted that absolute telomere measurements were not performed due to the complexity of the experiments, nonetheless RTL assays are commonly accepted as reliable and have been used in studies to detect telomere disorders [43], with short RTL measurements being used as a criteria for diagnosis in DC patients [44]. As to whether the telomere shortening could be a consequence and not cause of the disease, the symptoms of the patient manifesting phenotypes suggestive of telomere disorder lends support that the latter is probably the cause of disease.

4.5 Conclusion

In conclusion, the patient is likely to be suffering from a TBD since telomere shortening is evident, with the patient having a relatively shorter RTL of 0.11 (− 5.96 SD), which is below the 1st percentile (− 2.33 SD) with respect to age-matched normal subjects. Through comprehensive analysis of the patient's genome, we have shortlisted 8 variants in genes (*DKC1, ATXN3, PTPRQ, ABCB4, DIAPH3, TBP, PARP1*) associated with our patient's phenotype as potential candidates. Particularly, we identified a mutation in the *DKC1* gene (NM_001363.3:c.771+3A>G) near the intron-exon boundary, which could likely affect splicing and dyskerin function, resulting in reduced telomerase function [25]. However, upon curation of our shortlisted variants, all the variants were found to be of uncertain significance due to lack of supporting evidence in literature. Hence, the genetic cause of disease in our patient could not be confirmed. Nevertheless, these shortlisted variants are promising potential candidates, and their pathogenicity can be confirmed through further studies such as in vitro or in vivo functional studies if they are also found in other similar patients elsewhere.

References

1. Schoettler, M. L., & Nathan, D. G. (2018). The pathophysiology of acquired aplastic anemia. *Hematology Oncology Clinics of North America, 32*(4), 581–594.
2. Moore, C. A., & Krishnan, K. (2020). Aplastic anemia (updated 2020 Nov 23). In StatPearls [Internet]. Treasure Island (FL): StatPearls Publishing.
3. Vaht, K., Göransson, M., Carlson, K., Isaksson, C., Lenhoff, S., Sandstedt, A., et al. (2017). Incidence and outcome of acquired aplastic anemia: Real-world data from patients diagnosed in Sweden from 2000 to 2011. *Haematologica, 102*(10), 1683–1690.
4. Alzahrani, N., Ashor, N., Fathi, T., Bukhari, D., & Zaher, G. (2018). Idiopathic severe aplastic anemia with a delayed response to immunosuppressive therapy: A case report. *Clinical Case Reports, 6*(6), 1029–1032.
5. Calado, R. T., Regal, J. A., Kleiner, D. E., Schrump, D. S., Peterson, N. R., & Pons, V. et al. (2009). A spectrum of severe familial liver disorders associate with telomerase mutations. *PLoS One, 4*(11), e7926.
6. Gitto, L., Stoppacher, R., Richardson, T. E., & Serinelli, S. (2020). DC. *Autopsy and Case Reports, 10*(3).
7. Savage, S., & Bertuch, A. (2010). The genetics and clinical manifestations of telomere biology disorders. *Genetics in Medicine, 12*, 753–764.
8. Ball, S. E., Gibson, F. M., Rizzo, S., Tooze, J. A., Marsh, J. C., & Gordon-Smith, E. C. (1998). Progressive telomere shortening in aplastic anemia. *Blood, 91*(10), 3582–3592.
9. Gramatges, M. M., & Bertuch, A. A. (2013). Short telomeres: From dyskeratosis congenita to sporadic aplastic anemia and malignancy. *Translational Research: The Journal of Laboratory and Clinical Medicine, 162*(6).
10. Acquired Aplastic Anemia (n.d.). NORD (National Organization for Rare Disorders).
11. Hartung, H. D., Olson, T. S., & Bessler, M. (2013). Acquired aplastic anemia in children. *Pediatric Clinics of North America, 60*(6), 1311–1336.
12. Moore, C. A., & Krishnan, K. (2020). Bone marrow failure (updated 2020 Jul 13). In StatPearls [Internet]. Treasure Island (FL): StatPearls Publishing.

13. Balogh, E. P., Miller, B. T., & Ball, J. R., Committee on Diagnostic Error in Health Care; Board on Health Care Services & The National Academies of Sciences (2015). *The diagnostic process*. National Academies Press (US). www.ncbi.nlm.nih.gov
14. Fernández García, M. S., & Teruya-Feldstein, J. (2014). The diagnosis and treatment of dyskeratosis congenita: A review. *Journal of Blood Medicine, 5*, 157–167.
15. Nelson, A., Myers, K. (2021). Shwachman-diamond syndrome. 2008 Jul 17 (updated 2018 Oct 18). In M. P. Adam, H. H. Ardinger, R. A. Pagon, et al. (Eds.), GeneReviews® [Internet]. Seattle (WA): University of Washington, Seattle (1993–2021)
16. Joglekar, M. V., Satoor, S. N., Wong, W. K. M., Cheng, F., Ma, R. C. W., & Hardikar, A. A. (2020). An optimised step-by-step protocol for measuring relative telomere length. *Methods and Protocols, 3*(2), 27.
17. Cawthon, R. M. (2002). Telomere measurement by quantitative PCR. *Nucleic Acids Research, 30*(10), e47.
18. Alter, B. P., Rosenberg, P. S., Giri, N., Baerlocher, G. M., Lansdorp, P. M., & Savage, S. A. (2012). Telomere length is associated with disease severity and declines with age in dyskeratosis congenita. *Haematologica, 97*(3), 353–359.
19. Arias-Salgado, E. G., Galvez, E., Planas-Cerezales, L., Pintado-Berninches, L., Vallespin, E., Martinez, P., et al. (2019). Genetic analyses of aplastic anemia and idiopathic pulmonary fibrosis patients with short telomeres, possible implication of DNA-repair genes. *Orphanet Journal of Rare Diseases, 14*.
20. Savage, S. A. (2019, November 21). Dyskeratosis congenita. Seattle: Nih.Gov; University of Washington.
21. Rizvi, S., Raza, S. T., & Mahdi, F. (2015). Telomere length variations in aging and age-related diseases. *Current Aging Science, 7*(3), 161–167.
22. Shammas, M. A. (2011). Telomeres, lifestyle, cancer, and aging. *Current Opinion in Clinical Nutrition and Metabolic Care, 14*(1), 28–34.
23. Starkweather, A. R., Alhaeeri, A., Montpetit, A., Brumelle, J., Filler, K., Montpetit, M., Mohanraj, L., Lyon, D. E., & Jackson-Cook, C. K. (2014). An integrative review of factors associated with telomere length and implications for biobehavioral research. *Nursing Research, 63*(1), 36–50.
24. Anna, A., & Monika, G. (2018). Splicing mutations in human genetic disorders: Examples, detection, and confirmation. *Journal of Applied Genetics, 59*(3), 253–268.
25. OMIM Entry—*300126—DYSKERIN; DKC1 (n.d.).
26. Chalkoo, A. H., Kaul, V., & Wani, L. A. (2014). Zinsser-Cole-Engmann syndrome: A rare case report with literature review. *Journal of Clinical and Experimental Dentistry, 6*(3), e303–e306.
27. OMIM Entry—# 305000—DYSKERATOSIS CONGENITA, X-LINKED; DKCX (n.d.).
28. National Organization for Rare Disorders (2020). Dyskeratosis congenita.
29. OMIM Entry—*607047—ATAXIN 3; ATXN3 (n.d.).
30. OMIM Entry—*171060—ATP-BINDING CASSETTE, SUBFAMILY B, MEMBER 4; ABCB4 (n.d.).
31. OMIM Entry—*603317—PROTEIN-TYROSINE PHOSPHATASE, RECEPTOR-TYPE, Q; PTPRQ (n.d.).
32. OMIM Entry—*614567—DIAPHANOUS-RELATED FORMIN 3; DIAPH3 (n.d.).
33. OMIM Entry—*600075—TATA BOX-BINDING PROTEIN; TBP (n.d.).
34. Ramirez, M. H., Adelfalk, C., Kontou, M., Hirsch-Kauffmann, M., & Schweiger, M. (2003). The cellular control enzyme PolyADP Ribosyl Transferase is eliminated in cultured fanconi anemia fibroblasts at confluency. *Biological Chemistry, 384*(1), 169–174.
35. ATXN3 [Internet], Bethesda: National Library of Medicine (US), National Center for Biotechnology Information (2004)
36. Bettencourt, C., & Lima, M. (2011). Machado-Joseph disease: From first descriptions to new perspectives. *Orphanet Journal of Rare Diseases, 6*(1), 35.
37. Kawaguchi, Y., Okamoto, T., Taniwaki, M., Aizawa, M., Inoue, M., Katayama, S., Kawakami, H., et al. (1994). CAG expansions in a novel gene for Machado-Joseph disease at chromosome 14q32.1. *Nature Genetics, 8*(3), 221–228.

38. Stochmanski, S. J., Therrien, M., Laganière, J., Rochefort, D., Laurent, S., Karemera, L., et al. (2012). Expanded ATXN3 frameshifting events are toxic in Drosophila and mammalian neuron models. *Human Molecular Genetics, 21*(10), 2211–2218.
39. Stättermayer, A. F., Halilbasic, E., Wrba, F., Ferenci, P., & Trauner, M. (2020). Variants in ABCB4 across the spectrum of cholestatic liver diseases in adults. *Journal of Hepatology*.
40. Davit-Spraul, A., Gonzales, E., Baussan, C., & Jacquemin, E. (2010). The spectrum of liver diseases related to ABCB4 gene mutations: Pathophysiology and clinical aspects. *Seminars in Liver Disease, 30*(02), 134–146.
41. Vollrath, M. A., Kwan, K. Y., & Corey, D. P. (2007). The micromachinery of mechanotransduction in hair cells. *Annual Review of Neuroscience, 30*(1), 339–365.
42. Wu, X., Wang, S., Chen, S., Wen, Y., Liu, B., Xie, W., et al. (2018). Autosomal recessive congenital sensorineural hearing loss due to a novel compound heterozygous PTPRQ mutation in a Chinese family. *Neural Plasticity, 2018*, 1–6.
43. Montpetit, A. J., Alhareeri, A. A., Montpetit, M., Starkweather, A. R., Elmore, L. W., Filler, K., et al. (2014). Telomere length: a review of methods for measurement. *Nursing Research, 63*(4), 289–299.
44. Ratnasamy, V., Navaneethakrishnan, S., Sirisena, N., et al. (2018). Dyskeratosis congenita with a novel genetic variant in the DKC1 gene: A case report. *BMC Medical Genetics, 19*, 85 (2018).

Chapter 5
Topology and Geometry of 3-Band Models

Ching Hua Lee and Chien Hao Tan

Abstract Berry curvature is a property of N-band models which plays an analogous role of the magnetic field. The Majorana stellar representation (MSR) is a method of decomposing N-band states into multiple 2-band states, which paves way for a more intuitive geometric understanding of N-band models. We utilise the MSR to obtain a new formula for the Berry curvature of 3-band models in terms of individual contributions from each star and cross terms involving both stars, which could be insightful for investigating Berry curvature uniformity and topological behaviour of stars. We applied the MSR method to a model with uniform Berry curvature and investigated the cancellation of the divergences among three out of four of the terms to yield an overall non-divergent Berry curvature. In summary, the MSR approach aids the discovery of materials with uniform Berry curvature and is a powerful tool in the study of fractional Chern insulators (FCI).

5.1 Introduction

The study of topology in physics emerged in the late 1980s and has been an active field of study in recent literature. Historically, the concept of topological order was required to explain the chiral spin state in high-temperature superconductivity after it was discovered that Landau's spontaneous symmetry breaking was unable to do so. And more recently, the 2016 Nobel Prize has been awarded to David J. Thouless, J. Michael Kosterlitz and F. Duncan M. Haldane for "theoretical discoveries of topological phase transitions and topological phases of matter" [1].

Researchers study topological physics for their robustness to environmental noise and impurities. For instance, topological insulators, which we study in this paper, have potential applications for quantum computing. Imagine a thin film of electrons

C. H. Lee
Department of Physics, National University of Singapore, Singapore 117542, Singapore

C. H. Tan (✉)
1 Raffles Institution Ln,Bishan 575954, Singapore
e-mail: tanchienhao@gmail.com

H. Guo et al. (eds.), *IRC-SET 2021*, https://doi.org/10.1007/978-981-16-9869-9_5

Fig. 5.1 When a thin film of electrons is exposed to an external magnetic field, electrons in the middle will be in bound orbits, resulting in a current (red). While electrons at the edges will scatter off the edge while orbiting, resulting in an edge current (blue) that is robust to material impurities

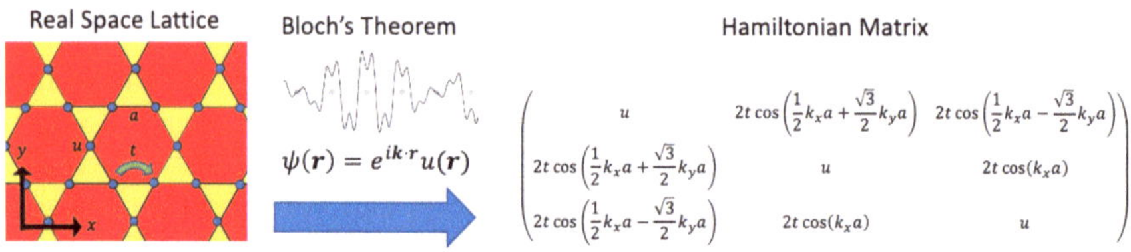

Fig. 5.2 An illustration of the Kagome lattice being converted into an 3-band model. As the lattice possesses translational symmetry, Bloch's theorem can be used to transform the real space tight-binding model into momentum space. The resultant 3-by-3 matrix is then diagonalised to obtain eigensolutions

in a strong magnetic field (refer to Fig. 5.1). The electrons in the middle (bulk) will be confined to circular orbits, but the electrons at the side will scatter off the edge and cause a current to flow around the edge of the material. The material's conductivity at the edge is not affected by small changes in the material, like impurities or imperfections and is therefore said to be protected by topology. In fact, this topologically protected edge conductivity is so robust that it is used to measure the fine structure constant.

5.1.1 Research Objectives

This paper investigates topological behaviour of 3-band models with the use of Majorana stellar representation (MSR). An N-band model is a theoretical model of condensed matter physics systems. Examples of condensed matter physics systems include graphene, superconductors, semiconductors, and generally solids with some crystalline or lattice structure. The value of N depends on the number of orbitals and sublattice sites one wishes to account for. Condensed matter systems, which often involve transition metals, typically involve many orbitals. Hence, the greater the value of N, the higher the degree of accuracy of the model, although this comes at the cost of increased complexity. These condensed matter systems can effectively be simplified to an N by N (hermitian) matrix by transforming into momentum space using the method of Bloch waves (refer to Fig. 5.2) [2]. The properties of the system (i.e. Berry curvature) are then investigated by calculating the eigenvectors (which correspond to the quantum states) and eigenvalues (which correspond to the (eigen-)energy of the states) of the square matrix. Unfortunately, each N-band eigenstate is a complex N-dimensional vector which is difficult to visualise. The Majorana stellar representation (MSR) approach promises to solve this problem.

In this paper, the goal of the MSR approach is to understand the flatness of the Berry curvature of 3-band models. Berry curvature plays the analagous role of a magnetic field in topological materials. A more uniform Berry curvature leads to better behaviour, and therefore, models with a flat Berry curvature are sought after. For 2-band models, there is a lower-bound for the uniformity due to the fact that there is no map from $\mathbb{T}^2$ to $\mathbb{S}^2$ that has a perfectly uniform Jacobian [3]. For higher-band models, a perfectly uniform Berry curvature is permitted. However, perfectly uniform models are unphysical as they require infinite-distance-hopping terms. Hence, the trimming of hopping terms that are too far would result in a trade-off between Berry curvature uniformity and experimental feasibility. An example of a relatively uniform 3-band model has been found in [3], and the Hamiltonian can be found in Appendix 1. We seek to understand the flatness of the Berry curvature of this model with the MSR approach, in hope of discovering other uniform models.

As a small but important digression, in most literature, "flat models" refer to models with a flat eigenenergy spectrum. A flat energy spectrum also has desirable properties, but it is very different from a flat Berry curvature. To avoid confusion, this paper refers to models with flat Berry curvature as uniform models from hereon. A flat energy spectrum is mathematically easier to achieve than a flat Berry curvature, as one can always add/subtract the identity matrix to the Hamiltonian to translate the energy. On the other hand, a flat Berry curvature is much harder to achieve mathematically, and in some cases, even impossible (such as in 2-band models). Additionally, to further achieve flat energy or flat curvature models in an experimental setup is an even greater feat, as one is constrained by physical materials in addition to mathematical difficulty. In this paper, we only focus on the mathematical aspects of uniform models.

It is also worth mentioning that although the techniques developed in this paper is most applicable to the study of topological insulators, the MSR approach could be applied to other areas such as topolectrical circuits [4] and high-spin systems (Fig. 5.3).

5.1.2 Berry Phase, Berry Curvature, and Chern Number

To further elaborate on what is meant by "topological behaviour", we will discuss the concept of Berry curvature in this paragraph. The concept of Berry phase

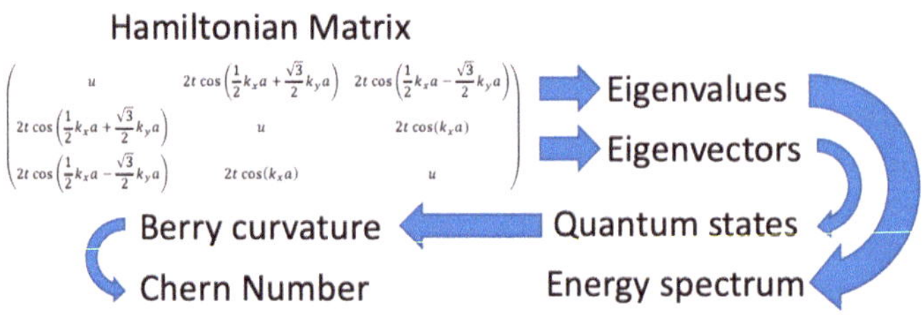

Fig. 5.3 The process of calculating the "topological behaviour" of a model

was developed in 1984 [5]. When a quantum eigenstate $|\Psi\rangle$ undergoes adiabatic cyclic evolution in parameter space (example in the proceeding paragraph), in addition to the dynamical phase $\exp(-\frac{i}{\hbar}\int_0^t \mathrm{d}t' E(t'))$, it acquires a geometrical phase $\exp(-\oint_{\mathcal{C}} \langle\Psi|\,|\mathrm{d}\Psi\rangle)$, called the Berry phase, which can be experimentally observed [5]. This also paves for the concept of a Berry connection and Berry curvature. The Berry connection is defined as $\mathcal{A} = -i\,\langle\Psi|\,|\mathrm{d}\Psi\rangle$ and the Berry curvature $\mathcal{F} = d\mathcal{A} = -i\,\langle\mathrm{d}\Psi| \wedge |\mathrm{d}\Psi\rangle$. Application of Stokes' theorem yields an expression of Berry phase in terms of a surface integral of Berry curvature, $\gamma(\mathcal{C}) = -\oint_{\mathcal{S}} \mathcal{F}$, where $\mathcal{C} = \partial\mathcal{S}$. The Berry curvature, which can have several physical consequences [6–8], is the focus of this paper.

5.1.2.1 Spin-1/2 Particle in Magnetic Field

We provide an example of an electron in an external magnetic field to aid in explaining Berry phase, connection, and curvature. An electron is a spin-1/2 particle, and its quantum behaviour is governed by the Hamiltonian $H \propto \boldsymbol{\sigma} \cdot \mathbf{B}$ and the Schrodinger equation $H\,|\Psi\rangle = i\hbar\frac{\partial}{\partial t}\,|\Psi\rangle$. This Hamiltonian is the potential energy of the electron in a magnetic field, and the quantum state of the electron is given by the eigenvectors to the Hamiltonian. We choose to present only the positive energy eigenstate for this paper as the method is easily applied to other eigenstates. Since the Hamiltonian is dependent on the direction of the magnetic field $\mathbf{B} = (B_x, B_y, B_z) \in \mathbb{R}^3$, the eigenvectors also depend on the magnetic field. The magnetic field is the parameter space for this example. A varying magnetic field would cause the eigenstate to vary, and we ascribe an eigenstate to each corresponding point in parameter space. Calculating the eigenstate for spin-1/2 particles is achieved by decomposing the 2×2 hermitian matrix H into a combination of pauli matrices.

$$H \propto B_x\sigma_1 + B_y\sigma_2 + B_z\sigma_3 \tag{5.1}$$

$$\sigma_1 = \begin{pmatrix} 0 & 1 \\ 1 & 0 \end{pmatrix} \quad \sigma_2 = \begin{pmatrix} 0 & -i \\ i & 0 \end{pmatrix} \quad \sigma_3 = \begin{pmatrix} 1 & 0 \\ 0 & -1 \end{pmatrix} \tag{5.2}$$

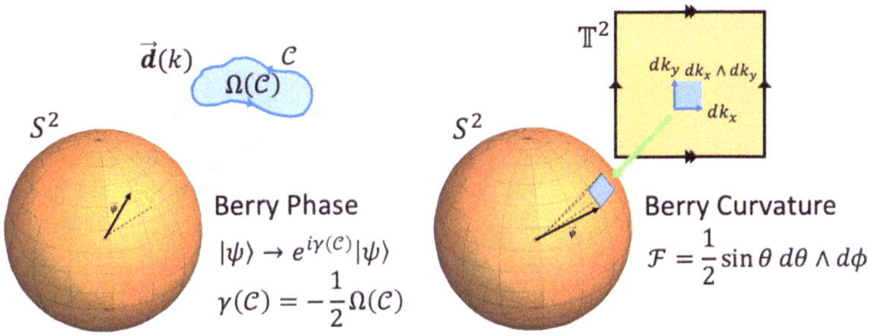

Fig. 5.4 Visualisations of the Berry phase and curvature of 2-band models. The Berry phase is acquired after a cyclic evolution in parameter space, and it can be thought of as (minus half) the solid angle of the path. The Berry curvature can be regarded as the Berry phase of a infinitesimal loop, and is hence (half) the differential solid angle (in spherical coordinates)

$$|\Psi(\mathbf{B})\rangle = \begin{pmatrix} B_z + |\mathbf{B}| \\ B_x + iB_y \end{pmatrix} \tag{5.3}$$

After an adiabatic evolution along a cyclic path $\mathcal{C}$ in parameter space, the eigenstate returns to its original value, but it picks up a phase factor $e^{i\alpha}$. This phase factor consists of two parts, the dynamical term $\exp(-\frac{i}{\hbar}\int_0^t \mathrm{d}t' E(\mathbf{B}(t')))$ and the geometrical term $\exp(i\gamma(\mathcal{C})) = \exp(-\oint_{\mathcal{C}} \langle\Psi|\,|\nabla_B\Psi\rangle \cdot dB)$. The geometrical term is only a function of the cyclic path traced out by the eigenstate and does not involve the time taken for cyclic evolution.

Moreover, we consider $\mathbf{B} = B_0 \begin{pmatrix} \sin\theta\cos\phi \\ \sin\theta\sin\phi \\ \cos\theta \end{pmatrix}$ in polar coordinates, then the eigenstate can be written as $|\Psi(\mathbf{B}(\theta,\phi))\rangle = \begin{pmatrix} \cos\theta/2 \\ e^{i\phi}\sin\theta/2 \end{pmatrix}$. The eigenstate $|\Psi\rangle$ is represented on the Bloch sphere as $\boldsymbol{\Psi} = \begin{pmatrix} \sin\theta\cos\phi \\ \sin\theta\sin\phi \\ \cos\theta \end{pmatrix} \in S^2$. As a mathematical interlude, the Bloch sphere representation owes its existence to the fact that the space of quantum states $\mathbb{C}^2/U(1) \sim \mathbb{CP}^1 \sim S^2$ is isomorphic to the 2-sphere topologically. The Bloch sphere representation is elegant because the Berry curvature of 2-band models can be interpreted as the differential solid angle on the Bloch sphere $\mathcal{F} = \mathrm{Im}\,\langle \mathrm{d}\Psi| \wedge |\mathrm{d}\Psi\rangle = \frac{1}{2}\sin\theta\,\mathrm{d}\theta \wedge \mathrm{d}\phi$. Therefore, the Berry phase of a cyclic evolution of the magnetic field is interpreted as the solid angle traced out by the cyclic path $\gamma(\mathcal{C}) = -\frac{1}{2}\Omega(\mathcal{C})$ (refer to Fig. 5.4).

5.1.2.2 *N*-Band Models

For condensed matter physics models, the parameter space is $(k_x, k_y) \in \mathbb{T}_2$ because (crystal) momentum is 2π-periodic. The Berry curvature is $\mathcal{F} = \mathrm{Im}\,\langle \mathrm{d}\Psi| \wedge |\mathrm{d}\Psi\rangle =$

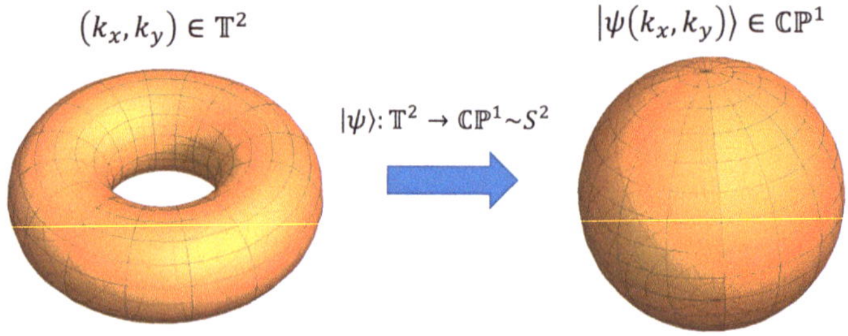

Fig. 5.5 The 2-band eigenstate is a map from the 2-torus to the 2-sphere. Maps from $\mathbb{T}^2$ to S^2 can be characterised by a winding number called the Chern number

$\mathrm{Im}(\langle\partial_x\Psi|\,|\partial_y\Psi\rangle - \langle\partial_y\Psi|\,|\partial_x\Psi\rangle)\mathrm{d}k_x \wedge \mathrm{d}k_y = F_{xy}\mathrm{d}k_x \wedge \mathrm{d}k_y$. And the Chern number of N-band models is defined as $C = 1/2\pi \int_{BZ} F_{xy} d^2k$. N-band condensed matter physics models are analogous to spin-$((N-1)/2)$ systems as the Hamiltonians of both systems are of the same size. 2-band models, which describe physical systems like polyacetylene (Su-Schrieffer-Heeger) [9], are analogous to spin-1/2 systems. One simply replaces the magnetic field $\mathbf{B}$ with the analogous vector $\mathbf{d}(k)$. Therefore, the calculations in the previous subsection on spin-1/2 particles carry over to 2-band models. The Chern number of 2-band models is thus quantised in integer values, since the total solid angle swept out by the eigenstate on the Bloch sphere is a multiple of 4π. As we have seen, the Bloch sphere representation of 2-band (or spin-1/2) states offers insights into its Berry curvature (refer to Fig. 5.5). The generalisation of the Bloch sphere representation for higher-band models is the goal of this paper. Specifically, we work on 3-band models where eigenstates live in $\mathbb{CP}^2 \sim (S^2 \times S^2)/\mathcal{S}_2$.

5.2 The Majorana Stellar Representation for N-Band Models

The Bloch sphere representation for the 2-band state is not immediately generalisable to higher-band models. However, the Majorana stellar representation (MSR) has potential to pave a geometrical intuition for higher-band models. The MSR has already been used to understand the Berry phase of higher-spin systems [10, 11], but the application of the MSR to Berry curvature has not been studied in-depth, and is therefore the goal of this paper. The MSR allows one to decompose an N-band state into $(N-1)$ entangled 2-band states (refer to Fig. 5.6). The Berry curvature, which we will derive later, will comprise of solid angles from the individual stars, as well as correlation terms. The contribution that only involve single stars have a simple geometrical interpretation, and therefore, the MSR has hope of allowing us to explore the geometry of higher-band models in a more intuitive manner.
From [12], an N-band state $|\Psi^{(N)}\rangle = \sum_{m=1}^{N} \Psi_m |m\rangle$ can be represented as $N-1$

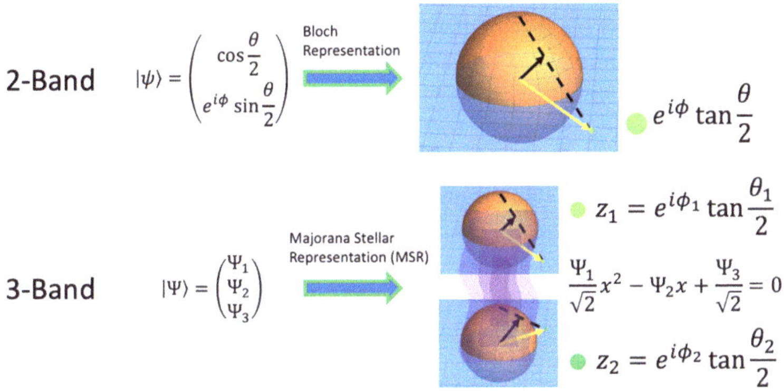

Fig. 5.6 2-band states can be projected onto the Bloch sphere. By MSR, 3-band states can be represented on 2 Bloch spheres as 2 entangled Majorana stars. The stars are given by roots to (5.4) followed by a stereographic projection from the complex plane to 2-sphere

stars on the Bloch sphere, $|u_i\rangle = \begin{pmatrix} \cos\theta_i/2 \\ e^{i\phi_i}\sin\theta_i/2 \end{pmatrix}$, with θ_i, ϕ_i given by the roots $x_i = e^{i\phi_i}\tan\frac{\theta_i}{2}$ to the polynomial $\sum_{k=1}^{N}\frac{(-1)^{k+1}\Psi_k}{\sqrt{(N-k)!(k-1)!}}x^{N-k} = 0$. Kindly refer to Appendix 2 for a more in-depth explanation.

For 3-band models, the 2 stars can be calculated by finding the roots to

$$\frac{\Psi_1}{\sqrt{2}}x^2 - \Psi_2 x + \frac{\Psi_3}{\sqrt{2}} = 0 \tag{5.4}$$

where Ψ_i are the components to the 3-component eigenstate $|\Psi\rangle$. The roots

$$x_{1,2} = \frac{\sqrt{2}\Psi_2 \pm \sqrt{2\Psi_2^2 - 4\Psi_1\Psi_3}}{2\Psi_1} \tag{5.5}$$

can be used to find the angles of the stars $|A\rangle$, $|B\rangle$ when represented on the Bloch sphere, $x_i = \tan\frac{\theta_i}{2}e^{i\phi_i}$, via stereographic projection from the complex plane to the 2-sphere (refer to Fig. 5.6). It is also important to note that due to the branch cuts of the square root function, star $|A\rangle$ and star $|B\rangle$ interchange along paths in the Brillouin zone. Kindly refer to Fig. 5.13 for a visual illustration. This exchange is later reflected in the calculation of Chern number for the individual stars.

5.3 Analysis of Uniform 3-Band Model with MSR

In this section, we investigate the uniform model obtained in [3]. The Hamiltonian can be found in Appendix 1.

An eigenstate of a 3-band model $|\Psi\rangle$ can be decomposed into 2 Majorana Stars $|A\rangle$, $|B\rangle$. The Berry curvature of 3-band models can be decomposed into 4 terms. The derivation is found in Appendix 3. The terms are grouped in the following way so as to allow for the most illuminating geometric interpretation. T_1 and T_2 have a simple geometric interpretation. T_1 and T_2 are half the differential solid angle subtended by stars A and B, respectively. However, a simple geometric interpretation for T_3 and T_4 has not been found. When we calculate the Chern number for T_1 and T_2 for the uniform model, we obtained 1.5 for both. If each star came from its own 2-band model, this Chern number would be an integer. However, the fact that it is not an integer is due to the mixing and entanglement of the 2 stars as mentioned earlier (refer to Fig. 5.13) in Sect. 5.2.

$$\begin{aligned}\mathrm{Im}\,\langle \mathrm{d}\Psi| \wedge |\mathrm{d}\Psi\rangle &= T_1 + T_2 + T_3 + T_4 \\ T_1 &= \mathrm{Im}\,\langle \mathrm{d}A| \wedge |\mathrm{d}A\rangle \\ T_2 &= \mathrm{Im}\,\langle \mathrm{d}B| \wedge |\mathrm{d}B\rangle \\ T_3 &= \frac{1}{4}\frac{d(A\times B)\wedge d(A-B)}{N_2^2} \\ T_4 &= \frac{1}{8}\frac{(A\times B)\cdot d(A-B)\wedge d(A\cdot B)}{N_2^4}\end{aligned} \tag{5.6}$$

We proceed to plot the T_1, T_2, T_3, T_4 terms separately as shown in Figs. 5.7, 5.8, 5.9 and 5.10. Figures 5.11 and 5.12 verify the correctness of the formula.

5.3.1 *Cancellation of Divergences in the Terms*

The total Berry curvature is relatively flat as we are calculating the Berry curvature for the uniform model. However, the individual terms T_1, T_2, T_3 are not flat as there are divergences (points where the value goes go infinity). It is observed that there are four points of divergence for T_1, T_2, T_3 each. These four divergences are symmetric about $k_x = \pi$ and $k_y = \pi$, and the location is numerically calculated to be around $k_x = k_y = 0.4219322\pi$, where $f(k_x, k_y) := 2\Psi_2^2 - 4\Psi_1\Psi_3 = 0$. Since the total Berry curvature does not have any divergences, the first 3 terms' (T_1, T_2, T_3) divergences must cancel to result in a non-divergent Berry curvature. We investigate this cancellation in the next paragraph.

To investigate the divergences near $k_x = k_y = 0.42\pi$ (which we call k_0 from hereon for convenience) for T_1, T_2, T_3, we proceed to plot $\Psi_1(k_x, k_y)$, $\Psi_2(k_x, k_y)$, $f(k_x, k_y)$ from $k_x = k_y = 0.41\pi$ to $k_x = k_y = 0.43\pi$. From the graphs (Figs. 5.14a, 5.15 and 5.16b), we observe that Ψ_1 and Ψ_2 are relatively smooth, whereas $f(k_x, k_y)$ can be approximated as $f \propto (k_x - k_0) + i(k_y - k_0)$.
Plugging this approximation for f into T_1, T_2 and T_3 explains the existence as well as the cancellation of divergences near $k_0 = 0.42\pi$. We shall begin this process.

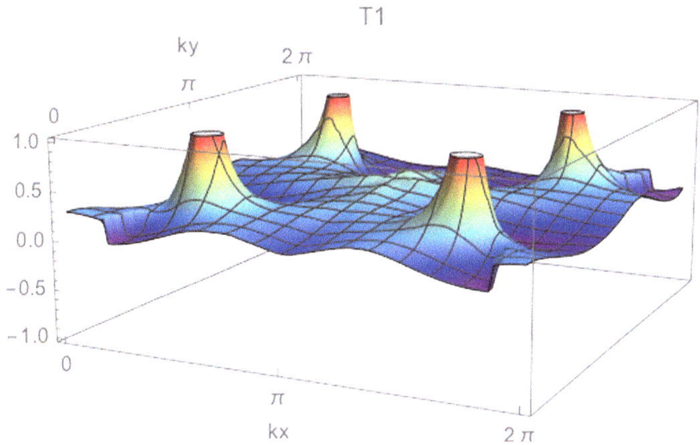

Fig. 5.7 Plot of T_1

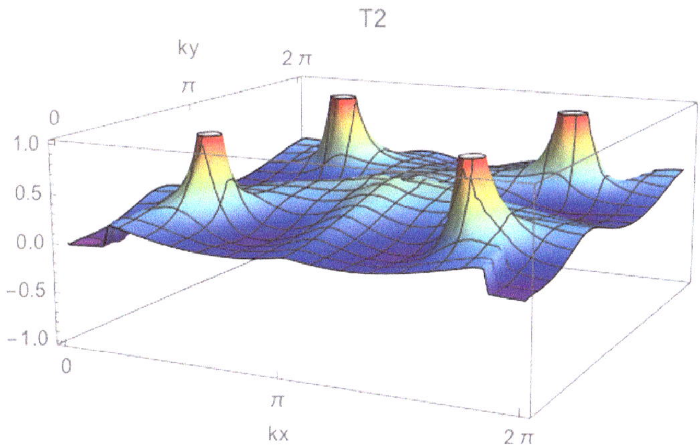

Fig. 5.8 Plot of T_2

Firstly, by stereographic projection of the Bloch sphere onto the complex plane, we obtain T_1, T_2 in terms of $\alpha_1 = \text{Re}(z_1), \alpha_2 = \text{Im}(z_1), \beta_1 = \text{Re}(z_2), \beta_2 = \text{Im}(z_2)$ [refer to Eqs. (5.3.1) and (5.3.1)]. Then, we look at the T_3 term, decomposing Eq. (5.9) into 6 pairwise ($\alpha_1, \alpha_2, \beta_1, \beta_2$, 4 choose 2) wedge products (refer to Eq. (5.44) in Appendix 4.

$$
\begin{aligned}
T_1 &= \text{Im}\,\langle \text{d}A| \wedge |\text{d}A\rangle \\
&= \frac{2}{(1+\alpha_1^2+\alpha_2^2)^2}\text{d}\alpha_1 \wedge \text{d}\alpha_2
\end{aligned}
\tag{5.7}
$$

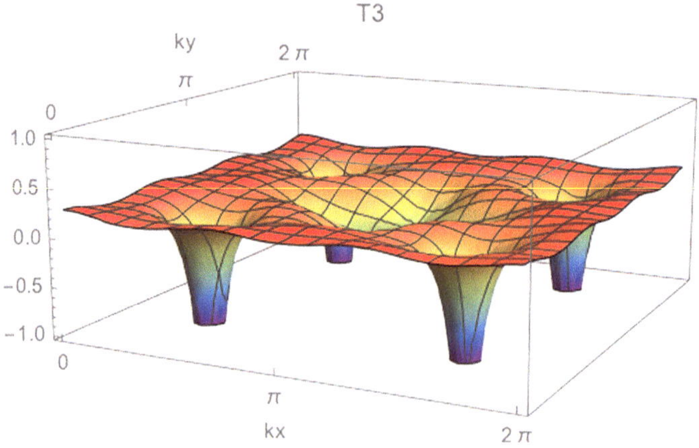

Fig. 5.9 Plot of T_3

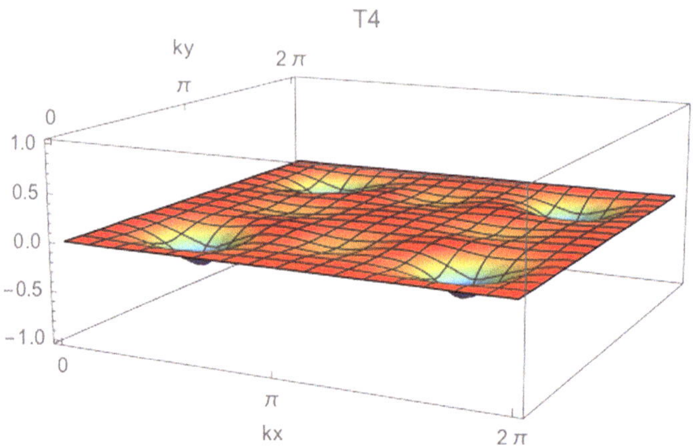

Fig. 5.10 Plot of T_4

$$T_2 = \operatorname{Im} \langle \mathrm{d}B | \wedge | \mathrm{d}B \rangle = \frac{2}{(1+\beta_1^2+\beta_2^2)^2} \mathrm{d}\beta_1 \wedge \mathrm{d}\beta_2 \tag{5.8}$$

$$T_3 = \frac{1}{4N_2^2} d(A \times B) \wedge d(A - B) \tag{5.9}$$

$$\text{where } N_2^2 = \frac{3}{2} + \frac{1}{2} A \cdot B \tag{5.10}$$

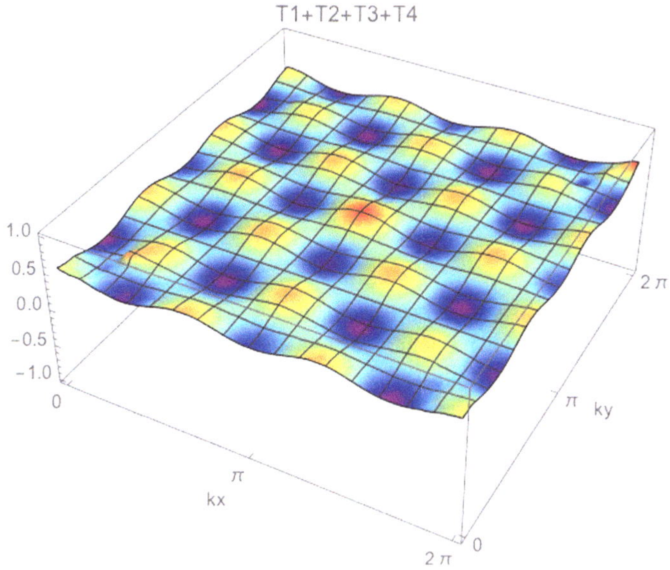

Fig. 5.11 Plot of $T_1 + T_2 + T_3 + T_4$

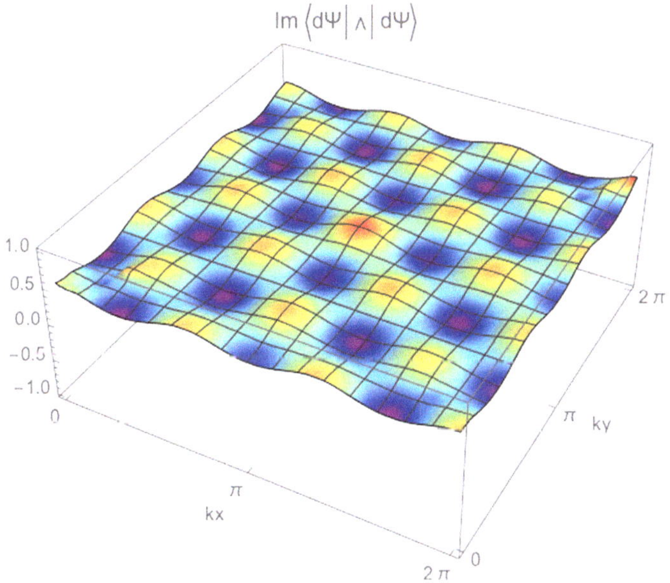

Fig. 5.12 Plot of Im $\langle \mathrm{d}\Psi | \wedge | \mathrm{d}\Psi \rangle$

Table 5.1 Decomposition of T_1, T_2, T_3 into wedge products. $\Gamma := 1 + \alpha_1^2 + \alpha_2^2$

Term/ coefficient of	$d\alpha_1 \wedge d\alpha_2$	$d\alpha_1 \wedge d\beta_1$	$d\alpha_1 \wedge d\beta_2$	$d\alpha_2 \wedge d\beta_1$	$d\alpha_2 \wedge d\beta_2$	$d\beta_1 \wedge d\beta_2$
T_1	$\frac{2}{\Gamma^2}$	0	0	0	0	0
T_2	0	0	0	0	0	$\frac{2}{\Gamma^2}$
T_3	$-\frac{1}{\Gamma^2}$	0	$\frac{1}{\Gamma^2}$	$-\frac{1}{\Gamma^2}$	0	$-\frac{1}{\Gamma^2}$

Evaluating the partial derivatives found in Eq. (5.44) is a straightforward but tedious process, which can be achieved with the help of symbolic Mathematica. Moreover, because we are focussed on $k_0 = 0.42\pi$ where $f = 0$ and therefore $z_1 = z_2$ ($\alpha_1 = \beta_1, \alpha_2 = \beta_2$), the massive expressions turn out to simplify greatly. We present the simplification of $\partial_{\alpha_1}(A \times B)\partial_{\alpha_2}(A - B) - \partial_{\alpha_2}(A \times B)\partial_{\alpha_1}(A - B)$ in detail [refer to Eqs. (5.45)–(5.52)] of Appendix 5 but provide only the result for the other 5 terms (refer to Table 5.1). We look at the divergent terms of $d\alpha_1 \wedge d\alpha_2, d\alpha_1 \wedge d\beta_2, d\alpha_2 \wedge d\beta_2, d\beta_1 \wedge d\beta_2$ (because they have non-zero coefficient in T_1, T_2, T_3). In order to evaluate the wedge products like $d\alpha_1 \wedge d\alpha_2$, we make use of the approximation for f near k_0.

$$f(k_x, k_y) \approx 1.85e^{-i0.7\pi}((k_x - k_0) + i(k_y - k_0)) \tag{5.11}$$

Evaluating partial derivatives and neglecting non-divergent terms (Appendix 6), we observe there are two types of terms responsible for divergence at k_0: (a) The $1/\sqrt{(k_x - k_0)^2 + (k_y - k_0)^2}$ term and (b) the $1/\sqrt{(k_x - k_0) - i(k_y - k_0)}$ term. Both need to have coefficients (from T_1, T_2, T_3) that cancel in order for the Berry curvature to be non-divergent.

$$D_1 := \frac{1.85}{16\Psi_1^2} \frac{1}{\sqrt{(k_x - k_0)^2 + (k_y - k_0)^2}} \tag{5.12}$$

$$D_2 := \mathrm{Re}\left[\frac{\sqrt{1.85e^{+i0.7\pi}}}{\sqrt{2}\Psi_1}\partial_z\left(\frac{\Psi_2}{\Psi_1}\right)\frac{1}{\sqrt{(k_x - k_0) - i(k_y - k_0)}}\right] \tag{5.13}$$

$$D_3 := \mathrm{Re}\left[\frac{\sqrt{1.85e^{+i0.7\pi}}}{\sqrt{2}\Psi_1}\partial_{\bar{z}}\left(\frac{\bar{\Psi}_2}{\Psi_1}\right)\frac{1}{\sqrt{(k_x - k_0) - i(k_y - k_0)}}\right] \tag{5.14}$$

From Table 5.2, the cancellation of divergent terms is observed. This concludes our discussion.

Table 5.2 Coefficients of divergent terms D_1, D_2, D_3 in T_1, T_2, T_3. We observe that all coefficients cancel when T_1, T_2, T_3 is added up. $\Gamma := 1 + \alpha_1^2 + \alpha_2^2$

		D_1	D_2	D_3
T_1	$d\alpha_1 \wedge d\alpha_2$	$\frac{2}{\Gamma^2}$	$\frac{2}{\Gamma^2}$	
T_2	$d\beta_1 \wedge d\beta_2$	$\frac{2}{\Gamma^2}$	$-\frac{2}{\Gamma^2}$	
T_3	$d\alpha_1 \wedge d\alpha_2$	$-\frac{1}{\Gamma^2}$	$-\frac{1}{\Gamma^2}$	
	$d\alpha_1 \wedge d\beta_2$	$-\frac{1}{\Gamma^2}$		$-\frac{1}{\Gamma^2}$
	$d\alpha_2 \wedge d\beta_1$	$-\frac{1}{\Gamma^2}$		$\frac{1}{\Gamma^2}$
	$d\beta_1 \wedge d\beta_2$	$-\frac{1}{\Gamma^2}$	$\frac{1}{\Gamma^2}$	
$T_1 + T_2 + T_3$		0	0	0

5.4 Conclusion

In this paper, we discussed concept of Berry curvature and explored a model with relatively uniform Berry curvature from [3] using the Majorana stellar representation (MSR). We observed and explained analytically the cancellations of divergences in the Berry curvatures T_1–T_3. Overall, the key result of this paper is the development of the MSR method (Eq. 5.6) and the cancellation of divergences when the MSR method is applied to the uniform model in [3].

5.5 Further Work

As the MSR method for Berry curvature has not been explored extensively, there is still plenty of room for further research. We are currently generalising the divergences analysis to any model and investigating the switching of the stars. The discriminant function $f(k_x, k_y)$ is also promising, and exceptional points (points where $f = 0$) hint at the possibility of Riemann surfaces and topological characterisation.

Appendix 1: Hamiltonian of Uniform 3-Band Model

$$H = \begin{pmatrix} \frac{51}{76} + \frac{2\cos k_x}{15} + \frac{2\cos k_y}{15} - \frac{2\cos k_x \cos k_y}{45} & -\frac{3\sin k_x}{38} + \frac{4\cos k_y \sin k_x}{31} + \left(\frac{3\sin k_y}{38} - \frac{4\cos k_x \sin k_y}{31}\right) i & -\frac{3\cos k_x}{25} + \frac{3\cos k_y}{25} + \frac{\sin k_x \sin k_y}{7} i \\ -\frac{3\sin k_x}{38} + \frac{4\cos k_y \sin k_x}{31} + \left(-\frac{3\sin k_y}{38} + \frac{4\cos k_x \sin k_y}{31}\right) i & \frac{63}{95} - \frac{4\cos k_x}{41} - \frac{4\cos k_y}{41} - \frac{2\cos k_x \cos k_y}{17} & -\frac{5\sin k_x}{33} - \frac{\cos k_y \sin k_x}{15} + \left(\frac{5\sin k_y}{33} + \frac{\cos k_x \sin k_y}{15}\right) i \\ -\frac{3\cos k_x}{25} + \frac{3\cos k_y}{25} - \frac{\sin k_x \sin k_y}{7} i & -\frac{5\sin k_x}{33} - \frac{\cos k_y \sin k_x}{15} + \left(-\frac{5\sin k_y}{33} - \frac{\cos k_x \sin k_y}{15}\right) i & \frac{2}{3} - \frac{\cos k_x}{28} - \frac{\cos k_y}{28} + \frac{4\cos k_x \cos k_y}{25} \end{pmatrix}$$

Appendix 2: Derivation of the Majorana Stellar Representation (MSR)

2.1 Schwinger Boson Representation

We begin by first discussing the Schwinger Boson Representation for spin-J quantum systems [13]. A spin-J Hilbert space is characterised by a basis and spin operators that act on the basis states. Namely, the $(2J + 1)$-dimensional space has basis states given by $|J, m\rangle$, $m \in \{-J, -J + 1, ..., J - 1, J\}$. Spin operators S^+, S^-, S^2, S_z act on these basis states in the following way.

$$S^+ |J, m\rangle = \sqrt{J(J+1) - m(m+1)}\, |J, m+1\rangle \tag{5.15}$$

$$S^- |J, m\rangle = \sqrt{J(J+1) - m(m-1)}\, |J, m-1\rangle \tag{5.16}$$

$$S^2 |J, m\rangle = J(J+1)\, |J, m\rangle \tag{5.17}$$

$$S_z |J, m\rangle = m\, |J, m\rangle \tag{5.18}$$

The spin operators also obey the $\mathfrak{su}(2)$ Lie algebra.

$$S_x := \frac{1}{2}(S^+ + S^-) \tag{5.19}$$

$$S_y := \frac{1}{2i}(S^+ - S^-) \tag{5.20}$$

$$[S_i, S_j] = i\epsilon_{ijk} S_k \tag{5.21}$$

It turns out, we can define two bosonic modes and use a clever definition of the spin operators to achieve the same commutation relations and basis states. If we define $[a, a^\dagger] = [b, b^\dagger] = 1$ (all other commutation relations vanish), and

$$S^+ = a^\dagger b \tag{5.22}$$

$$S^- = b^\dagger a \tag{5.23}$$

$$S_z = \frac{1}{2}(a^\dagger a - b^\dagger b) \tag{5.24}$$

then we can obtain the commutation relations

$$[S^+, S^-] = 2S_z \tag{5.25}$$

$$[S_z, S^+] = +S^+ \tag{5.26}$$

$$[S_z, S^-] = -S^- \tag{5.27}$$

which is exactly Eq. (5.21) after some manipulation.

Additionally, if we define the basis states as

$$|J, m\rangle = \frac{(a^\dagger)^{J+m}}{\sqrt{(J+m)!}} \frac{(b^\dagger)^{J-m}}{\sqrt{(J-m)!}} |\Omega\rangle \tag{5.28}$$

then one can check that it satisfies the following Eqs. (5.15)–(5.18) even with the new definitions (5.22)–(5.24). Equation (5.28) is the Schwinger Boson Representation for spin-J states in terms of 2 bosonic modes.

2.2. Majorana Stellar Representation

The Majorana Stellar Representation is simply a factorisation after converting a spin-J system to its Schwinger Boson Representation [12]. Let a spin-J quantum state be written in terms of the basis states $|J, m\rangle$ using Schwinger bosons.

$$|\Psi\rangle = \sum_{m=-J}^{J} C_m |J, m\rangle \tag{5.29}$$

$$= \sum_{m=-J}^{J} \frac{C_m (a^\dagger)^{J+m} (b^\dagger)^{J-m}}{\sqrt{(J+m)!(J-m)!}} |\Omega\rangle \tag{5.30}$$

We may factorise (5.30) in the following way (5.31). The $2J$ complex numbers z_i completely characterise the spin-J state. Moreover, when we stereographically project the complex numbers z_i onto the 2-sphere, we obtain $2J$ points in the Bloch sphere, which we call the Majorana Stars.

$$|\Psi\rangle = \frac{C_J}{\sqrt{(2J)!}} \prod_{i=1}^{2J} (a^\dagger + z_i b^\dagger) \, |\Omega\rangle \tag{5.31}$$

$$\begin{aligned} &= \frac{C_J}{\sqrt{(2J)!}} \left[(a^\dagger)^{2J} + (a^\dagger)^{2J-1} b^\dagger \left(\sum_{i=1}^{2J} z_i \right) \right. \\ &\left. + (a^\dagger)^{2J-2} (b^\dagger)^2 \left(\sum_{i<j} z_i z_j \right) + \ldots + (b^\dagger)^{2J} \left(\prod_i z_i \right) \right] \end{aligned} \tag{5.32}$$

Comparing coefficients of Eqs. (5.30) and (5.32).

$$m = J - 1 \colon \frac{C_{J-1}}{\sqrt{(2J-1)!}} = \frac{C_J}{\sqrt{(2J)!}} \sum_i z_i \tag{5.33}$$

$$m = J - 2 \colon \frac{C_{J-2}}{\sqrt{(2J-2)!2!}} = \frac{C_J}{\sqrt{(2J)!}} \sum_{i<j} z_i z_j \tag{5.34}$$

$$m = J - 3 \colon \frac{C_{J-3}}{\sqrt{(2J-3)!3!}} = \frac{C_J}{\sqrt{(2J)!}} \sum_{i<j<k} z_i z_j z_k \tag{5.35}$$

$$\ldots \tag{5.36}$$

$$m = -J \colon \frac{C_{-J}}{\sqrt{(2J)!}} = \frac{C_J}{\sqrt{(2J)!}} \prod_i z_i \tag{5.37}$$

Now, in order to find the values of z_i, $1 \leq i \leq 2J$ that satisfy Eqs. (5.33)–(5.37), we consider the polynomial equation with z_i as roots,

$$0 = \prod_{i=1}^{2J} (x - z_i) \tag{5.38}$$

$$\begin{aligned} &= x^{2J} - x^{2J-1} \left(\sum_i z_i \right) + x^{2J-2} \left(\sum_{i<j} z_i z_j \right) \\ &- x^{2J-3} \left(\sum_{i<j<k} z_i z_j z_k \right) + \cdots + (-1)^{2J} \left(\prod_i z_i \right) \end{aligned} \tag{5.39}$$

and substituting (5.33) to (5.37) into (5.39) yields

$$\sum_{k=0}^{2J} \frac{(-1)^k C_{J-k} x^{2J-k}}{\sqrt{(2J-k)!k!}} = 0 \tag{5.40}$$

Appendix 3: Derivation of Berry Curvature in Terms of Majorana Stars ($N = 3$)

We denote the stars for $N = 3$ models by $|A\rangle$ and $|B\rangle$. Starting from the expression from [10].

$$\mathrm{Im}\,\langle\Psi|\,|\mathrm{d}\Psi\rangle = \mathrm{Im}\,\langle A|\,|\mathrm{d}A\rangle + \mathrm{Im}\,\langle B|\,|\mathrm{d}B\rangle + \frac{1}{4}\frac{(A\times B)\cdot d(A-B)}{N_2^2}$$

where $N_2^2 = 1 + \langle A|\,|B\rangle\,\langle B|\,|A\rangle = \frac{3}{2} + \frac{1}{2}(A\cdot B)$. To find $\mathcal{F} = \mathrm{Im}\ \langle \mathrm{d}\Psi| \wedge |\mathrm{d}\Psi\rangle$, we take the exterior derivative of the Berry phase.

$$\mathcal{F} = \mathrm{Im}\ \langle \mathrm{d}\Psi| \wedge |\mathrm{d}\Psi\rangle \tag{5.41}$$

$$\mathcal{F} = \mathrm{Im}\,\mathrm{d}\,\langle\Psi|\,|\mathrm{d}\Psi\rangle \tag{5.42}$$

$$\begin{aligned}\mathcal{F} = {} & \mathrm{Im}\ \langle \mathrm{d}A| \wedge |\mathrm{d}A\rangle + \mathrm{Im}\ \langle \mathrm{d}B| \wedge |\mathrm{d}B\rangle \\ & + \frac{1}{4}\frac{d(A\times B)\wedge d(A-B)}{N_2^2} \\ & + \frac{1}{8}\frac{(A\times B)\cdot d(A-B)\wedge d(A\cdot B)}{N_2^4}\end{aligned} \tag{5.43}$$

Appendix 4: Decomposition of T_3 into Wedge Products

$$\begin{aligned}T_3 = {} & \frac{1}{4N_2^2} d(A\times B)\wedge d(A-B) \\ = {} & \frac{1}{4N_2^2}[\partial_{\alpha_1}(A\times B)\partial_{\alpha_2}(A-B) - \partial_{\alpha_2}(A\times B)\partial_{\alpha_1}(A-B)]\mathrm{d}\alpha_1\wedge\mathrm{d}\alpha_2 \\ & + \frac{1}{4N_2^2}[\partial_{\alpha_1}(A\times B)\partial_{\beta_1}(A-B) - \partial_{\beta_1}(A\times B)\partial_{\alpha_1}(A-B)]\mathrm{d}\alpha_1\wedge\mathrm{d}\beta_1 \\ & + \frac{1}{4N_2^2}[\partial_{\alpha_1}(A\times B)\partial_{\beta_2}(A-B) - \partial_{\beta_2}(A\times B)\partial_{\alpha_1}(A-B)]\mathrm{d}\alpha_1\wedge\mathrm{d}\beta_2 \\ & + \frac{1}{4N_2^2}[\partial_{\alpha_2}(A\times B)\partial_{\beta_1}(A-B) - \partial_{\beta_1}(A\times B)\partial_{\alpha_2}(A-B)]\mathrm{d}\alpha_2\wedge\mathrm{d}\beta_1 \\ & + \frac{1}{4N_2^2}[\partial_{\alpha_2}(A\times B)\partial_{\beta_2}(A-B) - \partial_{\beta_2}(A\times B)\partial_{\alpha_2}(A-B)]\mathrm{d}\alpha_2\wedge\mathrm{d}\beta_2 \\ & + \frac{1}{4N_2^2}[\partial_{\beta_1}(A\times B)\partial_{\beta_2}(A-B) - \partial_{\beta_2}(A\times B)\partial_{\beta_1}(A-B)]\mathrm{d}\beta_1\wedge\mathrm{d}\beta_2\end{aligned} \tag{5.44}$$

Appendix 5: Simplification of Coefficient of $\mathrm{d}\alpha_1 \wedge \mathrm{d}\alpha_2$ at (k_0, k_0)

$$\partial_{\alpha_1}(A \times B)\partial_{\alpha_2}(A - B) - \partial_{\alpha_2}(A \times B)\partial_{\alpha_1}(A - B)$$
$$= -\frac{8(1 + 4\alpha_1\beta_1 - \beta_1^2 + 4\alpha_2\beta_2 - \beta_2^2 + \alpha_1^2(-1 + \beta_1^2 + \beta_2^2) + \alpha_2^2(-1 + \beta_1^2 + \beta_2^2))}{((1 + \alpha_1^2 + \alpha_2^2)^3(1 + \beta_1^2 + \beta_2^2))} \quad (5.45)$$

$$\text{At } k_x = k_y = k_0, \quad (5.46)$$

$$f(k_0, k_0) = 0 \rightarrow \quad z_1 = z_2(\alpha_1 = \beta_1, \alpha_2 = \beta_2), \quad (5.47)$$

$$f = 0 \rightarrow A = B \rightarrow A \cdot B = 1 \rightarrow \quad N_2^2(k_0, k_0) = 3/2 + 1/2 = 2 \quad (5.48)$$

Substituting (47) into (45), it simplifies to (5.49)

$$\partial_{\alpha_1}(A \times B)\partial_{\alpha_2}(A - B) - \partial_{\alpha_2}(A \times B)\partial_{\alpha_1}(A - B) = -\frac{8}{(1 + \alpha_1^2 + \alpha_2^2)^2} \quad (5.50)$$

$$\text{Coefficient of } \mathrm{d}\alpha_1 \wedge \mathrm{d}\alpha_2 = \frac{1}{4N_2^2}$$
$$[\partial_{\alpha_1}(A \times B)\partial_{\alpha_2}(A - B) - \partial_{\alpha_2}(A \times B)\partial_{\alpha_1}(A - B)] \quad (5.51)$$

$$= -\frac{1}{(1 + \alpha_1^2 + \alpha_2^2)^2} \quad (5.52)$$

Appendix 6: Evaluation of Wedge Products

The 2 stars are given by complex roots (Eqs. 5.53 and 5.54). With the approximation of f near k_0 (Eq. 5.11), we can obtain partial derivatives $\partial_x = \frac{\partial}{\partial k_x}$, $\partial_y = \frac{\partial}{\partial k_y}$ of $z_1, z_2, \bar{z}_1, \bar{z}_2$. Then, by Eqs. (5.55)–(5.58), we obtain partial derivatives ∂_x, ∂_y of $\alpha_1 = \mathrm{Re}(z_1), \alpha_2 = \mathrm{Im}(z_1), \beta_1 = \mathrm{Re}(z_2), \beta_2 = \mathrm{Im}(z_2)$.

$$z_1 = \frac{\sqrt{2}\Psi_2 + \sqrt{f}}{2\Psi_1} \tag{5.53}$$

$$z_2 = \frac{\sqrt{2}\Psi_2 - \sqrt{f}}{2\Psi_1} \tag{5.54}$$

$$\alpha_1 = \frac{1}{2}(z_1 + \bar{z}_1) \tag{5.55}$$

$$\alpha_2 = \frac{1}{2i}(z_1 - \bar{z}_1) \tag{5.56}$$

$$\beta_1 = \frac{1}{2}(z_2 + \bar{z}_2) \tag{5.57}$$

$$\beta_2 = \frac{1}{2i}(z_2 - \bar{z}_2) \tag{5.58}$$

Defining $\partial_z = \frac{1}{2}(\partial_x - i\partial_y)$ and $\partial_{\bar{z}} = \frac{1}{2}(\partial_x + i\partial_y)$, listed below are the wedge products which have non-zero coefficient (refer to Table 5.1) in T_1, T_2, T_3.

$$\begin{aligned}\partial_x\alpha_1\partial_y\alpha_2 - \partial_y\alpha_1\partial_x\alpha_2 \approx{}& \frac{1.85}{16\Psi_1^2}\frac{1}{\sqrt{(k_x-k_0)^2+(k_y-k_0)^2}} \\ &+ \mathrm{Re}\left[\frac{\sqrt{1.85e^{+i0.7\pi}}}{\sqrt{2}\Psi_1}\partial_z\left(\frac{\Psi_2}{\Psi_1}\right)\frac{1}{\sqrt{(k_x-k_0)-i(k_y-k_0)}}\right]\end{aligned} \tag{5.59}$$

$$\begin{aligned}\partial_x\alpha_1\partial_y\beta_2 - \partial_y\alpha_1\partial_x\beta_2 \approx{}& -\frac{1.85}{16\Psi_1^2}\frac{1}{\sqrt{(k_x-k_0)^2+(k_y-k_0)^2}} \\ &- \mathrm{Re}\left[\frac{\sqrt{1.85e^{+i0.7\pi}}}{\sqrt{2}\Psi_1}\partial_z\left(\frac{\bar{\Psi}}{2\Psi_1}\right)\frac{1}{\sqrt{(k_x-k_0)-i(k_y-k_0)}}\right]\end{aligned} \tag{5.60}$$

$$\begin{aligned}\partial_x\alpha_2\partial_y\beta_1 - \partial_y\alpha_2\partial_x\beta_1 \approx{}& \frac{1.85}{16\Psi_1^2}\frac{1}{\sqrt{(k_x-k_0)^2+(k_y-k_0)^2}} \\ &- \mathrm{Re}\left[\frac{\sqrt{1.85e^{+i0.7\pi}}}{\sqrt{2}\Psi_1}\partial_z\left(\frac{\bar{\Psi}_2}{\Psi_1}\right)\frac{1}{\sqrt{(k_x-k_0)-i(k_y-k_0)}}\right]\end{aligned} \tag{5.61}$$

$$\begin{aligned}\partial_x\beta_1\partial_y\beta_2 - \partial_y\beta_1\partial_x\beta_2 \approx{}& \frac{1.85}{16\Psi_1^2}\frac{1}{\sqrt{(k_x-k_0)^2+(k_y-k_0)^2}} \\ &- \mathrm{Re}\left[\frac{\sqrt{1.85e^{+i0.7\pi}}}{\sqrt{2}\Psi_1}\partial_z\left(\frac{\Psi_2}{\Psi_1}\right)\frac{1}{\sqrt{(k_x-k_0)-i(k_y-k_0)}}\right]\end{aligned} \tag{5.62}$$

Appendix 7: Additional Figures

See Figs. 5.13, 5.14, 5.15, and 5.16.

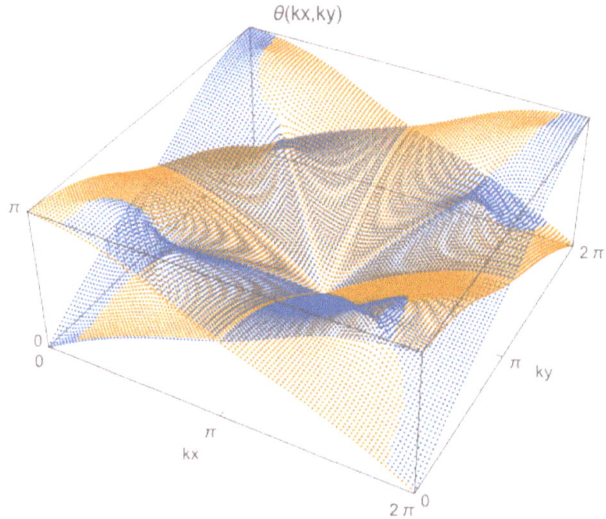

(a) Plot of $\theta(k_x, k_y)$

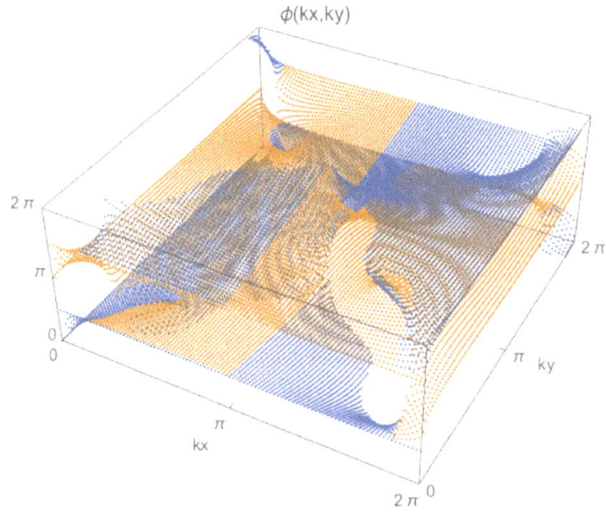

(b) Plot of $\phi(k_x, k_y)$

Fig. 5.13 Plot of θ and ϕ for both stars. The stars interchange positions as one traverses in the parameter space $\mathbb{T}^2$, and thus are said to be entangled

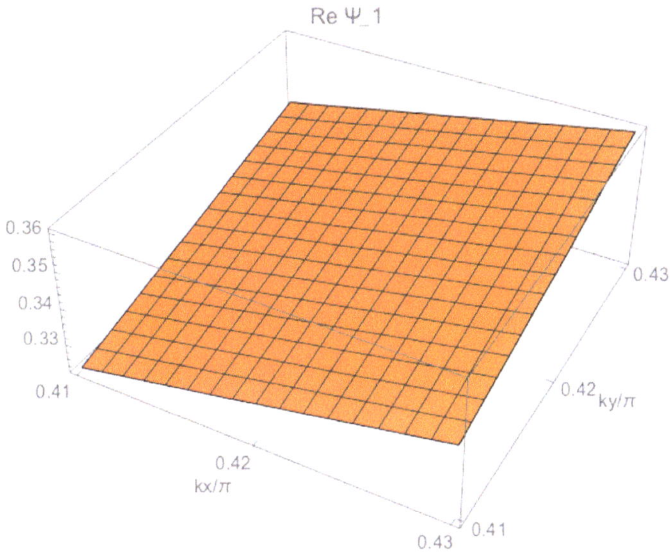

$\mathrm{Re}(\Psi_1)$ near k_0

Fig. 5.14 Plot of Ψ_1 near k_0

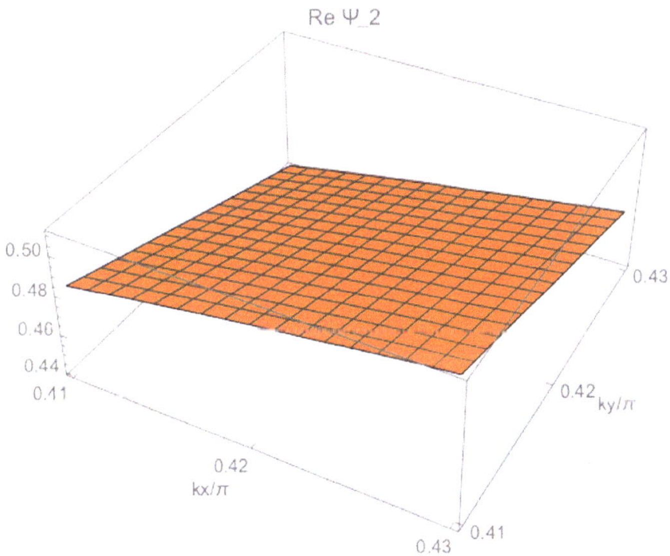

(a) $\mathrm{Re}(\Psi_2)$ near k_0

Fig. 5.15 Plot of Ψ_2 near k_0

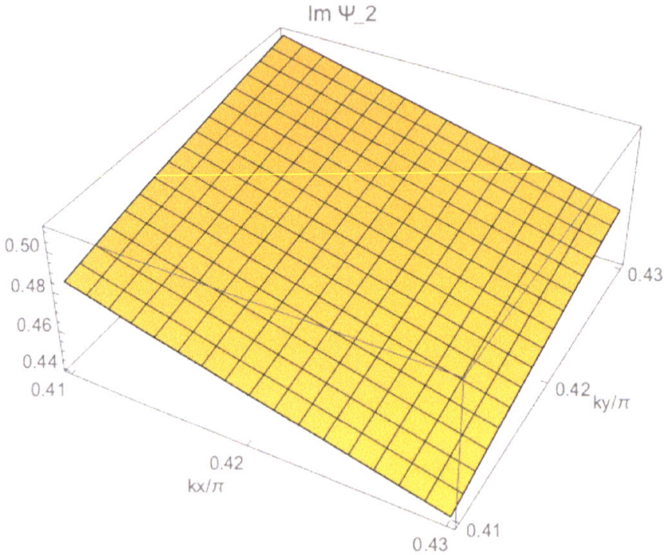

(b) $\mathrm{Im}(\Psi_2)$ near k_0

Fig. 5.15 (continued)

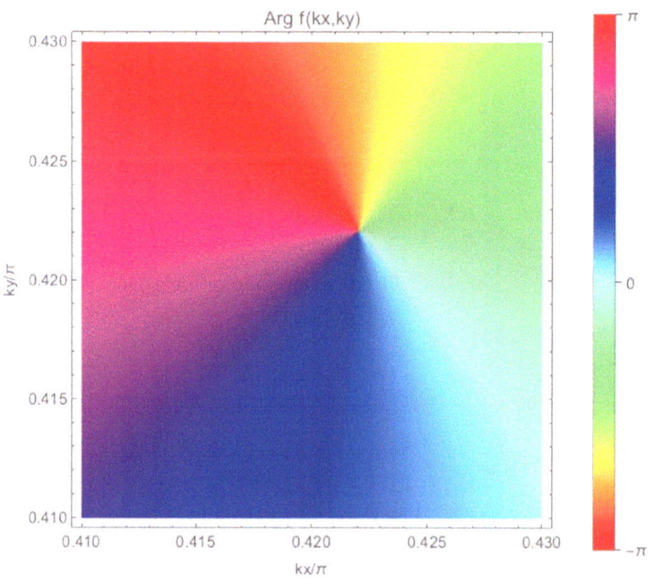

(a) $\mathrm{Arg}(f)$ near k_0

Fig. 5.16 Plot of f near k_0

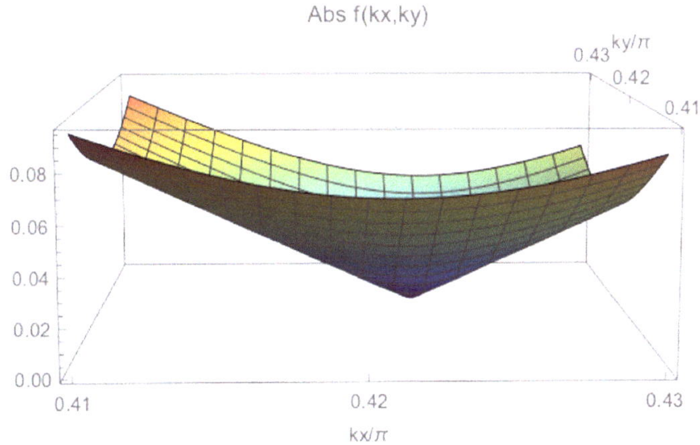

(b) Abs(f) near k_0

Fig. 5.16 (continued)

References

1. Topological Phase Transitions and Topological Phases of Matter (n.d.). Retrieved January 10, 2021, from https://www.nobelprize.org/uploads/2018/06/advanced-physicsprize2016-1.pdf
2. Kittel, C. (1996). *Introduction to solid state physics*. New York: John Wiley & Sons.
3. Lee, C. H., Claassen, M., & Thomale, R. (2017). Band structure engineering of ideal fractional Chern insulators. *Physical Review B, 96*(16). https://doi.org/10.1103/physrevb.96.165150
4. Lee, C. H., Imhof, S., Berger, C., Bayer, F., Brehm, J., Molenkamp, L. W., & Thomale, R. (2018). Topolectrical Circuits. *Communications Physics, 1*(1). https://doi.org/10.1038/s42005-018-0035-2
5. Changes, Quantal Phase, & Adiabatic, Factors Accompanying. (1984). Proceedings of the Royal Society of London. *A. Mathematical and Physical Sciences, 392*(1802), 45–57. https://doi.org/10.1098/rspa.1984.0023
6. Wimmer, M., Price, H. M., Carusotto, I., & Peschel, U. (2017). Experimental measurement of the Berry curvature from anomalous transport. *Nature Physics, 13*(6), 545–550. https://doi.org/10.1038/nphys4050
7. Virk, K. S., & Sipe, J. E. (2012). Anomalous THz emission from quantum wells with optically injected Berry curvature. In *Conference on lasers and electro-optics*. https://doi.org/10.1364/qels.2012.qtu1h.7
8. Haldane, F. D. (2004). Berry curvature on the fermi surface: Anomalous hall effect as a topological fermi-liquid property. *Physical Review Letters, 93*(20). https://doi.org/10.1103/physrevlett.93.206602
9. Asboth, J. K., Oroszlany, L., & Palyi, A. (2016). A short course on topological insulators. *Lecture Notes in Physics*. https://doi.org/10.1007/978-3-319-25607-8
10. Liu, H., & Fu, L. (2014). Representation of berry phase by the trajectories of Majorana stars. *Physical Review Letters, 113*(24). https://doi.org/10.1103/physrevlett.113.240403
11. Bruno, P. (2012). Quantum geometric phase in Majorana's stellar representation: Mapping onto a many-body Aharonov-Bohm phase. *Physical Review Letters, 108*(24). https://doi.org/10.1103/physrevlett.108.240402
12. Majorana, E. (1932). Atomi orientati in campo magnetico variabile. *Il Nuovo Cimento, 9*(2), 43–50. https://doi.org/10.1007/bf02960953
13. Schwinger, J. (1965). On angular momentum. In L. Biedenharn & H. Van Dam (Eds.), *Quantum theory of angular momentum* (p. 229). New York: Academic Press.

Chapter 6
Cross-Lingual Information Retrieval Random Shuffling Methodology

Alysa Lee Mynn

Abstract In this project, we evaluate the effectiveness of random shuffling in the cross-lingual information retrieval (CLIR) process. We extended the monolingual Word2Vec model to a multilingual one via the random shuffling process. We then evaluate the cross-lingual word embeddings (CLEs) in terms of retrieving parallel sentences, whereby the query sentence is in a source language and the parallel sentence is in some targeted language. Our experiments on three language pairs showed that models trained on a randomly shuffled dataset outperforms randomly initialized word embeddings substantially, despite its simplicity. We also explored smart shuffling, a more sophisticated CLIR technique which makes use of word alignment and bilingual dictionaries to guide the shuffling process. Due to the complexity of the implementation and unavailability of open source codes, we defer experimental comparisons to future work.

Keywords Cross-lingual information retrieval (CLIR) · Cross-lingual word embeddings · Random shuffling · Smart shuffling

6.1 Introduction

In monolingual information retrieval, the queries and answers for retrieval are in the same language. However, in cross-lingual information retrieval (CLIR), queries and answers are in different languages. This increases the difficulty of retrieving answers that are actually of interest to the user. For example, the user may enter a query in English, while the system's goal is to return a ranked list of documents in French that the user is interested in. Existing CLIR approaches include document translation, query translation, as well as the mapping of a both document and query to a third language or medium for comparison. Of the three, query translation is generally considered the most suitable due to its simplicity and effectiveness [1]. However, the main problem is dealing with translation ambiguity, which becomes

A. L. Mynn (✉)
Raffles Girls' School (Secondary) R.G.S., Singapore, Singapore
e-mail: Alysa.lee05@rafflesgirlssch.edu.sg

H. Guo et al. (eds.), *IRC-SET 2021*, https://doi.org/10.1007/978-981-16-9869-9_6

more pronounced when query sentences are shorter. With limited context, translations to related terms in the document's language would be difficult and inaccurate. In comparison, document translation is computationally more expensive and harder to scale as the entire document has to be translated to query language. However, its allure resides in its increased probability of correct translation to a synonymous query word, especially amongst more common query words. More recently, Vulic and Moens [2] introduced the concept of the cross-lingual word embeddings (CLEs) approach, which converts randomly shuffled parallel sentences into word embeddings for comparison. This method proved much more accurate than a model trained with a simpler baseline (translation of query before matching) for several reasons, one of which is due to its ability to efficiently convert parallel texts into dense vectors and map their proximity. The smart shuffling method introduced by Hamed, Sheikh and Allen in July 2020 [3] takes the CLE approach one step further with the help of a dictionary in the reordering process when shuffling the parallel sentences. In our exploration, we implemented our simplified interpretation of their algorithm and exemplified how it shuffles grouped words with similar definitions closer to each other. In traditional information retrieval, the queries and documents for retrieval are in the same language. However, in cross-lingual information retrieval (CLIR), queries and documents are in different languages. This increases the difficulty of retrieving documents that are actually of interest to the user. Generally, random shuffling is used in the CLIR process. Hence, we want to explore the effectiveness of this method.

6.2 Framework

6.2.1 Word2Vec Model and Its Parameters

For this experiment, we used Word2Vec, a popular method used to construct word embeddings from words in a document's vocabulary using a shallow neural network. It was developed by Tomas Mikolov in 2013 at Google [4]. The word embedding formed from each word is capable of capturing the context of a word in the document, as well as its semantic and syntactic similarity in relation to the other words. We chose to use the Skip-Gram model. According to Mikolov [5], this model has the ability to represent words well despite working with small amounts of data. Given a target word, the skip gram model tries to predict its context, i.e., the surrounding words. For each input word in the input layer, the input word is linearly transformed through a weight matrix to form its one-hot representation and activated with an activation function to create a hidden layer. Each word also goes through a backward pass (backpropagation) which re-calculates the input and output weight matrices. This process is repeated for every word in the training dataset in order to create word embeddings for using later on in the experiment. For our Word2Vec model, we use hyperparameters based on the default values. Our minimum count was 3, which meant that in the dataset, a word had to have a total frequency higher than 3 before

French sentence: *Appellant les knights des les d'Orient l'est*

English sentence: *Calling the knights of the Oriental East*

Shuffled Sentence: *de knights les of appellant calling east oriental knights les des the the d'Orient l'est*

Fig. 6.1 Randomly shuffled French–English sentence. In the shuffled sentence, shaded words are from the French sentence, while unshaded words are form the English sentence

it would be used for training the model. Our window size was 5 for our skip-gram model, which meant that the maximum distance between the current and predicted word in a sentence would be 5. We set the embedding size to 100.

6.2.2 *Random Shuffling*

For the random shuffling approach, each parallel sentence pair in the data set was tokenized. The word tokens of each sentence pairs are then randomly shuffled. Thereafter, the shuffled sentence will be used for training Word2Vec. The random shuffling creates a coarse-grained bilingual context for each word and enable the creation of a cross-lingual embedding space. Cross-lingual contexts allows the learned representations to capture cross-lingual relationships. While adjacent words in the shuffled sentences may not be correct translations, or may not approximate the original context closely, we hypothesize that if the sentences are short, random shuffling may still work adequately. Figure 6.1 illustrates random shuffling. For Word2Vec training, if the contextual window is set high enough, randomly shuffled words can still have a chance of forming useful cross-lingual associations. For example, the word "calling" would form connections with the words around it based on the window size. Due to this, the word "appellant", which is the corresponding French definition of "calling", would form a strong association with it as well. The random shuffling technique thus may be able to capture cross-lingual information despite its simplicity.

6.2.3 *Smart Shuffling*

In comparison, for the smart shuffling approach, word tokens are not shuffled randomly. Given a sentence in the source language, words from the parallel sentence in the targeted language are inserted as guided by the following procedure:

- Words with similar forms in both the source and target language are placed adjacent to each other in the shuffled sentence. For example in Fig. 6.2, the word "knight" is similar in both French and English.

French sentence: *Appellant les knights des les d'Orient l'est*

English sentence: *Calling the knights of the Oriental East*

Shuffled Sentence: *appellant les the knights knights of des les the d'Orient oriental l'est East*

Fig. 6.2 An example of a smartly shuffled sentence. The source language is French and the target language is English

- Looking up a cross-lingual dictionary which maps words in a source language to the words in the target language. If there are matches, the target word is inserted adjacent to the source word. For example in Fig. 6.2, the word "Appelant" maps to "Calling".
- If both above cases are not met, computing the character n-gram overlap between the dictionary translation of the source word and the target word, which is then used to compute a probability distribution. Given the source word, the target word is sampled. For example, "d'orient" and "oriental" overlaps substantially. Given "d'orient", "oriental" will be sampled with higher probability for placement beside "d'orient", as compared to other words in the English sentence. • If all above cases are not met, the target word is sampled from a uniform distribution, given the source word. Also, note that insertion of target words is adjacent to, but randomly before or after the source word.

We trained a Word2Vec model, utilising concepts such as random shuffling, cosine similarity, taking the mean reciprocal rank and calculating accuracy of test results.

6.3 Findings

We applied the random shuffling approach to datasets of parallel movie subtitles. We downloaded parallel move subtitles from OPUS [6] and pre-processed it to remove punctuation as well as lower all alphabets. This was so as to ensure that the Word2Vec model learns each word independent of the punctuation around it and does not mix up punctuations with words. Ensuring that all letters are in lowercase would lessen noise when processing and training the model as the model would not classify the capitalised and lowercase word as different words. For example, "America" and "america" would be classified as two different words though they are in actuality one word. Hence, the words should be converted to lowercase so as to minimise noise in the training process and prevent incorrect classification. Thereafter, we conducted five trials whereby for each trial, we randomly selected 1000 parallel sentence pairs as the test data set. For each trial, we evaluate the randomly initialized embeddings first on the test data set prior to training. Thereafter, it was trained on randomly shuffled data before being tested with the 1000 parallel sentence pairs to evaluate its accuracy. The aim of conducting multiple trials was to improve the estimate of the mean model performance. Arbitrarily selecting the randomly shuffled data for training also served as a less-biased representative of the overall data set [7].

6.3.1 Testing

For the testing process, we converted the words in the query and target sentences into word embeddings if they were in the vocabulary of the model. Each sentence was represented by a vector from the average of all the word embeddings it contained.

$$\text{Cosine Similarity}(A, B) = \frac{A \cdot B}{||A|| \times ||B||} = \frac{\sum_{i=1}^{n} A_i \times B_i}{\sqrt{\sum_{i=1}^{n} A_1^2} \times \sqrt{\sum_{i=1}^{n} B_1^2}} \tag{6.1}$$

Thereafter, we computed the cosine similarity between test vectors and candidate vectors. Equation 6.1 illustrates cosine similarity between vectors *A* and *B*, each of dimension n, and where ||.|| denotes the Euclidean norm of vectors. Cosine similarity ranges from -1 to 1, where -1 means that the results are perfectly dissimilar, whereas 1 is perfectly similar. A cosine value of 0 means that the two vectors are perpendicular to each other. For each test vector, the corresponding candidate vectors were ranked in descending order.

$$\text{MRR} = \frac{1}{|Q|} \sum_{i=1}^{n} \frac{1}{\text{rank}_i} \tag{6.2}$$

After the parallel sentence was sorted in order of descending cosine similarity, we then calculated the mean reciprocal rank (MRR) of the correct candidate vector with Eq. (6.2). For the *i*-th test sentence, the reciprocal rank is 1/ranki where ranki is the rank of the matching parallel sentence. For multiple queries, the reciprocal rank is the mean of the *N* reciprocal ranks. MRR is high if the correct corresponding candidate vectors are ranked high. We also computed the top 10 accuracy, whereby for each test sentence, we checked if the correct answer sentence was ranked within the top 10. Across all test sentences, we counted the number of times the correct candidate vector was within the top 10 candidate vectors and divided that by the number of test vectors to get our accuracy score. Before training, we calculated by probability that the parallel answer vector would have a 0.1% chance of being within the top 10 candidate vectors. We then hypothesised that, given our genism model consistently attained roughly 70% cosine-similarity score between the English word "house" and its French, Spanish and German synonyms ("maison", "casa" and "haus") after training, the top 10 accuracy score would be roughly 70%. In order to evaluate the trained model, the model was tested on 1000 parallel test sentences across three different language pairs. The mean reciprocal rank and accuracy were both derived from the averages of five trials for each language pair. For each trial, the model was tested twice, once before and once after training. “Pre-training” refers to the model before it was trained on the bilingual training data while “Post-training” refers to the same model after it has been trained.

6.3.2 Results

6.3.3 Equations

Table 6.1 represents our experimental results for the random shuffle data sets. On a whole, there was a significantly higher top-10 accuracy score after training. Using the English–French results as a benchmark, the top-10 accuracy score increased from 0.0838 to 0.673 after training. This means that after training, the corresponding parallel target sentence was within the top 10 candidate sentences almost 70% of the time. Similarly, the MRR score increased from 0.0585 pre-training to 0.547 after training. This signifies that the model has been trained properly. Another notable difference was that the English–Spanish experiment has a slightly higher accuracy as compared to the English–French and English–German experiments. This may be due to the comparatively larger number of cognates between Spanish and English words as compared to the other languages. Cognates are words with a common etymological origin. There are about 20 000 Spanish–English cognates, 1700 French–English cognates [8] and around 1000 German–English cognates [8]. Due to the high prevalence of cognates, it is plausible that cosine similarity between Spanish and English words would be higher, hence resulting in better classification and slighter greater accuracy. Nevertheless, having repeated each bilingual experiment on five test folds of data tabulating an average to be used in Table 6.1, we believe that our results are statistically significant.

Table 6.1 Results

		MRR	Top-10 accuracy
English–French	Pre-training	0.0585	0.0838
	Post-training	0.547	0.673
English–German	Pre-training	0.0387	0.0622
	Post-training	0.505	0.627
English–Spanish	Pre-training	0.0673	0.101
	Post-training	0.696	0.817

Pre-training refers to randomly initialised embeddings used prior to training. Post-training refers to the trained Word2Vec model based on random shuffling

6.4 Conclusion

We have explored a simple cross-lingual word embedding model based on random shuffling and it achieved almost 70% accuracy in matching short parallel sentences as compared to the 0.1% accuracy before training. This shows that despite its simplicity, random shuffling performs well when matching short non-complexed parallel sentences between romance languages. This model can thus be implemented in search engines to aid in bilingual query translation as well as information retrieval. However, one limitation of the model is its inability to recognise stylistic language. For example, should the idiom "let the cat out of the bag" be used, a search engine implemented with my model would search for words related to "cat" and "bag" despite the phrase having the connotation of "revealing facts previously hidden". As the skipgram model used only has a window size of 5 words, the idiom will not be considered in its totality. As a result, the meaning of the idiom may be distorted when translated. Another potential limitation is the translation gap. As my model has a tendency to search for words with similar embeddings to itself, the word chosen may not necessarily be precise. Hence, a translation gap arises. In order to circumvent that issue, Hamed, Sheikh and Allen proposed using the Smart Shuffling method which is able to bridge the translation gap. In future works, we hope to compare the effectiveness of the random shuffling method in relation to the smart shuffling method, as well as other more sophisticated models.

References

1. Ren, F., & Bracewell, D. (2009). Advanced information retrieval. *Electronic Notes in Theoretical Computer Science, 225*, 303–317. https://doi.org/10.1016/j.entcs.2008.12.082
2. Vulić, I., & Moens, M.-F. (2015). Monolingual and cross-lingual information retrieval models based on (Bilingual) word embeddings. In *Proceedings of the 38th international ACM SIGIR conference on research and development in information retrieval (SIGIR'15)* (pp. 363–372). New York, NY, USA: Association for Computing Machinery. https://doi.org/10.1145/2766462.2767752
3. Bonab, H., Sarwar, S. M., & Allan, J. (2020). Training effective neural CLIR by bridging the translation gap. In *Proceedings of the 43rd international ACM SIGIR conference on research and development in information retrieval (SIGIR'20)* (pp. 9–18). New York, NY, USA: Association for Computing Machinery. https://doi.org/10.1145/3397271.3401035
4. Mikolov, T., et al. (2013). Efficient estimation of word representations in vector space.
5. Mikolov, T. (n.d.). De-obfuscated python + question. Retrieved January 03, 2021, from https://groups.google.com/g/word2vec-toolkit/c/NLvYXU99cAM/m/E5ld8LcDxlAJ
6. Lison, P., & Tiedemann, J. (2016). OpenSubtitles2016: Extracting large parallel corpora from movie and TV subtitles. In *Proceedings of the 10th international conference on language resources and evaluation (LREC 2016).*
7. Facebookresearch (n.d.). Facebookresearch/MUSE. Retrieved January 03, 2021, from https://github.com/facebookresearch/MUSE
8. https://research.steinhardt.nyu.edu/scmsAdmin/media/users/xr1/glossaries/ELA/GlossaryCognatesFrenchUpdated5-5-2014.pdf

Chapter 7
Alternative to Chromatography: Purification of Therapeutic Protein/Peptides by Continuous Crystallization in Millifluidics

Rachel Chan Shurui, Wong Li Heng, and Ilangovan Harishiga

Abstract In the last decade, therapeutic proteins have emerged as an important class of pharmaceuticals due to their excellent target-specificity and low-toxicity profile, making them better equipped to treat certain diseases (e.g., cancer, enzyme deficiency, degenerative diseases) toward which conventional small-molecule drugs are not as effective. Recently, rapid technological advancements in protein/peptide synthesis have enabled pharmaceutical industries to increase production by more than 20-fold. However, this multifold increase in the upstream production capacity has created a bottleneck in the downstream processing, which heavily relies on chromatography. In this project, we aim to replace chromatography with continuous millifluidic crystallization as a more effective and efficient alternative to purify therapeutic protein/peptides in order to reduce the current stress in downstream processing.

Keywords Aspect ratio · Batch crystallization · Continuous crystallization · Reynolds number

7.1 Introduction

7.1.1 Background

I. Therapeutic proteins

Therapeutic proteins are proteins engineered in the laboratory for pharmaceutical use. They are highly effective in vivo and have revolutionized the treatment of diseases. Protein therapeutics permit an individualized treatment approach by supporting a specifically targeted therapeutic process through compensating the deficiency of an essential protein [1].

II. Continuous millifluidic crystallization

R. C. Shurui (✉) · W. L. Heng · I. Harishiga
Raffles Girls' School, Singapore, Singapore
e-mail: rachel.chan05@rafflesgirlssch.edu.sg

H. Guo et al. (eds.), *IRC-SET 2021*, https://doi.org/10.1007/978-981-16-9869-9_7

Continuous millifluidic crystallization can potentially provide improved control of crystal properties, improved process reproducibility, and reduced scale-up risk. Liquid and gas are introduced into one end of the tube at flow rates selected to spontaneously generate alternating slugs of liquid and gas that remain stable while cooling crystallization occurs in each liquid slug. Mixing within each stable self-circulating slug is maximized by controlling the slug aspect ratio through specification of liquid and gas flow rates [2].

7.1.2 Research Question and Objective

I. Research Question

How does aspect ratio affect the quality of lysozyme crystal products? In this experiment, the quality of crystals is taken to be their mass, length, and activity.

II. Research Objective

This project will study the effect of aspect ratio on lysozyme crystal products' quality. By tuning the air introducing pump (tuning injection interval to control water slug length, thus, water slug aspect ratio; tuning injection volume to control bubble length, thus, air slug aspect ratio), effects of different slug and bubble lengths on quality of crystal products can be investigated.

7.1.3 Literature Review

Aspects of continuous crystallization

I. Aspect ratio (AR)

The aspect ratio of the slug, referred to as the slug size, is the ratio between the length of slugs and the inner diameter of the tubing. Based on properly selected co-current flow rates of gas and liquid, stable alternating slugs of liquid and gas are spontaneously generated at aspect ratios close to 1, facilitating good mixing for enhanced heat and mass transport without mixing blades.

II. Slug flow

Slug flow in liquid–gas two-phase flow is a type of flow pattern. The slug usually refers to the heavier fluid, in this case the liquid, but it can also refer to the lighter fluid, in this case the gas.

Comparison between batch crystallization and continuous crystallization

Continuous crystallization results in less equipment footprint than batch crystallization, reducing + 20% in capital expenditure. As for process variability, batch crystallization displays comparably significant batch-to-batch variability regarding physical properties, making continuous crystallization more reproducible and reliable. Thus, continuous crystallization could be more feasible in the large-scale downstream processing of therapeutic proteins. However, continuous crystallization is not as well-researched as batch crystallization. Despite batch processes usually having higher yields in a once-through system, it is possible to achieve higher yields from continuous processes with appropriate recycling strategies. For material traceability, the tracing material process in batch processes is well understood from both an operational and a regulatory perspective while there is still a gap between both perspectives for a continuous process [3].

7.2 Materials and Methods

7.2.1 Materials

The chemicals used for our experiment include: Distilled water, Sodium acetate, Acetic acid, Hydrochloric acid, Sodium hydroxide, Lysozyme, Sodium chloride, Liquid nitrogen, 0.1 M Potassium phosphate, pH 7.0, Micrococcus lysodeikticus cells.

The apparatus used for our experiment include: Culture tubes, Centrifuge tubes, Mini Centrifuge tubes, Magnetic stirrer, Vortex machine, Centrifuge, Freeze dry machine, liquid nitrogen, Vacuum filter, Syringe filter, Pipette, Micropipette, UV-mini Spectrophotometer, UV-Visible Spectrophotometer, Weighing machine, Microscope.

7.2.2 Preparation of Buffer Solution

To obtain 0.5 L of buffer solution (pH 4.6), add 2.029 g of Sodium Acetate and 1.517 g of Acetic Acid to 400 mL of distilled water. Adjust solution to pH 4.6 using HCl or NaOH. Add distilled water until volume is 0.5 L.

7.2.3 Preparation of Lysozyme Solution

To obtain lysozyme solution, add a 1:1 ratio of lysozyme-buffer solution to NaCl solution in each test tube. The volume of lysozyme solution in each tube is dependent on the liquid aspect ratios (Table 7.1).

Table 7.1 Parameters of experiment

Inner diameter of tube (cm)	Aspect ratio	Total height of solution (cm)	Liquid volume (ml)	Lysozyme solution volume (ml)
1.15	0.25	4.60	4.64	2.32
1.15	0.5	2.30	2.35	1.18
1.15	1	1.15	1.16	0.58
1.15	2	0.58	0.46	0.23
1.15	3	0.38	0.36	0.18
1.15	4	0.29	0.16	0.08
1.15	5	0.23	0.10	0.05
1.15	6	0.19	0.06	0.03

7.2.4 Crystallization

The culture tubes were placed on 3 separate magnetic stirrers with stirring rates (SR) of 0 rpm, 100 rpm, and 250 rpm, respectively, for 24 h as crystallization took place. For every AR, two sample tubes were set up for each SR. After 24 h, the culture tubes were removed from the stirrers.

7.2.5 Freeze Drying

Transfer a fixed volume of lysozyme solution from each culture tube into small centrifuge tubes and centrifuge the tubes for 10 min. Separate the liquid from the solid crystals. Freeze the solid crystals in liquid nitrogen then freeze dry in the freeze dry machine for 24 h. After 24 h, remove the centrifuge tubes containing the freeze dried crystals from the freeze dry machine (Table 7.2).

7.2.6 Spectrophotometric Analysis

Add 1.98 mL of buffer solution to a cuvette and use that as a blank. Then add 0.02 mL of liquid from centrifuged lysozyme solution and mix well to obtain readings from the spectrophotometer.

Table 7.2 Volume of lysozyme solution (mL), total volume of crystal solution (mL) and volume of crystal solution transferred to small centrifuge tubes (mL) for each aspect ratio

Aspect ratio	Lysozyme solution volume (mL)	Liquid volume (mL)	Volume of liquid transferred (mL)
0.25	2.32	4.64	1
0.5	1.175	2.35	1
1	0.58	1.16	0.5
2	0.228	0.456	0.2
3	0.181	0.362	0.2
4	0.079	0.158	0.1
5	0.04875	0.0975	0.05
6	0.02905	0.0581	0.03

7.2.7 *Enzyme Assay*

2.0 mg of each freeze dried crystal sample was mixed with 2.0 mL of distilled water to prepare lysozyme crystal solutions of 1.0 mg/ml concentration. The solution was then diluted to 0.25 mg/mL by mixing 250 mL of each crystal sample with 750 mL of distilled water. A Micrococcus lysodeikticus cell suspension was prepared by suspending 12 mg of dried Micrococcus lysodeikticus cells in 40 mL of Potassium Phosphate (K_3PO_4) buffer of 0.1 M at pH 7.0, mixed thoroughly. A UV-Visible spectrophotometer was used for enzyme assay, adjusted to a wavelength of 450 nm and temperature of 25 °C. Using a micropipette, 1.9 mL of K_3PO_4 buffer and 1.0 mL of Micrococcus lysodeikticus cell suspension were added into a cuvette and incubated for 4-5 min to achieve temperature equilibration and establish blank rate. Then, 0.1 mL of lysozyme solution of 0.25 mg/mL concentration was added to the cuvette. Using the spectrophotometer, the change in A450/s from the steepest linear portion of the curve was recorded. All mixing was done thoroughly using a vortex machine.

7.2.8 *Crystal Analysis*

20 μL of the sample was diluted with 100 μL of saturated lysozyme. 10 μL of solution was then transferred to a glass slide and observed under a microscope. Pertinent images of the crystals were then saved onto a computer for further analysis. The images were then analyzed using a software, Image J, allowing us to obtain crystal lengths accurately to find the average lengths.

7.3 Results

See Tables 7.3, 7.4, and 7.5.

Table 7.3 Mass of crystals/g under varying aspect ratio and stirring rate conditions (yield)

	Mass/g								
	Aspect ratio								
Stirring rate/rpm	Sample number	0.25	0.5	1	2	3	4	5	6
0	1	0.023	0.028	0.012	0.0060	0.0031	0.0019	0.0017	0.0021
	2	0.037	0.028	0.011	0.0053	0.0036	0.0010	0.0018	0.00080
	Average mass/g	0.030	0.028	0.011	0.0057	0.0034	0.0015	0.0018	0.0015
100	1	0.041	0.019	0.013	0.0085	0.0058	0.0024	0.0023	0.0010
	2	0.042	0.013	0.0095	0.019	0.0092	0.0035	0.0017	0.0017
	Average mass/g	0.042	0.016	0.011	0.014	0.0075	0.0030	0.0020	0.0014
250	1	0.026	0.039	0.017	0.0016	0.0062	0.00090	0.0036	0.0012
	2	0.044	0.028	0.018	0.015	0.0041	0.0026	0.0038	0.0053
	Average mass/g	0.035	0.033	0.018	0.0084	0.0052	0.0018	0.0037	0.0033

Table 7.4 Average length of crystals under different stirring rates

	Stirring rate/rpm		
	0	100	250
Aspect ratio	Average crystal length/μm	Average crystal length/μm	Average crystal length/μm
0.25	7.39	4.34	6.85
0.5	6.64	4.54	4.17
1	5.52	4.09	4.10
2	4.42	3.69	3.45
3	6.35	3.64	3.98
4	5.22	3.03	3.68
5	6.41	3.48	3.35
6	5.56	2.82	3.61

Table 7.5 Average activity/units ml^{-1} under different stirring rates

	Stirring rate/rpm		
	0	100	250
Aspect ratio	Average crystal activity/units ml^{-1}	Average crystal activity/units ml^{-1}	Average crystal activity/units ml^{-1}
0.25	5864.04	7957.36	8042.64
0.5	9376.08	7920.00	7754.72
1	7256.04	7119.96	7872.00
2	7216.08	8512.80	8103.96
3	8527.92	9439.92	8280.00
4	5464.08	6735.96	7632.00
5	5896.08	8007.96	7708.45
6	6359.00	8391.96	7851.00

7.4 Discussion

Firstly, we will discuss batch crystallization as an introduction to continuous crystallization. In batch crystallization, the flow should neither be too turbulent nor too laminar to maximize crystal yield [4]. Laminar flow is denoted by a low Reynolds number (Re) of below approximately 2300 while turbulent flow is denoted by a high Re of approximately 3000 as seen in Fig. 7.1.

The Reynolds number can be calculated using the formula $Re = \rho u L/\mu$, where ρ is the density of the fluid (SI units: kg/m^3), u is the flow speed (m/s), L is the diameter

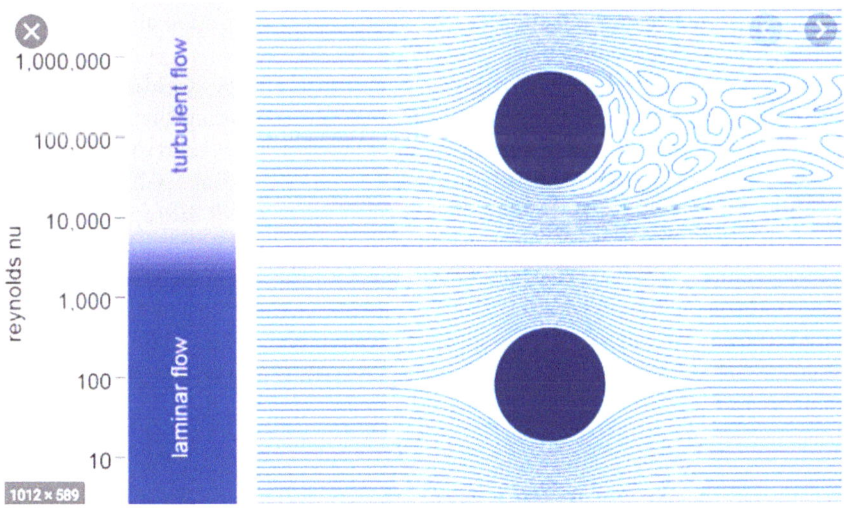

Fig. 7.1 Type of flow corresponding to values of Reynolds number [5]

of the crystallization tube (m) and μ is the dynamic viscosity of the fluid (Pa s). The density of crystal solution used is 1000 kg/m. For continuous crystallization, the flow speed is 0.00252 m/s (3 s.f.) while for batch crystallization at SR 0 rpm, 100 rpm, and 250 rpm, the flow speed is 0 m/s, 0.120 m/s (3 s.f.), and 0.301 m/s (3 s.f.), respectively. The diameter of the continuous crystallization tube is 1.59 mm, while that of the batch crystallization tube is 1.15 cm. The viscosity of the crystal solution is 0.001 Pa·s.

Using these values, the Re of continuous crystallization and batch crystallization at 0 rpm, 100 rpm, and 250 rpm are 6.34 (3 s.f.), 0, 1380 (3 s.f.), and 3460 (3 s.f.), respectively. Based on Fig. 7.1, batch crystallization under SR 0 rpm has a Re of 0 which is extremely laminar, while SR 250 rpm is highly turbulent with a Re of 3460. Among the three SR conditions for batch crystallization, 100 rpm with Re of 1380 is the most suitable for crystallization as it is neither too laminar, nor too turbulent.

Comparing the SRs and ARs under results Table 7.3, the condition with greatest yield in batch crystallization, with greatest average mass of 0.0419 (3 s.f.), is at SR 100 rpm and AR 0.25. Thus, we will compare batch crystallization in this condition with continuous crystallization.

For continuous crystallization, two types of flow ensure the best yield, namely, narrow bore laminar flow and turbulent tubular flow. For a long thin tube less than 1 mm in diameter, laminar flow is most favorable. Since the tube used was 1.59 mm in diameter, narrow bore laminar flow is not favorable. This leads to turbulent tubular flow, which is best for a very long tube, diameter unspecified [6]. Since the tube used is more than 1 mm in diameter, turbulent flow is best. However, according to our Reynolds number for continuous crystallization, the flow is more laminar. Therefore, continuous crystallization does not grant the best yield. Moreover, achieving proper conditions (e.g., tube diameter, flow type) for the best yield proves difficult in continuous crystallization (Fig. 7.2).

Therefore, we can conclude that yield wise, batch crystallization at AR 0.25 and SR 100 rpm is the best condition.

Now, on to the effect of AR on crystals in terms of mass (yield), crystal size (length), and activity for batch crystallization. When comparing the crystal yield against AR, the AR resulting in highest yields at SRs 0, 100, and 250 rpm are AR0.5 at 0.0277 g (3 s.f.), AR0.25 at 0.354 g (3 s.f.) and AR0.25 at 0.0295 g (3 s.f.), respectively. While the AR providing the greatest yields at SR 100 rpm is AR0.5, the results for AR0.25 are considerably close at 0.0274 g (3 s.f.).

When comparing crystal size against AR, the AR resulting in the greatest lengths at SRs 0, 100 and 250 rpm are AR 0.25 at 7.39 μm (3 s.f.), AR 0.5 at 4.54 μm (3 s.f.), and AR 0.25 at 6.85 μm (3 s.f.), respectively. While the AR providing the largest crystals at SR 100 rpm is AR0.5, the results for AR0.25 are considerably close at 4.34 μm (3 s.f.).

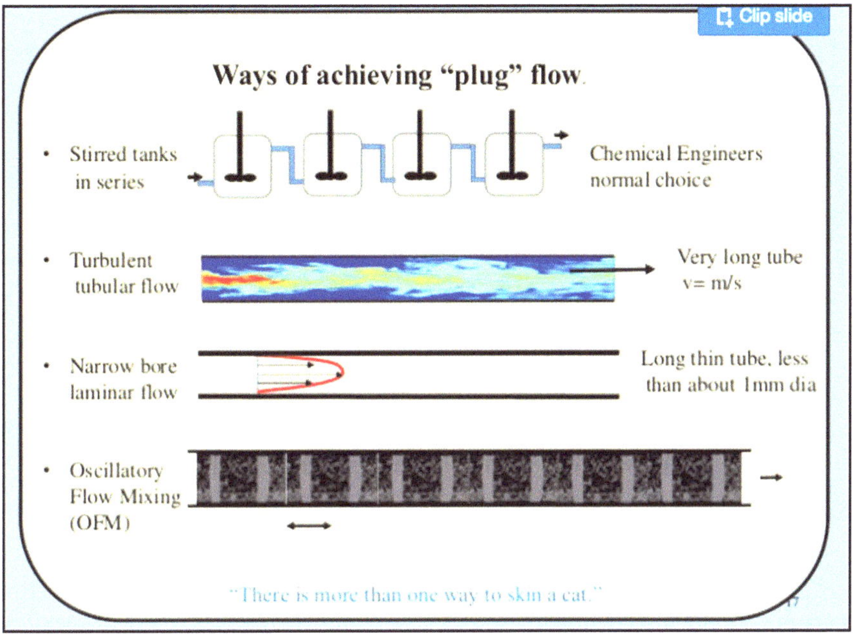

Fig. 7.2 Methods to achieve plug flow [6]

When comparing crystal activity against AR, the AR resulting in greatest activity at each SR is AR 0.5 at 9376.08 units ml^{-1}, with AR 3 next at 8527.92 units ml^{-1} at 0^{rpm}, 9439.92 units ml^{-1} at 100^{rpm}, and 8280 units ml^{-1} at 250 rpm. As seen, AR 0.25 is most effective in producing greatest crystal yields and sizes while AR 3 produces highest activity. Since AR 0.25 is most favorable for the majority of aspects determining crystal quality, AR 0.25 is most favorable in batch crystallization. In conclusion, batch crystallization proves to be more effective than continuous crystallization; and in batch crystallization, aspect ratio of 0.25 is most favorable, as it produces the best quality of crystals.

Appendix

See Figs. 7.3, 7.4, 7.5, 7.6, 7.7, 7.8, 7.9, 7.10, 7.11 and 7.12.

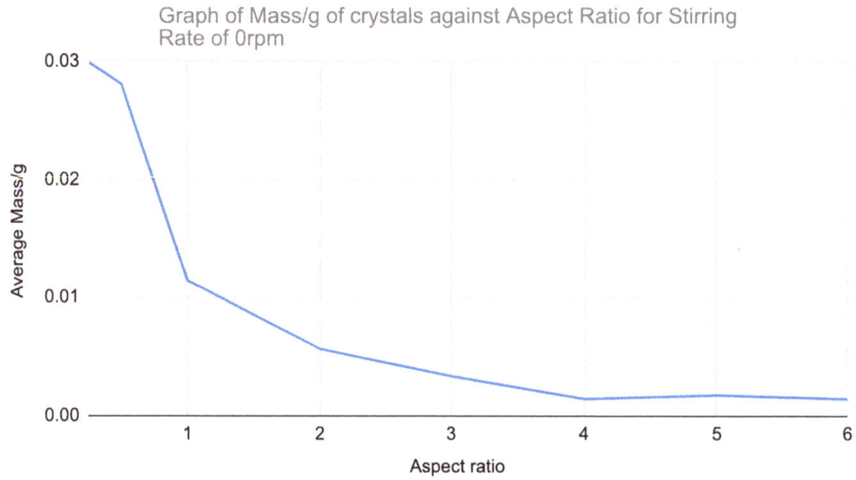

Fig. 7.3 Graph of mass/g of crystals against aspect ratio for stirring rate of 0 rpm

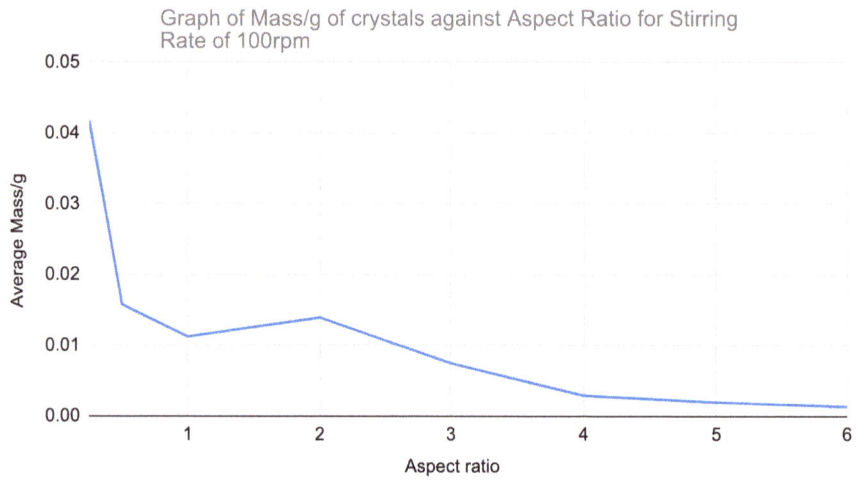

Fig. 7.4 Graph of mass/g of crystals against aspect ratio for stirring rate of 100 rpm

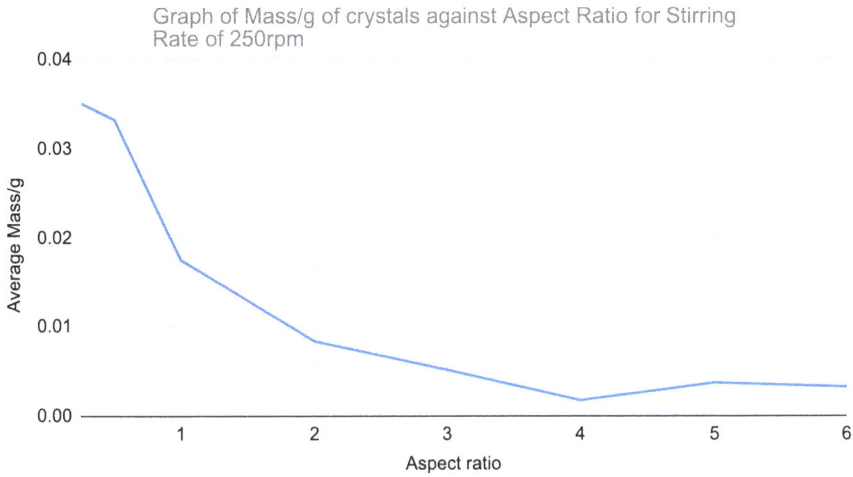

Fig. 7.5 Graph of mass/g of crystals against aspect ratio for stirring rate of 250 rpm

Fig. 7.6 Graph of crystal lengths/μm against aspect ratio for stirring rate 0/rpm

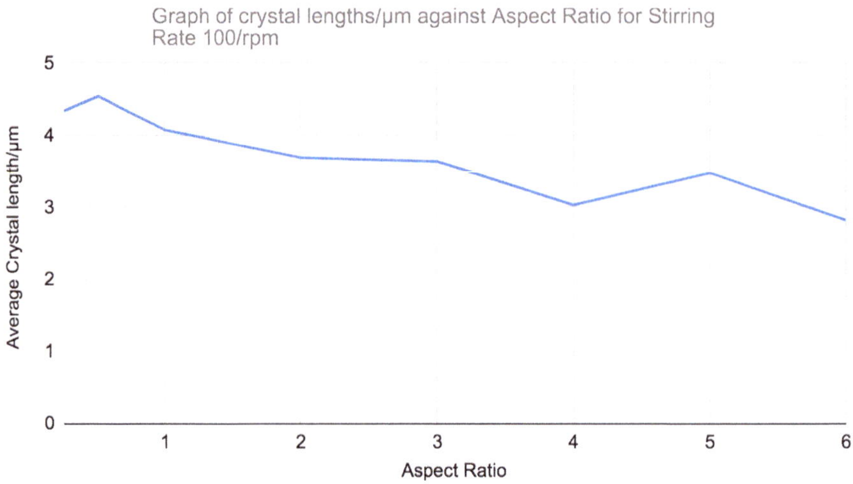

Fig. 7.7 Graph of crystal lengths/μm against aspect ratio for stirring rate 100/rpm

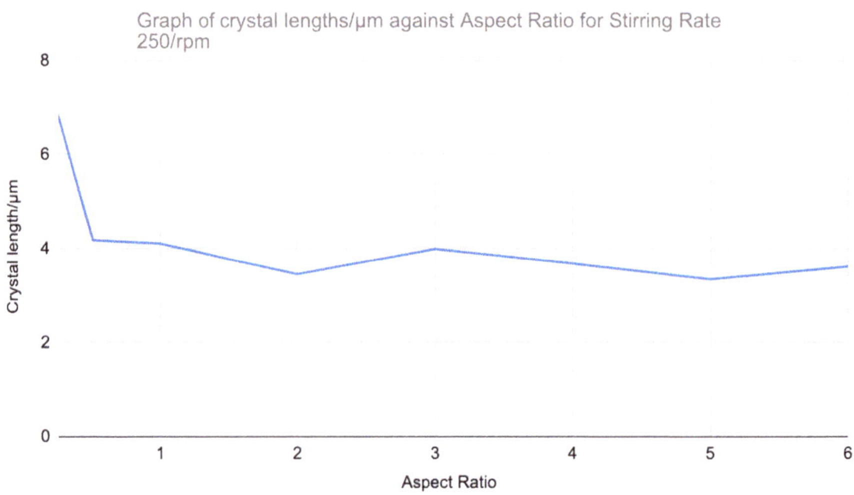

Fig. 7.8 Graph of crystal lengths/μm against aspect ratio for stirring rate 250/rpm

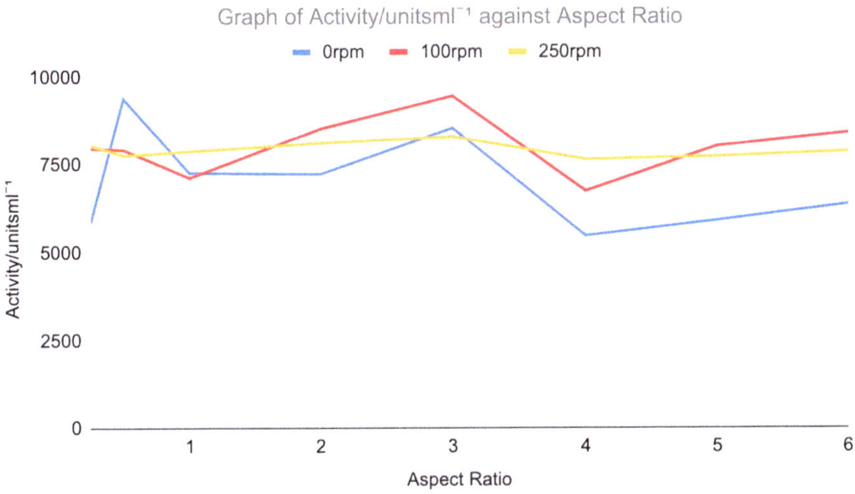

Fig. 7.9 Graph of activity/units ml^{-1} against aspect ratio

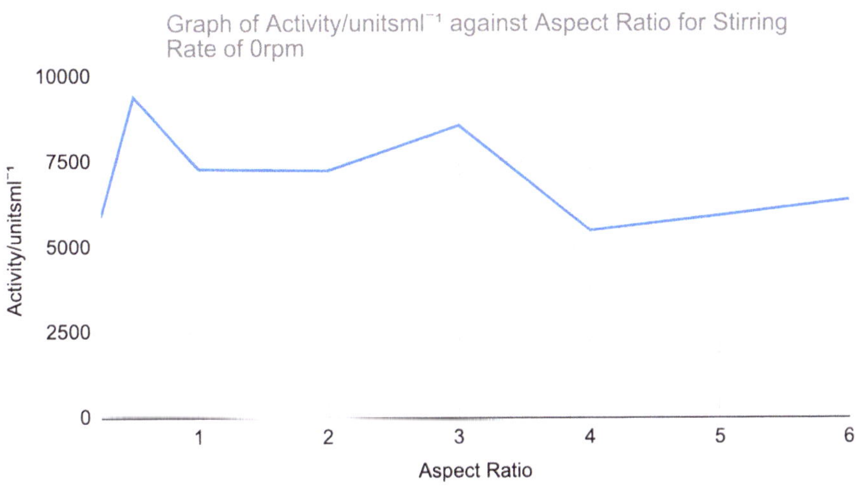

Fig. 7.10 Graph of activity/units ml^{-1} against aspect ratio for stirring rate of 0 rpm

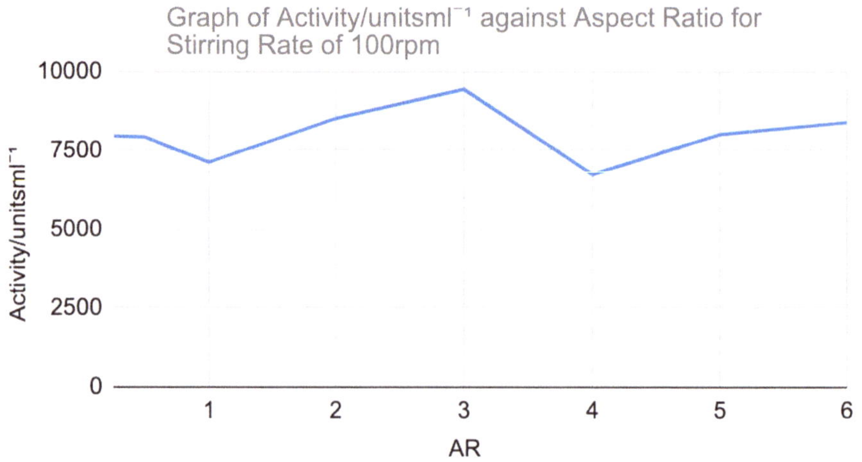

Fig. 7.11 Graph of activity/units ml^{-1} against aspect ratio for stirring rate of 100 rpm

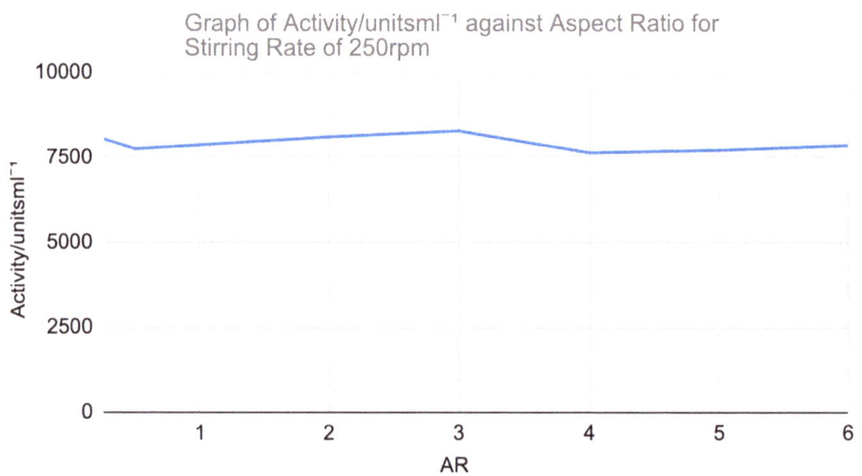

Fig. 7.12 Graph of activity/units ml^{-1} against aspect ratio for stirring rate of 250 rpm

References

1. Therapeutic Protein—Creative Biolabs (n.d.). Creative Biolabs. https://www.creativebiolabs.net/therapeutic-proteins_47.htm. Last retrieved, January 10, 2021
2. Pubs.acs.org. (2021). Microfluidic continuous seeded crystallization: Extraction of growth kinetics and impact of impurity on morphology. https://doi.org/10.1021/cg301538y?src=recsys. Last retrieved, January 10, 2021.
3. Myerson, A. S., et al. (2011). *Crystal Growth & Result, 11*, 887–895.
4. Ni, X. (n.d.). Continuous crystallisation with oscillatory baffled crystalliser technology (p. 3).

5. Laminar Flow versus Turbulent Flow (n.d.). Retrieved from https://www.reactor-physics.com/engineering/fluid-dynamics/laminar-flow-vs-turbulent-flow/. Last retrieved, January 10, 2021.
6. Malcolmmackley Follow (2014, May 24). Flow, crystallisation and continuous processing. Retrieved from https://www.slideshare.net/malcolmmackley/flow-crystallisation-and-continuous-processing. Last retrieved, January 10, 2021.

Chapter 8
Design of Underground Structures to Mitigate Effects of Soil Liquefaction

Kera Yong, Jovan Zhan Rui Mak, and Jia Hao Tan

Abstract Earthquakes have been a threat to mankind for aeons, with the earliest recorded earthquake occurring in 1177 BC. However, even with the harnessing of modern technology in plate tectonic studies, it is impossible to predict exactly when earthquakes will occur. Soil liquefaction, which is the suspension of saturated soil particles, is one of the most deadly hazards brought about by earthquakes and has claimed many lives thus far. However, there are few structural supports currently in the market to specifically combat this threat. Hence, this project aims to design and develop underground structures to mitigate the effects of soil liquefaction. There were two main families of designs: tap and fibrous, drawing inspiration from the complex root networks in plants. It was hypothesised that the fibrous designs would be more stable, and the further the structure's branches extended out, the more stable the design. Structural blueprints were designed in CAD and 3D printed, then tested in a soil liquefaction simulation. This was carried out via the fluidisation of 20 kg of sieved, fine sand. The accelerations in the X, Y and Z planes of the simulated building, in which a brick was used, were measured with an accelerometer, and after analysis of results, it was found that the fibrous root structures were generally more effective than the taproot structures at reducing the acceleration values of the simulated building. Overall, the full fibrous structure was the most suitable to be implemented in buildings in earthquake-prone countries. In the future, more investigations into the physical properties of the structures, as well as the mass and size of simulation buildings, can be conducted to assess the structural integrity of the structures more thoroughly.

Keywords Underground · Soil liquefaction · Earthquake

K. Yong (✉) · J. Z. R. Mak · J. H. Tan
National Junior College, Singapore, Singapore
e-mail: yong_rej_en_kera_ruth@njcgclassroom.com

J. Z. R. Mak
e-mail: mak_zhan_rui_jovan@njcgclassroom.com

J. H. Tan
e-mail: tan_jia_hao@njcgclassroom.com

H. Guo et al. (eds.), *IRC-SET 2021*, https://doi.org/10.1007/978-981-16-9869-9_8

8.1 Introduction

Earthquakes have long been a natural hazard for mankind, with the earliest recorded earthquake occurring in 1177 BC [1]. However, even with more recent studies into plate tectonics, there is still no way to predict exactly when earthquakes will occur. According to the United States Geological Survey, there are an estimated 500,000 detectable earthquakes in the world each year, out of which 100,000 can be felt and approximately 100 cause severe damage. Such a huge number of earthquakes would cause a colossal amount of damage, as can be seen from a study done in 2016 by the National Crime Records Bureau: an average of seven buildings collapse every day in India. Most of these buildings are residential and commercial buildings, with, respectively, an average of 4919 and 1610 fatalities for every incident [2].

Given that earthquakes can cause such large amounts of damage to cities and other communities, it is therefore imperative that precautionary measures against the decimating effects of earthquakes are taken. While the current primary concern that is dealt with is the great number of casualties, it is more important to address the root cause of building collapse, which is the building's inability to withstand soil liquefaction during earthquakes. Soil liquefaction occurs when saturated soil substantially loses strength and stiffness in response to an applied stress, which is the shaking during an earthquake [3]. The sudden change in stress condition fluidises the soil, leading it to exhibit properties of a liquid. It was found that the solutions to this issue primarily consists of additional structures that are above-ground, with only three underground structures to solve the issue. These structures are *Seishin* (structural reinforcement), *Taishin* (dampening) and *Menshin* (base isolation) [4] (Figs. 8.1, 8.2 and 8.3).

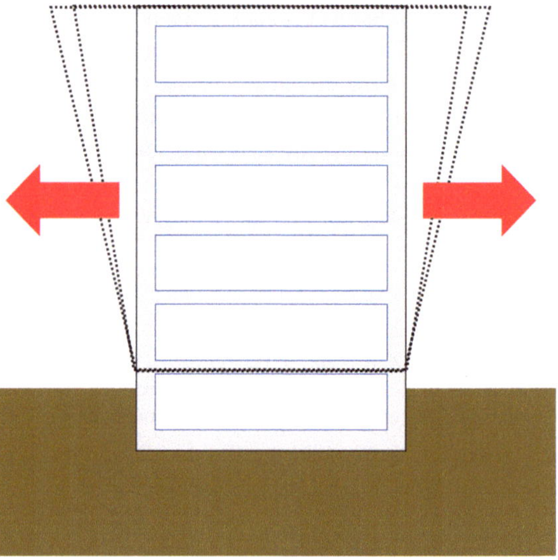

Fig. 8.1 Schematic diagram of Seishin (structural reinforcement)

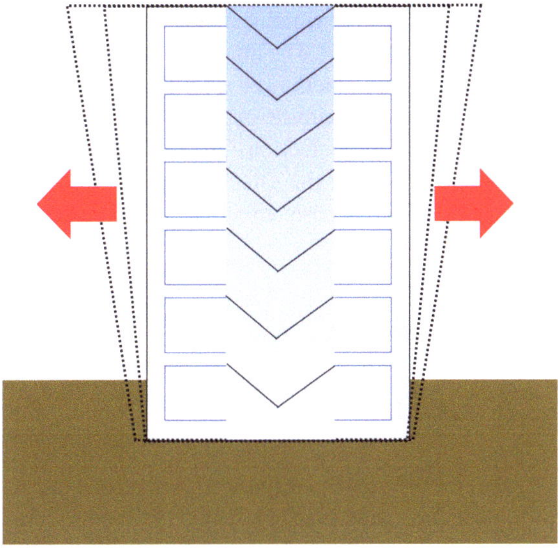

Fig. 8.2 Schematic diagram of Taishin (dampening)

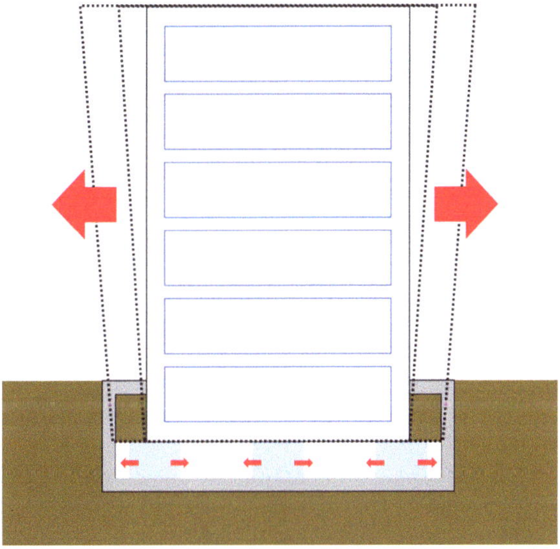

Fig. 8.3 Schematic diagram of Menshin (base isolation)

However, it is believed that underground structures are more effective than above-ground structures because they come into contact with the flowing soil during soil liquefaction more directly than above-ground structures. They also do not obstruct or affect the visible surroundings of the building. One such structure that occurs naturally is plant root systems, which has a primary function of anchoring the plant firmly to the ground, such that it will not be uprooted easily. The two most common root systems of plants are the tap and fibrous root systems. The taproot system has a distinctive main root which extends deep into the ground, with smaller peripheral roots branching out of it. The fibrous root system instead has no distinctive main root, but has many roots branching broadly across a large area rather than deep downwards.

Hence, this project aims to explore this area and engineer a structure that mitigates the effects of soil liquefaction. The aim is to investigate which design is the most effective in mitigating the effects of soil liquefaction, and also how the length of beams branching outwards affect its ability to keep the building above it stable.

8.2 Methodology

8.2.1 Structures

A total of six different underground structures were designed, by drawing inspiration from plant root structures [5]. Three structures were built according to the taproot system, where one main beam of length 20 cm branched out into three rows of four side beams arranged in a square planar around the main beam. The other three structures were built according to the fibrous root designs, where one thinner main beam of length 20 cm branched out into three rows of four side beams arranged in a square planar, angled downwards at a 45° angle. Additionally, thin downward beams were also added to simulate fibrous roots. The lengths of the side beams were varied by measuring out ½ the length, ¾ the length and the full length itself, which is 8 cm (taproot) and 6.5 cm (fibrous root) long (Figs. 8.4, 8.5, 8.6 and 8.7).

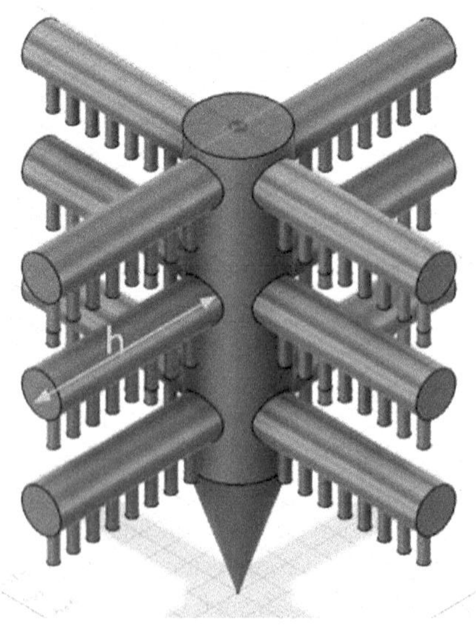

Fig. 8.4 Taproot design (top view)

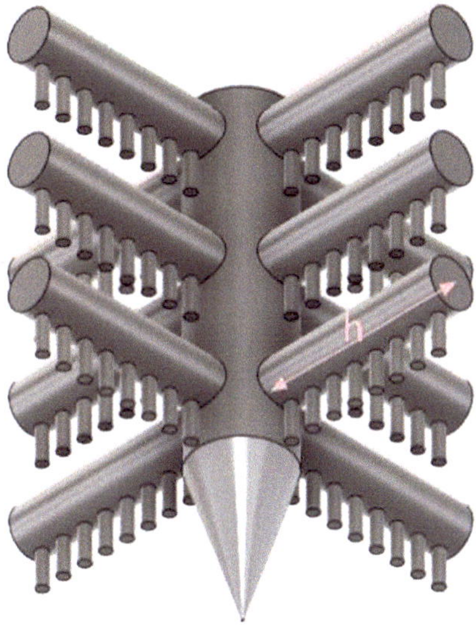

Fig. 8.5 Taproot design (bottom view)

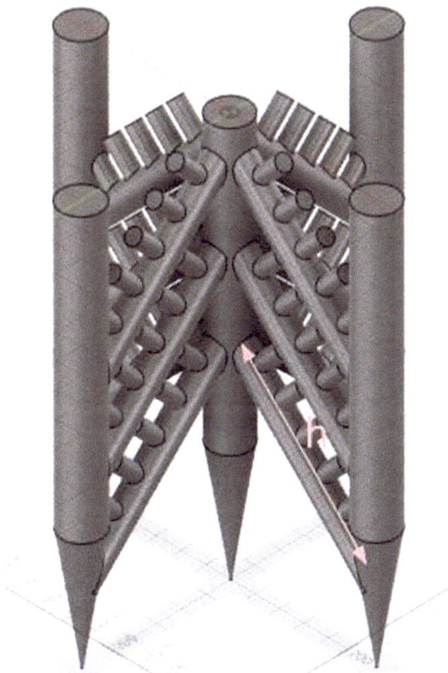

Fig. 8.6 Fibrous root design (top view)

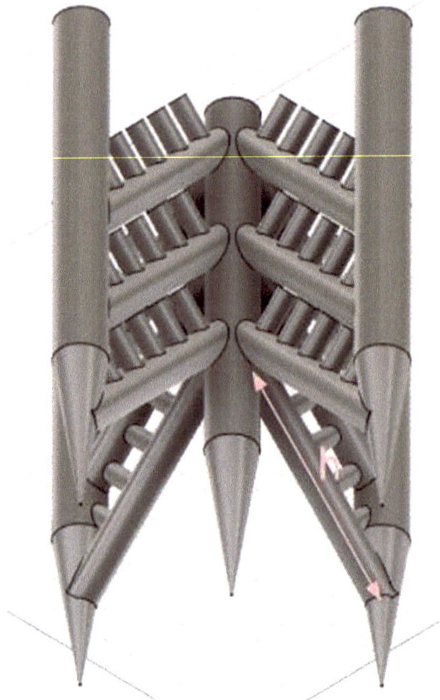

Fig. 8.7 Fibrous root design (bottom view)

Structures were designed on Fusion360, which allowed for three-dimensional digital viewing of all the blueprints from any angle. They were then 3D printed using a mixture of Acrylonitrile butadiene styrene (ABS) and polylactic acid (PLA).

8.2.2 Soil Liquefaction Simulation Setup

Polyvinyl chloride (PVC) pipes with a diameter of 0.5″ were cut according to the relevant dimensions of the 40 L container to be used, then fitted into the container just above the wheel cover. Two holes of diameter 1.0 mm were drilled in the PVC pipes at a 90° angle, at every 2.5 cm interval (Fig. 8.8).

20 kg of sand was filtered through two fine sieves with pores of 100 and 150 μm, and the fine sand was poured into the setup, above a cardboard fitting which represents hard bedrock deeper into the Earth's crust. An air compressor was then inserted into a centralised PVC pipe that fed into the other pipes and was secured with glue to ensure that there would be no leakage of air. The underground structure was placed in the sand with a 75 mm × 75 mm × 225 mm brick of 2.450 kg above the structure. The compressor was then turned on, pumping air out at 90 PSI (pounds per square inch). The air that escaped the small holes drilled into the PVC pipes allowed the sand in the container to flow like a fluid, as it escaped the holes at high speeds. The

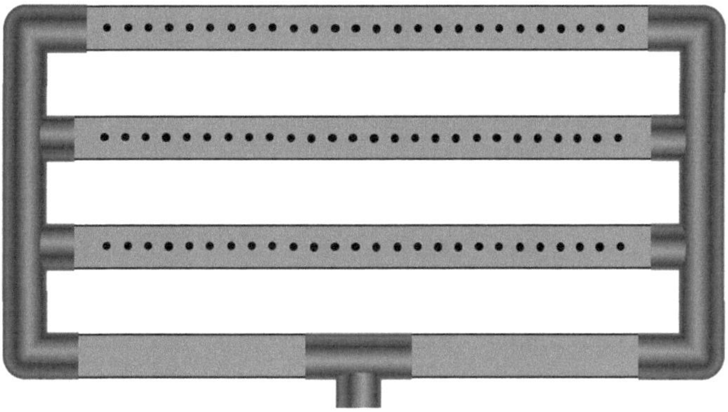

Fig. 8.8 Schematic diagram of PVC pipe network in 40L container

compressor was left on for 15 s, as that was the maximum time the air compressor could function at 90 PSI. An accelerometer will be placed on the brick to measure the acceleration of the brick above in the *X*, *Y* and *Z* directions, and the acceleration-time graphs will then be plotted. After conducting the experiment with a structure three times, the experiment was repeated with the other five structures individually.

This simulation is based on the concept of fluidisation, a process similar to soil liquefaction. When an upward-flowing gas exerts an upward force powerful enough to overcome the gravitational force acting on a particle, a particle becomes fluidised. When particles are fluidised, they exhibit fluid-like behaviour. This phenomena allows the soil liquefaction simulation to be carried out [6].

8.3 Results and Discussion

It was observed that all the structures sank into the sand upon activation of the liquefaction simulation. The brick collapsed in both the 50% taproot and 75% taproot setups, while the full taproot and 50% fibrous root structures caused the brick to be severely tilted, and the 75% and full fibrous root structures kept the building upright, with minimal tilting (Fig. 8.9).

As can be seen from Fig. 8.10, the fibrous root family had lower average accelerations overall for the *X* and *Y* axes as compared to their taproot counterparts. All the fibrous root designs had an average *Z* acceleration of 9.8 ms^{-2}, which is reflective of its initial sinking into the sand upon activation of soil liquefaction simulation, as well as that of the average pull of gravitational force on Earth acting on the setup. In contrast, the taproot family showed inconsistent average accelerations which were reflective of the inconsistent accelerations of the brick with the taproot structures.

Overall, the fibrous design was the better design as its three structures provided the lower average acceleration when compared to their taproot counterparts. The low acceleration readings show that the fibrous root designs, particularly the 75% fibrous

Results and Discussion

Structure	Min.X / ms^{-2}	Max.X / ms^{-2}	**Mean X / ms^{-2}**	Min.Y / ms^{-2}	Max.Y/ ms^{-2}	**Mean Y / ms^{-2}**	Min.Z / ms^{-2}	Max.Z / ms^{-2}	**Mean Z / ms.2**
Control	-14.4	0.9	**-3.2**	-3.6	9.7	**0.9**	-17.3	13.8	**5.6**
50% Tap	-23.8	5.6	**-4.7**	-3.3	7.7	**1.0**	1.2	21.2	**8.0**
75% Tap	-3.4	26.3	**4.0**	-58.2	8.9	**0.1**	-1.9	14.6	**5.4**
Full Tap	-2.9	17.1	**6.0**	-1.3	14.3	**1.0**	-6.5	3.5	**13.4**
50% Fibrous	-0.8	0.5	**-0.4**	0.0	0.3	**0.1**	6.9	11.3	**9.8**
75% Fibrous	-0.1	0.7	**0.5**	0.5	0.8	**0.6**	8.8	10.2	**9.8**
Full Fibrous	-0.4	0.3	**-0.2**	-0.1	0.7	**0.1**	9.0	11.9	**9.8**

Fig. 8.9 Table showing important acceleration values of all structures and control setup

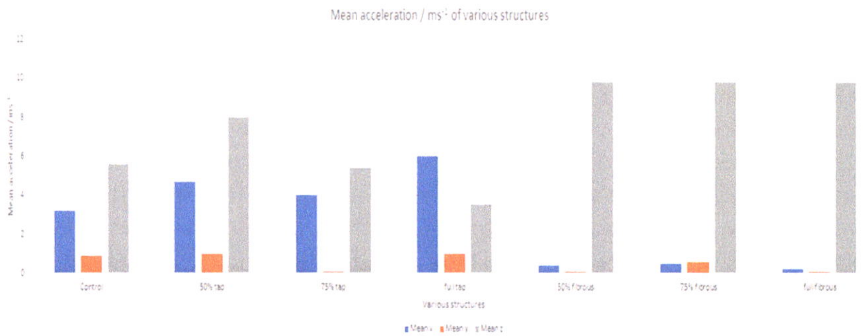

Fig. 8.10 Graph showing the comparison of mean accelerations for all setups

and full fibrous, would merely sink deeper into the ground upon activation of soil liquefaction and then remain in that position regardless of the speed or direction of the flow of sand.

As can be observed from Fig. 8.10, the longer the branches which extend outwards of the design, the lower its average acceleration values, when compared within the two families of designs (tap vs. fibrous). In particular, the mean *X* and *Y* acceleration values of the 75% fibrous structure is five times that of the full fibrous structure. This suggests a trend that the further out the branches of the designs expanded out, the more stable the structure.

The fibrous roots designs were completely undamaged after simulation, with neither loose nor broken pieces, while there was some damage to the taproot designs, suggesting that the taproot designs had engineering flaws. There were several loose and broken pieces, which were likely due to high stress and strain by external forces on them. This therefore suggests that the taproot designs in general were not as suitable to be applied in earthquake-prone countries as they would not be able to withstand soil liquefaction during earthquakes.

In particular, the most suitable structure that could possibly be implemented in the future in earthquake-prone countries is the full fibrous design. It had the lowest average acceleration for the *X* and *Y* axes as compared to all the other structural designs. This implies that it is the most structurally sound and has the least engineering flaws.

8.4 Conclusion

In conclusion, as seen from the results, the fibrous root designs are a better option than taproot designs in the designing of underground structures to mitigate the effect of soil liquefaction, as the fibrous root designs elicited much lower acceleration values in all three directions when compared to their taproot counterparts. Fibrous root designs were also found to be the least damaged after the simulation when compared to their taproot counterparts, most likely due to engineering flaws in the design that allowed the sand to apply a greater shear force which damages the structure. It is also noted that the greater the length of the branch extending outwards, the greater the surface area of the structure in contact with the sand, and the lower the overall acceleration values of the brick above. Thus, the full length fibrous root design is the most effective structure in mitigating the effects of soil liquefaction on the building above and can be implemented in earthquake-prone buildings in the future.

8.5 Future Work

Based on the results achieved from the experiment, further investigations can be done with regards to the physical properties of the structure itself. This would include the mass, material, stress and strain threshold of the structure. New designs can also be tested against current designs for its effectiveness in mitigating the effects of soil liquefaction on the building above, and buildings of different masses and sizes can be tested as well to assess the reliability of the structures even more thoroughly.

Annexe

See Figs. 8.11, 8.12, 8.13, 8.14, 8.15, 8.16 and 8.17.

Annexe

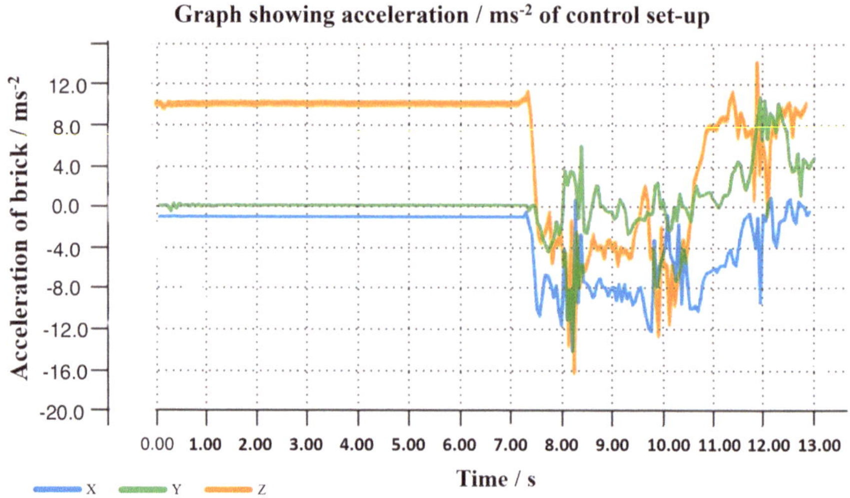

Fig. 8.11 Graph of mean acceleration of brick during control setup

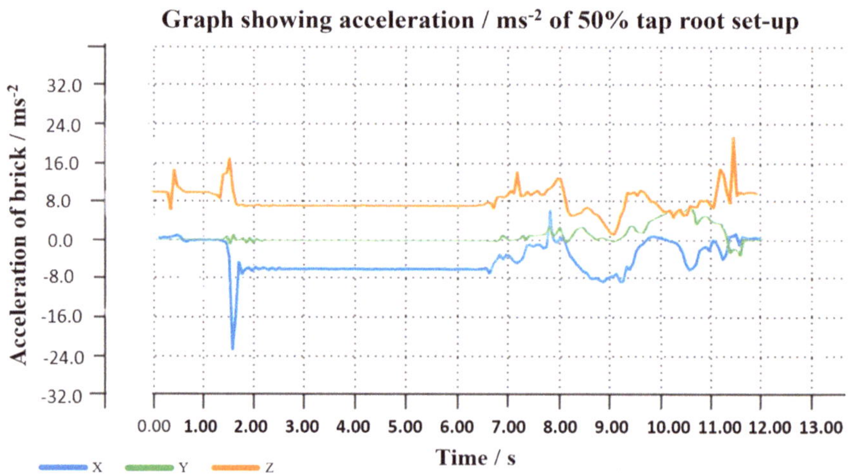

Fig. 8.12 Graph of mean acceleration of brick with 50% taproot structure

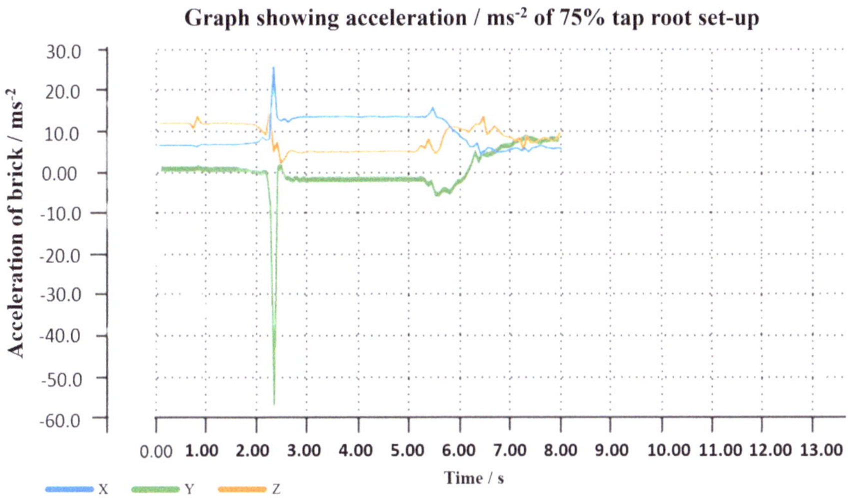

Fig. 8.13 Graph of mean acceleration of brick with 75% taproot structure

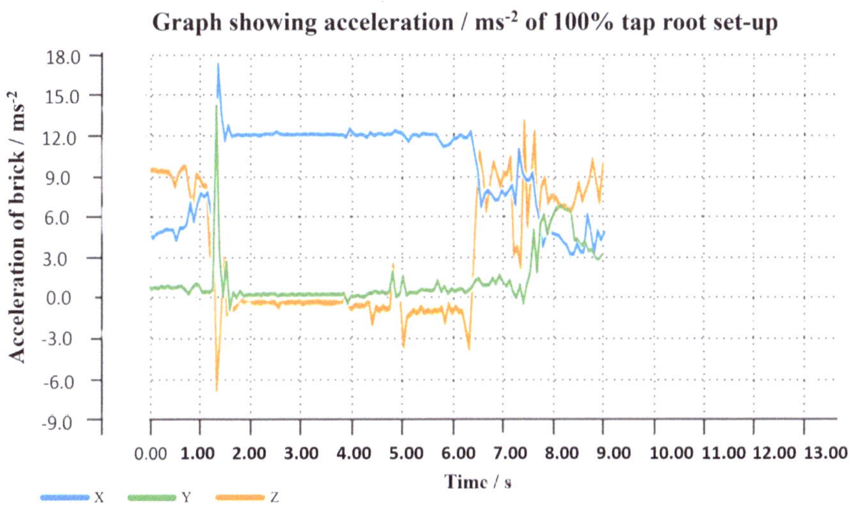

Fig. 8.14 Graph of mean acceleration of brick with full taproot structure

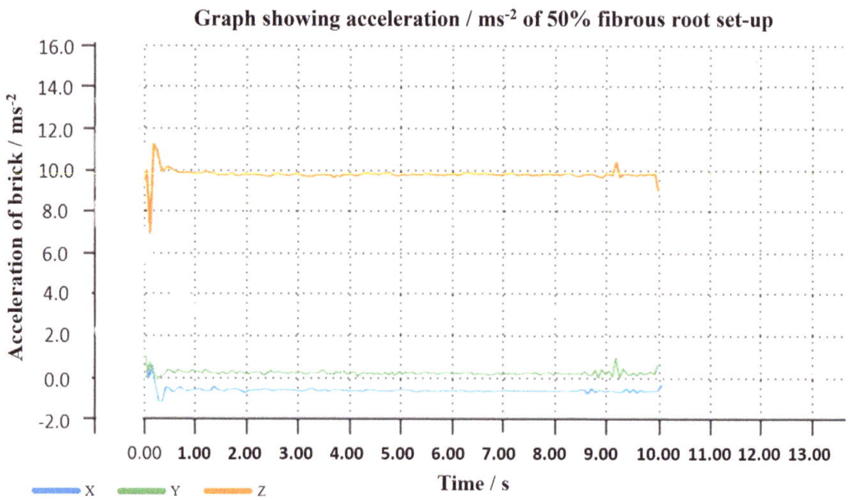

Fig. 8.15 Graph of mean acceleration of brick with 50% fibrous root structure

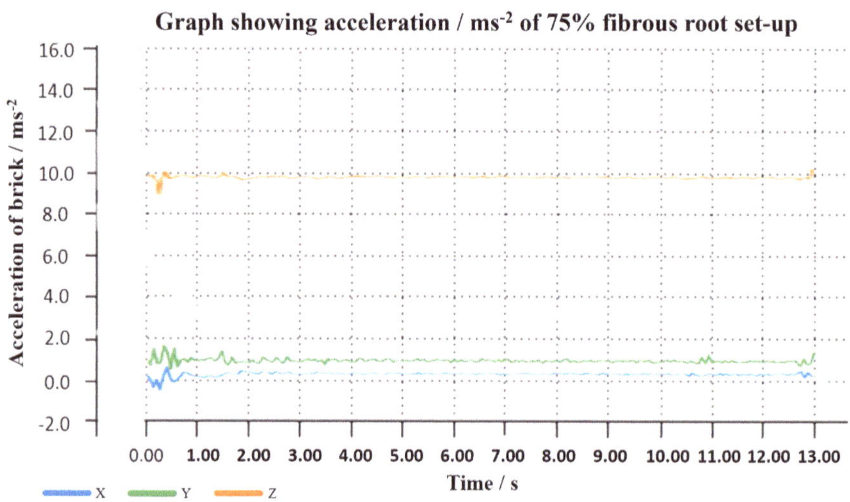

Fig. 8.16 Graph of mean acceleration of brick with 75% fibrous root structure

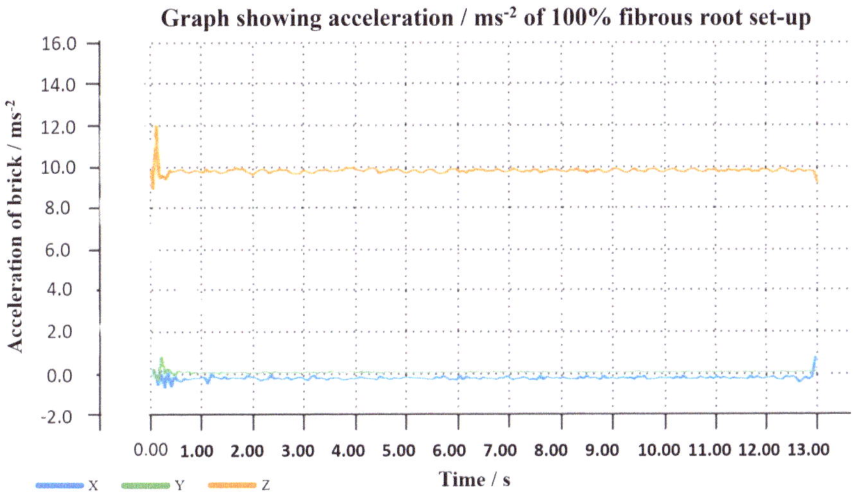

Fig. 8.17 Graph of mean acceleration of brick with full fibrous root structure

References

1. Earthquake Hazards (n.d.). Retrieved April 6, 2020, from https://earthquake.usgs.gov/learn/facts.php
2. Building Collapse: Some Precautions You Must Take (2016, October 15). Retrieved April 4, 2020, from https://www.icicilombard.com/insurance-information/home-insurance-info/article/building-collapse-some-precautions-you-must-take
3. Geography You and You (2018, May). *What is soil liquefaction? Causes, effects and measures*. What is soil liquefaction. https://geographyandyou.com/what-is-soil-liquefaction/
4. Real Estate Japan (2017, November 22). *Earthquake building codes & technology in Japan*. Resources—Real Estate Japan. https://resources.realestate.co.jp/buy/earthquake-building-codes-and-technology-in-japan/
5. McKenna, P. (2018, April 11). *Seismic cloak successfully deflects earthquake waves*. Retrieved January 21, 2021, from https://www.pbs.org/wgbh/nova/article/earthquake-shields/
6. American Institute of Chemical Engineers (2014, November). *Introduction to fluidization* (pp. 1–3). Retrieved from https://www.aiche.org/publications/cep

Chapter 9
Enhanced Real-Time Raw sEMG Signal Classification Through Bypass of Manual Feature Engineering and Extraction

Shriniket Subramanian, Narayanan Aravind, and Kane Ng Zheng Kang

Abstract The use of surface electromyographic (sEMG) signals obtained from the residual stump of an arm has been embraced for the purpose of developing prosthetics and bionic limbs. These sEMG signals are processed and classified into gestures. As the signals are stochastic and non-stationary, classification of these signals is a complex task. Conventional methods of feature recognition require handcrafted, manual feature extraction followed by classification. This paper explores the bypass of this preliminary step of feature extraction in order to reduce lag and computing complexity. Through data collected with the aid of an 8-channel Myoband developed by Thalmic Labs, our group trained and implemented three main raw data classifier models for real-time sEMG signal classification of nine gesture classes. The three methods presented are a raw naïve Bayes model implementation (81.55%), a novel raw sEMG ConvNet (96.88%), and finally, a novel ConvXGB model implementation (90.62%). The ConvXGB implemented is an adaptation of the new deep learning model developed in 2020. The significance of these novel models developed is the reduction in computational resources required for real-time classification of signals, leading to a reduction in lag time in real-time prosthetic devices, and the increased viability of embedded-systems classifications. The findings have the potential to also pioneer a new generation of more responsive prosthetic devices that are also easy to control.

Keywords Multi-channel time series classification · sEMG signals · Prosthetic devices · Bypassing feature extraction · ConvXGB model · GRU-FCN model · Naïve Bayes model

---

S. Subramanian (✉) · N. Aravind · K. N. Z. Kang
NUS High School of Math and Science, Singapore, Singapore
e-mail: H1610129@nushigh.edu.sg

H. Guo et al. (eds.), *IRC-SET 2021*, https://doi.org/10.1007/978-981-16-9869-9_9

9.1 Introduction

The loss of a forearm is a traumatic experience and severely impairs an individual's quality of life. Every year, around 50,000 people undergo amputation in the USA alone, and it is unimaginable as to how disrupted their life will become post-procedure. As such, biomedical research into prosthetics has gained traction over the past few years, and many prosthetics have been developed to attempt to achieve comparable degrees of freedom to the actual human hand. However, it has also increasingly come into question as to how natural control of the prosthetics is to be achieved. On top of degrees of freedom, main design goals of prosthetics would be to be lag free and be lightweight.

Many state-of-the-art prosthetics achieve control by detecting impulses in the muscles in the user's residual limb using mounted differential electrodes. These differential electrodes are invasive and need to be inserted under the skin of the user. While these electrodes provide more precise signals of greater amplitude and frequency, they are also inconvenient, intrusive, and introduce a chance of complications. The Thalmic Labs Myoband has thus been seen as a potential game changer by researchers for this purpose. The bands are worn on the subject, and collect data using non-invasive surface electrodes, that do not require prior preparation of the subject. The bands detect 8 channel surface electromyographic (sEMG) data from the muscles and wirelessly transmit the data to any device. These bands simplify the data collection process and are hence embraced by researchers. We have adopted this technology for acquiring data in this project. Electromyographic signals are complicated signals influenced by a range of physiological and anatomical properties such as muscle fatigue, muscle size, fiber type, orientation, electrode placement, impedance of skin, and more. While it is more convenient, sEMG signals are less precise and may carry more crosstalk than EMG signals from differential electrodes, the signal processing is further complicated. Being non-stationary stochastic signals with multiple channels of input [1, 2] and over 100 timesteps per gesture [3], sEMG signals tended to be too complicated for models to make sense of directly [4].

Sequence-based models such as (K-nearest neighbors) KNNs that can work with one input channel become too computationally expensive and impractical to implement in real time as the lag has to be kept within 200–300 ms [5] for the system to feel responsive. As such, manual feature extraction is often carried out to reduce the dimensionality of the data. These features are then fed into a classifier. Over time, accurate feature extraction from EMG signals has become the focus for accurate classification in both real-time and offline systems and is essential to the motion command identification. The performance of traditional feature-based methods depends on the quality of handcrafted features. However, it is difficult to design good features to capture all intrinsic properties of EMG to classify the various gestures accurately due to the signal's non-stationary nature. Thus, models are additionally limited by the feature extraction capabilities available. With manual feature extraction, the classification accuracy of nine classes in real time as of 2020 is currently 92.1% [6].

With current models having to extract over 50 features [6] in order to reliably classify nine classes of data, the lag time due to feature extraction is simply too much, and greatly exceeds the required 200–300 ms for a generally responsive prosthetic device. Inspired by recent advances in image classification, where automatic feature extraction occurs, this project investigates the feasibility of implementing a real-time classification system without manual feature extraction. Bypassing the computationally expensive feature extraction step allows for the lag time to be smaller. Automatic feature extraction might allow the model to be more accurate as well as the optimal feature set for each gesture is not fixed and can be learnt.

We thus developed several different artificial intelligence (AI) models capable of classifying this data and relaying the results to a mechanical hand.

9.2 Methodology (Data Collection)

One of the authors for the study was recruited for surface electromyography (sEMG) data collection.

Muscular activity was gathered using 2 Thalmic Myobands on the right forearm, each having eight non-invasive sEMG electrodes. This paper will refer to the band closer to the wrist as the lower band and the one further from the wrist as the upper band. Data was collected wearing two bands at the same time to reduce time spent collecting data, while ensuring both upper and lower placement data are not affected by the small inconsistencies between every repetition of each gesture. As a result, 8-channel data was collected for every movement at the maximal 200 Hz allowed by the Myobands. Nine classes of gestures were collected, with 30 repetitions in each class (see Fig. 9.1) [7, 8].

The Thalmic Myoband collects 200 data points every second, each expressed as an array of eight elements corresponding to the signal detected at each of the eight electrodes of the band. The data will then be stored as a text file with eight data points separated by spaces in each line, for each timestamp.

Two separate datasets were collected on different dates, one collecting 1 s of data per gesture (200 lines) and another collecting 2 s of data per gesture (400 lines). We will refer to the two datasets later on as the 200-line dataset and the 400-line dataset for clearer understanding [8].

The timer begins when the subject begins to form the gesture, from a neutral rest position. This method of data acquisition was practiced in place of recording 1 s of holding the gesture, because the nine gestures involve closely related muscles to hold the gesture, but obviously activate different muscles to perform the gesture from rest. For example, the index and middle gestures have similar sEMG oscilloscope traces and have roughly the same muscles activated, but the difference lies in the potentials fired at the start of the gesture. Rest was given as requested by the subject before the recording of each action to avoid fatigue.

The nine gestures were chosen primarily due to their prevalence in usage in both common literature and in the real world. Note that some of the gestures that we

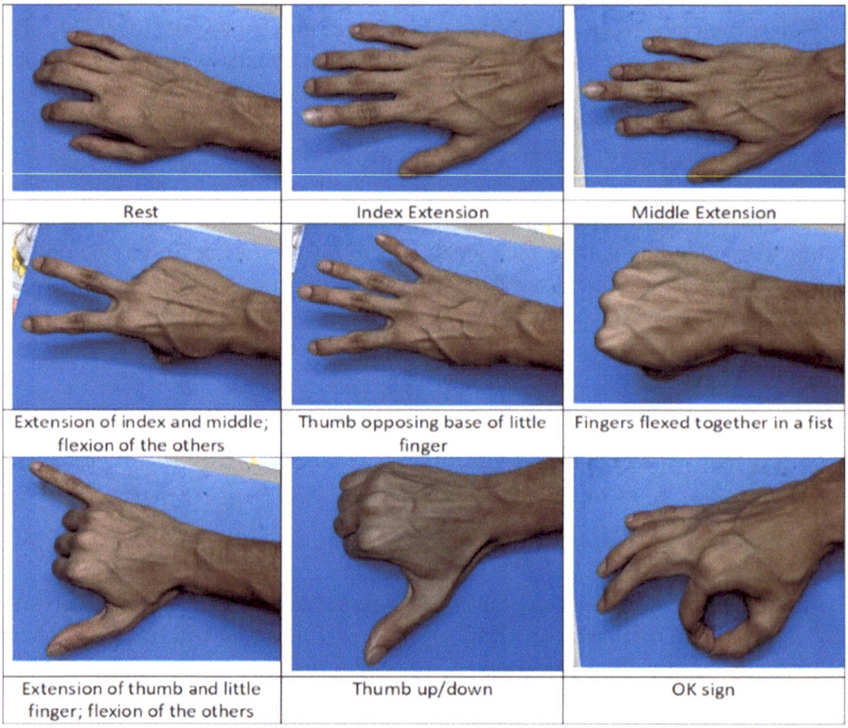

Fig. 9.1 Images of nine performed gestures

trained the classifier with are named after numbers, where the gesture represents the one-handed representation of a one-digit number. For example, to show the number four, all fingers bar the thumb are extended.

Four other plausible gestures such as large diameter grip, four gesture, three gesture, and seven gesture were trialed to form a 13-class dataset but were omitted due to lack of distinct data obtained from the gestures. Instead, a net 9-class dataset was used to train all the classifiers instead. In later parts, we will refer mainly to this 9-class dataset, and in certain parts will refer to the 13-class dataset as the **trial** classification dataset.

The data collected from each gesture was written into a text file by the Python program and saved separately. Following which, data was read from all the text files to save in arrays and NumPy objects for manipulation and training. Using the data collected, multiple datasets with variable number and combination of classes were created to test the efficacy of certain elements in the data that can contribute to the classification performance.

The data collected is represented as an (1, 8, 200) array for the 200-line dataset and (1, 8, 400) for the 400-line dataset. This arrangement of data is manipulated into various data formats for loading into the various models. Some strategies used by the team in manipulating the data are elaborated below. A form of data manipulation we trialed was filtering.

When passing through various tissues, EMG signals acquire noise. Noise is the unwanted electrical signal in an EMG signal. Noise is primarily from the ambient electromagnetic radiation, inherent noise in electrical devices and electrode impedance. Motion artifacts are introduced into the signal when the electrodes move relative to each other when a muscle flexes prominently. This noise and artifacts in the signal are serious issues to be considered, as this will adversely affect the quality of the signal. As these factors cannot be remedied easily without upgrading to a more complex electrode set up, we have decided to implement noise removal systems. As such, the team designed and adapted a Butterworth bandpass filter for this purpose to filter out noise that is not in the dominant frequency range of the band. We identified that the dominant frequency spectrum of the signals from the band was between 10 and 500 Hz through literature data [9]. A graph showing the difference between the noisy signal and the filtered signal is shown (Fig. 9.2). The impact of this designed filter will be discussed later.

For the naïve Bayes model which could not take the raw data in its multi-dimensional state as an input, the team carried out a flattening of the 3D (1 * 8 * 200) array into a 2D (1 * 1600) array.

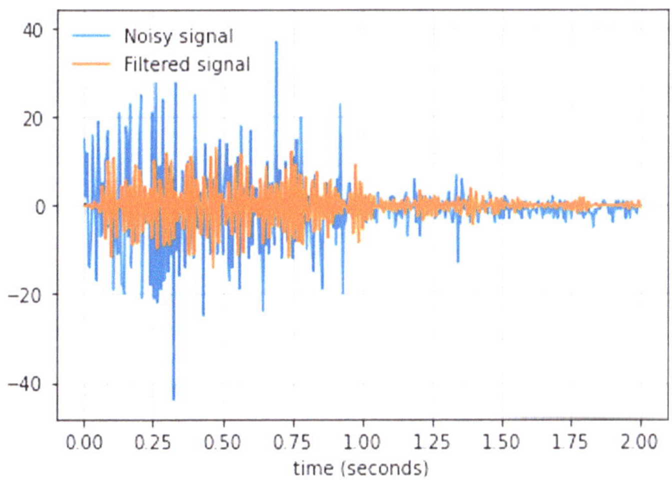

Fig. 9.2 sEMG trace of noisy versus filtered signal

9.3 Models Developed

9.3.1 Naïve-Bayes Classifier

The naïve Bayes model was trained using the 400-line dataset because it performed better with this dataset as compared to the other 200-line dataset.

We use this model, as it is a very commonly used model, albeit with extracted features. We can thus compare other solutions with this approach to check how much better other approaches fare when pitched against this model.

Raw data in the shape of (x, 8, 400) was flattened to form an x * 3200 array and input into a naïve Bayes model developed with the aid of Python's scikit-learn package. The naïve Bayes model was also optimized using grid search to find the optimal smoothing variable for best observed accuracy.

It is worth noting that such flattening of data, reduces the interdependency between the channels therefore potentially resulting in frequent misclassifications as the models will not be learning any properties specific to the interaction between the eight distinct channels of the Myoband.

This model developed is not state of the art; it was simply an elementary approach by which we can gauge the performance of the models in the absence of feature extraction. While such an approach compromises accuracy, it is fast.

Regardless, accuracy is still an important factor in determining the efficacy of the model, and it is not desirable to have the prosthetic device perform some other movement when given a gesture by the user.

Hence, we describe in later parts, other solutions to match, if not exceed the accuracies that are achieved by the current line of state-of-the-art models which practice feature extraction.

9.3.2 Gated-Recurrent-Unit, Fully-Connected-Network

A typical GRU cell consists of two gates: reset gate and update gate. The GRU FCN model was chosen due to GRUs being able to combat vanishing or exploding gradient problems due to its reset and update gates [10] and being easier to implement than LSTMs.

For the GRU, raw data was simply loaded as is with a data shape of (x, 8, 200) where x is the number of samples available. This model turned out to be competitive with state-of-the-art models, while showing minimal and consistent processing times. The model was also tested with 400-line data and 200-line data and was found to perform better with the 200-line dataset. This finding is important and is analyzed in later parts.

9.3.3 *Modified ConvXGB Classifier*

As an outstanding classifier and feature extractor, CNNs have achieved great success, especially in the field of image recognition [2]. Inspired by this advantage, we also exploit the GRU-FCN above as the feature extractor, and a new classifier XGBoost to replace the dense soft-max classifier. XGBoost is an efficient implementation of gradient boosted decision trees (GBDT) due to its block structure to support the parallelization of tree construction. In GBDT, gradient boosting refers to a kind of ensemble technique creating new models to predict the residuals or errors of prior models and making the final decision by summing up the predictions from all models. This model was chosen to be implemented due to its promise of strong feature extraction and accurate classification due to the blend of CNN and XGBoost. As such, we have implemented a novel GRU-CNN-XGB. In 2020, the general ConvXGB model was introduced [9]. Our implementation also includes a GRU for feature extraction and is the first of such models to be applied in this field, to our best knowledge [11].

9.4 Results and Discussion

Our team obtained four main datasets as a result of the manipulations described in data collection segment. There is a 200-line dataset and a 400-line dataset which are raw datasets derived directly from the Myoband. Secondly, through the designed Butterworth filter, we derived two filtered datasets for the 200-line and the 400-line, respectively.

We tried to optimize the model performance by identifying the dataset that led to the best accuracy for each of the models through trial and error.

It was found that the models performed better with the unfiltered datasets. It was further observed that the naïve Bayes model performed better with the 400-line dataset while the GRU and ConvXGB performed better with the 200-line dataset. As such, models were ultimately trained with the dataset they performed best with, and we obtained the following results.

The graph (Fig. 9.3) shows the highest accuracies that each of the three models described in our project has managed to yield. The orange line shows the current state-of-the-art accuracy, and it is worth noting that the GRU-FCN surpassed this standard while the ConvXGB is close to matching it. Unsurprisingly, the naïve Bayes did not perform as well.

The accuracy of the classical naïve Bayes model was 81.55%. The GRU-FCN and ConvXGB models were able to achieve accuracies that are well above that of the naïve Bayes model and very competitive with the conventional accuracy of 92.1% for nine classes of gestures. The results show that bypassing feature extraction is possible without compromising the accuracy classification. Further, the GRU-FCN and ConvXGB models are competitive alternatives that can be considered when

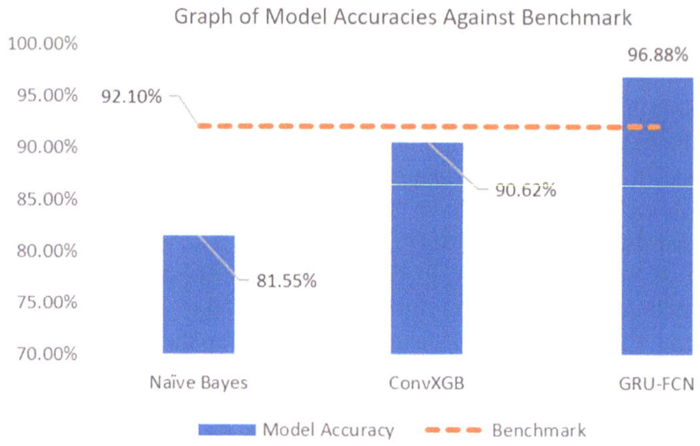

Fig. 9.3 Accuracies of models against testing data

designing prosthetic devices seeing as they are both faster and as accurate as state-of-the-art alternatives.

In addition to the above results, the data supports the conclusion that the lower placement of the band allows for more distinct data to be collected per class, leading to better classification performance. This result is consistently reflected in both the 9 class and **trial** 13 class classification (see Fig. 9.4) using the GRU-FCN model. This result can be explained biologically, with greater number of finer muscles fibers present as we progress down the forearm toward the wrist, which contribute to more distinct signals for each gesture when the band is placed lower along the forearm.

We also mentioned briefly earlier that the GRU-FCN performs better when the 200-line dataset is used. (see Fig. 9.5) a plausible explanation involves data density.

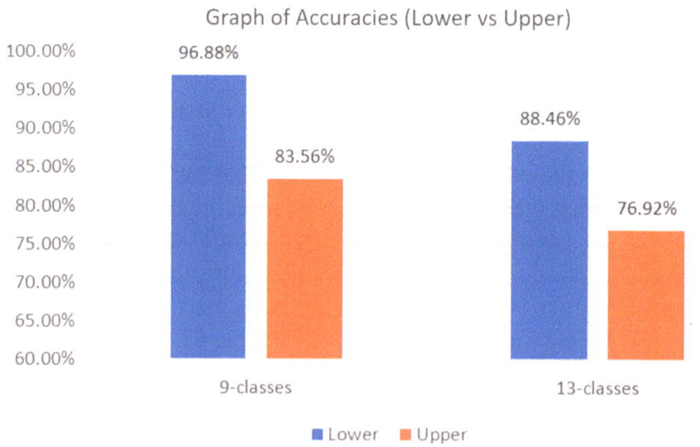

Fig. 9.4 Comparison of model performance between lower and upper band data for different training classes

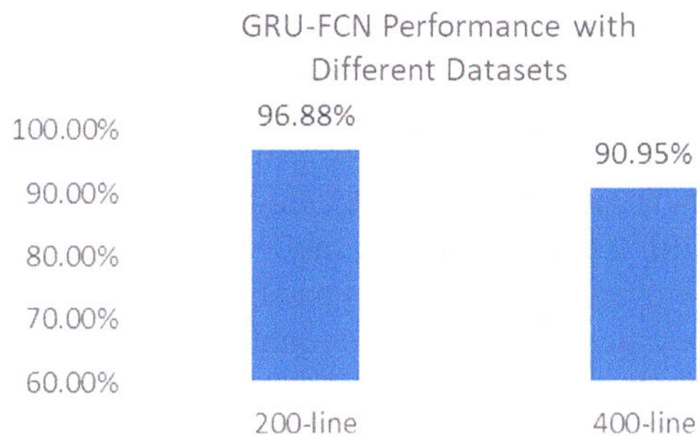

Fig. 9.5 GRU-FCN performance versus dataset types

The 400-line dataset involves a longer sequence to consider which results in more points being stored in short-term memory, leading to a greater chance of false associations and patterns being discovered, which reduces classification accuracy. This finding is useful and can be further explored in an attempt to boost model accuracies.

It is worth noting that the models developed did not perform too well with the filtered dataset. This was surprising as the filter that was developed with the goal of improving model accuracies; however, this was not observed and instead the models performed rather sub optimally instead in real time. After careful analysis of other studies involving sEMG signals and the physiological aspect of the electrodes, we have a plausible explanation. EMG electrodes are placed in the skin, closer to the muscles unlike sEMG electrodes which are on the surface. This means that the motion artifacts for EMG signals generally remain the same across gestures, while they differ between gestures sEMG data. That is to say, the noise generated is unique to each gesture and can enhance classification. Filtering reduces these motion artifacts and thus leading to a loss of important data. This results in a lowered model accuracy when fed filtered data. On the flip side, this means that using sEMG technology requires even lower computation, as filtering is not required.

Finally, yet most importantly, in terms of lag time, our models performed exceptionally well with the classifiers showing a remarkable 172.499 ms ($\pm$ 35.36977 ms) average lag time. Our models thus successfully fall under 300 ms of lag standard and thus fulfill the original objective of the implementation and bypass of the feature extraction step in order to improve lag times. Laboratory studies on pattern recognition usually report 300 ms of response lag [12]. Hence, the models developed are not only of comparable accuracies to existing alternatives but are also significantly faster. This fulfills the engineering goal of designing faster, yet accurate classifier models such that the responsiveness of prosthetic devices in real time can be improved [13].

9.5 Implications and Conclusions

This paper thus contributes three plausible novel implementations of raw sEMG signal classification models without feature extraction through exploration of models such as naïve Bayes, GRU-FCN, and GRU-ConvXGB. The GRU-ConvXGB and GRU-FCN models developed are also of comparable accuracy to state-of-the-art models that use feature extraction, and thus are equally as competitive and lucrative for prosthetics developers to consider.

The benefits of the models within this research paper include but are not limited to lightened computing complexity and reduced signal processing times when used to classify signals in real time. We highlight that reduced processing times are integral when it comes to mimicking the real hand with prosthetics. It would allow faster reaction of the prosthetic itself and thus would make it more cohesive and authentic. The novel approach explored by this paper, to the best of our knowledge, is one of the first of such endeavors in reducing lag times and can be further explored in future.

9.6 Limitations of Findings

However, there are certain limitations to the methods proposed and implemented. The data collected in this report is from an able-bodied participant. However, prosthetics largely concern users without a physical hand. As such, data derived from the residual stump of a hand—needs to be considered for a more realistic case scenario. Further, these gestures will need to be tried among a few human subjects to determine whether the EMG trace observed is user-specific or generally observed for any subject.

Further data can be collected from various subjects for training of classifiers to increase the generalizing abilities of the models as well.

Moreover, it was observed that there was some deviance in terms of data collected between different bands. This is something worth noting especially because most would assume that the data would be consistent across the same band product. As such, models may have to be retrained depending on the band being used for the prosthetic devices.

9.7 Future Work

Further optimization can be carried out to render the models to be more accurate as accuracy is highly important in real-time prosthetic devices. We believe that dataset sizes could be boosted to achieve this purpose such that the model can learn more detailed features of the data collected.

Further, we also recommend exploring variable timesteps of data recording. As mentioned earlier, we found deviations in the GRU-FCN's performance with 2 and

1-s timestep data. As such, this deviation can be explored further, and we could thus possibly improve the accuracy of the model further, by considering other possible timesteps of data collection.

Appendix

	Accuracies		
	GRU-FCN (%)	Naïve Bayes (%)	ConvXGB (%)
200 lines (1 s of data) (9 classes)	96.88	81.55	90.62

GRU-FCN	Lower		Upper	
	Loss	Accuracy (%)	Loss	Accuracy (%)
9 classes	0.4029	96.88	0.7225	83.56
13 classes	0.3854	88.46	1.1370	76.92

References

1. Lin, H. (2019, November 1). A multivariate time series classification method based on self-attention. https://doi.org/10.1007/978-981-15-3308-2_54
2. Zheng, Y., Liu, Q., Chen, E., Ge, Y., & Zhao, J. L. (2014, June). Time series classification using multi-channels deep convolutional neural networks. In *International conference on web-age information management* (pp. 298–310). Springer, Cham. http://staff.ustc.edu.cn/~cheneh/paper_pdf/2014/Yi-Zheng-WAIM2014.pdf
3. Reaz, M. B. I., Hussain, M. S., & Mohd-Yasin, F. (2006). Techniques of EMG signal analysis: Detection, processing, classification and applications. *Biological Procedures Online, 8*(1), 11–35.
4. Hatami, N., Gavet, Y., & Debayle, J. (2018, April). Classification of time-series images using deep convolutional neural networks. In *Tenth international conference on machine vision (ICMV 2017)* (Vol. 10696, p. 106960Y). International Society for Optics and Photonics.
5. Singh, R. E., Iqbal, K., Rasool, G., White, G., Wingate, S., & Hutchinson, H. Muscle synergy extraction from EMG (electro myogram) data: A comparison of algorithms (PCA, NNMF, pICA).
6. Scheme, K. E., Nurhazimah Nazmi, M. A., Tenore, A. R. F. V., Zecca, S. M., Guo, M. P. S., Phinyomark, P. P. A., & Kuiken, G. L. T. (2019, November 21). Evaluation of surface EMG-based recognition algorithms for decoding hand movements. Retrieved May, 2020, from https://doi.org/10.1007/s11517-019-02073-z
7. Balandinodidonato (n.d.). Balandinodidonato/MyoToolkit. Retrieved September, 2020, from https://github.com/balandinodidonato/MyoToolkit/blob/master/SoftwareforThalmic'sMyoarmband.md
8. Eliotxu (n.d.). Eliotxu/pyMyo. Retrieved August, 2020, from https://github.com/eliotxu/pyMyo

9. Englehart, K., & Hudgins, B. (2003). A robust, real-time control scheme for multifunction myoelectric control. *IEEE Transaction on Biomedical Engineering, 50*, 848–854.
10. Chung, J., Gulcehre, C., Cho, K., & Bengio, Y. (2014). Empirical evaluation of gated recurrent neural networks on sequence modeling. arXiv:1412.3555
11. Thongsuwan, S., Jaiyen, S., Padcharoen, A., & Agarwal, P. (2020, August 02). ConvXGB: A new deep learning model for classification problems based on CNN and XGBoost. Retrieved December, 2020, from https://www.sciencedirect.com/science/article/pii/S1738573319308587
12. Huang, Y., Englehart, K. B., Hudgins, B., & Chan, A. D. C. (2005). A Gaussian mixture model-based classification scheme for myoelectric control of powered upper limb prostheses. *IEEE Transactions on Biomedical Engineering, 52*, 1801–1811.
13. Chan, A. D. C., & Englehart, K. B. (2005). Continuous myoelectric control for powered prostheses using hidden Markov models. *IEEE Transactions on Biomedical Engineering, 52*, 121–124.

Chapter 10
Transforming and Solving Multi-channel Time Series Classification Problem Using Image Classification Techniques

Applying Knowledge From Maturing Image Classification Domain to Relatively Unexplored Problem of Complex Multi-channel Time Series Classification

Shriniket Subramanian, Narayanan Aravind, and Kane Ng Zheng Kang

Abstract The problem of multi-channel time series classification (MCTSC) is a widely unexplored one, with current state-of-the-art solutions, taking a precursor step of feature extraction to reduce the dimensionality of the data before being fed into classical classifiers. However, the problem with this approach would be the computationally expensive and labor-intensive feature extraction process, and the lag times associated with the process in real time. It is known that over 50 features ought to be extracted to reliably classify up to 9 classes of gestures using sEMG signals. This is quite inefficient although it may still achieve the end-goal of high classification accuracies. As such, the main solution to MCTSCs is to transform the longer sequence into a shorter one. This is a largely limiting technique. In stark contrast, the domain of image classification is a widely explored topic in recent times, with many innovative models and applications having been developed to fulfill the purpose of classifying images correctly with high accuracies quickly for real-time computer vision. Image classification is a largely mature field. This paper aims to show a proof-of-concept for translating the complicated MCTSC problem into a well-researched image classification problem such that we can better streamline the classification algorithm. This will allow us to tap on the capabilities developed in other fields well, such as dataset augmentation and advanced models. The paper overviews several approaches of translating raw, quantitative data into the form of images which will be fed into two specifically developed models (ImageNet model

S. Subramanian (✉) · N. Aravind · K. N. Z. Kang
NUS High School of Math and Science, Singapore, Singapore
e-mail: H1610129@nushigh.edu.sg

N. Aravind
e-mail: H1610092@nushigh.edu.sg

K. N. Z. Kang
e-mail: H1610046@nushigh.edu.sg

H. Guo et al. (eds.), *IRC-SET 2021*, https://doi.org/10.1007/978-981-16-9869-9_10

and convolutional neural networks), specifically designed for this purpose. We also overview some advanced image classification techniques that we also used to improve the accuracies of the classifiers in real time such as data augmentation.

Keywords Multi-Channel Time Series Classification · Image Data Classification · Prosthetic Devices · Convolutional Neural Network · Raw Data to Image Translation · ImageNet Classifier

10.1 Introduction

Multi-channel time series classification problems, specifically in the context of interpreting surface electromyographic muscle signals (sEMG), is useful in the design of artificial bionic limbs where the user can control a prosthetic device attached to his forearm, with wireless, non-invasive Myobands. These Myobands detect 8-channel sEMG signals from the arm, with distinct signals observed for different gestures. The data that can be obtained from the band is in the form of a $1 * 8 * x$ array with x being a variable depending on the amount of time the band is collecting data for. 200 lines of data is collected for 1-s timesteps, and 400 lines of data is collected for 2-s timesteps of data collection. The data in its raw form, however, is too complex to be given as input for training into classical models due to its multi-dimensional state. Nevertheless, instead of extracting features and reducing the dimensionality of the data, our team instead represents the data in the form of an image that can be given as input into an image classifier. The art of hiding data inside an image is not new, with steganography, a commonly practiced form of cryptography in modern times. However, instead of steganography, our team instead practices a novel approach of fabricating an image from scratch, for each sEMG (surface electromyogram) data point such that we can obtain an image dataset that can be used to train the image classifiers. In real time, the data can also similarly be converted into images and be classified accordingly. The advantage of doing this is the bypass of the manual feature extraction step as well as the transformation of a more complex multi-channel time series classification problem into a more generally well-known image classification problem. It also allows us to build up on years of research and use the various findings to better our results.

10.2 Data Collection

Data, for the purpose of training the classifiers, was collected through the recruitment of one subject to perform several common gestures with a MyoBand placed on his forearm. These bands are non-invasive and have eight sensors to detect sEMG signals from the forearm.

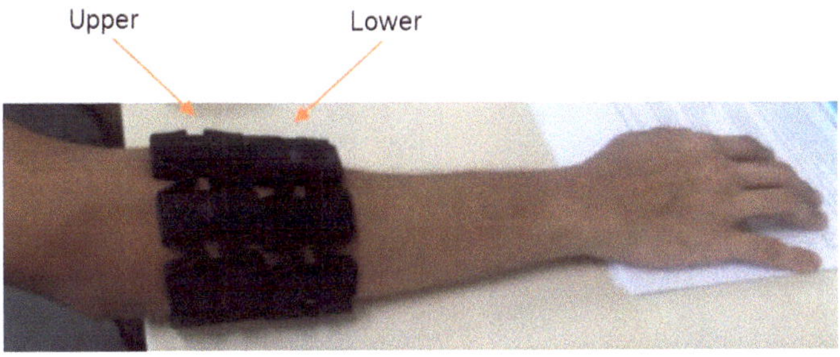

Fig. 10.1 Band positioning when collecting data

The data is saved into a text file and is later read and translated into images with methods described in the next part. An image dataset is assembled and two specifically developed artificial intelligence models are then trained with the datasets.

The data is also collected from two different band positions. We wanted to also investigate which band position generates better, distinct data for better classification. Two bands were worn by the subject, with one on the upper hand and one on the lower hand. This was done to reduce time collecting data and reduce inconsistencies when collecting data between upper and lower arm. The band positioning when collecting data is shown Fig. 10.1.

Note that two different image datasets are assembled with two different image translation approaches considered.

10.2.1 A Pixel for Every Point

As mentioned earlier, data in its raw form, as obtained from the MyoBand directly is in the form of a text file with 8 entries in each line for 400 lines or 200 lines depending on the timestep selected. This data can thus be encoded as an image by taking every value and matching a corresponding pixel with the value, thereby getting an image with 3200 or 1600 pixels. This image can then be fed into the image classifier for training.

As such, we took the value of every data point, added a weight of + 127 to it, to make sure every value is positive and formed a pixel with the corresponding value by assigning the R, G, and B channels of the pixel to be equal to that data point.

This image was initially represented as an 8×400 image but due to unsatisfactory results, was later adapted to be a 57×57 image with a few blank pixels such that the classifier can better learn the features of the image. A sample, generated image is shown below. Notice that you can even observe the visual differences with a naked eye (Fig. 10.2).

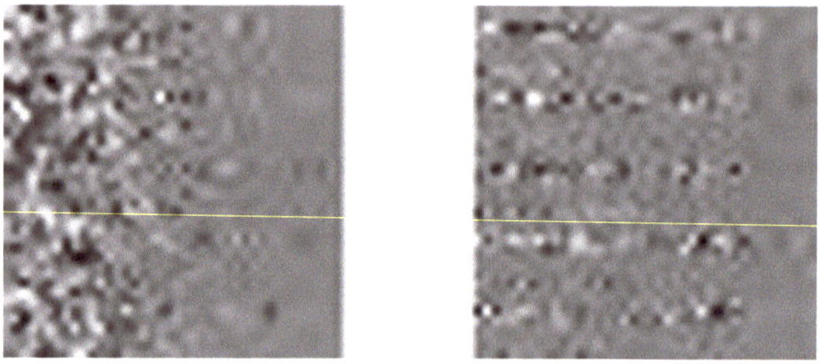

Fig. 10.2 Resized** example image for two different classes

10.2.2 Combinatorial Approach

However, by flattening the array into an image, the link between channels for the same timestep is reduced. However, the data between channels are indeed linked as they collect data at the same time, for the same gesture. It might be beneficial that the model learns this trait. As such, we introduced inter-dependency between channels into the image dataset using inspiration from the concept of linear combinations in linear algebra. Here is an image that explains the two methods.

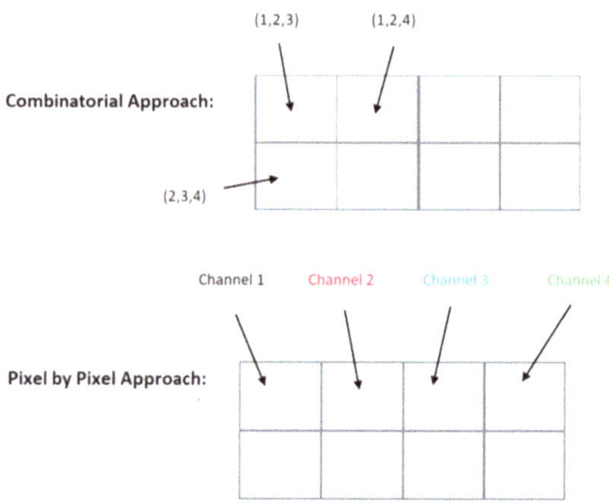

The image is formed by taking linear combinations of the eight channels in a completely random approach (i.e., by taking all possible combinations of three from a set of eight channels and performing linear combinations on them to form a pixel). We thus obtained a 56 × 400 image, and this image was then fed into a developed image classifier. An image output is shown below. A CNN model was developed and tweaked to handle the images translated with the parameters optimized for ideal classification of the following images (Fig. 10.3).

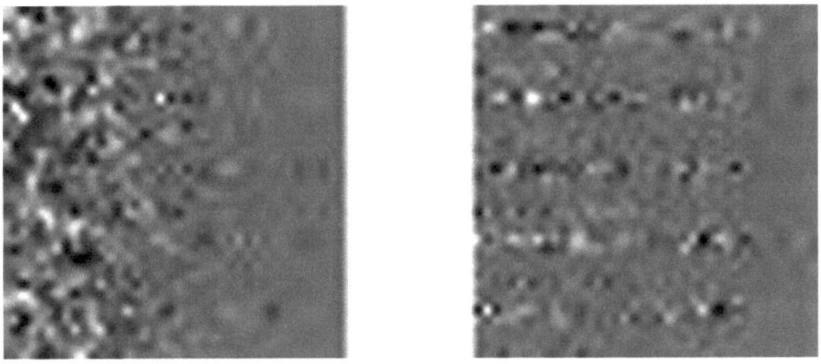

Fig. 10.3 Resized** example image output for combinatorial approach

10.2.3 Spectrograms

Spectrograms are a visual representation of the spectrum of frequencies of a signal as it varies with time. The fast Fourier transforms (FFT) is commonly applied in transformation of electromyography (EMG) signals from the time domain to the frequency domain. However, this technique has a limitation to provide the time–frequency information for EMG signals. This paper presents the analysis of EMG signal for contraction of muscle activity by using spectrogram. Spectrogram is one of the time–frequency representations (TFR) that represents the three-dimensional of the signal with respect to time and frequency in magnitude presentations.

Spectrogram provides high frequency resolution and can be calculated using Eq. (10.1):

$$S_x(t, f) = \left| \int_{-\infty}^{\infty} h(\tau) w(\tau - t) e^{-j2\pi f \tau} \mathrm{d}\tau \right|^2$$

where $h(\tau)$ is the input and $w(t)$ are the observation window. An EMG spectrogram looks like:

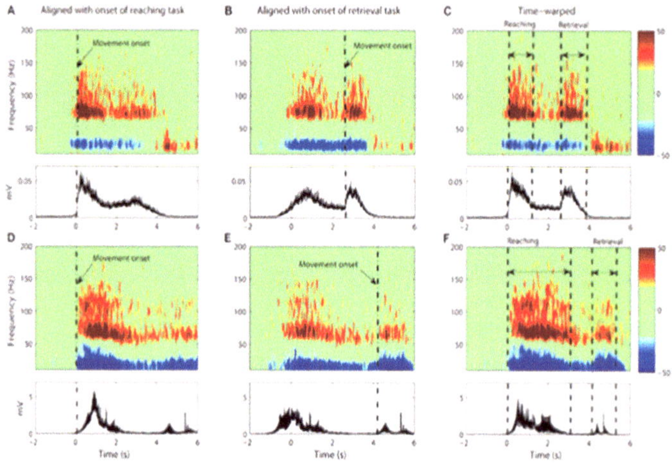

Firstly, generation of a spectrogram is really time- and computation-intensive process, which is not suitable for real-time classification. Secondly, spectrograms are very dense in information, but cannot represent multi-channel data well. Since we are using 8-channel data, this will be a big problem, resulting in significant data loss unless 8 spectrograms are made and overlapped which is imprecise and inaccurate. As such, this image creation technique should not be used.

10.3 Training the Models

As mentioned briefly earlier, two different image classifier models were developed to be trained with the image datasets.

The models developed are described below.

10.3.1 ImageNet Classifier

The collected raw data was encoded on a pixel-by-pixel approach as elaborated on earlier, for each of the 8-channel data to obtain a dataset of 40×40 images for the 200-line raw data and a dataset of 57×57 images for the 400-line raw data. To ensure that the data points were able to be translated into a pixel, a weight offset of $+\ 127$ was added to each of the data points to ensure that they fall in the [0,255] range for an RGB pixel. The resultant assembled image dataset was then fed into an ImageNet classifier to be trained. To prevent overfitting and greater generalization to work in real time, the dropout rate was tweaked to find the optimal rate. Further, batch normalization and image data augmentation were also carried out to facilitate generalization of the model, to make it more stable and improve training times. It was observed that the model had a significant improvement in classification accuracy when batch normalization and image data augmentation were performed. The techniques performed will be described in greater detail, later in this paper. However, these three techniques were primarily performed to improve the relevant accuracies of the ImageNet classifier in real time. By converting the sEMG data into images, we are effectively reframing a multi-channel time series classification problem into an image classification problem. The rationale behind such a method is to exploit the tools and techniques available in the image classification toolset [1]. This allowed us to use the ImageNet classifier and treat and optimize the problem using well-known techniques in image classification.

10.3.2 Convolutional Neural Network (CNN_56)

The resultant image dataset assembled from the combinatorial approach was then fed into a convolutional neural network to train and test. The CNN was adapted for the classification of sEMG signals through means of optimizing parameters of convolutional layers, number of dense layers as well as the dropout rate to find the optimal model in terms of performance. We were unable to use the ImageNet, as the image size of 56 * 400 is too large. Although we were unable to use pre-existing models, the increased familiarity and guidance to handle an image classification problem helped.

10.4 Image Classification Techniques

The purpose of translating the quantitative raw data into an image after all was to apply image data classification techniques that have been researched upon to improve the classifications. In this section, we elaborate on how translating the problem has enhanced our ability in the following points. The low classification accuracies could potentially be attributed to overfitting, due to the limited amount of data given to the model for training as well as the large disparities between the testing accuracies and the training accuracies. This means that the model is potentially memorizing the training data fed to it instead of learning them, resulting in poor classification accuracies in real time.

10.4.1 Tweaking Model Parameters

Rules and model optimization techniques were vague and not well defined in MCTCs. However, in image classification, such optimization techniques and manual algorithms to do so are quite well defined. This allows us to be more certain and precise with our models and confidently identify where the lapse is resulting in a lowered accuracy. Several model parameters were optimized to reduce potential overfitting of the training data.

Firstly, the dropout rate of the model was optimized with different values used, to find out which value trains the model the best. Dropout rate is a feature used to make the training data given, noisier such that the case where the neural net memorizes the small training dataset given to it, is minimized. Through random dropout of certain nodes, we found that the accuracies improved, confirming that the model did not perform well because of overfitting of the data points [2].

In addition, here are some general techniques used as well.

The activation functions of the classifier were trialed with functions like ReLU, Softmax and tanh used to determine which function worked best for training. Through trialing, we found that the Softmax activation function suited the training best with better testing accuracies observed, and thus, this activation function was adopted.

To combat the problem of overfitting, which was deemed to be the problem with the classification, we added the EarlyStopping function to cut off training when the model is starting to overfit. This ensured that the number of epochs for which the model was trained is optimized such that there is no overfitting or underfitting and such that the model is trained with the appropriate number of epochs.

10.4.2 Data Augmentation

Data augmentation is an advanced image classification technique which forces the model to learn different features of the image through expansion of the image dataset but with the same basic dataset. It artificially expands the training dataset and therefore aids the model in its learning as the training dataset is significantly expanded without exactly adding more data.

Augmentation techniques include rotating the image sideways, flipping the image, mirroring the image, zooming in, and many other techniques. This allows the model to see the image in a different light and therefore, learn more properties of the image that it may have missed out prior.

For this purpose, Keras image data generator was used in the model to dynamically the train the models with augmented images such that the model learns maximal properties of the image. Several augmenting techniques were trialed such as mirroring and flipping the image. Techniques such as zooming into the image was omitted due to unsatisfactory performance of the technique.

As such, by practicing such artificial expansion of the image dataset, the team was able to also improve classification accuracies by a marginal figure which nevertheless was useful in boosting the reliability of our model when put to the real-time classification test.

This is also an artificial expansion that lies unique to an image dataset, meaning that the transformation of the raw data into an image makes it easy for augmentation techniques to be practiced, which meets our goal of exploiting the existing image data classification toolkit to boost overall classification accuracies.

10.4.3 Pre-existing Models

The transformation of the data into images allows one to use image-based classification models. This means that instead of creating models from scratch or extensively modifying generic models, we are able to use transfer learning properties of trained models and harness existing architecture such as VGG, ImageNet, and more. This streamlines the classification process.

10.4.4 Generative Adversarial Networks

A Generative adversarial networks, or GAN, is a type of neural network architecture for generative modeling.

Generative modeling involves using a model to generate new examples that plausibly come from an existing distribution of samples, such as generating new photographs that are similar but specifically different from a dataset of existing photographs.

A GAN is a generative model that is trained using two neural network models. One model is called the "*generator*" or "*generative network*" model that learns to generate new plausible samples. The other model is called the "*discriminator*" or "*discriminative network*" and learns to differentiate generated examples from real examples. The two models are set up in a contest or a game (in a game theory sense) where the generator model seeks to fool the discriminator model, and the discriminator is provided with both examples of real and generated samples [3].

After training, the generative model can then be used to create new plausible samples on demand.

This model can boost datasets artificially by generating similar images. GANs can generate very complex images such as bedrooms, human faces, and other scenes as well. BigGAN from large-scale GAN training for high fidelity natural image synthesis in 2018 was able to produce hi-res images remarkably similar to photographs that were input. The point is that GANs are able to expand datasets of complex photographs and is thus logical to use in this scenario [4].

The rationale of this approach was to improvise on the lack of training data that our team had to train the models, and thus, we wanted to make use of existing data to generate more data and therefore better train our classifiers, as we know image classifiers require a lot of data. Artificial data generation is significantly better than data augmentation, as they create the most alike images.

The model was developed specifically to generate more data; however, the data was not satisfactory enough, and hence, the model was dropped. The reason for the

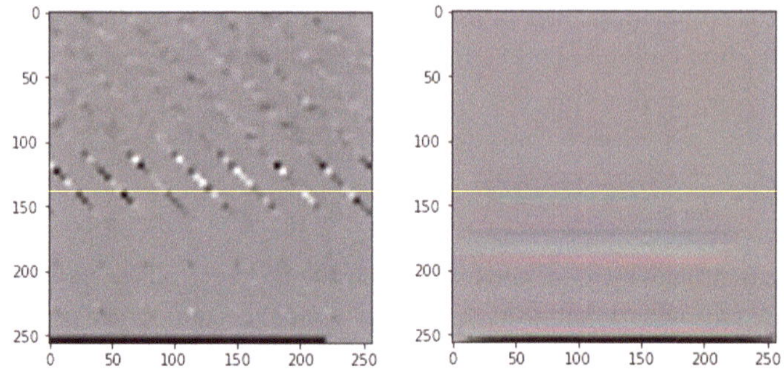

Fig. 10.4 Sample input and output from GAN)

unsatisfactory output from the model was due to an insufficient amount of training data to train the GAN network in the first place. Even top of the line pre trained GANs require at least 100 examples in each class to create satisfactory images. Although this project was unable to harness the GANs, we did test them out using different datasets like the MNIST fashion dataset and obtained satisfactory results [5].

We warn that one must be wary of a dataset that is too small to be artificially boosted at all, which is a constraint that we experienced when trialing this method (Fig. 10.4).

10.5 Results and Discussion

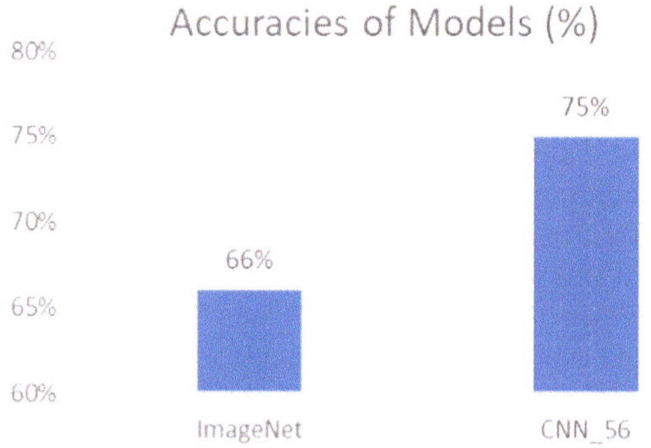

The above graph shows the relevant highest attained accuracies for each of the two models trained with the image datasets. The accuracies are relevant for nine classes of sEMG signal data translated into an image dataset, and the training process has also been optimized with various image data classification techniques.

The approach of having linear combinations between the eight channels of data yielded better accuracies than the other approach described in this paper.

Therefore, the combinatorial image dataset must have been more distinct as compared to the other dataset and the linear combinations must have accounted for this difference. As such, in terms of commercialization of such image data classification techniques, there must be an enforced dependency between the eight channels of data to get reliable and accurate classifications from the classifiers.

Moreover, the current benchmark for sEMG signal classification for nine classes is around 92% for models which practice feature extraction. Although the models that are described in this paper do not attain this high level of accuracy, it is worth noting that the training sets that were fed into the classifiers were limited and possible future expansion of this training data could prove vital in drastically improving classification accuracies. Further, these models are easy to implement, there was no need for any manual feature extraction to take place, and any transformation that was performed to the data was done automatically through translation of raw, quantitative data into image datasets using the methods prescribed earlier.

Finally, in terms of lag times of the models, the classifiers managed to accurately translate the quantitative data from the bands in real time with an average of 219 ms of lag. The benchmark for ideal lag to have responsive prosthetic devices is around 200–300 ms and since this lagtime falls in this range, it is worth noting that the image classifier models have the potential to be explored further in time to come.

10.6 Conclusion

The transformation of the problem of multi-channel time series classification into an image classification problem, in conclusion, not only allowed us to easily draft a solution by building upon years of research, but also allowed us to draft out quick, responsive solutions as well. We were able to build upon years of research to further build better solutions to the rather unexplored domain of multi-channel time series classification problem.

We also analyzed as to how we were able to exploit tools such as image data augmentation and batch normalization—tools that will never be available to us in raw data classification and therefore improve classification accuracies as well. This is therefore a possible avenue that one can investigate to solve relatively unknown problem—that is to transform it into a domain that we know better.

10.7 Future Work

Just as we saw that one of the methods to enforce a dependency between the eight channels of sEMG data was through the construction of linear combinations with equal weight attached to the three different randomly chosen channels, we could possibly also look deeper into each of the channels and identify which channel generates the most distinct data across all gestures trialed. This channel could thus have a larger weight attached to it, thereby generating even more distinct data and easily classifiable data.

Secondly, the main constraints of this paper, which resulted in rather suboptimal results was the lack of enough training data that resulted in overfitting and despite tackling the problem from different angles, the results only improved by around 10–20%. Therefore, since artificial dataset expansion can only go so far in improving accuracies, the expansion of training data through collection of even more data could be practiced to improve the accuracies of the relevant models. It is worth noting that image data classification requires substantial amounts of data to get reliable classifications from the image data classifier. As such, this is something that one should investigate and instead generate more data, if one is aiming to transform independent problems into the domain of image data classification specifically.

References

1. Brownlee, J. (2019, July 5). How to Configure Image Data Augmentation in Keras. Machine Learning Mastery. https://machinelearningmastery.com/how-to-configure-image-dataaugmentation-when-training-deep-learning-neural-networks/
2. Nandini, G. S., Kumar, A. P. S., & Chidananda, K. (2021, January 23). Dropout technique for image classification based on extreme learning machine. *Global Transitions Proceedings*. https://www.sciencedirect.com/science/article/pii/S2666285X21000157
3. Brownlee, J. (2019, July 12). 18 impressive applications of generative adversarial networks (GANs). *Machine Learning Mastery*. https://machinelearningmastery.com/impressive-applications-of-generative-adversarial-networks/
4. Brock, A., Donahue, J., & Simonyan, K. (2018). Large scale GAN training for high fidelity natural image synthesis. arXiv:1809.11096
5. Zhao, S., Liu, Z., Lin, J., Zhu, J. Y., & Han, S. (2020). Differentiable augmentation for data-efficient gan training. arXiv:2006.10738

Chapter 11
Catalytic Water Treatment

Aw Qi Zhen Shermane and Kristie Eliana Ramli

Abstract With the increasing lack of access to safe drinking water and the lack of proper sanitation services largely due to the presence of toxic Azo dyes in wastewater, generated from the textile industry, increasing research has been done to investigate better methods to treat wastewater. Though conventional methods are already present, such as filtration, photodegradation of organic pollutants has recently been the most widely investigated to solve this problem. Studies have shown titanium dioxide catalyst (TiO_2) is one of the most efficient methods in degrading Azo dyes. This is investigated through photocatalytic reactions of TiO_2 plates with Congo red solution. The effects of various operational parameters of TiO_2 plates, such as the different calcination temperatures and the concentration of Congo red solution used, are first determined to identify the final optimum reaction conditions needed to determine the optimum shape of the stainless steel reflector (flat, semicircle, right angle, parabolic). By comparing the results obtained for the percentage removal of Congo red dye, the most effective removal is when the TiO_2 plates are at a calcination temperature of 400 °C for 2 h, and the most effective shape for the stainless steel reflector shape is the parabolic one. Moreover, the ability of TiO_2 plates to be dye-sensitised into solar cells was also investigated using I_3/I^- electrolyte, and the voltage of the solar cell was measured. This ability of TiO_2 plates to be able to harvest solar energy into electrical energy is beneficial especially in the less developed and poorer countries.

11.1 Introduction

According to the World Health Organisation (WHO), in 2017, 2.2 billion people around the world do not have safely managed drinking water services, while 4.2 billion people do not have safely managed sanitation services [1].

The textile industry is one of the world's leading sources of water consumption and pollution. Processes involved in the production of clothing, including dyeing and

A. Q. Z. Shermane (✉) · K. E. Ramli
National Junior College, Singapore, Singapore
e-mail: shermaneaw8903@gmail.com

H. Guo et al. (eds.), *IRC-SET 2021*, https://doi.org/10.1007/978-981-16-9869-9_11

rinsing, release around 150 billion gallons of wastewater annually. This wastewater usually contains a large amount of Azo dyes with around 280,000 tons of dyes discharged in textile industrial effluent every year.

These disposal of toxic Azo dyes can cause extreme damage to the surrounding aquatic environment and the organisms living near the polluted water, due to the relatively high biochemical oxygen demand and chemical oxygen demand. These dyes are also carcinogenic for humans and difficult to degrade via natural means [2].

Although conventional methods such as chemical precipitation, filtration, electrodeposition, ion-exchange adsorption and membrane systems are already present to treat wastewater, these methods may not be very effective as it is often slow or non-destructive to certain persistent organic pollutants. Furthermore, these methods only serve to convert soluble pollutants into their solid phase, which would then require additional resources to remove [3, 4].

Currently, the use of titanium oxide photocatalysts is one of the most efficient methods of degrading Azo dyes. However, the use of the TiO_2 photocatalyst needs UV irradiation in order to degrade the dyes. This process is very expensive and inaccessible to those living in developing countries living near textile factories, where many lack access to clean drinking water. Thus, the use of photocatalyst is needed the most [5–8].

However, the metal sheets can be of different shapes, such as semicircle, flat or right angle. It is thus necessary to investigate the most efficient reflective shape for the best photocatalytic reaction of the TiO_2 plates. Past research has shown that the parabolic shape allows for the incident parallel rays to always converge at a single focal point no matter where on the surface of the mirror they actually strike, highlighting that the parabolic shape could be the most efficient shape [9]. As parabolic metal sheets are currently not present, the effectiveness of the parabolic shape of the metal sheet to act as a reflector will be investigated.

Additionally, in most of the developing countries, electricity is not easily accessible, while naturally sourced dyes may be easier to come by. One of the most commonly used and accessible fruits that are used in the making of the dye-synthesised solar cell is raspberry, which contains anthocyanin dye. Extensive research has been done on the efficacy of the TiO_2 solar cell to harvest light energy and its conversion to electrical energy. Due to the large band gap, suitable band edge levels for charge injection and extraction, long lifespan of excited electrons, exceptional resistance to photocorrosion, non-toxicity and low cost have made TiO_2 a popular material for solar energy applications. Hence, we plan to investigate the efficacy of the differently shaped plates on the generation of electricity via the dye-synthesised TiO_2 solar cell [10, 11].

11.1.1 Hypothesis

We hypothesise that the use of a parabolic surface will yield the most effective removal of CR dye as well as generate the highest current among all of the shapes.

11.2 Methodology

11.2.1 Preparation of TiO_2 Catalyst

The TiO_2 photocatalyst was prepared as follows: TiO_2 powder was grounded with acetic acid into a smooth paste using a mortar and pestle. A thin layer of the paste was then applied onto the conductive side of the fluorine-doped tin oxide (**FTO**) coated **glass** plates using a scalpel. The plates were then placed into an oven at 100 °C for 30 min to dry out the paste. The plates were then placed in the furnace at varying temperatures (200–600 °C) for 2 h to measure the effect on the temperature on the efficiency of dye removal.

11.2.2 Metal Sheet Preparation

Metal plates measuring 10 cm × 10 cm were hammered into four different shapes: straight, 90°, semicircle and a parabola.

11.2.3 Photocatalytic Experiments

11.2.3.1 Temperature

TiO_2-coated FTO glass plates were then placed in the furnace at different temperatures and were placed in a petri dish and filled with 40 ml of 20 mg/dl dye in the machine for 2 h, before they were taken out and the concentration was measured using a UV–vis spectrometer.

For the rest of the experiments, TiO_2-coated FTO plates were placed in the furnace for 2 $_h$ at 400 °C. A Hitachi TM3000 SEM machine was then used to scan the surface of the TiO_2 plates.

11.2.3.2 pH

To measure the optimum pH for solution of CR dye to use, three plates were placed in a petri dish and filled with 40 ml of 20 mg/dl dye of varying pH levels in a [UV] machine for 2 h, before they were taken out and the concentration was measured using a UV–vis spectrometer.

11.2.3.3 Shape of the Metal Sheets

In the UV machine, five dishes with the TiO_2 photocatalysts with 40 ml of 20 mg/dl CR dye were covered with the sheets of different shapes, one dish covered by one plate each, with one dish not covered with any plates. After 2 h, the concentration was measured using a UV–vis spectrometer.

11.2.4 TiO_2 Solar Cell

For the preparation of the raspberry dye, fresh raspberry was blended and filtered to obtain the dye. The TiO_2 plate was then soaked in a raspberry dye solution for 10 $_{min}$. It was then rinsed with ethanol to get rid of the excess dye. Next, another TiO_2-coated FTO plate, conducting side down, was passed through a candle flame to coat the conducting side with carbon. The I_3/I^- electrolyte solution was prepared using 0.5 M KI solution and 0.05 M I_2 in methanol. The electrolyte solution was sandwiched between a carbon-coated plate and a TiO_2-coated plate and bound together using binders. They were then placed on top of the different metal sheets, and the highest voltage recorded was taken using a multimeter. The experimentation was done in **Gelman** HLF **Class 100** Series Horizontal **Laminar Flow** workstation.

11.3 Results and Discussion

11.3.1 Concentration of Congo Red Solution

The rate of photocatalytic reaction is strongly influenced by concentration of the photocatalyst. Heterogeneous photocatalytic reactions are known to show proportional increase in photodegradation with catalyst loading. Generally, in any given photocatalytic application, the optimum catalyst concentration must be determined, in order to avoid excess catalyst and ensure total absorption of efficient photons. This is because an unfavourable light scattering and reduction of light penetration into the solution is observed with excess photocatalyst loading. The concentration of CR solution used in the experiment is 20 mg/dl.

11.3.2 Calcination Temperature of TiO_2 Plates

The increase in temperature enhances recombination of charge carriers and desorption process of adsorbed reactant species, resulting in decrease of photocatalytic

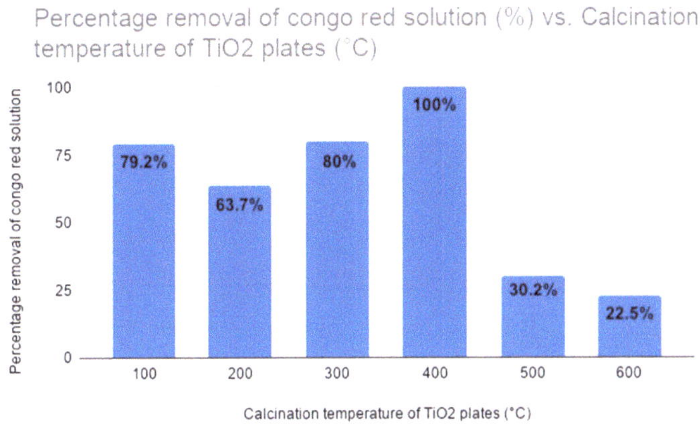

Fig. 11.1 Percentage removal of Congo red solution against calcination temperature of TiO_2 plates graph

activity. This is in conformity with Arrhenius equation, for which the apparent first-order rate constant Kapp should increase linearly with exponential $(-1/T)$. As shown in Fig. 11.1, the percentage removal of CR solution generally increases from 100 to 400 °C as products are able to desorb faster. Adsorption dominates here where desorption can be treated as the rate-limiting step. The optimum calcination temperature of TiO_2 plates is 400 C, with the 100% removal of the CR solution. Above 400 C, the TiO_2 plates become flaky and detached, and its photocatalytic activity is less effective. However, at high temperatures 500 and 600 °C, the reaction rate decreases with increasing temperature, because adsorption is decreasing with temperature in this range.

11.3.3 Analysis of the Optimum Shape of Stainless Steel Reflector

Figure 11.2 analyses the effectiveness of the different shapes of stainless steel reflector to increase light absorption. The most effective shape is the parabolic reflector as the percentage removal of the CR is the highest compared to the other shapes, at 68.8%. This is because the geometry of a parabola is effective in focusing light waves on a single location. The parabolic shape is such that incident parallel rays will converge at a single focal point no matter where on the surface of the mirror they actually strike. It can also be seen that with the stainless steel reflector, the percentage removal of CR doubles from 36.2 to 68.8%, highlighting how having a stainless steel reflector allows for a more efficient photocatalytic reaction.

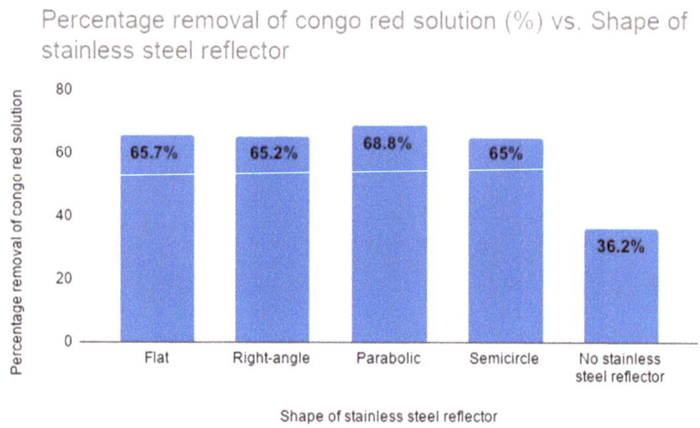

Fig. 11.2 Percentage removal of Congo red solution against the shape of stainless steel reflector graph

11.3.4 Titanium Dioxide Dye-Sensitised Solar Cell

The titanium dioxide (TiO_2) plates can act as a solar cell after being treated with raspberry dye. This takes place due to the condensation of alcoholic-bound protons with the hydroxyl groups that is present on the surface of the nano-structured TiO_2 plates. The plates generate electrons, allowing for electron transport, producing solar power. As seen from Fig. 11.3, this further highlights how the parabolic shape of the stainless steel reflector is most optimum for light absorption as the solar power generated by the dye-sensitised TiO_2 is the highest at 215.5 mV.

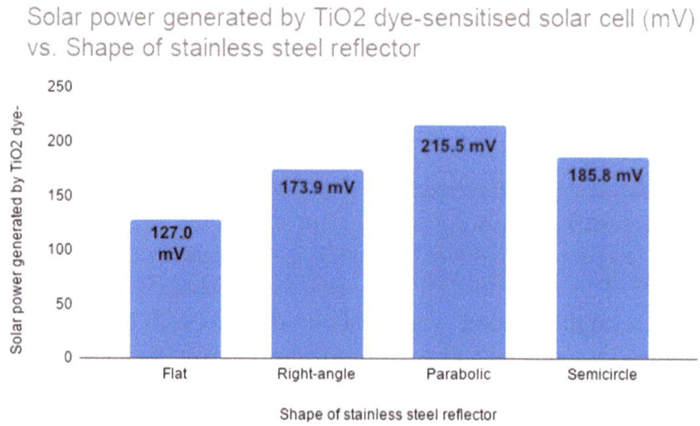

Fig. 11.3 Graph of solar power generated by TiO_2 dye-sensitised solar cell against the shape of stainless steel reflector used

11.4 Conclusion and Future Recommendations

This report addresses the non-strictly controlled discharge of industrial effluents, which has given rise to the current water contamination crisis. The photocatalytic degradation of these organic pollutants using titanium dioxide is the most widely studied method of overcoming the problem of water contamination by organic pollutants. Heterogeneous photocatalytic degradation using TiO_2 photocatalyst remains a viable alternative for the degradation of persistent organic contaminants in both air and water. Moreover, TiO_2 can also be dye-sensitised into a solar cell, harvesting light energy and converting it into electrical energy.

11.4.1 Future Works

The effectiveness of the TiO_2 photocatalyst can be investigated with more types of commonly used Azo dyes, like Chlorazol Blue B and Trypan blue. The metal sheets can also be bent into more shapes and angles, like 60°. The effectiveness of the TiO_2 substrate under constant flow of the dye solution can also be studied to provide information on the kinetics under actual application conditions. A pilot test can then be carried out. Moreover, the TiO_2-coated FTO plates can be placed in the furnace at more temperatures ranging from 400 to 500 °C.

Appendix 1. Calibration Curve

The absorbance of CR dyes of known concentrations (5–30 mg/dl) was measured using a UV–vis spectrometer to obtain the calibration curve.

Figure 11.4 allows for the understanding of the relationship between arbitrary units and concentration of the Congo red solution to aid in finding the percentage removal of the Congo red solution through the photocatalytic reaction of TiO_2.

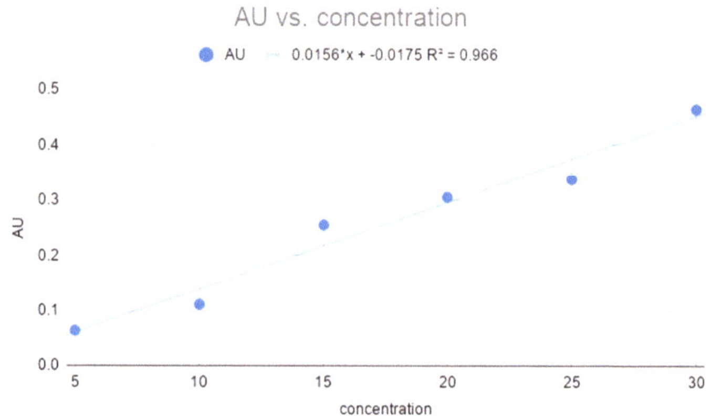

Fig. 11.4 Graph of arbitrary units (AU) against concentration

References

1. UNICEF, WHO (n.d.). 1 in 3 people globally do not have access to safe drinking water. Retrieved January 27, 2021, from https://www.who.int/news/item/18-06-2019-1-in-3-people-globally-do-not-have-access-to-safe-drinking-water-unicef-who
2. Lellis, B., Fávaro-Polonio, C. Z., Pamphile, J. A., & Polonio, J. C. (2019). Effects of textile dyes on health and the environment and bioremediation potential of living organisms. *Biotechnology Research and Innovation, 3*(2), 275–290.
3. Kraeutler, B., & Bard, A. J. (1978). Heterogeneous photocatalytic decomposition of saturated carboxylic acids on titanium dioxide powder. Decarboxylative route to alkanes. *Journal of the American Chemical Society, 100*(19), 5985–5992.
4. Evgenidou, E., Fytianos, K., & Poulios, I. (2005). Semiconductor-sensitized photodegradation of dichlorvos in water using TiO_2 and ZnO as catalysts. *Applied Catalysis B: Environmental, 59*(1–2), 81–89.
5. Soares, E. T, Lansarin, M. A., Moro, C. C., & Brazilian J. (2011). Styrene photocatalytic degradation reaction kinetics. *Journal of the Brazilian Chemical Society, 22*(10).
6. Chun, H., Yizhong, W., & Hongxiao, T. (2000). Destruction of phenol aqueous solution by photocatalysis or direct photolysis. *Chemosphere, 41*(8).
7. Mansilla, H. D., Bravoa, C., Ferreyra, R., Litter, M. I., Jardim, W. F., Lizama, C., & Freer, J. (2006). Photocatalytic EDTA degradation on suspended and immobilized TiO_2. *Journal of Photochemistry and Photobiology A: Chemistry, 181*(2–3), 188–194.
8. Krysa, J., Keppert, M., Jirkovsky, J., Stengl, V., & Subrt, J. (2008). Emanation thermal analysis study of N-doped titania photoactive powders. *Journal of Thermal Analysis and Calorimetry, 92*, 161–167.
9. Paraoblic Mirror: How It Works & Types (w/Examples) (2020, December 28). Sciencing. https://sciencing.com/parabolic-mirror-how-it-works-types-w-examples-diagram-13722364.html
10. Fernando, J. M. R. C., & Senadeera, G. K. R. (2008). Natural anthocyanins as photosensitizers for dye-sensitised solar devices. *Current Science Association, 95*, 663–666.
11. Sarwar, S., Ko, K., Han, J., Han, C., Jun, Y., & Hong, S. (2017, June 01). Improved long-term stability of dye-sensitized solar cell by zeolite additive in electrolyte. Retrieved January 27, 2021, from https://www.sciencedirect.com/science/article/pii/S0013468617312252

Chapter 12
Applying James–Stein Estimation to b-Bit Minwise Hashing

Jing En Daniel Toh, Rui Xian Matthew Kan, and Keegan Kang

Abstract b-bit minwise hashing (bBMWH) is an efficient hashing algorithm used in machine learning. The James–Stein (JS) estimator paradoxically produces a lower mean square error (MSE) than the traditional maximum likelihood estimator. Using 1000 documents from the Bag of Words Datasets (KOS) in the UCI Machine Learning Repository, we computed the pairwise resemblance for all documents in the dataset. We compared the performance of bBMWH with b from 1 to 4 bits with and without JS estimation, by calculating the precision, recall, F_1-score, and MSE in classifying pairs with resemblance $\geq R0$, with $R0$ from 0.30 to 0.60. Our results for $R0 = 0.30$ demonstrated that for $b = 4$ with JS estimation, the precision was high at 0.9 for a small sample size $k < 100$ and was maximized at 1.0 for higher k, while recall was decreased to 0.8. For $b = 3$, JS estimation improved precision without a significant drop in recall. JS estimation decreased the MSE of bBMWH for all b values investigated and especially for small k where MSE is higher. Our findings may be useful when precision is optimized over recall, e.g., spam detection. In cases where we want to estimate the pairwise resemblances for machine learning, bBMWH with JS estimation requires a smaller k to achieve the same MSE as bBMWH alone, thus saving computational time and storage space.

Keywords James–Stein · b-bit minwise hashing · Machine learning

J. E. D. Toh (✉) · R. X. M. Kan
NUS High School of Mathematics and Science, Singapore, Singapore
e-mail: h1710142@nushigh.edu.sg

R. X. M. Kan
e-mail: h1810065@nushigh.edu.sg

K. Kang
Engineering Systems and Design, Singapore University of Technology and Design, Singapore, Singapore
e-mail: keegan_kang@sutd.edu.sg

H. Guo et al. (eds.), *IRC-SET 2021*, https://doi.org/10.1007/978-981-16-9869-9_12

12.1 Introduction

Many machine learning applications are faced with very large and high-dimensional datasets, resulting in challenges in scaling up training algorithms and storing the data [1]. Hashing algorithms such as minwise hashing [2, 3] and random projections [4, 5] reduce storage requirements and improve computational efficiency, without compromising on estimation accuracy. b-bit minwise hashing (bBMWH) [6, 7] is a recent progress for efficiently (in both time and space) computing resemblances among extremely high-dimensional binary vectors. bBMWH can be seamlessly integrated [1] with linear support vector machine and logistic regression solvers.

In traditional statistical theory, no other estimation is uniformly better than the observed average when applied to observations. The paradoxical element in James–Stein estimation is that it contradicts traditional statistical theory elemental law if there are three or more sets of data, even when the three sets are completely unrelated [8, 9]. For example, the unrelated datasets of the estimates of the average price of HDB flats in Singapore, the chance of rain in London, and the average height of Americans can be combined to obtain an estimate better than computing the estimates individually in terms of mean squared error. When first proposed, the James–Stein estimator seemed counterintuitive and illogical. However, it has been proven to have lower mean squared error than the traditional maximum likelihood estimator, when there are at least three parameters of interest [8, 9].

12.2 Hypothesis

In this study, we hypothesized that adding James–Stein estimation to bBMWH improves the precision, recall, and $F1$-score and decreases the mean square error of the estimate from the hashing algorithm.

12.3 Materials and Methods

We briefly review the following: James–Stein estimation [9], minwise hashing [2, 3], and b-bit minwise hashing [7].

12.3.1 James–Stein Estimation

Given a random vector $z \tilde{} \mathcal{N}_N(\mu, I)$, the James–Stein estimator is defined to be

$$\hat{\mu}^{(JS)} = \left(1 - \frac{N-2}{S}\right) z \qquad (12.1)$$

where $S = |z|^2$.

N is the number of true means we want to estimate across datasets.

For $N = 3$, $\boldsymbol{\mu}$ could be a vector containing the true average price of HDB flats in Singapore, true chance of rain in London, and the true average height of Americans. Given some observations, we want to estimate $\boldsymbol{\mu}$ with $\hat{\boldsymbol{\mu}}$.

The maximum likelihood estimator (MLE) for $\boldsymbol{\mu}$, $\hat{\boldsymbol{\mu}}^{(MLE)}$ maximizes a likelihood function under an assumed statistical model, so that the observed data is most probable. The likelihood of a N-variate normal distribution has a closed form and thus can be maximized by using numerical methods such as the Newton–Raphson method to obtain the roots of its derivative.

The following theorem is taken from [9] and restated here:

Theorem 1 *For $N \geq 3$, the James–Stein estimator dominates the MLE $\boldsymbol{\mu}$ in terms of expected total squared error that is*

$$\boldsymbol{E}_{\mu}\left\{\hat{\boldsymbol{\mu}}^{(JS)} - \boldsymbol{\mu}^2\right\} < \boldsymbol{E}_{\mu}\left\{\hat{\boldsymbol{\mu}}^{(MLE)} - \boldsymbol{\mu}^2\right\} \tag{12.2}$$

for every choice of $\boldsymbol{\mu}$.

12.3.2 Minwise Hashing

Computing the size of set intersections is a fundamental problem in information retrieval, databases, and machine learning. Given two sets, S_1 and S_2, where

$$S_1, S_2 \subseteq \Omega = \{0, 1, 2, \ldots, D - 1\},$$

a basic task is to compute the joint size $a = |S_1 \cap S_2|$, which measures the (unnormalized) similarity between S_1 and S_2.

The Jaccard similarity or resemblance, denoted by R, provides a normalized similarity measure:

$$R = \frac{|S_1 \cap S_2|}{|S_1 \cup S_2|} = \frac{a}{f_1 + f_2 - a} \quad \text{where } f_1 = |S_1|,\ f_2 = |S_2|$$

Computation of all pairwise resemblances takes $\mathcal{O}(N^2 D)$ time, as one would need to iterate over all $\binom{N}{2}$ pairs of vectors and for each pair of vectors, over all D elements in the set.

In most cases, D is sufficiently big to make direct computation infeasible.

The original minwise hashing method [2, 3] has become a standard technique for estimating set similarity (e.g., resemblance). We briefly restate the algorithm here as follows:

Suppose a random permutation π is performed on Ω, i.e.,

$$\pi : \Omega \to \Omega, \quad \text{where } \Omega = \{0, 1, \ldots, D-1\}$$

A simple probability argument shows that

$$\mathbf{Pr}(\min(\pi(S_1)) = \min(\pi(S_2))) = \frac{|S_1 \cap S_2|}{|S_1 \cup S_2|} = R \tag{12.3}$$

After k minwise independent permutations, $\pi_1, \pi_2, \ldots, \pi_k$, one can estimate R without bias, as a binomial probability,

$$\hat{R}_M = \frac{1}{k}\sum_{j=1}^{k} 1\{\min(\pi_j(S_1)) = \min(\pi_j(S_2))\} \tag{12.4}$$

$$\text{Var}\left(\hat{R}_M\right) = \frac{1}{k}R(1-R). \tag{12.5}$$

This reduces the time complexity to $\mathcal{O}(N^2k)$ where k is the number of permutations, thus reducing the time taken while sacrificing some accuracy.

12.3.3 b-Bit Minwise Hashing

By only storing the lowest b bits of each (minwise) hashed value (e.g., $b = 1$ or 2), b-bit minwise hashing can gain substantial advantages in terms of computational efficiency and storage space [7].

The following theorem is taken from the paper on b-bit minwise hashing by Li and Konig [7], which we restate here as follows:

Theorem 2 *Define the minimum values under π to be z_i and z_2:*

$$z_1 = \min(\pi(S_1)), z_2 = \min(\pi(S_2)).$$

Define $e_{1,i}$ and $e_{2,i}$ to be the i th lowest bit of z_1 and z_2, respectively.

$$E_b = \boldsymbol{Pr}\left(\prod_{i=1}^{b} 1\{\, e_{1,i} = e_{2,i}\} = 1\right). \tag{12.6}$$

Assuming D is large,

$$\boldsymbol{Pr}\left(\prod_{i=1}^{b} 1\ \{\ e_{1,i} = e_{2,i}\} = 1\right) = C_{1,b} + \left(1 - C_{2,b}\right) R \tag{12.7}$$

where

$$r_1 = \frac{f_1}{D}, r_2 = \frac{f_2}{D}, \tag{12.8}$$

$$C_{1,b} = A_{1,b}\frac{r_2}{r_1 + r_2} + A_{2,b}\frac{r_1}{r_1 + r_2}, \tag{12.9}$$

$$C_{2,b} = A_{1,b}\frac{r_1}{r_1 + r_2} + A_{2,b}\frac{r_2}{r_1 + r_2} \tag{12.10}$$

$$A_{1,b} = \frac{r_1[1 - r_1]^{2^b - 1}}{1 - [1 - r_1]^{2^b}}, A_{2,b} = \frac{r_2[1 - r_2]^{2^b - 1}}{1 - [1 - r_2]^{2^b}} \tag{12.11}$$

$\hat{R}_b$ is an unbiased estimator of R:

$$\hat{R}_b = \frac{\hat{E}_b - C_{1,b}}{1 - C_{2,b}}, \tag{12.12}$$

$$\hat{E}_b = \frac{1}{k}\sum_{j=1}^{k}\left\{\prod_{i=1}^{b} 1\{\ e_{1,i} = e_{2,i}\} = 1\right\}, \tag{12.13}$$

where e_{1,i,π_j} and e_{2,i,π_j} denote the i th lowest bit of z_1, z_2 under the permutation π_j, respectively.

Following property of binomial distribution, we obtain

$$\begin{aligned}\mathrm{Var}\left(\hat{R}_b\right) &= \frac{\mathrm{Var}\left(\hat{E}_b\right)}{\left[1 - C_{2,b}\right]^2} = \frac{1}{k}\frac{E_b(1 - E_b)}{\left[1 - C_{2,b}\right]^2} \\ &= \frac{1}{k}\frac{[C_{1,b} + \left(1 - C_{2,b}\right)R][1 - C_{1,b} - \left(1 - C_{2,b}\right)R]}{\left[1 - C_{2,b}\right]^2}\end{aligned} \tag{12.14}$$

12.3.4 Experiment

We used Python 3.7.10 with vectorization to implement bBMWH and James–Stein estimation. We also used it to plot our graphs of the results. We computed the precision, recall, $F1$-score, and MSE at various R_0 values, using bBMWH with $b = 1,2,3,4$

Table 12.1 Document pairs in the dataset with $R \geq R_0$

R_0	Number of pairs
0.3	4230
0.4	2529
0.5	1020
0.6	353

bits with and without James–Stein estimation and also the original minwise hashing. We aimed to determine the smallest bit possible to save storage space and improve computational efficiency, while maintaining good levels of precision and recall.

Our experiment adopted a similar methodology as the Experiment 3 in the landmark b-bit minwise hashing paper by Li and Konig [7].

The word dataset used is a collection of the first 1000 documents (499,500 pairs) from the Bag of Words Dataset (KOS) in the UCI Machine Learning Repository [10].

We represented the i-th document as a binary vector X_i of size w, the total number of distinct words in the dataset. For this dataset, $w = 6906$. The j-th element of X_i will be 1 if the word occurs in the document and 0 otherwise.

We then computed the true pairwise resemblances for all documents in the dataset using the binary vectors and counted the number of pairs with $R \geq R_0$ (Table 12.1).

We conducted our experiment for $R_0 \in \{0.3, 0.4, 0.5, 0.6\}$ to represent the range covered in the abovementioned experiment. We ran bBMWH to compute the estimate of pairwise resemblances between vectors in X, represented as a square matrix. We then took the upper triangular portion of this matrix and flattened it to get a vector of $\binom{N}{2}$ elements *res*, representing the list of pairwise resemblances. We then ran James–Stein estimation on this vector which shrank the results toward 0, to obtain another vector *jsres*.

We compared these estimates to the true resemblances by selecting all elements with $R \geq R_0$ and computing the precision and recall of these estimates in identifying pairs with $R \geq R_0$.

Using the precision and recall, we calculated the $F1$-score. We also calculated the mean squared error (MSE) of these estimates with the true resemblance. This was done for $k = 500$ permutations and averaged over 100 iterations.

12.4 Results and Discussion

We present some findings for $R_0 = 0.30$ which are more significant here.

All experiments were done for $k = 500$ permutations and averaged over 100 iterations.

We plotted all graphs with error bars to represent the variance in our obtained values instead of relying on point estimates.

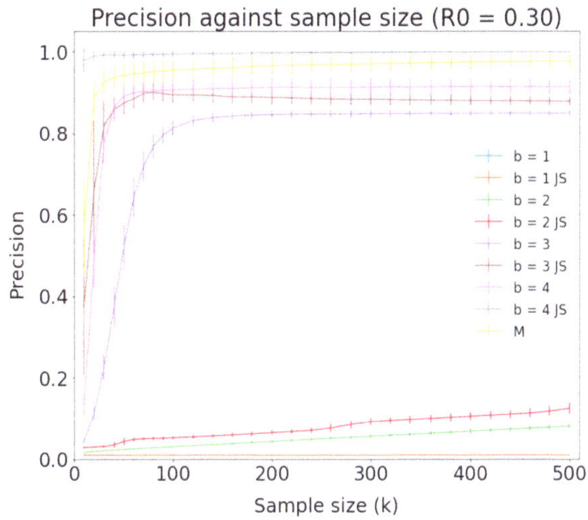

Fig. 12.1 Precision of bBMWH with and without James–Stein estimation with $R_0 = 0.30$ for $b = 1, 2, 3, 4$ and original minwise hashing

12.4.1 Precision

For $b \leq 2$, the precision for both bBMWH and bBMWH combined with James–Stein estimation was low at less than 0.2. This agrees with previous research [7] where using $b = 1$ bit per hashed value yields a sufficiently high precision only when $R_0 \geq 0.5$.

At $b = 3$, the precision for bBMWH increased to 0.8 even for a small $k = 100$. Adding James–Stein estimation to bBMWH further increased the precision.

At $b = 4$, the precision for bBMWH increased to 0.9 for $k < 100$. Adding James–Stein estimation to bBMWH further increased the precision to near 1.0 (Fig. 12.1).

12.4.2 Recall

A high recall of more than 0.9 was obtained for bBMWH alone and bBMWH with James–Stein estimation at the lowest bit of $b = 1$. This result is in accordance with previous research [7] where recall values are all very high even when very few bits are used and R_0 is low. The recall values are all very high and do not well differentiate between bBMWH with or without James–Stein estimation (except for low values of k).

However, for $b = 4$, the recall value decreased to 0.8 when James–Stein estimation was added. An increase in precision with James–Stein estimation results in a decrease in recall (Fig. 12.2).

Adding James–Stein estimation to bBMWH would be useful for applications where precision needs to be optimized over recall, for example, in spam detection where important emails wrongly classified as spam will not be seen by users.

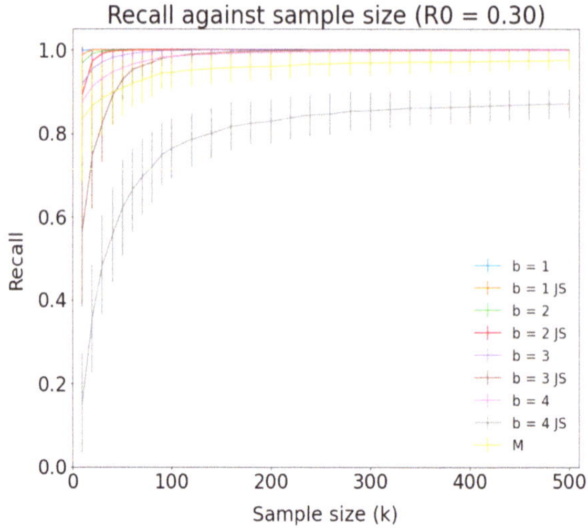

Fig. 12.2 Recall of bBMWH with and without James–Stein estimation with $R_0 = 0.30$ for b = 1, 2, 3, 4 and original minwise hashing

12.4.3 *Mean Square Error*

MSE is not affected by the threshold value R_0 chosen and is the same for all R_0 values.

Adding James–Stein estimation to bBMWH decreased the MSE for all b values. At small sample sizes of k, where MSE is higher, adding James–Stein estimation to bBMWH was especially useful in reducing the MSE as compared to bBMWH alone (Fig. 12.3).

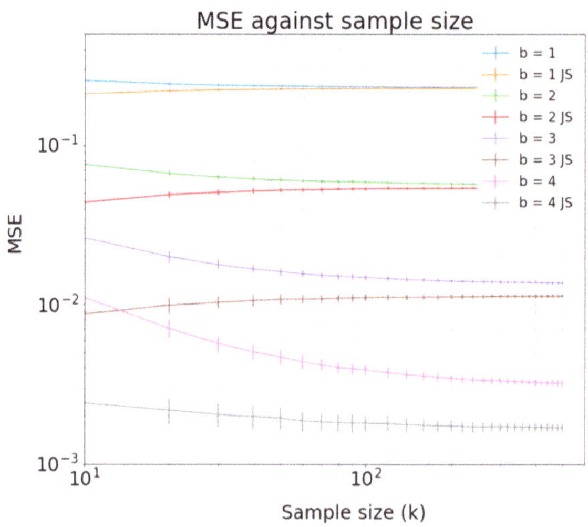

Fig. 12.3 MSE of bBMWH with and without James–Stein estimation for b = 1, 2, 3, 4

bBMWH (for small values of b) does require more permutations than the original minwise hashing; for example, by increasing k by a factor of 3 when using $b = 1$, the resemblance threshold is $R_0 = 0.5$. In the context of machine learning and b-bit minwise hashing, in some datasets, k has to be fairly large, e.g., $k = 500$ or even more, and this can be expensive [11]. This is because machine learning algorithms use all similarities, not just highly similar pairs.

Our results have potential applications in machine learning algorithms by achieving a low MSE without a need to increase the number of permutations k, thus saving on computational time and cost.

12.5 Conclusion and Implications

To our knowledge, we are the first to study the effect of adding James–Stein estimation to the b-bit minwise hashing algorithm.

At a low b-bit value of $b = 4$, the precision for bBMWH was high at 0.9 for a small sample size $k < 100$. Adding James–Stein estimation to bBMWH further increased the precision to 1.0 and decreased the recall to 0.8. For $b = 3$, James–Stein estimation increased the precision while not significantly decreasing the recall value. Adding James–Stein estimation to bBMWH would be useful for applications where precision needs to be optimized over recall, for example, in spam detection.

Detecting similar pairs of documents represented in a Bag of Words format is useful for search engines seeking to omit duplicate results in searches. For search engines, precision is important as a webpage wrongly marked as duplicate and omitted from search results will experience a decrease in site traffic, and users would be unable to obtain a complete search result. On the other hand, a small drop in recall that results in some duplicate pages not being omitted from search results will not significantly impact users' experience. Thus, improving the precision of classification of similar pairs while sacrificing slight recall is useful.

Adding James–Stein estimation to bBMWH decreased the MSE for all b values in the experiment. At small values of k, where MSE is typically higher, adding James–Stein estimation to bBMWH was especially useful in reducing the MSE. Our results have potential applications in machine learning algorithms by achieving a low MSE without a need to increase the number of permutations k, thus saving on computational time and cost.

References

1. Li, P., Shrivastava, A., Moore, J., & König, A. C. (2011). Hashing algorithms for large-scale learning. In *Proceedings of the 24th international conference on neural information processing systems* (pp. 2672–2680), December 2011.

2. Broder, A. Z. (1997). On the resemblance and containment of documents. In *Proceedings of the compression and complexity of sequences* (pp. 21–29), June 1997.
3. Broder, A. Z., Glassman, S. C., Manasse, M. S., & Zweig, G. (1997). Syntactic clustering of the web. *Computer Networks and ISDN Systems, 29*(8–13), 1157–1166.
4. Kang, K., & Hooker, G. (2017). Random projections with control variates. In *Proceedings of 6th international conference on pattern recognition application methods (ICPRAM)* (pp. 138–147).
5. Kang, K. (2017). Random projections with Bayesian priors. In *Proceedings of national CCF conference on natural language processing and Chinese computing (NLPCC)*, Dalian, China, pp. 170–182, Nov 2017.
6. Li, P., & König, A. C. (2011). Theory and applications of b-Bit minwise hashing. *Communications of The ACM—CACM, 54*(8), 101–109.
7. Li, P., & König, A. C. (2010). b-bit minwise hashing. In *Proceedings of 10th international conference on WWW*, Raleigh, NC, pp. 671–680, Apr 2010.
8. Efron, B., & Morris, C. (1977). Stein's paradox in statistics. *Scientific American, 236*(5), 119–127.
9. Efron, B. (2010). Empirical Bayes and the James—Stein Estimator. In *Large-scale inference: Empirical Bayes methods for estimation, testing, and prediction (Institute of Mathematical Statistics Monographs)* (pp. 1–14). Cambridge: Cambridge University Press.
10. UCI Machine Learning Repository Centre for Machine Learning and Intelligent Systems. Bag of Words Dataset (KOS). https://archive.ics.uci.edu/ml/datasets/bag+of+words. Irvine, CA: University of California, School of Information and Computer Science. Last retrieved, 2 January 2021.
11. Li, P., Shrivastava, A., & König, A. C. (2013). b-Bit minwise hashing in practice. In *Proceedings of 5th Asia-Pacific symposium on internetware* (no. 13, pp. 1–10), Oct 2013.

Chapter 13
Thermal Image Analysis of Singapore's Housing Infrastructure

Balakrishnan Naveen Mani Kumar, Mothiki Eswara Anirudh, Dhanabalan Jeevakaarthik, Clayton Miller, and Filip Biljecki

Abstract As a developed and densely populated city, Singapore proves to be a vulnerable victim of the effects of climate change. With Singapore heating up twice as fast as the rest of the world, it is highly important that factors contributing to it are identified and analysed to curb this issue and protect the well-being of the population and the environment. Firstly, with the many infrastructures that dot the city, Singapore falls prey to the urban heat island (UHI) effect. Secondly, with the use of air-conditioning accounting for 25% of the residential electrical consumption, cooling is still one of the largest energy demands for households and buildings. Hence, to obtain a better view as to how these factors come into play, thermal analysis of infrastructure was conducted. This was then coupled with simulations and analyses of thermal data of the flow of air around heated air-conditioning compressors. Results of these investigations revealed differences in temperature of the building with changes in altitude and also the pivotal role that the placements of air-conditioning compressors play in the dispersion of heat around compressors that are concurrently in use across multiple floors. This is crucial to investigate as this might affect the electrical consumption and electricity bills of households. This study also holds significance as it gives rise to a phenomenon known as the urban stack effect where temperature increases with increasing altitude due to the formation of a heat vortex. Moreover, with most modern infrastructures having a similar layout to the building studied in this investigation, this also allows for future studies to be conducted on the design of infrastructures that contribute towards the lessening of the UHI effect.

B. Naveen Mani Kumar · M. E. Anirudh · D. Jeevakaarthik
NUS High School of Math and Science, Singapore, Singapore

C. Miller
Department of the Built Environment, College of Design and Engineering, National University of Singapore, Singapore, Singapore

F. Biljecki (✉)
Department of Architecture, College of Design and Engineering, National University of Singapore, Singapore, Singapore
e-mail: filip@nus.edu.sg

H. Guo et al. (eds.), *IRC-SET 2021*, https://doi.org/10.1007/978-981-16-9869-9_13

13.1 Introduction

According to the Singapore Climate Research Centre, the city-state is heating up at a rate of 0.25 °C per decade with an average temperature increase of 1 °C since the 1950s [1, 2]. Climate research experts also believe that at the rate at which temperatures are rising, Singapore's daily maximum temperature could also reach up to 37–38 °C [3]. Coupled with the high humidity levels Singapore faces daily, Singaporeans are in danger of a myriad of health risks such as elevated heat stress, dehydration in the near future. This has led to a high level of importance and essentiality being channelled towards investigating ways in which such rises in temperatures can be countered. Hence, research is currently being done to understand the key contributors to such elevated temperatures and high rates of heating experienced in Singapore. Such research reveals one key factor that is bound to be one of the primary reasons for these elevated temperatures in densely populated tropical cities. This factor is the presence of the urban heat island (UHI) effect [4].

The UHI effect can be described as an effect that results in urban areas being significantly warmer than the rural areas. This difference is due to one of the fundamental characteristics that differ between urban and rural areas—contrast in building density. As the population in urban areas is significantly higher than that in rural areas, consequently, the building density in urban regions will be significantly higher as compared to the rural parts. Such infrastructural differences in urban areas as compared to rural areas give rise to the UHI effect. In Singapore, the key defining infrastructure of urban areas is the dense housing infrastructure. Demographic data report the presence of more than 1.37 million households with up to 80% of Singaporeans living in high-rise buildings (HRBs) such as public housing buildings or condominiums [5].

Consequently, a key contributor to the rapid heating of Singapore is the infrastructure of high-rise buildings. With more than 4.5 million inhabitants living in HRBs [6], they are the most common housing infrastructures in Singapore. This makes them a substantial contributor to the perpetuation of the UHI effect in Singapore. Analyses of the infrastructure and thermoregulation of HRBs will provide for effective identification of factors of HRBs that possibly contribute to the heating of the surrounding atmosphere. Hence, our study aims at analysing trends in the thermal signature of HRBs and infrastructural properties that affect heat absorption and temperature of the surface of HRBs. This will then allow us to gain an understanding of the heat flow of a HRB and the role it plays in the current urban warming issue. Such an understanding would further pave the path to finding the most suitable design of a HRB to combat the UHI effect that affects Singapore's weather to date.

This study focuses on one of the most crucial factors affecting the heating up of surroundings: air-conditioning units; more precisely, the compressors found on the exterior of the building. With the usage of air-conditioning units contributing to more than a quarter of the residential electrical consumption in Singapore, studying the changes in temperature of HRBs due to the use of air-conditioning units proves to be an integral part of our overall analysis of the heat flow of the building. In particular, our study aims to focus on the prevalence of the stack effect primarily due to the vertically stacked air-conditioning units. This stack effect refers to the effect a compressor has on the region above it in terms of temperature. The stack effect occurs due to the primary nature of hot air rising, which in this scenario is due to the hot air blown out of a compressor, as well as the surrounding air that gains heat from a compressor and rises. The significance of such a study is that firstly, analysing thermal images of the building would allow for detailed studies to be conducted regarding the design of buildings and the thermal flow within the building. Moreover, given that buildings are key contributors to the high rate of urban warming of Singapore, such a study also provides insight into infrastructural designs that can reduce the urban heat island effect. Finally, by understanding how heat flows within a building also opens up possibilities towards obtaining crucial information regarding possible economic disadvantages residents may face. Our study hypothesises that as a consequence of the stack effect, the rise of air heated by the compressors will result in the surrounding atmosphere of the air-conditioning compressors being progressively hotter with rising altitude. Such phenomena might also have an impact on varying indoor thermal comfort of apartments in a building.

13.2 Methodology

13.2.1 Case Study Object: A Condominium in Clementi

The measurements were conducted aiming at a condominium in Singapore, a high-rise development that was completed in 2019. The analysed buildings are 40-storey twin towers, each with 6–7 housing units per floor, with a total height of 129.5 m. The air-conditioning compressors of the corner units are found at the tails of the H-shaped towers. Meanwhile, the air-conditioning compressors of the centre units are found flush, in between the balconies.

13.2.2 Recreation of the Building in Simulation

Numerous studies involving the thermal analysis of infrastructures have made use of simulations to vividly identify and analyse the thermal flow of the building on a more descriptive and dynamic scale [7]. Hence, this investigation also adopts such a technique. Using real dimensions of the air-conditioning ledge and air-conditioning compressors, we modelled the air-conditioning ledge in a computational fluid dynamics (CFD) simulation using COMSOL Multiphysics. Such a simulation allowed for effective viewing of the air and thermal flow of the environment surrounding the air-conditioning ledge. Hence, this model was then used to simulate and investigate various scenarios of heat convection of running air-conditioning compressors to observe the thermal flow of the surrounding environment in such scenarios.

13.2.3 FLIR Camera Deployment and Measurement Setup

Analysis of thermoregulation of high-rise buildings can be effectively carried out through thermal imaging. To do so, this study made use of a forward-looking infrared (FLIR) camera. The FLIR camera can effectively and accurately capture temperature data of structures more than a kilometre away. In our case, the FLIR camera was set up on a high-rise building situated 400 m away. Infrared radiation that is captured through the camera is then digitalised and collated to form a thermal image of the object that is being captured, and in this scenario, the condominium building. For our study, the FLIR camera was placed for a few hours at a location that captures a major part of the building that is studied. This setup allows the thermal imaging of the façade of the building.

The thermal images captured were then analysed using a custom developed Python program that can display the temperature distribution in the exterior of the building as a colour map, thus allowing for a preliminary visual distinction of hotter and cooler regions of the building as seen in Fig. 13.1. The functionality of a FLIR camera is that every image captured by this device has a thermal signature on each of its pixels. This program can then extract temperature data stored in each pixel of the thermal image captured, which allows for extensive study of the overall heat flow of the infrastructure.

The measurements were largely taken during clear weather. The measurements were carried out for a total duration of 7 h from 5 in the evening till midnight over a span of a few days.

Fig. 13.1 Image on the left is captured by a thermal camera and the image on the right is captured by a conventional camera

13.2.4 Privacy Protections

Given the nature of our observational and measurement platform, it is of highest importance to ensure appropriate privacy protections for inhabitants of the HRB. The images that were taken for the subsequent analyses of thermal data were strictly limited in pixel resolution so that building interiors cannot be seen, thus abiding by the respective privacy restrictions. Moreover, no personally identifiable information was collected in the process [8].

13.2.5 Flow Diagram of Code Used for Data Analysis

As depicted in the flow diagram in Fig. 13.2, thermal images captured by the FLIR camera were the source of thermal data used for analysis. After running it through a series of programs written using the Python programming language, we obtain the thermal signature and profile of the image captured. The robust design of the program allows for modifications to be readily made to the code to achieve a myriad of functions, such as extraction and comparison of thermal signature [9] of various points, visual discernments of thermal flow of the infrastructure [10], and many more functionalities, which allows us to engage in a holistic analysis of the thermal data at hand.

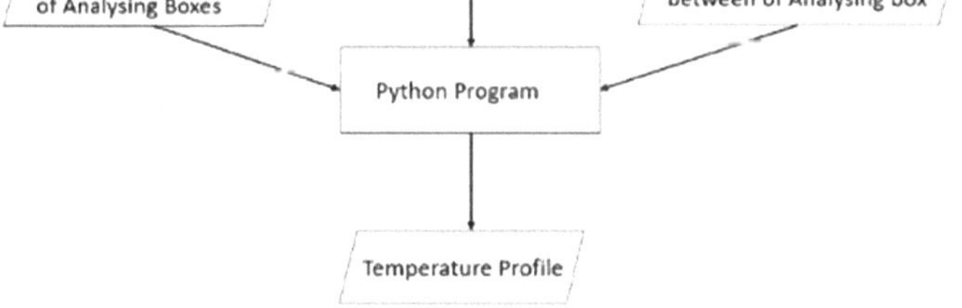

Fig. 13.2 Flow diagram of our python code

13.3 Simulations and Thermal Analysis of the Stack Effect

Our study investigates the concept of stack effect present in air-conditioning unit compressors. The stack effect [11] is a phenomenon that occurs primarily due to the nature of hot air rising because of its lower density. When an air-conditioning unit is in use, hot air, that gets heated due to the heat absorbed by the refrigerant, is blown out of a compressor. Air surrounding the compressor also gains this heat, and hence, resulting in an upward-rising mass of air. When this hot air rises and dissipates, it forms a vertical buoyancy-driven airflow [12]. This hot air might then get sucked into another compressor on a higher floor, causing a more shallow thermal conduction gradient between the compressor and the air which would result in even hotter air surrounding the compressor, longer cooling downtime, and higher electricity bill. With a higher number of floors [13] and extended utilisation of air-conditioning as a result of working from home during the recent period, the stack effect becomes increasingly significant. The cumulative effect of all the hot air that rises after gaining heat from the compressors leads to an increase in the temperature of the surrounding air with an increase in altitude. The significance of such an analysis was made possible through the insight obtained on the thermal flow of HRBs and the overall impact HRBs may have on the climate. This is because with the usage of air-conditioning units being significantly high among the Singapore resident population, possible accumulation and release of heat through the stack effect eventually could have drastic impacts on the global warming situation in and thermal signature of Singapore. Our investigation of stack effect comprises of two main aspects—conducting CFD simulations and analysing real-world data collected.

13.3.1 Simulations

The first study employs the use of computational fluid dynamics (CDF) simulations of the air-conditioning unit compressor stack located on the exterior of the condominium building.

Using a realistic model of the air-conditioning ledge, we simulated the heat flow from the compressor to the surrounding air as seen in Fig. 13.3. The red part indicates the heated compressor with a column of warm air above it. The result from the simulation establishes a firm physics grounding for our theory of vertical buoyancy-driven flow of hot air. Thus, we proceeded ahead to model a stack of such air-conditioning ledge similar to the ones found in the condominium building and ran simulations to find out the temperature and velocity of the surrounding air.

Through simulations of temperature and velocity distributions in Figs. 13.4, 13.5, 13.6 and 13.7, we note that there is significant air and heat flow in the upwards direction from the air-conditioning compressors. As explained by the stack effect, hot air that was surrounding the hotspot, which is the air-conditioning compressor, rises. Moreover, it is worth noting that the velocity of the air rising from the air-

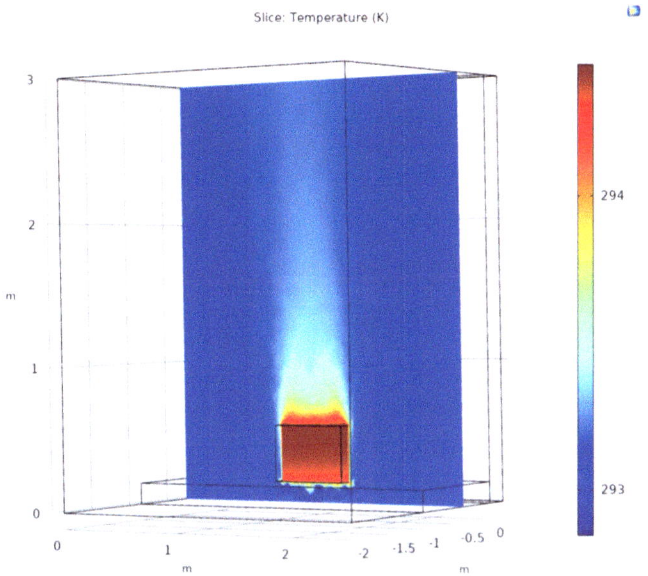

Fig. 13.3 Replication of the air-conditioning ledge of the condominium building in the simulation

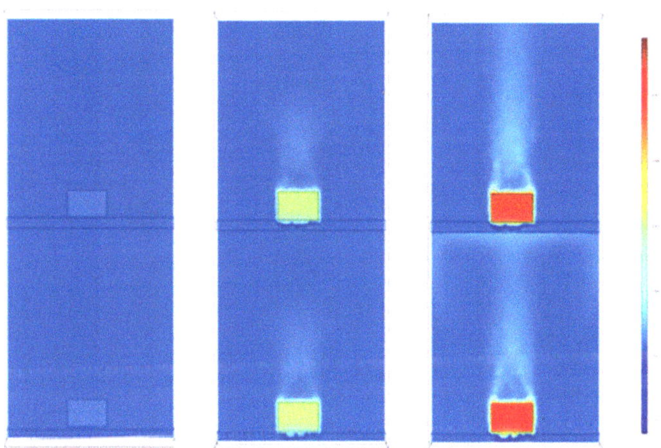

Fig. 13.4 Snapshots of the simulated temperature (in K) of the air surrounding the air-conditioning compressors across 2 floors

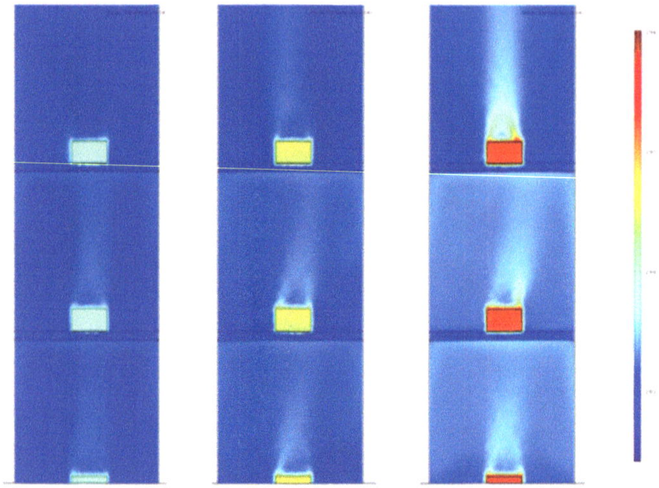

Fig. 13.5 Snapshots of the simulated temperature (in K) of the air surrounding the air-conditioning compressors across 3 floors

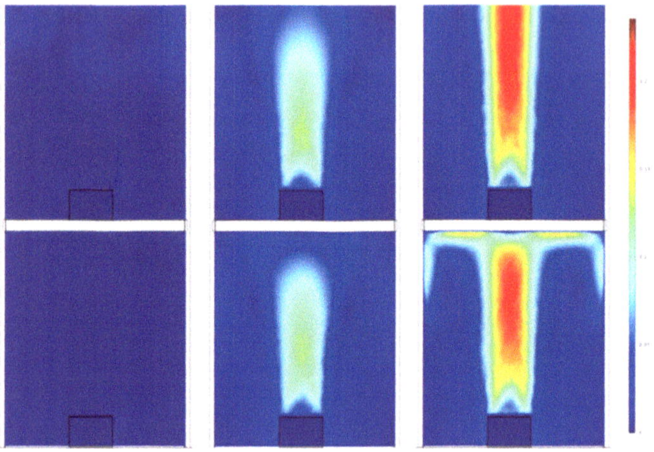

Fig. 13.6 Snapshots of the simulated velocity (in m/s) of the air surrounding the air-conditioning compressors across 2 floors

conditioning compressor on the higher levels is significantly higher than that of the level below it. As the hot air rises, it gains heat from the compressors above causing it to heat up further, resulting in a vortex of air which gains velocity coupled with heat as it rises, thus proving the existence of the stack effect.

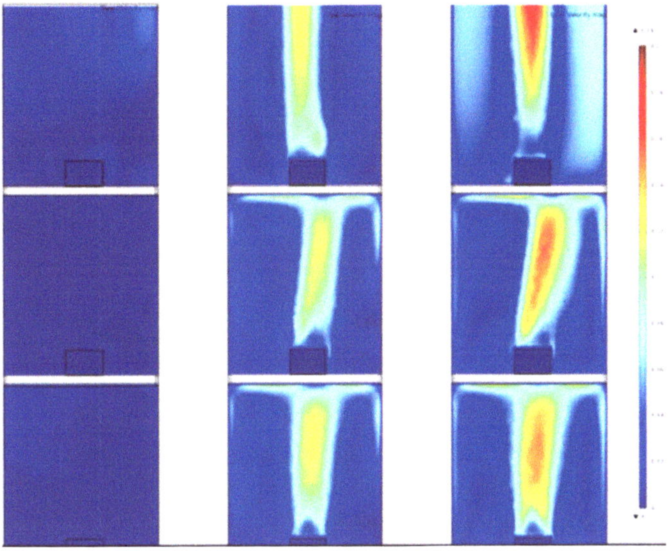

Fig. 13.7 Snapshots of the simulated velocity (in m/s) of the air surrounding the air-conditioning compressors across 3 floors

13.3.2 Analysis of Real-World Thermal Data

The second study analyses the temperature data of air-conditioning unit compressors found on the exterior of the condominium building. Using the Python program we designed, we analysed the thermal images to calculate the average temperature [15] of each of the air-conditioning compressors in the buildings. We then generated boxplots of the temperatures obtained for each level and matched these boxplots with their respective levels to display any trend in the temperature of compressors with increasing altitude, if any. A significant increase in the temperature of compressors with increasing altitude would indicate the prevalence of the stack effect.

In contrast to the expected results of a rising trend in the temperature as the altitude rises, Fig. 13.8 shows a decreasing trend in the recorded temperature with an increase in altitude, highlighting a possible absence or insignificant level of stack effect being present in this building.

Despite this overall observational absence of stack effect, after zooming in on each of the thermal images obtained, it was observed that on many occasions, there is indeed a stack effect being present, but only on the micro-scale. As seen in Fig. 13.9, there are two regions of the hotspot which show a running air-conditioning compressor and the temperature of the air above the air-conditioning compressor on the higher level is warmer than that of the lower year, thus indicating the presence of this effect in the micro-scale. CFD simulations of a functioning air-conditioning compressors substantiate this presence of stack effect on a micro-scale and the absence of it on a macro-scale.

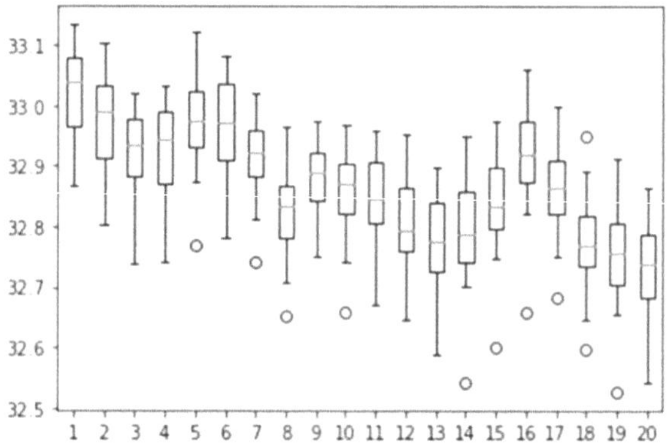

Fig. 13.8 Graph of temperature against compressors at the respective levels

As seen in Figs. 13.9 and 13.10 and the images of the building, each air-conditioning compressor is placed on a ledge. This ledge, which is made of concrete and has a thickness of about 15 cm, proves to be a crucial factor contributing to the absence of the stack effect. This behaviour is because, as seen in the simulations, the hot air from the exhaust of the air-conditioning compressor rises and encounters the ledge above. Once the hot air encounters the ledge, the air's velocity decreases, as the ledge serves as an obstruction of the vertical buoyancy-driven airflow.

Moreover, wind speed and direction due to the location's proximity to the coast and unpredictable weather patterns could have accelerated the rate of heat loss and dispersion of the hot air. Thus, contrary to the theory, such accumulation of heat through the rise of hot air and the stack effect is less significant. Additionally, the results obtained show the success of the strategic positioning of the air-conditioning unit compressors at the ends of the many HRBs constructed in Singapore allow for larger channels for heated air to escape through as well as higher exposure to wind, thus curbing the stack and consequently the UHI effect.

13.4 Conclusion

From the results obtained, it can be determined that in HRBs that follow the design of the studied condominium building, the stack effect is less prevalent with the results obtained, highlighting the converse of the hypothesised consequences had there been a prevalent stacking effect, in the building under study. Through extensive modelling of the air-conditioning units in the CFD simulations and analysing real-world thermal data of the building, it was concluded that the ledge present beneath each air-conditioning compressor obstructs the vertical buoyancy-driven airflow, hence preventing the primary contributor to the macro-scale stack effect phenomena, while also providing a significant level of insulation. As a result, the hot air cannot rise in purely vertical manner and is instead dispersed in other directions. This dispersion of the hot air released from every air-conditioning unit allows for it the inflow of

Fig. 13.9 A progressive zoom in of the snapshot of the colour-mapped thermal image

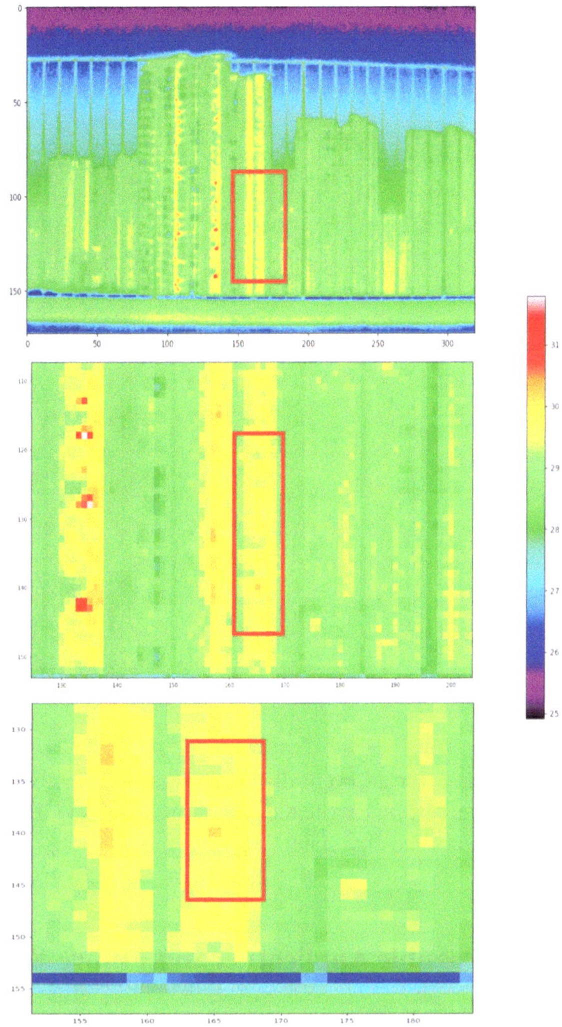

cooler air into the air-conditioning compressors located above. This, coupled with the phenomenon of temperature decreasing with increasing altitude due to several weather-based factors that accelerate the heat loss by allowing dispersion of hot air, caused a decreasing trend in the temperature of compressors.

Such an analysis reveals that one of the key infrastructural designs and properties that contribute towards the drastic reduction of the stack effect is the presence of various forms of insulation such as the ledges present beneath the air-conditioning compressors located on the exterior of the building. Hence, this investigation shows that placing compressors at an optimal vertical spacing from each other with barriers separating them allows for the reduction of the consequent urban stack effect which

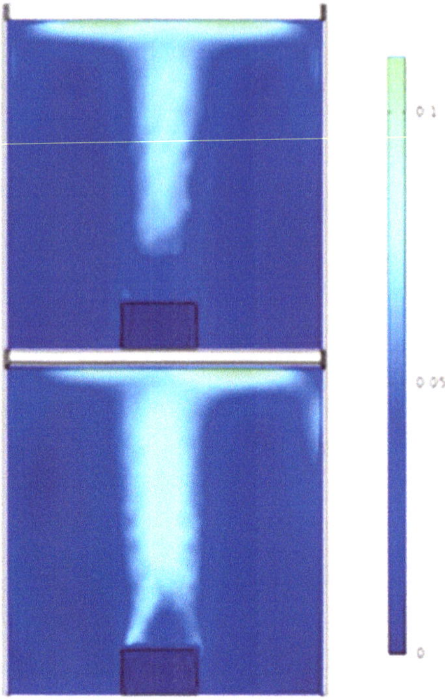

Fig. 13.10 A snapshot of the simulated velocity that shows micro-scale stack effect

would allow for an even, dispersive thermal flow within the building and the surrounding environment, henceforth mitigating the UHI effect and reducing the overall heating up of the city-state.

13.5 Possible Application and Future Work

For a thermal analysis that provides a more complete view, diverse types of infrastructures with varying designs can be examined and analysed to compare the stack effects present in different infrastructural designs. This study currently focuses on only one building due to the COVID-19 pandemic circumstances and restrictions, which during the experiments did not allow a more comprehensive survey. However, in the future, analysing a wider variety of buildings [16], with diverse design aspects that may affect the thermal flow within the building, would allow for a more exhaustive view of the optimal designs that mitigates the UHI effect and the heating of the environment. Analysis can also be done by considering other factors such as the façade colour, material, type, and placement of buildings, which could be also obtained from other sources such as street view imagery [17], and which would account for a greater number of factors that could possibly contribute to the UHI effect. Another plausible factor that could be incorporated into our analysis is electrical consumption data and electricity bills of the buildings under study, as these data would allow for a concrete analysis as to how significant the stack effect and other UHI effects are on the electrical consumption of residents living in such buildings.

The framework used in this study is to investigate the urban stack effect could also prove to be valuable in studies conducted on other environmental effects. For instance, this investigation has analysed the process and functionality of the urban stack effect through vivid and dynamic simulations that can accurately track air and thermal flow. Following this direction, insight was also obtained regarding a possible design—the presence of a ledge—that can be employed to ensure the reduction of this effect. The thermal imaging analysis conducted on the vertical thermal flow along the air-conditioning compressor column of the building also opens up opportunities for in-depth analysis to be done on the placements of air-conditioning compressors as well. This can be done to understand the best possible positioning of compressors to ensure the least effect on the surrounding environment. By obtaining such insight and opportunities for analysis, further research can be done in this area following a similar investigation framework to identify the most optimal infrastructural designs that can reduce the UHI effect that buildings have on the surrounding environment.

Acknowledgements We gratefully acknowledge NUS High School of Mathematics and Science for supporting this research project.

References

1. Ng, M. (2020). Nearly 1000 households in Tengah sign up for centralised cooling system. Retrieved January 20, 2021, from https://www.straitstimes.com/singapore/housing/nearly-1000-households-in-tengah-have-signed-up-for-centralised-cooling-system
2. Ng, D. (2020). Why Singapore is heating up twice as fast as the rest of the world. Retrieved January 20, 2021, from https://www.channelnewsasia.com/news/cnainsider/singapore-hot-weather-urban-heat-effect-temperature-humidity-11115384
3. Temperatures in Singapore could hit 40 °C as early as 2045: Scientists (n.d.). Retrieved January 20, 2021, from https://www.todayonline.com/singapore/singapore-could-swelter-through-40degc-days-2045-if-business-usual-emissions
4. Kardinal Jusuf, S., Wong, N. H., Hagen, E., Anggoro, R., & Hong, Y. (2007). The influence of land use on the urban heat island in Singapore. *Habitat International, 31*(2), 232–242.
5. Yuen, B. (2005, February). Romancing the high-rise in Singapore. Retrieved January 20, 2021, from https://www.researchgate.net/publication/222648100_Romancing_the_high-rise_in_Singapore
6. Entrop, A. G., & Vasenev, A. (2017). Infrared drones in the construction industry: Designing a protocol for building thermography procedures. *Energy procedia, 132*, 63–68.
7. Dobler, G., Bianco, F. B., Sharma, M. S., Karpf, A., Baur, J., Ghandehari, M., & Koonin, S. E. (2021). The urban observatory: A multi-modal imaging platform for the study of dynamics in complex urban systems. *Remote Sensing, 13*(8), 1426.
8. Lane, J., Stodden, V., Bender, S., & Nissenbaum, H. (Eds.). (2014). *Privacy, big data, and the public good: Frameworks for engagement*. Cambridge: Cambridge University Press.
9. Tanaka, H., & Lee, Y. (1988). Stack effect and building internal pressure. *Journal of Wind Engineering and Industrial Aerodynamics, 29*(1–3), 293–302.
10. Borrmann, D., Elseberg, J., & Nüchter, A. (2013). Thermal 3D mapping of building façades. In *In Intelligent autonomous systems* (Vol. 12, pp. 173-182). Berlin, Heidelberg: Springer.

11. Wong, N. H., & Heryanto, S. (2004). The study of active stack effect to enhance natural ventilation using wind tunnel and computational fluid dynamics (CFD) simulations. *Energy and Buildings, 36*(7), 668–678.
12. Bruelisauer, M., Meggers, F., Saber, E., Li, C., & Leibundgut, H. (2014). Stuck in a stack-temperature measurements of the microclimate around split type condensing units in a high rise building in Singapore. *Energy and Buildings, 71*, 28–37.
13. Lovatt, J. E., & Wilson, A. (1995). Stack effect in tall buildings. *In Fuel and Energy Abstracts, 4*(36), 289.
14. Barrett, R. E., & Locklin, D. W. (1968).Computer analysis of stack effect in high-rise buildings. *Ashrae Transactions*, 74(Part II), 155–169.
15. Bruelisauer, M., Meggers, F., Saber, E., Li, C., & Leibundgut, H. (2013, November 23). Stuck in a stack-temperature measurements of the microclimate around split type condensing units in a high rise building in Singapore. Retrieved January 22, 2021, from https://www.sciencedirect.com/science/article/pii/S0378778813007639
16. Lee, S., Moon, H., Choi, Y., & Yoon, D. (2018, February 14). Analyzing thermal characteristics of urban streets using a thermal imaging camera: A case study on commercial streets in Seoul, Korea. A case study on commercial streets in Seoul, Korea. Retrieved January 22, 2021, from https://www.mdpi.com/2071-1050/10/2/519
17. Biljecki, F., & Ito, K. (2021). Street view imagery in urban analytics and GIS: A review. *Landscape and Urban Planning, 215*, 104217.

Chapter 14
EEG-Based Evaluation of the Effect of Emotion on Relaxation Management

Xin Ru Ng, Maekayla Yen-C Lee, Xinyue Wang, and Phyo Wai Aung Aung

Abstract Emotion is a vital part of our lives in our interactions with the environment. With more hectic lifestyles, relaxation is critical to re-energising our mind. Emotional adjustment and relaxation bring benefits such as reduced anxiety and improved task performance. Studies have shown the relationship between emotions and relaxation in applying relaxation techniques to reduce negative emotional states, such as stress. In contrast, our study investigated, using EEG features, how different emotional states, fear, anxiety and happiness, affect one's ability to relax using EEG features. We collected EEG from 15 participants according to the experiment protocol comprising baseline, emotion-arousing "non-relaxed" tasks (happiness, fear and anxiety), each followed by a "relaxed task". We used six band-power features extracted from EEG signals applying statistical analysis methods to test research hypotheses. We applied band-power with statistical analysis methods to test our research hypotheses. Using paired samples *t*-test, individual band-power features (alpha, theta, beta) of "non-relaxed" tasks compared with that of baseline tasks showed a highly significant difference ($p < 0.001$). We then used supervised machine learning to test binary classification accuracy of relaxation state among different task pairs. The results showed lower 9.61% mean accuracy of modified tasks compared with baseline task. Baseline task pair achieved 82.98% while only obtained 74.54%, 73.52% and 72.06% accuracy for anxiety, fear and happiness non-relaxed and relaxed task pairs, respectively. Our results showed that emotions affect one's ability to relax and to varying degrees. These findings could allow a better understanding of what it is required to neutralise emotion effects and enhance relaxation.

Keywords Component · EEG · Relaxation

X. R. Ng · M. Y.-C. Lee · X. Wang (✉)
Nanyang Girls' High School, Singapore, Singapore
e-mail: w.xinyue9@gmail.com

X. R. Ng
e-mail: clairey.beary333@gmail.com

P. W. Aung Aung
Nanyang Technological University, Singapore, Singapore

H. Guo et al. (eds.), *IRC-SET 2021*, https://doi.org/10.1007/978-981-16-9869-9_14

14.1 Introduction

Across many studies, relaxation has consistently been shown to be extremely useful in people's daily lives, bringing significant health benefits such as reduced general depression, anxiety and stress [1, 2] and management of physical pain [3]. It also improves task performance, such as academic performance in students [4]. However, in our fast-paced society, many find it hard to relax and get such benefits, with a Cigna 360 Well-Being Survey showing that 92% of working Singaporeans report feeling stressed. This is 8% higher than the global average at 84% [5].

A common method used to determine the effectiveness of relaxation techniques is by using the brain–computer interface (BCI). We measure electrical activities of the brain through scalp electrodes using electroencephalogram (EEG). Analysis of EEG signals enables us to provide objective and quantifiable mental states, such as concentration and relaxation [6]. Past studies have been done, showing that conventional relaxation techniques improve BCI performance, such as mindfulness meditation [7].

Past studies have also shown how conventional relaxation methods affect one's emotions, such as meditation reducing negative mood. There is also previous research that has tried to find how relaxation affects fear, such as using relaxation as a therapy to overcome the fear of snakes or dentists [8, 9]. We found out various relaxation methods applied for stress relief and reducing anxiety with mixed outcomes [10]. However, there has been limited research on how different emotions affect one's ability to relax. There is some research that seems to suggest that humour allows one to relax more easily, but is not completely conclusive [11]. Our study aims to find the effect of three emotions, anxiety, fear and happiness, on relaxation as opposed to when emotions are not stimulated, and also compare the effects of the three emotions with each other.

14.2 Objectives and Hypotheses

We aim to find out if selected emotions (fear, anxiety and happiness) affect one's ability to relax and how the different emotions, when stimulated, would affect one's ability to relax as compared to when no emotions stimulated (during the baseline). We hypothesise that emotion induced relaxation would all hinder one's ability to relax with varying degrees of impact. We predict that in descending order, anxiety, fear and happiness would affect emotions the most, with happiness having a positive effect and allowing users to relax better than the baseline.

For our experiment, we aimed to find out how three different selected emotions, fear, anxiety and happiness, would affect one's ability to relax. We chose emotions that are more relevant to people's daily lives and are also easy to stimulate desirable emotional states with visual and audio stimuli. Our experiment stimulates these three emotions, each followed by the same relaxation task for each participant, to analyse the individual effect of the three emotions on the relaxation task. We collected EEG data of the participants through wearable Muse headband throughout the duration of the experiment. Then we could analyse to determine the relaxation states of the participants using machine learning and statistical analysis throughout the experiment. We also compare the effects of each stimulated emotion on each subject's relaxation.

14.3 Research Methods

Data collection and experiments started after obtaining ethical approval, and participants had obtained parental consent prior to their participation. The experiment strictly follows the standard safe management measures put in place due to the COVID-19 pandemic.

The experiment relies on audio and visual stimuli to induce different emotions in the subjects. Studies have shown that using audio and visual stimuli together is effective to improve relaxation [12]. There are three main blocks in the experiment, each comprising different emotion-inducing stimuli for each of the three emotions, and the same relaxation task for all three blocks. All emotion-inducing stimuli and relaxation tasks are 4 min each, red and blue colour boxes, as shown in Fig. 14.1. Before the experiment, participants also fill in a pre-experiment survey, where they would indicate their preferred choice of stimuli for certain tasks for personalised emotion stimulation.

Before showing emotion-inducing tasks, baseline data from each participant is first collected for comparison with emotionally induced tasks. Data at a non-relaxed state is first collected for 4 min, where participants are made to count the number of red-coloured dots while ignoring green-coloured dots on a screen. This is followed by a 15 s break. Next, participants are made to inhale and exhale as according to instructions on a screen for baseline relaxation. Participants take part in the emotion-inducing tasks after the baseline.

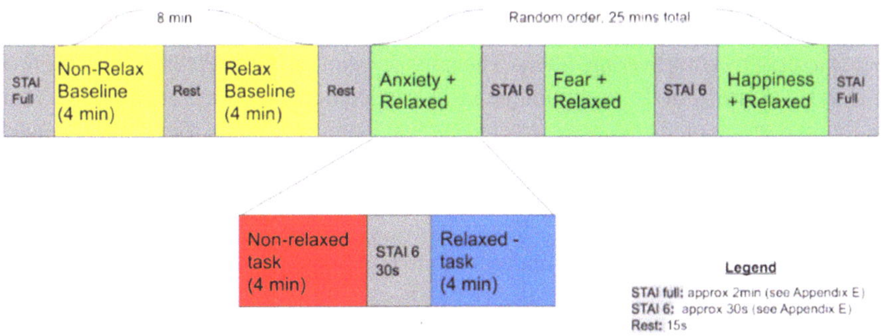

Fig. 14.1 Experiment protocol sequence and timing diagram

For the happiness- and fear-inducing tasks, all participants will watch videos with audio. Participants choose their choice of most emotion-inducing stimuli for these two emotions, happiness and fear, in the pre-experiment survey. Their personalised choice of stimuli will be used for the first 2 min of the task, while the last 2 min of stimuli is standardised and the same for all participants. For the anxiety task, the participants will be asked to answer multiple-choice math questions, while fast-paced, noisy music plays in the background. Math questions and multiple-choice questions mimic conditions in examinations, which is known to cause anxiety and stress, especially in students [13]. For the relaxation task, classical music is played, while participants choose their choice of most relaxing visual stimuli, either photographs of forests or oceans, through the pre-experiment survey. Classical music and photographs of nature scenes are used as both classical music [14, 15] and nature scenes [16, 17] have been proven to be effective relaxation methods.

After each stimulus, a short eight-question survey consisting of six questions of State-Trait Anxiety Inventory (STAI-6) (Appendix 3) and two questions regarding valence and arousal of the subject. This allows us to determine the emotional and relaxation states of the subject and is used in data analysis to help analysing EEG data. The full STAI survey and two valence–arousal questions are done at the start and end of the experiment.

We used OpenSesame software (Appendix 1) to design experiments that allow for graphical display, incorporating Python scripting, and allow us to run experiments with little to no coding experience. We explain full details of the experiment design in (Appendix 1).

14.4 Data Collection and Results

We collected EEG data from participants, as seen in Fig. 14.2, using a Muse 2016 headband. We also tracked eye gazes using a Tobii EyeX eye tracker and user key presses in response to particular visual stimuli presented (Appendix 2). We synchronised all data across different sensors and stimulus presentation events using world clock absolute timestamps.

We apply band-pass filters (0.5–45 Hz) to raw EEG signals and remove artefacts using a set of moving average filters. Then, we segmented the filtered data by synchronising timestamps of different unique tasks carried out during the experiment. With individual segmented data, we extracted six band-power features for relaxation states classification and statistical analysis as shown in Fig. 14.3.

We conducted analysis of variance (ANOVA) using band-power features on all tasks, non-relaxed and relaxed task pairs using band-power features. In our analysis, we only used four band-power features from Tp9, Tp10, Fp1 and Fp2 (theta, alpha, low beta, high beta) instead of all six bands as they can discriminate against the different relaxation states best. (Appendix 4). For all three comparisons, there was highly statistical significance ($p < 0.001$). Since there were generally significant differences for all the non-relaxed and relaxed tasks, we then moved on to paired

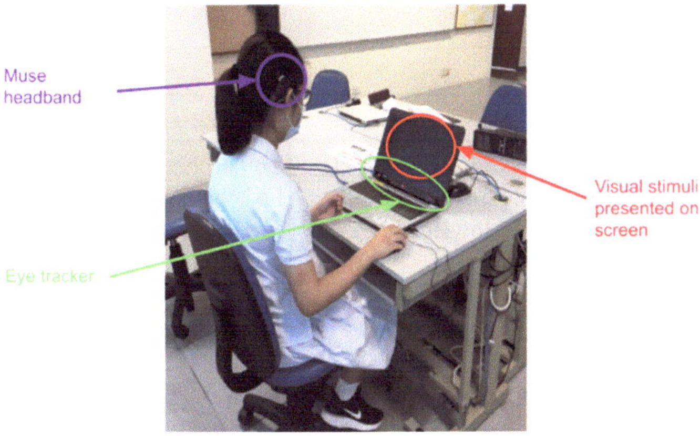

Fig. 14.2 Experiment set-up and subject performing the experiment task

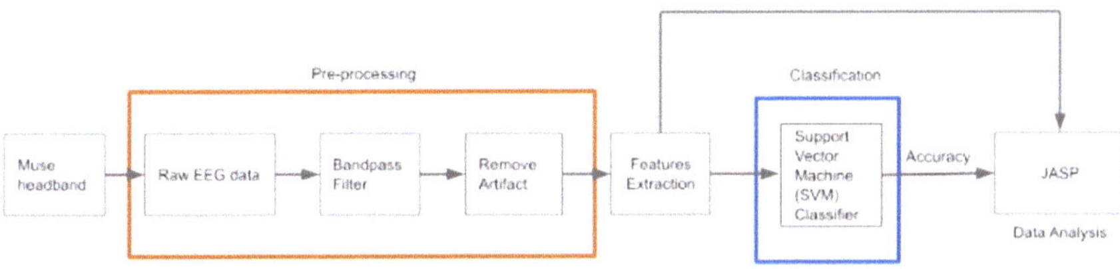

Fig. 14.3 Sequence diagram for data analysis

sample *t*-tests. We compared the individual non-relaxed tasks to the baseline and to each other to see if there is a significant difference between each pair as given in Table14.1.

Using ANOVA and *t*-tests, we had found that there is a significant difference between the different emotions. To see which non-relaxed task had a larger effect on relaxation, and whether it is a negative or positive effect, we then used supervised machine learning method to test binary classification accuracy of "relax" versus "non-relax" among different task pair (e.g. anxiety-relaxed versus anxiety-non-relaxed) using tenfold cross-validation as shown in Fig. 14.4.

We also used STAI questionnaire results in our analysis and compared it with the accuracy as seen in Fig. 14.5. Refer to (Appendix 3) for the calculation of the scores and more details.

From Fig. 14.5, participants' average STAI scores decreased the most in the relaxation task after the anxiety-inducing task, followed by after the fear-inducing task. The scores also increased slightly in the relaxation task after the happiness-inducing task.

Table 14.1 Results of paired sample *t*-tests

Non-relaxed tasks *p*-values*		
	Fear-nonrelaxed	Happiness-nonrelaxed
Anxiety-nonrelaxed	Tp9 highBeta: 0.981 Fp2 alpha: 0.055 Tp10 lowBeta: 0.658 Remaining: **<0.001**	Tp9 alpha: 0.264 Fp1 theta: 0.050 Tp10 lowBeta: **0.012** Remaining: **<0.001**
Fear-nonrelaxed		Fp1 lowBeta: 0.961 Fp2 alpha: **0.008 (<0.05)** Tp10 lowBeta: 0.003 Remaining: **<0.001**
Relaxed tasks *p*-values*		
	Fear-relaxed	Happiness-relaxed
Anxiety-relaxed	Tp9 theta: **0.009** Fp2 highBeta: 0.148 Remaining: **<0.001**	Tp9 alpha: **0.027** Fp1 theta: 0.230 Fp2 alpha: 0.288 Remaining:**<0.001**
Fear-relaxed		Tp10 theta: **0.004** Tp10 alpha: 0.834 Remaining: **<0.001**

* We use *p*-value thresholds for significant test as 0.05 and 0.001

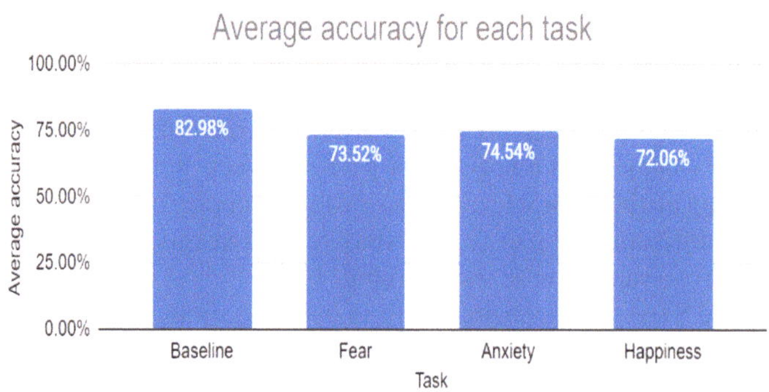

Fig. 14.4 Average classification accuracy of each task pair

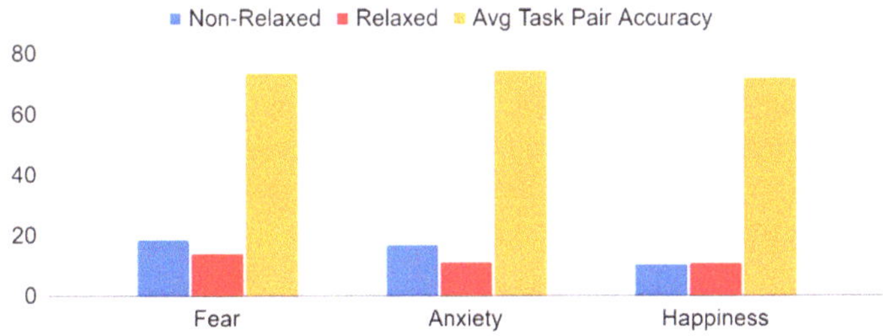

Fig. 14.5 Comparison of average STAI scores and average task pair accuracy

14.5 Discussion

We used statistical analysis using ANOVA and paired sample *t*-tests, and we found that there is generally high statistical significance ($p < 0.001$) for all non-relaxed tasks and between non-relaxed task pairs and relaxed task pairs. To find which emotion had the most impact on relaxation and to what extent, we used the accuracy from tenfold cross-validation to do so. Referring to Fig. 14.4, there is a large decrease in accuracy from the baseline pair to the non-relaxed pairs, with a 8.44%, 9.46% and 10.92% decrease in accuracy for anxiety, fear and happiness non-relaxed and relaxed task pairs, respectively, with an average of 9.61% decrease. This shows that all emotions do affect relaxation, and in order of ascending extent of effect, happiness, fear and anxiety affect relaxation the most. This was similar to what we have predicted.

STAI is used as a measure of anxiety, with higher scores indicating higher levels of anxiety. Referring to Fig. 14.5, our participants' average STAI scores showed the largest decrease in anxiety in the relaxation task after the anxiety-inducing task, followed by after the fear-inducing task. The scores also showed a slight increase in anxiety during the relaxation task after the happiness-inducing task. This supports our results from the accuracy. The average scores as seen in Fig. 14.5 also correctly predicted that anxiety and fear affected relaxation the most, while happiness affected relaxation the least. However, it did not correctly match whether relaxation was affected by fear or anxiety the most. This could be because STAI indicates a negative emotion or a positive emotion but not what kind of negative emotion it is.

Based on the results above, our main hypothesis has been proven correct; however, we have wrongly predicted the order of which emotions have the most impact on relaxation. Happiness affects relaxation the least, followed by fear and then by anxiety. We have wrongly predicted that happiness has a positive effect on relaxation. However, our hypothesis that STAI would largely correctly predict the effect on relaxation was correct. We could apply these findings to better consider one's emotional states when relaxing and a better understanding to neutralise emotional effects to enhance relaxation.

However, we did not take into consideration how the different emotion-inducing tasks, especially the happiness task, could have also caused participants to relax during the task itself, resulting in a lower-than-expected difference in relaxation state between the non-relaxed and relaxed tasks, which could have affected the accuracy found during the paired *t*-test. We also only had 15 participants, which was a small number, and larger sample size would have resulted in better representation. Future research could take a closer look at what causes the difference in the effect of relaxation states for the different emotions and what is the best way to mitigate the negative effects of these emotions on relaxation.

14.6 Conclusion

In our study, we aimed to find out the influences of three different emotions: fear, anxiety and happiness, on relaxation. After designing and implementing the experiment software, we collected EEG data from 15 participants according to the experiment protocol consisting of baseline, emotion-arousing (fear, anxiety and happiness) and relaxation tasks. We used signal processing and machine learning methods to compute and compare the mean accuracies of each non-relaxed and relaxed task pair. We also applied extracted band-power features with different statistical analysis methods to test our research hypotheses. Overall, our results showed that the different emotions affected relaxation to varying degrees. These findings could allow a better understanding of emotions' effects on relaxation, which could potentially be useful in areas such as stress management and psychological therapies. Our study is the first step in bringing notable interest and attention to applying different effects of individual emotions in relaxation and utilising particular emotions to improve the relaxation.

Appendix 1. Experiment Procedure

OpenSesame is a programme to create experiments for psychology, neuroscience and experimental economics and a graphical experiment builder for the social sciences. OpenSesame is free, open-source and cross-platform programme. It features a comprehensive and intuitive graphical user interface and supports Python scripting for complex tasks. Additional functionality, such as support for eye trackers, input devices and video playback, is available through plug-ins. OpenSesame can be used in combination with existing software for creating experiments. We used it to create our experiment. The below figures are the exact experiment design.

STAI - Full
Stimulus presentation
Instructions:
Question 1
1.
2.
3.
4.
Question 21
1 2 3 4 5 6 7 8 9
Instructions
STAI (20 questions)
Valence- arousal scale (1 to 9)
Timing
Approx 5s
Approx 60s
Approx 15s
User Response
Press any key
Answer questions by clicking appropriate response
Answer questions by clicking appropriate response
Event Logging
STAIfull_0
Keypress not recorded
STAIfull_1 to STAIfull_20
Keypress: STAIfull_x_response
STAIfull_20 to STAIfull_21
Keypress: STAIfull_x_response
Baseline (Non-relaxed)
Stimulus presentation
Instructions:
Instructions
Random no. of dots of random colour (red or green) appearing at random places on the screen
Blank screen
Timing
Approx 5s
Rand (0.5s to 2s)
Rand (3s to 4.5s)
User Response
Press any key
Count the number of red dots on the screen
Press the number of red dots from '5' to '9'
Event Logging
baseline_nonrelaxed_0
Keypress not recorded
baseline_nonrelaxed_1
Keypress not recorded
baseline_nonrelaxed_2
Keypress not recorded
Baseline (Relaxed)
Stimulus presentation
Instructions:
Instructions
expanding and shrinking circle with instructions to breathe in time (4 sec breathe in, 2 sec hold, 4 sec breathe out)
Timing
Approx 5s
total 210s (21 sets of breathing in, holding, and breathing out)
User Response
Press any key
Inhale, hold and exhale according to the instructions on-screen
Event Logging
baseline_relaxed_0
Keypress not recorded
baseline_relaxed_1
Keypress not recorded
Anxiety Non-relaxed Task
Stimulus presentation
Instructions:
Instructions
timed math questions with 4 options on screen to choose from
Timing
Approx 5s
10s each question, total 210s
User Response
Press any key
Answer questions by clicking one of the options
Event Logging
anxiety_0
Keypress not recorded
anxiety_1 to anxiety_17
Keypress: anxiety_x_response

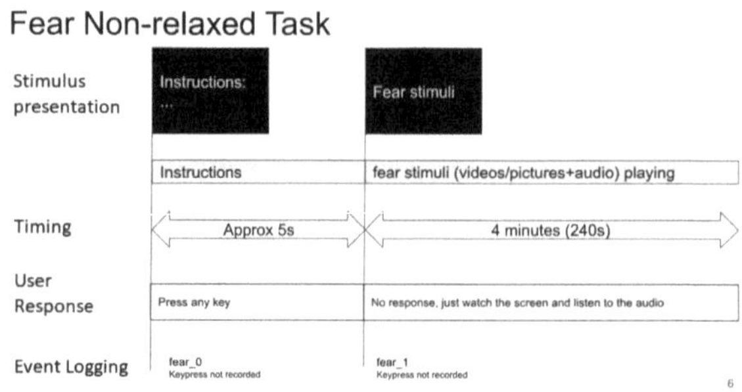
Fear Non-relaxed Task
Stimulus presentation
Instructions: ...
Fear stimuli
Instructions
fear stimuli (videos/pictures+audio) playing
Timing
Approx 5s
4 minutes (240s)
User Response
Press any key
No response, just watch the screen and listen to the audio
Event Logging
fear_0
Keypress not recorded
fear_1
Keypress not recorded
6

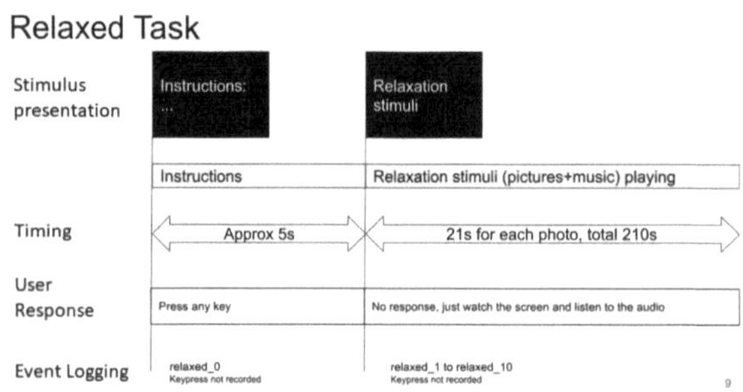
Relaxed Task
Stimulus presentation
Instructions: ...
Relaxation stimuli
Instructions
Relaxation stimuli (pictures+music) playing
Timing
Approx 5s
21s for each photo, total 210s
User Response
Press any key
No response, just watch the screen and listen to the audio
Event Logging
relaxed_0
Keypress not recorded
relaxed_1 to relaxed_10
Keypress not recorded
9

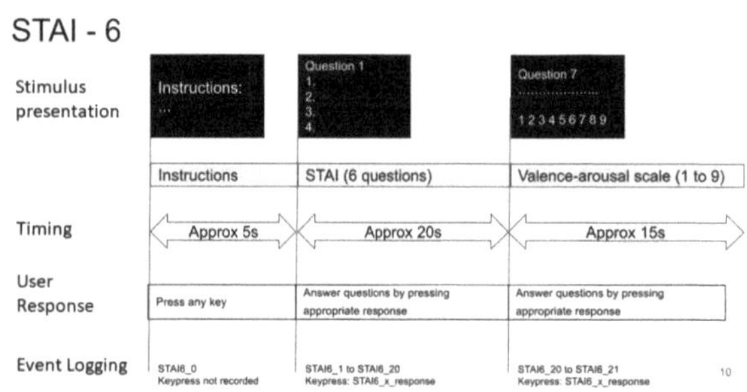
STAI - 6
Stimulus presentation
Instructions: ...
Question 1
1.
2.
3.
4.
Question 7
123456789
Instructions
STAI (6 questions)
Valence-arousal scale (1 to 9)
Timing
Approx 5s
Approx 20s
Approx 15s
User Response
Press any key
Answer questions by pressing appropriate response
Answer questions by pressing appropriate response
Event Logging
STAI6_0
Keypress not recorded
STAI6_1 to STAI6_20
Keypress: STAI6_x_response
STAI6_20 to STAI6_21
Keypress: STAI6_x_response
10

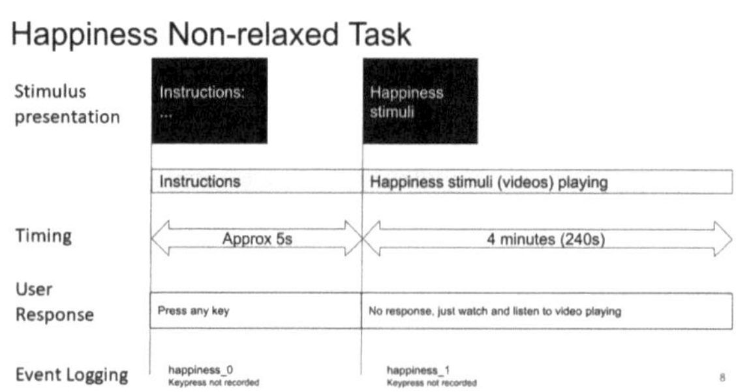
Happiness Non-relaxed Task
Stimulus presentation
Instructions: ...
Happiness stimuli
Instructions
Happiness stimuli (videos) playing
Timing
Approx 5s
4 minutes (240s)
User Response
Press any key
No response, just watch and listen to video playing
Event Logging
happiness_0
Keypress not recorded
happiness_1
Keypress not recorded
8

Appendix 2. Stimuli Used

Relaxation

Audio: "Venus, the Bringer of Peace", Gustav Holst.

Photographs: scenes of forests and oceans. Samples:

Fear: All: Jumpscares and scenes from horror movies.

Personalised: Jumpscares and scenes from horror movies, movie clips featuring snakes and movie clips featuring spiders.

Happiness:

All: Comedy video "Fastest Newscast Ever", Studio C.

Personalised: "Penguin chicks rescued by unlikely hero", BBC and clips from cartoons "Spongebob" and "Tom and Jerry".

Anxiety: Audio: "Threnody for the Victims of Hiroshima", Krzysztof Penderecki.

Questions: 30 mental arithmetic questions [e.g. Q: 3(4 + 46/2) (a) 81, (b) 73, (c) 78 and (d) 96].

To be consistent, we used both audio and visual stimuli for the relaxed task and all non-relaxed stimuli.

The order of the three blocks (of the non-relaxed stimuli each followed by a relaxation task) are randomised for each participant so as to ensure that the order of emotions induced does not affect the results.

Appendix 3. STAI Questionnaire

STAI stands for State-Trait Anxiety Inventory. It is a commonly used measure of trait and state anxiety. All 20 questions are rated on a four-point scale ("not at all", "somewhat", "moderately so" and "very much so"). Ten of the 20 items are anxiety-present, and are scored with one point for "not at all" and four points for "very much so", while the other ten items are anxiety-absent and are scored with one point for "very much so" and four points for "not at all". The total score ranges from 20 to 80, with higher scores indicating higher levels of anxiety symptoms. A cut-off score of 40 is commonly used to define probable clinical levels of anxiety. For the STAI-6, the total score is scaled down to range from 6 to 24, with a cut-off score of 12. The STAI has shown to be a valid and reliable measure.

Results of average STAI-6 scores and accuracy:

Average	Non-relaxed STAI-6 score	Relaxed STAI-6 score	Average task pair accuracy
Anxiety	16.93	10.93	74.543
Fear	18.33	13.87	73.51
Happiness	10.27	10.87	72.06

The results of average STAI-20 scores for the start of the experiment and for the end of the experiment are 39.93 and 48.33, respectively.

We analysed the participants' scores for each STAI-6 questionnaire done after each task, which has a total score ranging from 6 to 24. Overall, participants showed the highest levels of anxiety after the fear-inducing task, followed by the anxiety-inducing task and then the happiness-inducing task. The happiness-inducing task also resulted in a score of 10.27 which is below the cut-off point of 12 to define probable clinical levels of anxiety.

Participants' average STAI scores decreased the most in the relaxation task after the anxiety-inducing task, followed by after the fear-inducing task. However, the average score increased slightly in the relaxation task after the happiness task. Thus, the relaxation task after each non-relaxed task was most effective after the anxiety-inducing task, followed by after the fear-inducing task. The relaxation task was the least effective after the happiness-inducing task, and it also resulted in marginally higher levels of anxiety.

We also analysed participants' scores for the full STAI questionnaire done at the start and the end of the experiment, which has a total score ranging from 20 to 80. Most participants' scores increased by the end of the experiment, with an exception of two out of the 15 participants whose scores decreased, and one participant whose score remained the same. The mean of participants' scores increased significantly by the end of the experiment which showed that overall participants' levels of anxiety generally increased.

Appendix 4. Experiment Data

ANOVA results for all tasks.

Band-power feature	All tasks *p*-value*	All non-relaxed tasks *p*-value*	All relaxed tasks *p*-value*
Tp9 theta	< 0.001	< 0.001	< 0.001
Tp9 alpha	< 0.001	< 0.001	< 0.001
Tp9 lowBeta	< 0.001	< 0.001	< 0.001
Tp9 highBeta	< 0.001	< 0.001	< 0.001
Fp1 theta	< 0.001	< 0.001	< 0.001

(continued)

(continued)

Band-power feature	All tasks p-value*	All non-relaxed tasks p-value*	All relaxed tasks p-value*
Fp1 alpha	< 0.001	< 0.001	< 0.001
Fp1 lowBeta	< 0.001	< 0.001	< 0.001
Fp1 highBeta	< 0.001	< 0.001	< 0.001
Fp2 theta	< 0.001	< 0.001	< 0.001
Fp2 alpha	< 0.001	< 0.001	< 0.001
Fp2 lowBeta	< 0.001	< 0.001	< 0.001
Fp2 highBeta	< 0.001	< 0.001	< 0.001
Tp10 theta	< 0.001	< 0.001	< 0.001
Tp10 alpha	< 0.001	< 0.001	< 0.001
Tp10 lowBeta	< 0.001	0.006	< 0.001
Tp10 highBeta	< 0.001	< 0.001	< 0.001

* We use p-value thresholds for significant test as 0.05 and 0.001

References

1. Kashani, F., Babaei, S., Bahrami, M., & Valiani, M. (2012). The effects of relaxation on reducing depression, anxiety and stress in women who underwent mastectomy for breast cancer. *Iranian Journal of Nursing and Midwifery Research, 17*(1), 30–33.
2. Larson, H., Yoder, A., Brucker, S., Lee, J., Washburn, F., Perdieu, D., Polydore, C., & Rose, J. (2011). Effects of relaxation and deep-breathing on high school students: ACT prep. *Journal of Counseling in Illinois*, 16–27.
3. de Paula, A. A. D., de Carvalho, E. C., & dos Santos, C. B. (2002). The use of the 'progressive muscle relaxation' technique for pain relief in gynecology and obstetrics. *Revista Latino-Americana de Enfermagem, 10*(5), 654–659. https://doi.org/10.1590/S0104-11692002000500005
4. Aritzeta, A., Soroa, G., Balluerka, N., Muela, A., Gorostiaga, A., & Aliri, J. (2017). Reducing anxiety and improving academic performance through a biofeedback relaxation training program. *Applied Psychophysiology and Biofeedback, 42*(3), 193–202. https://doi.org/10.1007/s10484-017-9367-z
5. Cigna (2019, May 20). 2019 Cigna 360 well-being survey. Retrieved January 29, 2021, from https://www.cignaglobal.com/blog/healthcare/2019-cigna-wellbeing-survey
6. Bird, J. J., Manso, L. J., Ribeiro, E. P., Ekart, A., & Faria, D. R. (2018). A study on mental state classification using EEG-based brain-machine interface. *International Conference on Intelligent Systems (IS), 2018*, 795–800. https://doi.org/10.1109/IS.2018.8710576
7. Stieger, J. R., Engel, S., Jiang, H., Cline, C. C., Kreitzer, M. J., & He, B. (2020). *Mindfulness improves brain computer interface performance by increasing control over neural activity in the alpha band*. https://doi.org/10.1101/2020.04.13.039081
8. Berggren, U., Hakeberg, M., & Carlsson, S. G. (2000). Relaxation versus cognitively oriented therapies for dental fear. *Journal of Dental Research, 79*(9), 1645–1651. https://doi.org/10.1177/00220345000790090201
9. McGlynn, F. D., Moore, P. M., Rose, M. P., & Lazarte, A. (1995). Effects of relaxation training on fear and arousal during in vivo exposure to a caged snake among DSM-III-R simple (snake)

phobics. *Journal of Behavior Therapy and Experimental Psychiatry, 26*(1), 1–8. https://doi.org/10.1016/0005-7916(94)00075-W
10. Broota, A., Varma, R., & Singh, A. (1995). Role of relaxation in hypertension. *Journal of the Indian Academy of Applied Psychology, 21*(1), 29–36.
11. Gremigni, P. (2012). Is humor the best medicine? *Humor and Health Promotion*, 149–171.
12. Wang, F., He, Y., Pan, J., Xie, Q., Yu, R., Zhang, R., & Li, Y. (2015). A novel audiovisual brain–computer interface and its application in awareness detection. *Scientific Reports, 5*(1), 9962. https://doi.org/10.1038/srep09962
13. Caviola, S., Carey, E., Mammarella, I. C., & Szucs, D. (2017). Stress, time pressure, strategy selection and math anxiety in mathematics: A review of the literature. *Frontiers in Psychology, 8*, 1488. https://doi.org/10.3389/fpsyg.2017.01488
14. Alvarsson, J. J., Wiens, S., & Nilsson, M. E. (2010). Stress recovery during exposure to nature sound and environmental noise. *International Journal of Environmental Research and Public Health, 7*(3), 1036–1046. https://doi.org/10.3390/ijerph7031036
15. Huang, C., & Sheng, L. (2016). A pilot study on the portable EEG-based music effects. *Journal of Biomusical Engineering, 01*(S1). https://doi.org/10.4172/2090-2719.S1-002
16. Knight, W. E. J., & Rickard, N. S. (2001). Relaxing music prevents stress-induced increases in subjective anxiety, systolic blood pressure, and heart rate in healthy males and females. *Journal of Music Therapy, 38*(4), 254–272. https://doi.org/10.1093/jmt/38.4.254
17. Song, C., Ikei, H., & Miyazaki, Y. (2018). Physiological effects of visual stimulation with forest imagery. *International Journal of Environmental Research and Public Health, 15*(2), 213. https://doi.org/10.3390/ijerph15020213

Chapter 15
AI-Driven Data Analytics Assessing Methods to Localize Preexisting Natural Language-Processing Data

Sean Lim Shi-An, Karis Yuen Xin Er, Ryan Nathaniel Thesman, and Teow Loo Nin

Abstract BERT, which stands for bidirectional encoder representations from transformers, is a language representation model that has gained popularity lately for its novelty and uniqueness. This algorithm has already been used in many different fields which requires natural language processing (NLP), and because its bidirectional, unlike many other NLP models, the BERT model is able to learn the context of the word based on the sentence it is in instead of just the word itself. Because Singlish is normally spoken in an informal context and has deeper layers of meaning, hidden under sarcasm, we believed that the BERT model would be appropriate in analyzing Singlish text. Our group aims to exploit and adapt BERT for sentiment analysis of Singlish (Wikipedia Contributors in Singlish (2021), [1]) text. In this project, by applying a random forest classifier to sentence embeddings extracted from Singlish BERT, we obtained a 17% improvement in accuracy as compared to the original BERT, at 67% and 50% accuracies, respectively, on Singlish (Wikipedia Contributors in Singlish (2021), [1]) text.

Keywords BERT · Machine learning · Singlish · BLP

15.1 Introduction

Language model pretraining is effective in many fields which require natural language processing tasks. What such models effectively do are to imitate human being recounting from past experiences. Such models are given a dataset to train on, which will allow them to "remember" how natural language is like. This will then allow for the model to use this "remembered" knowledge on a testing set. BERT is one of the newer and more efficient versions of such models. Jacob Devlin and his colleagues

S. L. Shi-An (✉) · K. Y. X. Er · R. N. Thesman
NUS High School of Mathematics and Science, NUSH, 20 Clementi Ave 1, Singapore 129957, Singapore
e-mail: H1710121@nushigh.edu.sg

T. L. Nin
DSO National Laboratories, DSO, 12 Science Park Drive, Singapore 118225, Singapore
e-mail: tloonin@dso.org.sg

H. Guo et al. (eds.), *IRC-SET 2021*, https://doi.org/10.1007/978-981-16-9869-9_15

developed BERT at Google in 2018, and it is a state-of-the-art machine learning model used for NLP tasks. BERT's key technical innovation is applying the bidirectional training of transformer, a popular attention model, to language modeling. By being able to do so, most of such counterparts delve into the same field, utilizing only directional training. By being able to read the sentence as a whole unit, rather than individual words in a sentence in other directional models, it has improved natural language processing. This has inevitably provided new advances in such a field. However, it is seldom used in Singlish. This is to compare the effectiveness of our SingBERT [2] model as compared to the original one at reading Singlish sentences.

Singlish being one of the few main informal dialects found, especially in Singapore, has been adapted by many locals into their daily dialogues.

We presume that the original BERT would be ineffective in undertaking sentimental analysis as it would be unable to read the sentiments behind such Singlish words. This is especially so for sentences where the Singlish word will affect the sentiment of a sentence, for example, "This is swee swee," where in Singlish terms it is of positive sentiment, whereas the original BERT may be unable to recognize the Singlish word and may instead assign a neutral sentiment to this sentence. Therefore, we decided to investigate making a Singlish BERT such that sentences will be able to be labeled with the correct sentiment.

Our group has utilized BERT which has been fine-tuned to Singlish data. We hypothesize that a BERT model, which has been fine-tuned on Singlish data, can be utilized for sentiment analysis on Singlish text. We will test the accuracy of this fine-tuned BERT by using it for sentiment analysis on a labeled dataset, which has been compiled from multiple sources, including SGAG, reddit, YouTube as well as Mothership.sg. This dataset will be used as both the training as well as the testing dataset, with 80% used for training and 20% used for validation. We will also compare this fine-tuned BERT model as to one which has not be fine-tuned on Singlish.

15.2 Background

In the field of computer vision, researchers have repeatedly shown the value of transfer learning using the trained neural network as the basis of a new purpose-specific model. In recent years, research has also shown that a similar technique can be useful in many natural language tasks. BERT falls under this section and has made considerable improvements to this area.

A different approach, which is also popular in NLP tasks and exemplified in the recent ELMo paper, is feature-based training. In this approach, a pretrained neural network produces word embeddings which are then used as features in NLP models.

Sentiment analysis is a natural language processing technique that is used to determine the emotions behind data, whether it is positive, neutral, or negative. Sentiment analysis is often performed on textual data for various reasons such as to help businesses monitor brand and product sentiment in customer feedback and understand customer needs.

15.3 Methodology

Our idea is to utilize the concept of the original BERT, but with fine-tuning done to the BERT [3] model on Singlish. We would then compare the accuracy of this SingBERT [2] against the normal BERT [3] on analyzing sentiment in Singlish datasets.

15.3.1 BERT

To initialize BERT, we made use of a BERT tokenizer to tokenize all the sentences within our pre-made dataset, and thereafter map the tokens to their word IDs. We would afterward find the maximum sentence length inside the dataset that is not anomalies and set this number to be the padding number. The sentences would then be truncated to this number, which we found to be 45.

- *Masked Machine Learning (MLM)*

Before feeding word sequences into BERT, attention masks are being created for each sequence. This will allow the model to predict the original value of said masked words, based on the sentence given. This process is also known as MLM [4] or is also known as masked machine learning. Its purpose is to allow the AI to be able to learn how a sentence can fit together so that BERT is able to have bidirectional learning of the sentence by knowing the context of the word as well as the often-appearing adjacent words.

- *Preprocessing*

The dataset will then be split into the training set and testing set, which are around 80 and 20% each with similar ratios of neutral, positive, and negative statement in both datasets. The code will then convert all the input data and labels into torch [5] tensors. They will then be loaded in batches of 32, which in our case is to conduct sentimental analysis of Singlish comments. We will be utilizing a SingBERT [2] model which has been modified from the original hugging-face-BERT [3] model to be fine-tuned onto Singlish as well as Manglish. This was done by filling in the unused tokens in the original BERT model with several custom vocab tokens from Singlish words which did not already exist within the original BERT model [2]. We then "attach" a classifier on top of both models and train it on the labeled Singlish dataset, for sentiment analysis, and compare the accuracy of the models on a separate test set of Singlish statements.

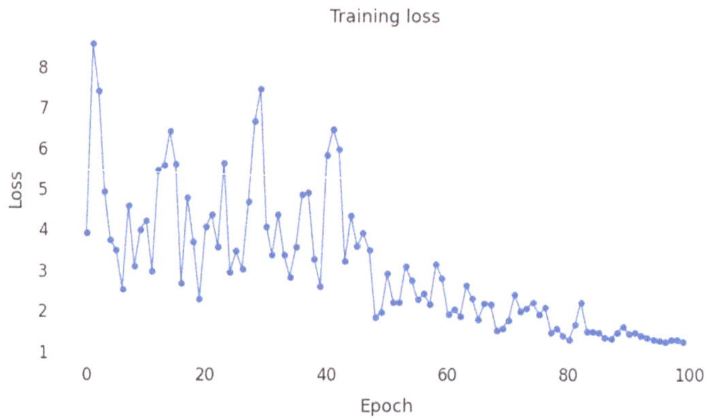

Fig. 15.1 Graph of loss per epoch

15.3.2 Classifier

- Using a linear classifier

To start, we tried training the models on the Singlish training set with a single linear classifier that was pre-built into the model [2]; however, after running the data over several epochs, it was noted that due to the complexity of the Singlish data, the model could not learn with just a single linear classifier resulting in the inability for the loss function to converge, and hence, we had to use a more complex classifier.

As we can see in Fig. 15.1, the loss function converges at approximately 1 and failed to converge at 0 after several epochs proving that this sentiment analysis task was too complex for linear classification, and the models could not accurately learn how to classify the statements in the dataset with just a linear classifier.

- Using random forest classifier

We decided to retrieve the sentence embeddings from the last hidden layer in the BERT models and train a random forest classifier with these embeddings. We would utilize the embedding at the first position, [CLS] tokens, as these were the sentence vectors that were fed into the original linear classifier. After training the two random forest classifiers on sentence embedding from the respective models, we would compare the accuracies of these classifiers on classifying sentence embeddings of the Singlish test dataset from the respective models.

Figure 15.2 shows the process of how we obtained the respective sentence embeddings from the two models before they passed through the classifier. We would input the Singlish sentences, from out dataset, into the respective tokenizers which out the respective word vectors. We would then input these word vectors into the respective BERT models. We would extract the sentence embeddings from the models at the last hidden layers for training of the random forest classifier. For both models, the

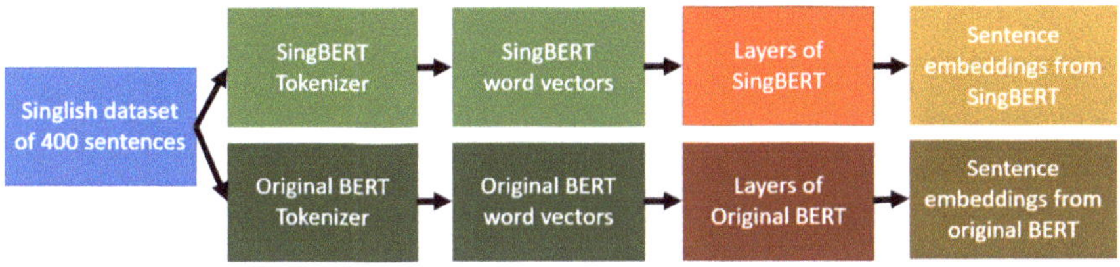

Fig. 15.2 How we obtained the sentence embeddings from the models

parameters we set were num_labels = 3 and output_hidden_states=True with the rest following the default parameters of the original BERT models.

Figure 15.3 detailed the training process. The sentence embeddings that we retrieve from the BERT models are in the form of a three-dimensional PyTorch [6] tensor. Form the tensor, we only extract the outputs for the [CLS] token and remove the outputs of the other 44 tokens. This is because the output for the [CLS] token is commonly regarded as the sentence embedding and is commonly used in classification. Along with the sentence embeddings, we also feed the correct labels for the training set into the random forest classifiers when training it.

- Process of obtaining accuracy of the two classifiers

Figure 15.4 shows how we obtained the predicted labels from the respective classifiers. We would feed the sentence embeddings of the 100 sentences in the test set into the random forest classifiers and get the predicted labels of the output of these sentences.

The classifiers attained a 99.75% accuracy on the training set of data meaning that the loss function converges to approximately 0 when using random forests as classifiers, showing that random forest is a suitable choice.

We also varied the number of trees we used in our random tree classifier to see which number optimal to produce the highest accuracy. We made use of scikit-learn [7] library to utilize these random tree classifiers and retained the default parameters.

- Creation of Singlish datasets for training and testing

Due to several time constraints, we had, we were unable to create a large dataset for our classifier to train and test on. This is also one of the reasons why we choose to

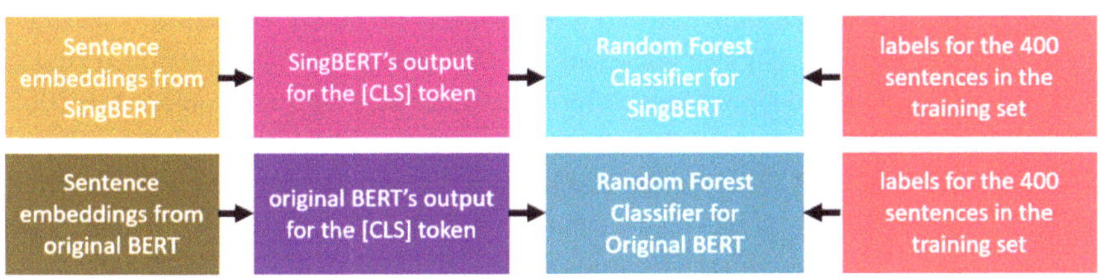

Fig. 15.3 How we trained the two random forest classifiers

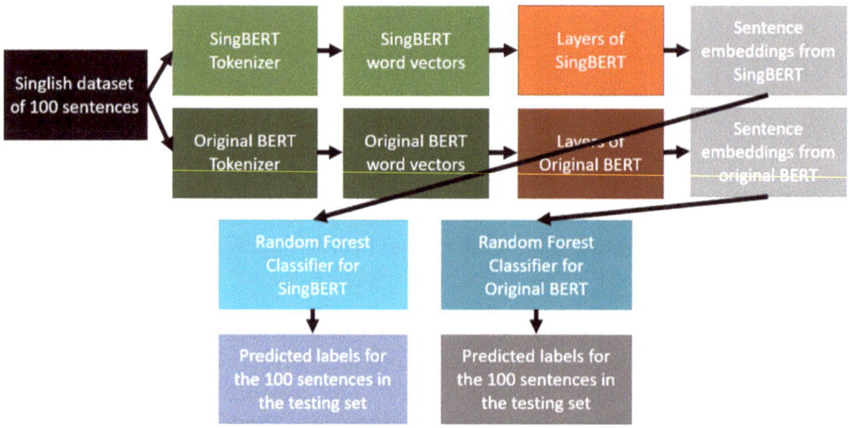

Fig. 15.4 How we obtained predicted labels from respective classifiers

use a random forest classifier in contrast to a neural network which requires a large dataset to be effective.

We took data from several sources namely

- YouTube comments from Channel News Asia (Local Singapore New Channel)
- Comments from SGAG (local satirical source)
- Statements from reddit, r/Singapore (social media filled with Singaporeans)
- Comments from Mothership sg (Local independent news source).

We split it up into two datasets—a training dataset with 400 statements and a test dataset with 100 statements. We labeled both datasets and ensured that both datasets had similar ratio of negative to positive to neutral sentiment statements, namely 2:1:2

15.4 Results

We ended the training when the model was able to achieve a 99.675% accuracy on the testing set for the SingBERT [2] model concluding that the loss function did converge, and that the random tree classifier was a suitable classifier for us to use for our sentiment analysis task. As seen from Table 15.2, without further fine-tuning, the SingBERT model with the random tree classifier obtained a maximum accuracy on the test set of 67%. This is achieved at 2000 trees. Afterward, when trying at 5000 trees, there was no significant difference in the accuracy as compared to 2000, showing that it has plateaued at that accuracy.

For the original BERT [3], the highest accuracy that it has attained is 50%, which was achieved at 500 trees, as can be seen from Table 15.1. There is no significant difference in the accuracy after 500 trees for the original BERT [3].

Table 15.1 Accuracy of results of the original BERT model on the testing dataset

Number of trees	Accuracy on testing dataset (%)
100	47
500	50
1000	50
2000	50
5000	49

Table 15.2 Accuracy of results of the SingBERT model on the testing dataset

Number of trees	Accuracy on testing dataset (%)
100	56
500	60
1000	64
2000	67
5000	66

As can be seen from the tables above, the BERT model that has been fine-tuned on Singlish is able to have a higher accuracy over all sets of trees, as compared to the BERT model which has not been fine-tuned on the Singlish dataset.

Furthermore, upon further checks, we found further insights on the way BERT's misclassifies Singlish statements which can be seen in Fig. 15.5.

From these results, it appears that the model tends to wrongly classify neutral statements as negative statements. We suspect this is probably because the training is not sufficiently large and. Another probably reason is because Singlish has elements

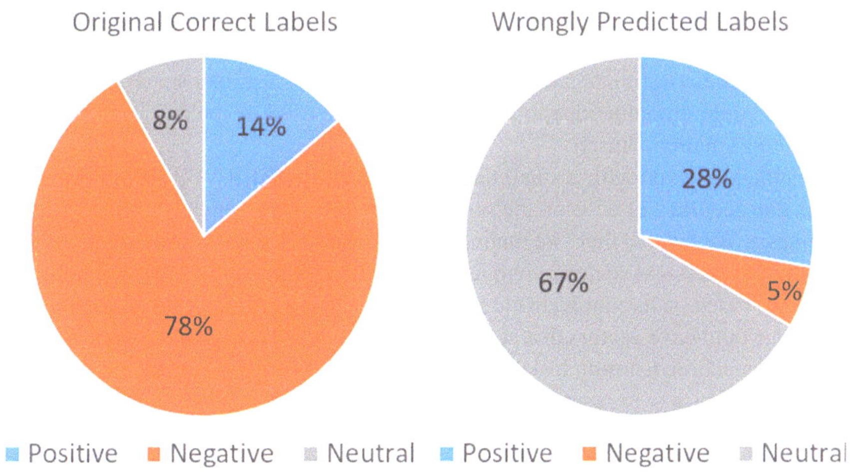

Fig. 15.5 More information on the wrongly predicted labels

of sarcasm which can create a deeper understanding of sentences depending on the context it is used in which the model is unable to learn.

15.5 Future Work and Discussions

We also feel that many Singlish comments that we have found online are often sarcastic, which may cause the model to return an even lower accuracy as opposed to what was expected. This can be seen in our result where a big proportion of statements wrongly labeled are statements that were negative but wrongly labeled as neutral due to the inability of the model to detect sarcasm.

This is something that we can work on as sarcasm is common in the Singlish context. As testing and training BERT on singular sentences may not suffice, we may try to implement BERT and alter it to be able to look across multiple sentences within a dialogue or conversation, thus allowing the model to learn the context meant by those sentences. This would allow us to be able to get the model to be able to detect sarcasm, thus allowing it to work better on more sentences at a dialogue level.

We may also want to try training on a larger dataset in the future using the SingBERT model [2], to test if the accuracy would increase. By having a bigger dataset, this would allow the model to learn from a larger variety of sentences, thus allowing it to have a better understanding on the context of words. The model would then be able to better assess the sentiment of the sentences based on the words, thus increasing the accuracy of the algorithm. The difference can be seen from an article on sentiment classification using BERT at GeeksforGeeks [7]. They have 50,000 sentences in their dataset, using half for testing and half for training, which is 100 times more than our dataset. They managed to get an 85% accuracy for their BERT model.

This shows how by having a larger dataset, the BERT model accuracy can change but due to time constraints and difficulty in sieving out Singlish statements as they were less common online due to them mainly being used in spoken words. We had to stick to a smaller dataset.

Nonetheless, even with a small dataset size, the SingBERT [2] still managed to achieve an accuracy of 67% on the sentiment analysis task which is in line with our hypothesis. We believe that one main reason attributing to this is because the BERT algorithm maps a text sentence into a numeric vector that captures the semantics of each word based on its context in the sentence such that words with similar contextual meaning would have vectors that are close to each other. This is the key property that facilitates machine learning tasks such as classification and clustering.

15.6 Conclusion

With these results, we can see that SingBERT [2] has been able to outperform normal BERT at reading and labeling a Singlish dataset for sentiment analysis. However, there are still several errors, and it has a strong tendency to wrongly classify neutral statements as negative statements.

References

1. Wikipedia Contributors (2021, January 3). Singlish. In *Wikipedia, the free encyclopedia.* Retrieved 15:15, January 5, 2021, from https://en.wikipedia.org/w/index.php?title=Singlish&oldid=998026810
2. Zanelim/singbert. Hugging Face (n.d.). Retrieved December 31, 2020, from https://huggingface.co/zanelim/singbert
3. BERT (n.d.). Retrieved December 31, 2020, from https://huggingface.co/transformers/model_doc/bert.html
4. Devlin, J., Chang, M.-W., Lee, K., Toutanova, K. (2019). BERT: Pre-training of deep bidirectional transformers for language understanding.
5. Paszke, A., Gross, S., Massa, F., Lerer, A., Bradbury, J., Chanan, G., Killeen, T., Lin, Z., Gimelshein, N., Antiga, L., & Desmaison, A. (2019). Pytorch: An imperative style, high-performance deep learning library. In *Advances in neural information processing systems* (pp. 8026–8037).
6. Pedregosa et al. (2011). Scikit-learn: Machine learning in python. *JMLR, 12*, 2825–2830.
7. Sentiment Classification using BERT (GeeksforGeeks). Retrieved December 31, 2020, from https://www.geeksforgeeks.org/sentiment-classification-using-bert/

Chapter 16
The Physics of a Bead Rolling on a Rotating Hoop—A Problem in Nonlinear Dynamics

Sun Xiaoqing

Abstract The problem of a bead sliding on a rotating hoop is a classical problem in physics texts. This paper develops a predictive model for a bead rolling in a grooved rotating hoop, by solving for the equation of motion using Newtonian mechanics with damping. Despite it being a common problem, such systems have rarely been investigated experimentally. Thus, in this paper, the theoretical model was verified with experimental results over a large range of parameters, proving the validity of the methods used. Intuitive explanations for the physics of the bead were provided, which may also be of pedagogical value.

Keywords Bead on rotating hoop · Nonlinear dynamics · Rolling resistance · Bifurcation

16.1 Introduction

The problem of a bead sliding on a rotating hoop is a classical problem in undergraduate physics, often used to illustrate the Lagrangian approach to mechanics, as well as the rich dynamics behind a seemingly simple system. It is a problem in nonlinear dynamics involving bifurcation and symmetry breaking. However, the equation of motion and behavior of this system has been given little experimental verification and intuitive explanation.

In this study, we extend the investigation to a bead rolling in a groove on a rotating hoop, with the relevant damping forces. Newtonian mechanics was used to solve for the motion of the bead, and experimental verification of these theoretical predictions was conducted for a more robust understanding of this phenomenon. An intuitive explanation for the phenomenon beyond the mathematical treatment common in physics texts was provided, in order to fully understand the relevant physics. This system may serve as a mechanical analog of other nonlinear systems with phase transitions [1]. Some ideas here can be extended for applications in mechanical

---

S. Xiaoqing (✉)
Raffles Institution, Bishan, Singapore
e-mail: xiaoqingsun004@gmail.com

H. Guo et al. (eds.), *IRC-SET 2021*, https://doi.org/10.1007/978-981-16-9869-9_16

systems with mobile masses, such as self-balancing systems [2], and those with spinning, such as drying/washing machines [3].

16.2 Theoretical Model

Consider a bead of radius R_{ball} and mass m, rolling in a grooved hoop of effective radius R_{CM}. The hoop is rotating about its vertical axis, with an angular velocity ω (Fig. 16.1).

In the theoretical model, Newton's 2nd law will first be used to obtain the forces on the bead, including damping. The equation of motion will thus be derived. The derivation of the equilibrium points and explanation of the bifurcation behavior of this nonlinear system will also be shown.

16.2.1 Using Newton's 2nd Law

Using $F = \text{ma}$, the tangential forces on the bead can be easily obtained and related to the angular displacement of the bead on the hoop (16.1).

$$F_{\text{centrifugal},t} - F_{\text{gravity},t} - F_f - F_{\text{damping}} = m R_{\text{CM}} \ddot{\theta} \tag{16.1}$$

of which we know the following (Fig. 16.2):

$$\begin{aligned} F_{\text{centrifugal},t} &= m\omega^2 R_{\text{CM}} \sin\theta \cos\theta \\ F_{\text{gravity},t} &= mg R_{\text{CM}} \sin\theta \end{aligned} \tag{16.2}$$

Fig. 16.1 Schematic of bead rolling in a grooved hoop

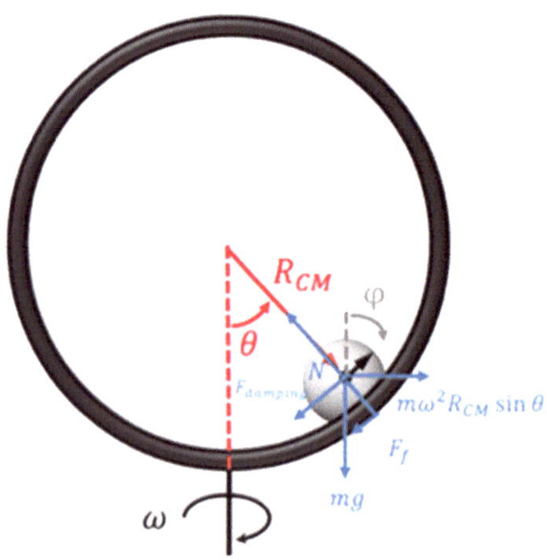

Fig. 16.2 Schematic of bead in the groove

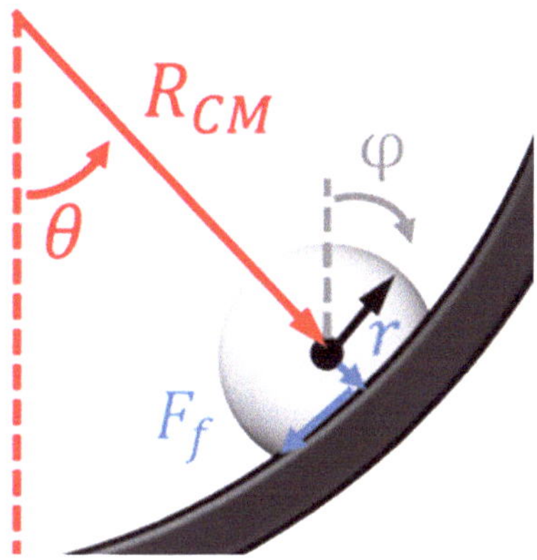

To find F_f, the static friction on the bead, notice that static friction is responsible for the torque on the bead that rotates the bead about its own axis (16.3). This rotation about its own axis can be related to the angular displacement along the hoop by Eq. (16.4) (which disagrees with the method proposed in [4][1]) Thus, the static friction force is given by (16.5).

$$I_{\text{ball}}\ddot{\varphi} = F_f r \tag{16.3}$$

$$r\dot{\varphi} = R_{\text{CM}}\dot{\theta} \tag{16.4}$$

$$F_f = \frac{2}{5} m R_{\text{ball}}^2 \frac{R_{\text{CM}}}{r^2}\ddot{\theta} \tag{16.5}$$

The final equation of motion is as follows:

$$\ddot{\theta} = \kappa\left[\omega^2 \sin\theta\cos\theta - \frac{g\sin\theta}{R_{\text{CM}}} - \frac{F_{\text{damping}}}{mR_{\text{CM}}}\right] \tag{16.6}$$

where $\kappa = \left(1 + \frac{2}{5}\frac{R_{\text{ball}}^2}{r^2}\right)^{-1}$.

The solution to (16.6) is not obvious. For $F_{\text{damping}} = 0$, it has been solved in literature using elliptic integrals [5]. However, we are interested in a realistic description of this system and thus have to consider the physical sources of damping.

[1] Equation (16.4) is the physically correct rolling condition as it relates the velocity of a point on the bead to the velocity of the centre of mass. The method proposed in [4] falls into the trap of directly equating contact distances without correcting for the curvature of the hoop surface; which is incorrect. This is a problem similar to the counterintuitive coin rotation paradox.

16.2.2 Sources of Damping

The possible sources of damping for a bead rolling in a groove would be rolling resistance and air resistance. Rolling resistance is a complex phenomenon caused by microscopic bumps and deformations on the bead and surface. It can be characterized by the following relation (for low speeds) [6]:

$$F_R = \mu N \tag{16.7}$$

where μ, the rolling resistance coefficient, has to be characterized experimentally.

Air resistance can be calculated using the following standard equation:

$$F_A = \frac{1}{2}\rho v^2 C_D A \tag{16.8}$$

where the dependence of C_D on the Reynolds number can be found in [7] (Figs. 16.3 and 16.4).

To determine which of these damping forces are more significant, the rolling resistance and air resistance models were compared with the experimental data, where a bead was allowed to oscillate at the bottom of a stationary hoop. It can thus be seen that air resistance is negligible. Henceforth, only, rolling resistance shall be considered.

The normal force N for a bead on a grooved rotating hoop has to balance the radial centrifugal and gravitational forces, the Coriolis force, as well as any static friction that points in the plane perpendicular to the path traced out by the bead. Since the directions and magnitudes of the static friction at the two points of contact are unknown, this is an indeterminate system. Thus, to solve for the normal force,

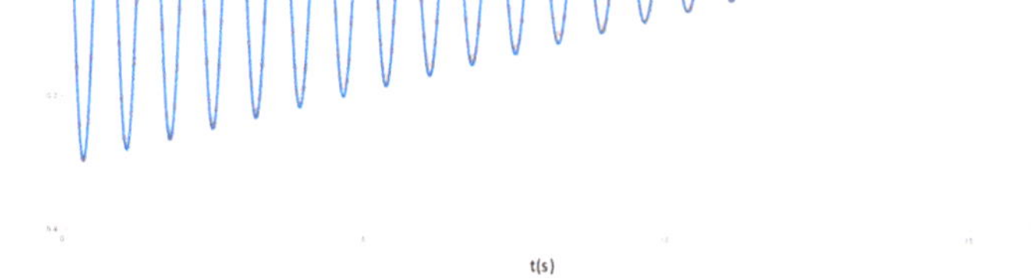

Fig. 16.3 Only, rolling resistance considered. μ was obtained by curve fitting to the equation of motion of bead on a stationary hoop with rolling resistance, obtaining $\mu = 0.0040$, which is of a reasonable order of magnitude

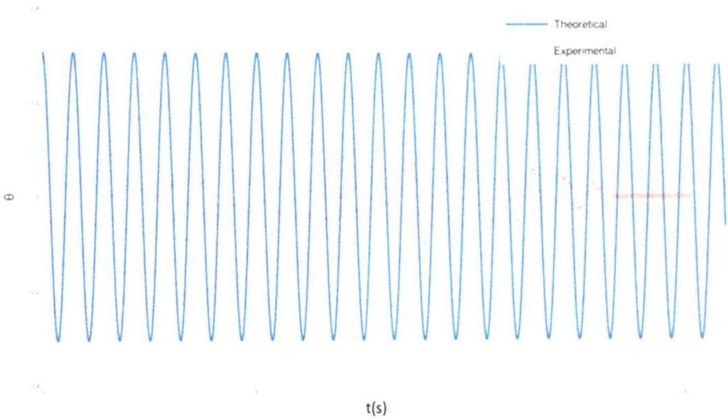

Fig. 16.4 Only air resistance considered. Constant terms were measured or calculated

the magnitude of static frictions have to be approximated to be negligible such that the normal force can be taken to balance only the radial centrifugal and gravitational forces, as well as the Coriolis force (Fig. 16.5).

$$N \approx mR_{\text{CM}}\sqrt{\left(\omega^2\sin^2\theta + \frac{g\cos\theta}{R_{\text{CM}}}\right)^2 + \left(2\omega\dot{\theta}\cos\theta\right)^2} \tag{16.9}$$

It was verified that the bead rolls without slipping. By considering the forces on the bead, it can be easily derived that $F_{\text{bead}} \leq \left(1 + \frac{5r^2}{2R_{\text{ball}}^2}\right)\mu_s N$ for rolling without slipping. This condition is fulfilled over the large range of θ and ω investigated; thus, the bead can be considered to roll without slipping. Rolling without slipping was also checked experimentally. Although slipping was not observed in these experiments,

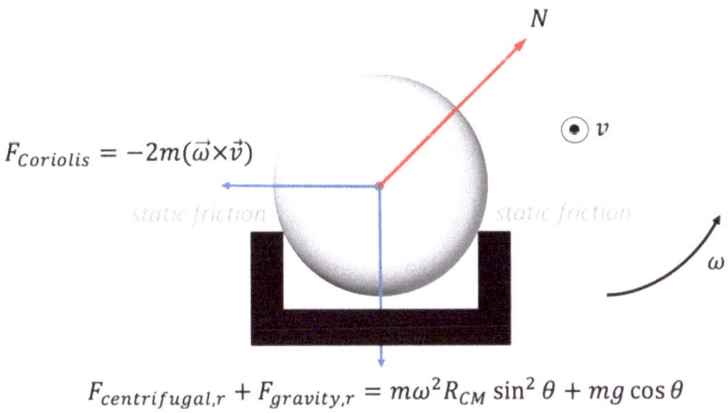

Fig. 16.5 Forced on bead in plane perpendicular to motion along hoop

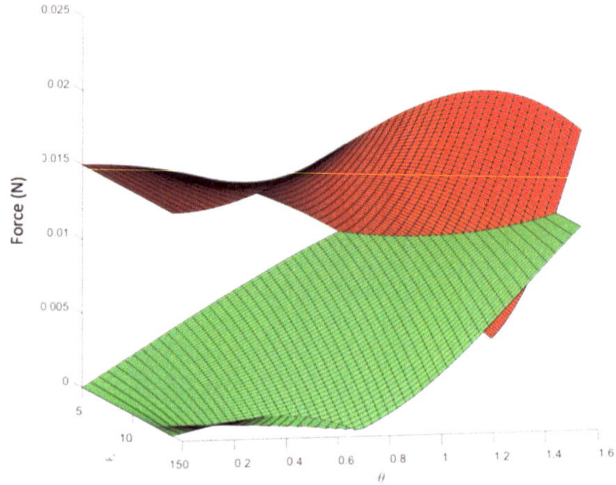

Fig. 16.6 Graph showing that force on bead is always smaller than maximum allowable force for rolling without slipping

the stick–slip transition can be easily added to the model for greater generalizability, with kinetic friction as another source of damping (Fig. 16.6).

16.2.3 Equilibrium Points and Bifurcation

The final equation of motion of the bead is as follows:

$$\ddot{\theta} = \kappa\left[\omega^2 \sin\theta\cos\theta - \frac{g\sin\theta}{R_{\mathrm{CM}}} - \mu\left(\mathrm{sign}\dot{\theta}\right)\sqrt{\left(\omega^2\sin^2\theta + \frac{g\cos\theta}{R_{\mathrm{CM}}}\right)^2 + \left(2\omega\dot{\theta}\cos\theta\right)^2}\right] \tag{16.10}$$

Setting $\ddot{\theta} = \dot{\theta} = 0$, the equilibrium points of this system are as follows:

$$\begin{aligned} &\theta = 0, \quad \text{for all}\,\theta \\ &\theta = \cos^{-1}\left(\frac{g}{\omega^2 R_{\mathrm{CM}}}\right), \quad \text{for}\,\frac{g}{\omega^2 R_{\mathrm{CM}}} \leq 1 \end{aligned} \tag{16.11}$$

The effective potential method is used to check the stability of equilibrium points, where the bead in the rotating frame is subject to effective potential $U_{\mathrm{eff}}(\theta)$. The local minima would be stable, and the local maxima would be unstable (Fig. 16.7).

$$\begin{aligned} U_{\mathrm{eff}} &= V_{\mathrm{gravity}} - T_{\mathrm{rotation\,of\,hoop}} \\ &= -mgR_{\mathrm{CM}}\cos\theta - \frac{1}{2}mR_{\mathrm{CM}}^2\omega^2\sin^2\theta \end{aligned} \tag{16.12}$$

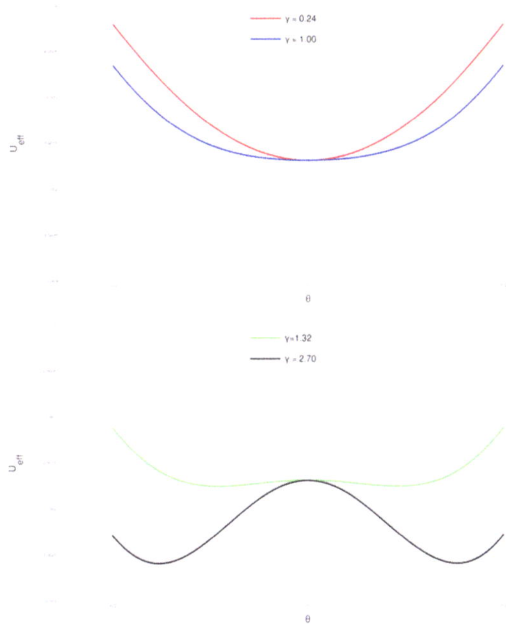

Fig. 16.7 Plot of U_{eff} against θ, for different values of γ where $\gamma = \frac{\omega^2 R_{\mathrm{CM}}}{g}$

Observe that below $\gamma = 1$, the system only has a stable equilibrium at $\theta = 0$. When $\gamma \geq 1$, $\theta = 0$ becomes unstable, and the system has equilibria at $\theta = \pm A$ where $A \neq 0$. This can be summarized in the bifurcation plot of this nonlinear system, which is a supercritical pitchfork bifurcation (Fig. 16.8).

A bifurcation happens when a small, smooth change made to the parameters leads to a large qualitative change in the behavior of the system. In this case, the stability of the point at the bottom of the hoop depends on the relative tangential magnitudes of the centrifugal and gravitational force.

While the gravitational force remains the same regardless of angular velocity, the centrifugal force increases with angular velocity, such that at near the equilibrium point, at low angular velocities, it is always smaller than gravitational force; but, at higher angular velocities, it is always larger than the gravitational force. Thus, it can

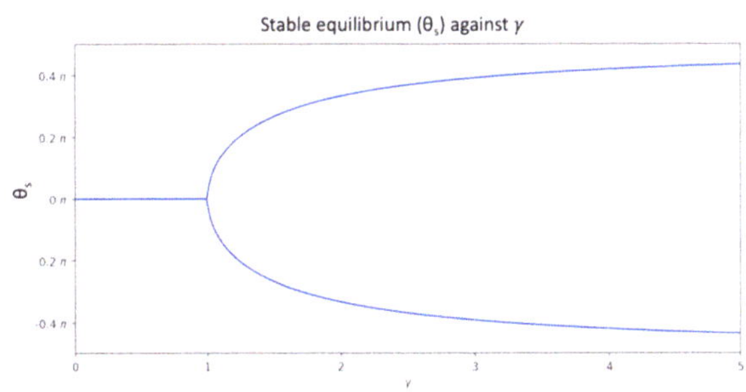

Fig. 16.8 Supercritical pitchfork bifurcation of system, showing the dependence of stable equilibrium angle on the parameters of the system

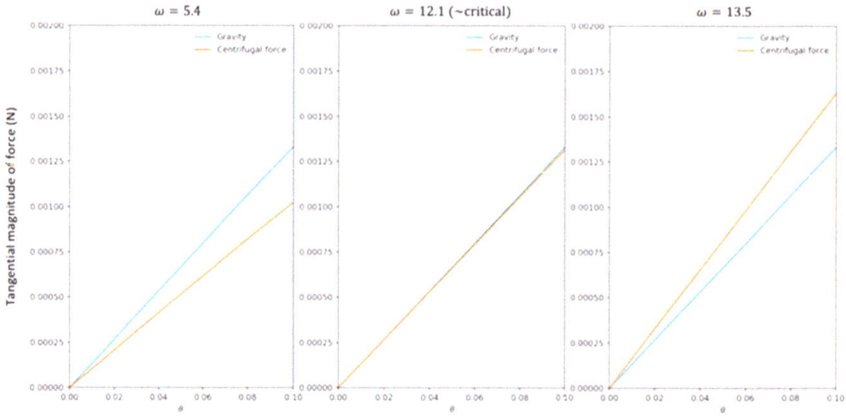

Fig. 16.9 Graphs showing tangential magnitudes of gravitational and centrifugal forces near the bottom of the hoop, for a few different angular velocities

be seen how a smooth change in angular velocity of the hoop physically leads to the change in stability of the equilibrium and thus a bifurcation (Fig. 16.9).

16.2.4 Solving Equation of Motion

Equation (16.10) cannot be solved analytically, due to the sign$\dot{\theta}$ term. Therefore, MATLAB ode45 (Runge–Kutta method) was used to solve it numerically and obtain how the θ term changes with time, as well as the phase plots of the system.

16.3 Experimental Results

16.3.1 Experimental Setup

The grooved hoop was constructed by joining two embroidery hoops together, and a uniform glass marble was used as the bead. A DC motor was used to power the grooved hoop. A phone camera was attached such that it rotates with the hoop. Another phone camera placed above the hoop recorded the angular velocity of hoop rotation. Using a spirit level, the hoop was fine-tuned to be vertical. The whole setup was held in place by multiple weighted retort stands, in order to reduce vibrations—however, some vibration was still present in the system due to the nature of the motor, which may have affected some experimental results. The videos were exported to the software Tracker, to find the position of the bead and angular velocity of hoop rotation (Fig. 16.10).

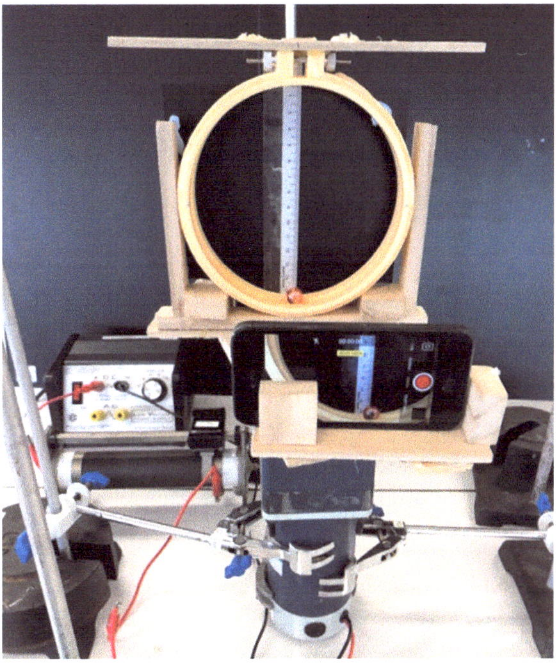

Fig. 16.10 Experimental setup of rotating grooved hoop

16.3.2 Bifurcation of System

There is a good agreement of values with the theoretical model. Since $\omega_{\text{crit}} = \sqrt{\frac{\gamma g}{R_{\text{CM}}}}$, ω_{crit} decreases when R_{CM} increases (Fig. 16.11).

16.3.3 Equilibrium Point $\theta = 0$

There is a good agreement of theoretical and experimental time domain motion, validating the method of obtaining the equation of motion. The following trends can be observed with increasing ω, and explained as follows (Fig. 16.12):

1. Amount of damping increases. This is due to increased centrifugal force that leads to increased normal force, and thus, damping increases (Fig. 16.13).
2. Period of oscillations increases. (There is also a dependence of period on the amplitude since the oscillation is not simple harmonic). This is because increased centrifugal force tangentially causes a decrease in restoring force $(F_{\text{gravity},t} - F_{\text{centrifugal},t})$ "downwards." The period increases greatly approaching ω_{crit} as this difference approaches 0 (Fig. 16.14).

The effect of rolling resistance coefficient was also investigated. This was achieved by changing the material lining the hoop, from plastic (lowest) to paper to masking

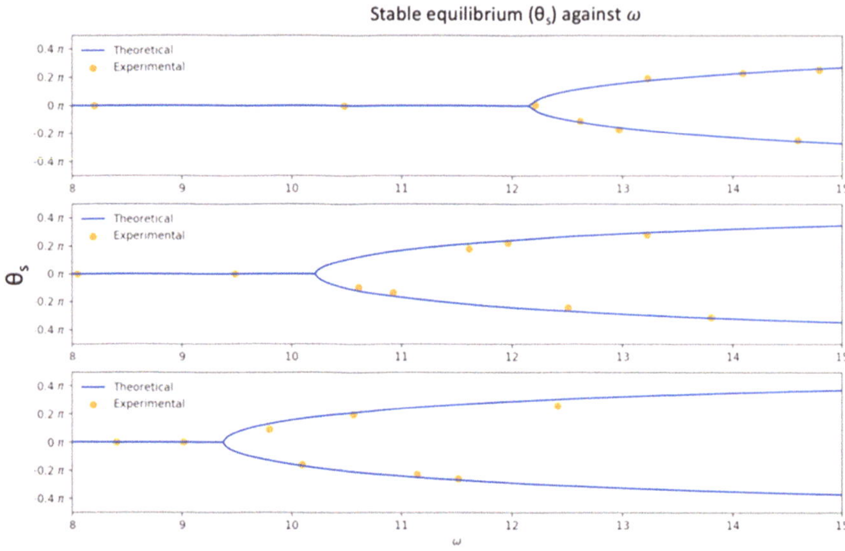

Fig. 16.11 Bifurcation curves for (from top) $R_{CM} = 6.62, 9.35$ and $11.10\,\text{cm}$

tape (highest). As rolling resistance coefficient increases, the amount of damping increases.

16.3.4 Equilibrium Point $\theta = \pm A$ Where $A \neq 0$

For the case of higher angular velocities, there is a good agreement of theoretical and experimental values except toward the end of the motion. This is likely due to the aforementioned vibration of the setup, which are most relatively significant at smallest amplitudes and lowest speeds of the bead. However, the good agreement elsewhere shows the theoretical model is still valid (Fig. 16.15).

The following trends can be observed with increasing ω:

1. Angle at stable equilibrium increases. This is expected from the solution of the equation of motion for the equilibrium points.
2. Number and amplitude of oscillations increase. This is due to the increase centrifugal force that pushes the bead further up the hoop (Fig. 16.16).
3. Period of oscillations decreases (instead of increasing as in the previous case). This is because the increasing centrifugal force increases the restoring force instead.

Notice too that oscillations are asymmetric about the stable equilibrium, with the bead spending more time below the stable point than above it. This is also expected

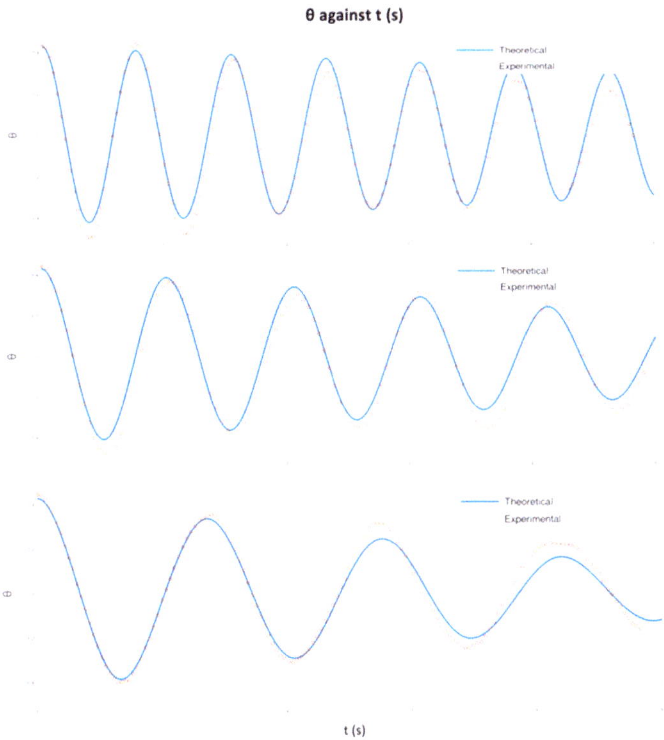

Fig. 16.12 θ against time, for arbitrary initial conditions $\theta = 0.44, \dot{\theta} = 0$. From top, $\omega = 5.4, 9.1$ and 10.7

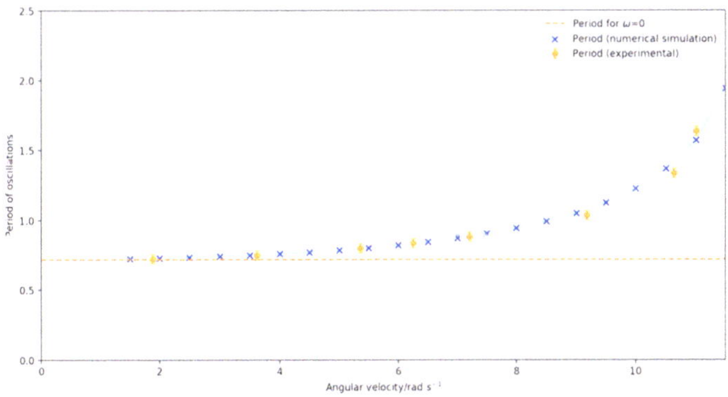

Fig. 16.13 Period of first oscillation against angular velocity of hoop

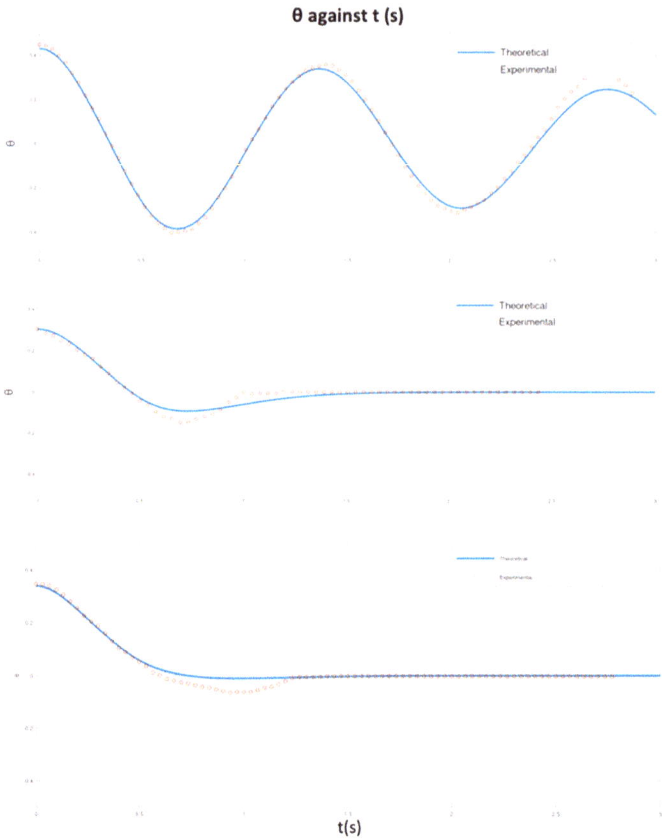

Fig. 16.14 θ against time, for rolling resistance coefficient (from top) $\mu = 0.0040, 0.021, 0.040$

from the effective potential plots, as the restoring force (difference between $F_{\text{gravity},t}$ and $F_{\text{centrifugal},t}$) is much smaller below the stable point than above it (Fig. 16.17).

Lastly, the effect of rolling resistance was also investigated. As before, increased rolling resistance causes the damping to increase.

16.3.5 Phase Portraits

It is interesting to study this system from the perspective of phase diagrams, for which experimental verification is presented for the first time. The phase plot consists of spirals (signifying underdamped motion) all approaching the equilibrium angle, which behaves like an attractor. The effect of increased angular velocity is in increasing the equilibrium angle, as well as increasing the number and size of the spirals (maximum velocities and displacements reached). The effect of increased

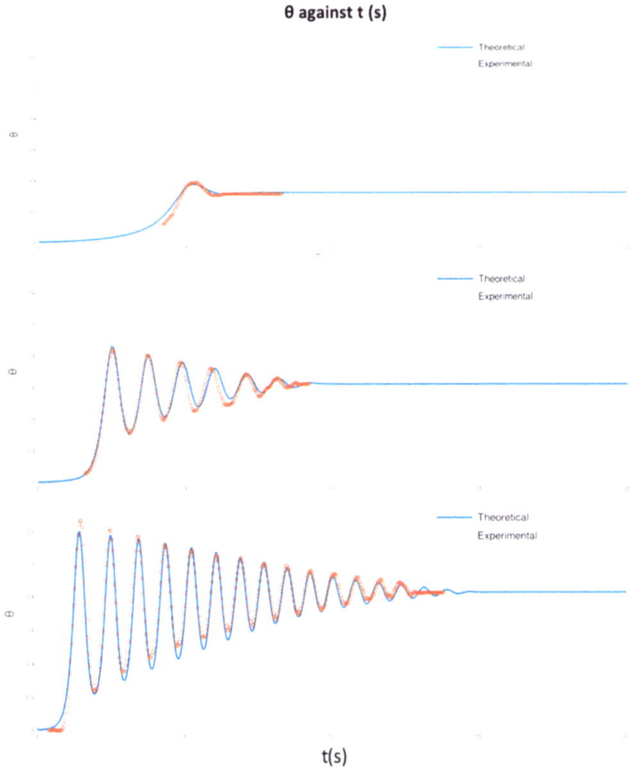

Fig. 16.15 θ against time. From top, $\omega = 12.9, 13.2,$ and 14.8

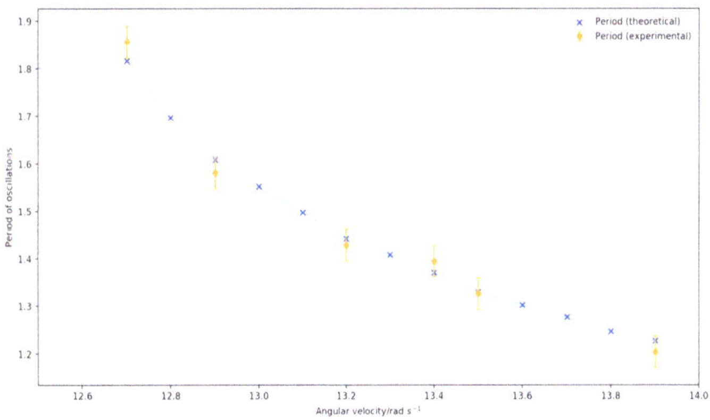

Fig. 16.16 Period of first oscillation against angular velocity of hoop

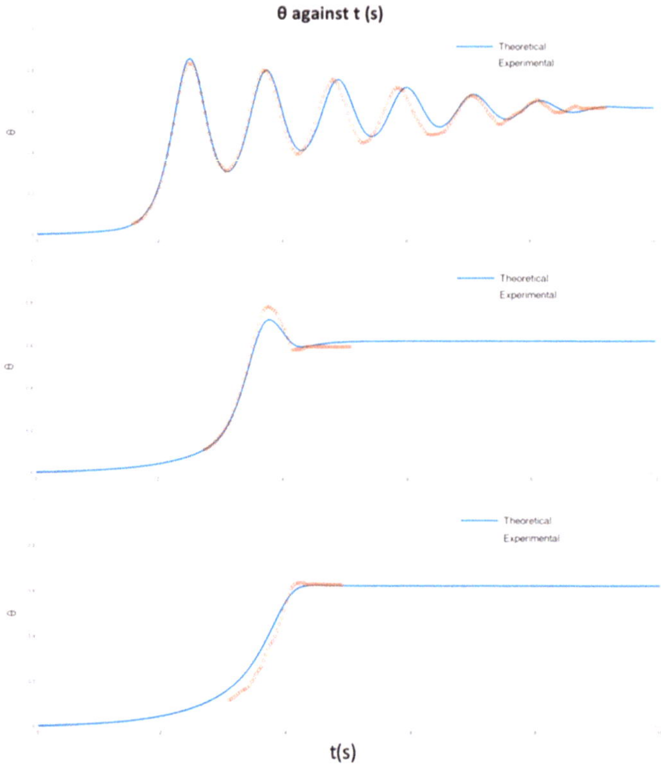

Fig. 16.17 θ against time, for rolling resistance coefficient (from top) $\mu = 0.0040, 0.021, 0.040$

rolling resistance is in decreasing the number of oscillations, which can be seen clearly in the lack of complete spirals for the highest resistance cases (Fig. 16.18).

16.4 Conclusion

In this paper, it was shown how Newtonian mechanics can be used to describe fully the motion of a bead on a rotating hoop, as seen in classical physics textbooks. The problem is extended to include a rolling bead, as well as the relevant form of damping, rolling resistance. This equation of motion is solved numerically to obtain a solution for the time domain motion of the bead, for any given set of initial conditions. The theoretical predictions were thoroughly experimentally verified for static and dynamic cases, with time domain and phase space illustrations used. The physics behind the bifurcation behavior of the bead, as well as its motion and oscillations, was also explained intuitively, which may be of pedagogical value.

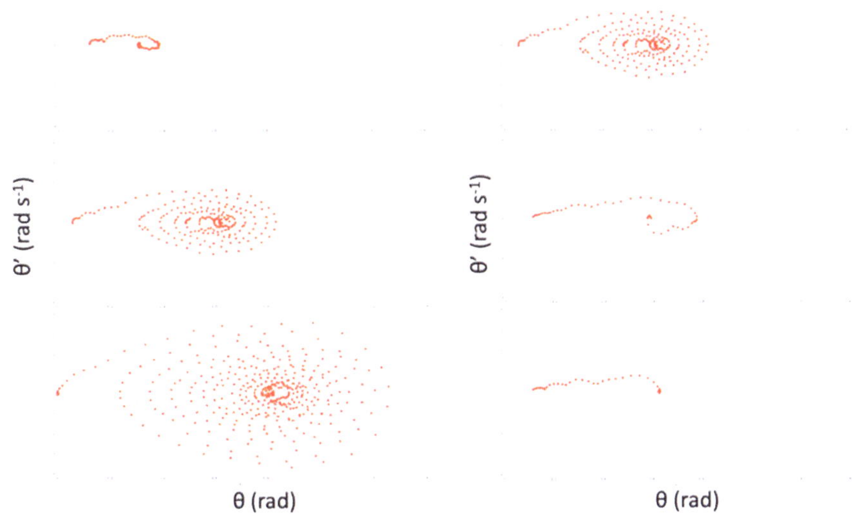

Fig. 16.18 Phase plots of $\dot{\theta}$ against θ, showing different amounts of damping for rolling resistance coefficient (from top) $\mu = 0.0040, 0.021, 0.040$

It may be interesting to extend the investigation into a tilted hoop or a hoop with its axis of rotation away from the line of symmetry; the effect of dry friction on the bifurcation and equilibria for a sliding bead setup may also be worth studying.

Acknowledgements I would like to thank my physics teachers and coaches at Raffles Girls' School and Raffles Institution for their support and guidance throughout the preparation for the Online International Young Physicists' Tournament for which I investigated this problem, as well as my teammates for supporting me throughout the competition.

References

1. Fletcher, G. (1997). A mechanical analog of first- and second-order phase transitions. *American Journal of Physics, 65*, 74.
2. Blekhman, I. I. (1981). *Synchronization in nature and technology*. Nauka.
3. Papadopoulos, E., & Papadimitriou, I. (2001) Modeling, design and control of a portable washing machine during the spinning cycle. In *Proceedings of the IEEE–ASME international conference on advanced intelligence mechatronics systems (AIM 2001)*, 8–11 July, pp. 899–904. Como, Italy
4. Raviola, L. A., Véliz, M. E., Salomone, H. D., Olivieri, N. A., & Rodríguez, E. E. (2017). The bead on a rotating hoop revisited: An unexpected resonance. *European Journal of Physics, 38*

5. Baker, T., & Bill, A. (2012). Jacobi elliptic functions and the complete solution to the bead on the hoop problem. *American Journal of Physics, 80*.
6. Vozdecký, L., Bartoš, J., & Musilova, J. (2014). Rolling friction—Models and experiment. An undergraduate student project. *European Journal of Physics, 35*.
7. Morrison, A. (2013). *An introduction to fluid mechanics*. Cambridge University Press.

Chapter 17
Development and Optimisation of a Rapid Paper-Based Test for the Detection of IL-6

Medha Shridharan, Patthara Kongsuphol, and Fernandez Kalpana

Abstract IL-6 serves as a key unique biomarker for the detection of various physiological malignancies, including tumour progression and immune conditions such as SIRS. IL-6 has many clinical applications in diagnosis. Current laboratory-based detection techniques for IL-6 (ELISA and Western Blot) require significant time, cost, machinery and specialist training. Current diagnostic tests also commonly use antibodies as binder proteins. These antibodies are unstable under heat, exhibit low performance and have long and expensive production processes, making them impractical for point-of-care (POC) clinical usage. Hence, we have identified an inherent need for a practical, inexpensive, rapid and durable diagnostic test for IL-6. We have developed and optimised a paper-based test that utilises unique engineered thermostable binder proteins which can detect IL-6 with high accuracy. The turnover time for the test is rapid, with results showing within 15 min, as compared to the 4-h turnover time for ELISA. The proteins used are also not prone to thermal denaturation and show activity over a broad pH range, making the test applicable in countless clinical contexts as it can be transported more easily than traditional methods. As minimal specialist training is required, the test is also more accessible. In this project, the engineering goal for this paper-based assay is achieved, as the test is optimised to a signal: noise ratio of 1.673 at 100 nM IL-6 and even higher at higher concentrations. Hence, this paper-based test we have optimised can detect IL-6 with high accuracy and sensitivity. With its rapid turnover time, simple methodology and high sensitivity, this paper-based assay has crucial and significant implications in improving clinical workflows.

M. Shridharan (✉) · F. Kalpana
NUS High School of Mathematics and Science, Singapore, Singapore
e-mail: h1810105@nushigh.edu.sg

F. Kalpana
e-mail: nhskalpa@nus.edu.sg

P. Kongsuphol
Singapore-MIT Alliance for Research and Technology, Antimicrobial Resistance IRG, Singapore, Singapore
e-mail: patthara@smart.mit.edu

H. Guo et al. (eds.), *IRC-SET 2021*, https://doi.org/10.1007/978-981-16-9869-9_17

Keywords Point-of-care diagnostics · Sandwich immunoassays · Paper-based diagnostic tests · IL-6 · Binder proteins · Thermostability · Rapid turnover time

17.1 Introduction

IL-6 is a potent inflammatory cytokine, playing a critical role in inflammation, the regulation of immune responses and hematopoiesis. Elevated levels of IL-6 are associated with tumour progression, making it a promising prognostic factor for malignancies such as colorectal cancer, breast cancer, gastric cancer, pancreatic cancer, endometrial cancer, non-small cell lung carcinoma, renal cell carcinoma and ovarian cancer [1].

Additionally, patients who meet Systemic Inflammatory Response Syndrome (SIRS) criteria and have elevated interleukin 6 (IL-6) levels have been shown to be at increased risk for complications such as pneumonia, multiple organ dysfunction syndrome (MODS), and death [2]. Higher levels of IL-6 in patient serum has also consistently been shown to correlate with mortality following severe injury and poor outcome from sepsis [3, 4]. Hence, IL-6 is a key unique biomarker for various physiological malignancies and is crucial in diagnosis.

However, currently, ELISA and Western Blots are the staple techniques for IL-6 detection. These laboratory-based techniques require significant time, cost, machinery and specialist training, hence making them impractical for general POC usage [5]. Current diagnostic tests commonly use newly-developed antibodies as binders for the biomarker, which are commonly unstable under heat, take much longer to optimise and have long and expensive production processes [6, 7].

Paper-based diagnostic immunoassays have been cited as a critical POC device with the potential to expedite testing times, reduce costs, and enable clinicians to make faster treatment decisions, thus improving patient healthcare outcomes [8]. The relative simplicity of these tests make them potentially transformative in the field of biomedical devices, as they have a simple methodology and readout and require low sample volumes [9].

In this project, we build on existing work from our institution by optimising a paper-based immunoassay in the context of the IL-6 biomarker using previously developed capture and receptor proteins that offer superior properties to antibodies. These proteins use the reduced charge Sso7d (rcSso7d) protein scaffold and are thus thermostable in addition to showing activity over a broad pH range, allowing the tests to be easily portable [10].

Developing and optimising a paper-based diagnostic test for IL-6 which allows for rapid, accurate and inexpensive testing has major implications in improving workflow in clinical settings. Additionally, making the test thermostable and pH-stable allows it to be transported and used more easily than traditional methods, with minimal specialist training required, making the test more accessible.

17.2 Engineering Goal and Objective

The aim of this project is to optimise the paper-based assay to achieve a S/N Ratio of at least 1.5 at 100 nM of IL-6. The S/N ratio refers to the measure of the magnitude of cyan intensity of the test when comparing the presence of 100 nM IL-6 (signal) and 0 nM (background), respectively. In many cases where reagents are varied, the measured colourimetric background fluctuates. Hence, the values are normalised and compared across various testing conditions by S/N ratio.

This allows the highest IL-6 signal as compared to the baseline signal to be identified to maximise assay sensitivity.

17.3 Materials and Methods

17.3.1 Mechanism of Assay

The IL-6 sandwich immunoassay employs the use of streptavidin/HRP-biotin chemistry and the oxidation of 3,3′,5,5′-Tetramethylbenzidine (TMB) substrate. The capture protein, 19 K, consists of the engineered rcSso7d-IL6 binder and the Cellulose Binding Domain (CBD), which allows the protein to bind to the cellulose paper. The reporter protein, 18 K, consists of another engineered rcSso7d-IL6 binder (conjugated to biotin) and horseradish peroxidase-conjugated streptavidin (SAHRP). The engineered 18 K binder binds to a different epitope of IL-6 than 19 K, hence preventing competitive binding and increasing assay sensitivity.

The streptavidin binds to the conjugated biotin. When TMB substrate is applied to the complex, it is oxidised by hydrogen peroxide (H_2O_2) in solution in the presence of horseradish peroxidase (HRP), causing it to turn blue. Hence, the cyan intensity of the test is used as a quantitative colorimetric measurement.

Beta Maltose Binding Protein (β-MBP) is used in 19 K as a spacer protein to improve the sensitivity of the assay. Without the spacer protein, biotin proves to be inaccessible for binding with streptavidin in in vivo biotinylation and shows reduced activity in in vitro chemical conjugation. The MBP-rcSso7d fusion protein conjugate yields high signal output by increasing biotin accessibility whilst preserving target-binding activity and protein stability [10] (Fig. 17.1).

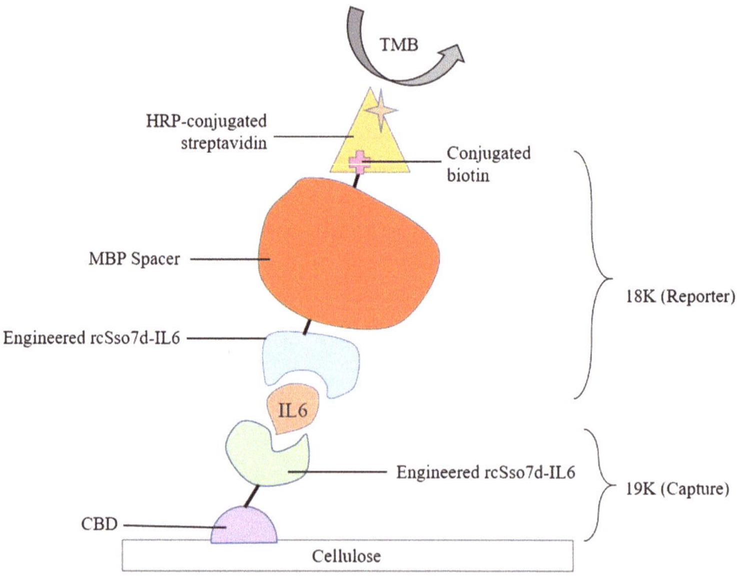

Fig. 17.1 Structure of paper-based IL-6 immunoassay

17.3.2 Key Differences Between Traditional ELISA and Paper-Based Assay

The key factors which allow for the rapid turnover times of the paper-based immunoassay are the short incubation times and lack of sequential reagent application as in traditional ELISA assays. The protein complex of 18, 19 K and IL-6 is formed in solution during incubation before being applied to the cellulose paper.

In the typical ELISA assay format, the capture protein is immobilised to the polystyrene surface, hence long periods of incubation are required to allow for effective binding and protein complex formation. However, in the paper-based assay, as binder and antigen proteins are in solution, there is a significantly higher rate of binding, hence long incubation is not required.

Additionally, traditional ELISA assays require sequential application of reagents and a washing step after each application to minimise cross-reactivity and background due to the high affinity of the polystyrene plate for all proteins. In the paper-based assay, only the CBD of 19 K exhibits high affinity for the cellulose, hence cross-reactivity is minimal. This allows for the complex to be formed in solution and applied to the paper with only a single wash step to minimise non-specific binding of unbound 18 and 19 K binders.

17.3.3 Sourcing of Materials

The proteins required for this assay are readily available in the laboratory. The 18 and 19 K proteins have been developed previously [11] and IL-6 can be purchased from protein manufacturers. Common laboratory buffers such as Phosphate-Buffered Saline (PBS) and Sodium Phosphate Buffer are used in the methodology. The cellulose papers are also treated with a blocking solution before experimentation to reduce background signal and improve assay sensitivity.

The blocking solution is prepared as follows:

- Sodium Phosphate Buffer base
- 1% Casein hydrolysate
- 5% Sucrose
- 0.1% between 20
- 10% glycerol
- 150 mM NaCl.

The methodology of each test is as follows:

1. 4 μL of Reagent A* + 36 μL of IL6 sample are added to Eppendorf tube
2. Solution from Step 1 is incubated on bench for x min^{+}
3. Solution is applied to cellulose paper
4. Spot is washed with 40 μL sodium phosphate buffer$^{\#}$
5. 40 μL TMB substrate is applied to paper
6. Cellulose paper is incubated on bench for 3 min
7. Picture of cellulose paper taken at 3 min
8. Image is analysed with ImageJ software and Microsoft Excel.

The key optimisation stages of the project are optimisation of SAHRP Concentration (0 pM to 200 pM), optimisation of Incubation Time (1–9 min) and optimisation of 18 K Binder (1–500 nM). A Dose Response Curve of IL-6 was also conducted to validate that the optimised values are effective.

*Reagent A is the mixture of 18, 19 K, SAHRP and IL-6. The concentrations of SAHRP and 18 K were optimised.

^{+}x min refers to the variation in incubation time as it was optimised.

$^{\#}$The assay is washed with a sodium phosphate buffer (pH 7.4) to minimise non-specific binding of unbound 18 and 19 K binders and improve signal: noise ratio of the assay.

17.4 Results and Discussion

17.4.1 Optimisation of SAHRP Concentration

The S/N ratio must be improved by minimising background whilst maximising signal, hence improving the "resolution" and sensitivity of the assay. Hence, various concentrations of SAHRP can be tested to achieve this goal, as shown in Fig. 17.2.

0, 37.5, 75, 100, 150 and 200 pM of SAHRP were tested against 0 and 100 nM of IL-6 and the S/N ratio was calculated (shown in Fig. 17.3). 0, 37.5 and 75 pM give high background signal, hence resulting in an S/N Ratio below 1.0 for these values. 100 pM SAHRP gives the highest S/N Ratio of 1.404 and is the optimal concentration of SAHRP.

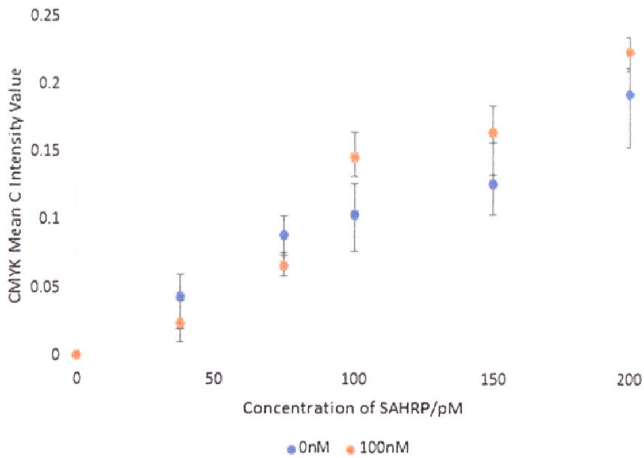

Fig. 17.2 CMYK mean C intensity value over various concentrations of SAHRP

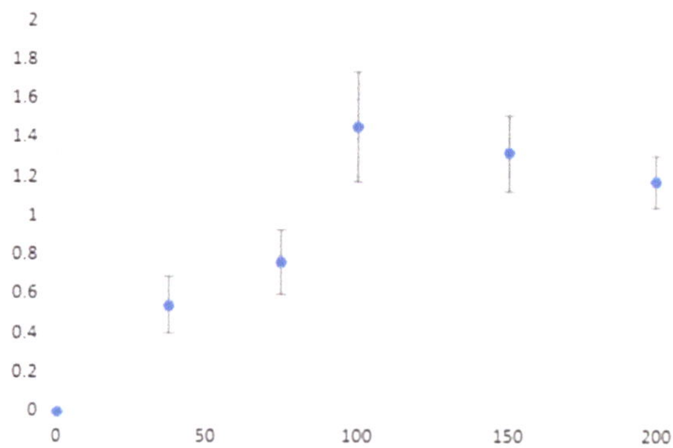

Fig. 17.3 Signal:noise ratio over various SAHRP concentrations

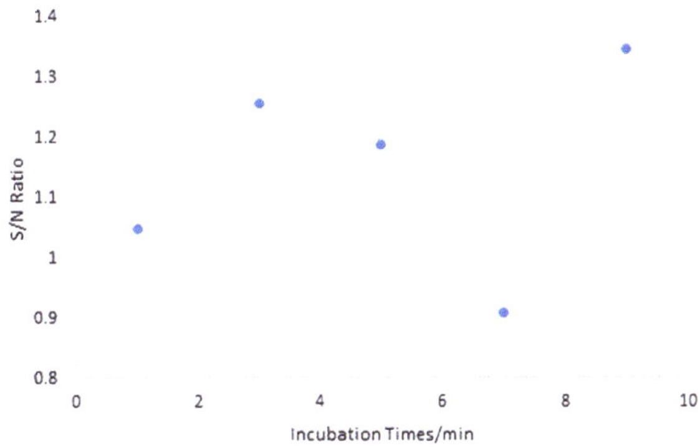

Fig. 17.4 S/N ratio over various incubation times (100 pM SAHRP)* [*7 min Incubation Time proves to be an outlier, in the upward trend, with a S/N Ratio of 0.909 (higher background than signal)]

17.4.2 Optimisation of Incubation Time

The next step to further improve the signal is to increase the incubation time of the solution. This allows for the formation of more protein complexes in solution before application to the paper, hence allowing for higher signal.

The protein binder mixture was incubated at room temperature with IL-6 for 1, 3, 5, 7 and 9 min with 100 pM SAHRP. Each value was tested with 0 and 100 nM of IL-6 and signal: noise ratio was calculated.

Figure 17.4 shows the effect of various incubation times on the S/N Ratio of the assay. From the results, 9 min of Incubation Time yields the highest S/N ratio of 1.346.

17.4.3 Optimisation of Concentration of 18 K Binder Protein

Subsequently, the signal can be further improved by optimising the concentration of 18 K reporter. A greater concentration of 18 K binder allows for more antigen complexes to be detected and reported when TMB substrate is applied, hence increasing the signal.

Various concentrations of 18 K binder, namely, 1, 5, 10, 50, 100, 500 nM, were tested against 0 and 100 nM of IL-6 to find signal: noise ratio. As shown in Fig. 17.5, the highest S/N ratio of 1.674 is achieved at 100 nM of 18 K binder.

17.4.4 Dose Response Curve

The assay was optimised at 100 pM SAHRP, 9 min incubation time and 100 nM 18 K binder in solution. To validate that these optimised values work well, a Dose Response Curve of IL-6 was conducted with 0, 1, 5, 10, 50, 100, 200, 400 and

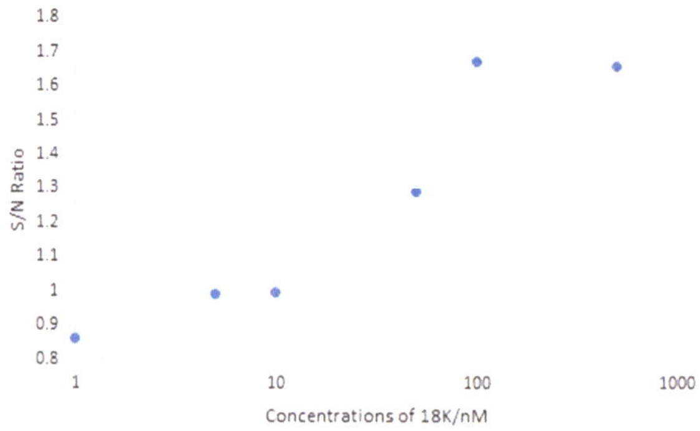

Fig. 17.5 S/N ratio over various concentrations of 18 K binder

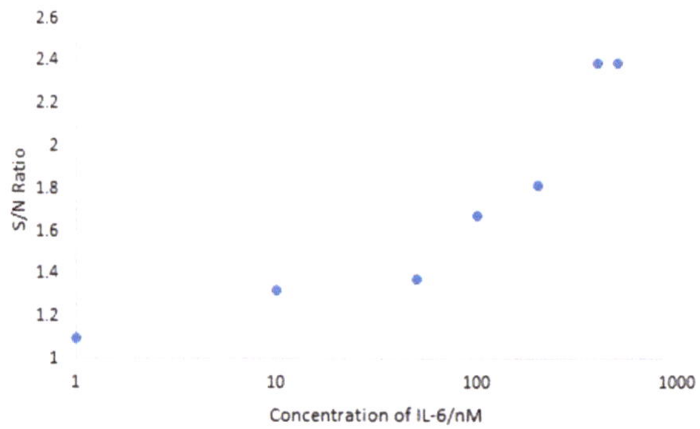

Fig. 17.6 Dose response curve of S/N ratio against various concentrations of IL-6

500 nM IL-6. This curve exemplifies the increase in signal as the concentration of IL-6 increases, proving the functionality of the assay.

The S/N Ratio for each concentration of IL-6 above 0 nM (signal) is compared to and divided by the background signal at 0 nM to obtain a S/N value for each datapoint, as shown in Fig. 17.6. There is a clear upward trend, indicating the rise in detection of IL-6.

The engineering goal for the assay has been achieved, with a S/N ratio of 1.673 at 100 nM IL-6 and even higher at higher concentrations. The assay appears to saturate at 400 nM, with minimal increase in detection from 400 to 500 nM of IL-6. Hence, the functionality and sensitivity of the assay is validated.

17.5 Conclusions and Implications

The paper-based test we have optimised can detect IL-6 with high accuracy and sensitivity. The unique engineered thermostable binder proteins used are not prone to thermal denaturation and show activity over a broad pH range, making the test easily portable and applicable in real-world clinical contexts. With its rapid turnover time, simple methodology and high sensitivity, this paper-based assay has crucial and significant implications in improving clinical workflow.

Acknowledgements I would like to thank Dr. Patthara Kongsuphol from the Singapore-MIT Alliance for Research and Technology (SMART Institute) and Mrs Kalpana Fernandez from NUS High School of Mathematics and Science for their guidance and support on this project.

References

1. Łukaszewicz, M. (2007, May/June). Clinical significance of interleukin-6 (IL-6) as a prognostic factor of cancer disease. Retrieved January 17, 2021, from https://pubmed.ncbi.nlm.nih.gov/18030875/
2. Boka, K., & Pinsky, M. R. (2020, November 12). What is the significance of increased interleukin 6 (IL-6) levels in systemic inflammatory response syndrome (SIRS)? Retrieved January 17, 2021, from https://www.medscape.com/answers/168943-41411/what-is-the-significance%20%20of-increased-interleukin-6-il-6-levels-in-systemic-inflammatory-response-syndrome-sirs
3. Dougnac, A., Riquelme, A., Calvo, M., Andresen, M., Magedzo, A., Eugenin, E., Marshall, G., & Gutiérrez, M. (2001). Estudio de la cinética de citoquinas en sepsis grave y su relación con mortalidad y score de disfunción orgánica [Study of cytokines kinetics in severe sepsis and its relationship with mortality and score of organic dysfunction]. *Revista medica de Chile, 129*(4), 347–358.
4. Mokart, D., Capo, C., Blache, J. L., Delpero, J. R., Houvenaeghel, G., Martin, C., & Mege, J. L. (2002). Early postoperative compensatory anti-inflammatory response syndrome is associated with septic complications after major surgical trauma in patients with cancer. *The British Journal of Surgery, 89*(11), 1450–1456. https://doi.org/10.1046/j.1365-2168.2002.02218.x
5. Huang, J., Harvey, J., Fam, W., Nimmo, M., & Tok, I. (2013, August 13). Novel biosensor for interleukin-6 detection. Retrieved January 17, 2021, from https://www.sciencedirect.com/science/article/pii/S1877705813010093X#:~:t%20xt=Currently%2C%20%20ELISA%20and%20western%20blot,cost%20yet%20sensitive%20protein%20detection
6. Miller, E. A., Traxlmayr, M. W., Shen, J., & Sikes, H. D. (2016). Activity-based assessment of an engineered hyperthermophilic protein as a capture agent in paper-based diagnostic tests. *Molecular Systems Design & Engineering, 1*(4), 377–381. https://doi.org/10.1039/C6ME00032K
7. Banta, S., Dooley, K., & Shur, O. (2013). Replacing antibodies: Engineering new binding proteins. *Annual Review of Biomedical Engineering, 15*, 93–113. https://doi.org/10.1146/annurev-bioeng-071812-152412
8. Yager, P., Edwards, T., Fu, E., Helton, K., Nelson, K., Tam, M. R., & Weigl, B. H. (2006). Microfluidic diagnostic technologies for global public health. *Nature, 442*(7101), 412–418. https://doi.org/10.1038/nature05064
9. Hristov, D. R., Rodriguez-Quijada, C., Gomez-Marquez, J., & Hamad-Schifferli, K. (2019). Designing paper-based immunoassays for biomedical applications. *Sensors (Basel, Switzerland), 19*(3), 554. https://doi.org/10.3390/s19030554
10. Sung, K.-J., Miller, E., & Sikes, H. (2018). Engineering hyperthermostable rcSso7d as reporter molecule for in vitro diagnostic tests. *Molecular Systems Design & Engineering, 3*, 877–882. https://doi.org/10.1039/C8ME00049B
11. Miller, E. A., Sung, K.-J., Kongsuphol, P., Baniya, S., Aw-Yong, H. Q., Tay, V., Tan, Y., Kabir, F. M., Pang-Yeo, K., Kaspriskie, I. G., & Sikes, H. D. (2019). Beyond epitope binning: Directed in vitro selection of complementary pairs of binding proteins. *ACS Combinatorial Science, 22*(1), 49–60. https://doi.org/10.1021/acscombsci.9b00176

Chapter 18
Determination of Trace Levels of Iron in *Spinacia Oleracea* Using Microcontroller-Based Photometer

Yunyin Xie, Jiayi Li, and Chyn Lynn Kate Lai

Abstract This work aims to investigate the trace levels of iron in different species of spinach using spectrophotometric analysis and a low-cost, simple and portable photometer built with a microcontroller. Experimental results show that iron reacts with 1,10-phenanthroline to produce an orange-red complex, which has a maximum absorption at a wavelength of 511 nm. The linear calibration graphs were obtained for 2×10^{-4} to 4×10^{-3} mg/ml of iron (II), with a correlation coefficient of 0.9994 for the iron(II)-phenanthroline complex. The molar absorption coefficient was found to be 8.642×10^3 L mol^{-1} cm^{-1}. The limit of detection is 1.1426×10^{-4} mg/ml. Nine different interference ions were investigated, and only copper and zinc, when added in excess, are found to interfere significantly with the absorbance of the complex. The iron(II)-phenanthroline complex is stable over 24 h. Under optimal conditions of temperature and pH, the concentration of iron in red and green spinach was successfully determined using our proposed method and achieved ±5% accuracy compared to a benchtop spectrophotometer which served as the reference.

18.1 Introduction

For many years, nutritional interest in iron was focused on its role in haemoglobin formation and oxygen in biological systems [1]. By determining the amount of iron present in various foods in our diet, we would be better able to plan our diets carefully and accurately to prevent iron deficiency and maintain a healthy average daily intake of iron. This knowledge would be especially important for people such as expectant mothers, who require a large amount of necessary daily iron intake.

Various volumetric and gravimetric methods are known for determination of iron [2]. However, chances of errors are great, especially in estimating trace quantities. In this study, spectrophotometric technique of analysis is chosen due to the very small amount of sample required, the reliability and easily reproducible results. Spectrophotometry is a method used to measure how much a chemical substance

Y. Xie · J. Li (✉) · C. L. K. Lai
Raffles Girls' School (Secondary), Singapore, Singapore
e-mail: 22YLIJI300G@student.ri.edu.sg

H. Guo et al. (eds.), *IRC-SET 2021*, https://doi.org/10.1007/978-981-16-9869-9_18

absorbs light, by measuring the intensity of light as a beam of light passes through a sample solution. According to Beer-Lambert's law, a substance's concentration and its absorbance are directly proportional, where $A = \varepsilon bc$. "A" is the Absorbance, "ε" is the molar extinction coefficient (M^{-1} cm^{-1}), b is the solution path length (cm), which is fixed at 10 *mm* for our investigations, and c is the concentration (moles $litre^{-1}$).

Although aqueous iron is nearly colourless, it can be determined spectrophotometrically from its intensely coloured complex with 1,10-phenanthroline in acidic (pH 2–3) solution. The orange-red coloured complex is formed by a stoichiometric reaction of ferrous ions with three molecules of 1,10-phenanthroline, as seen from the following reaction:

In order for this reaction to proceed, iron ions must first be reduced to its ferrous state. Hydroxylamine hydrochloride serves as a reducing agent and is able to rapidly reduce ferric ion to ferrous ion, as seen from the following reaction:

$$4Fe^{3+} + 2NH_2OH \rightarrow 4Fe^{2+} + N_2O + 4H^+ + H_2O$$

As trisodium citrate is a conjugate base of a weak acid (citric acid), it serves to adjust the solution to the required pH value by adding corresponding amounts of it.

In this present work, we proposed a laboratory-made, portable, microcontroller-based photometer for the determination of iron in *Spinacia oleracea*, and accessed its accuracy against a benchtop spectrophotometer. The laboratory-made photometer can be easily operated without much training. Furthermore, as the photometer is light and portable, it can be easily brought out of the laboratory for in situ analysis, unlike a benchtop spectrophotometer which can only be operated at a fixed position in a laboratory.

Spinacia oleracea was chosen as the sample for analysis of our field testing due to its high iron content. As iron is necessary for the production of chlorophyll, a greater concentration of chlorophyll pigments would tend to result in a higher amount of iron. Thus, green spinach was hypothesised to have a higher concentration of iron than red spinach. Similarly in the red spinach, the green part of the leaves were hypothesised to have a higher concentration of iron than the red part of the leaves.

18.2 Materials and Methods

18.2.1 Reagents and Solutions

The reagents and solutions used are of analytical grade and were used without further purification. They are as follows: Distilled water (Barnstead Water Purification System, Fisher Scientific Pte Ltd); Ammonium iron (II) sulphate hexahydrate, hydroxylamine hydrochloride and 1,10-phenanthroline monohydrate (Sigma

Aldrich); Sodium citrate (GCE Chemicals); Concentrated sulfuric acid and hydrochloric acid (Honeywell); 25% Ammonia solution (EMSURE).

18.2.2 Apparatus

Apparatus included volumetric flasks, beakers, measuring cylinders, micropipettes and droppers from Science Centre Singapore. Glass apparatus were submerged in a 10% hydrochloric acid bath overnight and rinsed several times with distilled water before usage. A pH probe (HANNA instruments) was used to determine the pH of the solutions.

18.2.3 Preparation of Solutions

Stock solution of concentration 0.04 mg/ml was prepared by dissolving 0.0281 g of ammonium iron(II) sulphate hexahydrate in distilled water with the addition of 1 ml of concentrated sulfuric acid. This was then transferred into a 100 ml volumetric flask and the volume was completed with distilled water. 1,10-phenanthroline solution of concentration 1 g/L was prepared by dissolving 0.1 g of 1,10-phenanthroline in 100 ml of distilled water. Since 1,10-phenanthroline is an organic compound of low solubility in water, this mixture was heated on a hotplate at around 80 °C for 20–25 min with constant stirring until the solid was fully dissolved. Hydroxylamine hydrochloride solution of concentration 100 g/L was prepared by dissolving 10 g of hydroxylamine hydrochloride in 100 ml of distilled water. Sodium citrate solution of concentration 50 g/L was prepared by dissolving 5.0 g of sodium citrate in 100 ml of distilled water. Standard solutions of concentrations ranging from 2×10^{-4} to 4×10^{-3} mg/ml of ferrous ions were prepared to obtain the calibration curve, by pipetting required volumes of the stock solution into beakers and diluting them in a 50 ml volumetric flasks. Required volumes of sodium citrate solution were added to adjust the pH of the solution to around pH 3, followed by 0.50 ml of hydroxylamine hydrochloride solution and 5.0 ml of 1,10-phenanthroline solution. Distilled water was used to complete the volume.

18.2.4 Lab-Made LED-Based Photometer

The microcontroller-based photometer was made using black acrylic, designed using LibreCAD and cut using Laser Cutter [Universal Laser System (USA)]. It has approximate dimensions of 18 × 6 × 13 cm. Our photometer consists of 3 main circuits, the RC low pass filter circuit, LED and resistor circuit, as well as button and resistor circuits. A low pass filter, constructed using simple resistor and capacitor connected

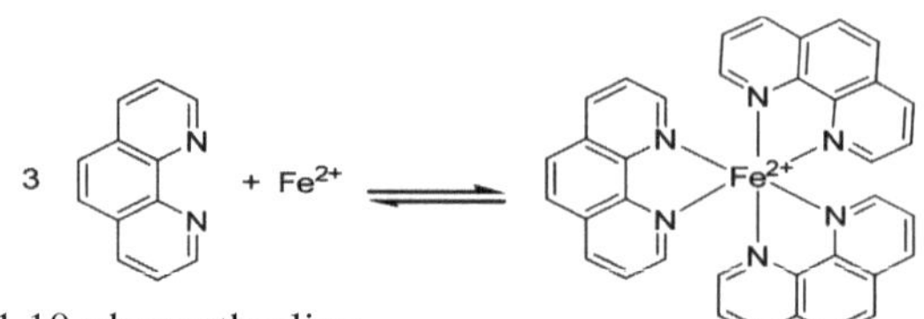

Fig. 18.1 Reaction between ferrous ions and 1,10-phenanthroline

in series, was used so as to filter out high-frequency signals whilst allowing low-frequency signal to pass through. This helps to prevent interference from external light sources, which has a high-frequency, whilst allowing the low-frequency light emitted from the LED to be detected by Arduino. For our circuit, a 2.2 μF capacitor and 100 kΩ resistor is used. This combination of resistor and capacitor allows maximum high-frequency signals to be filtered whilst not compromising the upper limit of detection of voltage [3]. Since the complex has a maximum absorption of 511 nm, as seen from the absorbance spectra in Fig. 18.3, an Ultrabright LED of dominant wavelength 510 nm (Element 14, Multicomp) was used as the light source. TSL257, which is a high-sensitivity low-noise light-to-voltage optical converter that combines a photodiode and a transimpedance amplifier on a single monolithic CMOS integrated circuit (Texas Advanced Optoelectronic Solutions (USA)), is used to detect and convert light intensity into voltage, which is transmitted to the microcontroller. The scheme shown in Fig. 18.1 indicates the photometer operation. A radiation beam is focused through the solution into the cuvette with 10 mm optical pathlength, occurring partial absorption. The transmitted radiation is detected by the photodiode in the TSL, and the analogue to digital conversion was performed by the microcontroller. Absorbance was then calculated and the results are then displayed on the Liquid Crystal Display (LCD). Cuvettes of 10 mm optical pathlength was employed for measurements. All measurements were recorded in triplicates for our investigations, with the exception of the time-based investigation.

18.2.5 Procedure for Determination of Iron in Spinacia Oleracea

After optimising the experimental conditions of the complex, the photometer was applied to find out how the amount of iron differs in different parts of the leaves of two species of *Spinacia oleracea,* namely, the green spinach and red spinach, both obtained from local supermarket. The amount of iron was investigated in the following: the leaves of the green spinach, both the green parts and red parts of the leaves of the red spinach.

Triplicate samples of 2.5000 g of each type of spinach leaf were prepared. The samples were heated in the oven at 170 °C for 25–30 min to obtain dry mass. The dry ashing technique was then applied to the samples by heating them on a hot plate

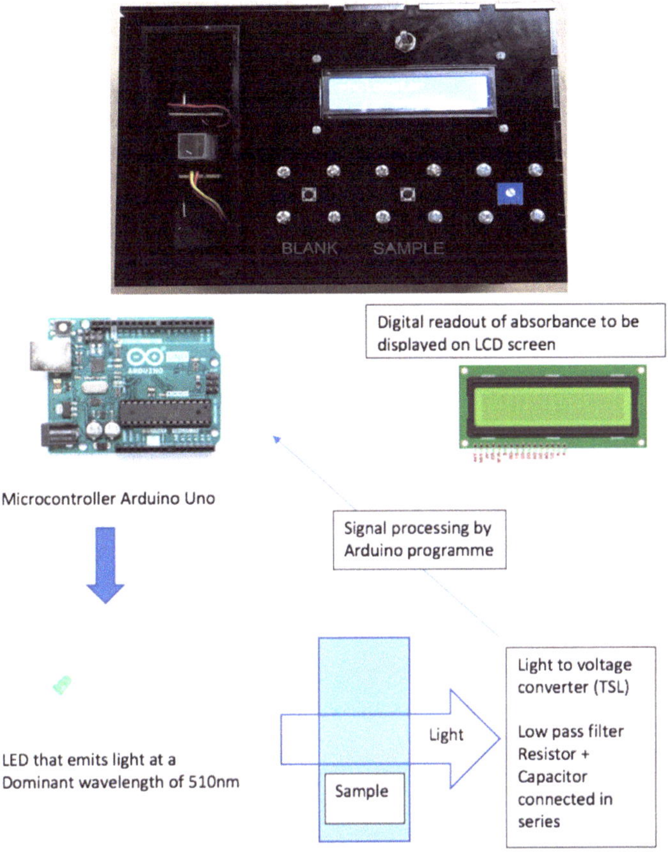

Fig. 18.2 Self-built microcontroller-based photometer

at around 450–500 °C. This is done so for around 7 h until all the samples turned into white ash. After cooling to room temperature, 5 ml of 4 M hydrochloric acid was added into each crucible containing the samples to dissolve and hydrolyse the ash so as to extract the iron ions. These solutions were then carefully transferred into centrifuge tubes and were centrifuged at 3900 rotations per second for 5 min, using Eppendorf Centrifuge 5804.

4 ml of each sample was micro pipetted into a beaker. The pH of the solutions were adjusted to around pH3 as before. 0.5 ml of hydroxylamine hydrochloride solution was then added, followed by 5 ml of 1,10-phenanthroline. The solutions were transferred into 50 ml volumetric flasks and the volumes were completed with distilled water. The solutions were left standing for around 21 h to allow for the development of the iron(II)-phenanthroline complex before the respective absorbances were measured.

18.2.6 Reference Method

The reference method for the determination of iron in *Spinacia Oleracea* was carried out by measuring the absorbance of different concentrations of iron(II)-phenanthroline complex in a benchtop spectrophotometer (Corona Electric, SH-1000).

18.3 Results

18.3.1 Absorption Spectra

The wavelength of maximum absorbance for the iron(II)-phenanthroline complex was determined by running a solution of 0.004 mg/ml iron in a benchtop spectrophotometer. The absorption spectra of the standard solution were recorded in the range 350–700 nm to determine the wavelength of maximum absorbance. The results obtained are shown in Fig. 18.3.

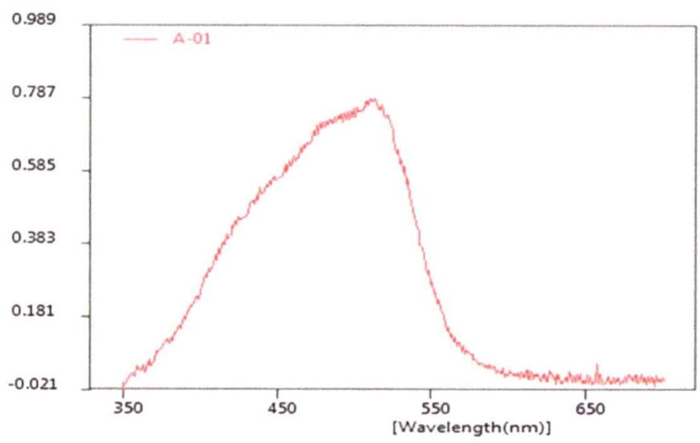

Fig. 18.3 Absorption spectra for iron(II)-phenanthroline complex of wavelength ranging from 350 to 700 nm

As seen from Fig. 18.3, the maximum absorbance peak has been identified at the 511 nm wavelength. Thus, the LED with the closest wavelength of 510 nm wavelength was chosen, so as to lessen the signal to noise ratio and ensure reliability.

In the next sections, we will analyse some factors affecting the absorbance of the iron(II)-phenanthroline complex.

18.3.2 Effect of pH

Iron is quantitatively complexed by 1,10-phenanthroline in the pH range of 2–9 [4]. The optimum pH value which allows for maximum absorbance of the complex was hence evaluated, using standard solutions of 0.004 mg/ml of iron. Sodium citrate solution was used to achieve the respective acidic pH values. Aqueous ammonia was added to achieve an alkaline pH.

Below pH 2, there is a high concentration of hydrogen ions and hence a strong competition between the hydrogen ions and the ferrous ions for reaction with the 1,10-phenanthroline molecule. Absorbance is hence reduced as not all the 1,10-phenanthroline molecules were able to react with ferrous ions to form the coloured complex. Between pH 2 and 7, there is less competition between the ferrous and hydrogen ions for the 1,10-phenanthroline molecule due to the decrease in hydrogen ions concentration. Hence, maximum absorbance of the complex is achieved within this range. Above pH 7, the possible formation of precipitates such as iron(II) hydroxides in alkaline solutions, as observed during the experiment, might have resulted in the drastic decrease in the absorbance of the complex.

As seen from Fig. 18.4, the maximum absorbance of the complex fell between the ranges of pH 2–3. We thus decided to adjust the pH of solution using aqueous sodium citrate to around 3 in future investigations.

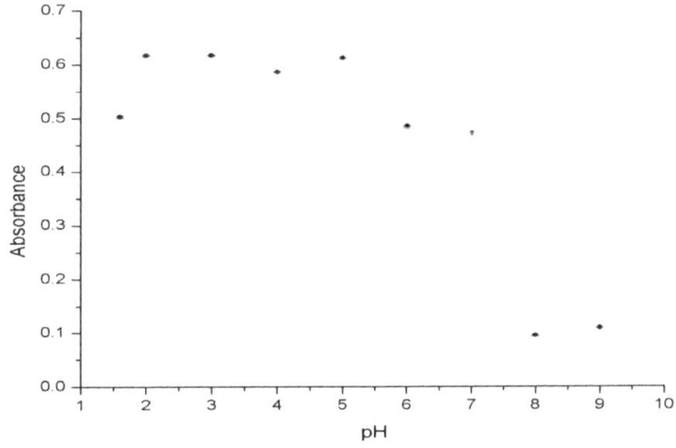

Fig. 18.4 Effect of pH on absorbance of iron(II)-phenanthroline complex

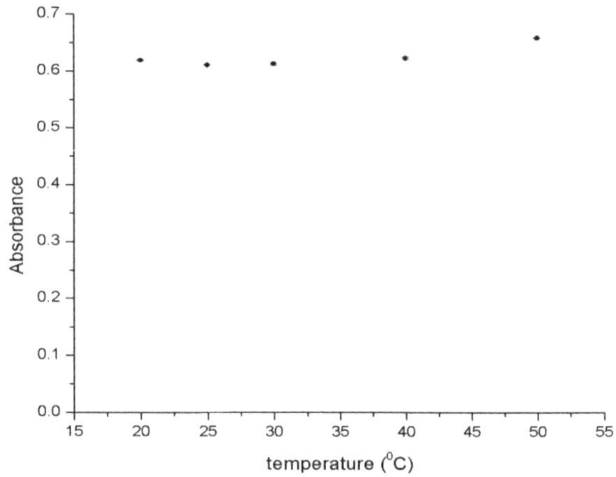

Fig. 18.5 Effect of temperature on absorbance of iron(II)-phenanthroline complex

18.3.3 Effect of Temperature

This investigation aims to find out how the absorbance of the complex varies over a range of temperatures from 20 to 50 °C, in order to decide on the optimum temperature to be used for our experiments. This was done so at regular intervals of 10 °C and compared with the absorbance of complex at 25 °C. Standard solution of 0.004 mg/ml of iron was heated or cooled from room temperature to the desired temperatures using a hot

plate and an ice bath, respectively.

As seen from Fig. 18.5, as the temperature increases from 20 to 50 °C, the absorbance of the complex increases slightly as well. However, the increase in absorbance as a result of the increase in temperature is not significant within this range. During experiments, it was observed that water vapour had started to condense on the inner surfaces of the beaker when the solution was heated to higher temperatures. The evaporation of water from the solution could hence have resulted in the solution becoming more concentrated, which might be a key reason to why the absorbance of the complex seemed to have an increasing trend. Considering that it is hard to maintain a high or lower temperature of the solution than room temperature, all solutions are measured at room temperature in further experiments.

18.3.4 Effect of Time

This investigation aims to find out whether the absorbance of a solution of 0.004 mg/ml of iron is able to remain stable over a given period of time. The length of investigation lasted for 4.5 h, and a total of 10 absorbance readings were taken at 30 min intervals throughout the investigation. Arduino was programmed such that it would automatically take one absorbance reading every 30 min.

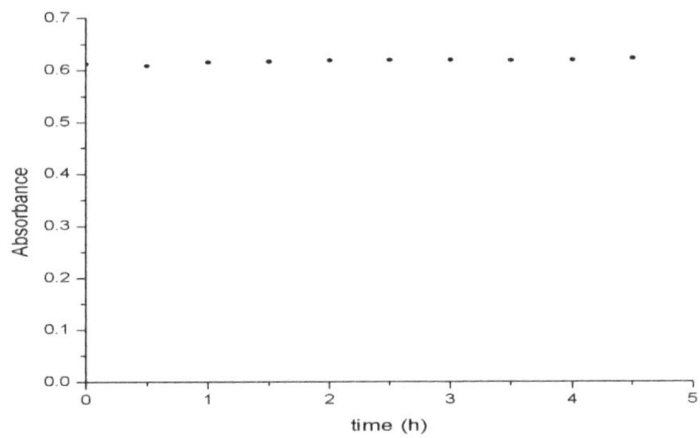

Fig. 18.6 Effect of time on absorbance of iron(II)-phenanthroline complex

As seen from Fig. 18.6, the plot reviewed does not show any oscillations of absorbance of the complex over the course of investigation of 4.5 h. It can hence be concluded that the absorbance of the complex remains relatively stable and measurements need not be taken immediately after the complex is formed. Therefore, all measurements were recorded within an hour of sample preparation, with the exception of spinach samples.

18.3.5 Effect of Interference Ions

The effect of some cations and anions on their interference with the production of quantitative colour reaction between 1,10-phenanthroline and ferrous ions were evaluated, by spiking the standard solution with the respective interference ions. This was done so according to the mass ratio of 1000:1 between that of the interference ion and ferrous ion. The criterion for an ion to be deemed as an interference was an absorbance value varying by more than ±5% from the expected value of 0.004 mg/ml of iron alone. [7]

As seen from Table 18.1, most of the ions were found to have no significant effect on the determination of iron. The ions which had the most significant interference were Cu(II) and Zn, and this was most likely due to their formation of a complex with 1,10-phenanthroline. The standard solution that was spiked with Cu(II) ions was light blue in colour, and this was due to the formation of the cupric-1,10-phenanthroline complex which has a maximum absorbance of 405 nm [5]. The standard solution that was spiked with Zn ions, however, was colourless. This could be due to the formation of the zinc-1,10-phenanthroline complex which shows UV-absorbance of maximum intensity at 270 nm [6]. The visible differences in colour of these two solutions as a result of the interference of the two ions hence resulted in the drastic deviation of the absorbance value at the wavelength of 510 nm.

Table 18.1 Effect of different ions on absorbance of iron(II)-phenanthroline complex

Interference ion (cation/anion)	Absorbance	Percentage interference (%)
Calcium, Ca^{2+}	0.561 ± 0.000436	+1.263
Magnesium, Mg^{2+}	0.555 ± 0.001	−4.391
Sodium, Na^{+}	0.557 ± 0.001	+0.505
Copper(II), Cu^{2+}	0.020 ± 0.001	-96.427
Potassium, K^{+}	0.548 ± 0.001	+1.101
Zinc, Zn^{2+}	0.022 ± 0.002	−96.263
Carbonate, $CO_3{}^{2-}$	0.568 ± 0.003	+2.508
Nitrate, $NO_3{}^{-}$	0.561 ± 0.000436	+1.263

$n = 3$, mean ± standard deviation

18.3.6 Calibration Curve

The absorbance of the standard iron(II) solutions were measured using both the photometer and the benchtop spectrophotometer. A linear graph was formed and the Beer-Lambert's law was observed from concentrations of 2×10^{-4} mg/ml to 4×10^{-3} mg/ml of iron.

As seen from Fig. 18.7 and Fig. 18.8, the absorbance of the iron(II)-phenanthroline complex was linear for 2×10^{-4} mg/ml to 4×10^{-3} mg/ml of iron at 510 nm. The equation of the straight line graph of best fit is $y = 154.75x + 0.0111$ from the photometer and $y = 219.62x - 0.0147$ from the benchtop spectrophotometer, with y being the absorbance, and x being the concentration of iron in mg/ml. The limit of detection of the photometer was found to be 1.1426×10^{-4} mg/ml, whilst that of the benchtop spectrophotometer was found to be 1.2331×10^{-4} mg/ml. The coefficient correlation obtained for the photometer was $R^2 = 0.9994$, whilst that for the benchtop spectrophotometer was $R^2 = 0.9993$. Both of these signified the reliability of the data points that are linearly fit. The molar absorption coefficient

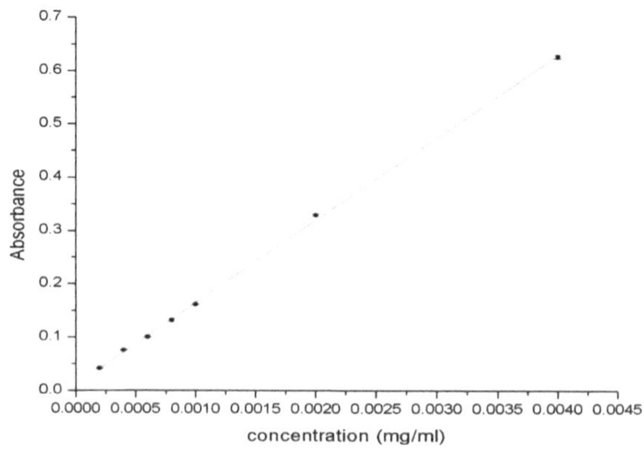

Fig. 18.7 Calibration curve plotted using results obtained from photometer

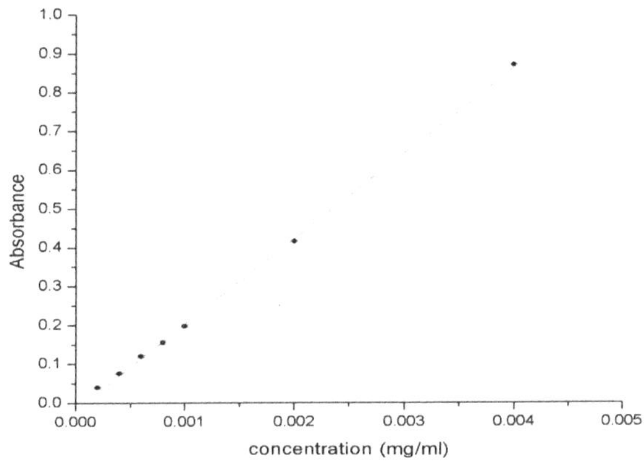

Fig. 18.8 Calibration curve plotted using results obtained from benchtop spectrophotometer

obtained from the photometer was $8.642 \times 10^{3\ M-1\ cm-1}$, whilst that obtained from the benchtop spectrophotometer was higher at $1.2265 \times 10^{4\ M-1\ cm-1}$ (refer to Annex B for methods of calculation). This shows that the spectrophotometer is more analytically sensitive than our self-built photometer. These two calibration curves were used to find the concentration of iron in spinach for our field testing, under the same experimental conditions.

18.3.7 Determination of Iron(II) in Spinacia Oleracea

The photometer developed was applied under the optimal conditions to determine the iron in two different species of spinach. Three samples of spinach leaves were analysed in triplicates. The three samples are: Red coloured leaves of red spinach (labelled as RR1, RR2 and RR3); Green coloured leaves of red spinach (labelled as RG1, RG2 and RG3); Green coloured leaves of green spinach (labelled as GG1, GG2 and GG3). The experiments were then repeated using the benchtop spectrophotometer as a reference method. The results obtained for the photometer are shown in Table 18.2, and the results obtained for the benchtop spectrophotometer are shown in Table 18.3. According to the U.S. Department of Agriculture, 100 g of raw spinach contains 2.71 mg of iron [8]. The mass of iron in 100 g of spinach leaves was calculated in our experiments so as to access the accuracy of our results by comparing them with this value.

18.4 Discussion

The mass of iron in 100 g of spinach leaves were calculated based on the calibration curve as constructed in Sect. 18.3.6. As seen from Table 18.2, the results obtained

Table 18.2 Determination of iron in different species of spinach leaves using photometer

Sample	Absorbance	Mass of iron in 2.5000 g of leaves/mg*	Mass of iron in 100 g of leaves/mg*
RR1	0.185 ± 0.002	0.0702	2.81
RR2	0.177 ± 0.001	0.0670	2.68
RR3	0.177 ± 0.001	0.0670	2.68
		Average: 0.0681	Average: 2.72
RG1	0.217 ± 0.002	0.0832	3.33
RG2	0.224 ± 0.001	0.0860	3.44
RG3	0.213 ± 0.002	0.0815	3.26
		Average: 0.0836	Average: 3.34
GG1	0.173 ± 0.001	0.0654	2.62
GG2	0.178 ± 0.001	0.0674	2.70
GG3	0.177 ± 0.002	0.0670	2.68
		Average: 0.0666	Average: 2.67

$n = 3$, mean ± standard deviation; *(Refer to annex B for explanations on the calculations)

Table 18.3 Determination of iron in different species of spinach leaves using benchtop spectrophotometer:

Sample	Absorbance	Mass of iron in 2.5000 g of leaves/mg*	Mass of iron in 100 g of leaves/mg*
RR1	0.246 ± 0.003	0.0742	2.97
RR2	0.221 ± 0.003	0.0671	2.68
RR3	0.243 ± 0.002	0.0733	2.93
		Average: 0.0715	Average: 2.86
RG1	0.278 ± 0.003	0.0833	3.33
RG2	0.293 ± 0.002	0.0876	3.50
RG3	0.277 ± 0.001	0.0830	3.32
		Average: 0.0846	Average: 3.38
GG1	0.232 ± 0.003	0.0702	2.81
GG2	0.241 ± 0.003	0.0728	2.91
GG3	0.221 ± 0.003	0.0671	2.68
		Average: 0.0700	Average: 2.80

$n = 3$, mean ± standard deviation; *(Refer to annex B for explanations on the calculations)

from our photometer showed that the green parts of the leaves in the red spinach had the highest concentration of iron, followed by the red parts of the leaves in the red spinach, and finally the green leaves of the green spinach. The average absorbances obtained for the photometer provides a relative error that does not exceed ±5% of the data obtained using the benchtop spectrophotometer. Hence for our field testing, which aimed to find out how the concentration of iron differs for different species of *Spinacia oleracea*, red spinach can be concluded to have a higher amount of iron compared to that of green spinach, whilst in the red spinach itself, the green parts of the leaves have a higher amount of iron compared to the red parts. The presence of a lower concentration of iron in the red parts of the red spinach could be due to the red parts containing a pigment called betacyanin, which is responsible for the red colour of the leaves. The presence of the Betacyanin pigment reduces the number of chlorophyll pigments [9]. Hence, this may lead to a lower concentration of iron in the red parts compared to the green parts within the red spinach.

Our data obtained from our field testing also proves that the accuracy of our photometer was in an acceptable range as compared to that of a benchtop spectrophotometer. Therefore, our LED-based photometer could serve as an advantageous alternative in the determination of iron due to its low-cost and portability. However, our photometer also has certain limitations. It is less sensitive compared to a benchtop spectrophotometer due to the 10 bit Arduino UNO used, causing it only able to measure up to 1024 levels of intensity. Our photometer can also measure only the intensity of solution

with a fixed wavelength from the LED, unlike the spectrophotometer which can measure the complex at different wavelengths. Hence our photometer will not be accurate in measuring other complexes that have a different wavelength of maximum absorbance than that of iron(II)-phenanthroline complex.

18.5 Conclusion

Trace concentrations of iron ions in different species of spinach have been studied in this work using a microcontroller-based photometer. From the experimental results, our hypothesis that green spinach had a higher amount of iron than red spinach is rejected, whilst our hypothesis that the green parts of the leaves of the red spinach has a higher amount of iron than that of the red parts is validated. Furthermore, the photometer built in the research offers many advantages such as portability, capability of in situ analysis, low-cost and having simple instrumentation as discussed previously, but it may not be as accurate and sensitive as the benchtop spectrophotometer, given its price and quality of materials used for construction.

18.6 Further Work

In the future, we would like to increase the resolution of the analogue to digital converter so that the photometer is able to detect a wider and nuanced range of values of light intensity that the photometer can detect. This can be done by replacing our current 10 bit Arduino UNO with a different microcontroller board that has a 16 bit analogue to digital converter, and hence enhancing the sensitivity of our photometer.

Table A.1 Effect of pH on absorbance of iron(II)-phenanthroline complex

pH	Absorbance/Au
1.6*	0.503 ± 0.002
2	0.617 ± 0.001
3	0.617 ± 0.001
4	0.586 ± 0.001
5	0.612 ± 0.002
6	0.486 ± 0.001
7	0.472 ± 0.001
8	0.096 ± 0.002
9	0.110 ± 0.001

$n = 3$, mean ± standard deviation
*Original pH with no addition of sodium citrate

Table A.2 Effect of temperature on absorbance of iron(II)-phenanthroline complex

Temperature/°C	Absorbance/Au
20	0.619 ± 0.001
25	0.611 ± 0.001
30	0.613 ± 0.002
40	0.623 ± 0.002
50	0.659 ± 0.001

We can also experiment with changing the wavelength of the LED to analyse other elements such as sodium and copper, or pH of solutions like seawater, testing and bringing this photometer into other relevant real-world contexts.

Acknowledgements We would like to express our sincerest gratitude to our mentor, Mr. Yeo Woei Ter, CRADL∑, Science Centre Singapore, as well as our teacher in charge, Ms. Cheong Szu Chuang, Raffles Girls' School.

Annex A

n = 3, mean ± standard deviation.

Annex B (Calculations)

Limit of detection.

3 times the Standard Deviation Over the Gradient of the Calibration Curve:

Table A.3 Effect of time on absorbance of iron(II)-phenanthroline complex

Time/h	Absorbance/Au
0	0.6116
0.5	0.6077
1	0.6154
1.5	0.616
2	0.6182
2.5	0.6193
3	0.6187
3.5	0.6183
4	0.6191
4.5	0.6218

For photometer, LOD = Standard deviation/154.75 × 3.3
For spectrophotometer, LOD = Standard deviation/219.62 × 3.3

Iron in 100 g of spinach.

Table 2: mg Fe / 2.5000 g of spinach = (Absorbance-0.0111)/154.75 × 50 × 5 / 4.

Table 2: mg Fe / 100 g of spinach = (Absorbance-0.0111)/154.75 × 50 × 5 / 4 / 2.5 × 100.

Table 3: mg Fe / 2.5000 g of spinach = (Absorbance + 0.0147)/219.62 × 50 × 5 / 4.

Table 3: mg Fe/ 100 g of spinach = (Absorbance + 0.0147)/219.62 × 50 × 5 / 4 / 2.5 × 100.

Molar absorption coefficient = gradient of the calibration curve multiplied by the relative atomic mass of iron.

References

1. Nazanin, A., Richard, H., & Roya, K. (2014). Review on iron and its importance for human health. *Journal of Research in Medical Sciences, 19*(2), 164–174.
2. Vogel (Ed.), *Vogel's Text of Qualitative Chemical Analysis*, 5th Ed Longman Book Society, Longman, UK, 1989
3. Irimia R., Gottschling M. (2016), Taxonomic revision of Rochefortia Sw. (Ehretiaceae, Boraginales). Biodiversity Data Journal 4: E7720.
4. Fortune W. B. and Mellon M. G. (1938), Determination of iron with o-phenanthroline; A spectrophotometric study, Ind. Eng. Chem. Anal. Ed.10, 60–64
5. Yuan, B., Zhu, Y., & Xu, K. (2015). The Study of 1,10-Phenanthroline-copper Complex By CV-Thin Layer Spectroelectrochemistry. *International Journal of Electrochemical Science, 10*, 4138–4145.
6. Lazici, D., Skundric, B., Jelena, P.-S., Sladojevic, S., Vasiljevici, L., Blagojevic, D., & Obrenovic, Z. (2010). Stability of tris-1,10-phenanthroline iron (II) complex in different composites. *Chemical Industry & Chemical Engineering Quarterly, 16*(2), 193–198.

7. Oktavia B, Lim L.W, and Takeuchi.T (2008), Simultaneous Determination of Fe(III) and Fe(II) Ions via Complexation with Salicylic Acid and 1,10-Phenanthroline in Microcolumn Ion Chromatography", Analytical Sciences, Vol.24
8. US Department of Agriculture, National Nutrient Database for Standard Reference Release 27 (2014). Retrieved from https://ndb.nal.usda.gov/ndb/foods/show/
9. Li, Y., Cui, L., Yao, X., Ding, X., Pan, X., Zhang, M., Li, W., & Kang, X. (2018). Trade-off between leaf chlorophyll and betacyanins in Suaeda salsa in the Liaohe estuary wetland in northeast China. *Journal of Plant Ecology, 11*(4), 569–575.

Chapter 19
Mining Next-Generation Sequencing Data to Identify Anti-nociceptive Signaling Pathways

Vanathi Pugalendhi and Hakkeem Nasira Farheen

Abstract Cisplatin is a platinum-based chemotherapy drug which causes damage to the peripheral nerves as a toxic side effect, resulting in cisplatin-induced peripheral neuropathy (CIPN). This diminishes patients' quality of life and possibly, their ability to work and live independently. A novel drug, CYM-5478, has been identified as a potential candidate which could ameliorate the pathology of peripheral neuropathy, although the mechanism in which it does this is not known in detail. Using RNA-Seq analysis, we identified genes in four pathways which have been shown to be significantly altered with cisplatin treatment—chemokine signaling pathway, cytokine-cytokine receptor interaction, NF-κB signaling pathway, and inflammatory medicated TRP channels. We then analyzed the genes identified in Rattus norvegicus treated with CYM-5478 to identify the pathways through which the drug exerts its protective effect. Through the characterization of genes using the data analysis method, t-test, we discovered multiple genes of interest, in particular, Gnb3 and Rac2. Finally, we validated observed pathway changes by evaluating homologous gene expression changes using an in vitro assay and performing RT-qPCR. Identified genes which were significantly altered in rats treated with cisplatin and with or without CYM-5478 did not show significant change in human astrocytes. However, we can better understand the mechanism through which CYM-5478 attenuates CIPN through the identification of these genes. This allows us to better understand the agonistic action of CYM-5478 on cisplatin toxicity, and the changes in their gene expression could be analyzed further in more vigorous future in vitro experiments involving human cells.

Keywords Cisplatin · CYM-5478 · Cisplatin-induced peripheral neuropathy (CIPN) · RNA-Seq · Anti-nociceptive · Volcano plots · Homologous gene expression · Immortalized normal human astrocyte (iNHA) cells

---

V. Pugalendhi (✉) · H. N. Farheen
NUS High School of Math and Science, Singapore, Singapore
e-mail: vanathipbd03@gmail.com

H. Guo et al. (eds.), *IRC-SET 2021*, https://doi.org/10.1007/978-981-16-9869-9_19

19.1 Introduction

Cisplatin, a platinum-based chemotherapy drug, has been shown to be effective in targeting a wide variety of cancers such as ovarian, testicular, bladder, head, and neck cancer. However, approximately 92% of chemotherapy patients using the drug cisplatin develop neurotoxic symptoms—characterized by tingling, numbness, and mechanical and thermal hyperalgesia—which cause patient discomfort and compromise cancer treatment. Most are dose-limiting, especially cisplatin-induced peripheral neuropathy (CIPN), which is characterized by sharp, stabbing pains, or tingling in the hands or feet caused by damage to the peripheral nerves. Symptoms progressively become more severe with continued treatment. A major cause of persistent pain in cancer survivors, CIPN is a toxic side effect which is often tolerated as an unfortunate consequence of life-saving chemotherapy. S1P is a lipid-based signaling molecule involved in neuroinflammatory processes through interactions with its five cognate receptors $S1P_{1\text{-}5}$, and S1P-related signaling has been identified previously as a potential target for treatment of CIPN. A specific sphingosine 1-phosphate receptor 2 (S1P2) receptor agonist, known as CYM-5478, has been identified as a potential candidate which could ameliorate the pathology of peripheral neuropathy, although the mechanism in which it does this is not known in detail [1]. The purpose of this project is to identify the pathways involved in CIPN that are affected by the action of CYM-5478 such as inflammation, the production of reactive oxygen species, and apoptotic pathways through the characterization of genes using research methods such as RNA-Seq. This will allow us to understand the changes in expression of specific genes in the S1P2 signaling pathway due to cisplatin and CYM-5478 treatment. We sieved out information from big data generated by previous experiments, with a narrower focus on genes involved in pathways that are changed by the action of cisplatin and our drug candidate. Genes in four pathways previously recognized to be dysregulated in patients undergoing chemotherapy with cisplatin as drug used with CIPN as a side effect were identified. Furthermore, we have evaluated whether these pathways have been changed by CYM-5478. We then singled out 2 genes—Gnb3 and Rac2 that were significantly changed and whose change was consistent with activation of sphingosine 1-phosphate receptor 2 (S1P2) by CYM-5478. Providing a more detailed explanation about how these genes are related and how CYM-5478 affects the expression of these genes to attenuate the effects of cisplatin toxicity will enable the trial of more drugs such as CYM-5478 to attenuate the effects of chemotherapy drugs, as well as help to identify targets to ameliorate the symptoms of CIPN.

We hypothesized that

1. CYM-5478 is able to attenuate toxic changes in gene expressions due to the action of cisplatin, by upregulating specific genes downregulated by cisplatin.
2. CYM-5478 is able to attenuate toxic changes in gene expressions due to the action of cisplatin, by downregulating specific genes upregulated by cisplatin.
3. The alteration of gene expression is not because of secondary response mechanisms within the body due to changes in chemical environment caused by the

action of cisplatin, but due to a direct effect on gene expression (e.g. transcription factors) by the drug CYM-5478 as it causes changes in cell composition. Any upregulation of inflammatory genes is due to the infiltration of immune cells and is not the direct effect of CYM-5478.

19.2 Methodology

A. Quantitative data analysis

Data analysis was conducted using mRNA expression levels obtained from a previous in vivo study done regarding measuring the expression levels of CYM-5478. Three groups of Rattus norvegicus were present, with one being the vehicle and the other two having been treated with cisplatin only and cisplatin and CYM-5478. Comparisons of cisplatin vs vehicle, cisplatin vs cisplatin and CYM-5478, and vehicle versus cisplatin and CYM-5478 were made. With the use of GraphPad Prism software, mRNA expression levels were analyzed using unpaired t-test with Welch's correction to create volcano plots. These plots helped to identify and ascertain which genes were significantly upregulated and downregulated. The genes were then grouped according to the pathways that are involved in. Using data already procured from the same study, four main pathways that CYM-5478 had known to affect were identified—chemokine signaling pathway, cytokine-cytokine receptor interaction, NF-κB signaling pathway, inflammatory medicated TRP channels. Among these, two genes were chosen—Gnb3 and Rac2 due to the previous studies showing that CYM-5478 has been known to activate the S1P2 receptor and affect these genes. The remaining genes identified show significant upregulation and downregulation purely due to the infiltration of immune cells and not due to the effect of CYM-5478 and therefore were not taken into consideration. MCP-1, which is a gene that is involved in producing inflammatory chemokines and Bax, an apoptotic marker was also chosen as they have been previously identified to be contributors to the pathology of CIPN and these genes would be downregulated in CIPN-related pathways.

B. Cell Culture

Immortalized normal human astrocytes (iNHA) cells were prepared with Dulbecco's modified eagle medium (DMEM) supplemented with 10% fetal bovine serum, 1% non-essential amino acids (Gibco, USA), 100 U/ml penicillin, 0.1 mg/ml streptomycin. They were incubated in 37 °C at 5% CO_2. 5×10^6 cells were cultured in 6-well plates overnight. The cells were subjected to four hour serum starvation before treatment began.

C. Treatment Conditions

iNHA cells were treated with the following conditions:

1. 10 μM CYM-5478,
2. vehicle control,

3. 20 μM cisplatin + 10 μM CYM-5478,
4. 20 μM cisplatin + vehicle control of CYM-5478 (10 μM).

D. RNA Extraction

1 ml of Trizol© Reagent was added, agitated using an orbital shaker for 5 min. 0.2 ml of chloroform was added per test tube. The cells were incubated at room temperature for two minutes and centrifuged at 12,000 × g for 15 min at 4 °C. The aqueous phase was removed and an equal amount of isopropanol was added and incubated for 10 min. It was then centrifuged at 12,000 × g for 10 min, and 75% ethanol was added to the resulting RNA pellet. The samples were centrifuged at 7500 × g for 10 min at 4 °C twice. The RNA pellets were then resuspended in 20 μl nuclease-free water, vortexed, and incubated in dry bath at 55 °C for 10–15 min. Yield and purity of RNA were quantified using Nanodrop.

E .Real-time Quantitative PCR (RT-qPCR)

Fast SYBR™ Green Master Mix (Thermo Scientific) was used to run RT-qPCR using the following primers: Bax, Rac2, Gnb3, and MCP-1. Beta-actin was used as a housekeeping gene. Applied Biosystems QuantStudio 6 Flex RT-qPCR machine was used to run a 3-step PCR (50 °C for 15 s, 95 °C for 20 s, 72 °C for 30 s) for 40 cycles.

19.3 Results

To identify the effects of CYM-5478 on gene expression, we generated volcano plots and identified genes that were significantly upregulated and downregulated (with significance achieved at p-value <0.05), which are highlighted in purple and are enlarged for clearer view (Figs. 19.1, 19.2, and 19.3). We primarily looked at four pathways—chemokine signaling pathway, cytokine-cytokine receptor interaction,

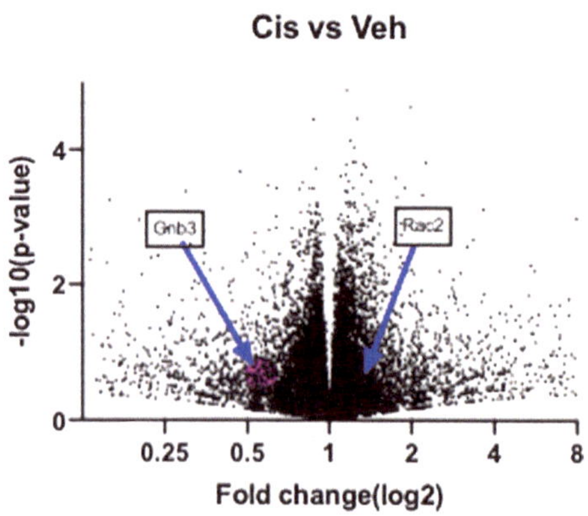

Fig. 19.1 Volcano plot depicting gene expression changes in in vivo analysis of rats treated with Cisplatin versus Vehicle

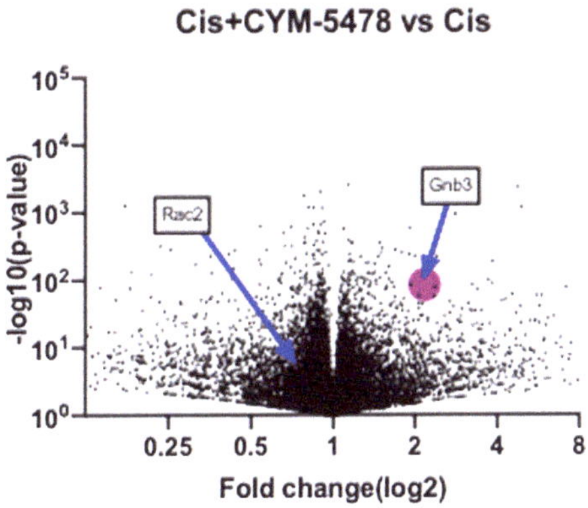

Fig. 19.2 Volcano plot depicting gene expression changes in in vivo analysis of rats treated with Cisplatin + CYM-5478 versus Cisplatin only

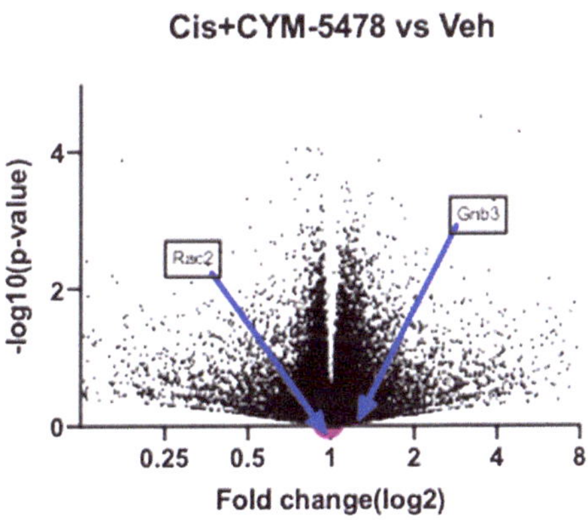

Fig. 19.3 Volcano plot depicting gene expression changes in in vivo analysis of rats treated with Cisplatin + CYM-5478 versus Vehicle

NF Kappa B signaling pathway, inflammatory medicated TRP channels that have been previously known to be primarily affected by CYM-5478 action. We sieved out genes based on the significance of regulation and whether they are involved in the 4 pathways that we had identified. These genes are displayed in Table 19.1, along with the p-values and q-values obtained from the volcano plots. The p-value refers to the probability of obtaining a change in gene expression at least as extreme as what is observed assuming that the null hypothesis (no change in gene expression observed) is correct. A p-value of less than 0.05 indicates a significant change in gene expression. The q-value is an adjusted p-value to control the false positive discovery rate.

In the RT-qPCR analysis, we measured the gene expression of 4 genes—*Bax, Rac2, MCP-1, and Gnb3*. *Rac2* and *Gnb3* were used from the array of genes that we identified from the volcano plot (Table 19.1). We used iNHA cell cultures exposed to 4 different types of treatments – vehicle, cisplatin only, CYM-5478 only and CYM-5478 and cisplatin and compared expression of the 4 genes across these treatment groups. *Bax* and *MCP-1* were used as means of comparison, as from the previous

Table 19.1 Table of p-values and q-values of genes of interest identified from the four pathways

Gene	CP versus V (p-value)	CP vs V (q-value)	CP versus CP + CYM-5478 (p-value)	CP versus CP + CYM-5478 (q-value)	Pathways involved
Dock2	0.0298	0.699	0.0491	0.838	Chemokine signaling, development HGF signaling, elevated platelet, and cytosolic Ca2+
Lyn	0.0382	0.764	0.0233	0.626	Chemokine signaling, NF-κB signaling, B cell receptor signaling, signaling by GPCR
Gnb3	0.0378	0.761	0.00585	0.305	Chemokine signaling, translation regulations by alpha-1 adrenergic receptors
Rac2	0.00665	0.704	0.018	0.559	Chemokine signaling, regulation of activated PAK-2p34, NGF pathway
Cntf	0.016	0.530	0.0016	0.144	Cytokine-cytokine receptor interactions, PEDF induced signaling, innate immune system
Camk2a	0.0488	0.840	0.0363	0.750	Inflammatory medicated TRP channels, translation regulations by alpha-1 adrenergic receptors
Trpm8	0.0000500	0.00994	0.0022	0.177	Inflammatory medicated TRP channels, CREB pathway, TRP channels

(continued)

Table 19.1 (continued)

Gene	CP versus V (p-value)	CP vs V (q-value)	CP versus CP + CYM-5478 (p-value)	CP versus CP + CYM-5478 (q-value)	Pathways involved
Asic1	0.007	0.339	0.0022	0.302	Inflammatory medicated TRP channels, transport of glucose and metal ions, ion channel transport
Asic3	0.00035	0.0472	0.0266	0.661	Inflammatory medicated TRP channels, transport of glucose and metal ions, ion channel transport
Syk	0.0465	0.823	0.0097	0.407	NF-κB signaling, B cell receptor signaling pathway, signaling by GPCR
Card11	0.00975	0.408	0.0142	0.498	NF-κB signaling, B cell receptor signaling pathway

literature, it has been established that they can be changed by the action of CYM-5478 action. As such, this gives us a means to check whether the expected increase of *Bax* and decrease of *MCP-1* is seen, thus allowing us to gauge the accuracy of our analysis.

The gene expression of the 4 genes was normalized to *β-actin*, and the results were plotted in the form of bar graphs (Figs. 19.4, 19.5, 19.6, and 19.7). However, the results of the analysis were inconclusive as the error bars charted in all 4 bar graphs overlapped across the 4 treatments, and hence, no definitive comparison could be made between the different treatments.

19.4 Discussion

For the sake of establishing the effect of CYM-5478 only, we used 2 comparisons only – cisplatin vs vehicle Fig. 19.1 and cisplatin vs cisplatin + CYM-5478 Fig. 19.2. The reason why these 2 comparisons were chosen is that they showed a clear contrast and change in gene expression of genes affected singularly by the introduction of the agonist into the treatment. We then further sieved out the genes that had contrasting regulations, as in upregulated in cisplatin vs vehicle and downregulated in cisplatin versus cisplatin + CYM-5478 and vice versa as we wanted to have a more focused and direct look on the effect of CYM-5478.

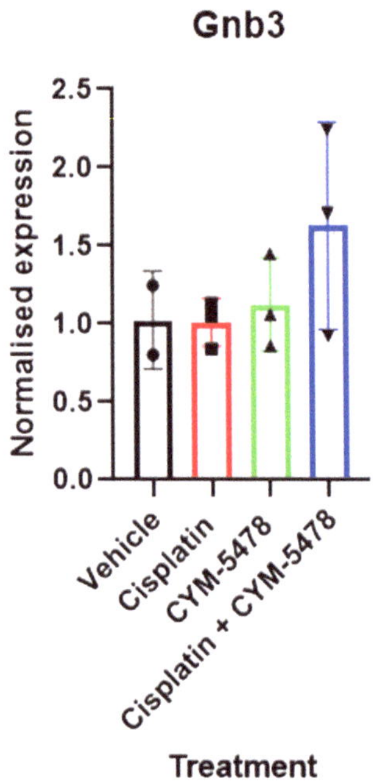

Fig. 19.4 RT-qPCR analysis of gene expression changes of *Gnb3* normalized to *β-actin* across the four treatments (n=3 per group)

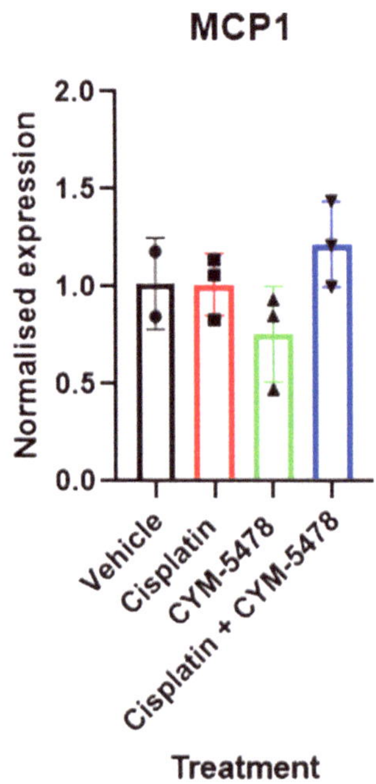

Fig. 19.5 RT-qPCR analysis of gene expression changes of *MCP-1* normalized to *β-actin* across the four treatments (n=3 per group)

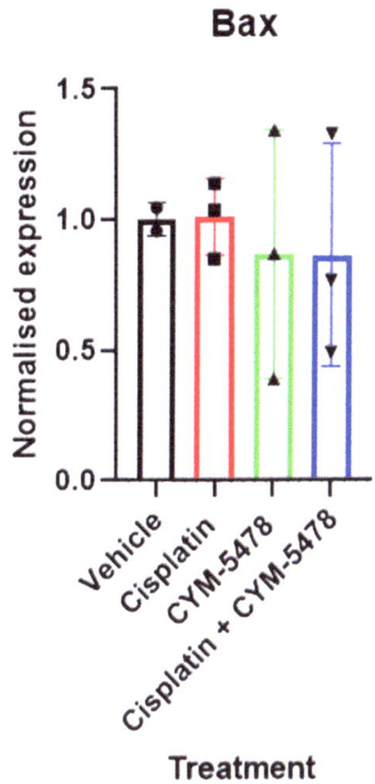

Fig. 19.6 RT-qPCR analysis of gene expression changes of *Bax* normalized to β-*actin* across the four treatments (n=3 per group)

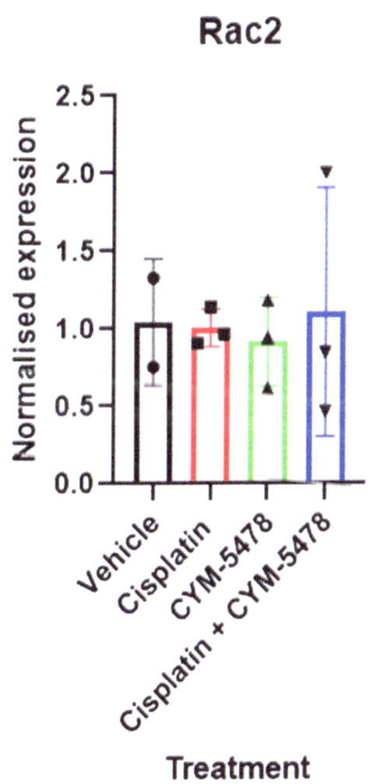

Fig. 19.7 RT-qPCR analysis of gene expression changes of *Rac2* normalized to β-*actin* across the four treatments (n=3 per group)

The panel of genes that we identified was *Dock2, Lyn, Gnb3, Rac2 (in chemokine signaling), Cntf (cytokine-cytokine receptor interaction), Camk2a, Trpm8, Asic1, Asic3* (inflammatory medicated TRP channels), *Syk, Card11,* and *Lyn* (NF Kappa B signaling pathway). Among these genes, with respect to possible directional gene changes evoked by CYM-5478 itself and secondary body responses such as immune responses triggered by cisplatin treatment, we singled out 2 genes—*Rac2* and *Gnb3*. *Rac2* is known to encode a member of the Ras superfamily of small guanosine triphosphate (GTP)-metabolizing proteins, which are involved in the generation of reactive oxygen species, consistent with a side effect commonly associated with cisplatin and it regulates processes, such as secretion, phagocytosis, and cell polarization. *Rac2* is also involved in the chemokine signaling pathway, which is associated with the action of CYM-5478. *Gnb3* codes for heterotrimeric guanine nucleotide-binding proteins (G proteins), which integrate signals between receptors and effector proteins and is involved in the chemokine signaling pathway.

After conducting RT-qPCR analysis with iNHA cells, we did not find any significant changes in gene expression across the different treatment groups. This could be due to the treatment time being too short for these genes to show any changes. This could also be due to the fact that the in vitro studies were conducted using human cells, unlike the in vivo study which involved rats, as such there might be significant differences in the gene expression pathways. Furthermore, only two genes were identified; therefore, this relatively small selection may not have included the target gene that we were looking for. The identification of these two genes may also be due to secondary body processes such as the possible mitochondrial damage caused by the action of cisplatin, which may have led to the marked upregulation and downregulation of the two genes identified.

19.5 Conclusion

We have identified genes whose expression was changed by the action of cisplatin, and then corrected by CYM-5478, in pathways which are known to be implicated in CIPN. No significant change in the gene expression of *Rac2* and *Gnb3* was observed. Hence, we identified genes that were changed through the cell-based assay, but they did not show the expected changes in gene expression in the in vitro conditions that we exposed them to.

19.6 Future Work

The optimization of treatment duration would allow for a more accurate picture of the effects of CYM-5478 in attenuating cisplatin toxicity and in ascertaining whether the changes in gene expression are still the same. This will allow us to better understand the mechanism of action of CYM-5478, which would be beneficial for the potential co-administration of CYM-5478 to relieve the toxic side effects of cisplatin. This would allow patients to have a better quality of life and a better treatment chance. Repeating the experiment with different cell types such as oligodendrocytes and skin cells could also reveal changes in gene expression not apparent in astrocytes. Moreover, exploring different genes of interest expressed in other cell types to prevent other side effects such as low blood count and ototoxicity is possible to aid in narrowing the dose-limiting side effects of chemotherapy. We can also identify if CYM-5478 is able to attenuate similar toxic effects of other widely used cytotoxic drugs such as oxaliplatin and vincristine, via identifying genes from affected pathways and evaluating homologous gene expression changes in in vitro assays.

Acknowledgements We would like to thank our research mentor Dr. Deron Raymond Herr, our teacher mentor Ms. Ruth Chan, as well as Ms. Brenda Lam and Ms Amelia Yam. Thank you for your constant guidance and support throughout our project.

References

1. Activation of sphingosine 1-phosphate receptor 2 ... —PubMed (2020, January 24). Retrieved on December 23, 2020, from https://pubmed.ncbi.nlm.nih.gov/31882542/

Chapter 20
Detection and Classification System for Rail Surface Defects Based on Deep Learning

Albert Ji, Yang Thee Quek, Eugene Wong, and Wai Lok Woo

Abstract The detection and classification of rail track surface defects is critical to the safety and productivity of urban transport operations. The requirement of fast and accurate detection and classification from the huge number of images taken drives for an automated solution. A method for detection and classification of rail surface anomalies and defects adopting deep learning approach based on the convolutional neural networks has been presented in this paper. The training and testing images are acquired from an automated video recording setup on the train. We proposed a convolutional neural network model trained to learn the features which will then be utilized to detect and classify defects from images taken by image acquisition device mounted on the train. The experimental results are promising and can be integrated to the current workflow of rail maintenance operations to improve the productivity.

Keywords Deep learning · Convolutional neural network · Rail surface defect · Defect detection · Defect classification

---

A. Ji (✉) · E. Wong
Newcastle University, Singapore Campus, Singapore, Singapore
e-mail: a.ji2@newcastle.ac.uk

E. Wong
e-mail: eugene.wong@ncl.ac.uk

Y. T. Quek
School of Engineering, Republic Polytechnic, Singapore, Singapore
e-mail: quek_yang_thee@rp.edu.sg

W. L. Woo
Northumbria University, Newcastle upon Tyne, UK
e-mail: wailok.woo@northumbria.ac.uk

H. Guo et al. (eds.), *IRC-SET 2021*, https://doi.org/10.1007/978-981-16-9869-9_20

20.1 Introduction

Heavy traffic, high-speed trains, braking and acceleration give severe stresses to rails which will show various types of defects roughly categorized into abrasion (loss of material through friction), deformation, fatigue and machining errors. Timely maintenance prolongs the rail's service life considerably and reduces the cost of maintaining the rail during its lifetime. Therefore, fast and accurate detection and classification of defects is critical [1]. Manual rail inspection through patrolling or reading images could be costly, delayed and time-consuming. Video cameras have been commonly adopted to monitor the health status of rails which produces huge number of rails images and only a small portion is unhealthy samples [2]. Therefore, a good defect detection and classification system should be able to identify as many defects without creating too many false alarms.

Deep neural networks have proved successful in many computer vision application areas and the technology has advanced rapidly. Deep learning of multi-layer neural network solves the issues of artificial feature extraction and optimization in traditional machine learning methods and preprocessing of images is less important to deep neural network approach than other hand-crafted features or spatial correlation statistics-based techniques. [3, 4] Convolutional neural network is based on this idea. Deep learning has been extensively adopted for surface defect detection and classification. The success will depend on the network's ability to learn efficiently the features. Large feature learning convolutional neural network might have large number of parameters and require large training datasets.

An application of deep convolutional neural networks to automatically detect and classify the rail surface defects is presented in this paper. Training images are fed into the model for the neural network to learn the features which will be used to predict the detection and classification results of the test images.

This paper is organized as follows. In Sect. 20.2, we describe the convolutional neural network structure and its operations to detect and classify the defects. In Sect. 20.5, we describe the hardware setup, the labeled image datasets and the proposed convolutional neural network. In Sect 20.10, we describe the results and conclude the paper.

20.2 Methods

20.2.1 Structure of Convolutional Neural Network

A deep convolutional neural network is a layered organization of neurons connected to other neurons. These neurons form a complex network to learn from the input signals with some feedback mechanism. The following diagram represents an "*N*" layered deep neural network (Fig. 20.1).

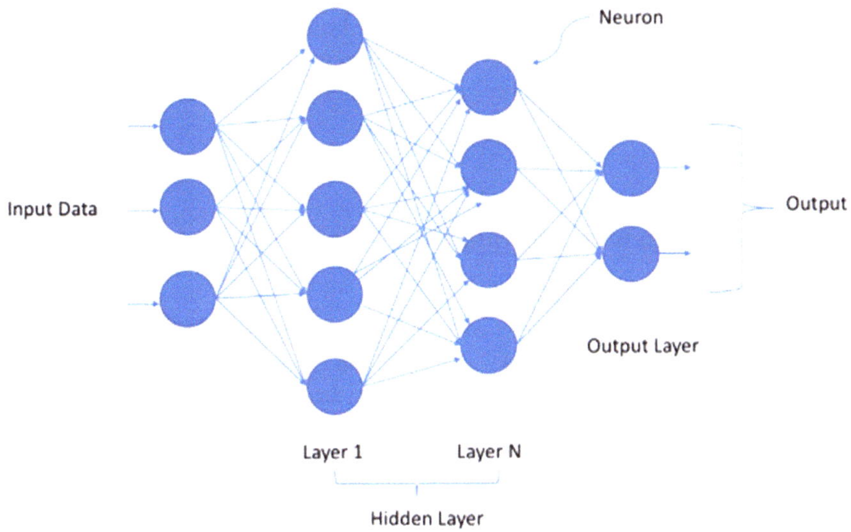

Fig. 20.1 A deep neural network with N hidden layers

Three main components exist in classical convolutional neural network [5].

(1) Convolution: The convolutional layer is central to the convolutional neural network and performs an operation called a "convolution" which involves the multiplication of a set of weights with the input. A two-dimensional array of weights, called a filter or a kernel, multiplies with an array of input data which is a dot product. The filter applies systematically across the entire input image to discover a particular feature that the filter is designed to detect. The filter is applied multiple times to the input array to generate a two-dimensional array of output values that represent a filtering of the input and is called a "feature map".

(2) Activation function: An activation function is a nonlinear function that takes in the output signal from the previous cell and converts it into some form that can be taken as input to the next cell. It is added into a neural network to help the network learn complex patterns for discriminating between image classes. Some examples of activation functions are Rectified Linear Activation (ReLU), Logistic (Sigmoid) and Hyperbolic Tangent (Tanh).

(3) Pooling: Feature map record the precise position of features in the input which is sensitive to small movements of the feature positions. A pooling layer is added after the convolutional layer and nonlinearity to create pooled feature map. Merging local information can reduce the dimensions of the feature maps and make the model invariant to local translation. Two common functions used in the pooling operation are: average pooling (calculate the average value for each patch on the feature map) and maximum pooling (calculate the maximum value for each patch of the feature map).

Convolutional and pooling layers are connected successively to create the convolutional neural network architecture. The use of many more hidden layers makes a neural network "deep" to learn more complex patterns which gives the name "deep learning."

20.2.2 Training with Convolutional Neural Network

A set of inputs is mapped to a set of outputs by a deep learning neural network through learning from training data. Learning can be regarded as an optimization problem where algorithms search and determine the model weights to predict the outputs. An optimization algorithm is used to train a neural network model and the backpropagation of error algorithm updates the model weights. The objective of training a neural network is to minimize the error and the objective function is therefore a loss or cost function whose calculated output can simply be called as "loss."

When adjusting model weights during training, cross-entropy losses (logarithmic loss) are used to minimize the loss, i.e., the smaller the loss the better the model [6]. Categorical cross-entropy is a loss function that is used in multi-class classification tasks.

Cross-entropy is defined as:

$$L_{\mathrm{CE}} = -\sum_{i=1}^{n} t_i \log(p_i) \text{ for } n \text{ classes},$$

Truth label as t_i and Softmax probability for the ith class as p_i.

The desired output (0 or 1) for each actual class is compared with its predicted class probability. The loss will then be calculated to penalize the predicted probability according to its difference from the actual value expected. The logarithmic penalty yields a small score for small differences tending to 0 and a large score for large differences close to 1.

For binary classification, we have binary cross-entropy defined as:

$$L = -\sum_{i=1}^{2} t_i \log(pi_i) = -\left[\mathrm{tlog}(p) + (1-t)\log(1-p)\right]$$

Stochastic gradient descent is extended to be Adam optimization which is based on adaptive estimation of first-order and second-order moments [7]. Adam is a combination of momentum update method and Root Mean Squared Propagation (RMSprop):

Require: α: Stepsize
Require: β1, β2 ∈ [0, 1): Exponential decay rates for the moment estimates
Require: f(θ): Stochastic objective function with parameters θ
Require: θ_0: Initial parameter vector
$m_0 \leftarrow 0$ (Initialize 1st moment vector)
$v_0 \leftarrow 0$ (Initialize 2nd moment vector)
$t \leftarrow 0$ (Initialize timestep)
while θ_t not converged **do**
$t \leftarrow t + 1$
$g_t \leftarrow \nabla_\theta f_t(\theta t-1)$ (Get gradients w.r.t. stochastic objective at timestep t)
$m_t \leftarrow \beta_1 \cdot m_{t-1} + (1 - \beta_1) \cdot g_t$ (Update biased first moment estimate)
$v_t \leftarrow \beta_2 \cdot v_{t-1} + (1 - \beta_2) \cdot g_t^2$ (Update biased second raw moment estimate)
$\hat{m}_t \leftarrow m_t/(1 - \beta_1^t)$ (Compute bias-corrected first moment estimate)
$\hat{v}_t \leftarrow v_t/(1 - \beta_2^t)$ (Compute bias-corrected second raw moment estimate)
$\theta_t \leftarrow \theta_{t-1} - \alpha \cdot \hat{m}_t /(\sqrt{\hat{v}_t} + \varepsilon)$ (Update parameters)
end while
return θ_t (Resulting parameters)

Momentum update method can be mathematically expressed as follows:

$$V_t = \gamma * V_{t-1} + \eta \nabla \theta * J(\theta)$$

$$\theta = \theta - V_t$$

The parameters of neural network (θ) are weights, biases or activations. The learning rate is η. We are to optimize the objective function (J). γ is momentum and a constant term. The past time-step is $V_{t\text{-}1}$ while the current is V_t.

Learning rate, or step size, is the amount that the weights are updates during training and has a small configurable positive value in the range between 0.0 and 1.0. Learning rate hyper parameter controls how quickly the model is adapted to the problem. Smaller learning rates require more training epochs. Too small learning rate cause stuck in the training process while too large causes suboptimal solution. To adjust learning rate during the training phase, adaptive learning rates reduce the learning rate to a pre-defined schedule such as RMSprop whose mathematical expression is shown as follows:

$$E[g^2]_t = 0.9 * E[g^2]_{t-1} + 0.1 * (g_t)^2$$

$$\Theta_{t+1} = \theta_t - -g_t * \frac{\eta}{\sqrt{E\left[g^2\right]_t + \varepsilon}}$$

The exponentially decaying average of squared gradients is represented as $E[g^2]_t$. γ is recommended to be 0.9 and the learning rate η has a default value 0.001.

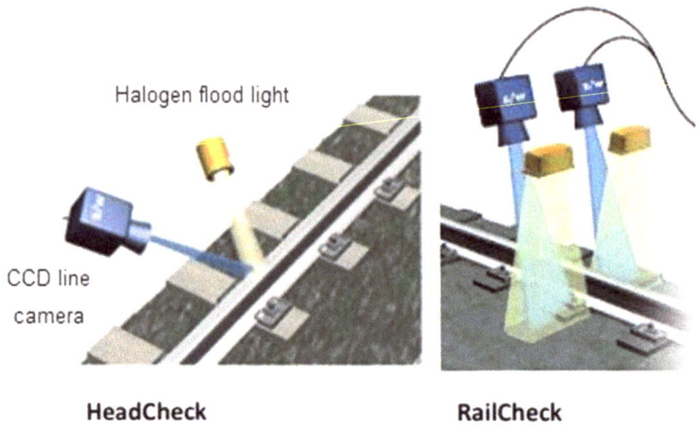

Fig. 20.2 Image acquisition setup

20.3 Implementation

20.3.1 Image Acquisition Setup

Rail vision systems are mounted on the rolling stocks to capture the video of the rail tracks when the rolling stocks are running. CCD line scan cameras are installed for both head check and rail check. Two halogen flood lights are installed for each camera. Maximum speed allowed for the trains is 100 km/h. Figure 20.2 illustrates the typical setup.

20.3.2 Data Description

Greyscale images are captured from the video taken by the CCD line cameras on top of the rail tracks. The images are manually labeled as "normal", "crackline," "chipoff," "corrugation," "missing," "pitting," "shelling," "transverse," "wheelburn," "watermark." The latter nine labels are regarded as defects for the detection problem. The original rail track images are adopted without segmenting the tracks from the ballast, the sleepers or other background textures around the rail tracks so as to minimize the image preprocessing efforts and maximize the productivity of operations. Figure 20.3 shows image samples of different defects. Some rail track anomalies are described in the following paragraphs.

Rail corrugation develops over time due to the rolling contact friction between the track and the train wheels. Waves, trough, and crests are left behind which may worsen in the long run. Heavily corrugated rails may experience concaving deformations on

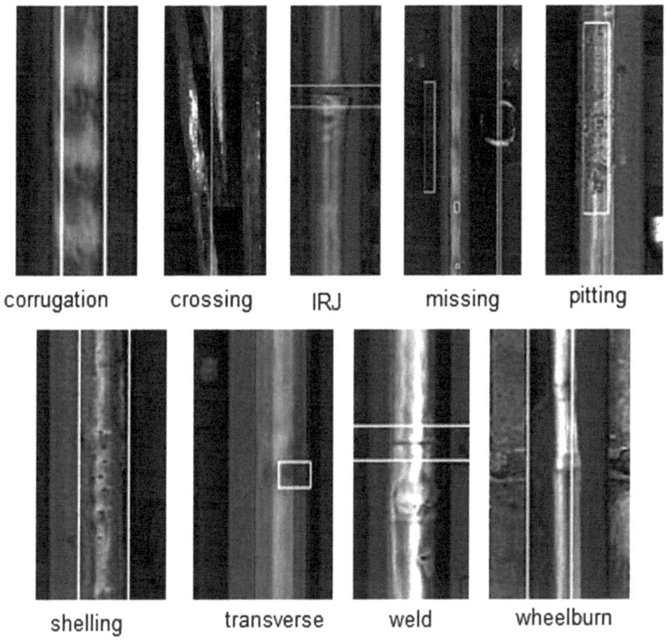

Fig. 20.3 Sample images of defects

the top of railroad tracks. Rail corrugation decreases the lifespan of the tracks and replacements must be made.

Most transverse cracks form in the area of faulty weld joints. Difference in weld material or a manufacturing flaw in the rail can be one of the causes of a weld error. Welding processes such as arc welding, pores, inclusions and misalignments can cause transverse cracks in the weld joints.

Shelling is caused by loss of material due to subsurface fatigue and occurs along the gage corner of rails. Shelling cracks develops on the inside of the rail of the gage corner and are usually noticeable as dark spots on the outside of the gage corner. In the early stages, the growth of the cracks may be fast and unforeseen with the failure appearing first before the detection of the defect.

The constant sliding of the railway wheels on the rails causes wheel burn defects. The sliding of the wheels raises the rails' surface to extreme temperatures. As a result of the rapid cooling, the rail material turns into the brittle martensite phase. Wheel burns could be seen on the surface of rails and may have similar appearances with squat defects. Wheel burns are usually found in pairs, opposite of one another on two rails.

The distribution of images from each category is summarized in the Table 20.1. As discussed previously, the number of normal images is much greater than those with defects for rail tracks. Therefore, the number of normal images is reduced to avoid the class imbalance problems in convolutional neural network.

Table 20.1 Distribution of image samples

Normal	57%
Corrugation	7%
Crossing	2%
IRJ	4%
Missing	1%
Pitting	1%
Shelling	7%
Transverse	2%
Weld	17%
Wheelburn	2%

20.3.3 Deep Learning Configurations

A computer with Intel Core i7-8700 CPU, Nvidia GeForce RTX 2080 Ti GPU and 64G RAM is configured for the deep learning experiment. Python 3.6, Nvidia CUDA Toolkit, cuDNN, TensorFlow with GPU support are installed on Windows 10 operating system to perform the deep learning experiments.

20.3.4 Neural Network Architecture

The deep neural network architecture in this paper is depicted in Figs. 20.4 and 20.5. The parameters of the neural network have been selected to build the model.

Input greyscale images are resized to 186 × 256 pixels before being fed into the neural network. For each category, 90% of the image samples are randomly reserved for training while the remaining 10% for testing.

The neural network architecture consists of four convolutional layers, two max-pooling layers and four fully connected layers. The kernel size of convolutional

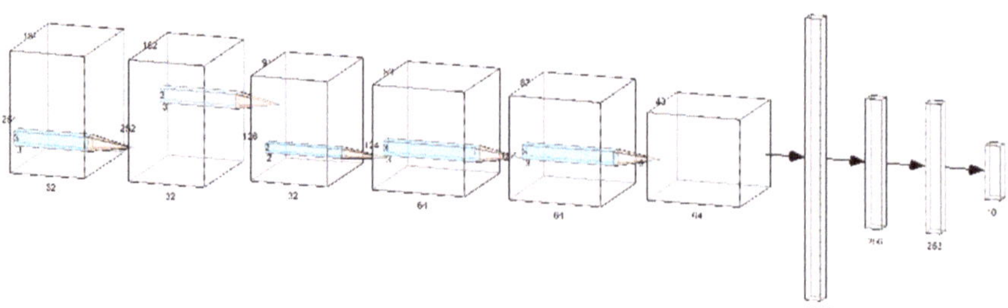

Fig. 20.4 Deep neural network architecture

Fig. 20.5 Sizes of tensors, kernels and vectors

Table 20.2 Confusion matrix of detection task

		Predicted	
		Normal (%)	Defective (%)
Actual	Normal	*95.00*	5.00
	Defective	20.11	*79.89*

layers is set as 3 × 3 pixels. We used max-pooling units of size 2 × 2 pixels. Rectified linear units (ReLU) was used as activation function for convolutional layers. We add a dropout layer as an effective regularization method after max-pooling to reduce overfitting by randomly dropping out nodes during training. After the convolutional and max-pooling layers, we use fully connected layers to perform high level reasoning in convolution neural network. The input images are categorized by the neural network into 2 output classes for detection task and 10 output classes for classification tasks.

20.4 Experimental Results

We run the deep convolutional neural network model for detection and classification tasks separately while keeping the training and testing images the same and only adjusting the number of output classes for the network classifiers at the end of the network. The learning rate of Adam optimizer is set as 0.001 for training the model. For both the binary classification of detection tasks and the multi-class classification of classification tasks, we count the number of true positives (TP), true negatives (TN), false positives (FP) and false negatives (FN).

20.4.1 Results

The confusion matrices of the detection task (binary) and the classification task (multiple-class) results trained on our deep convolutional neural network are presented in Table 20.2 and Table 20.3, respectively.

20.4.2 Discussions

The binary classification accuracy: (TP + TN)/(TP + TN + FP + FN) is calculated as 87.45%, and $F1$-score: 2TP/(2TP + FP + FN) is calculated as 88.33%. The performance is acceptable and has substantially improved the performance of the existing auto-detect method in operation [8].

Table 20.3 Confusion matrix of classification task

		Predicted									
		Normal (%)	Corrugation (%)	Crossing (%)	IRJ (%)	Missing (%)	Pitting (%)	Shelling (%)	Transverse (%)	Weld (%)	Wheelburn (%)
Actual	Normal	*95.83*	0.42	0.00	0.83	0.00	0.00	0.83	0.00	2.08	0.00
	Corrugation	12.90	*77.42*	0.00	0.00	0.00	0.00	0.00	0.00	6.45	3.23
	Crossing	28.57	14.29	*28.57*	0.00	0.00	14.29	14.29	0.00	0.00	0.00
	IRJ	11.11	5.56	0.00	*61.11*	0.00	0.00	5.56	0.00	16.67	0.00
	Missing	20.00	0.00	0.00	0.00	*0.00*	0.00	40.00	0.00	40.00	0.00
	Pitting	0.00	40.00	0.00	20.00	0.00	*20.00*	0.00	0.00	20.00	0.00
	Shelling	58.62	3.45	6.90	0.00	0.00	0.00	*10.34*	0.00	17.24	3.45
	Transverse	60.00	0.00	0.00	10.00	0.00	0.00	0.00	*20.00*	10.00	0.00
	Weld	25.33	9.33	0.00	0.00	0.0	0.00	2.67	0.00	*62.67*	0.00
	Wheelburn	44.44	22.22	11.11	0.00	0.00	0.00	0.00	0.00	22.22	*0.00*

The model performs well to correctly predict for normal images. First, this could be due to the higher percentage of normal images in data set which could be further affirmed by the classification task results. This potentially indicates a class imbalance problem which could be further investigated. Second, the parameters of the model such as convolutional kernel size and the dropout rate could be further tuned to improve the performances. Third, the size and number of layers can be expanded for the neural network model in this paper in order to enhance the performance, however, the training time will be increased.

For the classification task, the model performs satisfactorily to correctly detect some categories such as normal, corrugation, IRJ and weld. The performance with other defects is not as desired where categories such as missing and wheelburn are not detected correctly detected at all. We can first balance the percentage of the samples from different categories as there are few samples in low performance categories. The total number of image samples can be increased as well to enhance the training and improve the prediction performance. Besides, we can redefine the categories and merge some categories if it makes sense to the operations.

20.5 Conclusions

A deep learning approach for automatic rail surface defects detection and classification has been presented in this paper. We took advantage of the vast number of images generated from the rail operations by utilizing the learning ability of deep neural networks. Compared with traditional image processing approach, the deep learning approach reduces the preprocessing requirement and improves the operational productivity. The structure and the parameters of the convolutional neural network have been carefully analyzed and evaluated to build a deep convolutional neural network to perform the automatic detection and classification tasks.

The results for detection and classification tasks are promising with potentials improvement for certain types of defects. The current rail maintenance operations are carried out with much time and costs. The approach proposed in this paper can be integrated into the operations to further improve the productivity. In this regard, we will explore the methods to further enhance the accuracy level and incorporates mobile and IoT technologies to bring the deep learning approach onto the rail track which will support patrollers and engineers in identifying and locating the defects timely.

Acknowledgements This research is part of the project supported by SMRT Corporation Ltd.

References

1. Oregui, M., Molodova, M., Núñez, A., et al. (2015). Experimental investigation into the condition of insulated rail joints by impact excitation. *Experimental Mechanics, 55*, 1597–1612.
2. Li, Q., & Ren, S. (2012). A visual detection system for rail surface defects. *IEEE Transactions on Systems, Man, and Cybernetics Part C (Applications and Reviews), 42*(6), 1531–1542.
3. Gibert, X., Patel, V. M., & Chellappa, R. (2015). Robust fastener detection for autonomous visual railway track inspection. In *2015 IEEE winter conference on applications of computer vision* (pp. 694–701).
4. Marino, F., Distante, A., Mazzeo, P. L., & Stella, E. (2007). A real-time visual inspection system for railway maintenance: automatic hexagonal-headed bolts detection. *IEEE Transactions on Systems, Man, and Cybernetics Part C (Applications and Reviews), 37*(3), 418–428.
5. LeCun, Y., et al. (1989). Backpropagation applied to handwritten zip code recognition. *Neural Computation, 1*(4), 541–551.
6. Nasr, G. E., Badr, E. A., & Joun, C. (2002). Cross entropy error function in neural networks: forecasting gasoline demand. In *FLAIRS conference.*
7. Kingma, D. P., & Ba, J. (2015) Adam: A method for stochastic optimization. CoRR abs/1412.6980
8. Chicco, D., & Jurman, G. (2020). The advantages of the Matthews correlation coefficient (MCC) over F1 score and accuracy in binary classification evaluation. *BMC Genomics, 21*, 6.

Chapter 21
Analysing Gait Patterns of Parkinsons' Disease Patients to Predict Freezing of Gait (FoG) Using Machine Learning Algorithms

Ananya Nallapuraju, Chen Rui Ye, Prannaya Gupta, and Arthur Tay

Abstract Parkinson's disease (PD) is a neurodegenerative disorder that affects the dopamine producing neurons in the substantia nigra, an area of the brain, leading to shaking, stiffness, and difficulty walking. Parkinson's patients frequently exhibit the debilitating condition, freezing of gait (FoG), which is when patients cannot move their feet forward despite the intention to walk. Whilst the feet remain in place, the torso still has forward momentum, making falls very common. Several studies have reported methodologies to identify the PD patients at an early stage, such as PD scales and using inertial measurement units (IMU). However, these methods are ineffective as scales are not personalised, and the IMUs are generally big and bulky. In this study, acceleration data from the thigh of PD patients have been put through signal processing algorithms to postulate a generic freeze index, which is then run through multiple support vector machine models to analyse and therefore predict FoG events. The analysis found that vertical and horizontal lateral acceleration were the most accurate at predicting FoG. Ultimately, a small and lightweight prototype was created based on acceleration data from accelerometers placed on the thigh.

21.1 Introduction

Parkinson's disease (PD) is a progressive neurodegenerative disorder that affects the substantia nigra, an area of the brain, resulting in the progressive loss of dopamine neurotransmitters and poorer connection between the central nervous system and muscles [1]. This leads to many gait abnormalities, mainly tremors, bradykinesia, rigidity, and a loss of postural reflexes. Secondary motor symptoms that are evoked include stride length reduction, shuffling gait, step festination, and freezing [2, 3].

A. Nallapuraju · C. R. Ye · P. Gupta (✉)
School of Mathematics and Science, NUS High, Singapore 129957, Singapore
e-mail: h1810124@nushigh.edu.sg

A. Tay
NUS Faculty of Engineering, Singapore 117575, Singapore

H. Guo et al. (eds.), *IRC-SET 2021*, https://doi.org/10.1007/978-981-16-9869-9_21

Freezing of gait (FoG) is one of the most debilitating effects of PD and is defined as a brief, episodic absence, or marked reduction of forward progression of the feet despite the intention to walk [4]. It has serious social and clinical consequences for patients as it leads to falls [5] due to the torso still having forward momentum even though the feet remains in place, interferes with daily activities, and significantly impacts quality of life [6].

21.1.1 Aims and Objective

As the prevalence of PD increases in our ageing society, it is of increasing urgency that these elderlies are kept safe. This study aims to compare multiple machine learning models that can analyse public PD datasets and extract their motion patterns. From which, FoG can be more efficiently predicted. This study also aims to create a wearable prototype and apply the findings to real life.

21.1.2 Hypothesis

The support vector machine performs the best under the Gaussian kernel with all 3 parameters as training features.

21.2 Literature Review

21.2.1 FoG Characterization

Currently, FoG characterization is done using two main methods, such as gait tests and individual questionnaires. Gait tests include the timed up and go (TUG) test [7] and the Hoehn and Yahr (H&Y) scale [8].

In the second method, individual questionnaires are used. The freezing of gait questionnaire (FoG-Q) is a notable one, comprising 6 questions and utilising a 5-point scale to rank symptom severity [9].

However, both methods are highly inefficient in measuring FoG as it is highly sensitive to environmental triggers, medication, and the patient's mental state. Therefore, there has been research into using wearable inertial measurement units (IMUs) to display exactly the gait types of PD patients.

21.2.2 Utilisation of IMUs in Analysis

This section reports the previous studies which have explored the application of motion sensors on PD patients to accurately predict FoG. Ferster et al. [2] placed 9-axis IMUs (comprising 3D accelerometers, 3D gyroscopes, and 3D magnetometers) on both ankles of the subjects to extract gait features such as stride length and stride duration. Moreover, as FoG exhibits unique frequency ranges, they introduce and discuss frequency features such as dominant frequency, dominant frequency amplitude, and the inverse of the dominant frequency slope of the acceleration data to quantify changes in gait quality. Ferster was able to show specific changes in the stride duration, stride length, dominant frequency, and the inverse of the dominant frequency slope with up to four seconds prior to FoG on all subjects (Fig. 21.1).

Baechlin et al. [10] proposed placing accelerometers at three different parts of the body: the shank, thigh, and lower back, where the wearable computer is attached to.

Alam et al. [11] analysed the vertical ground reaction force using force insoles in patients' shoes to display gait cycles. Pinto et al. [12] again utilised accelerometers and gyroscopes to determine stride time, this time placing the accelerometer at the shank. In summary, a consolidated list of the main IMUs that were considered in this study is provided in Table 21.1.

Many works have utilised motion capture systems to annotate FoG events, synchronising sensor data, and computer analysis to make way for machine learning algorithms. Kuhner et al. [13] performed this experiment, setting up 12 cameras as

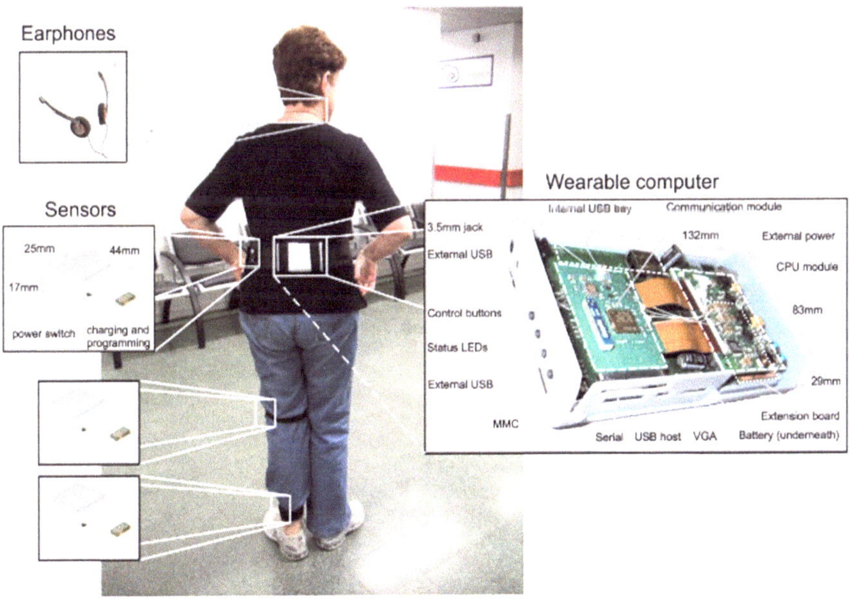

Fig. 21.1 Baechlin et al. 3-axial accelerometer wearable system

Table 21.1 Table of possible IMUs

IMU	Purpose	Measured parameter
Accelerometer	Measuring acceleration	Stride length, stride duration
Gyroscope	Measuring angular velocity	Step festination, gait asymmetry
Flexible goniometer	Measuring body joint angles	Flat foot strike
Force- sensitive insole	Measuring the tension and compression forces that act on the sensor	Gait cycle (not accurate for PD patients who suffer from flat footedness)

well as utilising an inertial measurement suit to create a 'live' system that reduces the latency of data processing.

The above literature confirms the success of utilising accelerometers, gyroscopes, and force insoles to effectively differentiate FoG from normal gait and has greatly helped in designing the proposed approach for this study.

21.3 Methodology

21.3.1 Daphnet Dataset

In the main investigation, the Daphnet freezing in users with Parkinson's disease (hereafter known as the Daphnet dataset) is employed and is the result of a study done by Baechlin et al. [10], carried out by the Laboratory for Gait and Neurodynamics, Department of Neurology, Tel Aviv Sourasky Medical Centre (TASMC). In this experiment, 17 samples were derived from 10 PD patients who were made to do various walking tasks, including walking back and forth in a straight line, doing several 180° turns, random walking including a series of initiated stops and 360° turns and walking simulating daily activities.[1]

Data were recorded using three 3D acceleration sensors attached to the shank, thigh, and the lower back of each subject. The sensors recorded at 64 Hz and transmitted the acceleration data via a Bluetooth link to a wearable computing system that was located at the lower back of the subjects (Fig. 21.2).

[1] Daily activities refer to entering and leaving rooms, getting something to drink, and returning to the starting room with the cup of water.

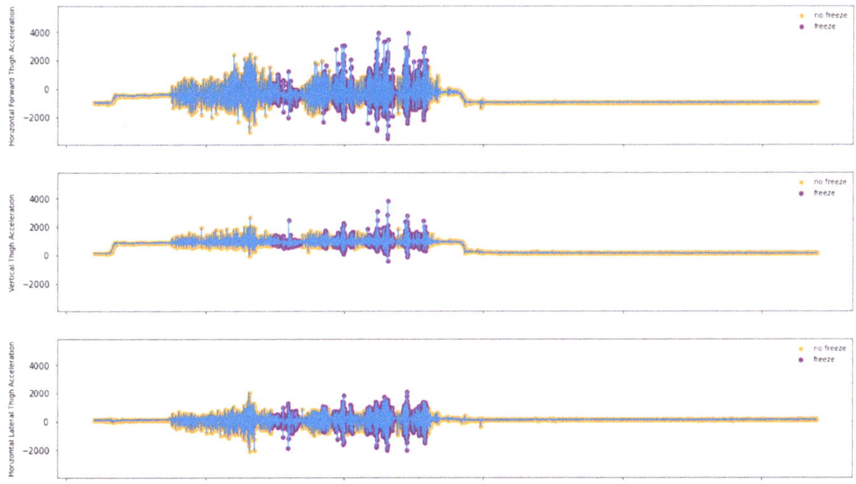

Fig. 21.2 Acceleration data from S03R02.txt. The blue line depicts the acceleration data provided

21.3.2 Signal Processing

This study uses signal processing algorithms to compute a postulated freeze index (FI). Moore et al. [14] defined the FI as the power ratio of freeze band (0.5–3.0 Hz) to locomotor band (3–8 Hz) derived from the frequency spectrum. The FI is postulated as a unitless magnitude of resistance experienced in the direction of acceleration from which it is tabulated. This approach was chosen instead of generic gait parameters because it was a lot more accurate and considered many other crucial factors in our investigation, like the rhythmic movement of the legs in normal and abnormal walking, gait velocity abnormalities, and others.

Moore et al. later identified FoG episodes at the time periods when FI exceeds a certain threshold. The code uses an updated form of Moore's algorithm, wherein it calculates the FI from data from a specific axis (horizontal forward, horizontal lateral, and vertical) from the sensors on the thigh by performing windowing of data with a length of 256 data points and steps up 32 data points. After performing a mean normalisation, the code finds the fast Fourier transform of the normalised window and uses that to compute the FI and supposed power spectral density (PSD). After this, it accounts for a standing case where the $PSD_{threshold} = 2^{11.5}$, and in cases where it is below this threshold, the FI becomes zero. This postulated FI was then considered.

In this experiment, a final function freeze (N) is defined,

where

Freeze (N) = FI postulated from the N-axis acceleration

For this investigation, the axes were defined as follows:

Axis	Axis with respect to the user
X	Horizontal forward
Y	Vertical
Z	Horizontal vertical

21.3.3 SVM Analysis

Unlike Baechlin et al. [10], where the FI with a threshold of 1.5 is used such that a FI greater than the threshold is considered FoG, SVM models have been tested against the derived FI values. This was in-line with a comprehensive review of four different types of machine learning algorithms (support vector machine (SVM); *k*-nearest neighbour (KNN); decision tree (DT); and Naïve Bayes (NB)) in classifying patients with FoG or no FoG done by Aich et al. [15]. They found that the SVM classifier with radial basis function provides the highest accuracy of 91.42% as well as the highest sensitivity and specificity of 90.89% and 91.21%, respectively.

Due to the large amount of data amassed from the signal processing algorithms, the Google Colaboratory tool [16] is used. It allows graphics processing nits (GPUs) of more powerful computers at Google to be utilised. This allows the SVM to run much faster such that results can be obtained faster for the investigation. The ThunderSVM utility [17] developed by the Xtra Computing Group at the NUS School of Computing (SoC) is also used to fully utilise the GPUs.

For the investigation, linear and Gaussian kernels have been employed and these SVMs are tested with either 2-axial freeze indices (i.e. taking freeze indices from 2 of the 3 axes defined) or 3-axial freeze indices (which took into account freeze (*X*), freeze (*Y*), and freeze (*Z*)).

21.3.4 Program

The programme has been developed using Arduino's programming language based on the C++ programming language. To get the best SVM model, a programme called sklearn-porter [18] developed by Darius Morawies, which converts the trained SVM model to *C* code, saved as a model.h file, has been utilised. This model is accessed by the Arduino programme, which also computes the freeze values. The Arduino Board is show below (Fig. 21.3).

This study utilises an Arduino Nano 33 BLE board that is attached to an elastic strap. It contains a 9-axial IMU, comprising a 3D accelerometer, 3D gyroscope, and 3D magnetometer. In the programme, upon a certain freeze event predicted by the SVM, the built-in light-emitting diode (LED) will turn on.

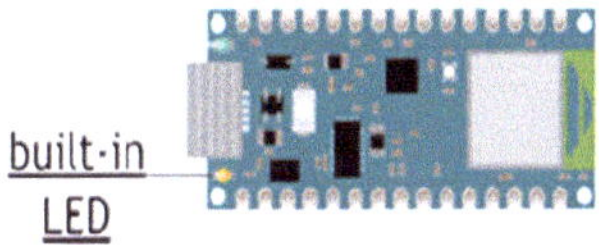

Fig. 21.3 The Arduino Nano 33 BLE pinout diagram. The built-in LED lights up upon detection of a freeze event

21.4 Results And Discussion

Upon performing the signal processing algorithms, the postulated function arrays for freeze (X), freeze (Y), and freeze (Z) are obtained. Below are the freeze indices based on the acceleration data shown in Fig. 21.4.

All 17 samples are merged and taken as the main argument X in the supervised machine learning structure. From here, a provided freeze value for each point given

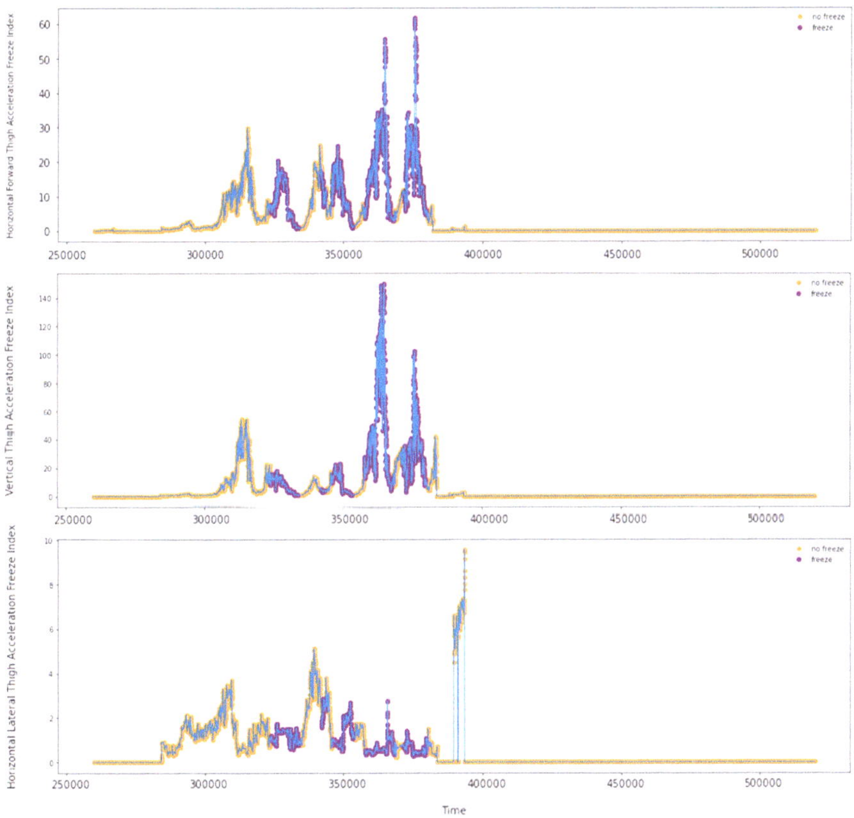

Fig. 21.4 FI tabulations from S03R02.txt

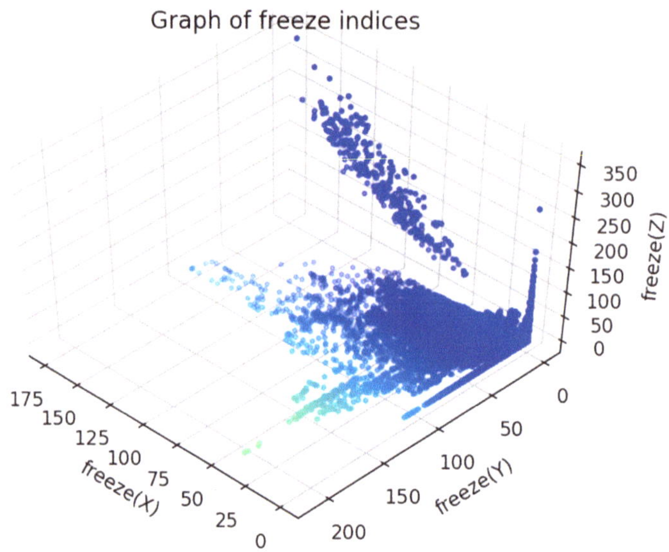

Fig. 21.5 Plot of freeze indices

by the dataset itself has been taken to function as the *y* argument. Below, all freeze indices are plotted against one another (Fig. 21.5).

Performing an SVM analysis using ThunderSVM, we gathered data (see Appendix) and found that the linear kernel with parameters freeze (Y) and freeze (Z) is the best since it is highly accurate, specific, and precise. After conducting an analysis, it was found that the weighted average value for precision and sensitivity/recall as 0.90 and weighted average F1 score as 0.86. The decision boundary function is shown (Fig. 21.6).

21.4.1 Prototype

In the end, a prototype (shown in Fig. 21.7) has been developed to verify the effectiveness of parameters based on public dataset analysis. It consists of an Arduino Nano 33 BLE board with a built-in 9-axial IMU consisting of a 3D accelerometer and 3D gyroscope attached to an elastic band. It is small, light, and very comfortable and is meant to be wrapped around the thigh.

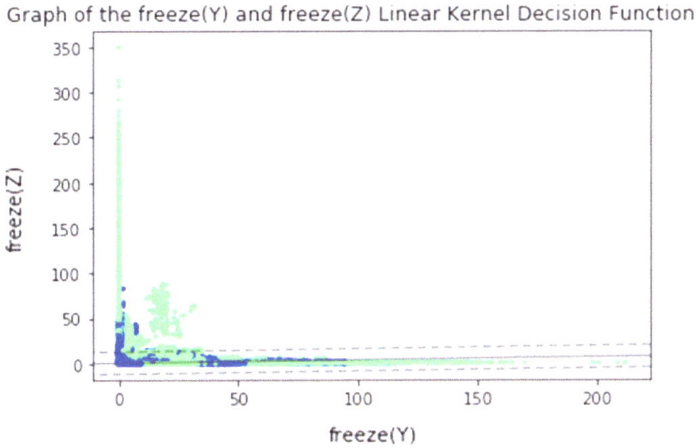

Fig. 21.6 Decision boundary of freeze (Y) and freeze (Z) SVM under linear kernel

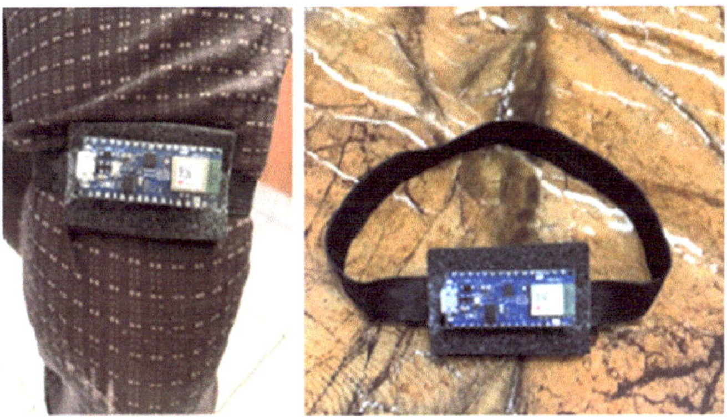

Fig. 21.7 The prototype

21.4.2 Final Application

The final product of this research project was an independent android application developed in Kotlin, which we titled 'Gait Analyser'. This application uses phone sensors to monitor these gait patterns and performs simplistic signal processing. There are two different systems, and users can sign in either as a patient or a caregiver. The patient has the option to activate a 'walk mode' such that when he walks, events such as freeze events or fall events will be detected, and appropriate action will be taken to counteract the effects of these events.

The application was created with the user interface taken into account. Since the majority demographic of the patients are elderly patients, it is ideal to make the application process as simplistic as possible. Buttons and TextViews were enlarged, and

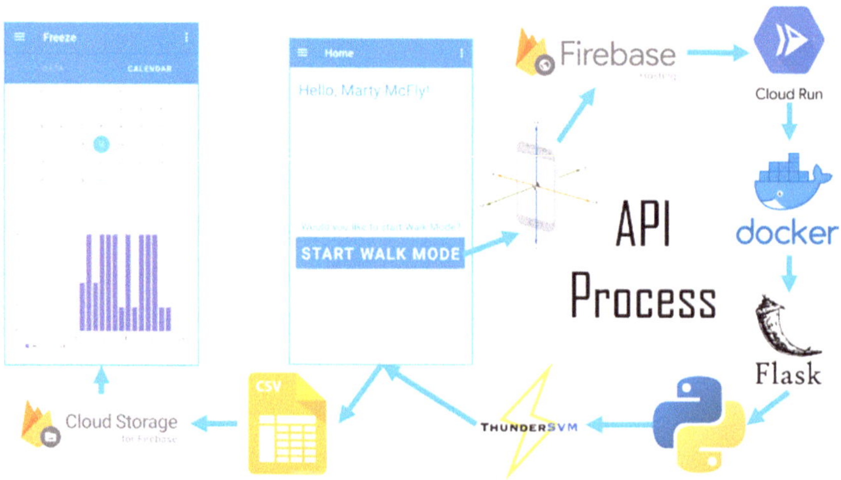

Fig. 21.8 The above details the process the android application takes

authentication processes were not complicated. The application was also connected to Google's Firebase's Cloud Firestore and Cloud Storage services, where user-specific authentication data and freeze data were stored, respectively. Database design was also taken into consideration, and authentication data were stored in a NoSQL database whilst freeze data were stored in the form of a CSV in Firebase Storage.

Users could also monitor the number of freeze events in separated tabs, whether in the form of a calendar widget or a master/detail flow. Bar graphs powered by the MPAndroidChart utility were utilised to show freeze data by hour. Information regarding the app was given in separate pages too (Fig. 21.8).

21.5 Conclusions

After analysing multiple machine learning models that have used acceleration data from accelerometers placed on the thigh, the most suitable parameters for classification are freeze (Y) and freeze (Z). The best model is the linear kernel model in terms of sensitivity. Furthermore, a prototype has been implemented using an Arduino Nano 33 BLE board.

As Parkinson's disease is becoming increasingly prevalent amongst the elderly population in Singapore, more elderlies are at the risk of falling and injuring themselves in cramped HDB spaces. Using the developed algorithm, prototype and mobile application, such falls can be predicted so that they can be mediated in time.

21.6 Future Work

There are a few improvements that could have been done to the project. Due to time limitations, only accelerometers were studied in this project, which could have limited the sensitivity of the device. Multiple sensors can be combined to achieve maximum accuracy including gyroscopes to find out rotational motion parameters as well. Additionally, only postulated freeze index values have been analysed based on the public datasets. Parameters such as stride length and stride duration have not been studied. In the future, such parameters can be taken into account when computing the general gait freeze moment. The sensitivity of the algorithm can also be improved. Since the prototype has already been built, the next phase is to test it in a laboratory with Parkinson's disease patients in order to determine its accuracy. Furthermore, by connecting this system to earbuds and implementing biofeedback via audio, the system can also mediate FoG.

Acknowledgements We would like to thank Professor Arthur Tay, Mr. Lim Yeow Heng and Mr. Lim Teck Choow for their support and encouragement in this study.

Appendix

See Table 21.2

Table 21.2 Table of algorithm evaluation functions with regards to kernels and parameters

Kernel	Parameters	Specificity	Sensitivity/recall	Precision	Accuracy	F_1 score
Linear kernel	Freeze (*X*) and freeze (*Y*)	1.000	0.000	0.000	0.903	0.000
	Freeze (*X*) and freeze (*Z*)	0.975	0.243	0.509	0.904	0.329
	Freeze (*Y*) and freeze (*Z*)	1.000	0.006	0.889	0.904	0.011
	All 3 parameters	1.000	0.002	0.867	0.904	0.005

(continued)

Table 21.2 (continued)

Kernel	Parameters	Specificity	Sensitivity/recall	Precision	Accuracy	F_1 score
Radial basis kernel	Freeze (X) and freeze (Y)	0.990	0.174	0.650	0.911	0.275
	Freeze (X) and freeze (Z)	0.990	0.183	0.664	0.912	0.286
	Freeze (Y) and freeze (Z)	0.991	0.163	0.665	0.911	0.262
	All 3 parameters	0.989	0.260	0.723	0.918	0.382

References

1. Braak, H., Ghebremedhin, E., Rüb, U., Bratzke, H., & Del Tredici, K. (2004). Stages in the development of Parkinson's disease-related pathology. *Cell and Tissue Research, 318*(1), 121–134. https://doi.org/10.1007/s00441-004-0956-9
2. Ferster, M. L., Mazilu, S., & Tröster, G. (2015). Gait parameters change prior to freezing in Parkinson's disease: A data-driven study with wearable inertial units. In: *Proceedings of the 10th EAI international conference on body area networks*. https://doi.org/10.4108/eai.28-9-2015.2261411
3. Nutt, J. G., Bloem, B. R., Giladi, N., Hallett, M., Horak, F. B., & Nieuwboer, A. (2011). Freezing of gait: Moving forward on a mysterious clinical phenomenon. *The Lancet Neurology, 10*(8), 734–744. https://doi.org/10.1016/s1474-4422(11)70143-0
4. Jankovic, J. (2008). Parkinson's disease: Clinical features and diagnosis. *Journal of Neurology, Neurosurgery & Psychiatry, 79*(4), 368–376. https://doi.org/10.1136/jnnp.2007.131045
5. Bloem, B. R., Hausdorff, J. M., Visser, J. E., & Giladi, N. (2004). Falls and freezing of gait in Parkinson's disease: A review of two interconnected, episodic phenomena. *Movement Disorders, 19*(8), 871–884. https://doi.org/10.1002/mds.20115
6. Moore, O., Peretz, C., & Giladi, N. (2007). Freezing of gait affects quality of life of peoples with Parkinson's disease beyond its relationships with mobility and gait. *Movement Disorders, 22*(15), 2192–2195. https://doi.org/10.1002/mds.21659
7. Podsiadlo, D., & Richardson, S. (1991). The timed "Up & Go": A test of basic functional mobility for frail elderly persons. *Journal of the American Geriatrics Society, 39*(2), 142–148. https://doi.org/10.1111/j.1532-5415.1991.tb01616.x
8. Hoehn, M. M., & Yahr, M. D. (1967). Parkinsonism: Onset, progression, and mortality. *Neurology, 17*(5), 427. https://doi.org/10.1212/wnl.17.5.427
9. Rozenfeld, S., Miskovic, A., Nitsch, K. P., & Ehrlich-Jones, L. (2017). Measurement characteristics and clinical utility of the freezing of gait questionnaire in individuals with Parkinson disease. *Archives of Physical Medicine and Rehabilitation, 98*(10), 2106–2107. https://doi.org/10.1016/j.apmr.2017.04.027
10. Bachlin, M., Plotnik, M., Roggen, D., Maidan, I., Hausdorff, J. M., Giladi, N., & Troster, G. (2010). Wearable assistant for Parkinson's disease patients with the freezing of gait symptom. *IEEE Transactions on Information Technology in Biomedicine, 14*(2), 436–446. https://doi.org/10.1109/titb.2009.2036165
11. Alam, M. N., Garg, A., Munia, T. T. K., Fazel-Rezai, R., & Tavakolian, K. (2017). Vertical ground reaction force marker for Parkinson's disease. *PLoS ONE, 12*(5), e0175951. https://doi.org/10.1371/journal.pone.0175951

12. Pinto, C., Schuch, C. P., Balbinot, G., Salazar, A. P., Hennig, E. M., Kleiner, A. F. R., & Pagnussat, A. S. (2019). Movement smoothness during a functional mobility task in subjects with Parkinson's disease and freezing of gait—An analysis using inertial measurement units. *Journal of NeuroEngineering and Rehabilitation, 16*(1). https://doi.org/10.1186/s12984-019-0579-8
13. Kuhner, A., Schubert, T., Maurer, C., & Burgard, W. (2017). An online system for tracking the performance of Parkinson's patients. In: 2017 *IEEE/RSJ international conference on intelligent robots and systems (IROS).* https://doi.org/10.1109/iros.2017.8205977
14. Moore, S. T., MacDougall, H. G., & Ondo, W. G. (2008). Ambulatory monitoring of freezing of gait in Parkinson's disease. *Journal of Neuroscience Methods, 167*(2), 340–348. https://doi.org/10.1016/j.jneumeth.2007.08.023
15. Aich, S., Pradhan, P., Park, J., Sethi, N., Vathsa, V., & Kim, H.-C. (2018). A validation study of freezing of gait (FoG) detection and machine-learning-based FoG prediction using estimated gait characteristics with a wearable accelerometer. *Sensors, 18*(10), 3287. https://doi.org/10.3390/s18103287
16. Bisong E. (2019) Google colaboratory. In *building machine learning and deep learning models on google cloud platform.* Apress, Berkeley, CA. https://doi.org/10.1007/978-1-4842-4470-8_7
17. Wen, Z., Shi, J., Li, Q., He, B., & Chen, J. (2018). ThunderSVM: A fast SVM library on GPUs and CPUs. *Journal of Machine Learning Research, 19*, 797–801.
18. Morawiec, D. (n.d.). Sklearn-porter—Transpile trained scikit-learn estimators to C, Java, JavaScript and others. https://github.com/nok/sklearn-porter

Chapter 22
An Overview of Transparent Insulation Materials and Its Future Developments

Abhishek Nautiyal, Prakhar Duggal, and Ravinder Kumar Tomar

Abstract TIMs or commonly known as transparent insulation material-based systems are among one of the most promising technologies for providing energy efficient housing along with solar transmission. This paper provides a thorough overview on some of the new potential; TIM-based projects discussing their details and place of their applications. One of the main goals of this paper is to provide a brief introduction for newcomers on the basic understanding of these systems by characterizing TI systems, on the basis of their geometric shape and type of material used. Other than these tasks, focus has also been showed on stating common barriers in enhancement and development of TIM-based systems, for further improvement of systems based on this technology. A comprehensive literature review has also been completed for the purpose of providing all information in a common place to save some valuable time.

Keywords Transparent insulation materials · Energy efficient housing · Transparent nanostructured adhesive film · Building glazing systems

22.1 Introduction

Since these past few decade's consequences of increase in earth's temperature are being felt worldwide, in January 2021, global ocean and land temperature were 0.80 °C above 20th century average, making it as the seventh warmest January in 142 years as per global records. If this is not enough to scare you, it is also specified that there is a 99% chance of 2021 being among the tenth warmest year on record, with a prediction of a further increase of temperature in future years.

Fortunately, because of several efforts, public awareness on important aspects like carbon footprint is increasing helping in achieving a significant change, and countries have also pledged to lower their emissions by several folds in upcoming years. Carbon footprint is regarded as a trend of carbon dioxide concentration in

A. Nautiyal (✉) · P. Duggal · R. K. Tomar
Department of Civil Engineering, Amity University, Noida, Uttar Pradesh, India
e-mail: abhishek.nautiyal@s.amity.edu

H. Guo et al. (eds.), *IRC-SET 2021*, https://doi.org/10.1007/978-981-16-9869-9_22

earth's atmosphere and is hence an important awareness to tackle above-mentioned problems. To address these issues, governments all around globe are encouraging and funding several promising projects, among these methods, TIMs are considered among most influential for bringing global change as they can capture and expeditiously retain solar heat energy by minimizing loss of heat and providing excellent thermal performance. But Tim's ability to abate thermal loss or providing solar transmission is highly dependent on its operating temperature, geometrical structure, or material types. Thus, in order to utilize maximum potential of TIM systems, these factors need to be discussed.

22.2 Tim Types

22.2.1 Glazing Parallel Structure

In this transparent insulation type of material, the absorber parallel structure irrespective of its main sub-types commonly referred to as multiple sheets or slab types are used as cover by placing them parallel with absorber. Furthermore, in this type of TIM-based system, heat losses are usually from radiative nature of heat transfers followed by those that are caused due to convection. But to reduce these heat losses, one would have to increase total number of parallel sheets which will certainly result in reduction of heat losses, but the major problem occurs when this action will simultaneously result in reduction of solar gains. Hence affirming the major drawback of using this method, this phenomenon is caused due to the increase in total no of optical reflections but will result in inability of this system for having lower heat type of loss and an overall high-solar gain simultaneously. Although this statement is true for all parallel glazing structures, but the effect varies on the selected sub-types, for example, multiple sheets system is further sub-divided into glass and polymer types. Although they belong to same sub-branch, their characteristics are somewhat different similarly the slab system also mostly consists of methylmethacrylate, but the effect of above-mentioned property is different than in the case of multiple sheets structure sub-types.

22.2.2 Glazing Perpendicular Structure

In this type of TIM system, the absorber perpendicular structure type is mainly separated in three main sub-types commonly known as capillaries, honeycombs, and V parallel slats. They are placed perpendicular to absorber so that they can provide forward type reflection of the incoming radiation resulting in minimum reflection losses with an overall reduced heat loss. What makes this system better than glazing parallel structure is that in this structure type, we can use an appropriate type of aspect ratio to efficiently censor the heat losses that are produced by convection. Unlike glazing parallel structure type, this method has many vastly used sub-types, mainly because of its property to efficiently suppress heat loss on basis of convection through an appropriate aspect ratio; this gives rise to vastly different sub-types. To properly explain above-mentioned sentence, let us investigate the various sub-types of this method in detail, the capillaries type of observer perpendicular system uses glass and polycarbonate-based individual system as in a system based on glass or on polycarbonate. Similarly, the honeycomb type of absorber perpendicular system uses polymethylmethacrylate, polyethylene, polyestercarbonate, polytetrafluorethylene, polyethesulfone, glass, polycarbonate, polypropylene, cellulose triacetate as its main constituent material to obtain dissimilar individual system. Lastly, the parallel and V slat type of absorbable perpendicular system uses glass or chlorinated ethylene propylene Teflon as a main constituent material to obtain individual systems (Fig. 22.1).

22.2.3 Cavity Structure

These types of transparent insulation material are mainly separated into two types, such as translucent foam type and multichannel duct plate type of materials. These structures are obtained by combining absorber perpendicular and absorber parallel structures. Along with that the main property of these structures as in the property that separates them from others is that they can censor convective heat loss more efficiently but even though they handles convective heat loss in a better way, these structures are subjected to a major setback regarding its high-reflection loss, basically meaning that execution of the system results in comparatively higher reflective surface, which results in the reduced transmittance, akin to that of a multiple film type of cover. Translucent foam kind mostly uses an acrylic base system, and multiple research is currently being done on this type of TIM system for further progress, but as of know, no significant progress is achieved in completely or significantly removing its main major disadvantage.

22.2.4 Quasi-homogeneous Structures

These types of TIMs have a homogeneous distribution off air in the material. The main types of these materials are fiber, granular aerogel, monolithic aerogel, and xerogel-based quasi-homogeneous structure. These types of structures have incredibly low value of reflectance, while the transmission of losses is mainly because of scattering and retention of incoming radiation. Heat loss by convection and radiation through air is minimum, heat conduction because of the solid is also very less in these quasi-homogeneous structures. It is also worth noting that they are among the only type of materials, which cannot be among single inclusive type. They are

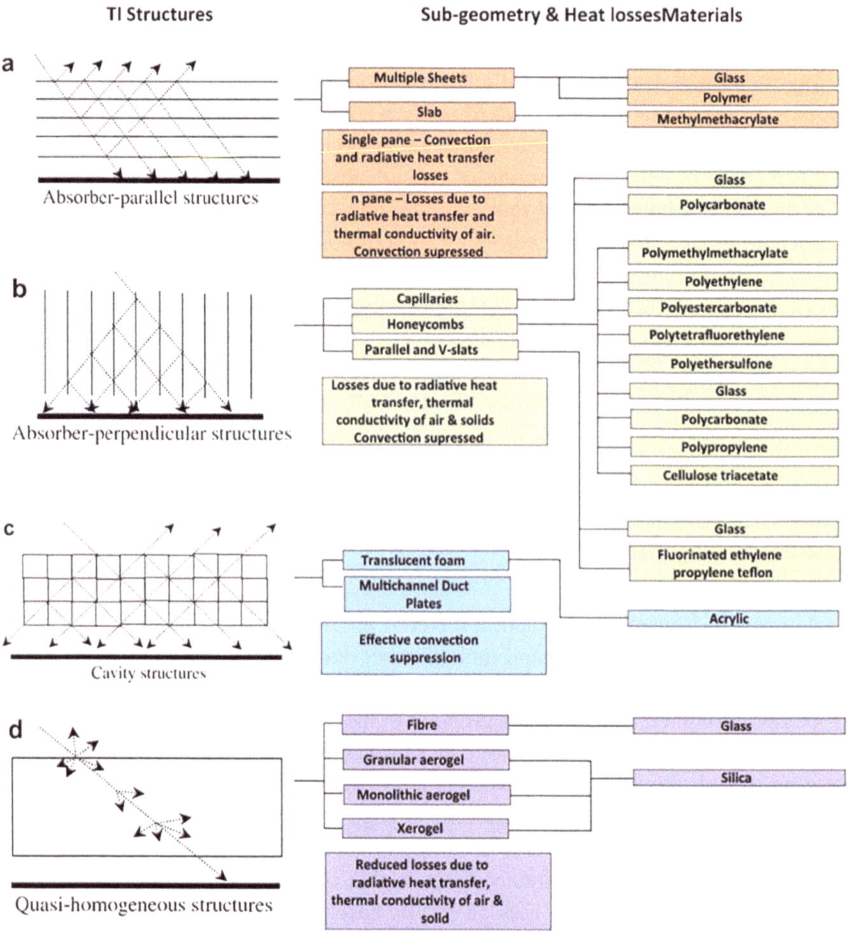

Fig. 22.1 Types of transparent insulation materials

getting a lot of attention these days, thanks to new research on silica aerogel, which is essentially a translucent material for filling the air gap in a double-glazed unit. The type of aerogel with a high porosity usually has an extremely low-heat conductivity, for example, it is in the range of (0.005–0.02 W/mK). In most popular types of aerogel, although monolithic silica aerogel is transparent enough to see through, its fragility E and greater cost limit its commercial adoption. Along with this, the granular aerogel has an extremely low transmittance for solar use, currently it is less than 0.50 due to the bulk scattering effect ultimately limiting most of its use in skylights only. Currently, aerogels and the glazing perpendicular type of transparent insulation material are among the most popular subjects for research and industry among all type of transparent insulation materials.

22.3 Applications of TIMs in Buildings

In general, there are three primary ways to use TIMs in buildings: the first, and probably the earliest, uses TIM under the front glass cover of a flat-plate solar collector; the second, which is most often referred to as a solar wall, uses TIM in front of a thermal mass wall or a thermal mass roof; and the third method uses TIMs in air cavity for flat-plate solar collector; and finally, TIM is used for air cavity of a window system or glazed facade in the third approach (Fig. 22.2).

22.3.1 Application of TIM for Flat-Plate Type Solar Collectors

The research on this sort of TIM system began in the 1960s, with the primary focus on use of honeycomb structures in solar absorbers to suppress convection. On begin with, only plastic/glass honeycombs were applied to flat-plate collectors. But recently, application of capillaries and silica aerogel-based systems is gaining popularity. They are used in space heating, water heating and are usually used to transform solar into useful heat predominating in producing heated water for industry, residential, and commercial applications. According to Wagner and Rommel (1992), it was mentioned that working temperatures between 40 and 80 °C are the only sustainable temperatures and that higher working temperature up to 260 °C can be achieved using glass honeycombs, but plastic-based honeycombs will melt at 120 °C. As the progress was made in its implementation researchers, experiments revealed that flat solar plate collectors with MSA integration could achieve efficiency of more than 60% at temperatures below 80 °C, bringing yearly energy-based gains to 41%. A TIM-based flat-plate collector has similar thermal-based efficiency, similar to a vacuum tube collector and its efficiency is projected to increase further as new research is being vastly funded in order to find a more energy efficient housing.

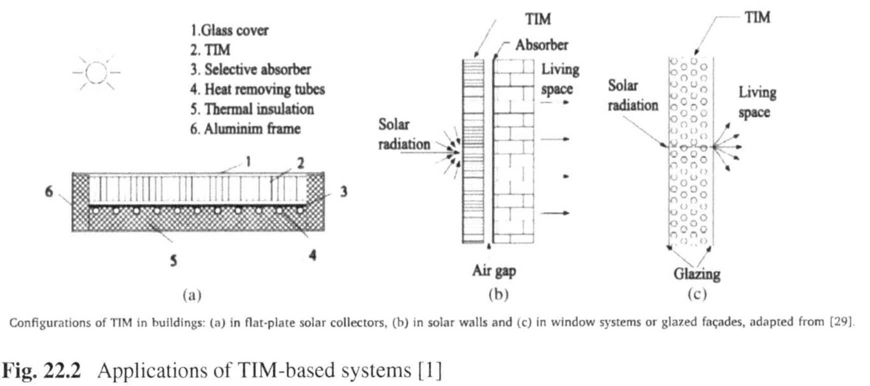

Fig. 22.2 Applications of TIM-based systems [1]

22.3.2 Application of TIMs in Solar Walls

When utilized in front of a masonry or concrete wall, TIM-based solutions and solar-based energy are always stored and transmitted in the thermal mass, while heat loss through the front surface to the external environment is reduced. If we take a roof system made up of square celled, air filled honeycomb material, laid over a blackened absorber along with concrete slab for thermal storage. Then for a honeycomb based upon 0.076 mm thick polycarbonate film, τ (θ) value was calculated as 0.9, and for team made up of 1 mm wide acrylic sheet, value was recorded as 0.60, these values are some of the initially obtained data from implementation of this setup. Furthermore, the IEA program on solar heating and cooling stated that the implementation of TIMs in solar wall is encouraged as a building renovation measure throughout Europe, further stating that forced ventilation and shading devices are to be used for solving problems of overheating in summer. Solar walls made of phase changing material (PCM) are being studied extensively in substitute of standard masonry or concrete walls. According to a study, 12 mm PMMA capillaries-based TIM, when applied to a double-glazed window in front of PCM in a glass container, only lost around 1% of the month's energy, showing the incredible benefits obtained by implementation of this system.

22.3.3 Application of TIMs for Window Systems or Glazed Facades

TIMs have shown tremendous potential to minimize overall heat loss while retaining solar and light transmittance when used in windows. Using TIMs can result in improved visual comfort level, because they have more homogeneous internal daylight distribution and its use result in reduction of both, space heating energy and lighting consumption. Application of these systems shows significant improvements in comparison with other type of insulation systems, for example, when a test on a residential home was conducted, then for a building with a 100 mm thick layer of honeycomb along with 3 mm square cells (encapsulated by an external glass layer), it is shown that in comparison with best performance practice of that time in winter timeframe for energy, the TIM glazing system required 40% less energy. However, application of this system was not much appreciated in 1994 as according to one researcher, the use of TIM glazing is usually restricted to skylights, industrial facades, commercial facades, and atriums. Reason given for this explanation was that geometric structure resulted in obstruction of clear view to the outside. In 1994, another statement was stated that for increasing visible transmittance for a TIM-based glazing, the capillary size needs to be increased while its thickness is to be reduced, maintaining distance was also suggested as one of the possible steps. From an application point of view in early 2000, it was suggested that the imperfections like rough or melted edges must be addressed by improving the production quality to make sure that the hindrance in clarity can be removed. But as of 2008, wide range application of transparent silica aerogel started to become popular for TIM glazing system, and since then, the craze has only increased even going as far as to use nanomaterial-based systems to further increase the quality which we will discuss at the end. According to an application of glazed aerogel facade in Dubai, it was revealed that when this glazing was paired with reflecting components (holographic optical elements) in glass laminate, aerogel glazed facade might assist reduce air conditioning demands by 17%, saving up to 26%.

22.4 Barriers in Development of TIM Systems

Overheating and lower working temperatures are one of the biggest issues for application of TIMs at higher temperature. The working temperature of plastic TIMs is 80 °C or less, whereas the working temperature for solar collectors can reach 250 °C, causing TIM melting and system failure. This is especially true if the TIM is spread across a large area of South faced building facades but to remove this problem, either we would require passive overheating protection devices or expensive shading would be required increasing to overall cost of the application.

Initially, problems like imperfections were a major issue for TIM application as the measured, and the estimated value of solar transmission was different because of this particular reason, but in 2003, profile extrusion process solved problems like improper glue dispensing and this process also helped cost reduction.

In one of most popularly researched topic of silica aerogels, barriers to it wider implementation are problems like long processing time, highly fragile nature, tendency to absorb water, and difficulties in production.

Another barrier would be that in monolithic silica aerogel, the optical transparency is moderately lacking in comparison with conventional glazing unit which leads to missed opportunity of its implementation instead of conventional glazing.

Lastly, another major setback is the extremely fragile nature of silica aerogel, as it is difficult to produce large cracks free pieces of monolithic-based silica type of aerogel, its sizes restricted to 0.6 m × 0.6 m as of 2015, because of which it is mainly used for research, even if this size would have increased with a few years, but still, its wide implementation would be restricted.

22.5 Literature Review

(1) ***Struhala* et al. *(2018)***

When compared to traditional insulation, translucent insulation has a reduced environmental impact. The highest achieved difference is 84% approx [2].

in favor of transparent insulation is achieved with South oriented facade assemblies that utilize the highest solar energy gains.

(2) ***Wong* et al. *(2007)***

Approximately, 100–200 kWh/m2 per year of solar gains have been measured or predicted for different TIM systems [3]. For a 10% discount, the simple payback term calculation for a given installation gives a payback period of fewer than 8 years.

(3) ***Braun* et al. *(1992)***

Simulation technique states that for vertical surfacing TI walls, they can be characterized by solar conversion efficiency of 30%, solar energy gains 100–200 kWh/m^2 with a large application potential [4].

(4) ***Cekon* et al. *(2017)***

By using a combination of selective absorber and transparent type of insulation materials, it is observed from a thermal standpoint that sorbent properties could help in improving the thermal properties significantly in comparison with concepts using high-emissivity absorber. From a thermal resistance point of view, this system provides comparable performance in comparison with conventional thermal insulation [5].

(5) ***Cekon* et al. *(2018)***

It compares two variants of solar absorbers (non-selective SA and selective SA) as combined with transparent insulation on developed TIF prototypes. The nSA concept showed almost constant thermal resistance between 1.20 and 1.24(m2) (K) /W, while depending on temperature gradient. In SA concept, a max of 1.73 (m2) (K) /W of thermal resistance is recorded at 10^{K} gradient and min value of 1.53 (m2) (K) /W at 40 K gradient [6].

(6) ***Čekon* et al. *(2018)***

Contemporary opaque facades have negligible solar heat gains, while the proposed TIF concept (first has one selective absorber (SA) with low-e coating, while the second also has a non-selective absorber (nSA) with high-emissivity level), has theoretical solar heat gains up to 178 kWh m^{-2} a^{-1} [7]. When compared to TIF with SA, TIF with nSA has 18% worse thermal performance and 3 percent higher solar heat gains.

(7) ***Osorioa* et al. *(2017)***

The efficiencies of conventional solar collectors decrease with an increment in tab (absorber's temperature) as a result of continued growth in heat losses. A slight reduction of efficiency for TIM-incorporated collectors is observed at low-absorber temperature, due to a solar radiation attenuation induced by TIM. At high tab, the TIM reduces the thermal losses leading to higher solar collectors' efficiencies. This behavior is important considering because high-operating temperatures are usually desired to obtain high-thermal cycles efficiencies [8].

(8) ***Paneri* et al. *(2019)***

According to a review of current TI systems, absorber perpendicular and quasi-homogeneous geometrical shapes are chosen for TIMs. Aerogel, a quasi-homogeneous geometrical structure, has emerged as one of the most effective TIMs, with a U value as low as 0.25 W/(m^2K) and a thickness of 70 mm [9].

(9) ***Zhu* et al. *(2007)***

Because of the low cost of the precursor, granulated aerogel is expected to be more useful in thermal energy engineering. Further research in ambient drying process led to discovery of xerogel which is a type of dense monolithic aerogel in which drying control chemical additives are added prior to drying [10].

(10) ***Jensen* et al. *(2004)***

When evacuated aerogel glazing of 55 × 55 cm^2 is made, the solar and daylight transmittance nature of aerogel glazing is optimized by the means of low-type iron glass covers with an anti-reflection coating. When exposed to direct non-perpendicular radiation, the U value is 0.66 Wm^{-2} K^{-1}, but after accounting for heat loss, it is 0.72 Wm^{-2} K^{-1}, as determined by hotbox measurements on a window with four aerogel panes linked in an interim frame [11].

(11) ***Cuce* et al. *(2014)***

A decrease of up to 90% in heat loss through external walls of residential buildings can be easily achieved by using a 20 mm aerogel insulation. Aerogel is 50% slimmer than polyurethane and 70% more than glass wool in construction. In glazing applications, monolithic aerogel performs better than granular aerogel because of its better light transmittance [12].

(12) ***Sun* et al. *(2015)***

When the aspect ratio of a PS-TIM structure's cell is less than 0.35, the free convection is fully suppressed despite temperature difference between two surface panes, design a suitable interval distance of the parallel slats in the air cavity to maintain the Nusselt number less than 1.2 to provide good reduction of convective heat transfer coefficient. The results of a convection, conduction, and radiation model show that a 3 mm-cell structure can reach 35%-46% decrease of thermal conductance while a 10 mm-cell structure can reduce it by 16%-18% [13].

(13) ***Osorioa* et al. *(2018)***

It was discovered that the integration of a TIM into parabolic trough collectors (PTCs) and central receiver (CR) collectors leads to a higher-thermal efficiency because of reduction in heat losses. It was also discovered that in typical concentrated solar collectors, high-concentration ratios are required to maintain high efficiency [14].

(14) ***Cekon* et al. *(2018)***

Long-term optical characterization of multi-wall panels utilizing co-extruded polycarbonate as a transparent insulation material and long-term solar radiation rate monitoring, its evaluation with three different multi-wall panels indicates that the internal part of structures and geometry currently used to produce polycarbonate-based materials may affect the solar transmittance functions of such components [15].

(15) ***Wallner* et al. *(2004)***

Cellulose acetate (CA) and cellulose triacetate (CTA) were identified as interesting alternatives for mainly used small, celled TI structures made of PC (polycarbonate). TI of $30m^2$ was produced and applied to the south oriented facade of an ultra-low-energy solar house which yielded excellent results as a wall system [16].

(16) ***Sun* et al. *(2018)***

An advance method that uses a matrix of bidirectional scatter distribution functions to characterize optical properties for complex window systems is regarded as more appropriate for predicting light transmission through glazing systems with TIMs [1].

22.6 Projects and Potential Future Developments

22.6.1 Insulating Particulate Coatings (2020)

The fundamental goal of the research is to create a thermal barrier made of silicon dioxide nanoparticles put on glass using a supersonic aerosol spraying procedure. The layer would reduce heat loss while remaining transparent, resulting in lower cost than conventional silica aerogels. This process will deposit a second layer that will reflect thermal radiation, as well as layers of thick polymers to limit adhesion and condensation on window panes. The life of this coating is 20 years, meanwhile this coating also claims to be resistant from damage by peeling, scratching, or by water vapor freezing process in pores of silica layer. https://arpa-e.energy.gov/technologies/projects/insulating-particulate-coatings [17].

22.6.2 Brightwall (2019)

It is a R&D project sponsored by the EC through seventh framework program. The major goal of this project was to create a completely new building material called transparent concrete, which comes in the shape of a concrete sandwich panel and has a lot of architectural and esthetic possibilities. The availability of natural light satisfies demands for energy-based efficiency and indoor environment. The finished product mixes high-strength concrete with optical fibers that are carefully inserted and may be utilized in both façade and weight bearing applications. The novel material's major use is in the construction of daylight-transferring external walls in new and renovated structures. Aside from lowering lighting costs and reducing emissions, the Brightwall allows for the delivery of high-quality natural light into buildings, which is beneficial to people's health. Brightwall addresses S&T constraints linked to the manufacture of prefabricated wall sandwich panels with embedded optical guides and thermal insulation from a conceptual standpoint, focusing on radical advances that enable dynamic control of daylight transmission using liquid crystals [18].

22.6.3 Engineer Polymer Film (2019)

The fundamental goal of the project was to create a transparent adhesive film which could easily be applied to single pane windows to reduce heat loss from heated rooms in the winter. This is accomplished by employing a revolutionary method based on a novel class of non-porous materials. Nanoparticles were employed to reflect heat and provide improved thermal insulation in this manner. Furthermore, because the transparent film does not hinder visible light, light can be transmitted through the window to brighten the interior. Along with this, soundproofing of the windows is also improved by implementation of this method. https://arpa-e.energy.gov/technologies/projects/engineered-polymer-film [19].

22.6.4 NanoSD (Nanobubble Thermal Barrier) (2019)

The purpose of this research is to create a nanostructured coating that is transparent and thermally insulating. This can help in reducing heat loss for single pane windows. This material is made by producing hollow polymer or ceramic nanobubbles, then employing heat and compression to consolidate them into a thick lattice structure. Because air makes majority of structure, the resulting nanobubble structure has excellent thermal barrier qualities. The film produced is transparent exhibiting flexible, lightweight, chemical/fire resistant, condensation resistant, and soundproofing properties. The goal of this project is to be highly cost-effective and reduced cost of window coating to less than \$5 per sq ft for this particular product. https://arpa-e.energy.gov/technologies/projects/nanobubble-thermal-barrier [20].

22.6.5 Nanoinsulate (2014)

This project successfully developed a durable cost-effective type; opaque transparent-based vacuum insulation panel (VIP's) using a new nanotech based on a combination of aerogel and nanofoams composites resulting in formation of high-barrier films that are very energy efficient. Aerogel have higher transparency than these composites, but they give three times higher-compressive strength. When a transparent panel was developed using this concept 9 mW/mK of thermal conductivity and U value of 0.450 W/m^{2}K at an overall thickness of 20 mm was observed as result [21].

22.6.6 Dynamic IR Window Film (2019)

The goal of this project is to develop nanomaterial-based, low-cost type of technology that can be used for flexible window frames to improve thermal insulation along with solar heat gains. The main nanomaterial for this project has incorporated two materials, and first thermal resistance is increased by using low-cost nanosheet, and secondly, a comparatively new kind of nanomaterial is used, this will allow infrared radiation for generating heat, which is allowed to pass through windows, when it is cold outside. In case of hot weather, the material will block solar infrared radiation from warming interior. This material also reflects thermal radiation, and its tunable emissivity helps it retain energy and provide insulation. In this project, the dynamic infrared radiation is passive in nature and only relies on environmental changes instead of electronic systems to achieve its results. https://arpa-e.energy.gov/technologies/projects/dynamic-ir-window-film [22].

22.6.7 Aerogel Insulated Pane (2020)

The goal of this project is to develop as silica aerogel-based insulation windowpane, for enhancing the efficiency of single pane windows. In this project, the silica aerogel has a combination of higher visible light transmittance, lower thermal conductivity, and lower haze. This task is accomplished by sandwiching an aerogel sheet between two glass panes to create a double-glazed pane structure. This process also employs a novel supercritical drying process that vastly reduces the time it takes for an aerogel to dry, hence improving productivity and lowering costs. Where weight and thickness prevent replacement with typical double pain units at a significantly reduced cost, Aspen aerogels window panes can be utilized to replace windows with single panes. https://arpa-e.energy.gov/technologies/projects/aerogel-insulated-pane [23].

22.6.8 *Thinner Coatings (2021)*

This research aims to use nanotechnology to create a transparent thermally insulating barrier with a thinner coating for reducing heat losses through single panes of glass, leaving behind a transparent, and thermally insulating porous oxide layer. Along with lowering heat loss and filtering damaging UV rays, the coating will help reducing water condensation on inside part of window surface. Another significant aspect of this project is the development of a scalable high-temperature spray-on technology for depositing coating onto glass at a low cost in the factory. https://arpa-e.energy.gov/technologies/projects/thinner-coatings [24].

22.7 Potential Developments (2021)

This covers few of the current developments incorporating uses of this material. Happold and Mark Dowson undertook an Engineering Doctorate, in collaboration with Brunel University, focusing on testing and proper development of a new building fabric technology, using translucent granular aerogel encapsulated inside the clear polycarbonate sheets for improving thermal performance of new and existing buildings. Three main types of technologies which are developed with promising initial results are

- Retrofitting polycarbonate panels filled with aerogel granules to existing windows for improving their thermal performance.
- Retrofitting of a translucent aerogel panel to the outside of a concrete wall, for the purpose of trapping solar energy to passively warm the buildings.
- An aerogel solar collector incorporating a translucent polycarbonate cover filled with granular aerogel insulation.

22.8 Conclusion

In this paper, a comprehensive review on transparent insulation materials (TIMs) and their new potential implementation in buildings has been discussed. Furthermore, development projects concerning use of nanostructured transparent adhesive film, using different principles are discussed in great detail for energy efficient housing. Along with these systems, system based on silica aerogel and translucent concrete sandwich panels are also being researched globally, for energy efficient housing. Moreover, transparent insulation materials are receiving major push all around world for reducing the global emissions and for also providing notable energy savings in a long run. The following points are remarkable for a brief overview of results from my literature review.

- To explore the optical features of sophisticated window systems, one of the most precise methods for estimating light transmission through glazing systems with TIMs is to use a bidirectional scatter distribution-based matrix.
- A research showed that around 90% reduction in heat losses through external walls of residential building can be achieved by 20 mm thick layer of aerogel insulation. In addition, the findings of this study show that monolithic aerogel performs better than granular aerogel in glazing applications, because of its better light transmittance.
- When a selective absorber and a transparent insulating material are used together, it is observed that with respect to thermal characteristics sorbent properties could result in an improvement of thermal properties significantly, compared to concepts using high-emissivity type of absorber
- In an overall comparison with normal insulation, it was discovered that transparent insulation material-based systems have lower environmental impact, achieving difference of 84% with south oriented facade for highest solar gains
- To date, studies of various TI systems have shown that absorber perpendicular and quasi-homogeneous geometrical shapes are favored for TIMs. With a thickness of 70 mm, quasi-homogeneous geometrical formations, particularly aerogel, have emerged as one of the most effective TIMs for obtaining U values as low as 0.25 W/m^2 K.
- When comparing selective absorber (SA) with low-e coating and non-selective absorber (nSA) with high-emissivity level. TIF concept with nSA has shown 18% worse thermal performance and 3% higher solar gains in comparison with SA type of TIF system. Overall, transparent insulation material-based systems are predicted to grow significantly in future providing a better and more energy efficient environment.

References

1. Sun, Y., Wilson, R., & Wu, Y. A review of transparent insulation material (TIM) for building energy saving and daylight comfort.
2. Struhala, K., Cekon, M., & Slavik, R. Life cycle assessment of solar façade concepts based on transparent insulation materials.
3. Wong, I. L., Eames, P. C., & Perera, P. S. A review of transparent insulation systems and the evaluation of payback period for building applications.
4. Braun, P. O., Goetzberger, A., Shmid, J., & Staul, W. Transparent insulation of building facades steps from research to commercial applications.
5. Cekon, M., & Curpek, J. A transparent insulation façade enhanced with a selective absorber: A cooling energy load and validated building energy performance prediction model.
6. Cekon, M., & Slavik, R. A Non-ventilated solar façade concept based on selective and transparent insulation material integration: An experimental study.
7. Cekon, M. Coupled transparent insulation system with low emissivity solar absorber: An experimentally validated building energy simulation study.
8. Osorioa, J. D., Rivera-Alvarez, A., Girurugwiroa, P., Yanga, S., Hovsapiand, R., & Ordonez, J. C. Integration of transparent insulation materials into solar collector devices.

9. Paneri, A., Wong, I. L., & Burek, S. Transparent insulation materials: An overview on past, present and future developments.
10. Zhu, Q., Li, Y., & Qiu, Z. Research progress on aerogels as transparent insulation materials.
11. Jenson, K. I., Schultz, J. M., & Kristiansen, F. H. Development of windows based on highly insulating aerogel glazing's.
12. Cuce, E., Cuce, P. M., Wood, C. J., & Riffat, S. B. Aerogel based thermal insulation superinsulation in buildings: A comprehensive review.
13. Sun, Y., Wilson, R., & Zhu, Z. Thermal and optical analysis of a novel glazing facade system with parallel slats transparent insulation material (PS-TIM).
14. Osorioa, J. D., Rivera-Alvarezc, A., & Ordoneze, J. C. Effect of the concentration ratio on energetic and exegetic performance of concentrating solar collectors with integrated transparent insulation materials.
15. Cekon, M., & Struhala, K. Thermal energy and life-cycle aspects of a transparent insulation façade: a case study.
16. Waller, G. M., Lang, R. W., Schobermayr, H., Hegedys, H., & Hausner, R. Development and application demonstration of a novel polymer film based transparent insulation wall heating system.
17. Insulating particulate coatings (2020). https://arpa-e.energy.gov/technologies/projects/insulating-particulate-coatings
18. Brightwall (2019). https://www.dti.dk/projects/brightwall/innovative-tailored-concrete-enabling-transmission-and-use-of-daylight-in-buildings/36685
19. Engineer polymer film (2019). https://arpa-e.energy.gov/technologies/projects/engineered-polymer-film
20. NanoSD (Nanobubble Thermal Barrier) (2019). https://arpa-e.energy.gov/technologies/projects/nanobubble-thermal-barrier
21. Nanoinsulate (2014). https://cordis.europa.eu/project/id/260086/reporting
22. Dynamic IR Window Film (2019).https://arpa-e.energy.gov/technologies/projects/dynamic-ir-window-film
23. Aerogel Insulated Pane (2020). https://arpa-e.energy.gov/technologies/projects/aerogel-insulated-pane
24. Thinner Coatings (2021). https://arpa-e.energy.gov/technologies/projects/thinner-coatings

Chapter 23
Study of Variation in Alccofine on Concrete Strength

Rahul Pandit, Prakhar Duggal, and R. K. Tomar

Abstract This paper comprises of various researches on Alccofine focussing on cements partial replacement in concrete. The main aim is to enhance, compressive strength along with its durability, to obtain in comparison with standard concrete mix, it has a much longer life span. Alccofine {GGBS} combination with fly ash is considered amongst extremely durable concrete-based composite. Alccofine is basically a supplementary cementitious material, it is an exceptionally fine particle in comparison with cement, so their binding properties show excellent performance, in some cases many times that of regular concrete. The 15% Alccofine is utilised as a cement substitute in a concrete mix to obtain our desired strength. The Alccofine is made from these elements like iron oxide, sulphur trioxide, silica, magnesia, alumina, and calcium oxide. It increases the compressive type of strength along with an overall increase in workability by using as a concrete substitute, and Alccofine is used. Furthermore, the workability is increased by increasing of Alccofine. The major aspect is that Alccofine acts as a partial replacement for cement, when it is used in concrete, it is beneficial for the construction such as durability, compressive strength, and workability.

Keywords Alccofine · Compressive strength · Performance · Concrete · Composite · Cementitious material

23.1 Introduction

As we all know, in the world any type of construction needs very regularly material is cement. Cement is a common building ingredient that is used to make concrete. Because it is malleable and resilient once set, concrete is a versatile building material used in construction.

India is a developing nation in the globe, hence the majority of the new structures are built there, but the development of new infrastructure is happening at a very high

---

R. Pandit (✉) · P. Duggal · R. K. Tomar
Department of Civil Engineering, Amity University, Noida, Uttar Pradesh, India
e-mail: rahul.pandit@s.amity.edu

H. Guo et al. (eds.), *IRC-SET 2021*, https://doi.org/10.1007/978-981-16-9869-9_23

pace. So, the concrete demand is increased day by day. As a result, in the long run, an increase in total concrete quality is required. By partially replacing cement with Alccofine, characteristics such as durability, workability, decreased water-cement ratio, heat of hydration, and overall resistance to corrosion/chemical-based assault can be improved Alccofine (GGBS).

First and foremost, we can obtain a longer lifespan by using Alccofine in concrete for increased durability. Its usage also aids in the attainment of higher compressive strength, as well as other characteristics such as tensile modulus and elastic strength.

23.1.1 Alccofine

Alccofine [1] is considered as one of them prominent futuristic material. The Alccofine is also known as a microfine, a type of additional cement mixers. This substance has significantly finer particle sizes than other hydraulic-based materials such as cement, fly ash, and silica. Now, Alccofine is a product made in India by the Counto Microfine Products Private Limited (CMPPL) (a joint venture with Goa-based Alcon developers and Ambuja Cement Limited).

23.1.2 Types of Alccofine

There is two type of Alccofine:

- Grouting Alccofine 1101
- SCM Alccofine 1203.

Alccofine 1101 (for grouting application) is a high-calcium silicate material, and which is basically a microfine type cementitious grouting-based material for rock anchoring and soil stabilisation.

In SCMs, Alccofine 1203 can be defined as a low-calcium-based silicate material. The fine, microfine, and ultrafine particle types with different sizes of Alccofine are represented by the Alccofine 1200 series (1201, 1202, and 1203). Use of Alccofine 1203 in any particular type of concrete mix design generally results in an increased strength with respect to normal concrete.

23.1.3 Application of Alccofine 1203

- For high-rise buildings.
- For bridges.
- For roads and airports.

23.2 Literature Review (Strength and Ductility Properties)

Soni et al. [2] focus on using 6.8% Alccofine to obtain an increase of strength in concrete with respect to traditional concrete. Suthar et al. [3] experimented and calculated a combination of 20–35% of fly ash with 4–14% of Alccofine/silica fume by cement weight to obtain at all ages, and concrete has a higher compressive strength than other concrete-based mixes. Sunil et al. [4] worked on a combination of Alccofine with 8–20% of fly ash to obtain increased compressive strength in concrete with respect to other concrete mixes at all ages.

Sourav et al. [5] worked on Alccofine (ultrafine slag) is better in workability and pump capacity of concrete by replacing some cement quantity with Alccofine. Also, hardened property of concrete is enhanced by adding 13% of Alccofine in a mix of traditional concrete. As the thesis's conclusion statement, they found that a 10% replacement of Alccofine for cement resulted in greater strength than conventional concrete. Siddharth et al. [6] experimented on combination of natural type sand, flash, and Alccofine, as well as M-sand, may be utilised as a cement substitute, resulting in a greater overall value change in compressive strength. It is also concluded that the highest compressive-based strength was obtained by replacement of cement with 30% fly ash and 10% Alccofine. The use of Alccofine increases the self-compatibility of concrete as well as its capacity to resist segregation.

Patel et al. [7] experimental work, the 30% glass powder and 10% Alccofine are partially replaced by cement to give a higher compressive strength under the 28 and 56 days on mainly the basis of their concrete mix design. Rajesh et al. [8] worked on the Alccofine has increased the strength in both compressions and flexure at 10% substitution with respect to cement. As a material Alccofine yields high workability and acts as a filler meanwhile to the concrete, it also increases strength rapidly at an early age to increase gradually. Sharma et al. [9] experimental work, they concluded the substitution of 15% Alccofine with cement and 10–45% of slag with fine aggregate to achieve high-strength concrete.

Reddy et al. [10] experimented on M40 grade concrete, a mixture of Alccofine and fly ash was used to replace cement. Cement is being replaced by a mixture of fly ash and Alccofine at (5–20%) with in increment of 5% at every attempt were used. They came to the conclusion that Alccofine has a higher overall performance in the production of long-term strength development properties. Chakravarthy et al. [11] experimented on the maximum replacement with Alccofine is at 16% for achieving compressive-based strength as 60.95 or 50.95% in comparison with traditional concrete after 7 days of curing and 28 days of curing interval on M25 grade of concrete.

23.3 Methodology

Manufacturing of Alccofine is done by Ambuja Cement in India. Abstraction of Alccofine is done by materials iron ore industries. Their chemical composition is the high percentage of alumina and silica content. So, this supplementary cementitious material is suitable for the replacement of silica fumes to obtained high-performance concrete. These all show the unique characteristics to modify concrete's performance in both hardened and fresh stages.

Table 23.1 shows the Alccofine chemical-based composition along with it physical type properties with their proportion, the chemical composition such as iron oxide, sulphur trioxide, silica, magnesia, Al_2O_3, and CaO. The table shows their physical characteristics, such as bulk density, specific surface areas, and specific gravity.

23.3.1 Materials

Cement:

For the study of Alccofine, we use the OPC 43 grade of cement.

Aggregate:

Alccofine uses fine aggregate such as river sand in accordance to IS: 383-1970 [12]. We also utilise coarse material that has been angularly crushed to a size of 20 mm.

Alccofine:

Alccofine is micro-supplementary cementitious materials (SCM), Counto Microfine Products Private Limited (CMPPL) (a joint venture between Ambuja Cement Limited and Goa-based Alcon developers) Mumbai.

Table 23.1 Chemical composition properties along with physical properties of Alccofine

Chemical composition		Physical properties	
constituents	Composition (wt.%)	Physical property	Results
Iron oxide (Fe_2O_3)	1.20	Bulk density (Kg/m^3)	680
Sulphur trioxide (SO_3)	0.13	Specific gravity	2.70
Silica (SiO_2)	35.30	Particle size distribution d10 (μm)	1.8
Magnesia (MgO)	8.20	d50	4.4
Alumina (Al_2O_3)	21.40	d90	8.9
Calcium oxide (CaO)	32.20	Specific surface area	12,000 cm^2/g

23.3.2 Test Methods

According to IS: 10,262-2009 code, [13] prepare M30 grade mix design concrete for a trial mix. The mix design ratio is 1 for cement, 1.55 for sand, and its 2.5 for aggregates, the water-to-cement ratio is set at 0.5. In M30 grade concrete, Alccofine is used as a partial cement substitute at various percentages (0, 5, 10, 15, 20, and 25%).

23.3.2.1 Compressive Strength

The most important role of Alccofine is its replacement with conventional concrete for higher compressive strength. To cast a specimen at 150 × 150 × 150 mm size and curing at this specimen 7, 14, and 28 days. At last, testing of the specimen is done on compression testing machine (CTM) in accordance to IS: 516-1956 [14] procedure.

The compressive strength = P/A in N/mm^2

Where, P = failure load in N,

A = cube area [150 × 150 mm^2].

23.3.2.2 Workability Test

Generally, workability is obtained by slump cone test on the fresh concrete [15]. Workability in concrete is calculated by filling cone with concrete which is later removed vertically to obtain the desired result. The horizontal displacement of unsupported concrete causes deduction in height also known as vertical settlement or slump.

23.3.2.3 Flexure Strength Test

This process is a bit similar to compressive strength, but it is originally dependent on beam and their size of specimen, 150 × 150 × 700 mm type was casted according to the IS: 516-1959 tested for 28 days. The flexural test gives the higher value of stresses just before the material yields. It is also known as the 'modulus of rupture'. A compression testing machine (CTM) is the major machine used for testing the specimen.

Bending strength is expressed.
$fb = (p \times l)/(b \times d^2)$ in N/mm^2
Where fb = Modulus of rupture (flexural concrete strength),
p = Failure load,

Table 23.2 Compressive strength of Alccofine

CUBE	7 days (N/mm^2)	14 days (N/mm^2)	28 days (N/mm^2)
0% Alccofine	18.37	20.34	22.66
5% Alccofine	19.56	21.33	26.67
10% Alccofine	19.87	22.87	28.5
15% Alccofine	21.35	25.33	32.4
20% Alccofine	20.78	22.66	30.8
25% Alccofine	19.44	21.3	29.65

l = length of specimen,
b = breadth of specimen
d^2 = Sq. of depth of specimen.

23.4 Results and Discussion

This research aids in learning more about how Alccofine is utilised to produce high-compressive strength by partially replacing cement. Compressive-based strength tested in accordance with fine aggregate and Alccofine (GGBS) along with M-sand and cement. When Alccofine is used for replacing some percentage of cement, its compressive strength shows an increment of value in comparison with traditional concrete.

So, we prepare concrete and cast their mould with size 150 × 150 × 150 mm. After 24 h, we demould the cast specimen and water curing after 7, 14, and 28 days, then tested on compression testing machine {CTM}. Basically, the specimen shape is cuboid with standard size 150 × 150 × 150 mm, cylindrical, and sometimes, prism is used for the calculating compressive strength.

Compressive strength = load/Area = $P/b \times d$ N/mm^2.

Table 23.2 shows the different quantities (0–25%) of Alccofine (GGBS) [16] which is replaced by cement in concrete. 15% of Alccofine gives a higher value of compressive-based strength in place of value below 15% and above 15% of Alccofine. So, the 15% Alccofine is replaced by cement to provide a higher value strength for 7, 14, and 28 days of curing period (Fig. 23.1).

23.5 Conclusion

- At 100% M-sand replacement and 15% Alccofine replacement for concrete, a higher value of compressive based strength is attained.

Fig. 23.1 Graph shows the compressive strength of Alccofine

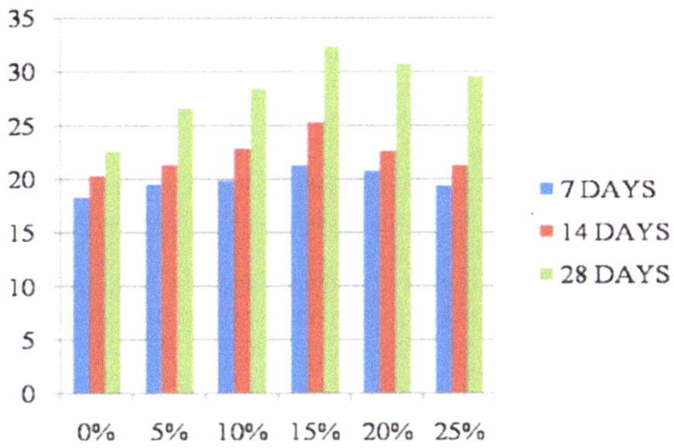

- The greater value of compressive strength is acquired at 15% of Alccofine used as a substitution of cement in concrete [17].
- Compressive strength is enhanced by 15% and gradually decreases when 15% Alccofine is added to the mix.
- Manufactured sand is used to completely replace natural sand in concrete mixes, allowing for maximum strength.
- The workability is increased by an addition in Alccofine content, even when the water-cement ratio is kept constant.
- The addition of Alccofine to concrete improves its fire resistance.
- Through the Alccofine, the overall durability of concrete is increased.

References

1. "Alccofine—Micro Materials," Counto Micro fine Products Pvt. Ltd. http://www.alccofine.com/
2. Soni, D. (2013). Experimental study on high performance concrete, with mixing of flyash and alccofine. *Indian Journal of Research, 3*(4).
3. Suthar, S. (2013). Study on effect of fly ash & alccofine addition, on the mechanical type of properties for high performance concrete. *International Journal for Scientific Research & Development, 1*(3).
4. Suthar Sunil, B. (2013). Study on strength development of high strength concrete containing fly-ash and alccofine. *Indian Journal of Research, 2*(3).
5. Saurav, relationship of concrete cylinder and concrete cube using ultrafine slag type of Alccofine (2014). *International Journal of Scientific & Engineering Research, 5*(5)
6. Upadhyay, S. P. (2014). Effect on compressive strength of high performance concrete incorporating alccofine and fly ash. *International Journal of Innovative Research & Development, 3*(2), 124–128.
7. Patel, M. Y. (2015). Study on mechanical properties of high performance concrete with alcoofine and waste glass powder. *International Journal of Science & Engineering Research, 6*(5), 105–110.
8. Rajesh Kumar, S., & Samanta, A. K. (2015). An experimental study on the mechanical properties of alccofine based high-grade concrete. *International Journal of Multidisciplinary Research and Development, 3*(4).

9. Devinder Sharma, S. S. (2016). Utilization of waste foundry slag and alccofine for developing high strength concrete. *International Journal of Electrochemical Science*, 3202–3204.
10. Vijaya Sekhar Reddy, M. (2016). Studies on eco-friendly concrete by partial replacement of cement with alccofine and fine fly ash. *ARPN Journal of Engineering and Applied Sciences, 11*(5), 3445–3448.
11. Kalyana Chakravarthy, P. R., & Raj, R. R. (2017). Analysis on compressive strength of concrete with partial replacement of cement with alccofine. *ARPN Journal of Engineering and Applied Sciences*, 2393–2395.
12. IS 383-1970, Bureau of Indian Standards, New Delhi, "Specification for Coarse and Fine Aggregates from Natural Sources for Concrete". Methods of Test for Aggregates for Concrete|| IS 2386-1963, Bureau of Indian standards, New Delhi.
13. IS 10262-2009, Bureau of Indian Standards, New Delhi, Recommended Guidelines for Concrete Mix Design.
14. IS 516-1959, Bureau of Indian Standards, New Delhi, Methods of Tests for Concrete Strength.
15. IS 456-2000, Bureau of Indian Standards, New Delhi, Plain and Reinforced Concrete-Code of Practice.
16. ASTM C 989-05, Standard Specification for Ground Granulated Blast—Furnace Slag for use in concrete and mortars.
17. ACI Committee 363. State of art report on high strength concrete, ACI 363–392.

Chapter 24
Linguistic Indicators of Success in Aviation Emergencies: A Cockpit Voice Recorder (CVR) Investigation

Yimin Huang and Andrew Prahl

Abstract Effective verbal communication is important in aviation safety. Miscommunication remains one key cause of aviation accidents. Hence, the main aim of this paper is to determine linguistic trends that are correlated with successful (i.e. safe landing, mitigated emergency) and unsuccessful (i.e. fatal crash) outcomes in aviation emergencies. This study analysed cockpit voice recorder (CVR) transcripts from a total of 57 aviation incidents with solvable or mitigatable causes using the Linguistic Inquiry and Word Count (LIWC) 2015 text analysis programme. Results revealed that more 'you' pronouns were used in successful flights, whilst more vision-based words were used in unsuccessful flights. In addition, we found significant differences in the use of tentative, differentiation, and discrepancy words in successful versus unsuccessful aviation outcomes. Results were interpreted in the context of various social and cognitive processes related to each linguistic category. Understanding the linguistic characteristics of aviation communication provides preliminary insight into linguistic patterns used by airline crews facing emergencies and highlights language use that is associated with successful outcomes.

Keywords Aviation communication · Linguistics · Cockpit voice recorder transcripts · Recommendations · LIWC

24.1 Introduction

The importance of communications in aviation safety was brought into the spotlight back in 1977 with the deadly Tenerife airport disaster. Two aeroplanes, the Pan Am and KLM Boeing 747, collided right on the runway, killing 583 people. In any aviation emergency, there may be many predisposing factors at play that makes it

Y. Huang (✉)
Raffles Institution, Singapore, Singapore

A. Prahl
Wee Kim Wee School of Communication and Information, Nanyang Technological University, Singapore, Singapore

H. Guo et al. (eds.), *IRC-SET 2021*, https://doi.org/10.1007/978-981-16-9869-9_24

difficult to pinpoint a single culprit. Still, thousands of deaths have been blamed on an issue that is seemingly easy to address: miscommunication.

However, miscommunication is not as straightforward as it seems. Factors such as pilots' workload, quality of audio signal, accent of pilot or controller, English language proficiency of operator, and failure to use standard phraseology are all thought to contribute to communication errors [1]. Such errors faced in communication have been an interest of the International Civil Aviation Organisation (ICAO), which has implemented international regulations on aviation terminology as well as language proficiency standards. Identifying specific measurable indicators in language use is thus useful in establishing a meaningful relationship between communication in the cockpit and the ultimate outcome of aviation emergencies.

Hence, the main aim of this paper is to determine linguistic trends that are associated with successful (i.e. safe landing, mitigated emergency) and unsuccessful (i.e. fatal crash) outcomes in aviation emergencies.

Current research centred around language use in groups proves useful to this research as most of the aviation incidents investigated in this study are flights involving a number of flight crew members. Thus, teamwork between members of the flight crew as well as their coordinated communication with air traffic controllers is crucial. One of the most challenging aspects in general aviation communication, as noted by pilots, was 'understanding other pilots' [1], since communication is very variable; there is more than one way to convey a message. For example, when manoeuvring the aircraft during routine flight, the pilot might say 'turn left' or specify some technical action to be taken in order to manoeuvre the plane to the left.

Another key finding about communication in groups is that discussing shared, or common information is not as productive in the decision-making process as discussing unique pieces of information [2]. Fischer, McDonnell, and Orasanu found that team success was associated with the extent to which team members shared task-critical information, equally participated and built on each other's contributions and showed agreement [3]. These findings suggest the need for both focus on the task at hand and building emotional rapport for effective collaboration in groups. This means that in the earlier example, the tone which the pilot uses in ordering a left turn also plays a role in the effectiveness of their communications.

In aviation emergencies, added external factors affect the pilots' communications. Under stressful circumstances, external stressors stemming from the surroundings and the task at hand may threaten the well-being of individual crewmembers and their functioning as a team. Pirzadeh and Pfaff's findings provide some good background on communication under stress. It was found that negative mood significantly correlates with an increase in the use of big words (more than six characters), which is negatively correlated with performance in the cockpit overall [4]. Sexton and Helmreich also found that big words are used more often because the information communicated is more technical, reflecting greater cognitive load [5].

This suggests some form of trade-off between the cognitive effort required to speak elaborately and the ability to maintain effective situation awareness. Such findings provide the basis for this study and are useful to note when interpreting the results. However, research of communications specific to the aviation setting is not commonly found, hence this study will provide a novel look into the role of communication in aviation.

This study analysed CVR transcripts from a total of 57 aviation incidents with solvable or mitigatable causes using the Linguistic Inquiry and Word Count (LIWC) 2015 text analysis programme.

This article first outlines the aims of this study, followed by the design and measurements taken under the research methodology. Results obtained are presented, then discussed and interpreted in relation to achieve our aims.

24.2 Aims/Objectives

The ultimate aims of this paper are (1) to determine linguistic trends that indicate successful and unsuccessful outcomes in aviation emergencies, and thus, (2) recommendations for aspects of language use that should be investigated further by aviation training professionals and group communication scholars.

The above aims will be achieved with the following objectives:

1. Create a database of comparable CVR transcripts.
2. Prepare transcripts by converting them into plain text form for analysis.
3. Extract data from LIWC to establish correlations between the linguistic categories tested and the success of aviation emergencies.

24.3 Methodology

24.3.1 Design

First, retrievable CVR transcripts from aviation emergencies were sourced from the following agencies and online databases, in no particular order:

- National Safety Board (NTSB) [6]
- Aviation Safety Network (ASN) [7]
- Cockpit Voice Recorder Database [8]
- Embry-Riddle Aeronautical University [9]
- Swedish Accident Investigation Authority (SHK) [10].

The NTSB and SHK are government agencies which maintain databases and archive of aviation incidents that they have investigated. These are credible sources as they follow stringent investigation procedures laid out by the standards and recommended practices in an international agreement Annex 13 of the Convention on International Civil Aviation [11]. The other organisations are also credible sources as they provide files of the accident reports from government agencies and/or their transcripts can be cross referenced with official investigation reports to verify authenticity.

The investigations are huge collaborative efforts between personnel from the country of crash, country where the aircraft is registered, country where it was assembled, etc. They analyse various technical data and produce their findings in an official report identifying causes and contributing factors, as well as safety recommendations. However, this paper focuses on the linguistic factors in aviation emergencies, hence only CVR transcripts were extracted and analysed for this study. Still, the implementation of the study must be taken in context of other technical issues pilots may have faced at the time.

Another note is that the amount of recorded information is greatly variable, depending on the flight duration. This variability in length of transcripts will be addressed in the next part of this section under measurements.

The transcripts were selected based on two criteria: presence of an emergency and solvability of the emergency.

The first criterion would often be easily fulfilled as transcripts from the above sources were made public only as part of an investigation report into an incident that had taken place. Under normal circumstances, the CVR continues recording over the old data.

From that large database of transcripts, only incidents with solvable causes can be extracted to be used for analysis, answering questions such as 'Were the pilots capable of solving the problem? Did the pilots have adequate time to respond to the problem'? Solvable causes are defined as incidents whereby pilots had sufficient time to communicate with each other to come up with a solution to solve the problem. Emergencies that were potentially solvable could result from various causes such as mechanical, structural failure, or pilot error that did not immediately lead to a crash. For example, a malfunction of one of the aircraft's engines that did not immediately lead to a crash would be a solvable incident as pilots can work around the situation and take appropriate actions to land safely.

On the other hand, unsolvable incidents would be those whereby once the problem was noticed, there was little the pilots could do to circumvent the disaster. Examples of such unsolvable incidents would be controlled flight into terrain (CFIT), and errors that could not have been avoided in time. Suicide crashes, hijacking, and loss of awareness would fall in this category as well.

Only solvable incidents were analysed for this study as we require a substantial conversation to analyse statistically. In solvable emergencies, pilots have the time to communicate with their co-pilots and other members of the flight crew to figure out how to avert the situation. That communication is what will be analysed to determine and what are the linguistic correlations with success in such emergencies.

If there was ambiguity in whether the emergencies were indeed solvable, we consulted the official accident report. Mention of pilot error or miscommunication as a cause of the accident would suggest that the problem would be solvable if the crew acted differently.

This criterion greatly reduces the number of transcripts that are used in this study, thus limiting sample size and this paper's applicability. However, this preliminary study serves to identify statistically significant associations between linguistic categories and aversion of aviation emergencies, which can be further expanded in the future studies.

After narrowing the database to comparable transcripts with solvable causes, the transcripts were further categorised into successful or unsuccessful outcomes, which would be the independent variables in our research. The dependent variable is all of the language categories analysed by LIWC.

A pilot study was first conducted to test the feasibility of the research design and is important in enhancing reliability and validity [12]. Its critical aspects are its study size, methods, and the content of the pilot study. The study comprised twenty transcripts, with eight successful and twelve unsuccessful transcripts. After showing some promise in producing statistically significant results (discussed in Sect. 24.4), the same experimental design is conducted on 37 more transcripts, with a combination of both successful and unsuccessful outcomes.

In effect, this research design mirrors that of many studies that investigate language use using LIWC, such as Fischer, McDonnell, and Orasanu's study of linguistic correlations of team functioning in space missions. They compared communication between members in teams that were consistently successful versus those that were consistently unsuccessful through analysis of verbatim transcripts using LIWC [3]. Features in team members' communications that reflected the quality of team functioning were identified. This is highly similar to this project's research methodology, given that aviation emergencies studied in this research is comparable to the complex critical thinking tasks experimented in the space missions. By clearly delineating success versus failure in the aviation incidents, specific linguistic categories can be identified as indicators for the ultimate outcome of the aviation emergencies. The communication style associated with the linguistic category can then be recommended as actionable suggestions for aviation professionals in aircraft operations during any future emergencies. Several other studies use LIWC in a similar fashion, to identify linguistic predictors of public approval of the US Congress [13], in describing hopes versus duties (discussed in Sect. 24.5) [14], and even in relationship goals in online dating [15].

24.3.2 Measurements

After identifying and categorising suitable transcripts, transcripts will be prepared for processing through linguistic analysis.

Each transcript was prepared into a group text file for all recorded speech, and individual text files collating the speeches of each individual. Words spoken at each timestamp were entered in a new line instead of in prose, since the sentences were not spoken continuously. All irrelevant information such as timestamps and transcribers' comments were removed, leaving only the verbatim text of what was spoken. This is to ensure accurate representation and translation of verbal communications in the cockpit to text form.

Within each group and individual text file, transcripts were further cut based on the moment of emergency (i.e. the time at which pilots become aware of an abnormal situation and the flights is no longer classified as routine), time of landing, and end of emergency. This is to narrow down to specifically pilots' reactions to the emergency at hand.

Transcripts were analysed using the LIWC software, which analyses text for attentional focus, emotions, social relationships, cognitive styles, and individual differences [16]. This software includes over eighty categories, including articles (a, an, the), self-references (I, me, my), cognitive, emotion word categories, et cetera. Linguistic categories that show statistical significance between successful and unsuccessful outcomes were identified and will be discussed in Sect. 24.4.

The LIWC programme has two central features—the processing component and the dictionaries. The processing feature is the programme itself, which opens a series of text files, in this case CVR transcripts, and compares each word in a given text file with the dictionaries which are collections of words that define a particular linguistic category.

LIWC was used by Fischer, McDonnell, and Orasanu in their study of communications in space teams [3], and by Dzindolet, Stover, and Pierce in groups' decision-making skills analysis [2]. In both studies, verbal discussions were transcribed into word files and were entered into the LIWC. Dzindolet measured the correlations between the relative frequency of each of the LIWC word categories used and the time spent on each of the group functions were computed, similar to our study of relation between frequency of linguistic categories and success. Results will finally be interpreted and analysed.

24.3.3 Limitations

The following limitations affect the accuracy and applicability of this research:

First, the small sample size of 57 CVR transcripts analysed makes it difficult to find significant relationships from the data, as statistical tests normally require a larger sample size to ensure a representative distribution of aviation emergencies around the world. Furthermore, the CVR transcripts are taken from limited sources, mainly NTSB, which is not representative of emergencies from other parts of the world. At different parts of the world, there could be different external factors that were not accounted for in this study. This small sample size also means there is a lack of reliable data that limits the scope of our analysis, thus presenting an obstacle in finding a trend or a meaningful relationship. Hence, future research in this direction may be needed in order to produce meaningful results.

Another limitation is that the method used to collect the data inhibited our ability to conduct a thorough analysis of the results. The only data accessible was online public databases where investigations are completed and complete reports are done. There was limited comparison with normal flights that did not meet any emergency, hence the gap is in comparing communications between incident flights and flights that went smoothly. Thus, there is a need for future researchers to revise a specific method for gathering data.

24.4 Results

Preliminary analysis using the LIWC software shows that more pronouns are used overall in unsuccessful outcomes ($M_{\text{diff}} = 1.7192$), but far more 'you' pronouns are used in successful ones ($M_{\text{diff}} = 0.8333$).

This result is reflected again in the subsequent test, where 'you' pronoun use is associated with a successful outcome ($M_{\text{diff}} = 0.5451$), but it is really used by the FOs that drives this result ($F = 4.160$, $p = 0.048$).

Significant results are also drawn around discrepancy, tentative, and differentiation words. Differentiation words are a class of words where people are distinguishing between entities, people, or ideas (e.g. 'but', 'without', 'else', 'has not'). Discrepancy words (e.g. 'should', 'would') tend to reflect what the person thought they should do or are descriptions of the ideal situation. Tentative words (e.g. 'may', 'can', 'likely') show uncertainty. In successful transcripts, all crew members used significantly more discrepancy ($M_{\text{diff}} = 0.5083$), tentative ($M_{\text{diff}} = 0.4710$), and differentiation words ($M_{\text{diff}} = 0.4904$) than in unsuccessful transcripts. The FO again plays a critical role in these results, ($F = 4.141, p = 0.048$; $F = 1.565, p = 0.218$; $F = 3.308, p = 0.076$).

More vision-based words (e.g. 'see') are used in unsuccessful flights ($M_{\text{diff}} = 0.5491$), and the captain plays a large role in this word use ($F = 4.095$, $p = 0.049$).

Table 24.1 highlights the results of ANOVA tests for significant linguistic categories for all crew members, showing the F-statistic and p-values. Similar data for subsequent tests for all crew members, captains, and first officers are shown in Tables 24.2, 24.3, and 24.4, respectively. No significant data were found for speech by the second officers, or flight engineers as the speech files were too short

Table 24.1 ANOVA results of preliminary study for all crew members

Linguistic process	*F*	*P*
Pronoun	2.976	0.102
You^	5.598	0.029*
Article	3.378	0.083
Interrog#	5.151	0.036*
Certain#	4.492	0.048*
Motion	2.766	0.114
Home	2.832	0.110
Relig	3.761	0.068

*Effects are significant at the $p = 0.05$ level
^Linguistic category was used more in successful outcomes
#Linguistic category was used more in unsuccessful outcomes

Table 24.2 ANOVA results for all crew members

Linguistic process	*F*	*p*
Discrep^	6.417	0.015*
Tone^	5.521	0.023*
WC^	5.236	0.027*
You^	4.891	0.032*
Conj^	4.748	0.035*
See#	4.735	0.035*
Tentat^	4.598	0.038*
Differ^	4.433	0.041*
Motion#	4.210	0.046*
Feel	3.308	0.076
Adj	0.077	0.077
Body	3.252	0.078
Posemo	3.202	0.080
Cogproc	3.184	0.081
Achieve	3.058	0.087

*Effects are significant at the $p = 0.05$ level
^Linguistic category was used more in successful outcomes
#Linguistic category was used more in unsuccessful outcomes

Table 24.3 ANOVA results for captains

Linguistic process	*F*	*p*
WC^	5.242	0.027*
See#	4.095	0.049*
Feel	3.408	0.072
Body	3.234	0.079

*Effects are significant at the $p = 0.05$ level
^Linguistic category was used more in successful outcomes
#Linguistic category was used more in unsuccessful outcomes

to analyse statistically. The focus of our experimentation would thus be on the interaction between pilot and co-pilot as they play more significant roles in the overall cockpit conversation.

24.5 Discussion

Success is connected with certain social dynamics of communication and status influence. Consistent with past research examining how status is reflected in language, and it is found that the use of pronouns influences processes like coordination in teams,

Table 24.4 ANOVA results for first officers

Linguistic process	*F*	*p*
Discrep^	4.141	0.048*
Tone^	6.647	0.013*
You^	4.160	0.048*
Conj^	6.547	0.014*
Differ	3.308	0.076
Motion	3.077	0.087
Feel	2.897	0.096
Posemo^	7.006	0.011*
Achieve#	4.897	0.032*
Affect	2.971	0.092

*Effects are significant at the $p = 0.05$ level
^Linguistic category was used more in successful outcomes
#Linguistic category was used more in unsuccessful outcomes

which have effects on outcomes like group effectiveness [17]. In successful outcomes, more first-person plural pronouns (e.g. 'we') and less differentiating pronouns (e.g. 'I') were used, reflecting collective identity [18] and fosters group attachments [19].

Unlike past research, however, more 'you' pronoun use is not observed to be counterproductive to group effectiveness [17]. In fact, more 'you' pronouns were used in successful outcomes, especially by FOs, which implies active participation of the FOs in the emergency response (i.e. 'you do this'). Problems with communication in the cockpit can arise due to both the overload of the cognitive system in situations of high-task load, and because of the specific social situation in the cockpit with one higher status, more experienced pilot and another less experienced crew member [20]. However, having members who equally contribute to the discussion subverts the status hierarchy, enabling success as compared to teams in which some member(s) talked more than others [3]. This can be seen in the case of the first officers in this study. In linguistic terms, FOs seem to be the key in emergency management. This means implications for training and safety.

In the light of past research that both discrepancy and differentiation words relate closely to words about cognitive processes [14], critical thinking seems to lead to success. In addition, whilst discrepancy words tend to correlate with hopefulness and differentiation words with obligations [14], we found that both discrepancy and differentiation words are associated with aviation success. This suggests that whilst distribution of tasks are vital in the situational environment on the flight deck, it is also helpful to discuss goals to maintain hopefulness.

Tentative words would suggest pilots are considering many options, whilst a lack of tentative words would indicate more 'tunnel vision' thinking. That sort of thought process may lead to overconfidence in a person's decision, associated with unsuccessful outcomes. The FO also plays a critical role in these results, implying that a more actively contributing, critically thinking, and analytical FO is associated with success.

On the other hand, vision words are associated with unsuccessful outcomes. This implies that more time spent on discussing their surroundings, the emergency situation, what they can and cannot see, leaves less time for actually analysing and solving the problem, less cognitive processing, and thus unsuccessful outcomes.

One recommendation we can make is that airlines and training institutions should consider tracking linguistic trends. In addition to training with simulations of emergency scenarios, institutions could consider tracking language use and analysing linguistic trends to better understand how the factors identified are truly affecting the problem-solving process on the flight deck. Moreover, implementing linguistic analysis at the training stage means that hundreds of iterations of the same emergency scenario with the same plane model and similar external factors can be run to eliminate the uncontrolled variables present in this real-world study.

24.6 Conclusion

The present study served to extend the previous research on communications in aviation flights, and results suggest some significant linguistic indicators relating to safety in aviation that merits further study and can be considered for implementation in pilot training. With consideration of such findings, future studies could benefit from narrowing the group variables by comparing between specific causes, such as mechanical error. Another direction to take could be to consider two-way communication: how using certain linguistic features associated with success can lead to more effective feedback from other interlocutor(s), in aviation or other settings. However, there exists the limitation that there could be many causes leading to an aviation emergency, and thus, each incident is unique as the conditions greatly vary. Unsolvable incidents could be looked into to undercover how mishaps can occur even with no signs of emergency. In terms of recommendations, this study provides the baseline that shows the significance of certain linguistic features and their psychological underpinnings in the cockpits. In sum, the results of this study showcase the importance of and need for future research into the role of linguistics in aviation.

Acknowledgements I would like to thank the Nanyang Research Programme at Nanyang Technological University for their support of this research and the guidance of my mentor, Asst. Prof. Andrew Prahl at Nanyang Technological University.

References

1. Molesworth, B. R. C., & Estival, D. (2015). Miscommunication in general aviation: The influence of external factors on communication errors. *Safety Science, 73*, 73–79. https://doi.org/

10.1016/j.ssci.2014.11.004

2. Dzindolet, M. T., Stover, A., & Pierce, L. G. (2005). Predicting group decision-making with a computerized text analysis tool. In *Proceedings of the human factors and ergonomics society 49th annual meeting—2005*.
3. Fischer, U., McDonnell, L., & Orasanu, J. (2007, June). Linguistic correlates of team performance: Toward a tool for monitoring team functioning during space missions. *Aviation, Space, and Environmental Medicine, 78*(5), Section II.
4. Pirzadeh, A., & Pfaff, M. S. (2012). Emotion expression under stress in instant messaging. In *Proceedings of the human factors and ergonomics society 56th annual meeting*.
5. Sexton, J. B., & Helmreich, R. L. (2000). Analyzing cockpit communications: The links between language, performance, error, and workload. *Human performance in extreme environments: The journal of the Society for Human Performance in Extreme Environments, 5*(1), 63–68. https://doi.org/10.7771/2327-2937.1007
6. NTSB Aviation Accident Database & Synopses. (2020). Retrieved from https://www.ntsb.gov/_layouts/ntsb.aviation/index.aspx
7. Aviation Safety Network. (2020). Retrieved from https://aviation-safety.net/investigation/cvr/transcripts/
8. Cockpit Voice Recorder Database. (2020). Retrieved from https://www.tailstrike.com/database.htm
9. Accident & Incident Information (2016). Embry-Riddle Aeronautical University Aviation Safety & Security Archives. Retrieved from http://archives.pr.erau.edu/resources/accidentinfo.html
10. Swedish Accident Investigation Authority. (2016). Retrieved from https://www.havkom.se/en/utredningar?categoryId=202&dateFrom=&dateTo=&reportDateFrom=&reportDateTo=&onGoingWithReport=1&q=cockpit+voice+recorder
11. Convention on International Civil Aviation—Doc 7300. (2019). Retrieved from https://www.icao.int/publications/pages/doc7300.aspx
12. Malmqvist, J., Hellberg, K., Möllås, G., Rose, R., & Shevlin, M. (2019). Conducting the pilot study: A neglected part of the research process? methodological findings supporting the importance of piloting in qualitative research studies. *International Journal of Qualitative Methods, 18*, 160940691987834. https://doi.org/10.1177/1609406919878341
13. Decter-Frain, A., & Frimer, J. A. (2016). Impressive words: Linguistic predictors of public approval of the U.S. Congress. *Frontiers in Psychology*. https://doi.org/10.3389/fpsyg.2016.00240
14. Vaughn, L. A. (2018). Contents of hopes and duties: A linguistic analysis. *Frontiers in Psychology*. https://doi.org/10.3389/fpsyg.2018.00757
15. van der Zanden, T., Schouten, A. P., Mos, M., van der Lee, C., & Krahmer, E. (2019). Effects of relationship goal on linguistic behavior in online dating profiles: A multi-method approach. *Frontiers in Communication*. https://doi.org/10.3389/fcomm.2019.00022
16. Pennebaker, J. W., Boyd, R. L., Jordan, K., & Blackburn, K. (2015). *The development and psychometric properties of LIWC2015*. University of Texas at Austin. https://doi.org/10.15781/T29G6Z
17. Van Swol, L. M., & Kane, A. A. (2019). Language and group processes: An integrative, interdisciplinary review. *Small Group Research, 50*(1), 3–38. https://doi.org/10.1177/1046496418785019
18. Ashmore, R. D., Deaux, K., & McLaughlin-Volpe, T. (2004). *An Organizing Framework for Collective Identity: Articulation and Significance of Multidimensionality*. https://doi.org/10.1037/0033-2909.130.1.80
19. Guerin, B. (2003). Language use as social strategy: A review and an analytic framework for the social sciences. *Review of General Psychology, 7*(3), 251–298. https://doi.org/10.1037/1089-2680.7.3.251
20. Krifka, M., Martens, S., & Schwarz, F. (2003). Group Interaction in the Cockpit: Some linguistic factors. Retrieved from https://www.researchgate.net/publication/255573424_Group_Interaction_in_the_Cockpit_Some_Linguistic_Factors1

Chapter 25
Gait Monitoring and Analysis for Parkinson's Disease Patients

Sanyukta Arunkumar, Eunice Tan Ee Wei, Arthur Tay, and Jerome Jayasundram

Abstract Parkinson's Disease (PD) is the second most common neurodegenerative disorder resulting in the degeneration of the central nervous system. PD patients usually experience ankle or foot stiffness, which affects their ankle joint movement, speed and stride length. Usually, patients with PD visit clinics every 3–6 months, where physicians conduct motor screening through exercises, such as the Timed Up and Go (TUG) test, to estimate the disease severity and medication effects. The TUG test is a performance-based measure of functional mobility that was initially developed to identify mobility and balance impairments in older adults. During our experimentation, the test requires the subject to rise from a chair, walk 7.0 m at a comfortable pace to a mark placed on the floor, turn around at the 7.0 m mark, walk back to the starting point, and return to sitting in the chair (Nocera et al, Using the timed up and go test in a clinical setting to predict falling in Parkinson's Disease, 2013). However, the disadvantage of this test is that the evaluation is subjective as a rating scale is often used, which may result in an unreliable diagnosis. Therefore, the aim of the project is to utilise wearable sensors, consisting of an accelerometer and a gyroscope, to monitor and analyse the gait of PD patients. A MinIMU-9 V5 was interfaced with a Bluno Beetle, which is the microcontroller of the device, to measure the various stride parameters and send the data to devices via Bluetooth Low Energy (BLE). Using the results obtained, the gait of PD patients can be analysed by reading the values of certain stride parameters at points of the graph where shuffling gait and freezing are prominent, which would indicate the degree of severity of the disease. From the results collected, it can be deduced that PD patients would take a longer time for each step of more than 1.08 s, have variable inconsistency between each step, a stride length much lower than 1.08 m, a range of acceleration in the x-axis much smaller than $-30\ 000 < \text{accX} < 20\ 000$, a lower tilt angle (less than 20°) during the TUG test and a tilt angle close to 0° when sitting down and standing

S. Arunkumar (✉) · E. Tan Ee Wei
Raffles Girls' School (Secondary), Singapore, Singapore
e-mail: sanyukta.arun05@rafflesgirlssch.edu.sg

A. Tay · J. Jayasundram
Department of Electrical and Computer Engineering, Faculty of Engineering, National University of Singapore, Singapore, Singapore

H. Guo et al. (eds.), *IRC-SET 2021*, https://doi.org/10.1007/978-981-16-9869-9_25

up. Moreover, the progression of the disease can be monitored by comparing their current gait patterns with the previous results.

Keywords Parkinson's disease (PD) · Timed up and go (TUG) test · Stride parameters · Rating scale · Accelerometer · Gyroscope · MinIMU-9 V5 · Bluno beetle · Bluetooth low energy (BLE) · Gait patterns

25.1 Introduction

Parkinson's Disease (PD) affects predominantly dopamine-producing neurons in a specific area of the brain called substantia nigra. PD usually affects the elderly and is a progressive disorder, where the symptoms develop gradually as the years pass [2]. It may start off with an almost unnoticeable tremor in a single hand and can also result in stiffness or slowing of movement. Hence, this shows the importance of gait monitoring and analysis for PD patients using wearable sensors. It enables doctors to accurately conclude how much the disease has progressed in a particular patient who has been diagnosed with PD. If a rating scale was used instead, the results may not be reliable and accurate, possibly resulting in the doctors either underestimating or overestimating the severity of the disease. Hence, using wearable sensors, they will be able to deduce how effective the current medications are and take the necessary actions to prevent the progression of the disease.

Gait analysis is the systematic study of locomotion using one's eyes and brain with the aid of measuring instruments. Gait analysis using 3-dimensional systems is currently the gold standard for measuring parameters including spatiotemporal variables and joint kinematics [3]. Quantitative gait analysis is essential for evaluating walking and running patterns for markers of pathology, injury, or other gait characteristics [4]. One of the most common methods of gait analysis is through the Timed Up and Go (TUG) test, which was used in this investigation as well.

It is expected that the portability, affordability, and applicability of wearable devices to many different populations will have contributed advancements in understanding the real-world gait patterns of people [4]. Data collection can be from various medical sensors attached to the patient. The data from medical sensors will be processed and transferred to the patient's smartphone and to the administrator's computer via communication network. After transferring and processing data in smartphone and administrator's computer, the required data will be sent to the clinic, assigned nurse, assigned doctor, emergency centre, and patient's family for decision-making about the patient [5].

Through gait monitoring and analysis of PD patients, the degree of severity of the disease can be determined by recording the values of stride parameters during the various gait phases, specifically dorsiflexion and plantarflexion. This is because due to the ankle or foot stiffness that PD patients usually experience, the ankle joint movement during the above mentioned phases would be the most affected. Therefore, gait monitoring and analysis is essential for the study of PD as it allows us to better understand how this disease affects kinematic movements. Symptoms of PD include freezing of gait, slowed movement (bradykinesia) and impaired balance.

Through gait analysis via a wearable device that is worn around the patient's ankle, we will be able to measure a patient's gait, inclusive of their walking mannerism, positions, and so on. As compared to constantly going to a research laboratory/medical facility, this device can be used anywhere, including from the comfort of the patient's own home, with the data conveniently sent to an IOS device with Arduino IDE via Bluetooth Low Energy (BLE), which will be further explained under materials and methods. The data collected by this device will then be translated onto a graph, which can then be analysed to monitor the severity of the patient's condition.

25.2 Materials And methods

In this project, we made a wearable sensor that would be attached to the PD patient's ankle. This device comprises a Bluno Beetle, a MiniMU-9 v5, a 3.7 V Lithium-Ion Rechargeable Battery 800mAH and a USB LI-Ion/Li-Poly Charger.

25.2.1 Bluno Beetle

The Bluno Beetle acts as the microcontroller of the device. It features extremely similar features to a standard UNO board, but utilises the CC2540 Bluetooth Chip for BLE Communications and has been designed to ensure that it is both compact and very affordable compared to the plethora of similar boards on the market [6].

25.2.2 The MiniMU-9 v5

The Pololu MinIMU-9 v5 is a compact (0.8" × 0.5") board that combines ST's LSM6DS33 3-axis gyroscope and 3-axis accelerometer and LIS3MDL 3-axis magnetometer to form an inertial measurement unit (IMU). These sensors operate at voltages below 3.6 V, which can make interfacing difficult for microcontrollers operating at 5 V. The MinIMU-9 v5 addresses these issues by incorporating additional electronics, including a voltage regulator and a level-shifting circuit, whilst keeping the overall size as compact as possible. The board ships fully populated with its SMD components, including the LSM6DS33 and LIS3MDL, as shown in the product picture.

Compared to the previous MinIMU-9 v3, the v5 version uses newer MEMS sensors that provide some increases in accuracy (lower noise and zero-rate offsets). The LSM6DS33 and LIS3MDL have many configurable options, including dynamically selectable sensitivities for the gyro, accelerometer, and magnetometer. Each sensor also has a choice of output data rates. The two integrated circuits (IC) can be accessed through a shared I^2C/TWI interface, allowing the sensors to be addressed individually via a single clock line and a single data line. Additionally, a slave address configuration pin allows users to change the sensors' I^2C addresses and have two MinIMUs connected on the same I^2C bus.

The nine independent rotation, acceleration, and magnetic readings (9DOF) provide all the data needed to make an attitude and heading reference system (AHRS). With an appropriate algorithm, a microcontroller or computer can use the data to calculate the orientation of the MinIMU board. The gyro can be used to very accurately track rotation on a short timescale, whilst the accelerometer and compass can help compensate for gyro drift over time by providing an absolute frame of reference. The respective axes of the two chips are aligned on the board to facilitate these sensor fusion calculations [7].

25.2.3 2.4 V Lithium-Ion Battery

As for the battery, we were initially planning to utilise a 2.4 V Lithium-Ion Battery Rechargeable (Secondary) 350 μAh, however, we were unable to find/purchase it anywhere. Hence, we eventually settled on a 3.7 V Lithium-Ion Rechargeable Battery 800 mAH.

25.2.4 USB Li-Ion/Li-Poly Charger

A USB Li-Ion/Li-Poly Charger was then used to charge the battery.

Both the battery and the charger were also soldered to the circuit board to enable it to be both portable and easily chargeable. The circuit diagram and circuit of our prototype are shown (Figs. 25.1 and 25.2).

25.2.5 Methods

In this wearable sensor, an IMU consisting of a gyroscope and an accelerometer is utilised to measure the gait of PD patients. In PD, IMUs have been employed to monitor gait patterns and stride parameters to further understand the intricacies of the Parkinsonian gait and to determine the efficacy of medications on the gait and other motor symptoms of PD patients.

The accelerometer was used to measure the average velocity, stride length and cadence of the PD patient. PD patients tend to have reduced walking velocity, stride length and cadence. Since PD patients tend to have a smaller tilt angle due to the

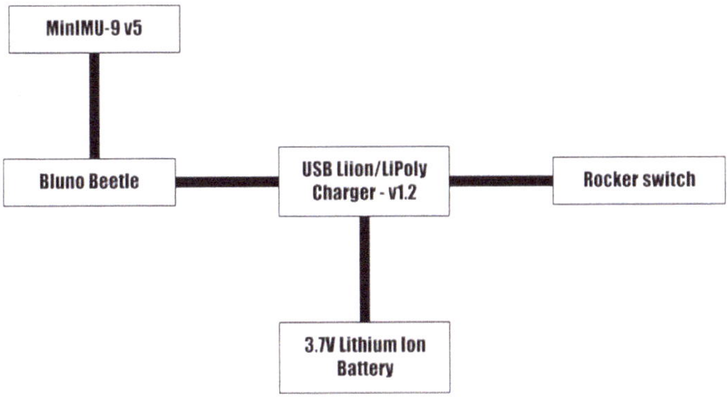

Fig. 25.1 Circuit diagram of prototype

Fig. 25.2 Circuit of the prototype

shuffling of gait caused by the rigidity of the feet, both the accelerometer and gyroscope were used to measure the tilt angle of the ankle of the PD patient. Moreover, the stiffness of the joints causes them to have a slower angular rotation of the ankle joint, which can be measured using the *x*-axis of a gyroscope. The *x*-axis was used because PD patients usually have ankle or foot stiffness, which can affect their ankle joint and how easy it is to bend their foot up and down [8]. When walking, this stiffness hinders the ankle joint movement that occurs during dorsiflexion and plantar flexion, thus slowing the movement made about the *x*-axis (roll). Hence, we will be using the tilt angle in the gyroscope *x*-axis and the acceleration in the accelerometer's *x*-axis to monitor and analyse the gait of PD patients.

Although it is possible to measure the tilt angle from either the accelerometer or gyroscope, we did not do so because the measurement from the accelerometer gets affected by sudden horizontal movements and the measurement from the gyroscope gradually drifts away from the actual value. In other words, the accelerometer reading gets affected by short duration signals and the gyroscope reading by long duration signals. These readings are, in a way, complementary to each other. The complementary filter is essentially a high pass filter acting on the gyroscope and a low pass filter acting on the accelerometer to filter out the drift and noise from the measurement [9]. Hence, by combining the measurements from the accelerometer and gyroscope using the complementary filter, the resultant angle will be stable and accurate.

Therefore, a code comprising the complementary filter formula was created to calculate the tilt angle accurately. The formula is currentAngle = α * (previousAngle + gyroAngle) + (1 − α) * (accAngle). The value of α was determined to be 0.987 using another formula $\alpha = \Gamma/(\Gamma + \mathrm{dt})$. Then, the graph printed in the Serial Plotter was used to identify the tilt angle of a PD patient at the various gait phases. The exact tilt angle values can be determined from the table in the Serial Monitor.

These parameters will be determined during a TUG test.

The wearable sensor will be worn around the ankle of the PD patient. The ankle was chosen because ankle stiffness is a common symptom in PD patients, causing the difference in tilt angle to be more prominent. Furthermore, ankle kinetics can be monitored and analysed more easily and accurately.

Firstly, a code was written in Arduino IDE to read data from the IMU and uploaded onto the Bluno Beetle. Since the sensor will be attached to the ankle, a BLE connection between the Bluno Beetle and a device needs to be established so that the device will be able to receive data wirelessly. BLE is used for applications that do not need to exchange large amounts of data, and can therefore run on battery power for years at a cheaper cost. As BLE devices are incompatible with standard Bluetooth devices, they will not inter-operate with them. Unlike classic Bluetooth, BLE remains in sleep mode constantly except for when a connection is initiated [10].

Therefore, an iOS application was created using *X*code. When the user first enters the application, the home screen will be displayed, as shown in Fig. 25.3. After he/she clicks the play button, the next screen will be displayed where the phone/tablet scans and connects to a BLE peripheral, which in this case is the Bluno Beetle, as shown in Fig. 25.4. After establishing the BLE connection using the app, data read by the Bluno Beetle, which consists of the tilt angle measured using the complementary filter formula and the acceleration in the x-axis, will be sent to the iPhone/iPad wirelessly. This data will be presented in the form of graphs in real-time, as shown in Fig. 25.5. Pictures of the app are shown.

25.3 Results

Since we were not allowed to test the prototype on a PD patient, it was tested on a healthy person instead. The graphs, reflecting the gait of a healthy person during a 7 m TUG test, are shown.

Fig. 25.3 Home screen

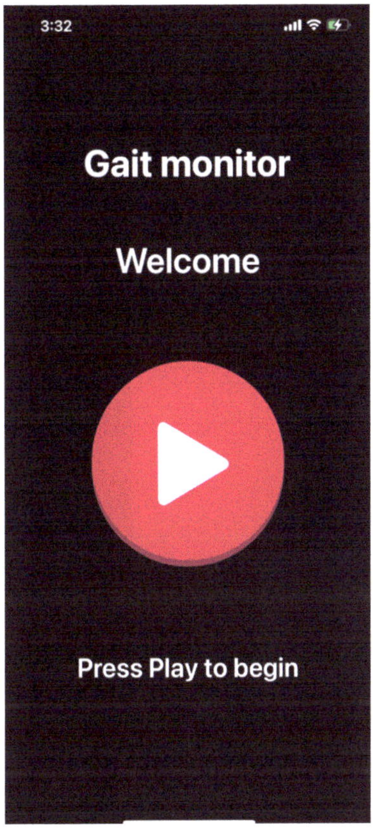

25.4 Discussion

In Fig. 25.6, each crest above 0 m/s^2 represents the healthy person lifting their foot up whilst walking, whilst each trough below $-10{,}000\ ^{m}/s^2$ represents the healthy person resting their foot on the ground. Each smaller peak and trough represents the small movement of the person's foot following the movement of their other foot. Fig. 25.6

The amount of time taken for the healthy person to walk the 7 m TUG is 14.07 s, whilst the average time taken for them to take each step is 1.08 s (obtained by finding the distance between two crests). The average length of a healthy person's step was calculated to be 1.08 m.

A PD patient is likely to experience slowed movement (also known as bradykinesia). Hence, it is likely that they would take a longer time for each step (>1.08 s). Another symptom of PD, shuffling or freezing of gait would also result in variable inconsistency between each step, as the patient would occasionally take a long pause in between each step. Moreover, the stride length will be much lower than 1.08 m. The range of acceleration in the x-axis will be much smaller than $-30{,}000 < \text{accX} < 20{,}000$.

Fig. 25.4 Connection status of Bluno Beetle

From approximately 0.0 s to 2.0 s and 11.0 s to 14.0 s, the healthy individual was likely to have been moving from a sitting to standing and standing to sitting position, respectively. This can be inferred based on the fact that the crests and troughs are less prominent than between 2.0 and 11.0 s, where the individual was walking. The person was likely to have turned at around 7.0 s, which was inferred based on the fact that the crests between 6.0 and 8.0 s were much lower (less than 0) as compared to the other crests. Moreover, the distance between the 2 crests at 7.0 and 7.2 s is much shorter than the other crests.

In Fig. 25.7 the tilt angle along the x-axis was measured, which helps us determine how PD affects the plantarflexion and dorsiflexion phases due to ankle stiffness. Thus, it would be expected that the PD patients would have a much lower tilt angle (less than 20°) during the TUG test and a tilt angle close to 0° when sitting down and standing up.

However, there are several limitations to this experiment. Firstly, we conducted this experiment with the assumption that the only differing factor between a healthy

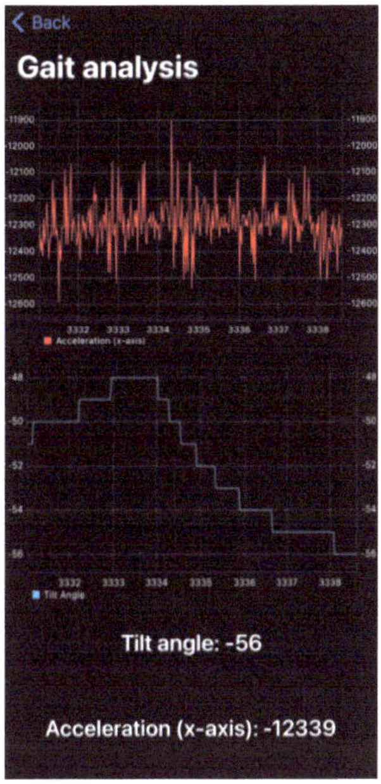

Fig. 25.5 Graphs displayed in real-time

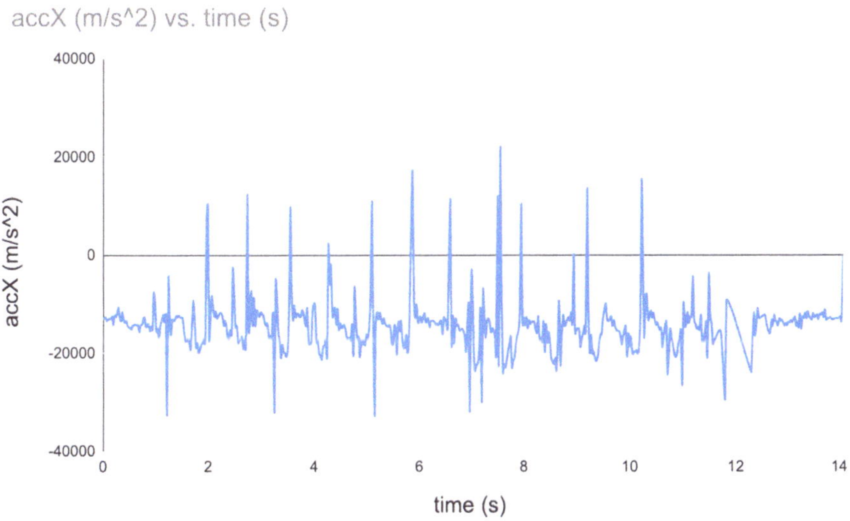

Fig. 25.6 Graph showing the *x*-axis accelerometer readings when the prototype is attached to the ankle of a healthy person during a TUG test

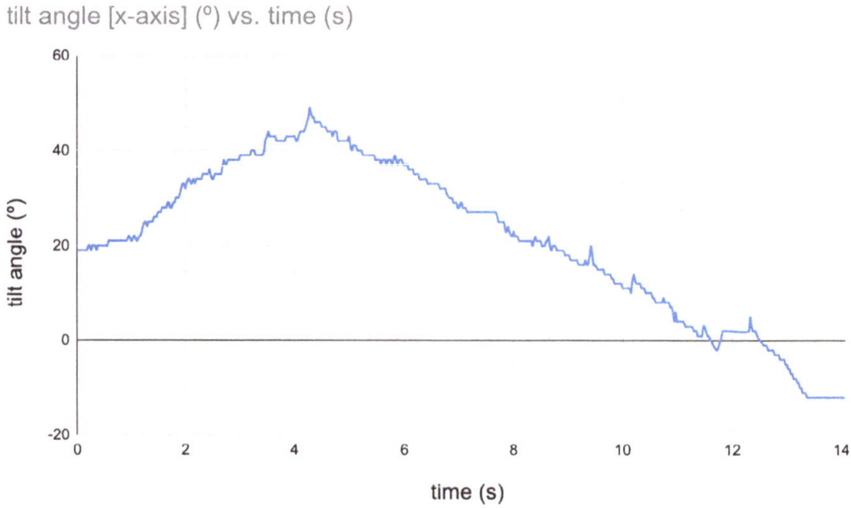

Fig. 25.7 Graph showing the tilt angle, which was calculated using the complementary filter formula, when the prototype was attached to the ankle of a healthy person during a TUG test

person and a PD patient would be that the PD patient has PD (and hence symptoms such as freezing of gait and bradykinesia) whilst the healthy person does not. There are other factors that could have influenced the person's gait, such as other health conditions (e.g. arthritis), fatigue, foot/leg numbness, stride length. Secondly, as mentioned previously, we did not get the opportunity to test out the device on an actual PD patient due to safety concerns. Hence, we were only able to test it out on a single healthy individual, which is an extremely small and limited sample size to base our results on. The gaits of other healthy individuals as well as PD patients are likely to differ.

If given the chance to further pursue this topic of research, we would work on getting a larger sample size of healthy patients as well as PD patients (who are preferably similar in terms of other factors such as other health conditions, age, gender, limb length) to conduct our experiment and base our results on, in order to make a fairer, more reliable comparison between the gaits of both parties. We would also like to expand on the idea of creating a mobile/desktop app on laptops/phone/tablet/iPad platforms that would be able to connect to our device via BLE and store the patients' data, which will then be transmitted to a professional for further analysis/diagnosis.

Acknowledgements We would like to thank our mentors, Assoc Prof Arthur Tay and Mr Jerome Jayasundram, our teacher in-charge, Ms Carmen Hoo, and the SMP community for giving us this opportunity and rendering us help whenever we needed it most.

References

1. Nocera, J. R., Stegemöller, E. L., Malaty, I. A., Okun, M. S., Marsiske, M., Hass, C. J., and National Parkinson foundation quality improvement initiative investigators. (2013). Using the timed up & go test in a clinical setting to predict falling in Parkinson's disease, 6 Mar, 2013. [Online serial]. Available: https://www.ncbi.nlm.nih.gov/pmc/articles/PMC4144326/#:~:text=As%20such%2C%20an%20accurate%20and,and%20sit%20down%20is%20timed [Accessed January 9, 2021].
2. Parkinson's Foundation, What is Parkinson's? [Online serial]. Available: https://www.parkinson.org/understanding-parkinsons/what-is-parkinsons [Accessed January 9, 2021].
3. Smith, R., & Seemungal, B. M. (2019). Diagnostic Approaches Techniques in Concussion/Mild Traumatic Brain Injury. In *Neurosensory disorders in mild traumatic brain injury*, [Online serial]. Available: https://www.sciencedirect.com/topics/neuroscience/gait-analysis [Accessed January 9, 2021].
4. Benson, L. C., Clermont, C. A., Bošnjak, E., & Ferber, R. (2018). The use of wearable devices for walking and running gait analysis outside of the lab: A systematic review, [Online serial]. Available: https://www.sciencedirect.com/science/article/abs/pii/S0966636218304715 [Accessed January 9, 2021].
5. Keikhosrokiani, P. (2020). Introduction to mobile medical information system (mMIS) development. *ScienceDirect* [Online serial]. Available: https://www.sciencedirect.com/topics/medicine-and-dentistry/wearable-sensor [Accessed January 9, 2021].
6. Bluno Beetle V1.1—Bluetooth 4.0 BLE Module, [Online document]. (n.d.), Available: https://www.diyelectronics.co.za/store/dfrobot/2072-bluno-beetle-v11-bluetooth-40-ble-module.html [Accessed January 9, 2021].
7. MinIMU-9 v5 Gyro, Accelerometer, and Compass (LSM6DS33 and LIS3MDL Carrier), [Online document]. (n.d.) Available: https://www.pololu.com/product/2738 [Accessed January 9, 2021].
8. Walker, R., & Lindop, F. (2018). Foot care and Parkinson's, [Online serial]. Available: https://www.parkinsons.org.uk/sites/default/files/2018-10/FS51%20Foot%20care%20and%20Parkinson%27s%20WEB.pdf [Accessed January 9, 2021].
9. Midhun, S. Arduino Self-Balancing Robot, [Online document]. Available: https://www.instructables.com/Arduino-Self-Balancing-Robot-1/ [Accessed January 9, 2021].
10. Khandere, (2019). Bluetooth Versus Bluetooth Low Energy: What's the Difference? [Online document]. Available: https://medium.com/@akash.kandhare/bluetooth-vs-bluetooth-low-energy-whats-the-difference-74687afcedb1. [Accessed January 9, 2021].

Chapter 26
An Eldercare Control System in IoT Cloud Environment

Leong Kai Ming Jeremy, Ching Chi Man Raymond, Yongqing Zhu, and Paul Horng Jyh Wu

Abstract Being one of the earlier countries to experience an increasingly aging population in Asia, Singapore will have to accommodate to the age structure in terms of policies planning and allowing the golden age to be self-sustainable alone. In this paper, we developed an eldercare control system based on Internet of things (IoT), named Merdeka-Pioneer monitoring system (MPMS), which provides care for silver and golden-aged citizens. The MPMS targets to monitor the air quality of the environment and detect motions and falling. It also incorporates the concept of IoT over the cloud environment, which allows caretakers or next of kin (NOK) to monitor data collected from the system remotely. The data collected can support preventive measures which can be adopted before any danger and misfortunes, especially in absence of the caretakers or NOK. In this paper, both hardware and software design and implementation are demonstrated in details. The prototype MPMS system has the ability to accurately detect a fall and changes in the environment and are able to send alerts and trigger alarms to the caretakers and NOK.

Keywords Eldercare · Real-time monitoring · IoT · Healthcare

26.1 Introduction

The demand for Internet of things (IoT) devices is on a rise nowadays due to the technology revolution. Many gadgets or devices we saw in movies back in the day can be fulfilled at the current stage. However, other than for personal usage and home environment, we neglected to address the seniors in Singapore. Siau [1] states that the elderly will make up half of Singapore's population by the year 2050. This is a challenge that Singapore will face inevitably due to the low birth rate in Singapore over the years. Devices and gadgets to monitor the well-being of the old aged should be in place to ensure that they are safe and sound. These devices can be simple detector of temperature, sensors to detect proximity and determine movement in the

L. K. M. Jeremy · C. C. M. Raymond · Y. Zhu (✉) · P. H. J. Wu
School of Science and Technology, Singapore University of Social Sciences, 463 Clementi Road, Singapore 599494, Singapore
e-mail: yqzhu@suss.edu.sg

H. Guo et al. (eds.), *IRC-SET 2021*, https://doi.org/10.1007/978-981-16-9869-9_26

house to ensure that the old-aged are doing well, and a camera to ensure that they are safe from harm.

This work targets to develop a Merdeka-Pioneer monitoring system (MPMS) to allow caretakers or NOK to monitor the Merdeka and Pioneer generation at home. Data analysis can be conducted on the data collected by MPMS, and the statistical results can be visualized for better management. Firstly, MPMS allows vital data about the indoor environment, e.g., air quality and proximity, to be collected in real-time automatically and sent to the cloud from the microcontroller via the Internet. The caretakers or NOK can retrieve the data from related app to access the data. A live dashboard will be able to demonstrate the real-time data being collected. Secondly, to ensure that the golden ager is free from harm, there will be a camera to be installed, either onboard or IP cameras which are readily available in the market. We will integrate the camera to our system by adding in the IP address of the camera. Lastly, wearables such as a watch with heart monitoring and/or blood pressure will allow NOK to monitor their status from time to time. This could possibly address the issue of the well-being of golden ager, ensuring they are safe and sound [2] and able to live at home by themselves. Their loved ones who need to work remotely are able to have peace of mind at work and can get notification when misfortune events happen. For example, when the data collected returned null over a period of time, pre-emptive measures would be triggered. A buzzer or notification would be sent out to ensure that help was provided in time when the golden ager faced difficulties.

MPMS includes a main device, wearable devices, and a backend IoT platform. For implementation, ESP8266 NodeMCU is used for both the main device and wearable devices, where the main device is installed in the flat, and wearable devices are carried by the elderly. The NodeMCU in main device is used for monitoring data such as air quality, motion in the room or home. The sensors and components, like motion sensor, gas sensor, buzzer, LED light, and camera, will be integrated with the mainboard. The wearable device is using a NodeMCU to detect falls, heart rate, and able to request assistance by pressing on a button.

Data collected from both devices will be sent to the cloud-based IoT platform, Blynk, which is adopted for this system to allow data storage, information collection, creating reports and triggers, and displaying of alerts and dashboard. NOK can view the information at the application to monitor any abnormalities that are happening to the golden ager. To further increase the accuracy of the fall detector, the motion sensor on the main device will be used to detect if there is any motion after the fall is being detected. An alert will be sent out triggering the buzzer and notifying the NOK accordingly.

The paper is organized as follows. Sect. 26.2 is the literature review. The detailed hardware design and software designs are illustrated in Sects. 26.3 and 26.4. The system testing is discussed in Sect. 26.5 and followed by conclusion in Sect. 26.6.

26.2 Literature Review

In the current market, there are not many IoT solutions that really allow NOK to monitor the golden age in Singapore while working remotely. One of the more relatable products is the elderly monitoring system that is selling of the shelf from M1 [3]. It has real-time tracking of elder activities at home using a motion sensor and a panic button to alert NOK with an SMS when activated. The statistics are all recorded, and the system is proven to be less intrusive compared to cameras and surveillance, but not actually using videos created by cameras.

Another one is assisted living platform (ALP), which is not readily available in the market as the initiative is by SMU [4]. Working similarly as M1 products, it actually does monitor the golden agers medication timing and their daily patterns.

One local example in Singapore is Home+ in 2018, where Aviva committed $135,000 for the deployment of the system and the services [5]. This benefits elders in Jurong GROs with the setup being two motion sensors. Instead of triggering the NOK, it will trigger the alarm to the Red Cross society. Despite the initiative is not as compelling as compared to the other two, this initiative is also a measure to address the objective of the aging society.

Some of the approaches in the market for elder care monitoring systems are mainly using proximity sensors to detect movement in a room or an area, which might have concerns about the accuracy to determine a fall. It is possible that the elderly might be sitting in the living room watching TV or fall asleep while the sensor detects no movement for a certain time and triggers a false alarm. To address this, cameras can be installed in the room, to ensure that the golden agers are doing well. Besides, it would be better to have a wearable device, like the one Health Promotion Board issued out for steps and heart monitoring [6], for the elderly to wear and detect the movement and fall. With the aid of real-time streaming data to ensure that every accident is real, the false alarm can be reduced. When the detections are accurate, it will trigger help to a buzzer installed outside of the flat to raise help from neighbors and send out an email notification and in-app notification to their NOK or caregiver to ensure that help is on the way.

Other than monitoring falls, our MPMS can also monitor the event of fire or gas leakages which might lead to death or cause the apartment to be on fire. For this, a camera and the sensor on the mainboard can address the real-time situation of the issue, increasing the credibility of the source as well as a quick response in the event of misfortune happened. In the event when the air quality rises above a threshold, a trigger to look into this matter will be sent out to ring the buzzer installed. These measures will allow help to reach in a short time and are also relatively more accurate and real-time.

26.3 Hardware Implementation

In this section, we start to introduce the detailed design and implementation of our MPMS system. Let us focus on the hardware first.

26.3.1 Hardware Testing

In order to ensure that all components, sensors, and microcontrollers are in working conditions before interfacing and integrating, there is a need to conduct pre-deployment testing and checks. Each component has been individually connected to NodeMCU for the integration test. This is to mitigate the risk of faulty hardware during the interfacing period to isolate any potential issue.

(a) *NodeMCU (ESP8266-12)*: With the inbuilt ESP8266 Wi-Fi module to NodeMCU, there is minimal integration needed. There are readily sample codes to test out with on Arduino's library.

(b) *Gas Sensor (MQ135)*: Figs. 26.1 and 26.2 show the connection that was done up for the gas sensor and using the code available in Arduino IDE to test for air quality. The gas sensor has 4 pins and is power by 5 V, we used only 3 pins

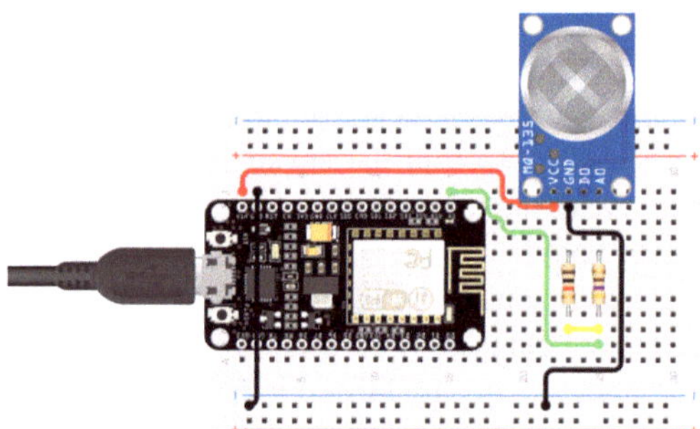

Fig. 26.1 MQ135 to NodeMCU

Fig. 26.2 MQ135 pin connection

NodeMCU	MQ135
VIN	VCC
GND	GND
-	D0
A0	A0

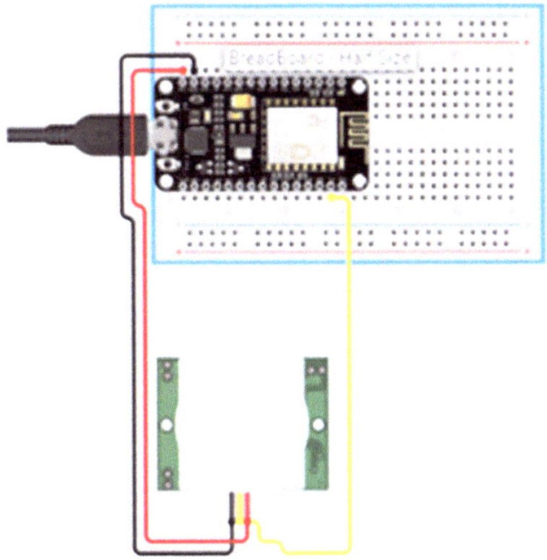

Fig. 26.3 HC-SR501 to NodeMCU

Fig. 26.4 HC-SR501 pin connection

NodeMCU	HC-SR501
VIN	VCC
GND	GND
D1	I/O

which are the VCC, ground, and analog pins. The input is then sent from the module to NodeMCU before sending to the Blynk application.

(c) *PIR Motion Sensor (HC-SR501)*: Figs. 26.3 and 26.4 show the connection that was done up for the motion sensor using the code available in Arduino IDE to test for motion. The PIR sensor is power by 5 V and has 3 input which is plus and minus for power and the blue line is for the signal. When detecting movement, it will display 1 and 0 for no movement. The input is then sent from the module to NodeMCU before sending to Blynk application.

(d) *6-Axis Accelerometer (MPU6050)*: Figs. 26.5 and 26.6 show the connection done up for 6 axis accelerometer and the sample code for testing. The MPU6050 is powered by VIN from NodeMCU. After powering on, the accelerometer will display 6 axis readings X, Y, Z and r yaw, pitch, row.

(e) *Pulse Oximeter and Heart Rate Sensor (MAX30102)*: Figs. 26.7 and 26.8 show the connection done up for pulse oximeter, heart rate sensor, and the sample code for testing. The MAX30102 is powered by VIN from NodeMCU,

Fig. 26.5 MPU6050 to NodeMCU

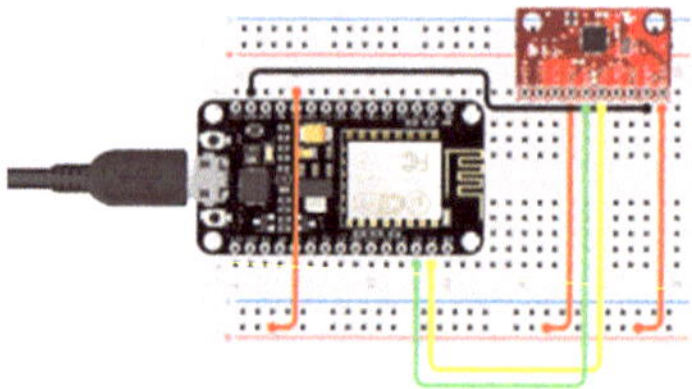

Fig. 26.6 MPU6050 pin connection

NodeMCU	MPU6050
VIN	VCC
GND	GND
D1	SCL
D2	SDA
D8	INT

Fig. 26.7 MAX30102 to NodeMCU

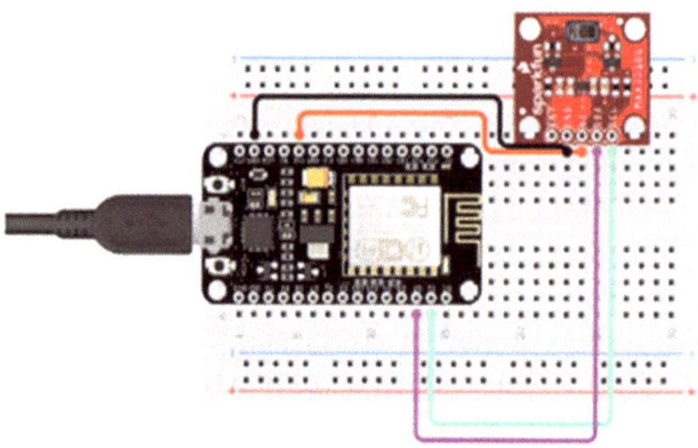

connected to the i2c pins on NodeMCU. After powering on, the accelerometer will display beats per minute (BPM) and the oxygen saturation (SpO2) readings.

(f) *ESP32-Cam*: Fig. 26.9 shows the standalone configuration for ESP32-CAM, the power will be drawn from NodeMCU but the connection is independent in this case as it is directly connected to Blynk.

Fig. 26.8 MAX30102 pin connection

NodeMCU	MAX30102
VIN	VCC
GND	GND
D1	SCL
D2	SDA

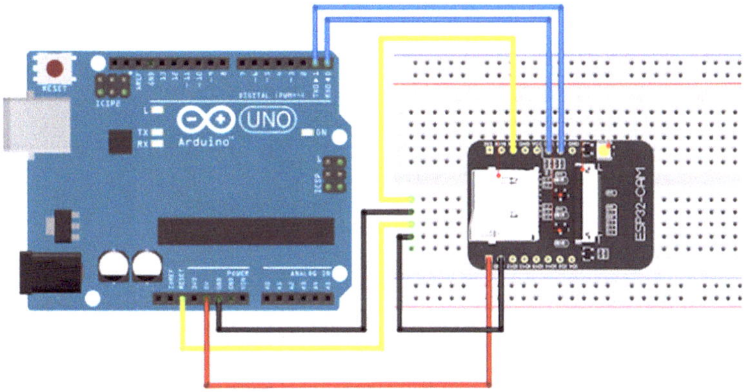

Fig. 26.9 ESP32-Cam

(g) *Buzzer Module*: Figs. 26.10 and 26.11 show the connection for the buzzer module done in Arduino IDE to test for the buzzer sound. The buzzer has 3 pins and is power by 5 V, the blue line is for the signal input for the buzzer.

(h) Button: Figs. 26.12 and 26.13 show the connection done up for the button and sample code used for testing. Connecting one end to the digital pin and the other end to ground. Once the button is pressed, it will go from low to high, lighting up the LED light, vice versa.

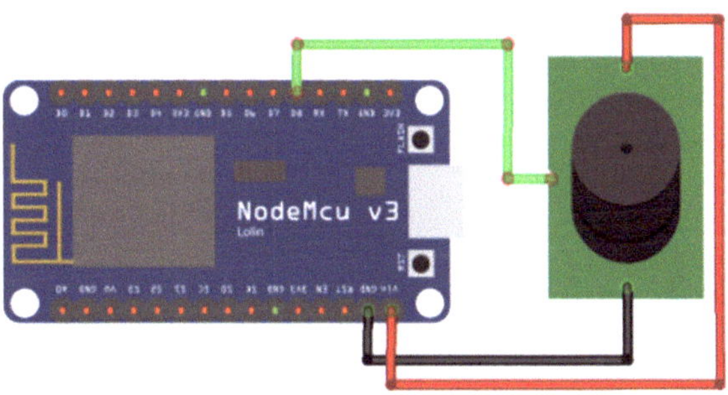

Fig. 26.10 Buzzer to NodeMCU

Fig. 26.11 Buzzer pin connection

NodeMCU	Buzzer
VIN	VCC
GND	GND
D0	I/O

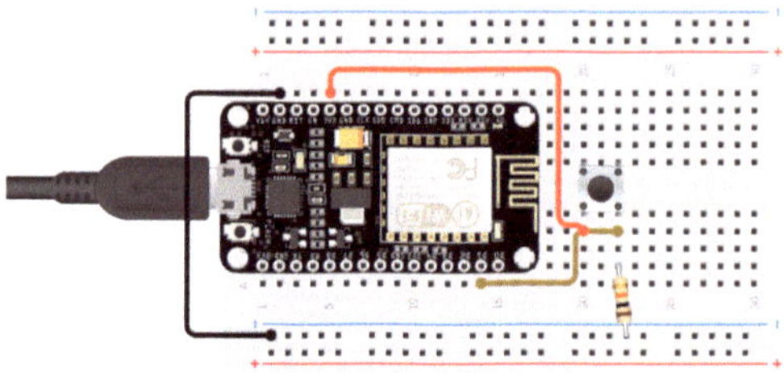

Fig. 26.12 Button to NodeMCU

Fig. 26.13 Button pin connection

NodeMCU	Button
GND	GND
D5	I/O

26.3.2 *Overall Sturcture of the Hardware Devices*

After testing each component with NodeMCU, the wearable device and main device for MPMS are integrated as shown in Figs. 26.14 and 26.15, respectively. Both devices use NodeMCU as the mother board. The main device is responsible for monitoring data such as air quality, motion in the room or home, while the wearable device aims to detect falls, heart rate, and able to request assistance by pressing on a button.

26.4 Software Implementation

Family members or caretakers usually do not have required technical knowledge of engineers to understand the complexity of the codes, binary numbers, or raw data

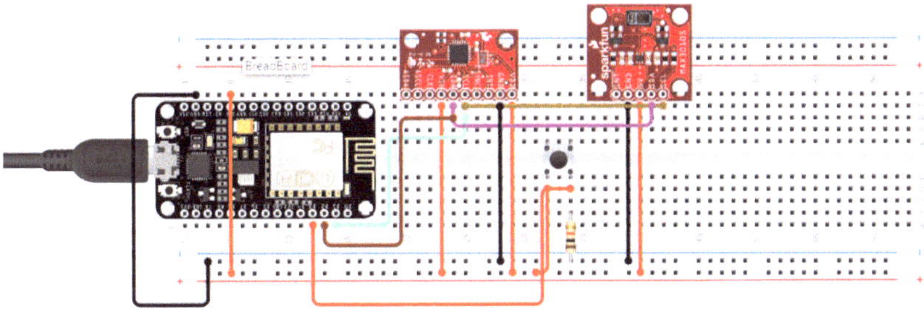

Fig. 26.14 Wearable device hardware design

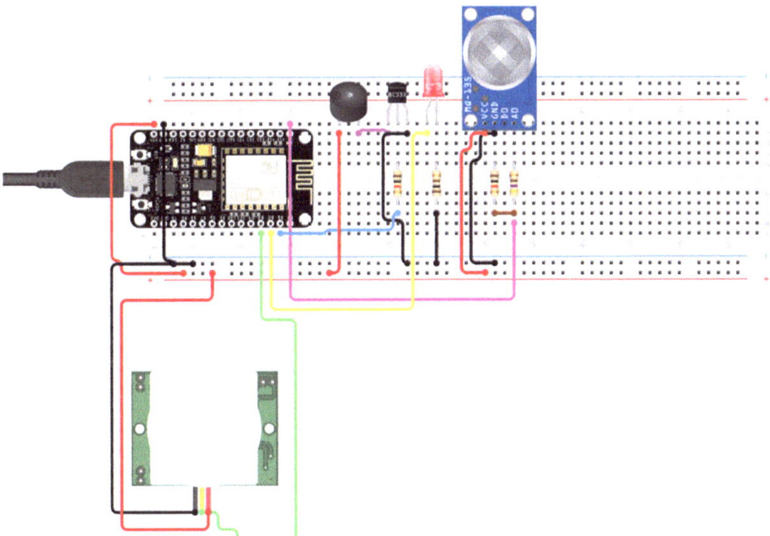

Fig. 26.15 Main device hardware design

before analyzing from sensors. Therefore, there is a need to come up with a user interface (UI) that allows to display, store, analyze data, and easily retrieve them for users to read. After much research, Blynk was chosen to be the platform, in terms of user experience and user interface (UI-UX) as it is available as a mobile application for users. The overall flowchart of communication protocols is shown in Fig. 26.16.

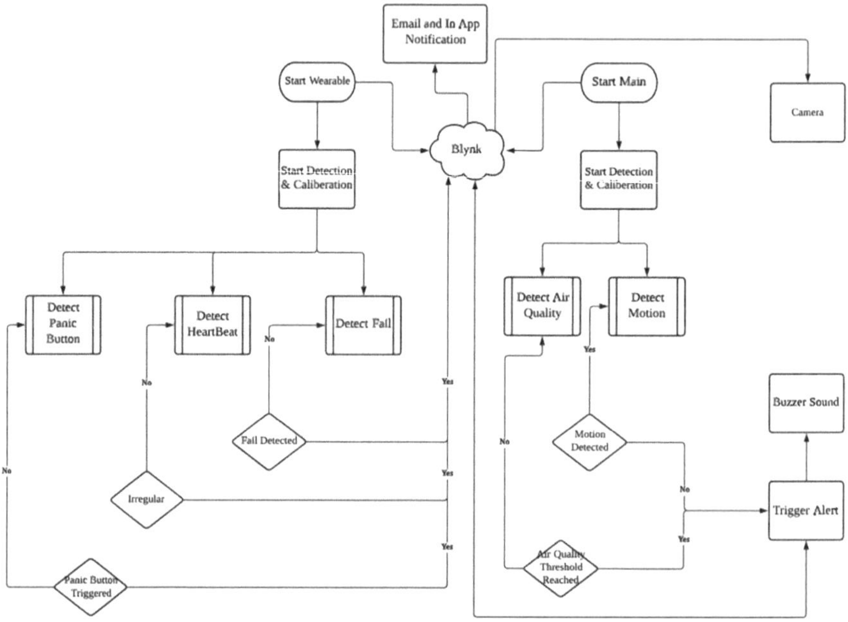

Fig. 26.16 Communication Protocol Flowchart

26.4.1 Blynk Application

Blynk is available on both IoS and android environments. It has a straightforward user interface that allows the user to customize the dashboard by dragging the widgets around. After creating a new project, users need to enter the board that they are using and a project name. For this project, NodeMCU is being selected as the board to be used.

Once done, the user can add widgets for the project. By clicking "+", a widget box will appear and select which widget that the projects require and customize the layout by dragging it around the page. For this project, the crucial 4 LED that was used was for fall detection, panic button, air quality, and pulse sensor. Click on the LED to change the color of the LED and naming for the LED. Remember the input such as V0 and V1, these will be hardcoded inside Arduino to trigger this button LED to light up. Next, we can add the rest of the button icons, email icons, and graphs needed for display. A super chart widget is being used to display the graph for motion, air quality, and beats per minute.

Once the super charge widget is up and running with the display setup, we can click on the settings to alter the information that we need such as the virtual pin to get details from. For email and in-app notification, drag both widgets and setup. It will send out the notifications when an alert is being triggered.

After setting up the last icon needed for the project, click on the play icon on the top right panel. It will start to run and populate the data and start reading and

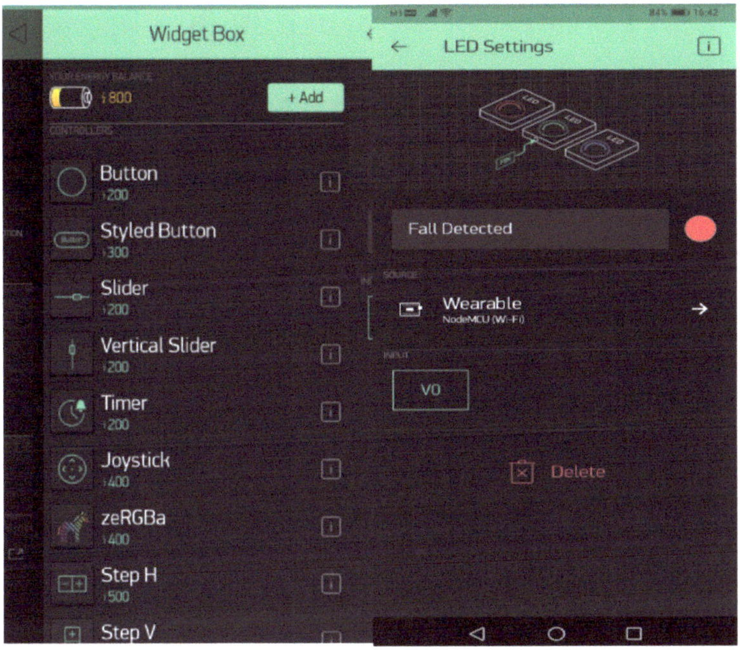

Fig. 26.17 Blynk new widget

displaying out on the dashboard. The play icon will turn into a stop icon even the user wishes to stop running the system. Selected screenshots of application dashboard are shown in the Figs. 26.17, 26.18, and Fig. 26.19.

26.5 Integration and Testing

In this section, we will exhibit the integration between the hardware and software for the IoT platform. Several testing are conducted and illustrated as follows.

26.5.1 Connections to Internet and IoT Platform

As NodeMCU has its own Wi-Fi module, the usage of the source code from the Blynk sample will allow the devices to connect to the Internet. The only changes that are needed were the SSID, password, and authentication code from Blynk to be input in the code to ensure successful onboarding of the device to the platform.

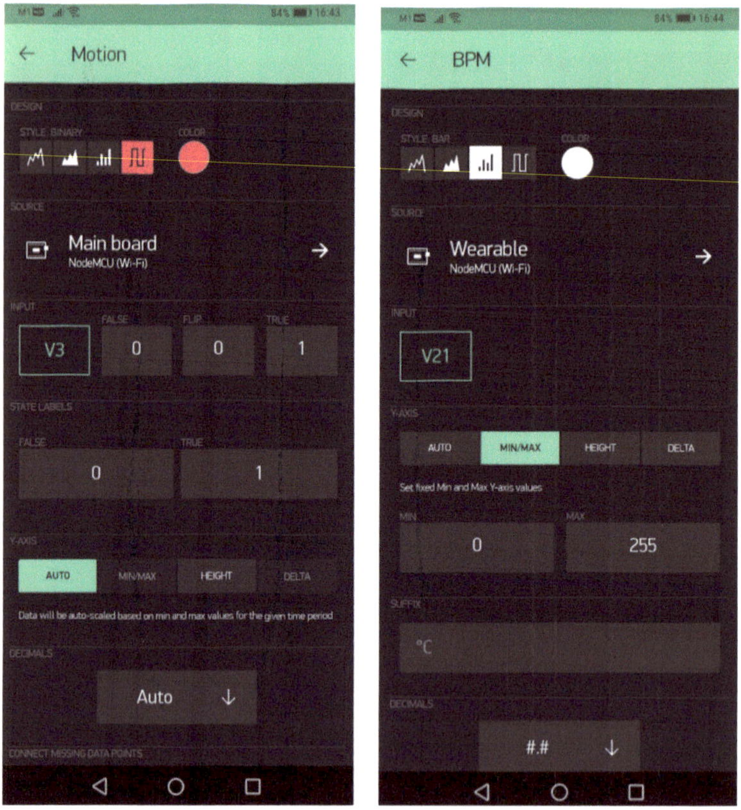

Fig. 26.18 UI graph setting for Blynk

26.5.2 Data Collection

Since boards on both devices are using NodeMCU, there is a need to include a link bridge to ensure synchronization between both boards on Blynk.

For the main device board, there are 2 sensors being interfaced to detect air quality and motion. The gas sensor can be read off the values using the library that was installed for MQ135, making it easier to use on Arduino IDE. For the motion sensor, we use a state to define either 1 or 0, meaning movement detected or no movement detected.

For the wearable device board, there are also 2 sensors being interfaced to the board where the 6-axis accelerometer will read the data from the sensor and display them through serial print. When a fall has been detected, a counter was used to trigger the alert once it hits the threshold. After a fall is being detected, it will need to be reset after acknowledged and someone that has attended to the golden ager. The user can either press the panic button on the wearable or on the app itself to cancel the alarm and a print will state that reset of fall detection is being executed. For the pulse

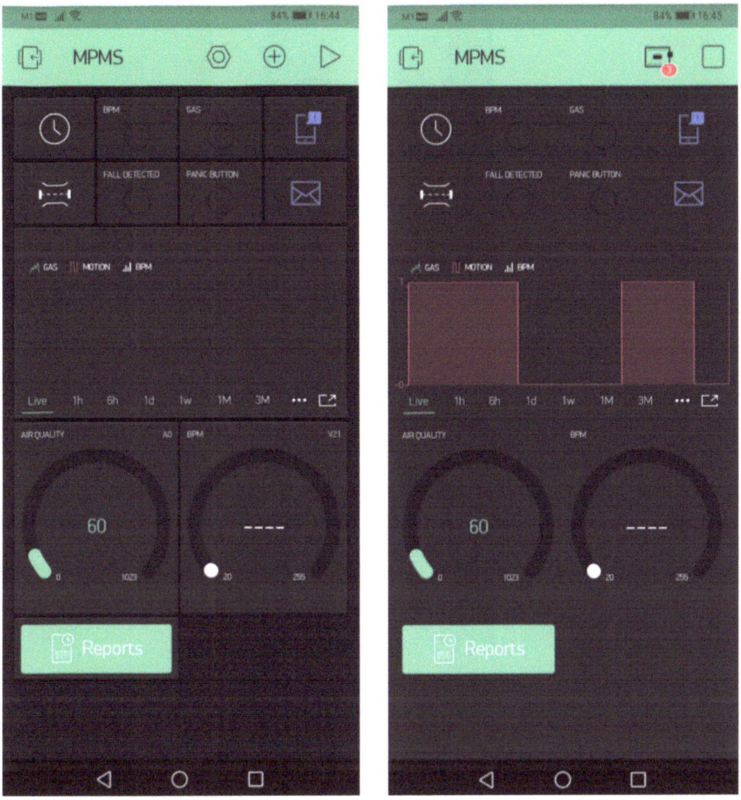

Fig. 26.19 Overview UI for Blynk

sensor, a minimal range is set, and when the unit falls off the range, it will alert the app about the abnormalities. Similar to fall detection, the alarm can be reset as well.

Lastly, on the wearable device, there is a panic button installed for golden ager to press for assistance. An alarm will trigger off once the button is pressed, alert NOK to look into it. A serial print will show that the button is being pressed. In order to reset, the user needs to press twice to cancel the alarm in the event of a false alert or triggering by accident.

26.5.3 Prototype Testing

For prototype testing, we used a USB Power Bank as wearable and a wall adapter USB to micro for NodeMCU. Once power-up, both would boot up and started to connect to the Wi-Fi and Blynk platform. Once the connection was established, the wearable device should be hanging around the wrist, but since it was still a prototype

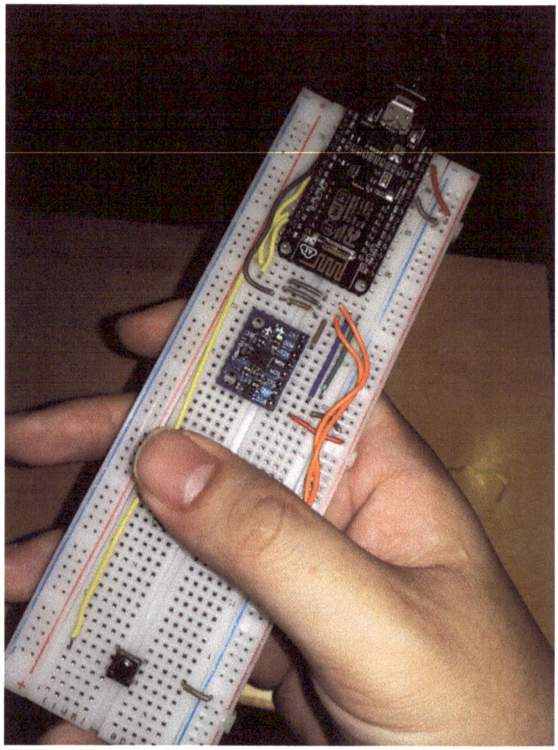

Fig. 26.20 Prototype wearable device

and had not been made into the desired look and feel, it was held on with hands. The prototype wearable device is shown in Fig. 26.20.

There are 4 scenarios being tested. Firstly, to stimulate a fall, the wearable was hold and fell onto the ground. This would trigger the red led on the Blynk application, when the motion sensor detected no movement, it would sound off the buzzer and sent notifications through the app and email as shown in Fig. 26.21. Canceling of alert could be done by pressing the panic button once or through the app.

We tested the scenario for the panic button pressed. Once it was pressed, the led would light up, and the trigger would sound off the buzzer and notification would be activated.

The third scenario being tested was the gas sensor. Once it was past the threshold, the led would light up, the trigger would sound off the buzzer and notification would be activated. Buzzer and LED would turn off when the air quality was back to normal or the user could overwrite by turning the button off on the app.

The last scenario being tested was the pulse sensor. Once it fell below average BPM or no pulse was detected, the led would light up, the trigger in-app notification would be activated.

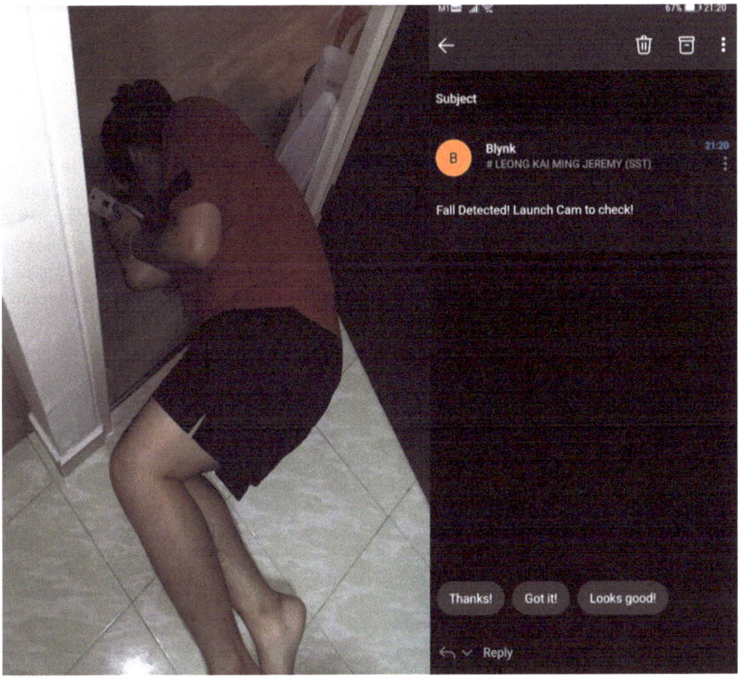

Fig. 26.21 Fall detection

26.6 Conclusion

The aging population in Singapore has always been an issue when Singapore's birth rate starts to lower, causing it to be an inverted triangle in terms of the population. In this paper, we have proposed a Merdeka-Pioneer monitoring system (MPMS) which can monitor elder falling and bad environment conditions in real-time. The MPMS collects multiple dimensional data based on various IoT sensors and manages data in the cloud. The hardware components have been tested and integrated, while software interfaces have been implemented. Prototype devices have been tested with both hardware and software integrated together. Multiple scenarios have been tested using the prototype devices to show the effectiveness of the proposed MPMS system.

References

1. Siau, M. (2020, March). Elderly to make up almost half of S'pore population by 2050: United Nations. [Online] Available at: https://www.todayonline.com/singapore/elderly-make-almost-half-spore-population-2050-united-nations
2. Alkhatib, S. (2020, January). 5 months' jail for maid who repeatedly abused elderly woman with dementia. [Online] Available at: https://www.straitstimes.com/singapore/courts-crime/5-

months-jail-for-maid-who-repeatedly-abused-elderly-woman-with-dementia
3. Ting, L. (2016, March). M1 launches home monitoring service targeted at the elderly. [Online] Available at: https://www.straitstimes.com/tech/m1-launches-home-monitoring-service-targeted-at-the-elderly
4. Choo, F. (2018, January). New tech for old folk to live at home safely. [Online] Available at: https://www.straitstimes.com/singapore/health/new-tech-for-old-folk-to-live-at-home-safely
5. Yahya, Y. (2018, July). New home monitoring system for elderly in Jurong Central alerts Red Cross in emergencies. [Online] Available at: https://www.straitstimes.com/singapore/new-home-monitoring-system-for-elderly-in-jurong-central-alerts-red-cross-in-emergencies
6. Choo, F. (2019, September). Silver challenge walking trails launched for seniors as part of national steps challenge. [Online] Available at: https://www.straitstimes.com/singapore/health/silver-challenge-walking-trails-launched-for-seniors-as-part-of-national-step

Chapter 27
Wideband Sinuous Antenna Design

Selina Peh Yuet Ning, Lim Zi Wei, and Ang Teng Wah

Abstract Wideband, frequency-independent sinuous antennas are useful in several applications due to their versatile polarisation capabilities and planar, compact structure. To ensure that the antenna performs optimally, its design should be optimised to achieve a consistent frequency response over its operating bandwidth. This project investigates the operating principles of the sinuous antenna to provide a comprehensive overview of how to best optimise its performance. Design characteristics of the sinuous antenna are investigated via parametric sweeps to observe how radiation patterns are affected. Mutual coupling in the sinuous arms, which decreases the consistency of frequency response and efficiency of antenna radiation, is observed for certain parameter values, and optimal values are recommended to avoid it. A linearly tapered microstrip balun and feed design are adapted for proper signal conversion and impedance matching, with discussion on possible interfering radiation affecting the overall performance of the antenna. Testing of the design prototype is done to evaluate the return losses and radiation patterns of the proposed design.

Keywords Sinuous antenna · Wideband · Frequency independent

27.1 Introduction

The sinuous antenna, patented by DuHamel [1], belongs to the class of wideband frequency-independent antennas, meaning that their geometry can be expressed entirely in terms of angles without specific dimensions, combining characteristics of spiral and log-periodic antenna structures [2]. What sets the sinuous antenna apart from other frequency-independent antennas is its dual linear and circular polarisation capabilities, despite its low-profile structure. Other low-profile frequency-independent antennas like spiral and certain log-periodic antennas cannot achieve

S. P. Y. Ning (✉)
Raffles Institution (Junior College), Singapore, Singapore
e-mail: 21YSELI857H@student.ri.edu.sg

L. Z. Wei · A. T. Wah
DSO National Laboratories, Singapore, Singapore

H. Guo et al. (eds.), *IRC-SET 2021*, https://doi.org/10.1007/978-981-16-9869-9_27

such orthogonal dual polarisation modes without making use of arrays or complicated modifications to their geometry or feed structures. For example, the spiral antenna can only undergo circular polarisation but not linear polarisation. Other frequency-independent antennas are capable of producing similar radiation patterns, such as quad-ridge horns or Vivaldi antennas, but they are high profile, three dimensional structures, compromising mobility [3].

Operation of a sinuous antenna can be explained using the radiating ring theory: current travels along sinuous curves of each antenna arm fed from the geometry's centre, propagating along the arms until it reaches the active region corresponding to $\lambda/2$, which forms a "ring" around the circumference of the antenna. Any residual currents attenuate significantly before they excite the next active region of the sinuous arms corresponding to lower frequencies [2].

In both linear and circular polarisation modes, the antenna radiation pattern consists of two rotationally symmetric bilateral beams. Linear polarisation can take place with just one excited sinuous arm pair, where opposite arms are driven in a balanced manner. At least two pairs of arms are required for circular polarisation, where the two arm pairs are excited with a 90° phase shift. A hybrid coupler can be used to attain the required phase shift, in which LHCP or RHCP can both be achieved. This paper will focus on the design for a four-arm sinuous antenna.

Sinuous antennas are utilised for many different purposes due to their polarisation capabilities and low profile, which makes them optimal for portable or compact systems. They are useful in direction finding where orthogonal polarisation modes can allow for good receiver–transmitter isolation [3]. In polarimetric radar remote sensing applications, higher accuracy in target classifications can be achieved with orthogonal polarisation modes by a sinuous antenna without compromising mobility [4]. They are also useful in radio astronomy applications, healthcare monitoring and several other ultra-wideband applications [4].

This paper investigates the radiation characteristics of the antenna, and identify how different design parameters affecting the geometry will impact these characteristics such that the design can be optimised for these applications. This entails consistency in radiation characteristics of the antenna, in terms of (a) gain flatness, where the frequency response of the antenna is predictable and consistent across its operating bandwidth and (b) directionality, in which the antenna's peak radiation direction or peak sensitivity to transmitted signals is in the same direction throughout the operating bandwidth and (c) input impedance, where impedance of the system remains consistent throughout its operating frequency range such that the same feeding and impedance matching structures can be used.

27.2 Sinuous Antenna Design

27.2.1 Generating the Sinuous Curve

Each sinuous trace is made up of p cells each, described by Eq. (27.1) in polar coordinates (Figs. 27.1 and 27.2)

$$\phi = -1^{p}\alpha \sin\left(\frac{\pi \ln\left(\frac{r}{R_p}\right)}{\ln \tau}\right) \pm \delta \tag{27.1}$$

α is the angular width of each cell. τ is the growth rate of successive cells, where $R_{p+1} = \tau R_{p.}$ δ determines the thickness of each sinuous arm. The cells of each arm are plotted from when $0 < \tau < 1$. In this case, R_p is the outer radius of the antenna. Some authors choose values of $\tau > 1$, in which R_p becomes the inner radius of the antenna, and the cells expand outward. These cells are then joined up and displaced across an angle to form each arm of the sinuous trace.

To close off the ends of each arm, parametric Eqs. (27.2) and (27.3) are used on the inner and outer radii, respectively. Each arm is then created by rotating the trace by 90° to form a four-arm sinuous trace [1].

$$x = R_{\text{inner}} \cos(\phi),\ y = R_{\text{inner}} \sin(\phi) \tag{27.2}$$

$$x = R_p \cos(\phi),\ y = R_p \sin(\phi) \tag{27.3}$$

Like many other frequency-independent antennas, the upper and lower frequency limits of the sinuous antenna are determined by the inner and outer radii of the sinuous trace in Eqs. (27.4) and (27.5).

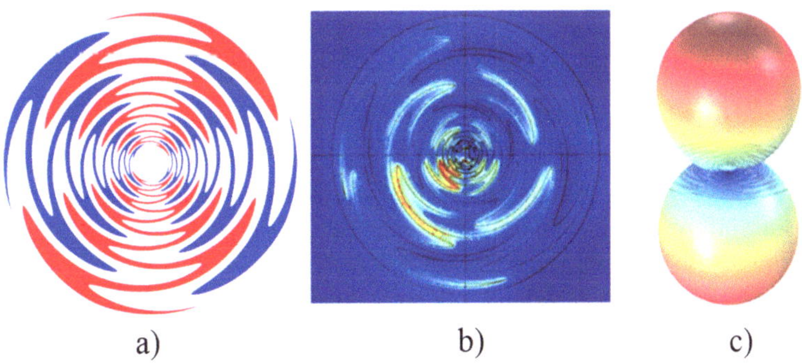

Fig. 27.1 **a** Example of four-armed sinuous antenna geometry, **b** electric field plot showing "radiating ring" formed by active region of the sinuous trace, **c** 3D radiation pattern plot of a planar sinuous antenna

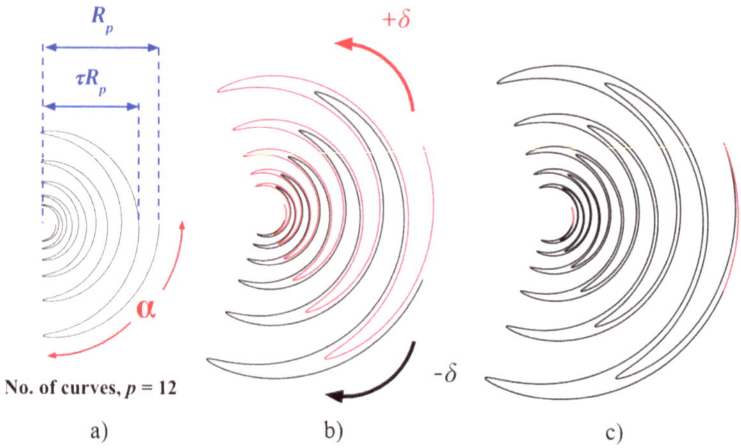

Fig. 27.2 Process of generating sinuous antenna geometry. **a** Generating sinuous curve cells according to 27.1, **b** creating sinuous arm thickness via rotation of sinuous curve cells, **c** closing off sinuous arm with parametric equations

$$R_{\text{inner}} = \frac{\lambda_{\text{upper}}}{4(\alpha + \delta)} \tag{27.4}$$

$$R_{\text{outer}} = \frac{\lambda_{\text{lower}}}{4(\alpha + \delta)} \tag{27.5}$$

However, antenna dimensions usually extend beyond these limits; the sudden truncation of the sinuous arm outer ends can cause unwanted distortion of the radiation patterns near the lower frequency limit if the outer radius corresponds too closely to that limit. As for the inner radius, interference from the feeding region can cause distortion of the overall radiation pattern as well [1]. Most authors design the inner radius to half the value calculated in (27.4) as a conservative estimate that ensures consistent radiation patterns, which is also done in this paper.

27.2.2 Evaluating Design Parameters

Specific values for each design parameter used to draw the sinuous trace must be chosen carefully in order to optimise antenna performance. Combinations of these design parameters can contribute toward different effects on the antenna's radiation characteristics and patterns. Simulations were conducted on COMSOL multiphysics to test the effects of these parameters before antenna fabrication. Arm thickness δ is usually maintained at 180°/2 N (N number of arms), specifically 22.5° for four-armed sinuous antennas, to maintain its self-complementary structure and consistent impedance across its operating bandwidth given a specific mode of operation. 22.5°

Fig. 27.3 Graph of far field patterns of sinuous antenna radiating at 3 GHz with varying α, illustrating difference in half-power (-3 dB) beamwidth, $\tau = 0.85$. Vertical dotted lines indicate half-power beamwidth

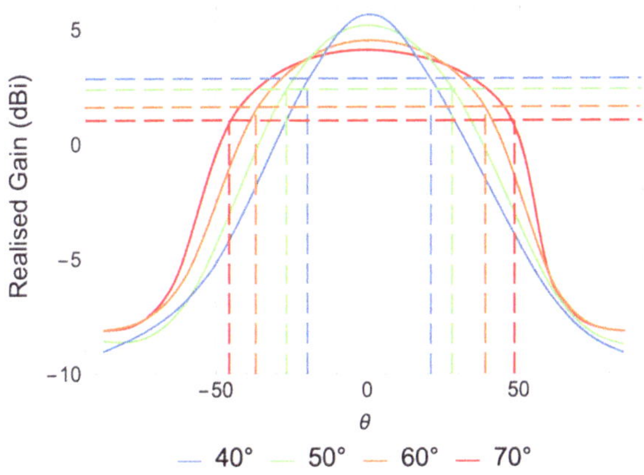

has also been found to provide sufficient arm thickness for ease of fabrication, whilst ensuring current flow is not hindered through the structure [5].

Maximising angular width α of the sinuous trace can increase the bandwidth of the antenna as specified by (27.4) since the outer and inner radius limits are determined by $(\alpha + \delta)$. Each active region supports a greater frequency range with the same antenna radius, minimising the required size to attain a certain operating bandwidth; increasing α from 40° to 70° can increase the antenna bandwidth by about 0.5 GHz when the outer radius = 10.0 cm.

In line with this, the antenna radiates with a greater beamwidth given each cell's increased curvature; beamwidth is inversely proportional to $(\alpha + \delta)$, where generally a 10° increase in $(\alpha + \delta)$ sees about 10° increase in beamwidth as simulated in Fig. 27.5, with some dependence on τ [1]. There is also a decrease in the maximum gain of the antenna with increased α, as the power radiated by the antenna is spread across a larger area (Fig. 27.3).

Larger values ($\alpha \gtrsim 80°$) are known to cause individual sinuous arm pairs to undergo circular polarisation, alternating between LHCP and RHCP depending on frequency of the signal and the orientation of the specific active regions and current flow, instead of being linearly polarised; thus, care must be taken in selecting values of α so as not to produce undesirable radiation characteristics [5].

Polarisation wobble of the sinuous antenna [6] has been highlighted as a concern when the sinuous antenna is linearly polarised. Due to the nature of the sinuous geometry, different areas of the sinuous geometry radiate at different frequencies. As a result, the radiation pattern will tilt varying with frequency. This mostly affects linear polarisation modes, where radiation is concentrated on a single plane, but these effects cancel out when the antenna is circularly polarised [6]. This phenomenon can lead to unfavourable "wobbling" of the direction of peak radiation across the operating bandwidth of the antenna, depending on which active regions of the antenna are excited at each frequency; one will observe an "oscillation" of the direction of peak radiation with variation in transmitting frequency of the antenna. With larger values of α and smaller values of growth rate τ, the angular range and radius in which each active region spans increases, thus increasing polarisation wobble. This is a common problem amongst log-periodic antennas, but it is relatively mild for a sinuous antenna, where variation can be less than $\pm 5^\circ$ [5] with proper selection of parameters (smaller α, greater τ). Our parametric sweep sees polarisation wobble less than $\pm 5^\circ$ up to $\alpha = 50^\circ$, with $\pm 10^\circ$ or greater deviation from the normal of the antenna plane only when α is close to 90°.

For a fixed operating bandwidth, growth rate τ determines the number of cells in the sinuous trace. τ is kept consistent to maintain the log-periodicity and self-complementary character of the antenna such that it maintains a consistent impedance and radiation characteristics throughout its operating bandwidth. In [1], a minimum value of $\tau = 0.65$ is recommended as it ensures sufficient attenuation through the specific active region excited in order to maintain rotationally symmetric, consistent radiation patterns throughout the antenna's operating bandwidth. Tending towards unity in the range of $0 < \tau < 1$ produces more uniform radiation patterns; with more available active regions within the same radius, frequencies within the operating bandwidth are better supported and produce smoother radiation patterns whilst at the same time increasing the mutual coupling between adjacent arms; these effects can be observed in the graphs in Fig. 27.5 [6]. However, it is of note that a larger growth rate can induce greater conductor losses and may make the antenna harder to fabricate due to its thinner cell traces. Some authors have varied τ for each cell in order to avoid excessively thin traces, but at the cost of forfeiting consistent radiation characteristics of the antenna as the geometry is no longer log-periodic [5].

Impedance of the sinuous antenna is consistent throughout its operation bandwidth due to its self-complementary geometry characteristics. By Deschamp's equation for *N* arm geometries in free space [6], input impedance to each sinuous arm is about 133.3 Ω. For a four-arm sinuous antenna, each balun connected to one arm pair will see an input resistance of about 267 Ω for both linear and circular polarisation modes, in which each arm pair is driven by independent sources at the ports of their baluns (see Sect. 27.4: *Feeding System* for connection of balun to sinuous arm pairs).

27.2.3 Mitigation of Unwanted Resonances in the Sinuous Antenna

Whilst it generally seems that larger α and τ leads to more wideband performance, simply, maximising the value of these parameters can produce inconsistencies in the frequency response of the antenna, aside from the polarisation wobble discussed before. Resonance occurs when the total length of an active region is close to $\lambda/2$ of a specific frequency, meaning that they are log-periodic according to growth factor τ. These resonant modes are usually unwanted as they can cause radiation pattern inconsistencies and additional ringing in the time domain when receiving signals at these specific frequencies [4], which would be undesirable for most applications.

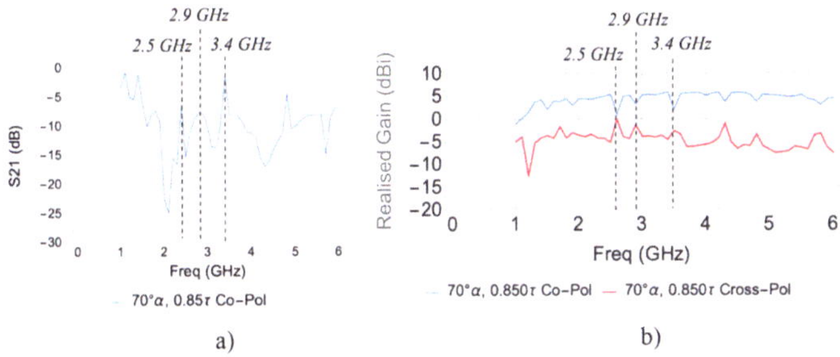

Fig. 27.4 Comparison of peaks in S21 and dips in co-pol gain of antenna's peak frequency response

The extent of which this resonance affects the consistency of the antenna gain patterns correlates with the extent of mutual coupling between the interleaving sinuous arms and adjacent cells [5]. With larger values of α and τ, greater interleaving and shorter distances between adjacent sinuous arms result in greater mutual coupling at these resonant frequencies and significant cross-polarisation effects. The amplification of such resonances can be attributed to mutual coupling between adjacent active regions; plotting the returns losses of the sinuous antenna shows significant spikes in $S21$ values at the same frequencies as the resonances spotted in the co-pol gain plots, suggesting that mutual coupling between the adjacent is responsible for power losses and the resulting changes in frequency response (Fig. 27.4).

To compare how these design parameters affect the extent of these resonances, a parametric study was conducted in which α was varied from 40° to 70°, whilst three values of τ were tested: 0.740, 0.825 and 0.850. Inner and outer radii of the sinuous trace were kept constant throughout. Using the frequency response graphs plotted in Fig. 27.5, we can compare the effects of α and τ on the radiation patterns of a sinuous antenna. We see that larger values of α cause sharper dips in the co-pol gain of the sinuous antenna. Resonant frequencies shift based on the values of both α and τ, affecting the length of the active regions. A maximum angle of $\alpha = 50°$ avoids significant drops in gain at resonant frequencies, and $\tau = 0.825$ and 0.850 generate smoother frequency responses overall given smaller α values.

27.2.4 Truncation of Sinuous Geometry

Truncation of the sinuous antenna can be carried out in order to mitigate the unwanted resonances [4], by making empirical modifications to cut off the sharp ends of the antenna arms at the ends of each active region. Fig. 27.6 shows the results of trimming 20° off a specific active region from a $\alpha = 70°$, $\tau = 0.825$ sinuous trace. By trimming only the ends of the outer six bends of the sinuous trace on each arm as in Fig. 27.6b, we mitigate only the corresponding resonances from each trimmed active region. This can be another solution to eliminate specific resonances, but it also comes at the cost of effectively decreasing the value of α. In the case of this project, selection of

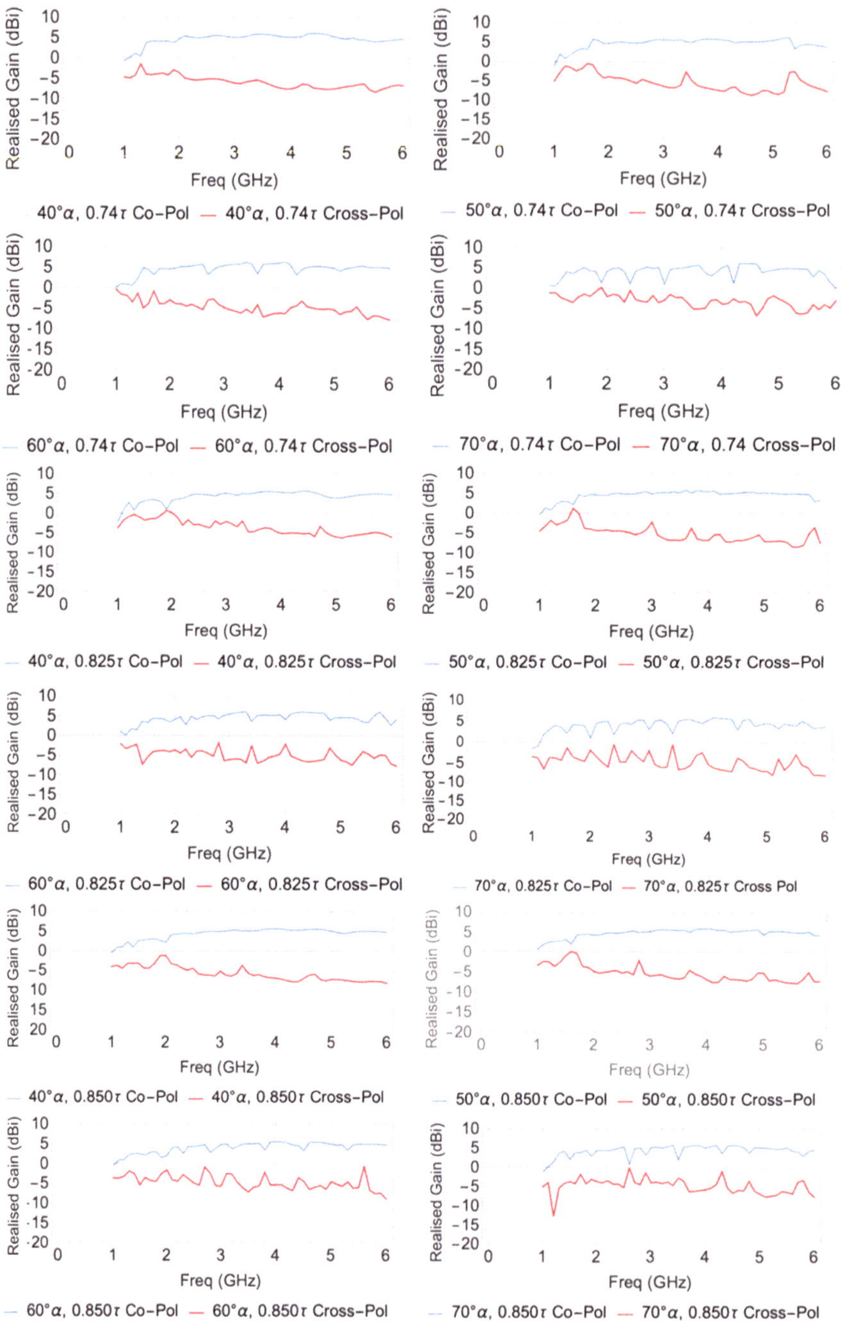

Fig. 27.5 Plots of peak frequency response with varying parameter values of α (from 40° to 70°) and τ (0.740, 0.825, 0.850), including dips in co-polarisation gain due to mutual coupling effects

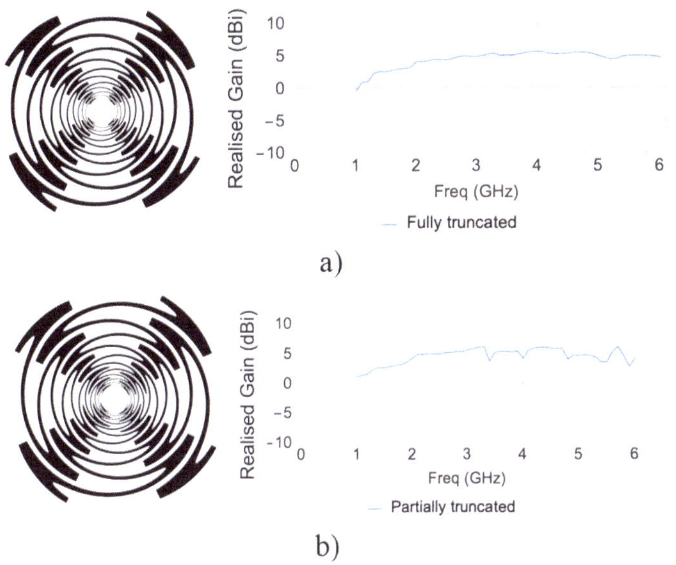

Fig. 27.6 Peak frequency response after **a** full truncation of sinuous arms by 20° and **b** partial truncation of outer six sinuous arm cells

design parameters is sufficient for the operating bandwidth, and no truncation was carried out. In literature, resonances caused by the traditional sharp end truncation of the sinuous antenna have also been mentioned. The same truncation method can be used to terminate the sharp ends and reduce the current reflections at the ends of each sinuous arm due to sharp truncation of the sinuous geometry [4] (Figs. 27.7, 27.8 and 27.9).

The selected parameter values for the fabricated antenna prototype are listed in Table 27.1. α and τ are picked in consideration of ensuring the frequency response of the antenna remains smooth across the operating bandwidth of 2–4 *GHz*. Truncation

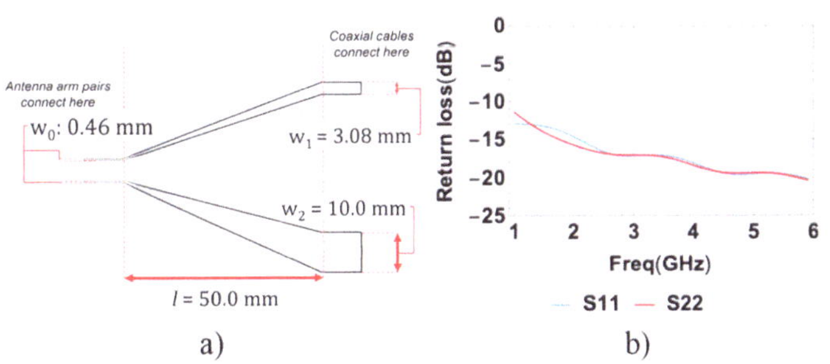

Fig. 27.7 **a** Design of linear balun and **b** return losses at each port (grey and black outlines in a) indicate opposite faces of substrate)

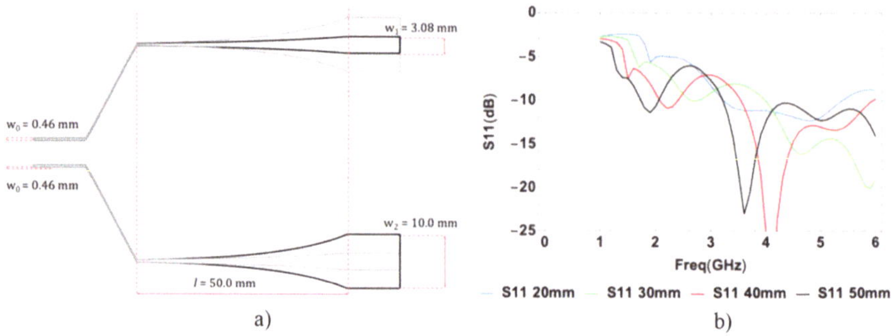

Fig. 27.8 **a** Design of exponential balun and **b** return losses at port 1 with increasing values of *l*(exponentially curved section of balun)

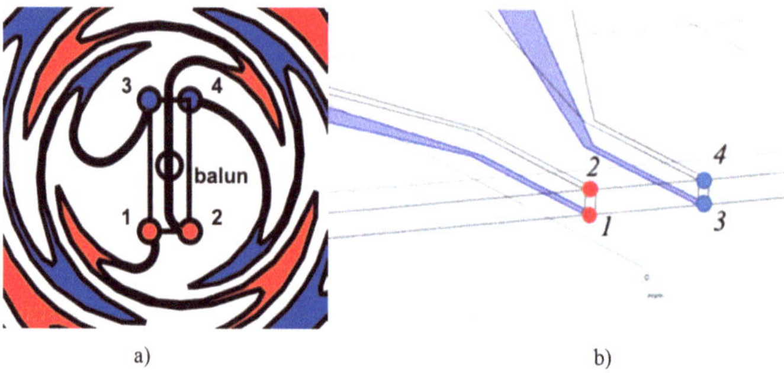

Fig. 27.9 **a** Top view of antenna feed and **b** connection points to balun

of the sinuous arms is not carried out as it produces similar effects to choosing smaller values for α of τ. See Fig. 27.10 for a visualisation of the final sinuous trace design.

27.3 Balun Design

Since the sinuous antenna is fed by 50 Ω coaxial cables (unbalanced transmission lines) at each port, impedance matching is required to minimise power losses and ensure efficient antenna performance. Similar to other frequency-independent antennas, a tapered microstrip balun can be used for impedance matching. The tapered transition across the microstrip transmission lines from the unbalanced end is gradually reshaped in terms of width until the impedance of the microstrip transmission line matches that of the antenna. This taper gradually decreases in width across a certain length (at least a quarter-wavelength, λ/4, for smooth transition).

A 60 mil substrate of dielectric constant $\varepsilon_R = 3.66$ was used to ensure that the balanced end of the microstrip transmission line could be fabricated easily. Widths of the microstrip lines at the two ends were determined using a microstrip calculator. At

Table 27.1 List of design parameter values used

Parameter	Effects		Chosen value
R_{inner}	Determines operating bandwidth (theoretically, 2 GHz: ~30 mm, 4 GHz: ~15 mm), requires adequate transition regions to avoid current reflection from sharp truncated ends (lower limit) and feed interference (upper limit)		6 mm
R_{outer}			50 mm
α (<90°)	Smaller values (<50°): decrease mutual coupling between arms to reduce unwanted resonances	Larger values ($50° < \alpha < 90°$): Increased frequency bandwidth, increased beamwidth, but $\alpha \geq 80°$ introduces alternating circular polarisation even in balanced excitation modes)	50°
τ ($0 < \tau < 1$)	Minimum value = 0.65 to ensure sufficient attenuation across active region	Maximum value = 1	0.825
	Smaller values: reduces conductor losses, easier to fabricate	Larger values: smooth frequency response, but thinner traces are harder to fabricate, induces greater current losses	
δ	22.5° to maintain self-complementary geometry of four-arm antenna		22.5°
No. of cells, p	Determined by growth rate τ given constraints of operating bandwidth		12

the unbalanced end, the width of the transmission line = 3.08 mm, which corresponds to an impedance of 50 Ω. At the balanced end, line width = 0.46 mm, which was calculated to correspond to the impedance of the sinuous arm itself at about 190 Ω, verified from simulation. Width of the ground plane was fixed at about three times the width of the unbalanced transmission line, making it 10 mm. The transmission and ground line of each balun are printed on opposite faces of the substrate such that the feed structure can be easily implemented, with a vertical section of consistent width at either end to connect the antenna and SMA connectors.

This planar structure consisting of two microstrip baluns was used to feed the two sinuous arm pairs separately via two discrete ports. The design was modified from [7] to avoid sharp turns in the diagonal sections of the transmission lines connecting the taper to the balanced twin lines, to arrive at the design in Fig. 27.7.

Different exponential taper lengths were simulated as an alternative balun design, as nonlinear tapers have been found to increase the operating bandwidth of the balun whilst keeping the overall structure more compact [4]. With an exponential taper, length of the balun can be decreased with somewhat comparable return losses by at least 10 mm (see Fig. 27.8b). However, the exponential taper requires sharp turns at the diagonal sections of the transmission lines connecting the taper to the balanced twin lines, which likely result in higher return losses.

Fig. 27.10 Top and side views of antenna prototype, showing the sinuous trace and accompanying balun

27.4 Feeding System

The feeding system shown in Fig. 27.9 is used to connect each arm pair to the corresponding balun trace. This configuration was used in order to connect opposite sinuous arms accordingly without any overlap within the feeding configuration [8]. Bow-tie feeds can be used as well [5], but this can introduce significant cross-pol radiation, especially at the higher end of the antenna's operating bandwidth. The proposed geometry also avoids having to mill each sinuous arm pair on opposite faces of the dielectric substrate, whilst ensuring that there is no overlap within the feeding system. There was an attempt to minimise the size of the feed system in order to reduce any cross-polarisation effects; the balun's balanced transmission lines were spaced about 5 mm apart with considerations towards better matching the inputs of the sinuous arms and to reduce any possible mutual coupling between the baluns.

27.5 Simulation and Experimental Results

The antenna and linear tapered balun prototype design from Fig. 27.11 was printed on 60 mil Rogers 4350b substrate using an LPKF PCB milling machine. The trace was connected to the balanced ends of the balun by copper wires, where each balun is then fed by a coaxial cable via an SMA connector. The structure was fixed in a 3D-printed mount to facilitate testing.

27.5.1 Vector Network Analysis

Vector network analysis (VNA) measurements were carried out to measure the return loss at each port of the fabricated antenna. When we compare the simulated and measured *S*-parameters, we see that there is a relatively good match between the two, with *S*11 and *S*22 values below −11.5 dB from 2–4 GHz (within the antenna's intended operating range). This indicates effective impedance matching for the sinuous antenna operation.

27.5.2 Radiation Pattern Measurements

Measurements of the antenna gain were conducted in an anechoic chamber, where the linear polarisation radiation patterns of the antenna were measured in the *E* and *H*-planes by exciting the vertical and horizontal arm pairs individually (*V* and *H* ports, respectively). Realised gain measurements within the antenna's operating bandwidth show that the antenna radiates with an acceptable efficiency, with similar patterns observed from both ports due to the self-complementary nature of the sinuous antenna

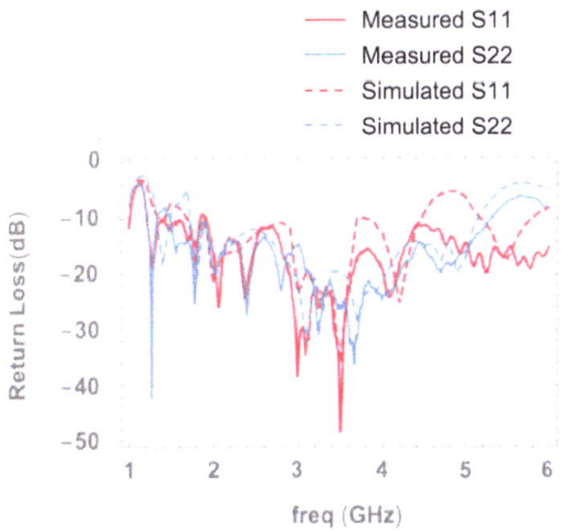

Fig. 27.11 Return losses of antenna prototype

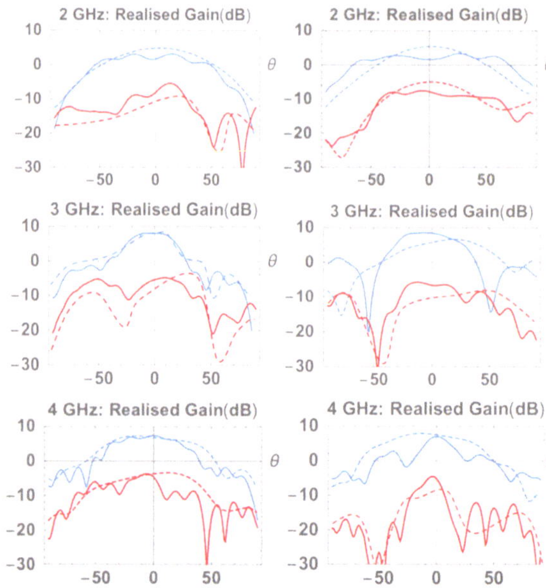

Fig. 27.12 H-port radiation patterns of antenna prototype (linear polarisation) across 2–4 GHz (*blue: co-pol, red: cross-pol, solid line: measurements, dashed: simulation*)

geometry. There is variation in beamwidth as the operating frequency increases from 2 to 4 GHz, with various smaller lobes accompanying the central lobe in the co-pol radiation pattern *(H-port and V-port radiation patterns are very similar due to their similar geometry; thus, only, H-port patterns are shown)* (Fig. 27.12).

27.6 Discussion

Measurement results show that the sinuous antenna and corresponding balun design functions relatively well across the intended frequency range of 2–4 GHz. Sufficient impedance matching throughout the intended frequency range (<-10 dB) and a relatively consistent frequency response and directionality across the operating bandwidth is observed (with the general exception of 3 GHz specifically). However, there are variations in the radiation pattern, affecting main-lobe beamwidth at higher frequencies. This is observed as flatter "fan-shaped" radiation patterns in contrast to the expected conventional bilateral radiation (see Fig. 27.1c). This "fan-shape" may be undesirable in certain applications that require highly directive radiation patterns such as direction finding. Maximum gain of the antenna is also somewhat higher than predicted at 3 GHz.

Inconsistencies in the radiation pattern are likely due to the feed design and balun design. Due to the feed's asymmetrical geometry, the radiation pattern at higher frequencies in the smaller active cells nearer the centre experience significant interference due to the radiation of the balun and feed area. This can destructively interfere with the antenna's radiation patterns, decreasing the antenna's beamwidth. Additionally, balun radiation can also interfere with the antenna's radiation pattern.

Overall, the antenna prototype is able to radiate at an acceptable efficiency across the intended operating bandwidth, whilst maintaining directionality of its radiation patterns.

27.7 Limitations/Further Work

Since the sinuous antenna is capable of radiating across much wider bandwidths, the feed design and design parameters can be further investigated to optimise antenna operation over larger bandwidths. Radiation patterns and response of the sinuous antenna during circular polarisation can be investigated as well.

A Marchand balun design can serve as more compact alternative to the tapered balun used as it can be attached parallel to the antenna plane. A symmetrical feed design can be devised to reduce inconsistencies in the antenna's radiation patterns. More investigation into combinations of sum and difference modes of the sinuous antenna to set up a working DF system can also be done. Unilateral radiation of the sinuous antenna is another desirable feature for several remote sensing applications. Different approaches to achieve this and improve radiation characteristics of the proposed antenna design can be investigated. Absorber cavities are commonly used to create a unilateral radiation pattern. The sinuous trace can also be superimposed onto a conical or pyramidal structure [1], which can increase co-pol gain and reduce back lobe radiation.

Acknowledgements The author would like to thank mentors Lim Zi Wei and Ang Teng Wah for their help and invaluable guidance through the course of this project.

References

1. DuHamel, R. H. (1987). Dual polarized sinuous antennas. *US Patent 4,658,262.*
2. Volakis, J. L. (2007). Antenna engineering handbook (3rd ed.).
3. Zhang, H. S., Xiao, K., Qiu, L., & Chai, S. L. (2014). Four-arm sinuous antenna for direction finding system. In *2014 IEEE international wireless symposium (IWS 2014).* https://doi.org/10.1109/ieee-iws.2014.6864258
4. Kang, Y., Kim, K., & Scott, W. R. (2015). Modification of sinuous antenna arms for UWB radar applications. *IEEE Transactions on Antennas and Propagation, 63*(11), 5229–5234.
5. Crocker, D. A., & Scott, W. R. (2019). Sinuous antenna design for UWB radar. In *2019 IEEE international symposium on antennas and propagation and USNC-URSI radio science meeting.* https://doi.org/10.1109/apusncursinrsm.2019.8888630
6. Cloete, J. H. & Sickel, T. (2012). The planar dual-polarized cavity backed sinuous antenna—A design summary. In *2012 IEEE-APS topical conference on antennas and propagation in wireless communications (APWC)*, pp. 1169–1172.

7. Zhang, X., Fu, G., & Zhang, Z. (2013). A planar sinuous antenna and the relevant balun. In *2013 international workshop on microwave and millimeter wave circuits and system technology*. https://doi.org/10.1109/mmwcst.2013.6814580
8. Gonnet, P., Sharaiha, A., Terret, C., & Skrivervik, A. (1999). Feeding networks for sinuous antennas. *Microwave and Optical Technology Letters, 1999*(20), 195–200.

Chapter 28
EEG-Based Evaluation of Classifying Attention States Between Single and Dual Tasks

Wang Yuting, Wang Yixuan, Portia Foo See Ern, and Aung Aung Phyo Wai

Abstract Everyone in their daily lives, occasionally, needs to juggle more than one task simultaneously. Although performing these tasks sequentially, one at a time, would result in optimal task performance, it is inevitable to multitask. Under such multitasking scenarios, a mixture of different attention paradigms is required to achieve the best performance outcome. Earlier studies have mainly focused on investigating their subjects' cognitive performance under dual tasks or single tasks separately. Not much research had been conducted comparing single tasks and dual tasks based on attention detection using electroencephalography (EEG). We designed an EEG experiment consisting three common cognitive tasks with single-tasking and dual-tasking paradigms to classify the attention levels of subjects. We collected data from twenty-five adolescents after seeking ethical approval and receiving parental consent. We used six bandpower features with machine learning and statistical analysis to evaluate attention detection performance among different task pairs. From our analysis, though there were less statistically significant differences between the mean *p*-value ($p = 0.21$) of accuracy between single tasks and dual tasks, it was also found that there was only 2% accuracy improvement obtained in dual tasks compared with respective single tasks in the subject-independent cross-validation of attention classification.

Keywords EEG · Attention · Cognitive tasks · Single task · Dual task

W. Yuting (✉) · W. Yixuan · P. F. S. Ern
Nanyang Girl's High School (NYGH), 2 Linden Dr, Singapore 288683, Singapore
e-mail: 7114180083@nygh.edu.sg

W. Yixuan
e-mail: 7114180223@nygh.edu.sg

P. F. S. Ern
e-mail: 711418000@nygh.edu.sg

A. A. P. Wai
School of Computer Science & Engineering, Nanyang Technological University (NTU), 62 Nanyang Dr, Singapore 637459, Singapore
e-mail: apwaung@ntu.edu.sg

H. Guo et al. (eds.), *IRC-SET 2021*, https://doi.org/10.1007/978-981-16-9869-9_28

28.1 Introduction

Everyone, in their daily lives, occasionally, needs to handle multiple tasks simultaneously. Although performing only one task at a time would result in more optimal performance, it may be inevitable for one to multitask sometimes [1]. For single-tasks, only, one task is performed at a time. For dual tasks, two tasks are performed simultaneously. An example of a dual task is watching television while holding a conversation. Generally, only, sustained attention is required for a single task, while a mixture of sustained with selective and/or divided attention is required in a multitasking scenario [2]. This is because a task with a dual-task paradigm requires two stimuli to be attended to at the same time, during which attentional resources are divided and/or selectively focused on a certain task [3].

Earlier studies have so far focused on the medical field, either investigating the performance of subjects performing a dual task, or exploring subjects' task performances on a single task [1]. Additionally, previous research has been done usually exclusively on specific target groups like patients with Alzheimer's disease and older adults [1]. Moreover, there are not many studies yet to evaluate attention performance between single tasks and dual tasks using well-known cognitive tests.

Attention is important for and positively correlated to learning, memory, and cognition capability. Attention affects everyone's performance in daily life, and that involves most of the tasks we perform which can be measured and quantified using electroencephalography (EEG). EEG measures electrical activities of the brain non-invasively with high temporal solution applied in various clinical and non-clinical applications [4–6]. There are several existing solutions to recognize user's attention levels so as to then assess, train, and improve their performance on tasks. Although EEG-based attention levels can be derived from spontaneous EEG bandpower features such as theta, alpha, and beta attention levels classified by another method—the supervised machine learning method, as training-based, generally outperforms the former method. In training-based attention classification, we need to train classifier models from EEG data collected through the calibration process. Generally, the standard cognitive tests like psychomotor vigilance task (PVT) and Stroop task are used as the attentive class in conjunction with opposite activities relevant to the inattentive class [7]. In this study, we also use three cognitive tasks, covering sustained and selective attention types, as baseline single-tasks and two variants of dual-task paradigm modified for each task. The goal is to evaluate the attention classification performance between single-tasking and dual-tasking paradigms of each attention calibration task.

The following Sect. 28.2 describes the details of research hypotheses that lead to experiment design including single-tasking and dual-tasking paradigms of three attention calibration tasks. Then, we explained about data collection procedure and data analysis steps in Sect. 28.3. Finally, we discussed the results obtained from data analysis and conclusion in Sects 28.4 and 28.5, respectively.

28.2 Hypotheses and Experiment Design

We leverage data-driven approach in interpreting EEG data to classify user's attentive levels ranging from 0 (inattentive) to 1 (attentive). We hypothesize that dual tasking improve attention performance as compared to single tasking of three cognitive tasks under consideration. With this performance improvement outcome, we can model better accurate machine learning models that can be trained from calibration EEG data for real-time attention detection [8].

To test our hypotheses, we design an experiment consisting of three common attention tasks in three variants: baseline single task and two modified dual tasks (*A* and *B*) on attention classification performance [9] as shown in Fig. 28.1a. During experiment, participants seated about forty to eighty cm away from the computer screen while wearing muse EEG headband. Additionally, we attached remote eye tracker below the screen to detect user's gaze coordinates on screen as shown in Fig. 28.1b.

Every subject goes through three main tasks, namely psychomotor vigilance task (PVT), Eriksen flanker task and Stroop task ordered in random sequence. Stroop task assesses a subject's selective attention where the subject is required to name the ink color of words [8]. Flanker task also tests a subject's selective attention where the subject is presented with a target stimulus that is surrounded by non-target stimuli, and the subject is required to respond to the target stimulus only [10]. PVT measures the speed with which subjects respond to a visual stimulus [11]. To minimize rapid reduction of subjects' attention level due to boredom and repetitiveness, we randomize the sequence of the abovementioned three tasks. Modified dual tasks differ from baseline task with the additional of distractor. The difference between modified dual tasks *A* and *B* is the type of distractor, visual, and audio stimulus, respectively, besides the primary visual attention task. The duration of each task was roughly five minutes, consisting of five blocks of attention and inattention trials with thirty seconds. Participants performed respective visual sustained or selective attention task during the attention trial and looked around the edges of the computer screen slowly without focusing on any point during inattention trial. The first block of every task was a practice block for participants to be acquainted with that particular task. Afterward, they would carry out four blocks continuously.

28.3 Data Collection and Analysis

We collected both EEG and eye tracker data from twenty-five teenagers aged (16.5 $\pm$ 2.5) years and in good health who took part in the experiment voluntarily. We had obtained ethical approval prior to the recruitment and seek parental consents on behalf of participants before the start of the experiment. Safe management measures to combat Covid-19 were also adhered precisely during the entire process of the experiment. EEG data are sampled at 256 Hz for four channels (Fp1, Fp2, Tp9, Tp10), and eye gaze data of 2D coordinates on the screen are sampled at 30 Hz.

As shown in Fig. 28.2, we process multi-channels EEG raw data through preprocessing and feature extraction steps to extract six bandpower features: namely, delta (1–4 Hz), theta (4–8 Hz), alpha (8–12 Hz), low beta (12–18 Hz) and high beta (18–30 Hz), and gamma (30–45 Hz) per channel. To extract these features, we applied fourth order Butterworth band-pass filters (0.5–45 Hz) to remove unwanted signal ranges and removed artifacts using set of moving average filters [12]. Further, we visually look through eye gaze data to remove the trials with high ocular artifacts. Then, the filtered and artifacts-removed signals were segmented with 2 s window size with step size of 200 ms. For all segments, we computed twenty-four bandpower features to classify between attention and inattention (binary classes) using support vector machine classifier [13, 14]. With bandpower features and accuracy outcomes, we test our research hypotheses using t-test and ANOVA statistical analysis using JASP software [15] as shown in Fig. 28.2.

To evaluate binary classification of attention performance, we consider different cross-validation (CV) methods such as tenfold, leave-1-trial-out (L1TO), and leave-1-subject-out (L1SO). The tenfold and L1TO CV methods result subject-specific attention classification model, whereas L1SO shows subject independent, model performance. As both training and testing datasets have data from the same subjects in the subject-specific CV method, it may have resulted in biased evaluation such that though a high accuracy was obtained, the classification model might not be able to predict unknown datasets very well. So, we also use subject-independent CV to see a less biased and consistent performance with unknown future EEG data. For fair comparison among different CV and cognitive tasks, we only use default support vector machine (SVM) parameters as outlined in [10] without optimizing different

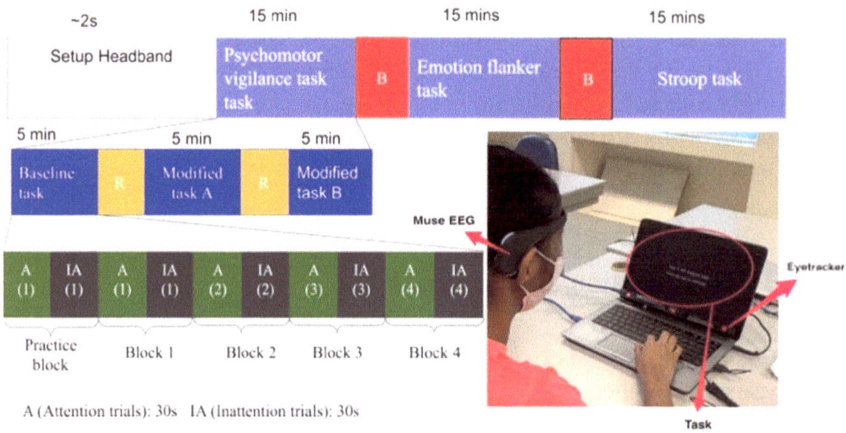

Fig. 28.1 Experiment design **a** experiment protocol **b** subject performing task in experiment

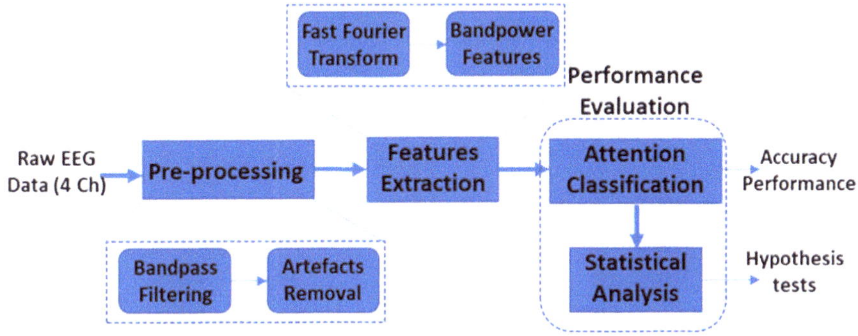

Fig. 28.2 EEG data analysis pipeline

hyper-parameters. SVM is a machine learning algorithm that analyzes data for classification and regression analysis, which sorts data into two different categories. It outputs a map of sorted data with the margins between the two as far apart as possible.

Figure 28.3 shows the mean accuracy of the three cross-validation methods, for baseline tasks, task *A* and task *B*, respectively, to determine overall accuracy differences among CV options. The accuracy comparison is done by comparing baseline task and their two respective dual tasks across three cognitive tasks. A one way between subjects repeated measures analysis of variance (ANOVA) was conducted to compare the effect of task difficulty on attention levels of participants under single-task and dual-task conditions. Due to differences in accuracy of each cognitive task across different CV options, the overall accuracies of both subject-specific

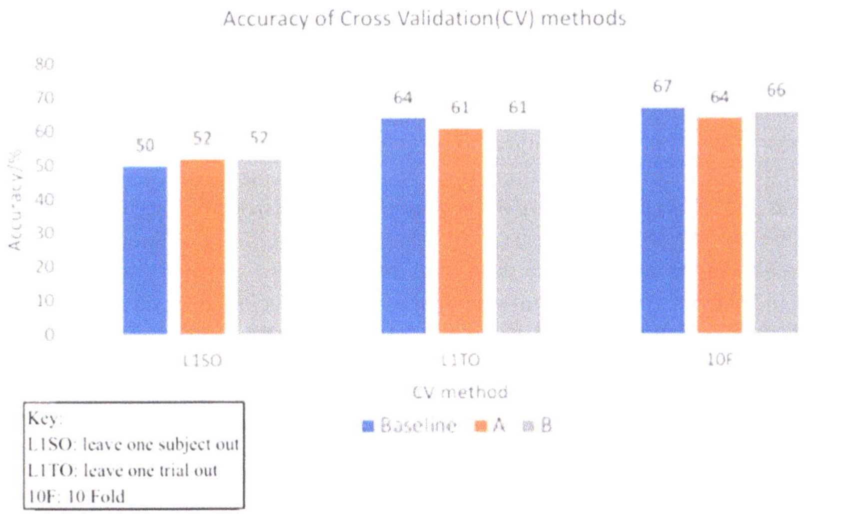

Fig. 28.3 Accuracy comparison of overall single-task and dual-task paradigms in three CV options

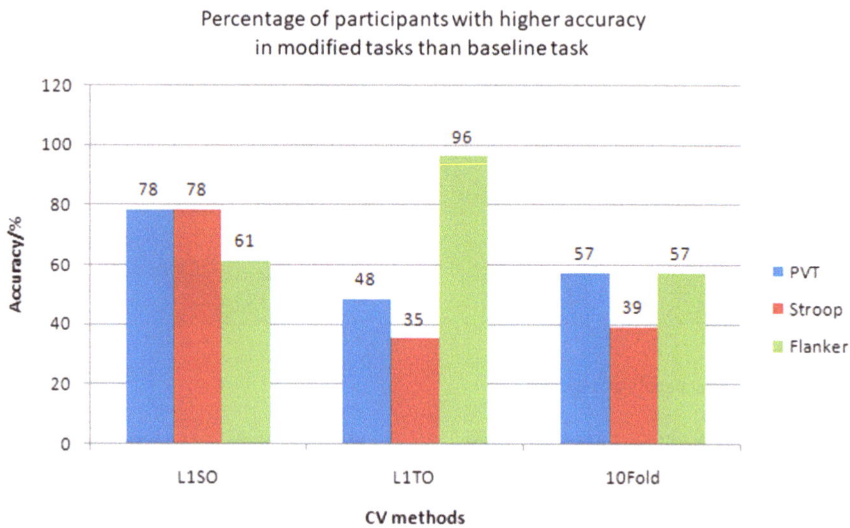

Fig. 28.4 Comparison of percentages of total subjects who achieve higher attention classification accuracy in dual tasks compared with single task among different CV options and cognitive tasks

CV show baseline single tasks have 2–3% higher accuracy than both dual-task variants. In contradictory, subject-independent CV results average 2% higher accuracy in dual tasks compared with baseline single task. As expected, the overall accuracy of subject-independent CV, L1SO is lower than both subject-independent L1TO and tenfold CV as shown in Fig. 28.3.

Figure 28.4 shows the percentage of participants that had a higher accuracy SVM model created from either of the modified tasks *A* or *B* than baseline task. Each subject's accuracy for each task was compared; then, the percentage of participants with higher accuracy was calculated. In total, there were twenty-four EEG bandpower features for classification and statistical analysis. The presence of each feature, from the theta wave, respectively, indicated focused attention [17], lack of engagement [18], active concentration with thinking [19], and engagement [20].

Table 28.1 shows the *p*-value of the accuracies obtained from the three CV methods for each task. The bolded values are tasks that have a significant difference between the modified tasks and baseline tasks. The mean *p*-value obtained was 0.210. As shown in Table 28.1, there was no significant effect of task difficulty on the attention

Table 28.1 Statistical significant test outcomes (*p*-value) comparison among single and dual tasks for each task and CV option (*: $p < 0.05$)

CV option	10 Fold	L1SO	L1TO
Flanker task	0.088	0.733	0.036*
PVT task	0.310	0.024*	0.219
Stroop task	0.042*	0.290	0.150

levels of participants at the threshold of p-value < 0.05 level for most of the tasks, namely tenfold flanker task, L1SO flanker task, tenfold PVT, L1TO PVT, L1TO Stroop task, and L1SO Stroop. At the same time, flanker task L1TO ($P = 0.036$), PVT L1SO ($P = 0.024$), and tenfold Stroop task ($P = 0.042$) showed significant effect of task difficulty on attention with a p-value < 0.05.

28.4 Results and Discussion

From Fig. 28.3, tenfold and L1TO methods do not show a significant difference in mean accuracy produced across all three tasks, with tenfold giving a slightly higher accuracy while the leave-one-subject-out method produced a lower mean accuracy across all three tasks. However, the L1SO method gave a higher mean accuracy of tasks *A* and *B* than the baseline task, while the tenfold and L1TO method produced higher accuracies in baseline task. The improvement in accuracy produced in the L1SO method indicates that a higher accuracy could be obtained in modified tasks than baseline tasks for anyone, without requiring subject-specific data training, using tenfold and L1SO CV methods.

Furthermore, with reference to Fig. 28.3, the L1SO method produced a higher percentage of participants with higher accuracies in modified tasks than baseline task across all three tasks, as compared to the other two methods. With an exception of L1TO method for flanker task, which gave a high percentage of 95.652% of participants that had a higher accuracy, the higher percentage further indicates that a higher accuracy could be obtained in modified tasks than baseline tasks for anyone.

Though the differences in accuracy of participants are not statistically very significant, our results do show an improved accuracy from baseline tasks to modified tasks. A higher classification accuracy implies that the difference in attention level of participants between the attention and inattention segment was greater, hence indicating better attention performance of participants. Thus, dual-task paradigm could improve attention performance of participants.

We experienced a number of limitations in our experiment. These include the lack of features used, with only bandpower features being used for data analysis. Other limitations also include the small number of participants, resulting in the amplification of any inaccuracies in a single participant's data. This may result in a greater deviation of the data from the true value. These could possibly result in the minimal improvement observed between baseline tasks and their modified tasks. It is also observed that there were greater improvements in attention in some participants but not the others. However, there seems to be a lack of common characteristics among these participants showing greater difference in attention levels. In our future work, we would like to analyze the common characteristics between participants of higher difference in attention level.

28.5 Conclusion

Our experiment has aimed to discuss the effect of task difficulty on attention level in the duration of the task and make use of a data-driven approach to analyze EEG data, at the same time applying supervised machine learning to classify between the two attention states of attentive and inattentive. From our experiment, we analyzed the EEG and eye-tracking data, from twenty-five participants to make a comparison of the attention level in participants under varying task difficulties. Our data analysis was also aided by SVM classifier with different CV options to evaluate the accuracies of this machine learning tool, obtaining *p*-values. We were able to conclude that though not statistically very significant, a dual-task paradigm leads to improved attention performance. Though we have yet to fully understand the effects of task difficulty on attention levels under single and dual-task conditions, we believe that our findings could be used to find the best balance between conducting single task versus dual task in the fast-paced world of today.

References

1. Silsupadol, P., et al. (2009). Effects of single-task versus dual-task training on balance performance in older adults: A double-blind, randomized controlled trial. *Archives of Physical Medicine and Rehabilitation, 90*(3), 381–387. https://doi.org/10.1016/j.apmr.2008.09.559
2. Yap, J. Y., & Lim, S. W. H. (2013). Media multitasking predicts unitary versus splitting visual focal attention. *Journal of Cognitive Psychology, 25*(7), 889–902. https://doi.org/10.1080/20445911.2013.835315
3. Lin, L., Cockerham, D., Chang, Z., & Natividad, G. (2015). Task speed and accuracy decrease when multitasking. *Technology, Knowledge and Learning*.
4. Lenartowicz, A., & Loo, S. K. (2014). Use of EEG to diagnose ADHD. *Current Psychiatry Reports, 16*, 498. https://doi.org/10.1007/s11920-014-0498-0
5. Ahani, A., et al. (2014). Quantitative change of EEG and respiration signals during mindfulness meditation. *Journal of NeuroEngineering and Rehab., 11*(1), 87. https://doi.org/10.1186/1743-0003-11-87
6. Niedermayer, E., & da Silva, F. L. (2005). *Electroencephalography: Basic Principles, Clinical Applications, and Related Fields*. 5th. Wolters Kluwer, [Google Scholar] A classical and comprehensive text, covering both basic and clinical aspects of EEG.
7. Wai, A. A. P., Dou, M., & Guan, C. (2020). Generalizability of EEG-based mental attention modeling with multiple cognitive tasks. In: *42nd IEEE EMBC*, pp. 2959–2962. https://doi.org/10.1109/EMBC44109.2020.9176346.
8. Lim, C. G., et al. (2019). A randomized controlled trial of a brain-computer interface based attention training program for ADHD. *PLoS ONE, 14*(5), e0216225. https://doi.org/10.1371/journal.pone.0216225
9. Adler, R., & Benbunan-Fich, R. (2014). The effects of task difficulty and multitasking on performance. *Interacting With Computers, 27*(4), 430–439. https://doi.org/10.1093/iwc/iwu005
10. McDermott, T. J., et al. (2017). Spatiotemporal oscillatory dynamics of visual selective attention during a flanker task. *NeuroImage, 156*, 277–285.
11. Jung, C. M., Ronda, J. M., Czeisler, C. A., & Wright, K. P., Jr. (2011). Comparison of sustained attention assessed by auditory and visual psychomotor vigilance tasks prior to and during sleep

deprivation. *Journal of Sleep Research, 20*(2), 348–355. https://doi.org/10.1111/j.1365-2869.2010.00877.x
12. Lotte F. (2014) A tutorial on EEG signal-processing techniques for mental-state recognition in brain–Computer interfaces. In: Miranda, E., Castet, J. (eds) *Guide to brain-computer music interfacing*. Springer, London. https://doi.org/10.1007/978-1-4471-6584-2_7
13. Lotte, F., et al. (2018). A Review of Classification Algorithms for EEG-based Brain-Computer Interfaces: A 10-year Update. *Journal of Neural Engineering, 15*(3), 55. https://doi.org/10.1088/1741-2552/aab2f2ff.ffhal-01846433f
14. Chang, C. C., & Lin, C. J. (2011). LIBSVM: A library for support vector machines. *ACM Transaction Intelligent System Technology, 2*(3), 27. https://doi.org/10.1145/1961189.1961199
15. Love, J. et al. (2019). JASP: Graphical statistical software for common statistical designs. *Journal of Statistical Software*, 88(2), 1–17. https://doi.org/10.18637/jss.v088.i02
16. JASP Team (2020). JASP (Version 0.14.1) [Computer software].
17. Gopher, D., Armony, L., & Greenspan, Y. (2000). Switching tasks and attention policies. *Journal of Experimental Psychology: General, 129*, 308–229.
18. Hillel, I., Gazit, E., Nieuwboer, A., Avanzino, L., Rochester, L., & Cereatti, A. et al (2019). Is every-day walking in older adults more analogous to dual-task walking or to usual walking? Elucidating the gaps between gait performance in the lab and during 24/7 monitoring. *European Review of Aging and Physical Activity*, 16(1). https://doi.org/10.1186/s11556-019-0214-5p
19. Meyer, D., & Kieras, D., (1997). A computational theory of executive cognitive processes and multiple-task performance: Part 2. Accounts of psychological refractory-period phenomena. *Psychological Review*, 104(4), 749–791.
20. Neural basis for brain responses to TV commercials: a high-resolution EEG study. *IEEE Transactions in Neural Systems and Rehabilitation Engineering*, 16(6), 522–531.

Chapter 29
Nanotechnology for Remediation of Water Environment

Pooja, Madhuri Kumari, and Ravinder Kumar Tomar

Abstract The application discussed in this paper is of nanotechnology in remediation and treatment of the water environment. Water treatment and remediation are considered one of the most promising environmental applications of nanotechnology, as various nanoparticles can purify water through a variety of mechanisms. Various types of nanoparticles are actively being researched and developed due to their particular activity against refractory pollutants and application versatility. Nanoparticles or nanomaterials are those which at least have one dimension in the 1–100 nm range. They have created a whole new generation of water purification technology. The application of advanced method of nanotechnology in comparison with contemporary processes such as sedimentation, chlorination, filtration, and disinfection creates the new potential for technological advancements in treatment of water and wastewater. The advantageous properties of nanoparticles in applications have been spotlighted in this paper. When in comparison with the traditional techniques, positive features of these materials as well as technical hurdles are also discussed in this paper. As best of my knowledge, not much developmental work has been completed out in this area. This paper is completely based on the literature review, and most measures mentioned in this paper are based on the theoretical or conceptual ideas, though their practical implications may behave as expected.

Keywords Nanotechnology · Nanoparticles · Environment · Nanomaterials · Sedimentation · Chlorination · Filtration · Disinfection

29.1 Introduction

Water is an inorganic, transparent, tasteless, odourless, and almost colourless chemical substance which is one of the main constituent of earth's hydrosphere and all known living organisms' fluids in which this acts as a solvent. It is necessary for all known forms of life, despite the fact that it contains no calories or organic ingredients.

Pooja (✉) · M. Kumari · R. K. Tomar
Department of Civil Engineering, Amity University, Noida, Uttar Pradesh, India
e-mail: 45poojaverma@gmail.com

H. Guo et al. (eds.), *IRC-SET 2021*, https://doi.org/10.1007/978-981-16-9869-9_29

Water covers about one-third of the earth's total surface. Most of the available water is saline and unfit for the human consumption. Out of total world's total available water on earth, fresh water makes up only 2.5% of the total. One of the most serious problems that people confront around the world is a lack of access to clean and safe water. The provision of safe and clean water has been become a basic human requirement.

Water contains various contaminants and the chemicals which are used whilst treatment of water includes chlorine (Cl2), ammonia (NH3), hydrochloric acid (HCl), ozone (O3), permanganate, ferric salts, coagulants, ion exchange resins, etc. All these compounds can damage freshwater resources to a larger extent. Contemporary water treatment technologies such as the coagulation, sedimentation, filtration, desalination, disinfection, and decontamination have been used since the early twentieth century. Chemically and operationally intensive, the foregoing processes necessitate big systems, infrastructure, and engineering skills, all of which are costly, making these processes inconvenient, unproductive, time- consuming ineffective, and expensive to use.

Membrane processes are significantly increasing in the water purification, as traditional water treatment procedures are unable to remove organic pollutants at the required levels. Nanofiltration (NF), microfiltration (MF), and ultrafiltration (UF) for the water and wastewater treatment and reverse osmosis (RO), nanofiltration (NF) for the desalination and water reclamation are examples of membrane techniques commonly employed in water purification. All traditional water purification systems have a number of disadvantages, whereasnanotechnology provides a framework for delivering speedy results. Nanotechnology has a number of intriguing applications to filter out organic and inorganic pollutants. Nanotechnology can efficiently remove various heavy metals such as cadmium, arsenic, lead, and mercury. It can also remove biological pollutants which can cause water-borne illnesses such as typhoid and cholera fever.

The study of manipulating matter on an atomic scale is known as nanotechnology. It refers to the micro-engineering and creation of functional systems. Nanomaterials and nanostructures have the nanoscale diameters ranging from 1 to 100 nm, and they frequently shows novel and drastically altered physical, chemical, and biological properties.

As nanoscale materials have a huge surface area compared to their volume, their reactivity in biological or chemical surface mediated reactions can be significantly increased when compared to the same substance at much larger sizes. They can be enhanced for specific applications to provide unique characteristics not present in micro or macroscale particles of the same substance. However, they can show changes in reaction rates that surface area alone cannot account for. These characteristics allow for higher interaction with contaminants leading in quick contaminant concentration reduction. Nanoscale materials can also penetrate very small spaces in sub-surface and remain suspended in groundwater provided proper coatings are applied, owing to their small size. Under the category of remediation and treatment, nanotechnology has huge potential in order to contribute to the long-term water quality, viability, and availability of water resources.

29.1.1 Classification of Nanoscale Materials

Nanoscale materials can be classified into majorly three classes such as natural, accidental, and artificial nanoscale materials.

Materials at nanoscale that can be obtained naturally or created artificially includes:

- Carbon nanotubes and fullerenes which exist as hollow spheres (bucky balls), ellipsoids, or tubes (nanotubes) made completely of carbon. They have great thermal and electrical conductivity. They are potent antioxidants, stable, little reactivity. Biomedical, super-capacitor, sensor, and photovoltaic applications all use them.
- Zinc oxide (ZnO), cerium oxide (CeO_2), iron oxide (Fe_3O_4) and titanium dioxide (TiO_2) are nanosized metal oxides that may block UV light. They are made up of hundreds or thousands of atoms in closely packed semiconductor crystals. Metal oxides are used in photo catalysts, pigments, medications to regulate release, medical diagnostics, and sunscreen.

Engineered nanoscale materials include:

- Zero-valent metals, for example, nanoscale zero-valent iron (nZVI), which has a high-surface reactivity and is used in the remediation of water, soil, and sediments.
- Dendrimers are defined as highly branched polymers with a wide variety of functional groups that can be designed and synthesized. Cones, spheres, and disc-like structures are common shapes of dendrimers. They are employed in medication delivery, chemical sensing, electrode modification, and DNA transfer agents.
- Two or more than two different nanoscale materials, or only one nanoscale substance mixed with bulk type materials, make up the composite nanoscale materials. Biological and synthetic molecules can be incorporated with the composite nanoscale materials to provide new electrical, catalytic, magnetic, mechanical, thermal, and imaging capabilities.

29.2 Methodology

As mentioned earlier, basis of this report is theoretical study and conceptual analysis. So, I have used various sources such as case studies on the applications of nanotechnology in remediation of water environment.

As nanotechnology encompasses a wide range of tools, techniques, and applications involving particles with diameters varying from a few to hundreds of nanometers and these size of particles has several unusual physicochemical and surface features that lend itself to new applications. Indeed, proponents of nanotechnology argue that this field of study might help solve some of the world's most pressing difficulties, such as providing a steady supply of safe drinking water for a growing population, as well as issues observed in medicine, energy, and agriculture.

Here, on the basis of properties, a comparative study between various types of nanoparticles has been done. I have considered some basic features that may affect the various properties of nanoparticles under variety of conditions, and then, the subsequent changes in the result of nanoparticles are mentioned. Using various case studies, I have also looked into the performance of nanoparticles in a variety of water environments. There is a critical comparison of the nanoparticles as not much developments are performed in this area yet. Hence, there is a dire requirement to consider the applications of nanotechnology in this area.

Although here lab testing was not done, the results obtained in this study are very much dependent on the theoretical verification and that forms the foundation for the future studies on the topic, as there are various other aspects which still have not been discussed as they are not in the scope of the study, such as the economical aspect and departmental willingness to adopt newer technologies and apply them on the existing circumstances.

29.3 Literature Review

29.3.1 Different Types and Applications of Nanoparticles

See Fig. 29.1.

29.3.2 Properties of Nanoparticles

All the bulk or heavy materials have diverse properties, such as mechanical, electrical, and optical qualities, which are further dependent on their structural properties like metals, insulators, and semiconductors. Nanoparticles have features that distinguish them from tiny molecules and their chemistry along with the manufacture can be considered to be complicated.

The ability of molecules to interact with nanoparticles on their surfaces and interchange with other molecules or particles indicates the chemistry of nanoparticles and how it relates to their fate in surface waters and sediments, as this is a critical factor in determining their ultimate fate. When one of the three spatial dimensions is comparable to or smaller than the wave length of de Broglie (λ) of the charge carrier of electrons and holes or the wavelength of light, periodic boundary conditions destroy crystalline materials or change the atomic density on the surface of amorphous materials. Many of the physical properties of the nanoparticles differ significantly from those of bulk materials as the result of this trait, resulting in a wide range of novel uses.

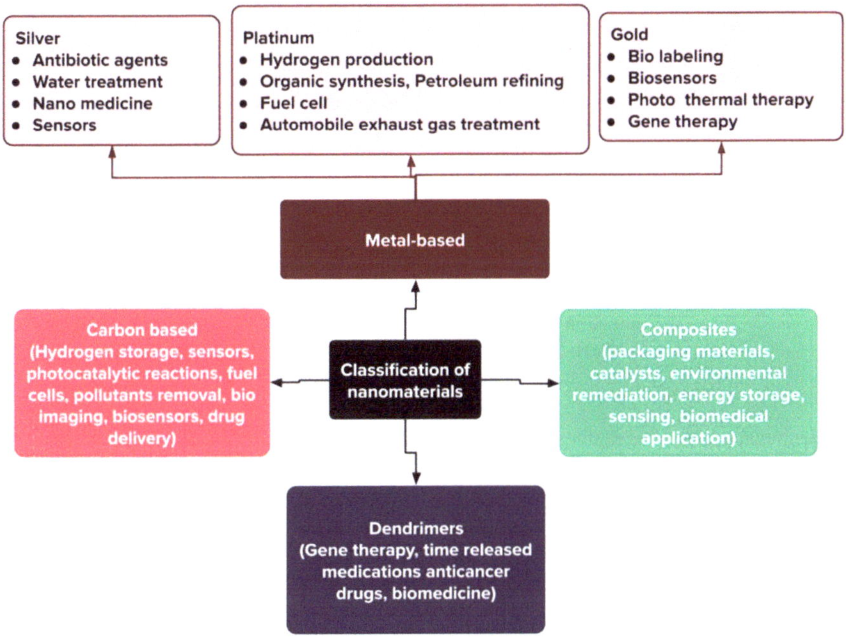

Fig. 29.1 Classification as well as the broad range of utilization of various types of nanoparticles

29.3.2.1 Adsorption

Adsorption is defined as the property of all solid state substances to attract molecules of liquids or gases to their surfaces when they are in close proximity. Adsorbents are solids which are used in order to adsorb the gases or dissolved chemicals, and the molecules which are adsorbed are referred to as the adsorbate collectively. Nanoadsorbents have a far higher rate for the adsorption of organic chemicals than granular or powdered activated carbon because of their large specific surface area.

They hold significant promise for developing new, more effective, and faster decontamination techniques for inorganic and organic pollutants such as the heavy metals and micro-pollutants. The increased process efficacy allows for the development of even more compact water and wastewater treatment devices having smaller footprints specifically for the decentralized applications and of point-of-use systems, in addition to saving adsorbent materials (Table 29.1).

Out of the following types of nanomaterials mentioned above, nanoadsorbents are the subject of current research:

- nanoadsorbents made of carbon such CNTs
- nanoadsorbents made of metals such as ZnO, TiO_2
- polymeric nanoadsorbents
- zeolites

Table 29.1 Overview of types of nanoparticles with properties and respective applications

Nanomaterial	Positive properties	Negative properties	Applications
1. Nanoadsorbents	High-specific surface, higher adsorption rates, small footprint, short intra-particle diffusion distance	High-production costs	Point-of-use, removal of organics, heavy metals, bacteria
2. Nanometals and nanometal oxides	short intra-particle diffusion distance, compressible, abrasion-resistant, magnetic, photocatalytic (WO_3, TiO_2)	Less reusable	Removal of heavy metals (arsenic) and radionuclides, media filters, slurry reactors, powders, pellets
3. Membranes and membrane processes	Reliable, largely automated process	Relative high-energy demand	All fields of water and wastewater treatment processes

(i) **Carbon nanotubes (CNTs)**

Carbon nanotubes (CNTs) are carbon allotropes with a cylindrical nanostructure. CNTs are classified as the single walled nanotubes or multi-walled nanotubes depending upon their manufacturing procedure. CNTs have a large specific surface area, as well as highly assessable adsorption sites and a chemistry that may be adjusted. CNTs also have antibacterial characteristics because they cause oxidative stress in bacteria and damage cell membranes. Despite of the fact that chemical oxidation takes place, no harmful by-products are formed, which is a significant benefit over traditional disinfection methods like chlorination and ozonation.

Adsorption-based techniques are simple and easy to use for point-of-use water purification devices, but their capacity to remove salts is limited. Traditional desalination methods are energy-intensive and technically demanding. Despite their major benefits over activated carbon, CNTs are unlikely to be used on a broad scale for big municipal wastewater and water treatment plants in the coming future due to high-production costs.

(ii) **Nanoadsorbents made of metals**

Nanoscale metal oxides are one of the potential adsorbents for heavy metals and for the radionuclides which are effective alternatives to activated carbon. They have a huge amount of specific surface area, a short intra-particle diffusion distance which can be compressed without losing much of their available surface area. Because some of these tiny metal oxides are superparamagnetic, they can be separated and recovered using a low-gradient magnetic field. Adsorptive media filters and slurry reactors can both benefit from them.

Nanometals and nanometal oxides are typically compacted into porous pellets or used as powders in industrial applications (Table 29.2).

Table 29.2 Metal-based nanoadsorbents with properties and respective applications

Nanometals and nanometal oxides	Positive properties	Negative properties	Applications	Novel approaches
Nanosilver and nano-TiO_2	Nanosilver: bactericidal, low-human toxicity Nano-TiO_2: high-chemical stability, very long life time	Nanosilver: limited durability Nano-TiO_2: requires ultraviolet activation	Point-of-use, water disinfection, anti-biofouling surfaces, decontamination of organic compounds	TiO_2 modification for activation by visible light, TiO_2 nanotubes
Magnetic nanoparticles	Simple recovery by magnetic field	Stabilization is required	Groundwater remediation	Forward osmosis
Nano zero-valent iron	Highly reactive	Stabilization is required (surface modification)	Groundwater remediation (chlorinated hydrocarbon, perchlorates)	Entrapment in polymeric matrices for stabilization

(iii) **Polymeric nanoadsorbents**

Dendrimers which are repetitively branched molecules are the polymeric nanoadsorbents that can be used to remove organics and heavy metals. The interior of hydrophobic shells can adsorb organic substances, whilst the specialized outer branches can adsorb heavy metals.

Dendrimers are used in an ultrafiltration device to extract copper from the water. Using this combined dendrimer-ultrafiltration technology, nearly all the copper ions were collected. A pH adjustment is all it takes in order to renew the adsorbent. They are able to remove up to the count of 99% of some dyes.

However, because to the complicated multistage synthesis of dendrimers, no commercial providers have emerged so far.

(iv) **Zeolites**

Since the early 1980s, zeolites in conjunction with the silver atoms or particles have been well known. Nanoparticles such as silver atoms can be incorporated in the structure of porous zeolite. They are then exchanged with the other cations present in solution and freed from zeolite matrix. When a metallic surface comes into contact with liquids, a considerable small number of silver ions is released.

For the elimination of heavy metals such as arsenic, both CNTs and nanometals are particularly effective nanoadsorbents. Nanometals and zeolites are benefitted from their cost-effectiveness and their compatibility with the current water treatment systems in this application field because they are capable of being used in pellets and in beads for the fixed absorbers. In terms of ecotoxicity, the nanometals, carbon

Table 29.3 Overview of various types of nanoadsorbents with their properties and their respective applications

Nanoadsorbents	Positive properties	Negative properties	Applications	Novel approaches
Carbon nanotubes	Highly assessable sorption sides, bactericidal, reusable	High-production costs, possibly health risk	Point-of-use, heavily degradable contaminants (pharmaceuticals, antibiotics)	Ultra-long carbon nanotubes with extremely high-specific salt adsorption
Nanoadsorbents made of metals	Highly reactive	Stabilization is required (surface modification)	Groundwater remediation (chlorinated hydrocarbon, perchlorates)	Entrapment in polymeric matrices for stabilization
Polymeric nanoadsorbents	Bi-functional (inner shell adsorbs organics, outer branches adsorb heavy metals), reusable	Complex multistage production process	Removal of organics and heavy metals	Biodegradable, biocompatible, nontoxic bio-adsorbent (combination of chitosan and dendrites)
Zeolites	Controlled release of nanosilver, bactericidal	Reduced active surface through immobilization of nanosilver particles	Disinfection processes	Nanozeolites by laser induced fragmentation

nanotubes, and zeolites discussed here are made up of well-characterized fundamental materials that are found in nature and are categorized as harmless. As a result, the potential toxicity of nanoadsorbents is largely determined by their size and form, as well as chemical stabilizers and surface modifications. Because the ecotoxicity of each new alteration of a recognized nanomaterial must be re-evaluated, a general assessment of nanoadsorbents in terms of their toxicity potential is impossible (Table 29.3).

29.3.3 Nanomaterials and the Water Environment

In recent years, researchers have looked into the performance of nanoparticles in a wide variety of aqueous environments, including the ocean, river, lake, and wetlands. Environmental research groups have made it a priority to find environmentally friendly remediating agents in order to achieve the goal of restoring damaged soil and water bodies in order to begin an environmentally friendly economy.

Several recent analyzes have raised concerns about electro membrane (separation) technologies for removing charged components from solutions, such as producing fresh water from brackish water. Currently, nanoparticles are commonly used for treatment of sewage, and nano-photocatalytic technology is regularly used for oxidizing various organic compounds present in water by producing the hydroxide ions and superoxide ions with excellent activity and strong oxidation. Different nanostructured materials are currently employed in capacitive deionization (CDI), a new water treatment method.

In aqueous environments, the nanoparticles exhibit a variety of characteristics, with some nanomaterials demonstrating greater potential for water contamination remediation (Fig. 29.2).

Nanofiltration (NF) membrane technology, which uses pressure-driven membranes to separate the chemicals found in sewage, is also available. Furthermore, the hydroxyl groups on the nanoparticle surface bind to specific cations, causing certain nanomaterials to have an obvious adsorption impact on organics or metal ions. Carbon nanoparticles, for example, offer higher stability in acidic or alkaline environments and have a considerable large specific surface area (SSA), porous structure, and specialized binding sites, when compared to the traditional

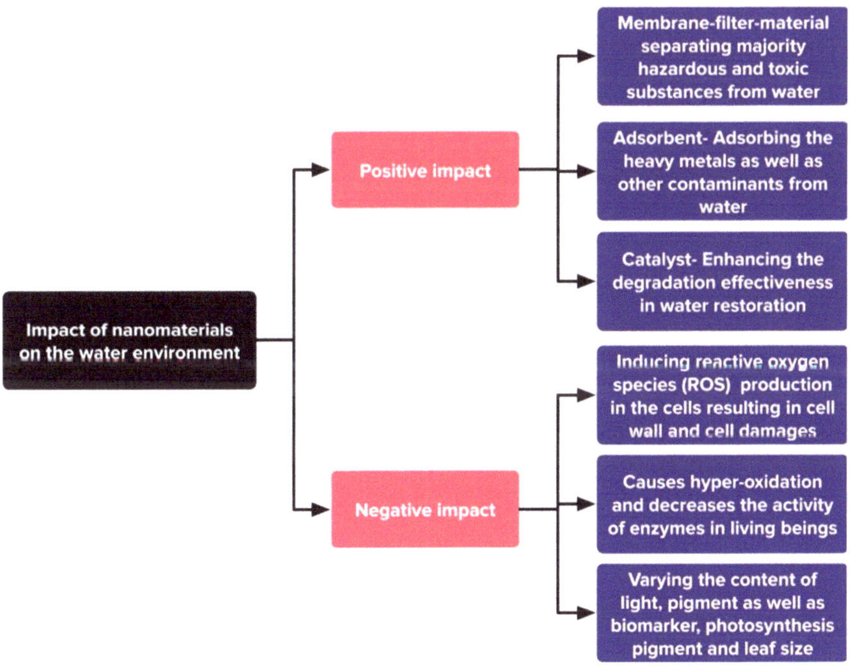

Fig. 29.2 Environmental impacts of variety of nanoparticles on water environment

sewage treatment chemicals. As a result, using nanoparticles for wastewater, treatment overcomes the shortcomings of traditional technology whilst simultaneously demonstrating good remedial efficacy.

29.3.4 Applications of Various Nanomaterials in the Water Environment

Various nanomaterials such as TiO_2 nanoparticles, silver nanoparticles, nano zero-valent iron (nZVI), zinc oxide (ZnO) nanoparticles, nanoscale graphene, nanofiber materials, carbon nanotubes, and nanofiltration membrane materials can help to clean up the contaminated water. Table 29.4 shows the most often used manufacturing procedures, operational principles and pollution cleans up for the aforementioned nanomaterials.

Mechanical milling is a nanomaterial fabrication technology which was developed in the 1970s as a method for creating sophisticated alloys and phase mixtures from the powder particles. The aforementioned technique can overcome the quantity limitations of nanocrystalline synthesis, allowing for large-scale production of nanocrystalline powders. One notable advantage is that it can operate at lower temperatures, allowing the grains to develop more slowly. The plasma process can be divided into two types: plasma spray synthesis and microwave plasma synthesis.

The ability to generate unagglomerated particles, a narrow particle size distribution, and better production rates is all advantages of this approach. Chemical reactions between the gaseous precursor and the substrate surface are stimulated in the chemical vapour deposition (CVD) method to deposit a thin solid coating onto the substrate. Because of its simplicity of scale-up, high-production yields, and low set-up cost, the aforementioned approach is a widely utilized material-processing technology. Laser ablation is a manufacturing technology that entails the use of a laser beam with an optical focussing system and a feeding apparatus.

Because of its lower yield and higher operational expense, the laser ablation process is not widely used, especially on a large scale. The centrifugal force can be used to create ultrafine fibres and nanoscale fibres utilizing centrifugal spinning technology. As a substitute approach for generating nanofibers, a solution blow spinning process combines components of both melt blowing and electrospinning technologies.

29.3.5 Limitations of Nanoparticles in Remediation for Water Applications

The influence of nanoparticles for water and wastewater technology on the water environment is critical to their commercialization. To examine the health concerns

Table 29.4 Use of different nanomaterials in treatment of polluted water

S. No.	Nanomaterials	Mechanism	Remediation	References
1	Titanium dioxide (TiO_2) nanoparticles	Ultraviolet light has the potential to excite TiO_2 nanoparticles and produces free radicals with high-catalytic activity, resulting in increased photo-oxidation and reduction capacity	Different inorganic and organic substances like formaldehyde	[13]
2	Silver nanoparticles	Bacteria will dry up and die as a result of the interaction with their metabolic enzymes. The smaller the particle, better the germicidal effectiveness	Role of water photocatalyst and the antibacterial agent	[14]
3	Zero-valent iron (nZVI)	Pollutant adsorption could be aided by the thin iron oxide core's surface complexation and electrostatic interaction	Azodyes, heavy metals, chlorinated aromatic compounds, organo-chlorine and nitro-aromatic compounds	[15]
4	Graphene nanomaterials	Porous features of the nanomaterial adsorb pollutants from the water	Dyes and heavy metals	[16]
5	Nanofiber material	Membranes of different pore widths are used to filter particles of varied dimensions	Different contaminants present in eutrophicated landscape water	[17]

of nanoparticles, numerous research has been conducted including toxicity tests, life cycle analysis, technology assessment, and routes of distribution of nanoparticles in water bodies.

The findings of these studies have led to a better understanding of the behaviour of nanoparticles such as carbon nanotubes (CNTs), titanium dioxide (TiO_2), and silver nanoparticles in aqueous systems which have aided stakeholders from government, politics, and industry in developing new laws and regulations or amending existing ones. However, because no uniform standards and circumstances for experimental tests and measurements have been established, numerous research has produced contradicting results, slowing down the crucial decision-making processes.

29.4 Discussions and Conclusion

This paper has looked at how nanomaterials have recently been used in water remediation for environmental cleanup. We also talked about the detrimental effects of nanoparticles in water. Carbon nanotubes (CNTs), graphene-based nanomaterials, silver nanoparticles, zinc oxide (ZnO), titanium dioxide (TiO_2), and nano zero-valent iron were amongst the nanomaterials studied.

According to the observations, graphene nanoparticles have a high-adsorption capability for surfactants, antibiotics, and other hazardous pollutants. These nanoparticles can also successfully remove heavy metal ions in water at very low concentrations. Despite the fact that the manufacturing of graphene oxide-based nanomaterials and their usage in environmental pollution purification are fraught with problems, it is expected that graphene oxide-based composites will be used in the field of wastewater treatment in the coming years as a result of the various steps taken by scientists. Furthermore, carbon nanotubes also have the potential to operate as excellent purifiers as they are capable of separating biological, inorganic, and organic impurities from water. Furthermore, the functionalized carbon nanotubes have a remarkable ability to absorb heavy metals from water and eliminate organic dyes.

Their unregulated discharge, on the other hand, has the potential to threaten biota in a variety of environmental domains, including soil and water systems. Though the nanotechnology for water remediation is new and its application, more lab or even field investigations are still needed to better understand the possible risks of nanomaterials.

References

1. Yuan, L., Richardson, C. J., Ho, W., Wesley Willis, C., Colman, B. P., & Wiesner, M. R. *Stress Responses of Aquatic Plants to Silver Nanoparticles.*
2. Hu, A., & Apblett, A. *Nanotechnology for Water Treatment and Purification.*
3. Kumar, S., Ahlawat, W., Bhanjana, G., Heydarifard, S., Nazhad, M. M., & Dilbaghi, N. *Nanotechnology-based water treatment strategies.*
4. Patanjali, P., Singh, R., Kumar, A., & Chaudhary, P. (2019). Nanotechnology for water treatment: A green approach. In *Green Synthesis, Characterization and Applications of Nanoparticles*, (pp. 485–512).
5. Theron, J., Walker, J. A., & Cloete, T. E. *Nanotechnology and water treatment: Applications and emerging opportunities.*
6. Lu, H., Wang, J., Stoller, M., Wang, T., Bao, Y., & Hao, H. *An overview of nanomaterials for water and wastewater treatment.*
7. Batley, G. E., Kirby, J. K., & McLaughlin, M. J. Fate and risks of nanomaterials in aquatic and terrestrial environments.
8. Ghadimi, M., Zangenehtabar, S., & Homaeigohar, S. An overview of the water remediation potential of nanomaterials and their ecotoxicological impacts.
9. Ghasemzadeh, G., Momenpour, M., Omidi, M., Hosseini, M. R., Ahani, M., & Barzegari, A. Applications of nanomaterials in water treatment and environmental remediation.

10. Bhattacharya, S., Saha, I., Mukhopadhay, A., Chattopadhyay, D., Ghosh, U. C., & Chatterjee, D. Role of nanotechnology in water treatment and purification: Potential applications and implications.
11. Amin, M. T., Alazba, A., & Manzoor, A. A review of removal of pollutants from water/wastewater using different types of nanomaterials.
12. Anjum, M., Miandad, R., Waqas, M., Gehany, F., & Barakat, M. A. Remediation of wastewater using various nano-materials.
13. Singh, R., & Dutta, S. (2018). A review on H2 production through photocatalytic reactions using TiO2/TiO2-assisted catalysts. *Fuel 220*, 607–620.
14. Rtimi, S., Dionysiou, D. D., Pillai, S. C., & Kiwi, J. (2019). Advances in catalytic/photocatalytic bacterial inactivation by nano Ag and Cu coated surfaces and medical devices. *Application of Catalyst B Environment, 240*, 291–318.
15. Huang, X. Y., Ling, L., & Zhang, W. X. (2018). Nanoencapsulation of hexavalent chromium with nanoscale zero-valent iron: High resolution chemical mapping of the passivation layer. *Journal of Environmental Science, 67*, 4–13.
16. Lim, J. Y., Mubarak, N. M., Abdullah, E. C., Nizamuddin, S., & Khalid, M. (2018) Recent trends in the synthesis of graphene and graphene oxide based nanomaterials for removal of heavy metals—A review. *Journal of Industrial Engineering Chemistry, 66*, 29–44.
17. Lim, C. T. (2017) Nanofiber technology: Current status and emerging developments. *Progress in Polymer Science, 70*, 1–17.
18. Tlili, I., & Alkanhal, T. A. Nanotechnology for water purification: Electrospun nanofibrous membrane in water and wastewater treatment.
19. Baby, R., Saifullah, B., & Hussein, M. Z. *Carbon nanomaterials for the treatment of heavy metal-contaminated water and environmental remediation.*
20. Zhang, W., Zhang, D., & Liang, Y. Nanotechnology in remediation of water contaminated by poly- and perfluoroalkyl substances.
21. Adeleye, A. S., Conway, J. R., Garner, J., Huang, Y., Su, Y., & Keller, A. A. Engineered nanomaterials for water treatment and remediation: Costs, benefits, and applicability.
22. Karn, B., Kuiken, T., & Otto, M. Nanotechnology and in situ remediation: A review of the benefits and potential risks.

Chapter 30
Elucidating the Role of MAMDC2 in Head and Neck Squamous Cancer

Sheryl Ke Ying Tay, Rachel Shiyi Lu, Kah Yee Goh, and Darren Wan-Teck Lim

Abstract The tumour matrisome comprises proteins within the extracellular matrix of the tumour microenvironment that regulate cancer growth and development. We had previously published a pan-cancer 29 gene signature that predicts for cancer prognosis. One of the genes in this signature is *mamdc2*. It encodes for MAM domain-containing protein 2 (MAMDC2), of which the biological functions are still uncertain. We found that *mamdc2* was highly expressed in five head and neck squamous cell carcinoma (HNSCC) lines derived from patients with distant metastasis but underexpressed in a HNSCC line derived from a patient without distant metastasis. This supports a correlation between the upregulation of *mamdc2* and distant metastasis. By performing several functional assays, overexpression of *mamdc2* was shown to promote cell proliferation in vitro. Immunofluorescence also demonstrated the intra-cellular localisation of MAMDC2 in the cytoplasm and nucleus, which suggests that MAMDC2 may be involved in intra-cellular signalling or transcription and may interact with pathways regulating cell proliferation. These findings support further investigation into the role of *mamdc2* in cancer development. Additionally, *mamdc2* could be a candidate for drug development and anti-cancer therapy.

S. K. Y. Tay (✉) · R. S. Lu
Raffles Girls' School (Secondary), Singapore, Singapore
e-mail: sheryl.tay05@rafflesgirlssch.edu.sg

R. S. Lu
e-mail: rachel.lu05@rafflesgirlssch.edu.sg

K. Y. Goh
Division of Medical Oncology, National Cancer Centre Singapore, Singapore, Singapore
e-mail: goh.kah.yee@nccs.com.sg

D. W.-T. Lim
Division of Medical Oncology, National Cancer Centre Singapore, Office of Academic and Clinical Development, Duke-NUS Medical School, Institute of Molecular and Cell Biology, A*STAR, Singapore, Singapore
e-mail: darren.lim.w.t@singhealth.com.sg

H. Guo et al. (eds.), *IRC-SET 2021*, https://doi.org/10.1007/978-981-16-9869-9_30

30.1 Introduction

The extracellular matrix (ECM) is a complex network of macromolecules, with distinctive physical, biochemical, and biomechanical properties. Dysregulation of the ECM occurs commonly in cancer and can directly promote metastasis and cellular transformation. The ECM is a major component of a tumour's local microenvironment or the *niche*; hence, it plays an important role in the development of cancer [1].

The ECM proteome, also known as the "matrisome," of normal and diseased tissues has been further characterised because of its importance to cancer development [2]. In 2017 and 2019, Lim et al. [3, 4] used a bioinformatics approach to characterise the genome of the tumour matrisome and devised the tumour matrisome index (TMI), a 29-matrisome-gene expression-based classifier for prognosis and adjuvant therapy response across 11 major cancer types, including early-stage non-small cell lung cancer (NSCLC), breast cancer, and colorectal cancer. It has the potential to distinguish cancers from normal tissues and predict patient survival.

Squamous cell carcinoma is the most common type of head and neck cancer in Singapore, accounting for approximately 300 cases per year [5]. Up to 15% of HNSCC patients eventually develop distant metastasis, which led to dismal prognosis [6]. Despite the potential applications of the TMI, the TMI was not specifically interrogated in HNSCC. Hence, we sought to test if those 29 genes were differentially expressed in head and neck tissue. We found that the majority of the 29 genes were indeed differentially expressed across six patient-derived HNSCC lines relative to a normal squamous epithelial cell line.

From the 29 genes, *mamdc2* was identified as the candidate gene in this project. *mamdc2* is a gene on human chromosome 9 that encodes MAM domain-containing protein 2. The protein contains 686 amino acids, and its quaternary structure and biological function are unknown. It was found that *mamdc2* was highly expressed in HNSCC lines derived from patients with distant metastasis but underexpressed in a HNSCC line derived from a patient without distant metastasis.

Previous works [7, 8] have also shown that MAMDC2 is overexpressed in breast cancer cells and chronic myeloid leukaemia. *mamdc2* was also found to be downregulated in NSCLC [3]. Despite the differential expression of *mamdc2*, the biological functions of *mamdc2* are still unclear and very much unexplored. Hence, we hypothesised that overexpression of MAMDC2 in HNSCC promotes cellular proliferation and distant metastasis.

30.2 Materials and Methods

30.2.1 Plasmids

An expression plasmid containing *mamdc2* fused to flag epitope was created for transfection into HN26 cell line for *mamdc2* overexpression studies. To generate the *mamdc2flag* insert, the total RNA of HN19 cells was extracted and reverse transcribed into cDNA using oligo dT primer (Invitrogen Superscript IV cDNA synthesis kit). The resulting cDNA was amplified with primers (primer sequences are listed in Table 30.1 in the Appendix) to generate a dsDNA that contained the *mamdc2* isoform 1 gene fused to a 5xglycine polylinker and 1xflag epitope at the C terminal. The dsDNA was directly inserted into an intermediate TOPO vector (Invitrogen Zero Blunt TOPO PCR Cloning Kit) and transformed into competent *E. coli* cells. Next, the *mamdc2flag* insert was excised from the TOPO plasmid (refer to Fig. 30.7 in the Appendix) and cloned into pcDNA3.1 expression vector using BamH1 and EcoRI restriction enzymes. The integrity of *mamdc2flag* insert was verified by DNA sequencing (Axil Scientific) and a test digest (refer to Fig. 30.8 in the Appendix). An overview of the plasmid construction workflow can be found in the Appendix (Fig. 30.6).

30.2.2 RNA Isolation and Quantitative Real-Time PCR (qRT-PCR)

Qiagen RNeasy Mini Kit was used to extract total RNA from a panel of HNSCC cell lines (HN26, HN19, HN43, HN64, HN90, and HN1) and Het1A (normal squamous epithelial cell line control). The same kit was also used to extract total RNA from HN26 cells transfected with the indicated plasmids. DNase treatment was performed during RNA extraction to improve the purity of extracted RNA. qRT-PCR was performed using PowerUp SYBR Green Master Mix to measure the expression of *mamdc2* in HNSCC lines on Techne Prime Pro 48 real-time qPCR machine. Forward and reverse primers specific to *mamdc2* total (isoform 1 and isoform 2), *mamdc2* isoform 1 and *beta-actin* were used (sequences in Appendix Table 30.1). *Beta-actin* was used for normalisation. Data were analysed using the comparative Ct method ($2^{-\Delta\Delta C_t}$).

30.2.3 Cell Culture and Transfection

A patient-derived HN26 HNSCC line from Dr Gopalakrishna Iyer's laboratory in NCCS was used for overexpression of MAMDC2. They were cultured in Roswell Park Memorial Institute 1640 Medium (RPMI 1640) supplemented with 10% foetal bovine serum (FBS) and 1% L-glutamine. The cells were grown in a 37 °C incubator with 5% CO_2.

To set up the overexpression experiment, 400 K of HN26 cells was seeded on a 6-well plate 24 h prior to transfection. HN26 cells were transfected with 1.25 μg of pcDNA3.1::mamdc2flag plasmid or pcDNA3.1 control plasmid using Invitrogen Lipofactamine 3000 Kit following the manufacturer's instructions. Cells were harvested at 48 h post-transfection for functional assays (wound healing assay, Transwell migration assay, and Matrigel invasion assay) and immunofluorescence. The overexpression of *mamdc2* was confirmed using qRT-PCR (refer to Appendix Table 30.2).

30.2.4 Cell Proliferation Assay

HN26 cells were seeded at 8 K/well in a flat-bottom 96-well plate and subsequently transfected with the indicated plasmids. The proliferation assay was done according to the quick protocol for Vybrant MTT Cell Proliferation Assay Kit (Thermo Scientific) with slight modifications. Instead of the recommended 50 μl, 100 μl of dimethyl sulfoxide (DMSO) was used to dissolve the MTT formazan product. The absorbance was measured at 540 nm with Hidex Sense plate reader. The absorbance readings were taken every 24 h after transfection, up to 72 h.

30.2.5 Wound Healing Assay

Wound healing was used to indirectly measure migration and proliferation. Transfected HN26 cells were seeded at 30 K/well on the Ibidi Culture-Insert in 35 mm μ-Dish. The wound healing assay was conducted according to manufacturer's instructions. The wound was imaged at 0 h, 8 h and 24 h, and 4–5 random fields were taken per sample.

30.2.6 Transwell Migration and Matrigel Invasion Assays

For Transwell migration assay, Corning 6.5 mm Transwell with 8.0 μm Pore Polycarbonate Membrane Insert in 24-well plates were used and for Matrigel invasion assay, Corning BioCoat Matrigel Invasion Chambers with 8.0 μm PET Membrane in 24-well plates were used. Transfected HN26 cells were starved overnight with basal RPMI. The next morning, the cells were harvested, and 100 K and 125 K cells were seeded into the upper chambers (Transwell insert). RPMI containing 10% FBS was added to the bottom chambers of the Transwell plate as a chemoattractant. The cells were incubated at 37 °C for 24 h for migration assay and 48 h for invasion assay. Thereafter, the Transwell inserts were stained with 0.5% crystal violet in 25% methanol. The cells were then imaged, taking five random fields for each sample.

30.2.7 *Immunofluorescence*

Immunofluorescence was conducted to determine the subcellular localisation of MAMDC2. At 48 h post-transfection, 50 K of transfected HN26 cells were harvested and spun down onto a glass slide with a cytospin funnel. The cells were fixed in 4% paraformaldehyde (PFA) and permeabilized with phosphate buffered saline (PBS) containing 0.3% Triton X-100. The cells were then blocked with 3% bovine serum albumin (BSA) with 0.2% Triton. After blocking, the cells were incubated with primary anti-FLAG antibodies (1:200; Sigma) for 2 h in blocking buffer. The secondary antibody used was Alexa Fluor 488-conjugated anti-mouse (1:1000; Life Technologies). The cells were then incubated with Hoescht (1:400; Life Technologies) in PBS as the nuclear stain. After staining, the glass slide was mounted with Merck Millipore FluorSave Reagent to prevent photobleaching and viewed under Zeiss Axioimager Z1 fluorescence microscope.

30.2.8 *Statistical Analysis*

ImageJ software was used to measure the wound area and the number of cells that migrated and invaded in Transwell assays. Student's t-test was used to calculate the *p*-values (Graphpad Prism software). Statistical significance was set at $P < 0.05$.

30.3 Results

30.3.1 mamdc2 *Overexpression is Possibly Associated with Distant Metastasis*

To confirm the preliminary findings that *mamdc2* was differentially expressed in HNSCC lines, we conducted qRT-PCR analysis of cDNA derived from a panel of six patient-derived HNSCC lines. The relative fold changes of *mamdc2* mRNA expression in each cell line were compared to Het1A (normal squamous epithelial control). The results (Fig. 30.1) showed that compared to Het1A, *mamdc2* was downregulated in HN26 and upregulated in HN19, HN43, HN64, HN90 and HN1. Since HN26 was derived from a patient without distant metastasis, whilst HN19, HN43, HN64, HN90 and HN1 were derived from patients with distant metastasis, higher expression of

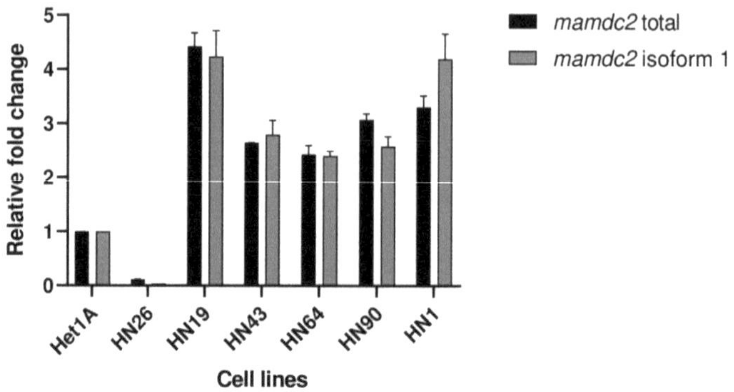

Fig. 30.1 Endogenous expression of *mamdc2* total and *mamdc2* isoform 1. Relative fold change in *mamdc2* mRNA expression in each line was normalised to *beta-actin* and compared to Het1A. Error bars represent standard error mean (SEM)

MAMDC2 may be correlated with enhanced propensity for distant metastasis of cancer cells.

There are two reported isoforms of *mamdc2*, isoform 1 and isoform 2. As seen in Fig. 30.1, the expression levels of *mamdc2* isoform 1 were also not significantly different from the expression levels of *mamdc2* total (i.e. isoform 1 and isoform 2). Therefore, *mamdc2* isoform 1 appears to be the predominant isoform, and hence, we built an expression plasmid specific to *mamdc2* isoform 1.

30.3.2 Effect of MAMDC2 Overexpression on Cell Activities

To test whether high expression of *mamdc2* may correlate with metastasis, *mamdc2* isoform 1 was overexpressed in HN26 cell line, and the effects on cell proliferation, migration, invasion and wound healing were examined. The cells transfected with pcDNA3.1 were named as Empty vector, and the cells transfected with pcDNA3.1:mamdc2flag were named as MAMDC2flag.

30.3.3 mamdc2 *Overexpression had No Obvious Effects on Cell Migration and Invasion*

To determine the effects of *mamdc2* on cellular migration and invasion, a Transwell migration assay and a Matrigel invasion assay were performed, respectively. The ability of the cells to migrate was assessed by the ability of the cells to pass through a polycarbonate membrane, whilst the invasive potential of the cells in the Transwell assay was determined by the ability of the cells to invade a Matrigel matrix barrier, which mimics the basement membrane according to the manufacturer. The results (Fig. 30.2) showed that there was no significant difference in the number of migrated

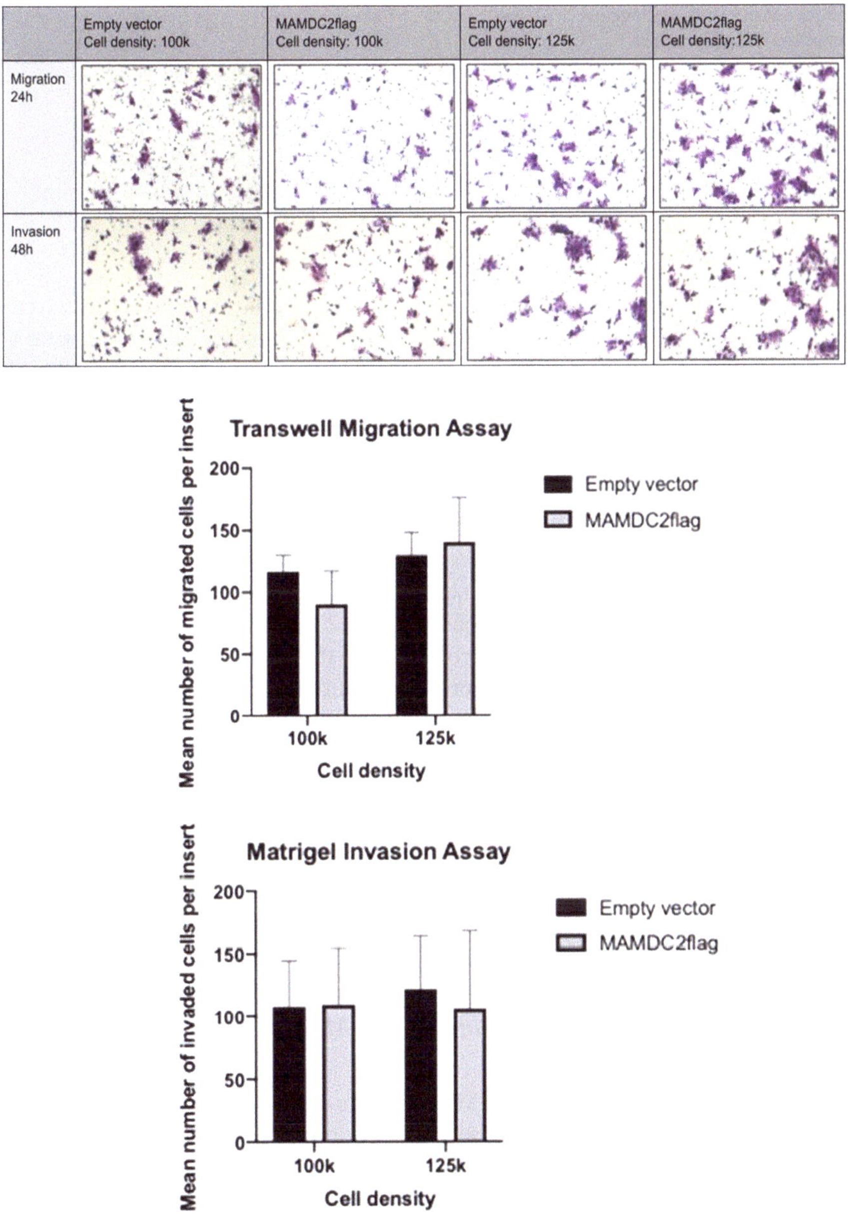

Fig. 30.2 *mamdc2* overexpression had no obvious effect on migration and invasion. Error bars represent standard deviation (SD)

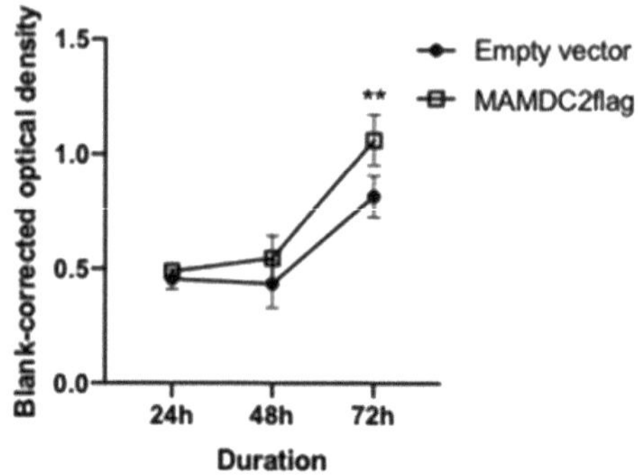

Fig. 30.3 Overexpression of *mamdc2* increased the rate of proliferation relative to control. The blank-corrected optical density is an average taken from 2 independent MTT assays. Error bars represent standard deviation (SD). (**$P < 0.01$)

and invaded cells between *mamdc2* overexpression and control group at the two cell densities (100 and 125 *k*). Hence, high expression of *mamdc2* does not appear to impact cellular migration or invasion.

30.3.4 mamdc2 *Overexpression Promotes Cell Proliferation*

To test the effect of *mamdc2* overexpression on cell proliferation, the MTT proliferation assay, which measures cellular metabolic activity, was performed. The results (Fig. 30.3) showed that overexpression of *mamdc2* significantly increased the rate of proliferation activities in the cells compared to the control group. This suggests that high expression of *mamdc2* may promote proliferation.

Apart from the MTT assay, a wound healing assay was also conducted, which indirectly measures cell proliferation and migration. As shown in Fig. 30.4, overexpression of *mamdc2* significantly increased the rate of wound closure (i.e. a smaller wound area) compared to the control. This higher rate of wound closure could be due to a higher rate of cell proliferation or migration. Based on our data, we believe *mamdc2* overexpression is more likely to affect proliferation rather than migration since the results of MTT assay also supported an increase in proliferation.

30.3.5 Subcellular Localisation of MAMDC2

Knowledge of the localisation of MAMDC2 within the cell may help in determining the function and interactions of the protein. Using immunofluorescence techniques, we found that the intra-cellular localisation of MAMDC2 in transfected HN26 cells was in the cytoplasm and the nucleus (Fig. 30.5). Therefore, MAMDC2 may be involved in intra-cellular signalling or transcription, and may interact with pathways regulating proliferation.

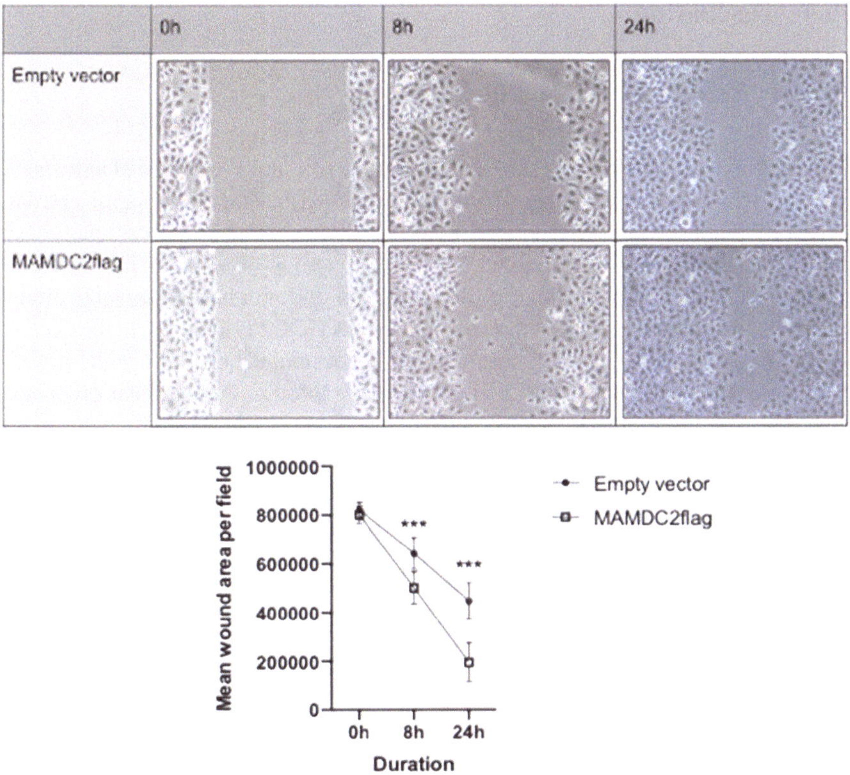

Fig. 30.4 Overexpression of *mamdc2* led to higher rate of wound closure compared to the control. Error bars represent standard deviation (SD). (***$P < 0.001$)

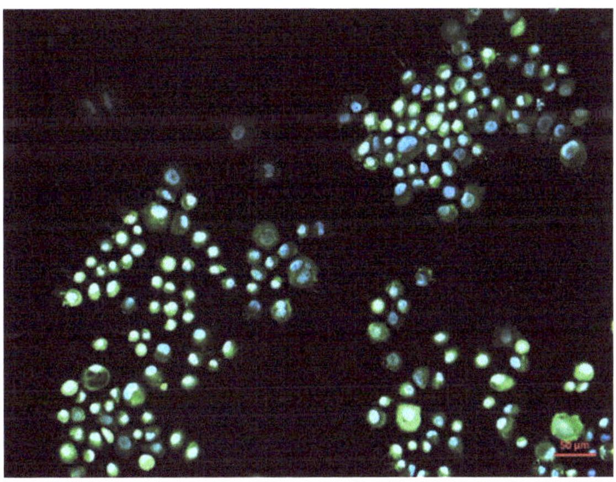

Fig. 30.5 Representative image of HN26 cells transfected with pcDNA3.1::mamdc2flag and stained at 48 h post-transfection using Hoechst (blue; nuclear stain) and Alexa488 (green; binds to anti-FLAG antibodies). Image taken under 20 × objective

30.4 Discussion

30.4.1 Role of MAMDC2 in Cancer

Our experiment results provide new insight into the role and expression of *mamdc2* in HNSCC, which there has been no previously published data on. Our unpublished data and qRT-PCR data support the possible association of *mamdc2* with distant metastasis in HNSCC, and our results revealed that overexpression of *mamdc2* in transfected HN26 cells does promote cellular proliferation but does not have observable effects on cellular migration. Our work has shown that MAMDC2 is localised in the nucleus and cytoplasm within the cell. Since the ECM is not studied here, our results suggest that MAMDC2 is likely to affect cell proliferation through intra-cellular processes. However, how this contributes to metastasis still requires further study.

However, our findings are contradistinct to the data from two breast cancer studies [9, 10] which showed that *mamdc2* is downregulated in breast cancer. Furthermore, *mamdc2* is downregulated in oral squamous cell carcinoma mouse model [11], but our study showed that *mamdc2* is consistently upregulated in HNSCC lines derived from patients with distant metastasis (5 tumour cell lines) and downregulated in HN26, a HNSCC line derived from a patient without distant metastasis. The underexpression of *mamdc2* in HN26 could be a possible reason for a lack of distant metastasis in the patient. MAMDC2 might have different functions in different cancer tissues, which may explain the differences seen. Differences in experimental methodology may also account for the observed differences.

30.4.2 Limitations and Future Studies

Whilst *mamdc2* overexpression was mainly used in our study, this may present a problem as the effects of gene overexpression in cells may be considered as non-physiological. Hence, to verify the results obtained from our overexpression studies, we propose that the knockdown of *mamdc2* can be done in other HNSCC cell lines with high endogenous expression of *mamdc2* (HN19, HN1) to determine if the results are consistent with the overexpression studies.

Regarding areas for future study, more research is needed to investigate how *mamdc2* increases cell proliferation, as our current data suggest that high expression of *mamdc2* promotes proliferation. Future studies can look into pathways regulating proliferation, such as the epidermal growth factor receptor (EGFR) signalling pathway and determine whether *mamdc2* affects any components of this pathway.

To further confirm that *mamdc2* overexpression does indeed promote cell proliferation, staining of the Ki-67 antigen and BrdU labelling can be carried out in future studies, as these tests are more sensitive to cell proliferation than the MTT assay used. The Ki-67 antigen is strongly associated with cancer cell proliferation, whilst BrdU labelling works by incorporating BrdU into replicating DNA.

Additionally, immunofluorescence with an appropriate membrane stain can be performed to observe if MAMDC2 is also localised on the cell membrane, as a faint green outline can be seen surrounding the cells in the representative image shown (Fig. 30.5). If this was to be true, it can be speculated that MAMDC2 possesses an additional function of being a membrane protein that may transmit extracellular signals to the cell interior. As reported by Avilés-Vázquez et al. [7], *mamdc2* encodes for cell membrane proteins in chronic myeloid leukaemia primitive cells, which further supports this theory.

30.4.3 Conclusion

In conclusion, high expression of *mamdc2* promotes cell proliferation and possibly distant metastasis. This observation may be useful for further studies on the role of *mamdc2* in cancer development. Furthermore, *mamdc2* could be a candidate for drug development and anti-cancer therapy.

Acknowledgements We would like to express our deep gratitude to our research supervisor, Dr. Goh Kah Yee, for teaching us valuable research skills and knowledge, her continuous guidance and immense patience throughout our research journey. We would also like to offer our special thanks to our mentor, Associate Professor Dr Darren Lim, for all his guidance and the great opportunity to conduct research at the IMCB-NCCS-MPI-Singapore OncoGenome (INMSOG) Laboratory, Institute of Molecular and Cell Biology. In addition, we would like to thank Mr Terence Cheng for assisting us in performing the immunofluorescence and Ms Tham Su Chin for showing us important experimental techniques. We would like to acknowledge Dr Gopalakrishna Iyer's laboratory in NCCS for providing us with the necessary HNSCC cell lines for our study. We thank our teacher–mentor, Ms Carmen Hoo, for supporting and coordinating this whole project.

Appendix

See Table 30.1, Figs. 30.6, 30.7, 30.8 and Table 30.2.

Table 30.1 Primers used in PCR/qRT-PCR

Primer	Orientation	Sequence	Purpose/remarks
MAMDC2-cDNA-F1 upstream primer	Upstream primer (Forward)	ccaaGGATCCATGCTGTTAAGGGGCGTCCTCCTGG	Generate pcDNA3.1::mamdc2flag plasmid Restriction sites: EcoRI, BamHI Purchased from Integrated DNA Technologies (IDT)
MAMDC2-cDNA-R1 (specific to isoform 1)	Downstream primer (Reverse)	CTTGTAGTCTCCTCCGCCTCCTCCATACTCAATTTCATTTAAGTCCTCA	
MAMDC2-cDNA-R2 (specific to isoform 1)	Downstream primer (Reverse)	ccttGAATTCTTACTTGTCATCGTCGTCCTTGTAGTCTCCTCCGCCTCCTCC	
MAMDC2-F	Upstream primer (Forward)	TGCCTACCAAGGTGGTTTTCA	Target-specific genes for qRT-PCR Purchased from Integrated DNA Technologies (IDT)
MAMDC2-R	Downstream primer (Reverse)	TCCAGGTGGAGGTGGAAGTT	
MAMDC2-isoform 1-F	Upstream primer (Forward)	ACACTACTGGGGTAGGCTACTA	
MAMDC2-isoform 1-R	Downstream primer (Reverse)	TTCCAGAGACTCCTCGCAGA	
ACTB-F	Upstream primer (Forward)	GGCACCCAGCACAATGAAG	
ACTB-R	Downstream primer (Reverse)	GCCGATCCACACGGAGTACT	

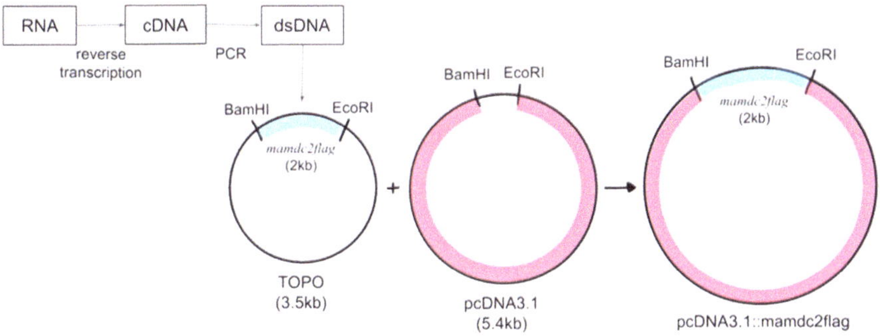

Fig. 30.6 Overview of pcDNA3.1::mamdc2flag plasmid construction

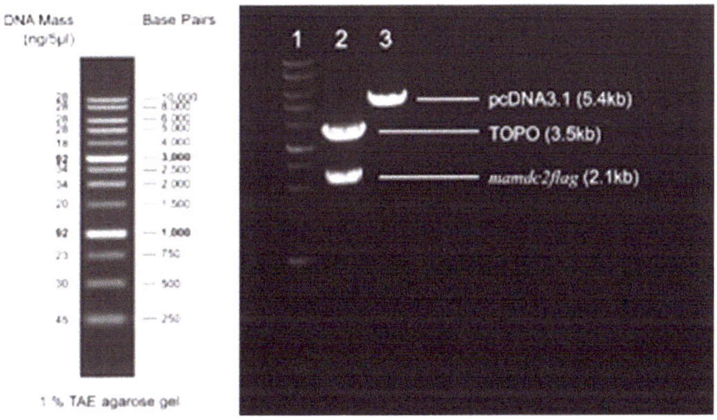

Fig. 30.7 DNA gel electrophoresis image of BamHI and EcoRI restriction digest to obtain the *mamdc2flag* insert. Lane 1: DNA marker. Lane 2: BamHI/EcoRI digest of TOPO::mamdc2flag. Lane 3: BamHI/EcoRI digest of pcDNA3.1 empty vector. The *mamdc2flag* insert from lane 2 and pcDNA3.1 vector from lane 3 were extracted, purified and ligated as shown in Fig. 30.6

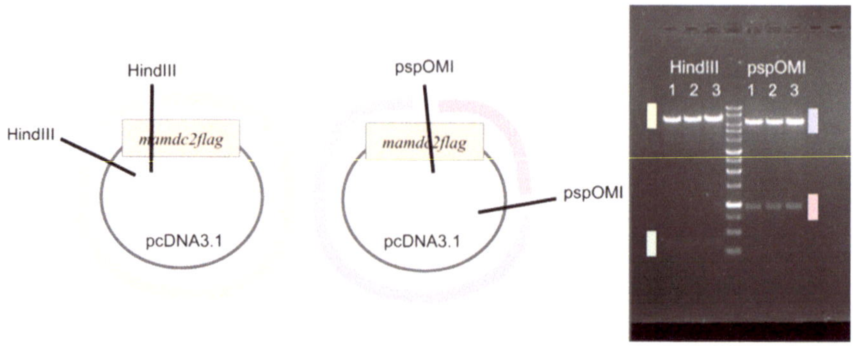

Fig. 30.8 Verification of bacterial clones that carry the correct *mamdc2flag* insert in pcDNA3.1 vector. After transformation, three putative bacterial clones containing pcDNA3.1::mamdc2flag plasmid were selected. HindIII and PspOMI restriction enzymes were used to verify the presence of the correct *mamdc2flag* insert and pcDNA3.1 vector. The results of restriction digest were visualised by DNA gel electrophoresis. The coloured areas show the DNA fragments obtained after the test digest

Table 30.2 Verification of the overexpression of *mamdc2* in transfected HN26 cells

Sample name		Assay name	Cq mean	Relative fold change in mRNA expression normalised to beta-actin and compared to empty vector	Remarks
Cell line	Plasmid transfected				
HN26	Empty vector (pcDNA3.1)	*mamdc2* total	31.81	1	-
		mamdc2 isoform 1	32.20	1	-
		beta-actin (control)	18.26	–	-
	pcDNA3. 1::mamd c2flag	*mamdc2* total	15.93	42,864.82	Upregulated
		mamdc2 isoform 1	15.96	55,443.25	Upregulated
		beta-actin (control)	17.78	–	-

Relative fold change in mRNA expression was calculated using this formula: $2^{(-\Delta\Delta Cq)}$

References

1. Lu, P., Weaver, V. M., & Werb, Z. (2012). The extracellular matrix: A dynamic niche in cancer progression. *Journal of Cell Biology, 196*(4), 395–406. https://doi.org/10.1083/jcb.201102147. PMID:22351925;PMCID:PMC3283993
2. Naba, A., Clauser, K. R., Ding, H., Whittaker, C. A., Carr, S. A., & Hynes, R. O. (2016). The extracellular matrix: Tools and insights for the "omics" era. *Matrix Biology, 49*, :10–24. https://doi.org/10.1016/j.matbio.2015.06.003. Epub 2015 Jul 8. PMID: 26163349; PMCID: PMC5013529
3. Lim, S. B., Tan, S. J., Lim, W. T., & Lim, C. T. (2017). An extracellular matrix-related prognostic and predictive indicator for early-stage non-small cell lung cancer. *Nature Communications, 8*(1), 1734. https://doi.org/10.1038/s41467-017-01430-6.PMID:29170406;PMCID: PMC5700969
4. Lim, S. B., Chua, M. L. K., Yeong, J. P. S., Tan, S. J., Lim, W. T., & Lim, C. T. (2019). Pan-cancer analysis connects tumor matrisome to immune response. *NPJ Precision Oncology, 22*(3), 15. https://doi.org/10.1038/s41698-019-0087-0.PMID:31123708;PMCID:PMC6531473
5. Wong, T. H., Skanthakumar, T., Nadkarni, N., Nguyen, H. V., & Iyer, N. G. (2017). Survival of patients with head and neck squamous cell carcinoma by housing subsidy in a tiered public housing system. *Cancer, 123*(11), 1998–2005. https://doi.org/10.1002/cncr.30557 Epub 2017 Jan 30 PMID: 28135397.
6. Duprez, F., Berwouts, D., De Neve, W., Bonte, K., Boterberg, T., Deron, P., Huvenne, W., Rottey, S., & Mareel, M. (2017). Distant metastases in head and neck cancer. *Head and Neck, 39*(9), 1733–1743. https://doi.org/10.1002/hed.24687 Epub 2017 Jun 26 PMID: 28650113.
7. Avilés-Vázquez, S., Chávez-González, A., Hidalgo-Miranda, A., Moreno-Lorenzana, D., Arriaga-Pizano, L., Sandoval-Esquivel, M. Á., Ayala-Sánchez, M., Aguilar, R., Alfaro-Ruiz, L., & Mayani, H. (2017). Global gene expression profiles of hematopoietic stem and progenitor cells from patients with chronic myeloid leukemia: the effect of in vitro culture with or without imatinib. *Cancer Medicine, 6*(12), :2942–2956. https://doi.org/10.1002/cam4.1187. Epub 2017 Oct 13. PMID: 29030909; PMCID: PMC5727298
8. Denisov, E. V., Skryabin, N. A., Gerashchenko, T. S., Tashireva, L. A., Wilhelm, J., Buldakov, M. A., Sleptcov, A. A., Lebedev, I. N., Vtorushin, S. V., Zavyalova, M. V., Cherdyntseva, N. V., & Perelmuter, V. M. (2017). Clinically relevant morphological structures in breast cancer represent transcriptionally distinct tumor cell populations with varied degrees of epithelial-mesenchymal transition and CD44+CD24- stemness. *Oncotarget, 8*(37), 61163–61180. https://doi.org/10.18632/oncotarget.18022.PMID:28977854;PMCID:PMC5617414
9. Lee, H., Park, B. C., Soon Kang, J., Cheon, Y., Lee, S., & Jae Maeng, P. (2020). MAM domain containing 2 is a potential breast cancer biomarker that exhibits tumour-suppressive activity. *Cell Proliferation, 53*(9), e12883. https://doi.org/10.1111/cpr.12883. Epub 2020 Jul 24. PMID: 32707597; PMCID: PMC7507446
10. Sultan, G., Zubair, S., Tayubi, I. A., Dahms, H. U., & Madar, I. H. (2019). Towards the early detection of ductal carcinoma (a common type of breast cancer) using biomarkers linked to the PPAR(γ) signaling pathway. *Bioinformation, 15*(11), 799–805. https://doi.org/10.6026/97320630015799.PMID:31902979;PMCID:PMC6936658
11. Wu, L., Jiang, Y., Zheng, Z., Li, H., Cai, M., Pathak, J. L., Li, Z., Huang, L., Zeng, M., Zheng, H., Ouyang, K., & Gao, J. (2020). mRNA and P-element-induced wimpy testis-interacting RNA profile in chemical-induced oral squamous cell carcinoma mice model. *Experimental Animal, 69*(2), 168–177. https://doi.org/10.1538/expanim.19-0042. Epub 2019 Nov 19. PMID: 31748426; PMCID: PMC7220707

Chapter 31
Predicting In Vivo Drug–Drug Interactions Between Rivaroxaban and Tyrosine Kinase Inhibitors Arising from Mechanism-Based Inactivation of Cytochrome P450 3A4

Ng Wei Shuen, Hou Weihan, and Choo Jun An Jerry

Abstract Cancer-associated venous thromboembolism (CA-VTE)—a highly debilitating cardiovascular comorbidity predominantly elicited by Virchow's triad that complicates neoplastic disease management—is amongst the principal causes of death in ambulatory cancer patients. In CA-VTE prophylaxis, direct oral anticoagulant rivaroxaban is becoming a viable alternative as compared to the clinically prescribed intravenously administered low-molecular-weight heparin. Concurrently, tyrosine kinase inhibitors (TKIs) represent a novel paradigm of anticancer therapies which dysregulate key signal transduction cascades indispensable for tumour cell proliferation, consequently halting carcinogenesis. However, hepatic and intestinal cytochrome P450 3A4 enzymes (CYP3A4) are known to catalyse TKI bioactivation, generating reactive metabolites which render CYP3A4 non-functional through mechanism-based inactivation. A possibility of detrimental drug–drug interaction (DDI) surfaces owing to the concomitant consumption of rivaroxaban (victim) and TKI (perpetrator), where the former elevates to supratherapeutic levels, predisposing CA-VTE patients to major bleeding risk. This provided a strong impetus towards predicting the extent of DDI to pre-emptively abate the potentially lethal ramifications of DDI-induced haemorrhage. A mechanistic static model was harnessed to predict the likelihood of major internal bleeding induced by synchronous co-administration of rivaroxaban and four archetypal TKIs namely, imatinib, nilotinib, crizotinib and pazopanib. Premised on interactions via CYP3A4 metabolism, our results predicted the comparatively low risk of DDI-induced major bleeding between rivaroxaban and the four TKIs. Stemming from rivaroxaban's narrow therapeutic index and its elimination via other hepatic enzymes and renal active transporters, the potential DDI between rivaroxaban and anticancer drugs should be further screened based on these additional potential interactions.

N. W. Shuen (✉) · H. Weihan · C. J. A. Jerry
Hwa Chong Institution (Junior College), Singapore, Singapore
e-mail: 161756U@student.hci.edu.sg

H. Guo et al. (eds.), *IRC-SET 2021*, https://doi.org/10.1007/978-981-16-9869-9_31

Keywords Cancer-associated venous thromboembolism · Rivaroxaban · Tyrosine kinase inhibitors · CYP3A4 · Mechanism-based inactivation · Drug–drug interactions · Mechanistic static modelling · Major bleeding

31.1 Introduction

Cancer-associated venous thromboembolism (CA-VTE)—comprising deep vein thrombosis and pulmonary embolism [6] is a common comorbidity [11]. Cancer patients are predisposed to acquiring VTE by 4–7 times as compared to their non-cancer counterparts [1], owing to Virchow's triad (i.e. venous stasis, endothelial injury and hypercoagulability) [14]. The occurrence of CA-VTE complicates neoplastic disease management, resulting in poorer clinical outcomes. In fact, CA-VTE is amongst the leading causes of death in ambulatory cancer-afflicted patients, whereby CA-VTE patients often face a sixfold decrease in survival rate as compared to VTE patients not afflicted by cancer [9].

In recent years, direct oral anticoagulants (DOACs) such as rivaroxaban (Fig. 31.1), epitomises twenty-first century therapeutics for VTE-afflicted patients. DOACs have gradually overshadowed conventional anticoagulant treatment that involved the administration of vitamin K antagonists and low-molecular-weight heparin due to their direct inhibition of Factor Xa in the coagulation cascade, a more selective and target-specific treatment. Key advantages of DOACs include more rapid onset of action [5], fewer dietary considerations and less extensive monitoring due to their linear pharmacokinetic properties, as well as greater efficiency in reducing coagulation in patients afflicted with CA-VTE.

Tyrosine kinase inhibitors (TKI) represent a new paradigm of cancer therapeutics that function as target-specific moieties which pinpoint key molecular pathways instrumental to the proliferation and survival of cancerous cells. This is in contrast to conventional chemotherapeutic agents which indiscriminately exert cytotoxicity to all cells—including non-cancerous cells [8]. The four TKIs investigated in this study are utilised in different and distinct types of cancers (Fig. 31.2); crizotinib is used in first-line treatment of non-small cell lung cancer involving tumours caused

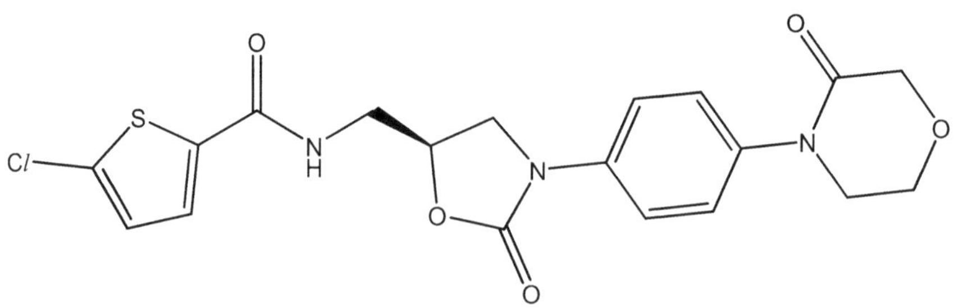

Fig. 31.1 Chemical structure of rivaroxaban

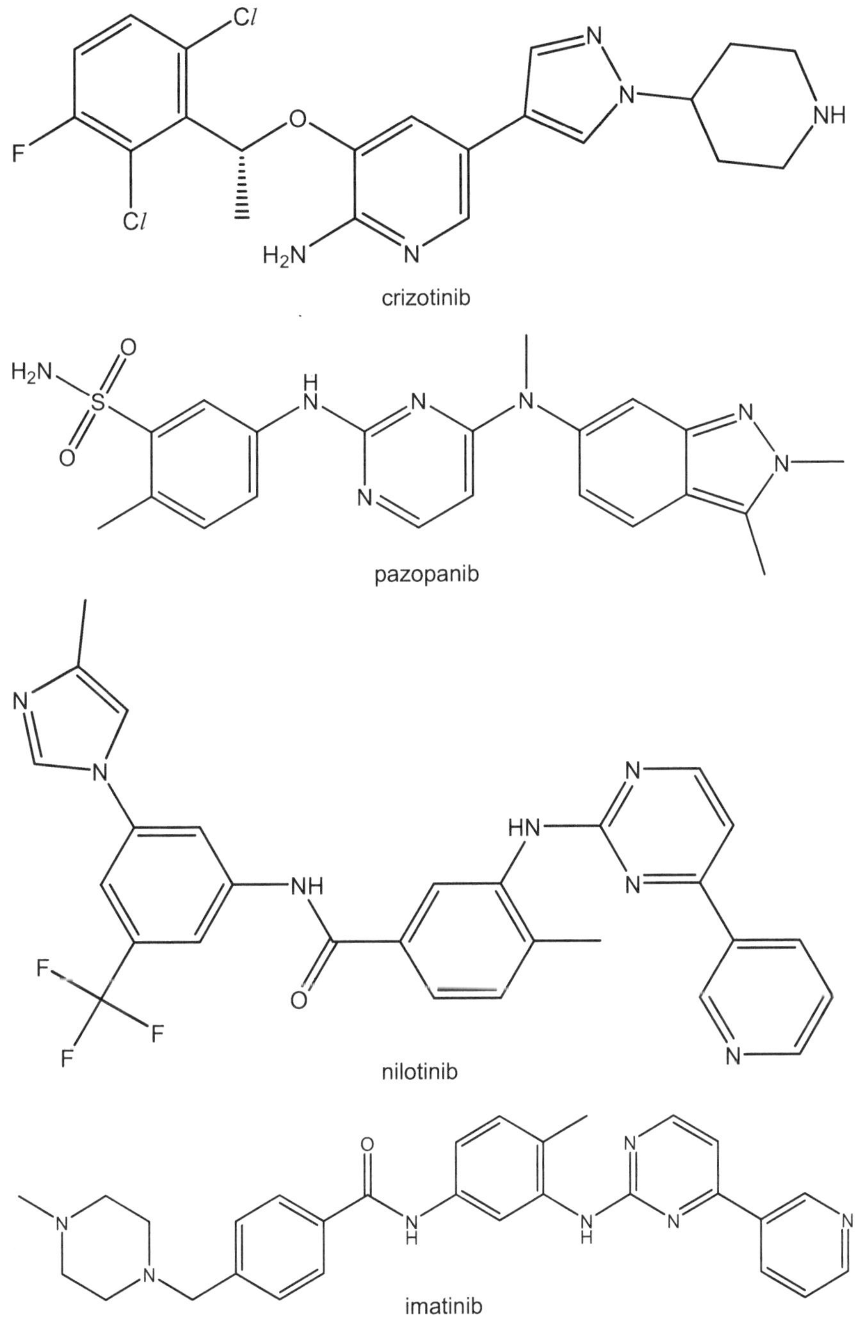

Fig. 31.2 Chemical structures of the four tyrosine kinase inhibitors of interest

by a mutated *ALK* gene. It inhibits ALK phosphorylation, restricting kinase activity. Pazopanib is used in first-line treatment of advanced renal cell carcinoma and soft tissue sarcoma [10]. It functions by targeting kinases of vascular endothelial growth factor receptors, inhibiting angiogenesis [7]. Nilotinib and imatinib are used in first- and second-line treatment respectively of acute lymphoblastic leukaemia as well as chronic myeloid leukaemia. They inhibit the mutated BCR-ABL1 kinase created by the Philadelphia chromosome abnormality, preventing carcinogenesis.

Cytochrome P450 3A4 (CYP3A4), a predominant isoform of the cytochrome P450 superfamily (CYP450), is responsible for metabolism of ~60% of clinically available drugs [17]. Several CYP3A4-mediated bioactivation pathways can initiate reactive metabolite (RM) formation from metabolic conversion of various drug moieties [3, 4, 15].

A key consequence of RM formation is mechanism-based inactivation (MBI), a unique form of quasi-irreversible or irreversible enzyme inhibition where the RM formed is sequestered onto the CYP3A4 enzymes, rendering them non-functional. MBI of CYP3A4 elicited by a perpetrator drug could block the metabolic pathways of the victim drug, leading to its accumulation beyond therapeutic threshold levels (Fig. 31.3). Such an interaction is also termed as a drug–drug interaction (DDI) and could precipitate dose-dependent toxicities.

Potential MBIs are greatly scrutinised in the study of DDIs, where multiple different substrates are co-administered to patients with comorbidities. Such DDIs

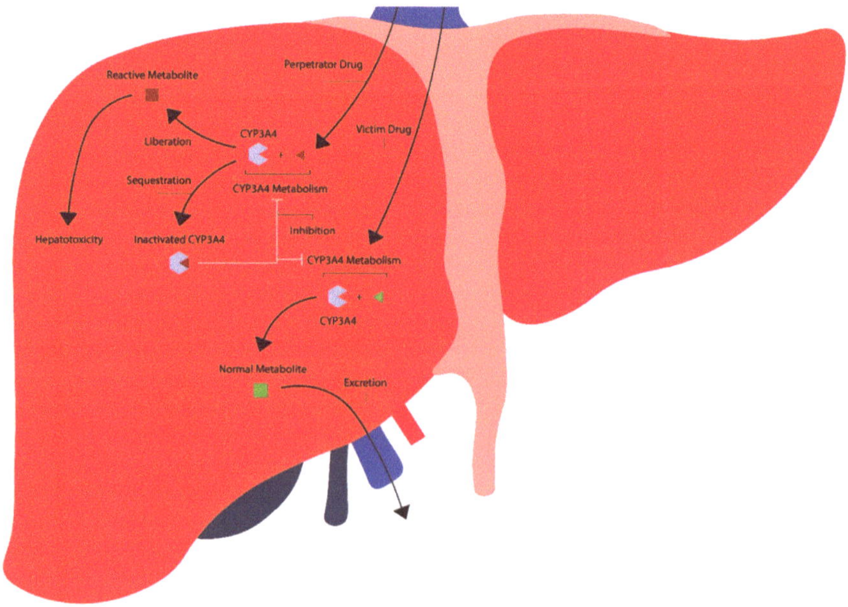

Fig. 31.3 MBI of CYP3A4 and drug–drug interactions which cause accumulation of victim drug

can significantly alter the victim drug's plasma concentrations to reach supratherapeutic levels, causing excessive bleeding in the case of rivaroxaban. As rivaroxaban is a substrate of CYP3A4 and crizotinib, pazopanib, nilotinib and imatinib are known mechanism-based inactivators of CYP3A4, a question arises as to whether such a clinically-relevant drug combination could potentially result in DDI, and if so, to what extent?

31.2 Objectives and Hypothesis

In this study, four TKIs were investigated, namely crizotinib, pazopanib, nilotinib and imatinib. Rivaroxaban may be potentially prescribed along with these TKIs for the prophylaxis of CA-VTE in various cancer types. Thereafter, relevant pharmacokinetic parameters and methodologies obtained from a plethora of scientific literature will be systematically incorporated into a mechanistic model to accurately evaluate the significance of DDI for each corresponding TKI.

Here, we hypothesise that the concomitant consumption of rivaroxaban, which is a substrate of CYP3A4, and the four TKIs which are known as mechanism-based inactivators of CYP3A4 can precipitate in varying degrees of DDI. These DDIs may subsequently predispose a patient to an elevated risk of major internal bleeding as the concentration of rivaroxaban increases to supratherapeutic levels.

31.3 Methods

31.3.1 Data Collection

Inactivator concentration at half-maximum inactivation rate constant (K_I), maximum inactivation rate constant (k_{inact}), oral absorption rate constant (k_{a}), fraction of inactivator dose absorbed into the gut wall (f_a), total daily oral dose of inactivator (D), maximum steady-state concentration ($C_{\text{ss,max}}$) and fraction of the drug unbound within the blood ($f_{\text{u,b}}$) for each of the four selected TKIs (i.e. crizotinib, pazopanib, nilotinib and imatinib), as well as fraction of the drug metabolised by CYP3A4 ($f_{m,\text{CYP3A4}}$) and oral bioavailability (F_{oral}) for rivaroxaban, degradation rate constant for CYP3A4 in the liver or the gut ($k_{\text{deg,H/G}}$), and enterocytic blood flow (Q_G) were selectively recorded from Filppula et al. and Cheong et al.

31.3.2 Estimating the Extent of Metabolic Drug–Drug Interaction Arising from CYP3A4 Inactivation Using a Mechanistic Static Model

The in vivo concentrations of the selected drugs available to CYP3A4 in the liver were estimated based on $C_{\mathrm{ss,max}}$ corrected for plasma protein binding, represented by $[I]_{\mathrm{H}}$ (Eq. 31.1).

$$[I]_H = f_{u,b} \times C_{ss,max} \tag{31.1}$$

The in vivo concentrations of the selected drugs available to CYP3A4 in the gut were estimated based on dose, rate of absorption and fraction absorbed, represented by $[I]_{\mathrm{G}}$ (Eq. 31.1). Enterocytic blood flow (Q_G) corresponded to 248 mL min^{-1}.

$$[I]_G = D \times k_a \times \frac{t_a}{Q_G} \tag{31.2}$$

Observed time-dependent inactivation of CYP3A4 in the liver (A) represents observed time-dependent inactivation of CYP3A4 in the liver, with subscript H denoting liver values. $k_{\mathrm{deg,H}}$ corresponded to 0.000146 min^{-1}. A is calculated with Equation.

$$A = \frac{k_{\deg,H}}{k_{\deg,H} + \frac{k_{\mathrm{inact}} \times [I]_H}{K_I + [I]_H}} \tag{31.3}$$

Observed time-dependent inactivation of CYP3A4 in the gut (X) represents observed time-dependent inactivation of CYP3A4 in the gut, with subscript G denoting gut values. $k_{\mathrm{deg,G}}$ corresponded to 0.000481 min^{-1}. X is calculated with Equation.

$$X = \frac{k_{\deg,G}}{k_{\deg,G} + \frac{k_{\mathrm{inact}} \times [I]_G}{K_I + [I]_G}} \tag{31.4}$$

A modified mechanistic static model that only took into account time-dependent inhibition was used to predict the extent of metabolic DDI (Eq. 31.5). AUCR, the area under curve ratio, represents the area under the plasma concentration–time curve of the affected drug in the presence and absence of the inhibitor.$f_{m,\mathrm{CYP3A4}}$ of rivaroxaban (victim drug) corresponds to 0.18. The fraction of rivaroxaban escaping intestinal extraction (F_G) was previously calculated to be 0.89.

$$\mathrm{AUCR} = \frac{1}{f_{m,\mathrm{CYPA4}} \times A + (1 - f_{m,\mathrm{CYPA4}})} \times \frac{1}{(1 - F_G) + F_G} \tag{31.5}$$

31.3.3 Predicting the Risk of Major Bleeding

The risk of major bleeding occurring as a result of accumulation of the victim drug is given by Eq. 31.6. The area under the curve at steady state (AUC_{SS}) corresponds to 3.164 mg h L^{-1}.

$$\text{Risk of Major Bleeding } (\%) = 1.79 \times 1.32^{AUC_{SS} \times AUCR} \quad (31.6)$$

31.4 Results and Discussion

31.4.1 Static Modelling of Metabolic DDI

A mechanistic static model modified from Artusson et al. was utilised with the time-dependent inhibition parameters of four target pharmacokinetic time-dependent CYP3A4 inactivators to determine their corresponding AUCR values. Data sourced from literature review was subsequently applied to Eq. (31.6) to comprehensively predict the risk of major bleeding, as shown in Tables 31.1 and 31.2.

Table 31.1 Pharmacokinetic parameters of TKI obtained from Filppula et al.

Drug	K_I (μM)	K_{inact} (min^{-1})	$fu_{,p}$	B: P	$C_{ss,max}$ (μM.L^{-1})	D (mg)	k_a	fa
Crizotinib	0.78 ± 0.130	0.054 ± 0.030	0.0930	1.10	0.0007	1000	0.8	0.4
Imatinib	3.07 ± 0.710	0.020 ± 0.001	0.0683	0.73	0.0053	400	1.1	1.0
Nilotinib	1.29 ± 0.230	0.031 ± 0.001	0.0013	0.76	0.0030	800	6	0.8
Pazopanib	1.88 ± 0.290	0.016 ± 0.001	0.0013	0.76	0.1029	800	6	0.8

Table 31.2 Calculated parameters for static modelling

Drug	$f_{u,b}$	$[I]_H$ (μM)	$[I]_G$ (μM)	A
Crizotinib	0.0845	0.0000614	48.76	0.648
Imatinib	0.0933	0.000491	58.70	0.477
Nilotinib	0.00173	0.00000522	487.4	0.973
Pazopanib	0.00173	0.000178	589.8	0.607

31.4.2 Discussion

Crizotinib, imatinib, nilotinib and pazopanib were all predicted to increase the AUC values of all drugs by approximately 1.2-fold. Imatinib had the greatest AUCR value of 1.24, while nilotinib had the smallest AUCR value of 1.13. Correspondingly, the risk of major bleeding stemming from the co-administration of rivaroxaban with each of the TKIs was predicted to be approximately 5%, with imatinib displaying the highest risk of major bleeding at 5.32% and nilotinib displaying the lowest risk of major bleeding at 4.83%. This increase from the baseline risk of major bleeding when rivaroxaban is administered independently, 4.31%, was calculated using Equation by taking its AUCR value to be 1.

Area under plasma concentration–time curve (AUC) measures the total systemic exposure of a given drug. According to the FDA guidelines, a strong drug inhibitor can increase the AUC by ≥fivefold; moderate inhibitor by 2 to <fivefold; weak inhibitor by <twofold. When CYP3A4 inhibitors are concomitantly consumed with other substrates, the metabolic and excretion pathways for the substrates are blocked, increasing the substrate residence time in the blood plasma, as represented by AUCR—the multi-fold increase in AUC (FDA, 2020). From Table 31.3, it is evident that all the four TKIs yielded an average of 1.2-fold increase in AUC of rivaroxaban, therefore suggesting that the magnitude of DDI when subjected to in vivo conditions is relatively small. Hence, it can be deduced that the co-administration of each TKI with rivaroxaban exerts a relatively small effect on the systemic exposure of rivaroxaban.

The risk of bleeding stems from accumulation of DOACs such as rivaroxaban which could cause blood thinning and potentially major internal bleeding. Given that rivaroxaban is metabolised by CYP3A4, co-administered CYP3A4 inactivators have the potential to disrupt rivaroxaban metabolism and culminate in its accumulation. Stronger inactivators inhibit rivaroxaban metabolism to a greater extent, leading to greater accumulation and hence a much greater risk of major bleeding, whereas weaker inactivators can be expected to give rise to a less significant increase in the risk of major bleeding.

As calculated earlier, for patients with VTE who have been independently administered with rivaroxaban only, the risk of major bleeding amounts to 4.31%. This value can be referred to as the baseline risk of major bleeding, with the extent of increase in risk of major bleeding being utilised as the benchmark that can be referenced

Table 31.3 Calculated AUCR values and risk of major bleeding

Drug	AUCR	AUC_{SS} (mg h L^{-1})	Risk of bleeding (%)	Net increase in risk of bleeding (%)
Crizotinib	1.20	3.164	5.13	0.82
Imatinib	1.24	3.164	5.32	1.01
Nilotinib	1.13	3.164	4.83	0.52
Pazopanib	1.21	3.164	5.18	0.87

against to ascertain the clinical significance of co-administration of rivaroxaban with the various TKIs of interest.

Imatinib, the strongest CYP3A4 inactivator, imposed a risk of 5.32% of causing major bleeding when co-administered with rivaroxaban, whereas nilotinib, the weakest CYP3A4 inhibitor, imposed a slightly lower risk of 4.83% in giving rise to major bleeding. Therefore, co-administering imatinib with CYP3A4 poses the greatest risk out of the four TKIs of interest. Nevertheless, the increase in propensity towards major bleeding was relatively insignificant for each drug, with the maximum percentage increase in major bleeding being 1.01%; in contrast, the average net increase in the likelihood of major bleeding for individuals consuming anticoagulants is 1.88% [16]. Intuitively, the co-administration of rivaroxaban with each of the four TKIs predisposes individuals to major bleeding to a relatively small extent.

However, patients with CA-VTE complications often suffer from thrombocytopenia [1], a common symptom that is characterised by abnormally low platelet count found in the patient's bloodstream. As platelets are a paramount component essential in blood clotting, patients with abnormally low platelet count often face difficulty in wound closure due to ineffective plugging of the site on injury and thus experience prolonged healing time. The marginal increased systemic exposure of rivaroxaban due to the MDI-induced DDI might therefore still potentially augment the risk of major internal bleeding due to underlying thrombocytopenia.

Although the AUCR value for each of the four TKIs shown in Table 31.3 shows that the extent of DDI between rivaroxaban and the four drugs is relatively small, their inhibitory effects play a pivotal role in the customisation of patient treatment, owing to rivaroxaban's narrow therapeutic index. The therapeutic index is the specific range of drug dosage at which the drug exhibits therapeutic effects, and a drug administered outside of its therapeutic window would culminate in ineffective therapy or adverse pharmacodynamic reactions. Rivaroxaban's narrow therapeutic index [2] makes it difficult to regulate its anticoagulation effects when rivaroxaban is concomitantly administered and consumed with either of the four TKIs of interest. Pharmacokinetic DDIs of both the perpetrator and victim drugs can thus precipitate the accumulation of rivaroxaban, the victim drug, in the bloodstream, potentially elevating its blood plasma concentrations to supratherapeutic levels, ultimately resulting in blood thinning and predisposing one to unintended and potentially lethal haemorrhage.

This marks the cornerstone of DDIs, where even a small magnitude of DDI between rivaroxaban and the four weak CYP3A4 inactivators can possibly snowball into a cascade of deleterious responses which can engender major internal bleeding, despite strong consensus have long been established regarding DOACs' superiority over conventional anticoagulants in monitoring requirements and rate of absorption. Considering the different severities of CA-VTE from patient to patient, the dosage of rivaroxaban for each patient would also vary accordingly, thereby positioning patients with greater rivaroxaban dosages at a greater major bleeding risk due to higher drug plasma concentration, which could be further augmented by the four TKIs. Therefore, considering the variation in dosage of rivaroxaban from individual to individual, dosages of the four TKIs should be carefully adjusted.

A major assumption for mechanistic static modelling resides in the vast number of molecular pathways involved in the metabolism and elimination of a drug. The static modelling equations provides a simplistic overview of the process in which a victim drug's pharmacokinetic parameters are affected by the perpetrator drug but lacks the intricacies of all the potential metabolic pathways which could be undetected. Subsequently, this could lead to an under-prediction or over-prediction of the calculated AUCR values, thereby possibly complicating patient treatment [13].

Furthermore, other factors such as drug dosage, possible undocumented alternative pathways for drug metabolism and excretion as well as patient-to-patient variances have not been taken into account. Moreover, our study of MBI only targeted inactivation but not reversible inhibition, which then could lead to multiple output results. Additionally, CYP3A4 was mainly focused in this study due to the key role it plays in metabolising a large proportion of xenobiotics, but there are other hepatic CYP450 enzymes and renal transporters that account for the elimination of rivaroxaban, hence adding on to the limitations of static modelling and dampening its efficacy in translating experimental data into clinical application.

To better reflect in vivo conditions and realistic settings, one should consider the usage of patient-specific modelling in addition to mechanistic static modelling. Patient-specific modelling refers to the formulation of human pathophysiological models using patient-specific data, thereby customising experimental findings to each individual rather than deriving an average set of data applicable to a wide group of patients [12], overcoming the limitations of mechanistic static modelling.

31.5 Conclusion and Recommendations for Future Work

In this study, the AUCR values of the four target TKIs were calculated using static modelling-based equations, which served as a proxy indicator of the potency of the DDIs between rivaroxaban and the four TKIs. Our results have displayed that the four target TKIs result in a small magnitude of DDI, which lowers the risk of DDIs if co-administered to patients afflicted with CA-VTE. Nevertheless, one should take a precautious stance on administering rivaroxaban with other drugs to avoid pharmaceutical complications in cancer management due to major internal bleeding.

An application of our results is its potential use in the biomedical field. Our study has demonstrated that each of these TKIs poses a relatively small risk of major bleeding when co-administered with rivaroxaban, thus making them potentially compatible for co-administration in most cases of future clinical prescription. Predicting such risks is paramount given the absence of large trials with alternative treatment strategies for co-administration of potentially incompatible pharmaceutical drugs.

An unexplored but equivocally significant aspect that can be worked on as future work would be the MBI- and DDI-mediated toxicities elicited by other hepatic enzymes and renal active transporters. After sourcing from a myriad of search engines and literature libraries, pharmacokinetic data from only the four TKIs studied above could be found. If more drugs not limited to the current four drugs investigated were to be studied, the possibility for the drugs to precipitate into hepatotoxicity presents a new challenge in the treatment of VTE. By quantifying the K_I, K_{inact}, $[I]_H$ and $[I]_G$, the corresponding AUCR value and risk of major bleeding can be easily computed using our highly applicable and efficient quantitative method.

In addition, another possible detrimental outcome of CYP3A4-mediated MBI, and thus, DDI is idiosyncratic hepatoxicity, an under researched field where the severity of liver damage varies drastically from patient to patient independent of drug dosage, and whereby the exact mechanism for such a peculiar phenomenon has been unelucidated hitherto. Therefore, postulated mechanisms that can lead to such idiosyncratic hepatotoxicity like the hapten theory and danger theory could be explored [3].

References

1. Bauersachs, R., Khorana, A. A., Lee, A. Y., & Soff, G. (2020). Cancer-associated venous thromboembolism: Treatment and prevention with rivaroxaban. *Research and Practice in Thrombosis and Haemostasis, 4*(4), 532–549.
2. Ding, S., Wang, L., Xie, L., Zhou, S., Chen, J., Zhao, Y., & Shao, F. (2020). Bioequivalence study of 2 formulations of rivaroxaban, a narrow-therapeutic-index drug, in healthy chinese subjects under fasting and fed conditions. *Clinical Pharmacology in Drug Development, 9*(3), 346–352.
3. Guengerich, F. P. (2010). Mechanisms of drug toxicity and relevance to pharmaceutical development. *Drug Metabolism and Pharmacokinetics*, 1010210090–1010210090.
4. Irving, R. M., & Elfarra, A. A. (2012). Role of reactive metabolites in the circulation in extrahepatic toxicity. *Expert Opinion on Drug Metabolism & Toxicology, 8*(9), 1157–1172.
5. Julia, S., & James, U. (2017). Direct oral anticoagulants: A quick guide. *European Cardiology Review, 12*(1), 40.
6. Jun, M., Lix, L. M., Durand, M., Dahl, M., Paterson, J. M., Dormuth, C. R., & Wu, C. (2017). Comparative safety of direct oral anticoagulants and warfarin in venous thromboembolism: multicentre, population based, observational study. *bmj, 359*, j4323.
7. Keisner, S. V., & Shah, S. R. (2011). Pazopanib. *Drugs, 71*(4), 443–454.
8. Kojima, K., & Burger, J. A. (2020). Treatment algorithm for Japanese patients with chronic lymphocytic leukemia in the era of novel targeted therapies. *Journal of Clinical and Experimental Hematopathology*, 20002.
9. Lorga, R. A., Bratu, O. G., Marcu, R. D., Constantin, T., Mischianu, D. L. D., Socea, B., & Diaconu, C. C. (2019). Venous thromboembolism in cancer patients: Still looking for answers. *Experimental and Therapeutic Medicine, 18*(6), 5026–5032.
10. Maruzzo, M., Martin-Liberal, J., Messiou, C., Miah, A., Thway, K., Alvarado, R., & Benson, C. (2015). Pazopanib as first line treatment for solitary fibrous tumours: The Royal Marsden Hospital experience. *Clinical Sarcoma Research, 5*(1), 5.
11. Moik, F., & Ay, C. (2020). How I manage cancer-associated thrombosis. *Hämostaseologie, 40*(01), 038–046.
12. Neal, M. L., & Kerckhoffs, R. (2010). Current progress in patient-specific modeling. *Briefings in Bioinformatics, 11*(1), 111–126.
13. Peters, S. A., Schroeder, P. E., Giri, N., & Dolgos, H. (2012). Evaluation of the use of static and dynamic models to predict drug-drug interaction and its associated variability: Impact on drug discovery and early development. *Drug Metabolism and Disposition, 40*(8), 1495–1507.

14. Renni, M. J. P., Araujo, M. L. C., Trugilho, I., Bergmann, A., & e Coura, C. P. D. M. (2017). Rivaroxaban used in the treatment patients with gynecologic cancer and venous thromboembolism: The experience of Instituto Nacional de Câncer—Rio de Janeiro, Brazil. *International Journal of Gynecologic Cancer, 27*(5).
15. Sasaki, E., & Yokoi, T. (2018). Role of cytochrome P450-mediated metabolism and involvement of reactive metabolite formations on antiepileptic drug-induced liver injuries. *Journal of Toxicological Sciences, 43*(2), 75–87.
16. Shoeb, M., & Fang, M. C. (2013). Assessing bleeding risk in patients taking anticoagulants. *Journal of Thrombosis and Thrombolysis, 35*(3), 312–319.
17. Yan, L., Wang, Y., Liu, J., Nie, Y., Zhong, X. B., Kan, Q., & Zhang, L. (2017). Alterations of histone modifications contribute to pregnane X receptor-mediated induction of CYP3A4 by rifampicin. *Molecular Pharmacology, 92*(2), 113–123.

Chapter 32
Low-Reynolds Number Airfoil Shape Optimization Using Genetic Algorithm with Mode Parameterization

Yu Shuhuai

Abstract Genetic algorithm (GA) has been widely used in airfoil optimization. However, GA is limited in high-dimensional optimization problems due to the curse of dimensionality. In the past decades, various dimension reduction methods have been proposed in airfoil parameterization to enhance convergence and performance of optimization algorithms. To improve the efficiency of using GA in airfoil optimization, a machine learning technique called singular-value decomposition (SVD) is used in this paper. Mode parameterization of airfoil is performed by SVD, which can significantly reduce the number of variables required in the optimization process. Moreover, airfoils generated by SVD are more realistic compared to the traditional points controlling method. The results show that the implementation of SVD can generate airfoils with better performance and increase the optimization efficiency.

Keywords Airfoil optimization · Genetic algorithm · Mode parameterization · Low-Reynolds number

32.1 Introduction

The performance of aircraft largely depends on its airfoil, which generates lift forces and extracts energy. Thus, the airfoil design is crucial in the entire aircraft design, making airfoil optimization an important field in aerodynamic research. Traditionally, the optimization of airfoil is based on wind tunnel experiment. Using wind tunnel is accurate in determining airfoil performance, but it is costly, time consuming, and ineffective. In the past decades, computational fluid dynamics (CFD) has become an alternative for aerodynamic performance analysis. The advent of CFD method enables batch optimization on a large scale so classic optimization algorithm like genetic algorithm (GA) can be used. However, optimization of airfoil is a multi-objective and therefore complicated problem. CFD usually requires huge computing

Y. Shuhuai (✉)
Hwa Chong Institution, Singapore, Singapore
e-mail: yushhci@gmail.com

H. Guo et al. (eds.), *IRC-SET 2021*, https://doi.org/10.1007/978-981-16-9869-9_32

power, hindering the efficiency of GA. As such, various works have been done to decrease the number of evaluations needed in GA.

Zhang and Wang [1] proposed an airfoil optimization method using the multi-island genetic algorithm based on nonlinear programming by quadratic Lagrangian optimization method. Their results showed that the optimization process was enhanced by the parameterization of airfoils, and airfoils with better performance were obtained. Jones [2] developed a parallel genetic algorithm methodology to address rotorcraft aerodynamics and aeroacoustics concerns. Obayashi [3] combined an inverse optimization method with a multi-objective genetic algorithm, and the method had been shown to produce superior aerodynamic performance. Liu [4] utilized the controlled elitist NSGA-2 coupling with CST to obtain airfoils with excellent aerodynamic performance at ultra-low-Reynolds number.

Various dimension reduction techniques are combined with GA to improve the efficiency of airfoil optimization. One way to perform dimension reduction is through shape parameterization. Masters et al. [5] compared seven different airfoil parameterization methods, including class function/shape transformation (CST) [6], B-spline, Hicks-Henne bump function [7], singular-value decomposition (SVD) [8], Bézier surfaces, and the PARSEC method. They concluded that the airfoil mode parameterization was the most efficient approach.

In this paper, a mode parameterization method using SVD is utilized with GA to generate airfoils with better performance. Compared with GA using one-point controlling method, GA with SVD generates better result and doubles the efficiency.

32.2 Airfoil Data Parameterization

The airfoil database used in this research is taken from the University of Illinois at Urbana-Champaign (UIUC) airfoil database with 1552 existing airfoil shapes. Each airfoil data contains several plane coordinates, and the shape of the airfoil can be obtained by connecting those coordinates. An open-source B-spline interpolation Python script by Chen [9] is used to interpolate the raw airfoil data, and smoothing is applied. After some selection, 1154 airfoils (the performance of the rest airfoils cannot be effectively evaluated) are selected and used in the following optimization process.

As proposed by Li [10], using mode shapes in a design space is better than using control points. Mode parameterization is able to significantly reduce the number of parameters while keep underlying airfoil features. It offers a way to exclude abnormal airfoil shapes, like shapes with uneven surfaces, to increase the efficiency of optimization. In most optimization problems, fewer parameters usually result in a decrease of the design space, lowering the chance of finding the global optimum [5]. Therefore, a suitable number of modes should be chosen to ensure both optimization efficiency and airfoil geometry accuracy.

The modes of the airfoils are derived through singular-value decomposition (SVD) [10]. SVD is a factorization process which generalizes the eigen decomposition of a square normal matrix to any matrix. It is assumed that the modes can be calculated by observing universal variation about a shape. Through the previous B-spline interpolation, each airfoil can be represented by 64 surface points. So, the y coordinates of the selected 1154 airfoils form a snapshot matrix:

$$A = \begin{bmatrix} y_{(1,1)} & y_{(2,1)} & \cdots & y_{(1154,1)} \\ y_{(1,2)} & y_{(2,2)} & \cdots & y_{(1154,2)} \\ \vdots & \vdots & \ddots & \vdots \\ y_{(1,64)} & y_{(2,64)} & \cdots & y_{(1154,64)} \end{bmatrix} \tag{32.1}$$

Using SVD,

$$A = U \sum V^T \tag{32.2}$$

where each column in U are airfoil modes, $\sum$ is a diagonal matrix, and V is a matrix whose columns are the right singular vectors. The first 20 modes obtained are displayed as shown in Fig. 32.1.

Twenty airfoil modes are ranked according to their explained variance ratio, which is the ratio between their respective eigenvalue and the sum of all eigenvalues. For example, the first airfoil mode contains 92.8% of the original airfoil information. As shown in Fig. 32.2, green lines represent the original airfoils, while red lines represent the airfoils re-plotted using SVD. It is difficult to differentiate between the re-plotted airfoil and the original airfoil when more than five modes are used. Therefore, it is sufficient to just use the first five modes in the following evaluation. Let $\boldsymbol{M}_5$ be the first five modes, and vector $a_n = (c_1, c_2, c_3, c_4, c_5), n = 1, 2, \ldots, 1154$ be the five coefficients for their respective mode for each airfoil, and the original y coordinates of airfoil can be calculated as follows:

$$Y_n = M_5 a_n^T, \quad n = 1, 2, \ldots, 1154 \tag{32.3}$$

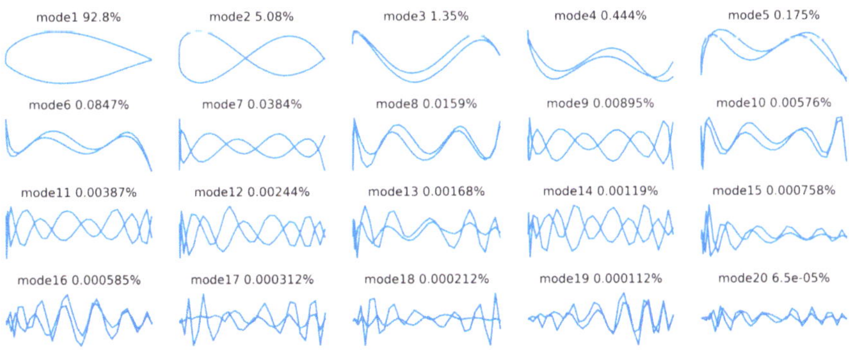

Fig. 32.1 Ranked airfoil modes derived from SVD

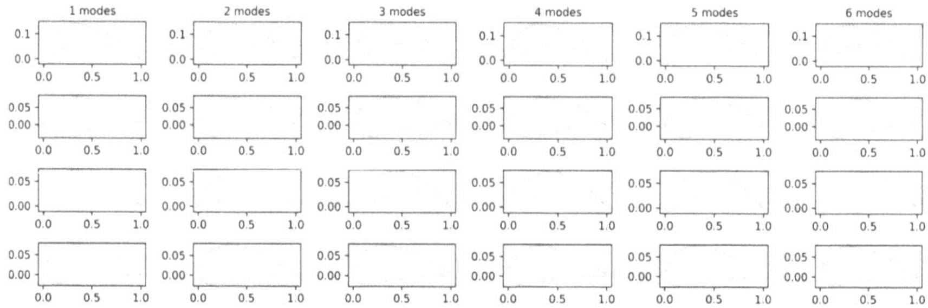

Fig. 32.2 SVD representation of airfoils using different number of modes (green line: original airfoil shape; red line: re-plotted airfoil shape)

32.3 Airfoil Evaluation

For the past decades, airfoils are increasingly evaluated using CFD method, an analysis of the fluid flow using numerical solution methods. Various models, like Reynolds-averaged Navier–Stokes equations and Favre-averaged Navier–Stokes equations, are used in CFD to perform aerodynamic analysis. To ensure accuracy of the calculation, CFD usually needs high density mesh. However, the computation time increases exponentially as the mesh density increases, and it may take several hours to perform one evaluation on a personal computer. An alternative for CFD method is to use XFOIL [11] software, which solves the boundary layer and transition equations simultaneously.

Morgado [12] showed that when the Reynolds number is low, and XFOIL gave the overall best prediction results compared to various CFD methods. Thus, XFOIL is used in this paper to perform airfoil performance evaluation because its ability to analyze low-Reynolds number airfoil flows rapidly and accurately.

The evaluation process is automated by using a Python script. Usually, airfoils are evaluated with different angles of attack at various Reynolds number and Mach number. Due to limited computing power, in this paper, airfoils are only evaluated when angle of attack α is 3°, Reynolds number Re at $5 \times 10_5$ and Mach number M at zero. The design angle of airfoil is usually within 3°–5°, so the evaluation at 3° can still reflect the general performance of airfoil.

32.4 Genetic Algorithm

Genetic algorithm (GA) has been widely used to find global optimum for complex multi-objective problems. Inspired by the natural evolutionary, GA is based on natural selection and genetics, including mutation, crossover, and reproduction. Unlike conventional algorithm which starts optimization from a single point, GA searches the design space through a population to find the global optimum with higher efficiency. In this paper, GA is applied to optimize the performance of airfoil. The fitness function F is defined as follows:

$$F = \frac{C_L}{1.5} + \frac{1}{100}\frac{C_L}{C_D} \tag{32.4}$$

The objectives are to maximize both the lift coefficient C_L and the lift drag ratio $C_L \backslash C_D$. One of the advantages of GA is that the fitness function can be defined freely to meet various designing goals. Therefore, the coefficients in (32.4) can be adjusted in order to meet different design goals. The tournament selection is implemented to ensure airfoil with the best performance is selected [12]. Then, all airfoils are given random perturb, so the algorithm is less likely to stuck in local minimum. For better convergence, adaptive mutation rate is applied such that the mutation rate is adjusted based on the rate of change of the fitness value. As proved by Ebrahimi [13], using adaptive mutation rate could enhance both the efficiency and outcomes of the optimization. A one-point crossover operator is used with 80 percent probability of combination, as Tes [14] showed GA performance is compromised when a smaller value is used. The reproduction is done by linearly combining y coordinates of two airfoils with randomly generated ratio. Newly generated airfoils then replace airfoils with low fitness value in the original population to maintain a stable population number.

32.4.1 SVDGA

The mechanism of SVDGA is similar to the one of traditional GA. But unlike traditional GA which directly manipulates the geometry of airfoil, SVDGA manipulates the SVD coefficient vector a_n. This ensures that most of the airfoils created by SVDGA are authentic by avoiding random oscillation on the surface of the airfoil (Fig. 32.5).

32.5 Results

The results obtained by GA and SVDGA are compared in Fig. 32.4. A direct observation from the graph is that SVDGA is better in searching the design space, while GA is more likely to converge in a local minimum. Fitness of the airfoils in GA converges quickly to a value around 1, while it undergoes huge fluctuates in SVDGA. The fitness of the best airfoil generated by SVDGA doubles compared to the original best airfoil, while it is around 20% increase in GA. The result confirms the hypothesis that SVDGA, with fewer dimension parameters, can increase the optimization efficiency and obtain better result.

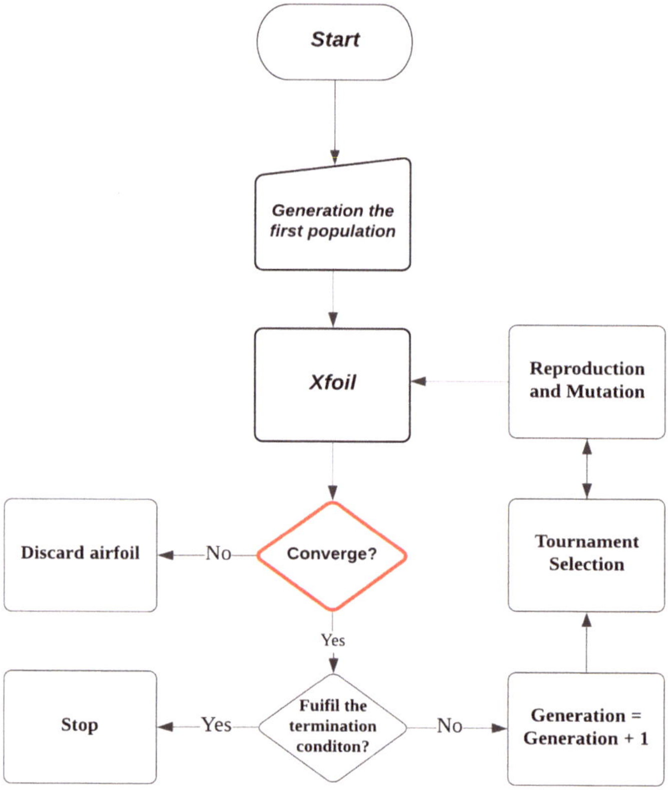

Fig. 32.3 Flowchart for GA optimization

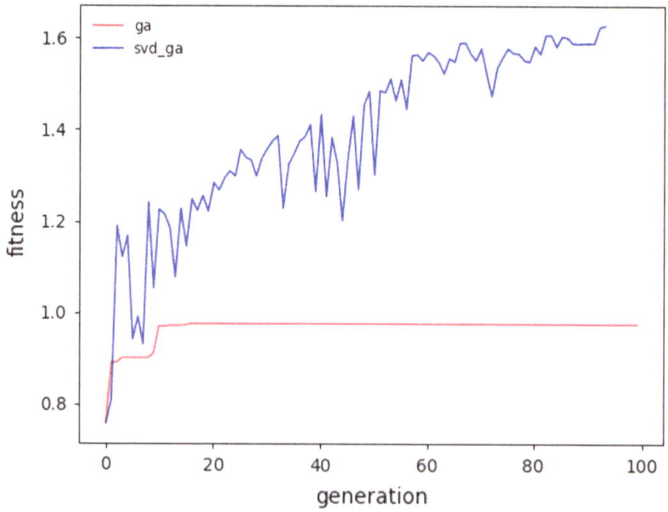

Fig. 32.4 Fitness score comparison

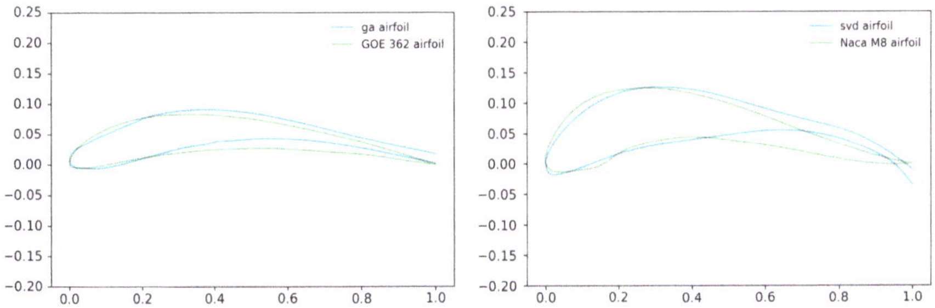

Fig. 32.5 Optimized airfoils obtained by GA and SVDGA

Table 32.1 Performance comparison

Airfoil name	Table column head		
	C_L	C_D	C_L/C_D
GOE 362 (baseline)	0.785(+0%)	0.0278(+0%)	28.2(+0%)
GA Foil	0.866(+12.9%)	0.0217(-21.9%)	39.9(+41.5%)
Naca M8 (baseline)	0.554(+0%)	0.0715(+0%)	7.76(+0%)
SVDGA foil	1.596(+188%)	0.0772(+8.3%)	20.7(+167%)

*Number in bracket denotes the percentage increase/decrease compared to the baseline airfoil

The final optimized airfoils are plotted in Fig. 32.5. In order to compare the performance of two algorithms objectively, two airfoils with similar shapes are found in the original airfoil database. Airfoil obtained by GA is compared with GOE 362 while airfoil obtained by SVDGA is compared with Naca M8 airfoil. It is observed that airfoil generated by SVDGA is more realistic and smoother because some discontinuity is observed on the surface of the airfoil generated by GA.

Next, performance of each airfoil is evaluated using XFOIL. C_L, $C_{D,}$ and C_L/C_D values for each airfoil are listed in Table 32.1, and percentage improvement is calculated between generated airfoils and comparison airfoils. The results demonstrate that SVDGA is far better than GA in terms of the performance of the final optimized airfoil. Except SVDGA foil, none of the other airfoils has a C_L larger than 1, while CL of SVDGA foil reaches nearly 1.6. SVDGA almost triples the C_L of the original airfoil, while the CD of SVDGA foil does not increase much. This leads to a 167% increase in the C_L, C_D ratio, but this increase is only around 40% in GA.

32.6 Conclusion

In this work, a mode parameterization method is introduced by using SVD. This parameterization method is then combined with traditional GA method in airfoil optimization. The whole airfoil optimization process is automated using Python and XFOIL, where airfoils with better aerodynamic performance are obtained. Results obtained by GA and SVDGA are further compared, demonstrating SVDGA's ability to generate more realistic and smoother airfoil. This work has also confirmed the importance of using parameterization in airfoil optimization and showcases that SVD, as one of the important machine learning techniques, can be successfully implemented in GA to obtain better results.

Acknowledgements The author gratefully acknowledges the supports by Hwa Chong Institution and the CenTaD research program. The author also thanks Mr. Ma Zhou for his continuous support and guidance.

References

1. JZhang, T.-T., Huang, W., Wang, Z.-G., & Yan, L. (2016). A study of airfoil parameterization, modeling, and optimization based on the computational fluid dynamics method. *Journal of Zhejiang University-SCIENCE A, 17*(8), 632–645. https://doi.org/10.1631/jzus.a1500308.
2. Jones, B. R., Crossley, W. A., & Lyrintzis, A. S. (2000). Aerodynamic and aeroacoustic optimization of rotorcraft airfoils via a parallel genetic algorithm. *Journal of Aircraft, 37*(6), 1088–1096. https://doi.org/10.2514/2.2717
3. Obayashi, S., Takahashi, S., & Fejtek, I. (1998). Transonic wing design by inverse optimization using MOGA. In *Sixth annual conference of the computational fluid dynamics society of Canada.*
4. Liu, Z., Dong, L., Moschetta, J.-M., Zhao, J., & Yan, G. (2014). Optimization of nano-rotor blade airfoil using controlled elitist NSGA-II. *International Journal of Micro Air Vehicles, 6*(1), 29–42. https://doi.org/10.1260/1756-8293.6.1.29
5. Masters, D. A., Taylor, N. J., Rendall, T., Allen, C. B., & Poole, D. J. (2016). A geometric comparison of aerofoil shape parameterisation methods. In *54th AIAA aerospace sciences meeting*. https://doi.org/10.2514/6.2016-0558.
6. Kulfan, B. M. (2008). Universal parametric geometry representation method. *Journal of Aircraft, 45*(1), 142–158. https://doi.org/10.2514/1.29958
7. Hicks, R. M., & Henne, P. A. (1978). Wing design by numerical optimization. *Journal of Aircraft, 15*(7), 407–412. https://doi.org/10.2514/3.58379
8. Toal, D. J. J., Bressloff, N. W., Keane, A. J., & Holden, C. M. E. (2010). Geometric filtration using proper orthogonal decomposition for aerodynamic design optimization. *AIAA Journal, 48*(5), 916–928. https://doi.org/10.2514/1.41420
9. Chen, W., Chiu, K., & Fuge, M. (2019). Aerodynamic design optimization and shape exploration using generative adversarial networks. In *AIAA SciTech Forum*, AIAA, San Diego, USA.
10. Li, J., Bouhlel, M. A., & Martins, J. R. R. A. (2019). Data-based approach for fast airfoil analysis and optimization. *AIAA Journal, 57*(2), 581–596. https://doi.org/10.2514/1.j057129
11. Drela, M. (1989). XFOIL: an analysis and design system for low reynolds number airfoils. In *Lecture notes in engineering low reynolds number aerodynamics* (pp. 1–12). https://doi.org/10.1007/978-3-642-84010-4_1.

12. Deb, K. (2011). Multi-objective optimisation using evolutionary algorithms: an introduction. In *Multi-objective evolutionary optimisation for product design and manufacturing* (pp. 3–34). https://doi.org/10.1007/978-0-85729-652-8_1.
13. Ebrahimi, M., & Jahangirian, A. R. (2017). Airfoil shape optimization with adaptive mutation genetic algorithm. *Journal of Aerospace Science and Technology*, pp. 47–59.
14. Tes, D., & Chan, Y., (2000). Multi-point design of airfoil by genetic algorithm. In *8th Annual Conference of the CFD Society of Canada*, Montreal, Canada.

Chapter 33
Personalized Guided Perceptual Learning: A Proof of Concept Study

Jin Xiaoxuan and Lee Wai Yee

Abstract Conventional training often uses fixed-intensity programs which can lead to plateaus in learning trajectories and sub-optimal outcomes. Hence, individualized programs are adopted to achieve a highly personalized outcome, which can be applicable to cognitive training, specifically perceptual training in our case. Artificial intelligence (AI), in this case, can identify single subject (N-of-1) learning profiles and enhance learning trajectories by designing personalized training regimens. The objective of this study is a proof of concept that aims to test whether personalized guided learning programs are warranted and to assess if CURATE.AI can be used in perceptual learning training. Participants were required to train under both adaptive training programs and CURATE.AI training programs. Their perceptual learning ability was assessed based on their performance in terms of accuracy and difficulty level across the sessions. Perceptual learning ability of each participant has improved in varying degrees and even worsened in some cases under the adaptive training. CURATE.AI was harnessed for personalized training. Results from traditional adaptive approach highlight high variance across participants and indicate that individualized approaches are warranted. In our proof of concept, we were able to generate a CURATE.AI personalized perceptual learning profile. Future work is warranted to test the efficacy of CURATE.AI-guided perceptual learning training compared to traditional adaptive learning and fixed-intensity training methods.

Keywords Artificial intelligence · CURATE.AI · Single subject profile · Personalized training · Perceptual learning

33.1 Introduction

Perceptual learning is defined as performance improvement after practice or experience with stimuli related to the tasks [1]. It is important to study as it is an inherent part of our perceptual systems and plays a crucial role in our ability to complete

J. Xiaoxuan · L. W. Yee (✉)
Hwa Chong Institution, Singapore, Singapore
e-mail: waiyeeks@gmail.com

H. Guo et al. (eds.), *IRC-SET 2021*, https://doi.org/10.1007/978-981-16-9869-9_33

daily tasks. An essential role of perceptual learning is its contribution to our development. As many aspects of our body change as we age, perceptual learning helps us adapt to these changes in order for our body to function normally [2]. Perceptual learning ability can also affect performance to a great extent, leading to an increase in accuracy from purely by chance to more than 90% correct in many two-choice tasks [3].

Different from higher-order forms of learning, perceptual learning includes improved sensitivity without the effects of cognitive, motor, or other non-perceptual factors. Thus, decreases in the strength, quality, or duration of a stimulus required to achieve a certain level of accuracy can be used to indicate perceptual learning [4].

Moreover, there are a wide range of perceptual learnings which occur under a variety of conditions that probably indicate a diverse set of changes in neurons [4]. Examples of perceptual learning include developing an ability to distinguish between different odors, musical pitches, or shades of colors. As such, perceptual learning in different sensory modalities has been studied, which include changes in respective sensory pathways. In our report, we will narrow the scope and focus on visual perceptual learning.

Perceptual learning occurs for numerous types of visual tasks that demand different levels of ability to assess the situation. Tasks including orientation [5], spatial frequency [6], phase [7], contrast [8], and acuity [9] have shown the improvements in the capability of detection or discrimination for single features. Tasks including compound stimuli [10], textures [11], and motion [12] help to enhance pattern discrimination. In our report, we will conduct the perceptual learning task in the area of orientation discrimination, which is the ability to tell the difference between two stimuli that are rotationally offset from each other.

Repetitive practice has been proven as effective in vernier acuity in adults with visual deficiencies [13]. Other studies have also used fixed intensity of stimuli to train visual perceptual learning [14]. Conventional interventions were often delivered at a fixed dose or intensity for all individuals. Such a strategy might result in sub-optimal outcomes due to plateaus in learning trajectories. To try and improve this, adaptive learning approaches have been adopted. Adaptive learning is usually technology-based which will vary its training programs based on previous interactions [15]. Thus, instead of the same intensity of training being provided every time, the intensity will change depending on factors like how well the participant is performing. However, the change in intensity still follows a fixed pattern. For instance, the 3up1down adaptive learning method follows a pattern where the intensity is increased by one level when participants get three consecutive rights and lower by one level when participants get one wrong.

It is unknown if further enhanced learning results can be achieved using personalizing training programs by adjusting intensity according to the subject's performance instead of fixed patterns. In this case, artificial intelligence (AI) might be a great solution for personalization. Using AI, it is plausible that we will be able to identify N-of-1 (single subject) training profiles, and produce individualized training through modifying training features such as duration and intensity. In light of this, we intend to use CURATE.AI [16, 17], an AI platform, to help identify the N-of-1 learning profiles dynamically, which will allow one's own profile to be subjected to change during the course of training. Hence, CURATE.AI has the advantage of truly personalizing a perceptual learning intervention in a continual way.

33.2 Research Objective

To our knowledge, there are no existing papers on the effects of AI -guided perceptual learning training. Thus, in this report, our objective is to examine if personalized learning can be achieved using CURATE.AI. Our research question(s) are: (1) Is personalized perceptual learning training warranted? and (2) Can CURATE.AI be used to guide personalized perceptual learning? If personalized perceptual learning training is warranted and CURATE.AI guided training can be done, we hope to discover a more efficient way of improving one's cognitive skills.

33.3 Hypothesis

We hypothesize that personalized perceptual learning is needed. In light of the success of CURATE.AI in single-drug optimization [18], combination treatment [19] and other clinical settings, we hypothesize that perceptual learning can be guided by CURATE.AI.

33.4 Methodology

This study is part of the larger study and received full ethical approval from the Institutional Review Board (refer to IRB submitted separately).

33.4.1 Perceptual Learning Task

For our experiment, we used Gabor patches to measure perceptual learning. Gabor patches are a sinusoid convolved with a Gaussian. During the task, the stimuli was a Gabor patch and each trial featured two Gabor patches with slight different orientations. The difference in rotation was controlled by training procedure. The general objective for the task was to train the ability of the participants to differentiate the direction of rotation between two patches with a minor rotational difference as efficiently as possible. The outcome of perceptual learning was measured by the change in performance of individuals, which is their accuracy of responding to the task, calculated by number of correct answers divided by the number of total answers.

33.4.2 Procedure

Data were remotely collected from volunteer participants. Participants performed the perceptual training using a bespoke program hosted by Amazon Mechanical Turk using their own computers or tablets. They participated in four testing sessions across a week at the same time each day and another four sessions under CURATE.AI approach.

33.4.3 Standard Learning Approach

The traditional standard learning approach is adaptive training. We used 3-down, 1-up (3D1U) staircase approach, which we increased the task difficulty level (reduce the difference between the stimuli) when the participant was able to get three correct answers in a row, while getting one wrong answer decreased the task difficulty level (increase the difference between the stimuli) [20].Change in difficulty level was recorded down as one reversal and each session had 10 reversals or end when participants achieved the highest level of difficulty with a 0.3° difference.

33.4.4 AI Approach

CURATE.AI [16, 17] is an AI-derived and mechanism-independent platform that plots the correlation between an intervention intensity (input) and a phenotypic result (output), according to a specific individual's data. It is developed based on a finding that a quadratic polynomial equation can be utilized to establish the relationship between a changing input and a measurable output. Hence, the data collected will be used to calibrate the platform and determine the person's personalized CURATE.AI profile. The profile will be used to recommend the following training intensity that AI deems to provide the best result. As the participant's performance changes overtime, so does the CURATE.AI profile. In other words, the profile is dynamically recalibrated along with the progress of the training. With the learning trajectory profiles for individuals identified, CURATE.AI seeks to find the optimal personalized training program. CURATE.AI profiles were developed in this study as follows: (1) Session 1 is rotation thresholding to get optimal rotation difficulty; (2) Session 2 is noise thresholding to get optimal rotation difficulty; (3) Session 3 is developing a 7×7 matrix from the two values obtained from Session 1 and Session 2 (minus 3 levels of difficulty). These combinations are randomly ordered and duplicated and tested; (4) Session 4 is same as session 3. CURATE.AI profiles are created.

33.4.5 Data Analysis

This is a proof of concept study, so analysis is exploratory. Data quality were assessed first. Individual and group performance across sessions and tasks are plotted to identify trends. Data were cleaned, and only data from participants with complete datasets were included in analysis. Accuracy and difficulty over time were explored. CURATE.AI profiles were created. This analysis will inform future larger-scale studies.

33.4.6 Statistical Analysis

As analysis is exploratory, only descriptive statistical analysis was performed.

33.5 Results

33.5.1 Data Quality

3U1D: 19 datasets were collected. 68.4% of datasets were full ($n = 13$). 31.6% were missing data and subsequently not used in analysis ($n = 6$). Four participants were missing data after session 1, one participant was missing data after session 2, and one participant was missing data after session 3.

CURATE.AI: Due to technical difficulties, only one complete dataset was collected.

33.5.2 3U1D

Individual and group mean accuracy across sessions are plotted in Fig. 33.1a, b, respectively. Accuracy across sessions is highly variable across participants. The overall group trend shows participants to slightly decrease in accuracy across sessions; however, accuracy still remains high (around 78%). Individual and group mean difficulty levels across sessions are plotted in Fig. 33.2a, b, respectively. Achieved difficulty level across sessions varied between participants.

33.5.3 CURATE.AI

Figure 33.3 depicts the generated personalized CURATE.AI profile for the collected

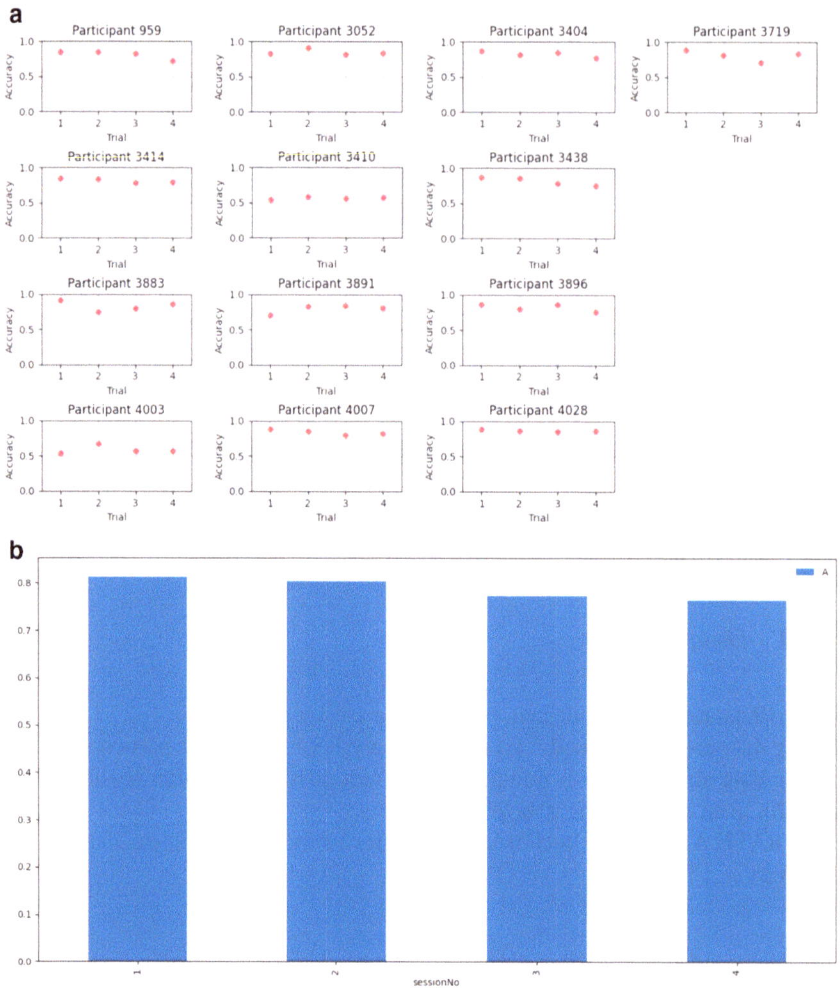

Fig. 33.1 **a**, **b** Individual and group accuracy across sessions

dataset. For this individual, optimal noise and rotation difficulty achieved about 75% accuracy (point 0). This surface predicts how this individual will perform 3 levels below and above their identified optimal threshold. For this individual, they will perform best at the lowest difficulty for noise and rotation and perform worst at the highest level of noise and difficulty. Difficulty level appears to influence accuracy more than noise for this individual.

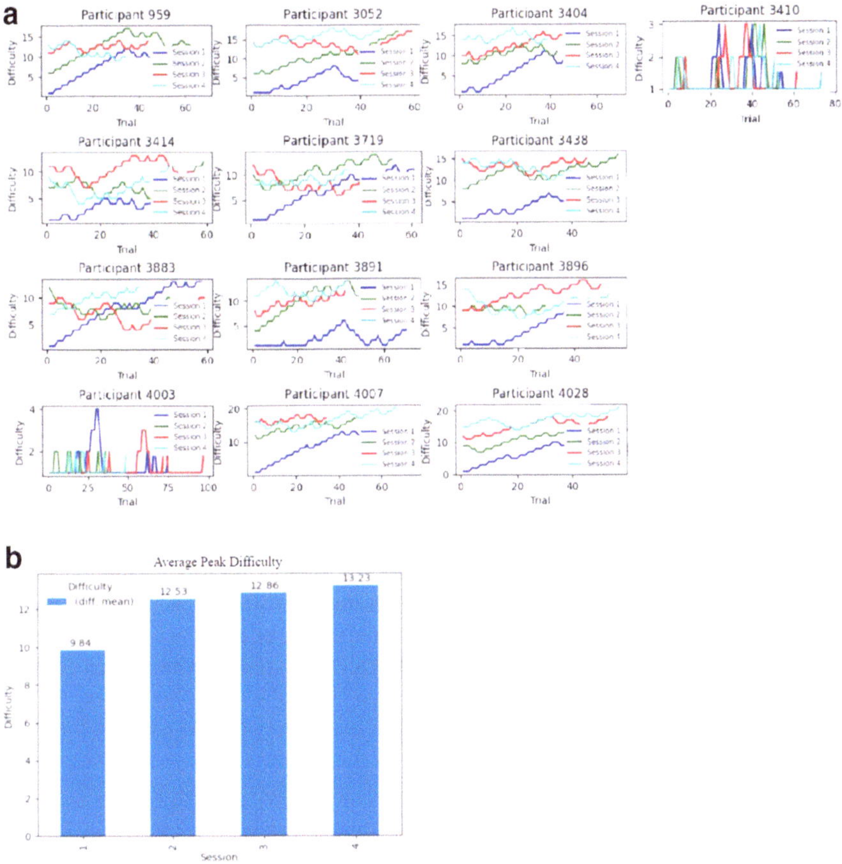

Fig. 33.2 **a**, **b** Individual and group difficulty level across sessions

33.6 Discussion

This is a proof of concept report which aimed to explore the potential of conducting personalized perceptual learning experiments online. We explored the feasibility of implementing two personalized perceptual learning approaches, traditional adaptive learning and CURATE.AI-guided learning. Proof of concept exploratory analysis indicates that perceptual learning is highly individualized and personalized learning approaches are appropriate perceptual learning optimization. CURATE.AI can be used to establish personalized perceptual learning trajectories.

According to Fig. 33.1, there was improvement in participants' perceptual learning ability, as group accuracy stayed nearly the same while the difficulty level/peak difficulty level actually rose. However, on an individual level, participants' performance varies after being through the training. For example, Participant 4028 has improved throughout the experiment, whereas Participant 3438 worsened (Fig. 33.1). In light

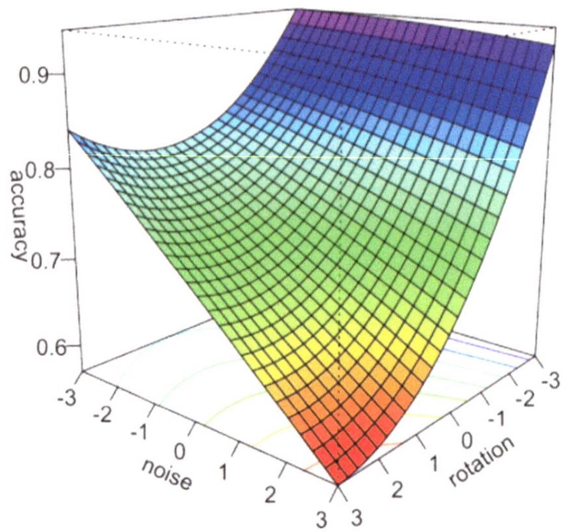

Fig. 33.3 CURATE.AI profile

of this individual variation, personalized programs powered by AI can be utilized. In this case, we managed to generate the personalized profile for perceptual learning for one participant, showing that indeed CURATE.AI has its potential to enhance one's learning trajectory according to one's own strengths and weaknesses.

33.6.1 Limitations

This study is not without limitations. This study was initially planned to occur in person in a controlled laboratory setting. However due to COVID-19, we were forced to move experimentation online. Development of the tool took longer than expected and with online development comes an increased need for user testing and pilot testing. As the online testing platform is still in developmental stages, datasets from CURATE.AI testing sessions were lost from numerous participants, and only one was saved. In spite of this, we were still able to demonstrate that CURATE.AI profiles can be created to be used to guide perceptual learning training. This also informs on technical aspects of online testing implementation.

33.6.2 Future Work

An experiment that includes more participants with diverse backgrounds (e.g., age, race groups) can be carried out to test its effectiveness in a more general demographic population. Additionally, a larger sample size will show differences in perceptual learning across individuals. Finally, now that it is established that CURATE.AI profiles can be generated from perceptual training, future work should test the efficacy of CURATE.AI-guided perceptual learning training compared to traditional adaptive learning and fixed-intensity training. Moreover, the visual perceptual learning ability is essential as it forms important foundations of complex cognitive processes and the ability for perceptual learning is retained throughout life [21]. Hence, if the CURATE.AI program is successfully developed, it will be much beneficial to youngsters whose brains are still in their developing stage.

33.7 Conclusion

Our analysis excluded participants who did not complete four sessions. 31.6% dropout rate also exemplifies the necessity of this study as a proof of concept before launching in a larger scale since it may face multiple obstacles with larger opportunity cost. However, results from the traditional adaptive approach highlight high variance across participants and indicate that individualized approaches are warranted. In our proof of concept, we were able to generate a CURATE.AI personalized perceptual learning profile. Future work is warranted to test the efficacy of CURATE.AI-guided perceptual learning training compared to traditional adaptive learning and fixed-intensity training methods.

Acknowledgements The authors gratefully acknowledge the support by Hwa Chong Institution, teacher mentor Mr. Kay Siang Low and the guidance from Dr. Alexandria Remus under the NUS N.1 Institute for health leading Project CURATE.

References

1. Deveau, J., & Seitz, A. (2014, September 25). *Applying perceptual learning to achieve practical changes in vision*. Retrieved August 12, 2020, from https://www.frontiersin.org/articles/https://doi.org/10.3389/fpsyg.2014.01166/full
2. Seitz, A. (2017, July 10). *Perceptual learning*. Retrieved August 12, 2020, from https://www.sciencedirect.com/science/article/pii/S096098221730619X
3. Ball, K., & Sekuler, R. (n.d.). *A specific and enduring improvement in visual motion discrimination*. Retrieved August 12, 2020, from https://pubmed.ncbi.nlm.nih.gov/7134968/
4. Gold, J., & Watanabe, T. (2010, January 26). Perceptual learning. Retrieved August 12, 2020, from https://www.ncbi.nlm.nih.gov/pmc/articles/PMC3821996/

5. Dosher, B., & Lu, Z. (1998, November 10). *Perceptual learning reflects external noise filtering and internal noise reduction through channel reweighting*. Retrieved August 12, 2020, from https://www.ncbi.nlm.nih.gov/pmc/articles/PMC25004/
6. Bennefl, R. G., & Westheimer, G. (n.d.). *The effect of training on visual alignment discrimination and grating resolution*. Retrieved August 12, 2020, from https://pubmed.ncbi.nlm.nih.gov/1857628/
7. Fiorentini, A., & Berardi, N. (n.d.). *Perceptual learning specific for orientation and spatial frequency*. Retrieved August 12, 2020, from https://pubmed.ncbi.nlm.nih.gov/7412873/
8. Adini, Y., Wilkonsky, A., Haspel, R., Tsodyks, M., Sagi, D. (n.d.). *Perceptual learning in contrast discrimination: The effect of contrast uncertainty*. Retrieved August 12, 2020, from https://pubmed.ncbi.nlm.nih.gov/15669907/
9. Westheimer, G. (n.d.). *Is peripheral visual acuity susceptible to perceptual learning in the adult?* Retrieved August 12, 2020, from https://pubmed.ncbi.nlm.nih.gov/11163615/
10. Fiorentini, A., & Berardi, N. (n.d.). Learning in grating waveform discrimination: Specificity for orientation and spatial frequency. Retrieved August 12, 2020, from https://pubmed.ncbi.nlm.nih.gov/7314493/
11. Ahissar, M., & Hochstein, S. (1993, June 15). *Attentional control of early perceptual learning*. Retrieved August 12, 2020, from https://www.ncbi.nlm.nih.gov/pmc/articles/PMC46793/
12. Lu, Z., Chu, W., Dosher, B., & Lee, S. (2005, April 12). *Independent perceptual learning in monocular and binocular motion systems*. Retrieved August 12, 2020, from https://www.ncbi.nlm.nih.gov/pmc/articles/PMC556263/
13. Polat, U. (2009, June 09). *Making perceptual learning practical to improve visual functions*. Retrieved August 12, 2020, from https://www.sciencedirect.com/science/article/pii/S004269890900282X
14. Shao-Chin Hung and Aaron R. Seitz (2014 June 18) Prolonged Training at Threshold Promotes Robust Retinotopic Specificity in Perceptual Learning. Retrieved from https://www.jneurosci.org/content/34/25/8423.full
15. Murray, M. C., & Pérez, J. (2015). Informing and performing: A study comparing adaptive learning to traditional learning. *Informing Science: The International Journal of an Emerging Transdiscipline, 18*, 111–125. Retrieved from http://www.inform.nu/Articles/Vol18/ISJv18p111-125Murray1572.pdf
16. Blasiak, A., Khong, J., & Kee, T. (2019). CURATE.AI: Optimizing Personalized Medicine with Artificial Intelligence. *SLAS TECHNOLOGY: Translating Life Sciences Innovation, 25*(2), 95–105. Retrieved May 10, 2020 from https://doi.org/10.1177/2472630319890316
17. Kee, T. (2019, September 1). *Harnessing CURATE.AI as a digital therapeutics platform by identifying N-of-1 learning trajectory profiles*. Wiley Online Library. Retrieved May 10, 2020 from https://onlinelibrary.wiley.com/doi/abs/https://doi.org/10.1002/adtp.201900023
18. Zarrinpar, A., Lee, D.K., Silva, A., Datta, N., Kee, T., Eriksen, C., Weigle, K., Agopian, V., Kaldas, F., Farmer, D., Wang, S. E., Busuttil, R., Ho, C. M., & Ho, D. (n.d.). *Individualizing liver transplant immunosuppression using a phenotypic personalized medicine platform*. Retrieved August 12, 2020, from https://pubmed.ncbi.nlm.nih.gov/27053773/
19. Lee, D. K., Chang, V. Y., Kee, T., Ho, C. M., Ho, D. (n.d.). *Optimizing combination therapy for acute lymphoblastic leukemia using a phenotypic personalized medicine digital health platform: Retrospective optimization individualizes patient regimens to maximize efficacy and safety*. Retrieved August 12, 2020, from https://pubmed.ncbi.nlm.nih.gov/27920397/
20. Leek, M. R. (2001, November 1). Adaptive procedures in psychophysical research. *Attention, Perception, & Psychophysics*. Retrieved November 11, 2020 from https://link.springer.com/article/https://doi.org/10.3758/BF03194543?error=cookies_not_supported&code=34a4fece-e5d3-45eb-bd2b-41a099185ff0
21. Kellman, P. J., & Garrigan, P. (2009, June). *Perceptual learning and human expertise*. Retrieved from https://www.ncbi.nlm.nih.gov/pmc/articles/PMC6198797/

Chapter 34
Investigating the Public's Experiences and Knowledge of Communicating with People with Dementia

Chen Mo, Clarissa Ng Ming Yi, and Yu Jing Rong

Abstract In a rapidly ageing society, a significant number of the elderly are susceptible to dementia, a condition that results in varying degrees of communication impairments. Yet, current research shows that many Singaporeans lack effective communication skills with patients with dementia, contributing to the increased loneliness felt by patients with dementia. In order to understand the severity of the issue mentioned, this project designed an online survey via Qualtrics which was distributed through snowball sampling to potential parties on various platforms such as social media and emails. With a total of 259 participants, our results show that Singaporeans generally lack technical communication skills when conversing with patients with dementia. Yet, many have had experiences communicating with patients with dementia. Results of this study add to previous research done on society's treatment towards patients with dementia and also offer possible ways Singapore can be more inclusive towards the dementia community.

Keywords Singapore · Dementia · Communication · Knowledge · Experiences

34.1 Introduction

Dementia is an overall term used to describe symptoms that impact memory, performance of daily activities and communication abilities [1]. Alzheimer's disease is the most common type of dementia, and dementia gets worse with time [1]. Common symptoms of dementia include a deterioration in memory, thinking, behaviour and the ability to perform everyday activities [2]. Worldwide, around 50 million have dementia with nearly 10 million new cases every year [2]. This trend is also seen in

C. Mo (✉)
National Institute of Education, Nanyang Technological University, Singapore, Singapore
e-mail: mo.chen@nie.edu.sg

C. N. M. Yi
Hwa Chong Institution, Singapore, Singapore

Y. J. Rong
River Valley High School, Singapore, Singapore

H. Guo et al. (eds.), *IRC-SET 2021*, https://doi.org/10.1007/978-981-16-9869-9_34

Singapore where number of people with dementia is increasing annually, and one in 10 people aged 60 and above in Singapore may have dementia [3]. The number of patients is expected to exceed 100,000 in a few years' time [4], and Singapore is projected to be the fifth oldest worldwide by 2050 [5], thus a significant proportion of the population in Singapore faces the threat of getting dementia.

Despite the prevalence of dementia, many Singaporeans lack experiences interacting with people with dementia and lack knowledge on how to communicate and engage those with this illness. For example, 56% of Singaporeans surveyed rated themselves low in dementia knowledge while 75% of patients with dementia feeling rejected and lonely [6]. There is a lack of evidence on Singaporeans' knowledge of communication strategies with people with dementia, further highlighted under 'Literature Review'. Thus, there is an urgent and pertinent need to investigate the degree to which Singaporeans today know about this condition as well as how accurate their knowledge is. Specific research questions include:

(a) What is the current status of Singaporeans' knowledge of dementia and experiences with people with dementia? (b) How much experience do Singaporeans have when it comes to communicating with patients with dementia? (c) To what degree do Singaporeans know about communication techniques in interacting with people with dementia?

34.2 Literature Review

Currently, there is limited research that examines how to effectively communicate with patients with dementia. Ripich et al. (1995) developed the FOCUSED program, a seven-step program used to enhance communication between patients with dementia and people. FOCUSED represents functional and face to face, orientation to topic, continuity of topic-concrete topics, unstick and communication blocks, structure with yes/no questions, exchange conversation and encouraging interaction, and short, simple sentences [7].

In Singapore, there are a few studies that examined individuals' perceptions of people with dementia. Chin et al. (2019) [8] conducted an ethnography study to explore the experiences of nursing staff communicating with Singaporean Chinese with dementia in Singapore. Their findings highlighted the need for nurses to understand cultural characteristics among Singaporean Chinese with dementia on emotional expressions. In addition, Tan et al. [9] investigated the factor structure of the Dementia Knowledge Assessment Scale among a sample of 282 informal caregivers for people with dementia in Singapore, as well as how their knowledge correlated with their sociodemographic variables. The results showed that there is a need for more dementia literacy intervention for caregivers of patients.

Lastly, a national survey was conducted by Singapore Management University and Singapore Alzheimer's Disease Association in 2019 [10] to understand Singaporeans' attitudes and awareness towards Alzheimer's disease and dementia. The survey was given to more than 5600 people across Singapore, and the results showed that more needed to be done to raise dementia awareness.

Despite the informative findings from the aforementioned studies, to our best knowledge, there is yet no research study that has specifically looked into Singaporeans' experiences and knowledge in terms of communicating with patients with dementia. The project adds to the current research literature by presenting an analysis of the interactions Singaporeans have with these patients. Through understanding the current landscape of Singaporeans' interactions with patients, we can identify obstacles that hinder interaction and develop suitable solutions to optimise interactions.

34.3 Methodology

Ethical approval was obtained from National Technological University's Institutional Review Board prior to the commencement of the study. Participants qualified for this study had to be (a) at least 21 years of age, (b) have lived in Singapore for more than 5 years and (c) currently reside in Singapore. The invitation letter was then sent to eligible participants through WhatsApp, Facebook and Telegram. We also reached out to various social enterprises Jamiyah Nursing home, Dementia Friendly Singapore and National Library Singapore to invite participants. Additionally, the invitation letters were also pasted around NTU's campus. Out of 259 participants, 223 participants completed all items on the survey that was adapted from researchers at the University of Minnesota Duluth [11, 12]. Demographics of the participants are shown in Table 34.1.

Table 34.1 Demographic of respondents

Question no.	Demographic		
	Question (n = total respondents)	*n*	%
1	*Age group (n = 223)*		
	21–25	19	8.5
	26–35	23	10.3
	36–45	40	17.9
	46–55	101	45.3
	56 and above	40	17.9
2	*Gender (n = 223)*		
	Male	70	31.4
	Female	153	68.6
3	*Occupation (n = 223)*		
	Student	16	7.2
	Working adult	182	81.6
	Retired adult/pioneer	25	11.2

34.4 Results

34.4.1 *Singaporeans'* Knowledge *of* Dementia *and* Experiences *with* People *with* Dementia

Table 34.2 shows the results regarding Singaporean's knowledge about dementia. 65% of the sample (155/239) rated themselves unconfident in differentiating dementia from other related conditions, and only 57% (136/239) managed to identify

Table 34.2 Singaporeans' knowledge of dementia and experiences with people with dementia

Question no.	Knowledge of dementia and experiences with people with dementia		
	Question (n = total respondents)	*n*	%
4	*Confident in describing the difference between dementia and related conditions such as Alzheimer's disease? (n = 239)*		
	Yes	83	34.7
	No	156	65.3
5	*What is dementia? (n = 239)*		
	Significant loss of intellectual functioning and adaptive behaviour associated with advancing age	89	37.2
	Group of symptoms including memory loss and cognitive impairment	137	57.3
	A natural part of ageing	5	2.1
	Disease associated with senility and a rapid loss of communication	3	1.3
	Not sure what it is	5	2.1
6	*Know anybody personally who has dementia? (n = 236)*		
	Yes	115	48.7
	No	121	51.3
7	*If yes, how was this person related to you? Select all that apply (n = 123)*		
	Family member	66	52.9
	Relative	32	26
	Neighbour	6	4.9
	Colleague at workplace	3	2.4
	Friend	11	8.9
	Others (please specify):	6	4.9
	Friend's parent	3	2.4
	Neighbour's mother	1	< 1
	Client	1	< 1

the correct definition of dementia. Therefore, most Singaporeans lack basic information about this condition despite its prevalence in society. On the aspect of personal experiences, Table 34.2 also shows that 49% (115/236) know someone with dementia, and most of these interactions came from familial ties.

34.4.2 Singaporeans' Knowledge of Communication Techniques for People with Dementia

Table 34.3 shows that only a small minority (15%, 35/235) rated themselves very comfortable when interacting with patients with dementia while most chose somewhat uncomfortable (38%, 88/235) or somewhat comfortable (45%, 105/235). The lack of knowledge further applied to the awareness of communication methods, where an overwhelming majority (79%, 183/233) rated themselves as being unaware of techniques to communicate with someone with dementia.

Despite the majority lacking concrete knowledge on various aspects of dementia, respondents reached a general consensus that eye contact should be maintained (168/233), and one should present one idea at a time whenever communicating with persons with dementia (163/233). The least popular options chosen were—asking about activities done earlier that day (71/233), asking open-ended questions (20/233) and avoiding pronouns (18/233). Table 34.3 also shows that relating to the patient was the most popular response (106/233) when someone heard a patient mention something incorrect, while kindly correcting them emerged as the second most popular response (58/233).

34.4.3 What Needs to be Done?

Table 34.4 indicates that when asked if enough has been done to let Singaporeans learn more about the dementia community, an overwhelming 93% (212/227) chose 'No', and 95% (214/225) felt that they would benefit from some training on how to communicate with individuals with dementia. The most common responses when respondents were asked to give suggestions on future work regarding dementia awareness included presenting accurate media coverage on dementia beyond stereotypes in shows, more seminars in the community and integrating dementia awareness programmes to schools as part of the school syllabus.

Table 34.3 Results showing Singaporeans knowledge on communication technique

Question no.	Knowledge on communication techniques		
	Question (n = total respondents)	*n*	%
8	*How comfortable do you feel interacting with someone who has dementia? (n = 235)*		
	Very uncomfortable	7	3.0
	Somewhat uncomfortable	88	37.5
	Somewhat comfortable	105	44.7
	Very comfortable	35	14.9
9	*Do you feel that you have the proper skills to communicate effectively with patients with dementia? (n = 233)*		
	Yes	49	21.0
	No	184	79.0
10	*Do you know any techniques to communicate with patients with dementia? (n = 233)*		
	Yes	50	21.5
	No	183	78.5
11	*If yes, where did you learn those technique(s) to communicate with patients with dementia? Select all that apply. (n = 88)*		
	Reading books/magazines etc.	15	17.1
	Attending talk sessions or events	15	17.1
	Reading resources online.	30	34.1
	Getting advice from specialists or doctors.	18	20.5
	Others (please specify): Personal experience/ From family members in medical profession/ Personal logic	9	10.2
12	*Which techniques do you think are appropriate in communicating with patients with dementia? Select all that apply (n = 1275)*		
	Asking closed-ended questions	159	12.5
	Presenting most important information first and again at the end	124	9.7
	Provide written information during the discussion	69	5.4
	Asking open-ended questions	20	1.6
	Maintaining eye contact	168	13.2
	Facing the individual	163	13.2
	Presenting one idea at a time	174	13.4
	Asking about activities done earlier that day	71	5.6
	Avoiding pronouns	18	1.4

(continued)

Table 34.3 (continued)

Question no.	Knowledge on communication techniques		
	Question (n = total respondents)	*n*	%
	Validating the individual's responses	132	10.4
	Offer simple choices	167	13.1
	Others (please specify): Write notes/provide helping words when patient is stuck/make links to long-term memory/show concern, give attention/offer only 1 option/use pictures and gestures/not sure	8	0.63
13	*Likely response when patient says something incorrect? (n = 227)*		
	Kindly correct them	58	25.6
	Think about their underlying intention and address that intention	106	46.7
	Ignore the comment	19	8.4
	Change the topic	17	7.5
	Not sure how I would respond	27	11.9
14	*Enough done to let Singaporeans learn more about the dementia community? (n = 227)*		
	Yes	15	6.6
	No	212	93.4
15	*Do you think you would benefit from training on how to communicate with patients with dementia? (n = 225)*		
	Yes	214	95.1
	No	11	4.9

34.5 Discussion

34.5.1 What is the Current Status of Singaporeans' Knowledge of Dementia and Experiences with People with Dementia?

Many Singaporeans (48.7%, 115/236) have experience with patients, and such a result was expected given the increasing number of patients with dementia in Singapore. Yet a significant number of Singaporeans (65%, 156/239) reported lacking knowledge of what dementia is.

Table 34.4. Results showing future work to be done

Question no.	Future work		
	Question	*n*	%
1	*What do you think can be done? (n = 195)*		
	Education in terms of talks/workshops/campaigns to raise awareness – integrating lesson in school curriculum – educating relatives of patients with dementia	118	60.5
	More training and support for caregivers	16	8.20
	Utilising the media – TV shows/short film ads – Accurate media coverage on dementia	19	9.74
	Not sure	14	7.18
	Others – Community sharing – Publish dementia-related guidebooks – Government intervention in removing taboo around topic – Roadshows – More newspaper articles	28	14.4

34.5.2 How much Experience do Singaporeans have when it comes to Communicating with Patients with Dementia?

Singaporeans are generally low in experience when it comes to communicating with patients with dementia. From the results, only 25% (66/236) of participants have family members who have dementia. This would mean that they are caregivers (formal/informal) of people with dementia and may have high levels of experience. However, rest of the participants (75%, 170/236) either have varying degrees of experience communicating with people with dementia or none at all since they may not be direct caregivers.

Despite having low levels of experience in communicating with people with dementia, majority (105/235) feel somewhat comfortable communicating with patients with dementia. This suggests that Singaporeans are open to the idea of communicating with people with dementia, potentially allowing for higher levels of communication between Singaporeans and people with dementia.

However, higher levels of communication between Singaporeans and patients with dementia may not necessarily be good if most Singaporeans lack communication experience and skills. Such worry is evident as 79% (184/233) feel that they are lacking in communication skills. This may result in negative communication experiences with patients due to unsuccessful communication attempts between both parties.

34.5.3 To What Degree do Singaporeans Know About Communication Techniques in Interacting with Patients with Dementia?

From the results, it can be inferred that the majority of Singaporeans today are only subconsciously aware of proper techniques to interact with patients with dementia. Despite 79% indicating that they are unaware of specific communication techniques, the top three communication methods chosen by respondents were the recommended ones. This suggests that many are unknowingly aware of some communication techniques, despite not having learnt them formally. However, the option that received the least votes (avoiding pronouns) is actually a strategy advocated by professionals, as they encourage people to avoid pronouns and identify individuals and things by name when communicating with patients with dementia [13]. Thus, while Singaporeans know the most commonly encouraged and discouraged techniques, more could be done to increase awareness for other techniques.

Though a significant number of participants may be more concerned with dementia-related issues (than the general population) since the project uses snowball sampling by forwarding research invitations to organisations that have closer ties to the elderly, such a method is well reasoned. These participants could have greater contact with patients with dementia (caregiving roles, relatives etc.), hence a more urgent need to investigate their level of knowledge and experience with communication with patients with dementia.

34.5.4 Limitations

When asked to rate their level of comfort when interacting with people with dementia, 88 respondents indicated they were 'Somewhat Uncomfortable' while another seven indicated that they were 'Very Uncomfortable'. However, reasons to identify the root cause for this discomfort were not identified. Identifying them could allow future solutions rolled out to target these root causes to fundamentally change the mindset of

Singaporeans and dispel possible stereotypes that led to the aforementioned discomfort. Furthermore, our study's target group was the general population, but Singaporeans under the age of 21 were not included. This may restrict our results to be representative of the adult population rather than general public. Further analysis can also be narrowed down to various demographics (race, age, gender, caregivers and non-caregivers) to further understand the knowledge and communication gaps of specific groups.

34.6 Conclusion

In conclusion, the project found out that the Singapore public generally lacks meaningful communication experiences with the dementia community and concrete knowledge on suitable communication strategies. However, the survey also revealed that the public hoped that with the cooperation of various stakeholders. More teaching on how to communicate with patients with dementia can and should be provided for the general public, so as to make us a more dementia-friendly nation.

Acknowledgements We sincerely thank our supervisor Dr. Chen for her invaluable guidance and mentoring throughout the entire research process. We would also like to thank all the organisations who accepted our request and facilitated the dissemination of our survey, namely the National Library Singapore, Jamiyah nursing home and Dementia Friendly Singapore.

References

1. Healthline. (2020). *Dementia and Alzheimer's: What are the differences?* https://www.healthline.com/health/alzheimers-disease/difference-dementia-alzheimes
2. World Health Organisation. (2019). *Dementia.* Retrieved from https://www.who.int/news-room/fact-sheets/detail/dementia
3. Alzheimer's Disease Association, Singapore. (2020). *Dementia in Singapore.* Retrieved from https://alz.org.sg/dementia/singapore/#:~:text=Locally%2C%20according%20to%20the%20Well,in%20a%20few%20years'%20time
4. Health Xchange. *Dementia in Singapore.* Retrieved from https://www.healthxchange.sg/seniors/ageing-concerns/dementia-singapore. Last retrieved July 1,. 2020.
5. United Nations. (2015). *World population ageing.* Retrieved from https://www.un.org/en/development/desa/population/publications/pdf/ageing/WorldPopulationAgeing2019-Report.pdf
6. Straits Times. 30 April 2019. *3 in 4 with dementia feel lonely, rejected: Survey.* Retrieved from https://www.straitstimes.com/singapore/3-in-4-with-dementia-feel-lonely-rejected-survey
7. Weirather, R. R. (2010). *Communication strategies to assist comprehension in dementia.* Retrieved from https://www.ncbi.nlm.nih.gov/pmc/articles/PMC3104619/
8. Chin, S. Y., Lopez, V., Tan, M. L., & Goh, Y. S. (2019). *'I Would Like to Be Heard.' Communicating with Singaporean Chinese patients with dementia: A focused ethnography study.* Retrieved from https://pubmed.ncbi.nlm.nih.gov/30227768/

9. Tan, G. T. H., Yuan, Q., Devi, F., Wang, P., Ng, L. L., Goveas, R., Chong, S. A., & Subramaniam, M. (2020). *Dementia knowledge and its demographic correlates amongst informal dementia caregivers in Singapore, Aging & Mental Health.* Retrieved from https://doi.org/10.1080/13607863.2020.1740914
10. Singapore Management University. (2019). *Remember. For. Me.* Retrieved from http://www.screeningstatistics.com/alzheimers/
11. Habben, R., Cyr, R., Hyppa-Martin, J., Mizuko, M., & Collins, D. (2014). *SLPs & other professionals who work with individuals with dementia: a need to Collaborate?* Presentation at American-Speech-Language-Hearing Association Convention. Orlando, FL, USA
12. Olson, C., Leukuma, B., & Hyppa-Martin, J. (2018). *Supporting inclusion for individuals with dementia: SLP & financial professionals' evolving opportunities for interprofessional collaboration.* Presentation at American-Speech-Language-Hearing Association Convention; Boston, MA, USA
13. Hyppa-Martin, J., & Hofmann, R. (2018). *Sensitive stewardship and donors with dementia.* The NonProfit Times. Retrieved from https://www.thenonprofittimes.com/npt_articles/sensitivestewardship-donors-dementia/

Chapter 35
The Study of Epidemic Model in the Context of Singapore: Modelling the Transmission and Vaccination Mechanisms of COVID-19 in Singapore

Lukang Guo, Jiquan Gao, and Linxuan Feng

Abstract As countries throughout the globe are battling against the novel coronavirus disease 2019 (COVID-19), the modelling of infectious diseases—to study the mechanisms by which diseases spread, to predict the future course of an outbreak, and to evaluate strategies to control an epidemic—has become increasingly important. In our research, we investigated the SIR epidemic model and applied it to model the event of COVID-19 within the context of Singapore, using basic assumptions, mathematical manipulation, and programming. In the course of our model construction, we divided our project into three stages. Firstly, we collected and processed publicly available data in Singapore from *Worldometer* and *Statista*. In our second stage, we used Python codes to implement the SIR model, and adopted particle swarm optimization (PSO) algorithm to determine the infection rates β and recovery rates γ of COVID-19 over time. We found that the initial basic reproductive numbers R_0 was 4.137212757 in Singapore. Subsequently, we analysed the effectiveness of the preventive measures adopted by Singapore during different phases, and concluded that they have effectively reduced active cases by 99.11%. In our final stage, we adopted the SIR-V model to investigate the vaccination rate required to keep the epidemic under control, which we evaluated to be 0.5533%.

Keywords COVID-19 · The SIR model · Particle Swarm Optimization (PSO) · SIR model with vaccination

L. Guo (✉) · J. Gao · L. Feng
Hwa Chong Institution, Singapore, Singapore
e-mail: 162307b@student.hci.edu.sg

J. Gao
e-mail: 161909u@student.hci.edu.sg

L. Feng
e-mail: 162418h@student.hci.edu.sg

H. Guo et al. (eds.), *IRC-SET 2021*, https://doi.org/10.1007/978-981-16-9869-9_35

35.1 Introduction

35.1.1 Background

In the field of applied mathematics, mathematical modelling of infectious diseases is a useful tool to study the mechanisms by which diseases spread, to predict the future course of an outbreak, and to evaluate the effectiveness of different preventive strategies to control an epidemic. The study of epidemiology has become significantly pivotal, especially in light of the recent development of novel coronavirus disease 2019 (COVID-19) that has plagued numerous countries around the world since the end of 2019 and the start of 2020.

35.1.2 Objectives

Our modelling process incorporated multiple goals to thoroughly study the epidemic diseases:

- Investigate the rationales of the SIR model in disease simulation and apply the model to determine the various parameters (infection rate β, recovery rate γ, and reproduction number R_0) that represent the nature of COVID-19 in Singapore's context.
- Use our results to evaluate the effectiveness of various preventive measures during the different stages of Singapore.
- Investigate the minimum vaccination rate α required to keep an epidemic under control in Singapore, and the relationship between the vaccination rate and the level of infectiousness (R_0) of a virus.

35.2 Assumptions and Justification

Assumption 1: We assumed the total population in Singapore, N, was constant at 5.6 million throughout the period of model simulation.

Justification: The active infected cases, I, of COVID-19 was only 0.37% relative to the total population, and the yearly population growth rate in Singapore is merely 0.79% in 2020 [1]. Hence, the effect of change in total population is negligible to our model.

Assumption 2: There is a homogeneous mixing of the infected I and the susceptible S populations.

Justification: The social network in Singapore is dynamic. Considering that COVID-19 has a certain latency period [2], people who are infected but have yet to show symptoms may still be mixing around with others.

Assumption 3: We assumed all the infection and recovery cases reported have a delay of 6–7 days from their onset.

Justification: According to a study led by *Abbott S, Hellewell J, Thompson RN* et al., the distribution of the delay in reporting infectious cases has an average 6.5 days [3]. We modified the domain of our simulation for each phase of Singapore based on this delay.

Assumption 4: We defined the infected population I as people who are capable of spreading the disease to the susceptible people, even if they are asymptomatic or quarantined. We defined the removed population R as those who no longer possess the ability to spread the disease, either because they are immune or dead.

Justification: We defined I as such because it has the same meaning as the reported active cases. We did not distinguish between the diseased and the recovered for population R because both are not affected by the epidemic anymore.

35.3 Model Development

35.3.1 The SIR Model

In the SIR epidemic model, the total population N is divided into three groups, namely the susceptible S, the infected I, and the recovered R. With the assumption that total population is constant, we have:

$$S + I + R = N \tag{35.1}$$

The model comprises a system of differential equations. The rate of change of S is directly proportional to the susceptible population, the population already infected, and the amount of contact between these two populations. Hence, it is given by:

$$\frac{\mathrm{d}S}{\mathrm{d}t} = \frac{\beta SI}{N} \tag{35.2}$$

where β, the infection rate, is the average number of contacts per person per unit time.

The rate of change of R is directly proportional to the infected population and is given by:

$$\frac{\mathrm{d}R}{\mathrm{d}t} = \gamma I \tag{35.3}$$

where γ, the recovery rate, is the fraction of the population group that will recover in any day.

From Eq. (35.1), we can deduce that $\frac{dS}{dt} + \frac{dI}{dt} + \frac{dR}{dt} = 0$. Thus, the rate of change of I is given by:

$$\frac{dI}{dt} = \frac{\beta SI}{N} - \gamma I \tag{35.4}$$

There is no explicit analytical solution to this system of differential equations, so we employed the *odeint* ordinary differential equation solver from the *scipy* module, which helps us derive numerical solutions using the Runge–Kutta methods.

Here, it is necessary to introduce the basic reproduction number, R_0, which is calculated as $R_0 = \frac{\beta}{\gamma}$. We used R_0 to indicate how contagious an infectious disease is, as it denotes the expected number of people an infected person can successfully spread the disease to [4]. $R_0 = 1$ is the threshold to trigger an epidemic outbreak:

- If $R_0 < 1 \Leftrightarrow \beta < \gamma$, people recover faster than getting infected, and the disease will die out.
- If $R_0 > 1 \Leftrightarrow \beta > \gamma$, people are infected at a higher rate, and there may be an outbreak.

35.3.2 Model Fitting in Singapore's Context

Having expounded the rationales of the SIR model, we then applied this theoretical framework based on practical COVID-19 data in Singapore. Our aim in this section is to use the particle swarm optimization (PSO) algorithm to accurately determine the initial parameters β and γ of the COVID-19 pandemic in Singapore.

35.3.2.1 Data Collection

To thoroughly investigate the trend of COVID-19 development, we collected publicly available data across a 341-day duration (starting from 23 January 2020, when the first case was reported, to 28 December 2020), including:

- Daily active cases, $I_{\text{real}}(t)$ (total number of people currently infected on day t), from *Worldometer* [5].
- Total number of infected cases in Singapore, from *Statista* [6] and *Worldometer*.
- Total recovered cases, $R_{\text{real}}(t)$, calculated by subtracting daily active cases from the corresponding total number of infected cases.

The trends of collected data I_{real} and $R_{\text{real}}(t)$ are displayed in Fig. 35.1.

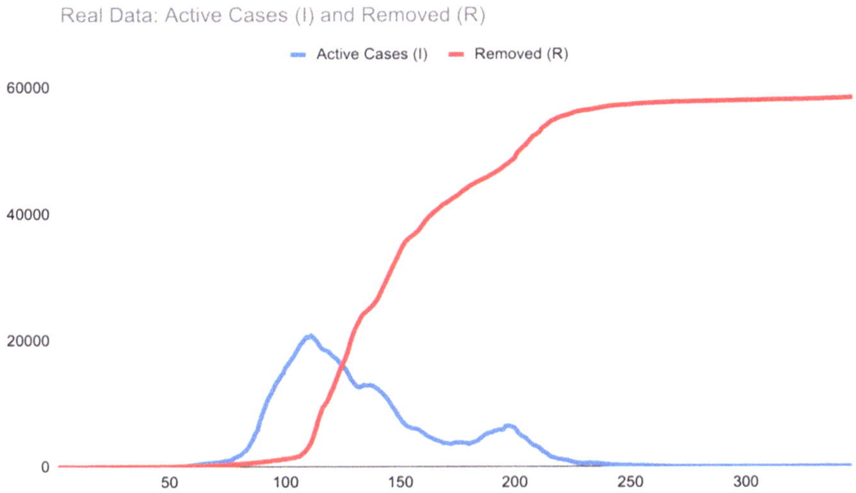

Fig. 35.1 Actual data: active infection cases (blue) and removed cases (red)

35.3.2.2 Particle Swarm Optimization (PSO) Algorithms

Theoretically, each pair of vectorized parameters $\theta = (\beta, \gamma)$ in the SIR model will produce a distinct set of solution curves for $S(\theta, t)$, $I(\theta, t)$, and $R(\theta, t)$. The parameter vector θ with the best fit to the real data $S_{\text{real}}(t)$, $I_{\text{real}}(t)$, and $R_{\text{real}}(t)$ is the one that minimizes the ordinary least squares between the simulated data and the real data. Since we assumed the total population N is constant, and $S = N - I - R$, we could ignore the susceptible population S when calculating the ordinary least squares. Here, we defined the ordinary least squares residual loss function:

$$E(\theta) = \frac{1}{\max(I_{\text{real}})} \sum_{t=1}^{n} [I(\theta,\ t) - I_{\text{real}}(t)]^2 + \frac{1}{\max(R_{\text{real}})} \sum_{t=1}^{n} [R(\theta,\ t) - R_{\text{real}}(t)]^2 \tag{35.5}$$

where we equalized the weight of I and R through dividing their least squares sum by their maximum true values, respectively. The upper bound n is the last day of a simulation. Hence, the best fit parameter vector $\theta = (\beta, \gamma)$ is the one that minimizes $E(\theta)$.

To this end, we employed the particle swarm optimization (PSO) algorithm. A set of 50 potential solutions, represented by particles with x-coordinate β and y-coordinate γ were randomly scattered around the search-space $0 \leq \beta; \gamma \leq 1$. Their

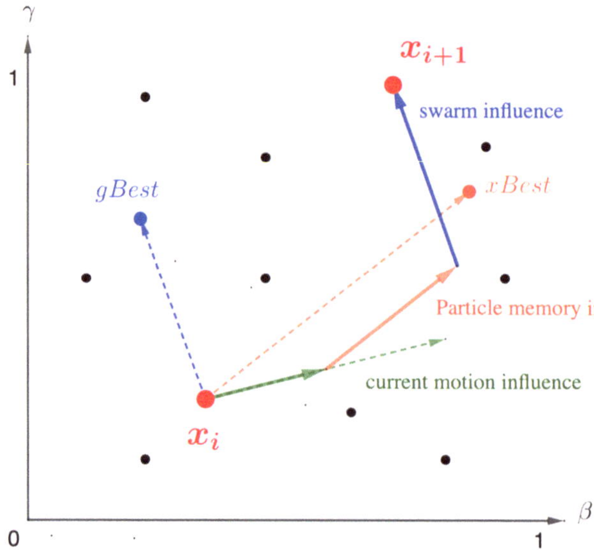

Fig. 35.2 Movement of a single particle

respective z-coordinates were thus $E(\beta, \gamma)$. In each iteration, each particle's movement was influenced by its own inertia, its prior best-known solution $x\,Best$ (cognitive component), and the global best known solution $g\,Best$(social component). More specifically, the velocity v^i_{k+1} and displacement x^i_{k+1} of the ith particle in the $(k+1)$th iteration were given by the formulae:

$$v^i_{k+1} = wv^i_k + c_1 r_1 \left(x\ Best^i_k - x^i_k\right) + c_2 r_2 \left(g\ Best^i_k - x^i_k\right)$$
$$\text{and } x^i_{k+1} = x^i_k + v^i_k k \tag{35.6}$$

where we set the parameters inertia $w = 0.9$, cognitive rate $c1 = 0.5$, social rate $c2 = 0.3$. r_1 and r_2 were random real numbers within (0,1). We illustrated the movement of a single particle in Fig. 35.2. The iteration was then carried out for 500 times, and eventually, some particles converged to the best-fit solution for β and γ where $E(\beta, \gamma)$ was minimum. The PSO was implemented using the package *PySwarm*.

35.3.2.3 Initial Parameters and Reproduction Number, R_0

The initial parameters of COVID-19 must be obtained before any preventive measures are taken, so as to most accurately reflect the contagiousness of the virus itself. We assumed that no major preventive measures had been taken prior to the Circuit Breaker, a nationwide lockdown which began on 7 April (Day 76) [7]. Considering

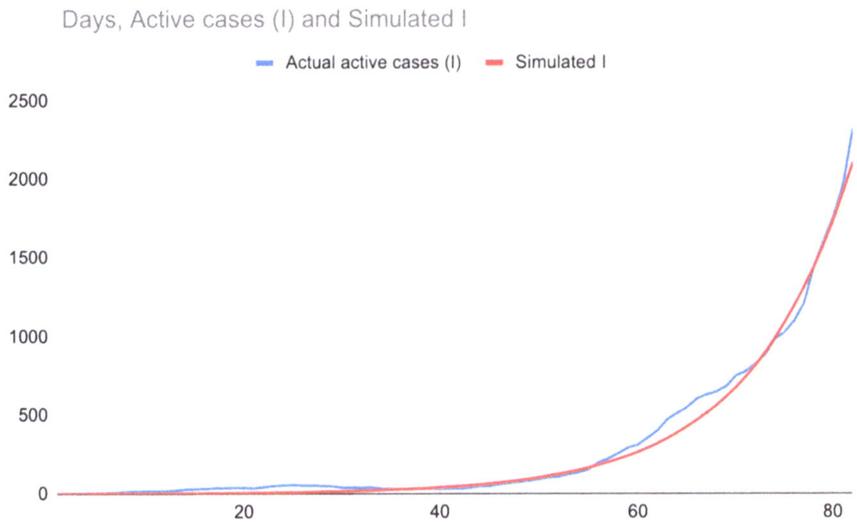

Fig. 35.3 A comparison between our simulated trend (red) and actual data (blue)

that the average delay of reporting of the infected cases is 6 to 7 days (by assumption 3), we extended the initial simulation period to Day 82. Hence, in Eq. (35.5), we set the upper bound of sigma notation to $n = 82$.

Using the PSO algorithm, we obtained the infection rate and recovery rate of COVID-19 at its onset as:

$$\beta = 0.124600646, \ \gamma = 0.03011705061$$

The initial basic reproduction number R_0 was given by:

$$R_0 = \frac{\beta}{\gamma} = 4.137212757$$

This suggested that the nature of COVID-19 was highly infectious, given its high initial R_0 value. Figure 35.3 illustrates our simulation of the initial viral transmission in Singapore.

35.3.3 Simulation on Subsequent Preventive Measures

As Singapore gradually implemented various preventive strategies, the infection rate and recovery rate changed accordingly. Hence, we divided the 341-day duration into five stages based on the several major regulations carried out by the authorities and

Table 35.1 SIR parameters of Singapore's different stages

Stage	Days (range of t)	Infection rate, β	Recovery rate, γ	R_0
Pre CB	1–82	0.124600646	0.03011705061	4.137212757
CB part 1	83–107	0.09968657734	0.005784718782	17.23274391
CB part 2	108–136	0.03542533256	0.04750462208	0.745723910
Phase 1	137–154	0.02887605357	0.06167369544	0.468206961
Phase 2	155–341	0.0520287245	0.06974407237	0.745994931

used PSO to fit the best parameters in each stage. The results and evaluations are shown in Table 35.1 next page.

In the first part of the Circuit Breaker (CB part 1), the immediate effect of lockdown was reflected in the reduction of infection rate. However, the recovery rate was lower, possibly due to the lack of capacity of local hospitals to handle the influx of COVID-19 patients [8].

In the second part of the Circuit Breaker, the reproduction number became less than 1, meaning that the virus was under control and more people were recovering. This showcased the lockdown was effective to contain the virus, and the healthcare capacity had been improved [9].

In Phase 1 (1 June to 17 June), strict safety management measures (SMM) such as social distancing and compulsory mask-wearing were effective in maintaining the low viral transmission rate, as seen by the low β and R_0 values.

In Phase 2 (18 June to 28 December), the ease of regulations incurred higher infection rate, but the continuation of SMM kept R_0 below 1 so that the virus was still safely subdued [10].

To evaluate the effectiveness of these measures, we predicted the trend of I when no preventive measures were implemented using only the initial β and γ values throughout the 341 days. The result, plotted in logarithmic scale in Fig. 35.4, was staggering: the peak of infection could easily have reached about 100 times of the actual peak, and this zenith would have occurred in late-July (around Day 180). Our data suggested that the preventive measures in Singapore had reduced the peak of infection cases by 99.11%.

35.4 The SIR Model with Vaccination

35.4.1 Minimum Vaccination Rate Required to Control COVID-19

Beyond investigating the SIR mechanism, we were also interested in investigating the vaccination rate that is required for a country like Singapore to keep an epidemic under control. To this end, we derived an extended SIR model with the vaccinated

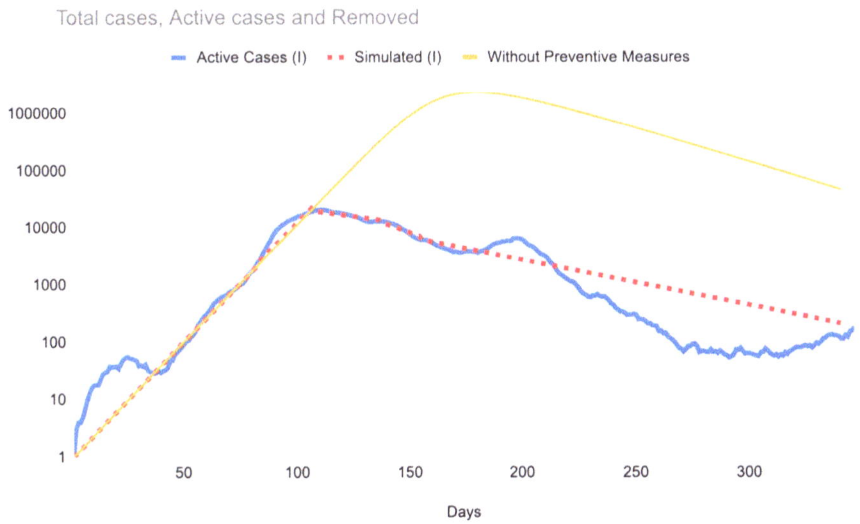

Fig. 35.4 Simulated infected population with and without preventive interventions in logarithmic scale

population taken into account, with a few more assumptions: firstly, the entire population is categorized into four groups: the susceptible, the infected, the recovered, and the fully vaccinated. In our model, we took the vaccinated population as those who have been fully vaccinated, i.e. taken the second jab. Secondly, still holding the total population constant, we also assumed that the majority of Singaporeans took or would take the Pfizer-BioNTech COVID-19 vaccine [11] whose vaccination efficacy remained constant. Thirdly, we assumed that there is no viral mutation that might render the vaccine ineffective.

Our enhanced SIR model with the vaccination process is demonstrated in the system of differential equations (35.7):

$$\begin{cases} S + I + R + V = N \\ \dfrac{\mathrm{d}S}{\mathrm{d}t} = -\dfrac{\beta S I}{N} - \alpha S + \dfrac{\sigma \beta V I}{N} \\ \dfrac{\mathrm{d}I}{\mathrm{d}t} = \dfrac{\beta S I}{N} - \gamma N \\ \dfrac{\mathrm{d}R}{\mathrm{d}t} = \gamma \mathrm{I} \\ \dfrac{\mathrm{d}V}{\mathrm{d}t} = \alpha S - \dfrac{\sigma \beta V I}{N} \end{cases} \tag{35.7}$$

where S, I, R, V, N denote the susceptible, infected, recovered, fully vaccinated, and total population in Singapore, respectively; $\alpha, \beta, \gamma, \sigma$ denote the vaccination rate (as the proportion of the total population vaccinated per unit time, $0 \leq \alpha \leq 1$),

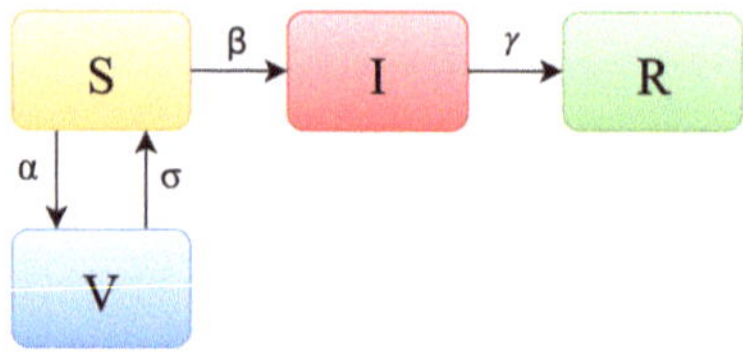

Fig. 35.5 Schematic diagram of our SIR-V model

transmission rate, recovery rate, and vaccine inefficacy, respectively. We took the initial transmission and recovery rate of COVID-19 modelled in Sect. 35.3.2 (i.e.$\beta = 0.124600$, $\gamma = 0.030117$). For vaccine inefficacy σ $(0 \leq \sigma \leq 1)$, the lower the σ, the highest the ability a vaccine can protect a vaccinated person from immunity relapse [12]. Hence, in the case of the Pfizer-BioNTech COVID-19 vaccine, we assumed it is $\sigma = 0.09$ since the official report declares its efficacy is around 91% [13]. Therefore, a probability of 0.09 existed for a vaccinated person to suffer from immunity loss and become a member of the susceptible population again. The relationship of our SIR-V model can be illustrated using the schematic diagram in Fig. 35.5.

Our objective was to determine the minimum vaccination rate, α, required to keep COVID-19 under control, which we defined as keeping the total number of infections below 1% of the total population in the long run, i.e.

$$\lim_{n\to\infty} [I(n) + R(n)] < 0.01N \tag{35.8}$$

Using the parameters specified above, we computed the minimum vaccination rate required, as well as the number of Singapore residents that must be vaccinated per day to keep the total infection below the 1% population threshold as:

$$\text{Vaccination rate: } \alpha = 0.005533$$

$$\text{Number of people vaccinated per day: } \alpha N \approx 30.985$$

Subsequent report showed that Singapore's daily vaccination rate had exceeded our figure [14]. As Singapore continues improving her vaccination effort, it is generally safe to conclude that Singapore will be able to keep the pandemic under control, assuming that the virus does not mutate. However, it is notable that we were using the initial parameters (β, γ) to model the vaccination rate rather than those with safety management regulations. This is because the purpose of vaccination is to allow a society to resume its original lifestyle where people have multiple contacts per day without having to worry about the disease in general. Hence, if Singapore could vaccinate at a rate that outstrips the spread of COVID-19 with high initial R_0, there is no worry that if R_0 were to drop further after regulations, we can still keep the transmission under control. We also assumed the vaccination starts at Day 1 because our SIR-V model to compute the minimum α was hypothetical, i.e. we were not finding the parameters based on real data anymore.

35.4.2 Vaccination Rate for Epidemics with Different R_0

Having evaluated the minimum vaccination rate, α, required to keep COVID-19 under control, we extended our investigation to predict the α for diseases with differing levels of infectiousness, and in different countries. For instance, what is the vaccination rate required when each Singaporean is having 5 contacts per day versus that when people are having 20 contacts per day?

Since the frequency of contacts is reflected in the infection rate β, we used different $R_0 = \frac{\beta}{\gamma}$ to solve the differential equations (35.7) and obtained the corresponding minimum vaccination rate α required to keep the epidemic under control, and the number of people vaccinated per day, αN. The relationship between R_0 and αN is shown in Fig. 35.6.

The data suggested that a positive linear relation was a good approximation for α and R_0 when $R_0 \geq 2.5$ and is not overly large. The trend lines from our simulation were obtained as:

$$\alpha = 2.6 \times 10^{-3} R_0 - 5.01 \times 10^{-3}$$
$$\alpha N = 14{,}566\, R_0 - 28{,}083$$

suggesting that for every increase of 1 in R_0, additional 14,566 citizens have to be vaccinated every day in Singapore to keep COVID-19 under control. For an epidemic

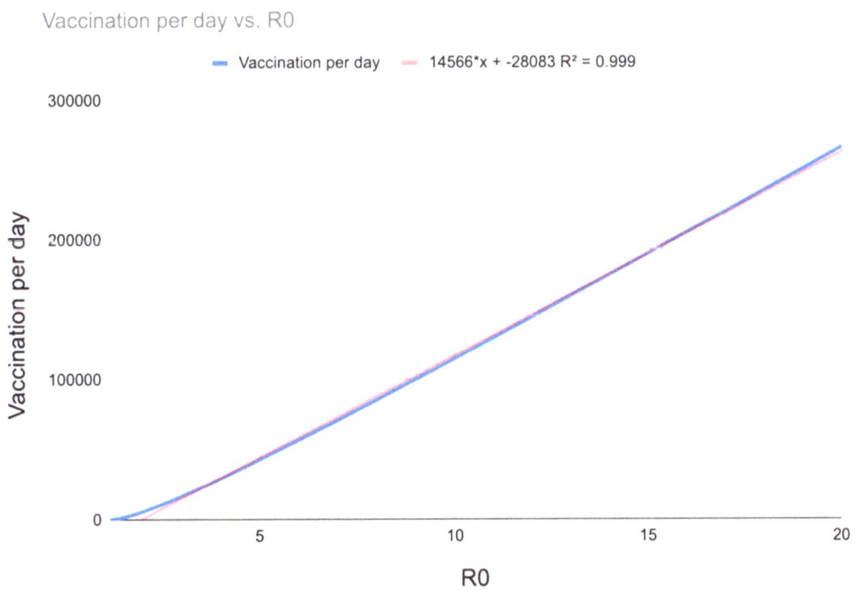

Fig. 35.6 Number of people that need to be vaccinated per day (αN) against the infectiousness (R_0) in Singapore

with $1 < R_0 < 2.5$, it is still preferred to carry out the computation by solving equations (35.7) as the runtime will be relatively fast.

We believed that a similar approach could be adapted to other countries with different population sizes and different vaccine efficacy (σ), to predict the vaccination rate required to contain the virus in their societies.

35.5 Discussion

35.5.1 Strengths and Limitations of Our SIR Model Simulation

Strength:

- Our SIR model could accurately and reliably compute the parameters values β and γ of COVID-19 during Singapore's different stages, demonstrating the epidemic mechanism.
- Our results of these parameters could be rightfully explained by the various preventive measures in these phases (refer to Sect. 35.3.3). Our results were also able to reflect the effectiveness of these measures.
- The implementation of the SIR and SIR-V models was fast and straightforward and entailed low computational costs [15].
- Our findings on the vaccination rates were socially beneficial as societies can leverage them to predict the vaccination rates to protect their citizens.

Limitations:

- The result of simulation based on the system of differential equations in the SIR model was non-random, thus may not be fully realistic.
- Our SIR model was not able to predict fluctuations in the active cases I, such as the second peak from Day 185 to Day 210 (refer to Fig. 35.1). Such fluctuations might be due to factors not directly related to the social dynamics, such as the increase in testing intensity for the foreign migrant workers during the final stage of dormitory clearance [16].
- Our SIR model could not evaluate the effectiveness of individual strategies because most measures were implemented simultaneously.
- Our SIR-V model assumed that there is no viral mutation, which might not hold true in the dynamically changing social environment.

35.5.2 Implications of Our Model

- Our research successfully demonstrated the mechanism by which the COVID-19 spread in Singapore, and potentially provided an insight into how the society's preventive interventions could assist in clamping down the viral transmission.
- Our algorithms could be easily adapted to model the COVID-19 developments and vaccination rates in countries with large populations such as China, the USA, Brazil, and many European countries. They could also model the spread of previous epidemic diseases, such as the severe acute respiratory syndrome (SARS) and the Ebola virus outbreaks.
- Our model could be harnessed to predict future events. For instance, we may predict if Singapore's Phase 3 measures (from Day 342 onward) can still subdue the virus by maintaining the reproduction number below 1, and when the pandemic would end in other countries with different vaccination rates.

35.6 Conclusion

In conclusion, our research thoroughly investigated the SIR epidemic model and applied it to the novel coronavirus disease 2019 (COVID-19) transmission in Singapore. Based on the publicly available data from *Worldometer* and *Statista*, we successfully employed particle swarm optimization (PSO) algorithm to compute the infection rate β and recovery rate γ of COVID-19 during different stages (refer to Sect. 35.3.3). Our results showed that COVID-19 was highly contagious, with its initial basic reproduction number R_0 as 4.137212757. As preventive measures such as Circuit Breaker lockdown, social distancing, and compulsory mask-wearing followed up, the infection rate and recovery rate varied, and R_0 eventually dropped below the transmission threshold of 1. Our results also suggested that these measures have effectively reduced the maximum active cases by 99.11%.

As Singapore stepped up to implement nationwide vaccination, we enhanced our simulation using the SIR-V model and computed the minimum vaccination rate, α, required to keep COVID-19 under control (infect less than 1% of total population in the long run) is 0.005533, which means that about 30,985 people have to be vaccinated per day to restrain viral transmission effectively. We also modelled the relationship between α and R_0 as a linear trend (Sect. 35.4.1) which can be used to predict the required vaccination rate.

While our model was limited in monitoring the fluctuations and evaluating individual preventive measures in close detail, it was nevertheless holistic and reliable in demonstrating the transmission mechanism of various epidemic diseases in different locations, and possessed the potential to predict future trends and recommend suitable vaccination rates for societies to effectively restrain viral development.

Acknowledgements We thank our mentor, Dr. Fedor Duzhin from the Nanyang Technology University, for his pertinent guidance and advice in assisting us with our research and modelling processes.

Format Rectification

Equations

(1) $S + I + R = N$

(2) $\frac{dS}{dt} = -\frac{\beta SI}{N}$

(3) $\frac{dR}{dt} = \gamma I$

$$\frac{dS}{dt} + \frac{dI}{dt} + \frac{dR}{dt} = 0$$

(4) $\frac{dI}{dt} = \frac{\beta SI}{N} - \gamma I$

(5) $E(\theta) = \frac{1}{\max(I_{\text{real}})} \sum_{t=1}^{n} [I(\theta, t) - I_{\text{real}}(t)]^2 + \frac{1}{\max(R_{\text{real}})} \sum_{t=1}^{n} [R(\theta, t) - R_{\text{real}}(t)]^2$

(6) $v_{k+1}^{i} = w v_k^i + c_1 r_1 (x\text{Best}_k^i - x_k^i) + c_2 r_2 (g\text{Best}_k^i - x_k^i)$

$$x_{k+1}^{i} = x_k^i + v_k^i k$$

$$\beta = 0.124600646, \gamma = 0.0301170561$$

$$R_0 = \frac{\beta}{\gamma} = 4.137212757$$

(7) $S + I + R + V = N$

$$\frac{dS}{dt} = -\frac{\beta SI}{N} - \alpha S + \frac{\sigma \beta VI}{N}$$

$$\frac{dI}{dt} = \frac{\beta SI}{N} - \gamma N$$

$$\frac{dR}{dt} = \gamma I$$

$$\frac{dV}{dt} = \alpha S - \frac{\sigma \beta VI}{N}$$

(8) $\lim_{n\to\infty}[I(n)+R(n)] < 0.01N$

$$\text{Vaccination rate: } \alpha = 0.005533$$

$$\text{Number of people vaccinated per day: } \alpha N \approx 30,985$$

$$\alpha = 2.6\times 10^{-3}R_0 - 5.01\times 10^{-3}$$

$$\alpha N = 14{,}566R_0 - 28{,}083.$$

References

1. Singapore Population (LIVE). Worldometer. [Online]. Available: https://www.worldometers.info/world-population/singapore-population/. Accessed February 17 2021.
2. Transmission of SARS-CoV-2: implications for infection prevention precautions.," World Health Organization. [Online]. Available: https://www.who.int/news-room/commentaries/detail/transmission-of-sars-cov-2-implications-for-infection-prevention-precautions. Accessed February 17 2021.
3. Abbott, S., Hellewell, J., Thompson, R. N., et al. (2020). Estimating the time-varying reproduction number of SARS-CoV-2 using national and subnational case counts. *Wellcome Open Research*, December 8 -2020. [Online]. Available: https://wellcomeopenresearch.org/articles/5-112. Accessed February 17 2021.
4. Admin. Epidemic theory & techniques for analysis of infectious disease data. *Health Knowledge*, July 8 2019. [Online]. Available: https://www.healthknowledge.org.uk/public-health-textbook/research-methods/1a-epidemiology/epidemic-theory. Accessed February 17 2021.
5. Worldometer's covid-19 data. [Online]. Available: https://www.worldometers.info/coronavirus/country/singapore/. Accessed February 17 2021.
6. Hirschmann, R. (2021). Singapore: COVID-19 recoveries by day. *Statista*, April 7 2021. [Online]. Available: https://www.statista.com/statistics/1101490/singapore-number-of-patients-recovered-from-covid-19/. Accessed February 17 2021.
7. COVID-19 circuit breaker: Closure of workplace premises. MCI—Gov.SG. [Online]. Available: https://www.gov.sg/article/covid-19-circuit-breaker-closure-of-workplace-premises. Accessed February 17 2021.
8. Khalik, S. (2021). Surge in coronavirus infections strains healthcare system. *The Straits Times*, May 24 2021. [Online]. Available: https://www.straitstimes.com/singapore/surge-in-virus-infections-strains-healthcare-system. Accessed February 17 2021.
9. Baker, J. A. (2020). *Stretched but coping: How Singapore's healthcare system has cranked up efforts to deal with COVID-19*. CNA, May 4, 2020. [Online]. Available: https://www.channelnewsasia.com/news/singapore/covid-19-singapore-health-capacity-hospitals-treatment-12698282. Accessed February 17 2021.
10. Ending circuit breaker: phased approach to resuming activities safely. [Online]. Available: https://www.gov.sg/article/ending-circuit-breaker-phased-approach-to-resuming-activities-safely. Accessed February 17 2021.

11. More in S'pore opting for Pfizer Covid-19 vaccine despite longer wait than for Moderna. *The Straits Time*. [Online] Available: https://www.straitstimes.com/singapore/covid-19-more-opting-for-pfizer-vaccine-despite-longer-wait-than-for-moderna. Accessed July 17 2021.
12. Ghostine, R., Gharamti, M., Hassrouny, S., & Hoteit, I. (2021). An extended SEIR model with vaccination for forecasting the Covid-19 pandemic in Saudi Arabia using an Ensemble Kalman filter. *Mathematics, 9*(6), 636. https://doi.org/10.3390/math9060636
13. Pfizer and BioNTech CONFIRM high efficacy. Pfizer and BioNTech Confirm High Efficacy. (n.d.). Available: https://www.pfizer.com/news/press-release/press-release-detail/pfizer-and-biontech-confirm-high-efficacy-and-no-serious.
14. Singapore aims to vaccinate 2 in 3 residents against Covid-19 by National Day: Are we on track? *The Straits Time*. [Online] Available: https://www.straitstimes.com/multimedia/graphics/2021/06/singapore-covid-vaccination-tracker/index.html. Accessed July 17 2021.
15. Modelling epidemics lecture 7: Stochastic individual based epidemiological models. Available: https://jvanderw.une.edu.au/L7_ModellingEpidemics3.pdf. Accessed February 17 2021.
16. Iau, J. (2020). 908 new Covid-19 cases in Singapore, nearly all are migrant workers in dorms. *The Straits Times*, August 5, 2020. [Online]. Available:https://www.straitstimes.com/singapore/health/908-new-covid-19-cases-in-singapore-including-4-in-community-and-4-imported. Accessed February 17 2021.

Chapter 36
From Phenotypes to Endotypes: Deciphering the Immune Signatures Associated with Atopic Dermatitis (AD)

Alina Wee, Yeo Rong Quan, and Justin Ng

Abstract Atopic dermatitis (AD) affects 13.1% of the local population and is the most common skin disorder in Singapore. A complex disease like AD is not just defined by its phenotype or observable characteristics as the underlying pathophysiology may be different. Hence, we investigated endotypes which are subtypes of a disease condition, often defined by distinct pathophysiological mechanisms at the molecular level. In this study, we investigated dermotype in AD patients, which further stratified them into two distinct dermotypes—dermotype A and dermotype B. It was found that the two dermotypes within AD display varying properties. However, the immune signatures associated with these two dermotypes remain unknown. Therefore, this study identified the underlying immune phenotypes associated with dermotype-specific patients. It proved that patients with dermotype B displayed higher expression of IGHE genes which correlates with the higher IgE levels observed, and elevated levels of CCL13 and CCL17 biomarkers, thus reflecting type 2 inflammation. On the other hand, higher expression of PTCH1 gene in dermotype A confers greater skin hydration and barrier function. Based on our findings, it is recommended that treatments targeting type 2 inflammations can be used to effectively treat AD patients with dermotype B, while patients with dermotype A can be administered with low dosage of drugs tackling IGHE genes or type 2 inflammations. Understanding of the underlying immune differences between both dermotypes is crucial in enabling a more specialized treatment of AD at the molecular level.

36.1 Introduction

Atopic dermatitis (AD) affects 13.1% of the local population and is the most common skin disorder in Singapore [1]. AD is characterized by dry, itchy, and inflamed skin that causes pain, distress, and loss of sleep. AD arises from the complex interaction

Mentored by Dr. Anand Kumar Andiappan (A*STAR SIgN).

A. Wee (✉) · Y. R. Quan · J. Ng
Hwa Chong Institution & A*STAR, Singapore, Singapore
e-mail: 162242e@student.hci.edu.sg

H. Guo et al. (eds.), *IRC-SET 2021*, https://doi.org/10.1007/978-981-16-9869-9_36

of genetics and the environment of patients and results in a dysregulated immune response.

Loss-of-function mutations in the filaggrin (FLG) gene have been proven to be a strong genetic causative agent of AD. [2]. FLG mutation increases risk of skin barrier dysfunction which increases the skin's susceptibility to environmental triggers such as *Staphylococcus aureus*, inducing high levels of immunoglobulin E (IgE) antibodies [3].

Biomarkers found in blood, including IgE and Th2 cytokines, are involved in the immune response of AD. An immune response of AD would first require exposure to allergens in the environment, resulting in production of allergen-specific IgE antibodies and would bring about allergen sensitization. Cytokines are the inflammatory mediators responsible for most biological effects in the immune system, such as cell mediated immunity and allergic type reactions. Th2 cells mostly produce IL4, 5, and 13, which are associated with the promotion of IgE and eosinophilic responses in atopy [4]. Therefore, there is an overexpression of Th2 cytokines observed in patients with AD [5].

Hitherto, there is no known treatment to cure AD. Current treatments only address AD at a physiological level and bring about side effects such as skin thinning, stretch marks, and bruising susceptibility. There is a need to go further and address AD at a molecular level so that we can provide targeted, personalized treatments to patients instead of the current generic treatment protocol.

AD patients can be stratified into endotypes, which are subtypes of a disease condition, often defined by distinct pathophysiological mechanisms at the molecular level. Past research has shown that food allergy is an endotype of AD (Leung et al., 2019). Recent studies have found that AD patients can be further stratified into two distinct skin dermotypes—dermotype A and dermotype B. Dermotypes A and B display differences in skin hydration levels and transepidermal water loss [6]. Therefore, it is evident that the two dermotypes within the AD population display varying properties.

Since the microbe–host interactions on skin can influence immune response and contribute to AD, skin microbiomes can be analyzed to find biomarkers present, which can then be used to monitor AD progression. Hence, this study seeks to find the immune signatures associated with these two dermotypes.

Hypothesis

FLG and *S. aureus*—representing genetic and environmental factors, respectively—trigger immune dysregulation, and dermotypes will have different IgE and TH2 cytokine levels.

36.2 Objectives

The objective of this study was to determine underlying immune phenotypes associated with dermotype-specific patients.

36.3 Methodology

Blood phenotyping

IgE levels were detected using ImmunoCAP, immune markers using Luminex technology, and filaggrin genotyping using a customized sequencing panel.

36.4 Isolation of Peripheral Blood Mononuclear Cells (PBMC) and Gene Expression

Transcriptional profile of isolated PBMC was determined using next generation sequencing by estimating reads per million transcript. Total RNA was extracted following the double extraction protocol: RNA isolation by acid guanidinium thiocyanate-phenol–chloroform extraction (TRIzol, Thermo Fisher Scientific, Waltham, MA, USA) followed by a Qiagen RNeasy Micro-clean-up procedure (Qiagen, Hilden, Germany). Gene expression was determined using next generation sequencing by the SMARTSeq v2 protocol using an indexed paired-end sequencing run of 2×151 cycles on an Illumina HiSeq 4000 system (Illumina) (25 samples/lane).

36.5 Dermotype Definition

Previously published data were used to define the dermotype [6] using whole genome metagenomics of skin tape samples from the study cohort.

36.6 Data Analysis

(1) The Mann–Whitney U test (for 2 setups only) or Kruskal–Wallis test (for 3 or more setups) was carried out to find out if there are significant differences in immune markers found in dermotype A and B. The t-test (for 2 setups with more than 15 samples) is carried out to determine if there are significant differences in the means of test setups.

(2) R-squared test was conducted to statistically measure the association between two categorical datasets. R value represents the proportion of the variance for a dependent variable, explained by an independent variable or variables in a regression model. We used the R-squared test to correlate immune markers with dermotype A and B.

(3) False discovery rate (FDR) was used to determine the genes that were significantly expressed in dermotypes. FDR refers to the ratio of number of false-positive results to the number of total positive test results and is used in experiments where multiple comparisons are tested to correct for random events that falsely appear significant.

36.7 Results and Discussion

(A) There was a significant difference in IgE levels between dermotype A and dermotype B. Patients with dermotype B had significantly higher IgE levels as compared to patients with dermotype A (Fig. 36.1).

n	20
Median for dermotype A	776
Median for dermotype B	11,726

(B) Of the 56 immune markers analyzed, chemokine ligand 13 (CCL13/MCP-4) and chemokine ligand 17 (CCL17/TARC) displayed a significant difference between dermotype A and dermotype B. For both immune markers, dermotype B had a higher median level as compared to dermotype A (Figs. 36.2 and 36.3).

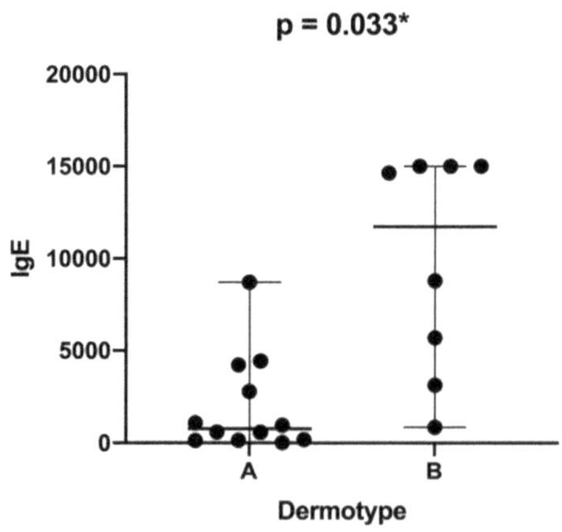

Fig. 36.1 Significant differences in IgE levels between dermotypes A and B

n	20
Median for dermotype A	42.895
Median for dermotype B	48.84

n	20
Median for dermotype A	82.605
Median for dermotype B	140.2

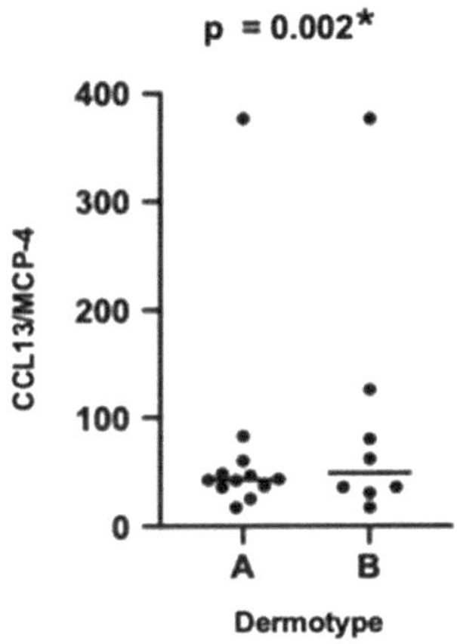

Fig. 36.2 Significant difference in CCL13 and level between dermotype A and dermotype B

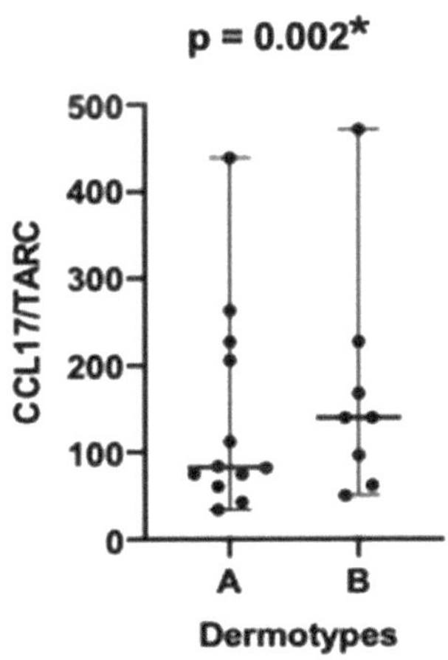

Fig. 36.3 Significant difference in CCL17 level between dermotype A and dermotype B

There were elevated levels of CCL13 and CCL17 biomarkers found in dermotype B. Since these biomarkers are produced by TH2 cells, this suggests that immune signatures associated with dermotype B have strong inflammatory response and Th2 response, thereby reflecting a type 2 inflammation.

(III) Among the 33,000 genes screened, only, PTCH1, IGHE, NLGN4Y, and AC021127.1 were significantly expressed across both dermotypes. Given that NLGN4Y is only present in the Y chromosome and that AC021127.1 is a non-coding gene; the genes selected for analysis were narrowed down to protein patched homolog 1 (PTCH1) and immunoglobulin heavy constant epsilon (IGHE)*n* (Fig. 36.4).

Higher expressions of PTCH1 gene were found in patients with dermotype A. PTCH1 produces patched-1 proteins which brings about an increase in regulatory T cells. AD pathology is reduced via the sonic hedgehog pathway which regulates epidermal homeostasis. Additionally, [7] has found that the sonic hedgehog pathway maintains epidermal stem cells which in turn promotes differential epidermal tissue repair. Evidently, higher expression of PTCH1 gene in dermotype A confers greater skin hydration and barrier function, as previously observed [6] (Fig. 36.5).

Patients with dermotype B displayed higher expression of IGHE genes which correlates with higher IgE levels and elevated levels of CCL13 and CCL17 biomarkers previously observed in section **(A)** and **(B)**. Since these biomarkers are produced by TH2 cells, this further supports the fact that immune signatures associated with dermotype B have a strong inflammatory response and Th2 response, reflecting a type 2 inflammation.

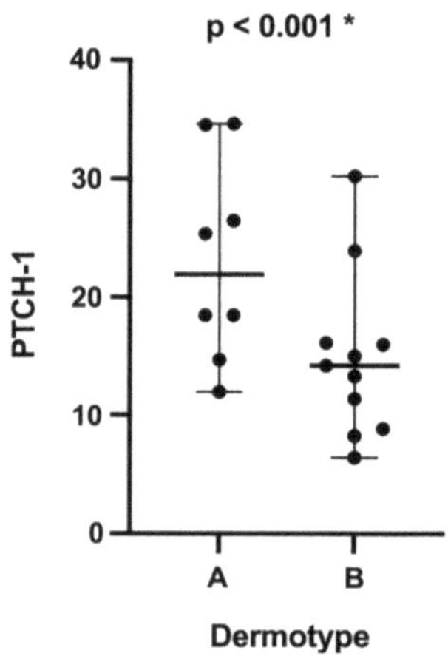

Fig. 36.4 Dermotype A patients **(median = 21.9143)** exhibited higher expression of PTCH1 genes as compared to dermotype B **(median = 14.2142)**

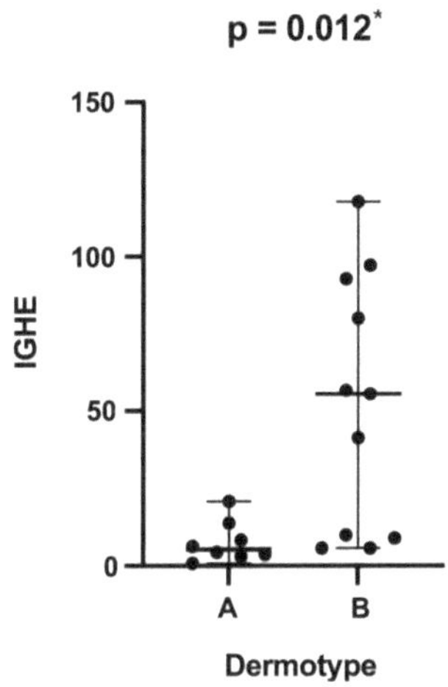

Fig. 36.5 Higher levels of IGHE were observed in dermotype B patients **(median = 55.5569)** than those of dermotype A **(median = 5.32781)**

36.8 Conclusion

A stratified approach is needed to understand the underlying phenotypes and endotypes of a clinically heterogeneous disease such as AD, which will be useful in defining treatment strategies. The endotype that was investigated in this study was dermotypes, and it was found that distinct skin microbiome configurations found on skin dermotypes can be used to stratify AD patients, and there were unique immune signatures found on AD patients with different skin dermotypes.

While there have been challenges standardizing the size and ethnicity of samples, this study conclusively demonstrates that AD phenotype has different underlying immune signatures. AD patients can be stratified into two distinct groups based on their skin dermotype—A or B. Immune signatures associated with dermotype B reflect type 2 inflammation, whereas higher expression of PTCH1 gene in dermotype A confers greater skin hydration and barrier function, as previously observed [6].

Dermotype B has shown to promote type 2 inflammatory responses, which are associated with the promotion of IgE and eosinophilic responses in atopy [4]. Based on our findings, we recommend that treatments targeting type 2 inflammations be used to effectively treat AD patients with dermotype B, while patients with dermotype A can be administered with low dosage of drugs tackling IGHE genes or type 2 inflammations. This specialized treatment not only addresses AD at the physical level but also at the molecular immune level.

For future work, eosinophil count can be obtained to check if there is a correlation between eosinophils and skin dermotype. Additionally, the systemic inflammation component in AD may be different in AD patients with and without food allergy. Hence, this can be evaluated in further studies.

References

1. Tay, Y. K., et al. (2020). Atopic dermatitis microbiomes stratify into ecological dermotypes enabling microbial virulence and disease severity. *Journal of Allergy and Clinical Immunology*. https://doi.org/10.1016/j.jaci.2020.09.031
2. Gao et al. (2009). Filaggrin mutations that confer risk of atopic dermatitis confer greater risk for eczema herpeticum. *Journal of Allergy and Clinical Immunology, 124*(3), 507–513. https://doi.org/10.1016/j.jaci.2009.07.034
3. Seite, S., & Bieber, T. (2015). Barrier function and microbiotic dysbiosis in atopic dermatitis. *Clinical, Cosmetic and Investigational Dermatology, 2015*(8), 479–483. https://doi.org/10.2147/CCID.S91521
4. Bojuniewicz et al. (2018). Atopic dermatitis yardstick: Practical recommendations for an evolving therapeutic landscape. *Annals of Allergy, Asthma & Immunology*. https://doi.org/10.1016/j.anai.2017.10.039
5. Brunner, et al. (2017). The immunology of AD and its reversibility with broad spectrum and targeted therapies. *The Journal of Allergy and Clinical Immunology*. https://doi.org/10.1016/j.jaci.2017.01.011
6. Angeline, S., Tay, L., et al. (2020). Atopic dermatitis microbiomes stratify into ecologic dermotypes enabling microbial virulence and disease severity. *Journal of Allergy and Clinical Immunology, 147*, 1329–1340. https://doi.org/10.1016/j.jaci.2020.09.031
7. Abe, Y., & Tanaka, N. (2017). Roles of the hedgehog signalling pathway in epidermal and hair follicle development, homeostasis, and cancer. *Journal of Developmental Biology*. https://doi.org/10.3390/jdb5040012

Chapter 37
Synthesis and Modification of LDHs for the Adsorption of Cationic and Anionic Dyes and Phosphates

Lucas Koh Eu Jen, Tan Yan Zhong, and Yeo Tze Zhuan

Abstract Phosphates and dyes are two common water pollutants. Excess phosphates discharged cause eutrophication due to the hyper-fertilization of phosphorus, while the dyes discharged render water unsafe for consumption. Thus, it is necessary to adsorb both phosphates and dyes from water bodies. We proposed layered double hydroxides (LDHs) as a solution to this. This study proposes the modification of Zn-Al LDH with sodium dodecyl sulfate (SDS), significantly increasing the adsorption capacity of the LDH for cationic dyes, namely methylene blue. The Zn-Al LDH synthesized by co-precipitation involved NaOH with $ZnCl_2$ and $AlCl_3$ salts in a 2:1 ratio. Subsequently, Zn-Al LDH was reconstructed with SDS solution to intercalate the dodecyl sulfate anions. LDHs synthesized were characterized using XRD. The effects of initial methyl orange, methylene blue, phosphate concentration, and pH on the adsorption by the LDHs were studied. Reusability tests were conducted for methyl orange to evaluate the efficacy of the Zn-Al LDH over continuous cycles. Zn-Al LDH has a maximum adsorption capacity of 1110 mgg^{-1} and 262 mgg^{-1} on methyl orange and phosphate, respectively, outperforming several other LDHs. Modifying the Zn-Al LDH with SDS enhances its maximum adsorption capacity on methylene blue, attaining a maximum adsorption capacity of 149 mgg^{-1}, outperforming several other LDHs. The Zn-Al LDH can remove close to 95% of dye over multiple cycles.

37.1 Introduction

Rapid population growth and industrialization have resulted in the large-scale pollution of water bodies with various pollutants via wastewater discharge [1]. Dyes such as methyl orange and methylene blue which are widely used in textile, paper, and pharmaceutical industries are often discharged into water bodies [2]. Wastewater containing these dyes is toxic, carcinogenic, and mutagenic to life forms mainly due to carcinogens like benzedrine, naphthalene, and other aromatic compounds [3].

L. K. E. Jen (✉) · T. Y. Zhong · Y. T. Zhuan
Hwa Chong Institution, Singapore, Singapore
e-mail: 171424L@student.hci.edu.sg

H. Guo et al. (eds.), *IRC-SET 2021*, https://doi.org/10.1007/978-981-16-9869-9_37

Eutrophication of water bodies is another major environmental problem. Water bodies are polluted with phosphate ions due to sewage discharge and fertilizer runoffs [4]. Phosphorus is a key element causing eutrophication, which leads to the excessive growth of algae [5]. This directly affects water quality and harms aquatic life through oxygen depletion. Consequently, the removal of phosphates from surface waters is necessary.

The conventional methods for treating dye-containing wastewater are coagulation, flocculation, reverse osmosis, and adsorption on various adsorbents. Adsorption by activated carbon is the preferred method because of its efficiency, capacity, and scalability for commercial usage [6]. However, activated carbon is expensive, which limits its application. Major treatment processes to remove phosphate from water include ion exchange, chemical precipitation, and biological treatment [7]. However, chemical precipitation results in the problem of sludge handling, while biological treatment is slow and ineffective due to variability in the temperature of wastewater that would make the execution of this process inaccessible for wastewater treatment [8].

This project proposes layered double hydroxide (LDH) as an alternative to conventional adsorbents. Layered double hydroxides (LDHs) are naturally occurring and synthetic anionic clays, in which divalent cations within brucite-like layers are replaced with trivalent cations. The general chemical composition of LDHs can be described by the following formula [9]: $\left[M^{2+}_{(1-\alpha)}N^{3+}(OH)_2\right]^{\alpha+}\left[A^{n-}\right]_{\alpha/n} \cdot mH_2O$, where M^{2+} is the divalent cation, N^{3+} is the trivalent cation, A^{n-} is the interlayer anion, and α is the molar ratio of$N^{3+}/\left(M^{2+} + N^{3+}\right)$. The resulting positive charge is balanced by hydrated anions found in the interlayer space between the brucite sheets [10]. LDHs can be used as sorbents for the removal of anions from aqueous solutions [11], due to their flexibility, low operational costs, ease of operation, and their potential to recover phosphate for use in biological fertilizers, as researchers have already pointed out that the world's phosphate reserves may be depleted before 2050 [11].

While many studies have used $CO_3{}^{2-}$ as their intercalated anion, few have used Cl^- ions [12]. Conventional LDHs synthesized by co-precipitation are hydrophilic and have a low affinity for organic pollutants including dyes [13]. Thus, this study proposes the use of Cl^- intercalated Zn-Al LDH as an effective method to adsorb both dyes and phosphates. In order for the LDH to adsorb cationic dyes, they will be modified with sodium dodecyl sulfate (SDS).

Hypothesis

This study hypothesizes that the Zn-Al LDH intercalated with Cl^- ions can be modified with SDS. Zn-Al LDH is comparable to other studies in the removal of methyl orange and phosphates. SDS-Zn-Al LDH is comparable to other studies in the removal of methylene blue. Zn-Al LDH can be reused for at least three cycles without significant drop in effectiveness.

37.2 Materials and Methods

37.2.1 Materials

Zinc chloride, aluminum chloride, sodium hydroxide, and sodium dihydrogen phosphate were procured from GCE Chemicals. Sodium dodecyl sulfate (SDS) was purchased from Sigma-Aldrich.

37.2.2 Synthesis of Zn-Al LDH

A solution (150 ml) containing 0.2 mol of $ZnCl_2$ and 0.1 $_{mol}$ of $AlCl_3$ was slowly added to a second solution containing 300 ml of 2.0 mol dm^{-3} NaOH by a burette for 150 min under constant vigorous stirring at 250 rpm at room temperature. After the mixing, a thick slurry was obtained. It was then aged at 85°C for 2 h using a hot plate. After the aging process, the precipitates were centrifuged, filtered, washed with deionized water, and dried in an oven until constant mass. The Zn-Al LDH obtained was then crushed into a powder.

37.2.3 Modification of Zn-Al LDH with SDS

Zn-Al LDH adsorbs anionic but not cationic species. To allow LDH to adsorb cationic dyes, it was modified with SDS ($CH_3(CH_2)_{11}SO_4Na$) by reconstruction method. Zn-Al LDHs were calcined at 450 °C for 2 h to destroy the layered structure. With stirring, 1 g of calcined LDH was added to 50 ml of 0.500 mol dm^{-3} SDS. SO_4^{2-} anions in the SDS were exchanged with the Cl^- anions to be intercalated between layers of cations.

37.2.4 Batch Adsorption Studies

20 ml of dye or phosphate of varying concentrations (from 200 to 1200 ppm) were stirred with 0.020 g of LDH/SDS-LDH for 24 h. The mixture was centrifuged, and the supernatant was obtained. The remaining concentration of the dye was determined using a UV–VIS spectrophotometer (Shimadzu UV1800) at 664 nm (for methylene blue) and 465 nm (for methyl orange). The remaining concentration of phosphate was determined using a colorimeter (HACH DR 890). The adsorption capacity was then calculated. To determine the adsorption mechanism and maximum adsorption capacity, the equilibrium concentration data was fitted into two most common isotherms—Freundlich and Langmuir isotherms.

37.2.5 *Effect of pH*

20 ml of 50 ppm concentration of dyes/phosphates at pH 2, 4, 6, 8, and 10 were mixed with 0.020 g of LDH for 24 h. pH was adjusted using HCl and NaOH. Adsorption was carried out as described in Sect. 2.4. The remaining concentration of dye or phosphate was then determined, and percentage removal (%) was calculated.

37.2.6 *Reusability Tests*

Zn-Al LDH was regenerated through calcination at 400 °C. The LDH was then reused to adsorb 50 ppm of methyl orange. Remaining concentration of the dye was then determined, and percentage removal was calculated. A total of four cycles of regeneration were carried out.

37.3 Results and Discussions

37.3.1 *Characterization of Chloride Intercalated Zn-Al LDH*

37.3.1.1 By X-Ray Diffraction (XRD)

The XRD pattern of Zn-Al LDH (Fig. 37.1) reveals two theta peaks at 11.3°, 22.6°, and 34.4°, which corresponded with the symmetrical basal reflections of the (003), (006), and (009) planes, respectively. On the other hand, two theta peaks at 38.8° and 46.0° corresponded with the asymmetrical basal reflection of the (015) and (018) planes, respectively, which are characteristic of the rhombohedral structure exhibited by the LDHs. The XRD patterns of Zn-Al LDH were similar to those reported in literature [14], confirming the successful synthesis of Zn-Al LDH.

Figure 37.2 reveals two theta peaks at 11.3°, 13.7°, 16.4°, and 21.4° which corre-

Fig. 37.1 Zn-Al LDH XRD

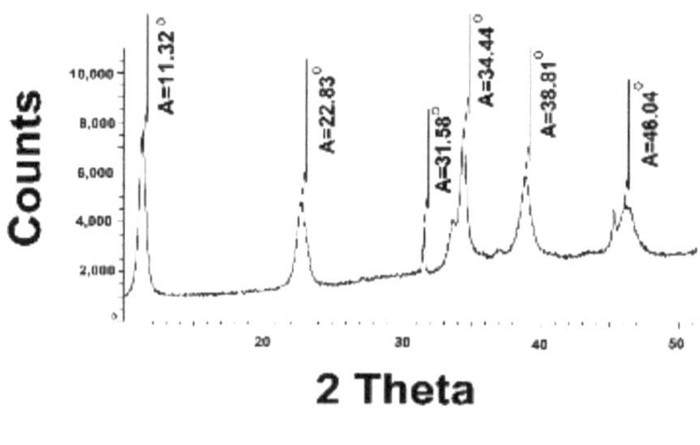

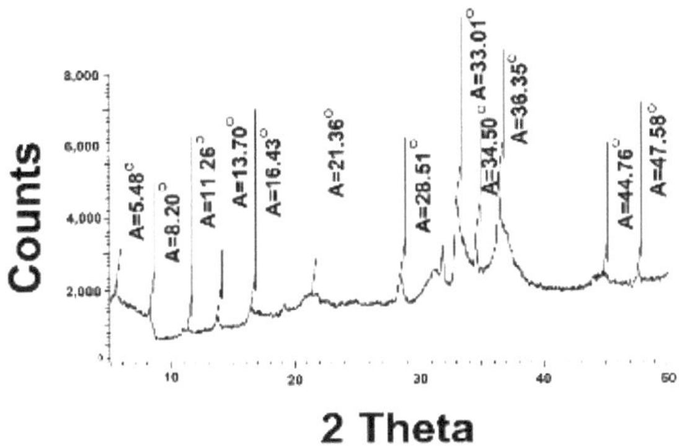

Fig. 37.2 SDS-Zn-Al LDH XRD

spond with the formation of a super lattice with ordered dodecyl sulfate anions intercalated between layers [15], confirming the successful synthesis of SDS-LDH by reconstruction.

37.3.1.2 By Scanning Electron Microscope (SEM)

The surface morphologies of the LDHs were similar to those reported by Qu et al. [14]. Both LDHs have flake-like structures which form agglomerates. The particles of SDS-Zn-Al LDH are bigger than that of Zn-Al LDH (Figs. 37.3 and 37.4). This could have been due to the calcination process undergone by the SDS-LDH, resulting in it being regenerated to form larger particles.

Fig. 37.3 SEM of Zn-Al LDH

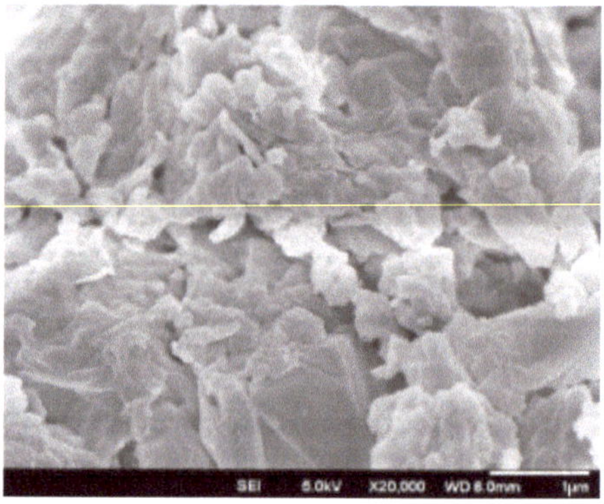

Fig. 37.4 SEM of SDS-Zn-Al LDH

37.3.2 Adsorption Studies

37.3.2.1 Effect of Initial Concentration of Methyl Orange (MO) and Phosphate Ions on the Adsorption Capacity of Zn-Al LDH

Figures 37.5 and 37.6 show the effect of initial concentration of methyl orange and phosphate on the adsorption capacity on Zn-Al LDH, respectively.

The general trend observed is that adsorption capacity increases as the orange and phosphates increase. This could be due to the fact that as phosphate and methyl orange concentration increases, more phosphate ions and methyl orange molecules are available for sorption by the active sites of LDH.

LDH adsorbed methyl orange and phosphate ions through electrostatic forces of attraction as the metal cations in the LDH attract the SO_3^{2-} anions in methyl orange and the phosphate anions. Phosphate anions and methyl orange containing SO_3^{2-} can also be exchanged with the Cl^- ions intercalated within the LDH via ion exchange [16]. Thus, there are various pathways for adsorption to occur, increasing the efficacy of Zn-Al LDH.

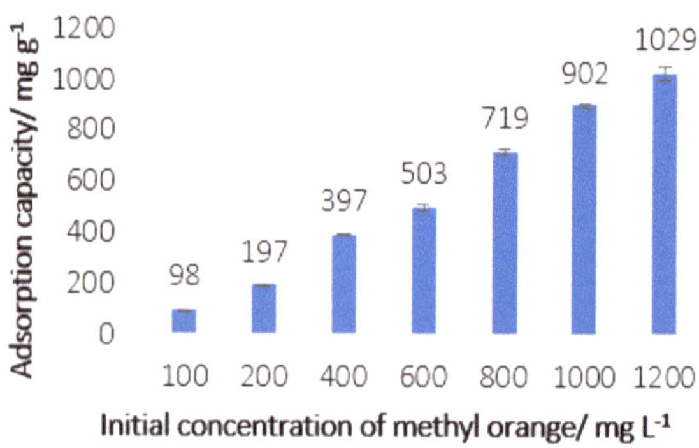

Fig. 37.5 Effect of initial concentration of methyl orange on adsorption capacity of Zn-Al LDH

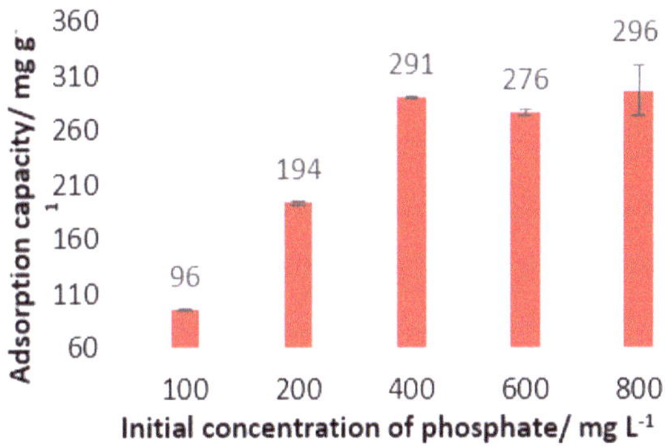

Fig. 37.6 Effect of initial concentration of phosphate on adsorption capacity of Zn-Al LDH

Equilibrium concentration data for methyl orange and phosphate was fitted into Langmuir and Freundlich isotherm. Langmuir isotherm was determined to be a better fit for the adsorption of both methyl orange and phosphate, with an R^2 value of 0.970 and 0.998, respectively. The results suggest that the adsorption of methyl orange and phosphates by Zn-Al LDH is monolayer. The maximum adsorption capacities were then derived from the gradient of the linear plots of Langmuir isotherm and were compared with the maximum adsorption capacity of other adsorbents.

For both methyl orange and phosphate, Zn-Al LDH synthesized in this study outperforms other forms of LDH and other adsorbents. Zn-Al LDH intercalated with Cl^- ions is more effective than Zn-Al LDH intercalated with CO_3^{2-} ions at adsorbing methyl orange (Table 37.1). Similarly, Zn-Al LDH intercalated with Cl^- ions is more effective than Zn-Al LDH intercalated with NO_3^- ions at adsorbing phosphates (Table 37.2). This could be due to the Cl^- ions having a smaller ionic radius, decreasing the interlayer spacing [17], allowing more Cl^- ions to be intercalated between the layers, and increasing the ion exchange capabilities of the LDH.

As shown in Table 37.2, Zn-Al LDH is more effective at adsorbing phosphates compared to Ni–Al and Mg–Al LDH. This could be due to the divalent ions having affinity to phosphate in the following order $Zn^{2+} > Cu^{2+} > Mg^{2+} > Ni^{2+}$ [24]. Thus, Zn-Al LDH intercalated with Cl^- ions is the most effective at adsorbing both methyl orange and phosphates.

Table 37.1 Maximum adsorption capacity (q_{max}) of different adsorbents on methyl orange

Type of adsorbent	q_{max}/(mg/g)	References
Zn-Al LDH intercalated with chloride ions	1110	This study
Zn-Al LDH intercalated with carbonate ions	684	[18]
Zn-Al LDO	182	[19]
Modified chitosan magnetic composite	758	[20]

LDO layered double oxide

Table 37.2 Maximum adsorption capacity (q_{max}) of different adsorbents on phosphate

Type of adsorbent	q_{max}/(mg/g)	References
Zn-Al LDH intercalated with chloride ions	263	This study
Zn-Al LDH intercalated with nitrate ions	78	[20]
Ni–Al LDH	27	[21]
Mg–Al LDH	128	[22]
Fe_3O_4-SiO_2-LDH	57	[23]

37.3.2.2 Effect of Initial Concentration of Methylene Blue (MB) on the Adsorption Capacity of SDS-Zn-Al LDH

Figure 37.7 shows the effect of initial concentration of methylene blue on the adsorption capacity on SDS-Zn-Al LDH. The general trend observed is that adsorption capacity of SDS-Zn-Al LDH increases as the initial concentration of methylene blue increases.

The dodecyl sulfate anions occupied the interlayer regions and resulted in the formation of hydrophobic interactions between the alkyl chains of SDS and methylene blue, increasing the adsorption capacity of SDS-Zn-Al LDH [25].

Equilibrium concentration data of methylene blue was fitted into Langmuir and Freundlich isotherm. Langmuir isotherm was a better fit with an R^2 value of 0.947. This shows that the adsorption of methylene blue was monolayer. The maximum adsorption capacity was derived and compared with the maximum adsorption capacity of other researchers (Table 37.3).

The modification of LDH with SDS enhances the adsorption capacity of the LDH due to the increased hydrophobic interactions [25] as compared to Zn-Al LDH without modification. Zn-Al LDH modified with SDS also outperforms LDHs synthesized by other researchers (Table 37.3).

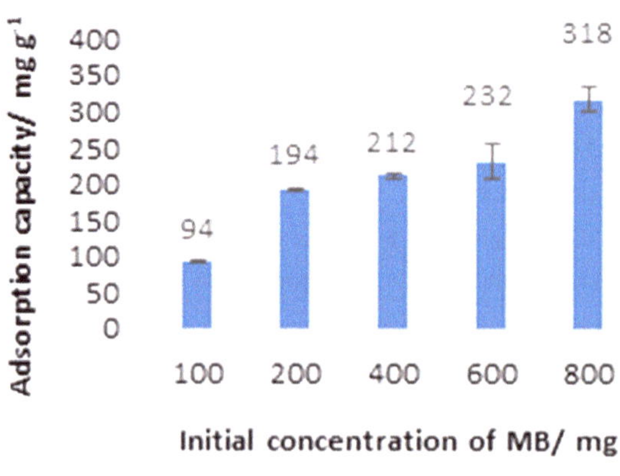

Fig. 37.7 Effect of initial concentration of MB on adsorption capacity of SDS-Zn-Al LDH

Table 37.3 Maximum adsorption capacity (q_{max}) of different LDHs on methylene blue

Type of adsorbent	q_{max}/(mg/g)	References
SDS-Zn-Al LDH	149	This study
Zn-Al LDH	31	[26]
SDS Mg–Al LDH with iron oxide	110	[27]
Mg–Al LDH	44	[28]

37.3.2.3 Effect of pH on Adsorption of Methylene Blue (MB) and Phosphate by Zn-Al LDH

As shown in Fig. 37.8, percentage removal of phosphate is lowest at pH 2. From pH 4–10, the percentage removal of phosphate remains relatively constant. There is a significant difference in percentage removal of phosphate when pH varies (*p*-value of Kruskal–Wallis test = 0.01208 < 0.05). This could be due to the disordering of the LDH structure at low pH, resulting in the leaching of Al^{3+} and Zn^{2+} ions into the solution [29], reducing the efficacy of the LDH as positively charged Zn^{2+} and Al^{3+} ions attract the negatively charged PO_4^{3-} ions through electrostatic forces of attraction.

A similar trend was observed in Fig. 37.9. Percentage removal of methylene blue was also lowest at pH 2. The difference in percentage removal when pH varies is not significant (*p*-value of Kruskal–Wallis test = 0.4067 > 0.05). This could be due to low pH promoting hydrophilic interaction rather than hydrophobic interactions, reducing the efficacy of the SDS-LDH in adsorbing methylene blue. Additionally, large amounts of H_3O^+ ions exist at low pH which can compete with methylene blue, a cationic dye for active sites of the SDS-LDH [30].

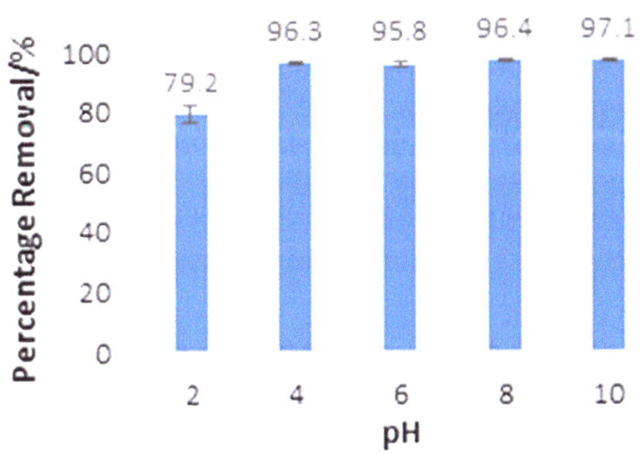

Fig. 37.8 Effect of pH on adsorption of phosphate

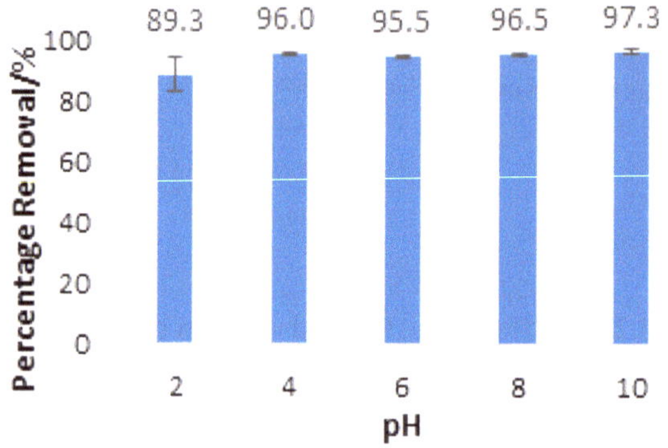

Fig. 37.9 Effect of pH on adsorption of methylene blue

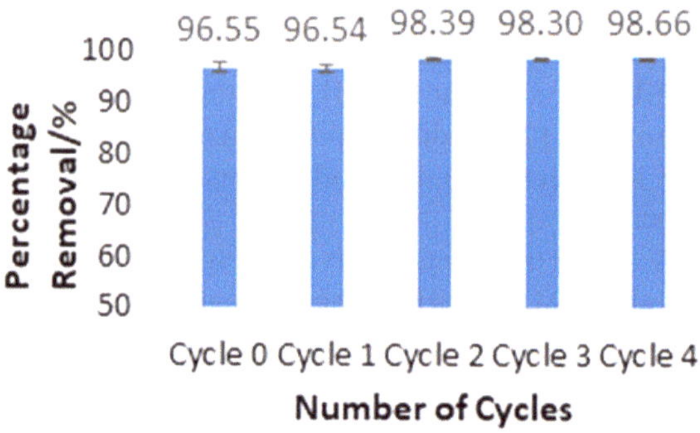

Fig. 37.10 Reusability of Zn-Al LDH in the adsorption of methyl orange

37.3.3 Reusability Studies

As shown in Fig. 37.10, at fourth cycle of adsorption, Zn-Al LDH is still able to adsorb methyl orange effectively without an observable decrease in percentage removal. This is because calcination at 400 °C degrades the methyl orange intercalated between layers of the LDH. Although the high temperature breaks down the layered structure of the LDH, forming layered double oxides (LDOs), due to its "memory effect," after being added to the dyes, it underwent structural reconstruction and formed LDH again [31], thus it can be regenerated and can sustain its ability to adsorb dye.

37.4 Conclusion and Future Work

Zn-Al LDH has been successfully synthesized via co-precipitation reaction involving sodium hydroxide with $ZnCl_2$ and $AlCl_3$ salts in a 2:1 mol ratio. Zn-Al LDH has also been successfully modified with SDS via reconstruction method to obtain SDS-Zn-Al LDH. The adsorption of methyl orange, phosphate, and methylene blue fits the Langmuir isotherm model, suggesting that the adsorption is monolayer. Zn-Al LDH has a greater maximum adsorption capacity on methyl orange dye and phosphates as compared to several other forms of LDH and adsorbents. Modifying the Zn-Al LDH with SDS enhances its maximum adsorption capacity on methylene blue. SDS-Zn-Al LDH also has a greater maximum adsorption capacity on methylene blue as compared to other forms of LDH. Zn-Al LDH is able to remove more than 95% of both methyl orange and phosphate at pH 4 and above. In addition, Zn-Al LDH can be modified easily with SDS to adsorb cationic dyes, and thus, it is a convenient and promising alternative to conventional methods of removing dyes. Zn-Al LDH can be reused multiple times to adsorb methyl orange, rendering it a suitable and eco-friendly alternative to conventional adsorbents.

In the future, the desorption and recyclability of phosphate can be studied as phosphate reserves in the world are being depleted. In real life, wastewater contains more than one type of pollutant. Hence, it would be interesting to find out if the presence of other pollutants would affect the ability of Zn-Al LDH in adsorbing dyes and phosphates. Finally, both the Zn-Al LDH and SDS-Zn-Al LDH can be incorporated into filters to remove both cationic dyes, anionic dyes and anions such as phosphate from wastewater.

References

1. Yaseen, D. A., & Scholz, M. (2018). Textile dye wastewater characteristics and constituents of synthetic effluents: a critical review. *International Journal of Environmental Science and Technology*, 1–32. https://doi.org/10.1007/s13762-018-2130-z
2. Edris, B., Zarei, A. A., Nadi, H., & Mohammad, A. Z. (2014). Adsorptive removal of methyl orange and reactive red dyes by *Morninga peregrina* ash. *Indian Journal of Chemical Technology, 21*(2), 105–113. Retrieved from https://pdfs.semanticscholar.org/320b/f585ee82a282 22b925f7031fd0f1d215d73d.pdf
3. Carmen, Z., & Daniel, S. (2012). Textile organic dyes—characteristics, polluting effects and separation/elimination procedures from industrial effluents—a critical overview. *Organic Pollutants Ten Years After the Stockholm Convention - Environmental and Analytical Update.* https://doi.org/10.5772/32373
4. Maryna, S., et al. (2016). Alarming nutrient pollution of Chinese rivers as a result of agricultural transition. *Environmental Research Letters, 11*, 014–024. https://doi.org/10.1088/1748-9326/11/2/024014
5. Subha, E., Sasikala, S., & Muthuraman, G. (2015). Removal of Phosphate from wastewater using natural adsorbents. *International Journal of ChemTech Research, 7.*
6. Mattson, J., & Mark, H. (1971). *Activated carbon: Surface chemistry and adsorption from solution.* M. Dekker.
7. Jiang, J. Q., & Wu, L. (2010). Preliminary study of calcium silicate hydrate (tobermorite) as crystal material to recovery phosphate from wastewater. *Desalination and Water Treatment, 23*(1–3), 49–54.
8. Long, F., Gong, J.-L., Zeng, G.-M., Chen, L., Wang, X.-Y., Jiu-Hua Deng, J.-H., Niu, Q.-Y., Zhang, H.-Y., & Zhang, X.-R. (2011). Removal of phosphate from aqueous solution by magnetic Fe–Zr binary oxide. *Chemical Engineering Journal, 171*, 448–455.
9. Goh, K. H., Lim, T. T., & Dong, Z. (2007). Application of layered double hydroxides for removal of oxyanions: A review. *Water Research, 42*(6–7), 1343–1368.
10. Cai, P., Zheng, H., Wang, C., Ma, H., Hu, J., Pu, Y., & Liang, P. (2012). Competitive adsorption characteristics of fluoride and phosphate on calcined Mg–Al–CO_3 layered double hydroxides.

Journal of Hazardous Materials, 213–214, 100–108. https://doi.org/10.1016/j.jhazmat.2012.01.069
11. Theiss, F. L. (2012). *Synthesis and characterisation of layered double hydroxides and their application for water purification, Introduction to Layered Double Hydroxides*. Master thesis, Queensland University of Technology, Australia. Retrieved from https://eprints.qut.edu.au/54620/1/Frederick_Thesis.pdf
12. Forano, C. (2004). Environmental remediation involving layered double hydroxides. In F. Wypych, & K. G. Satyanarayana (Eds.) *Clay surfaces: Fundamentals and applications* (pp. 425–458). Elsevier Academic Press.
13. Fernandez, L., Borzecka, W., Lin, Z., Schneider, R. J., Huvaere, K., Esteves, V. I., Cunha, A., & Tome, J. P. C. (2017). Nanomagnet-photosensitizer hybrid materials for the degradation of E2 in batch and flow modes. *Dyes and Pigments, 142*, 535–543.
14. Qu, J., He, X., Li, X., Ai, Z., Li, Y., Zhang, Q., & Liu, X. (2017). Precursor preparation of Zn–Al layered double hydroxide by ball milling for enhancing adsorption and photocatalytic decoloration of methyl orange. *RSC Advances, 7*(50), 31466–31474. https://doi.org/10.1039/c7ra05316a
15. Starukh, G., Rozovik, O., & Oranska, O. (2016). Organo/Zn-Al LDH Nanocomposites for Cationic Dye Removal from Aqueous Media. *Nanoscale Research Letters, 11*(1). https://doi.org/10.1186/s11671-016-1402-0
16. Almoisheer, N., Alseroury, F. A., Kumar, R., Aslam, M., & Barakat, M. A. (2019). Adsorption and anion exchange insight of indigo carmine onto CuAl-LDH/SWCNTs nanocomposite: Kinetic, thermodynamic and isotherm analysis. *RSC Advances, 9*(1), 560–568. https://doi.org/10.1039/c8ra09562k
17. Mahjoubi, F. Z., Khalidi, A., Cherkaoui, O., Elmoubarki, R., Abdennouri, M., & Barka, N. (2016). Treatment of textile effluents by chloride-intercalated Zn-, Mg- and Ni-Al layered double hydroxides. *Journal of Water Reuse and Desalination, 7*(3), 307–318. https://doi.org/10.2166/wrd.2016.041
18. Ni, Z., Xia, S., Wang, L., Xing, F., & Pan, G. (2007). Treatment of methyl orange by calcined layered double hydroxides in aqueous solution: Adsorption property and kinetic studies. *Journal of Colloid and Interface Science, 316*(2), 284–291. https://doi.org/10.1016/j.jcis.2007.07.045
19. Yang, D., Qiu, L., & Yang, Y. (2016). Efficient adsorption of methyl orange using a modified chitosan magnetic composite adsorbent. *Journal of Chemical & Engineering Data, 61*(11), 3933–3940. https://doi.org/10.1021/acs.jced.6b00706
20. Zhou, J., Yang, S., Yu, J., & Shu, Z. (2011). Novel hollow microspheres of hierarchical zinc–aluminum layered double hydroxides and their enhanced adsorption capacity for phosphate in water. *Journal of Hazardous Materials, 192*(3), 1114–1121. https://doi.org/10.1016/j.jhazmat.2011.06.013
21. Yan, L., Yang, K., Shan, R., Yan, T., Wei, J., Yu, S. J., Yu, H. Q., & Du, B. (2015). Kinetic, isotherm and thermodynamic investigations of phosphate adsorption onto core–shell Fe_3O_4@LDHs composites with easy magnetic separation assistance. *Journal of Colloid and Interface Science, 448*, 508–516. https://doi.org/10.1016/j.jcis.2015.02.048
22. Wang, Y., & Li, G. (2015). Adsorption behavior of phosphate on Mg–Al layered double hydroxide/montmorillonite composite. *Desalination and Water Treatment, 57*(38), 17963–17972. https://doi.org/10.1080/19443994.2015.1088803
23. Li, F., Jin, J., Ji, H., & Yang, M. (2020). Removal and Recovery of Phosphate and Fluoride from Water with Reusable Mesoporous Fe_3O_4@mSiO_2@mLDH Composites as Sorbents. *Journal of Hazardous Materials, 388*, 121734. https://doi.org/10.26434/chemrxiv.9876635
24. Almojil, S. F., & Othman, M. A. (2019). Screening different divalent and trivalent metals containing binary and ternary layered double hydroxides for optimum phosphate uptake. *Scientific Reports, 9*(1). https://doi.org/10.1038/s41598-019-52031-w
25. Kong, Y., Huang, Y., Meng, C., & Zhang, Z. (2018). Sodium dodecyl sulfate-layered double hydroxide and its use in the adsorption of 17β-estradiol in wastewater. *RSC Advances, 8*, 31440–31454. https://doi.org/10.1039/C8RA05726E

26. Bi, B., Xu, L., Xu, B., & Liu, X. (2011). Heteropoly blue-intercalated layered double hydroxides for cationic dye removal from aqueous media. *Applied Clay Science, 54*(3–4), 242–247. https://doi.org/10.1016/j.clay.2011.09.003
27. Miranda, L. D., Bellato, C. R., Fontes, M. P., Almeida, M. F. D., Milagres, J. L., & Minim, L. A. (2014). Preparation and evaluation of hydrotalcite-iron oxide magnetic organocomposite intercalated with surfactants for cationic methylene blue dye removal. *Chemical Engineering Journal, 254*, 88–97. https://doi.org/10.1016/j.cej.2014.05.094
28. Aguiar, J. E., Bezerra, B. T. C., Braga, B. D. M., Lima, P. D. D. S., Nogueira, R. E. F. Q., Lucena, S. M. P. D., & Silva, I. J. D. (2013). Adsorption of anionic and cationic dyes from aqueous solution on non-calcined Mg-Al layered double hydroxide: Experimental and theoretical study. *Separation Science and Technology, 48*(15), 2307–2316. https://doi.org/10.1080/01496395.2013.804837
29. Barrado, E. (2015). *Layered double hydroxides for application in catalysis and electroluminescent devices*. Doctoral dissertation, Universitat Rovira i Virgili, Spain. Retrieved from https://www.tdx.cat/handle/10803/285961#page=1
30. Tang, R., Dai, C., Li, C., Liu, W., Gao, S., & Wang, C. (2017). Removal of methylene blue from aqueous solution using agricultural residue walnut shell: equilibrium, kinetic, and thermodynamic studies. *Journal of Chemistry*. Retrieved from https://doi.org/10.1155/2017/8404965
31. Wang, H. (2019). *Novel hybrid regeneration process for adsorbent used in wastewater treatment*. Retrieved from https://www.watertank.se/projects/novel-hybrid-regeneration-process-for-adsorbent-used-in-wastewater-treatment/

Chapter 38
Decitabine as a Latency Perturbing Agent in Epstein-Barr Virus (EBV) Positive Natural Killer/T-Cell Lymphoma (NKTL)

Wang Liang Wei, Chin Jung Ee, Lee Jie Ying, and Li Miaomiao

Abstract Natural killer/T-cell lymphoma (NKTL) is an aggressive form of cancer that occurs at relatively high frequencies, and it is strongly associated with infection by Epstein-Barr virus (EBV). EBV is a ubiquitous virus that persists asymptomatically in healthy human hosts, often for life. Infected cells are able to achieve a state of latency by minimizing viral protein production and thus evading immune detection. Methylation of the viral promoters on the EBV genome, including C promoter (Cp) and LMP promoter (LMPp), downregulates viral protein transcription such as EBV-encoded nuclear antigens (EBNAs) 2, 3A, 3B, 3C, and latent membrane protein (LMP) 1 and 2. Conversely, demethylation activates viral promoters to increase EBV's immunogenicity, enabling immune clearance. Decitabine, a hypomethylating agent, is hypothesized to demethylate viral promoters and increase viral protein expression, thereby increasing immunogenicity. Immunoblot analysis showed that NK-YS cells treated with decitabine upregulated LMP1 and LMP2A expression, while EBNA3A expression remained undetectable, suggesting decitabine's more potent effect on LMPp than on Cp. Higher levels of LMP1 and LMP2A were also observed at post-decitabine treatment with lower initial seeding cell densities. Furthermore, we found that decitabine upregulated LMP2A to a greater extent than LMP1, implying the stabilizing effect of LMP2A on LMP1. Thus, decitabine may be useful in stimulating NKTL immunogenicity via LMP upregulation. Specific targeting of LMP2A may also synergize with existing immunotherapies to enhance efficacy of clinical treatments.

W. L. Wei · L. Miaomiao
Singapore Immunology Network (SIgN)—A*STAR, Singapore, Singapore
e-mail: wang_liang_wei@immunol.a-star.edu.sg

L. Miaomiao
e-mail: 162089y@student.hci.edu.sg

C. J. Ee · L. J. Ying (✉)
Hwa Chong Institution, Singapore, Singapore
e-mail: 162074w@student.hci.edu.sg

C. J. Ee
e-mail: 162385x@student.hci.edu.sg

H. Guo et al. (eds.), *IRC-SET 2021*, https://doi.org/10.1007/978-981-16-9869-9_38

Keywords Natural killer/T-cell lymphoma · Epstein-Barr virus (EBV) · Latency · Methylation · Decitabine · Cp · LMPp · EBNA

38.1 Introduction

38.1.1 Epstein-Barr Virus (EBV) in Natural Killer/T-Cell (NKTL)

NKTL is an aggressive form of cancer that occurs with relatively higher frequencies in parts of Asia and South America [1], and it is strongly associated with infection by EBV [2]. EBV is a ubiquitous virus that is usually able to persist asymptomatically for life in healthy human hosts [3]. Infected cells are usually able to achieve a state of latency and immune evasion by producing viral proteins such as EBV-encoded nuclear antigens (EBNAs) 2, 3A, 3B, and 3C, and latent membrane proteins (LMPs) 1 and 2 minimally. In the four known latency programs of EBV [4], the aforementioned proteins are fully expressed in latency III, which is hence the most immunogenic of the four, while partially expressed in the other latency programs which are hence less immunogenic [5]. Although the majority of EBV infections are "silent," it is still collectively responsible for over 200,000 new cancer cases worldwide each year, and NKTL is one of the EBV-related cancer types [6]. However, NKTL still has a poor prognosis today and has yet to have targeted therapy [7]. Insights into the pathways that maintain EBV latency in NKTL are critical in paving the way for more targeted clinical therapeutics and better disease prognosis.

38.1.2 Methylation and the Maintenance of Latency Programs in NKTL

While many studies investigated EBV latency in B cells, the mechanism of EBV latency in natural killer/T cells remains largely unknown. Key EBV proteins, LMP1 and LMP2A, are expressed in EBV-positive NKTL that provide the survival and proliferation signals to infected B cells [8].

DNA methylation has a role to play in the establishment and maintenance of restricted EBV latency programs, by decreasing the expression of LMPs. Incoming EBV viral DNA, initially linear and free of any epigenetic modifications [9], circularizes upon entering the host cell and is quickly methylated [10]. Upon initial infection, DNA methyltransferase (DNMT) 3A and 3B of the host cell form de novo DNA methylation patterns on latency gene promoters, the C, W, and Q promoters (Cp, Wp, and Qp, respectively) which are responsible for promoting the expression of EBNAs. In B cells, the LMP promoter is enhanced by EBNA2 through the latter's interactions with binding factors RBP-Jk and PU.1, although this is not strictly necessary [11]. EBV-encoded LMPs, in turn, seemingly increase the expression levels of all three cellular DNMTs, DNMT1, 3A, and 3B, causing hypermethylation [12]. Tellingly, cells without DNMT3B showed a reduction in DNA re-methylation after treatment with decitabine, an inhibitor of DNMT, thereby highlighting its function in forming new methyl marks on DNA [13]. DNMT1 has a maintenance role in DNA methylation; loss of DNMT1 in mouse embryonic cells leads to genome-wide demethylation [14].

38.1.3 Decitabine as a Possible Drug Candidate for NKTL

Decitabine (5-aza-2′-deoxycytidine) is a well-known irreversible inhibitor of DNA methyltransferases. Once inside a cell, decitabine is phosphorylated and activated by the enzyme deoxycytidine kinase to its triphosphate form, aza-dCTP [15]. Aza-dCTP inhibits methylation of the promoter by forming a covalent bond with DNMT1, thus rendering DNMT1 unable to copy the pattern of DNA methylation of the parental cell to the newly synthesized DNA strand during cell replication [16]. Furthermore, when decitabine is incorporated into DNA, ubiquitin-E3 ligase activity is induced which results in the proteasomal degradation of DNMT1 [10].

Thus, blocking methylation through the use of decitabine is hypothesized to result in an increased expression of LMP1 and LMP2A. With an increased expression of LMPs, immune detection of NKTL may occur more efficiently to result in remission.

In the present study, decitabine will be investigated as a potential latency perturbing agent in NKTL cells.

38.2 Hypothesis

Decitabine increases latency gene expression in EBV-positive NKTL cells.

38.3 Materials and Method

38.3.1 Cell Culture

NK-YS, a specific EBV-positive NKTL cell line, was chosen among others as it showed the lowest background expression of EBNAs. Any perturbation that induced the re-expression of EBNAs would thus be readily observed. These cells were treated with decitabine (5 μM) or DMSO. Cells were maintained at 37 °C and 5% CO_2 in HyClone™ RPMI-1640 (Roswell Park Memorial Institute) medium supplemented with 10% fetal bovine serum (FBS), penicillin/streptomycin and 100U/ml interleukin-2 (IL-2) for two days before media replacement and drug refreshment.

AMAI cells, which originated from an EBV-transformed B lymphoblastoid cell line (LCL), were grown in HyClone™ RPMI-1640 medium supplemented with 10% fetal bovine serum (FBS) and penicillin/streptomycin and at 37 °C with 5% CO_2. Cell viability was determined using INCYTO C-Chip DHC-N01 and trypan blue staining. For experiments, NK-YS cells were seeded at 100 k and 250 k cells per ml.

38.3.2 Cell Lysis

For the optimization phase, 6 sets of AMAI cell samples were lysed using BIO-RAD Laemmli buffer containing BME. Thereafter, each sample was separately heated for 15 min, 30 min, 45 min, or 60 min at 70 °C, or for 5 min or 10 min at 95 °C. After optimization, NK-YS cells were subject to lysis in BIO-RAD Laemmli buffer containing BME and heating for 45 min at 70 °C.

38.3.3 Immunoblotting

10 μl of protein ladder (BIO-RAD Precision Plus Protein™ Dual Color Standards, USA) and 25 μl of cell lysates from NK-YS treated with decitabine and DMSO were resolved by 10% SDS-PAGE stain-free gel (BIO-RAD Laboratories, USA) electrophoresis and transferred to a polyvinylidene difluoride (PVDF) immobilon membranes (Merck Millipore) in a BIO-RAD trans-blot SD semi-dry transfer cell. To prevent non-specific reactivity, the membrane was blocked in 0.05% Tris-buffered saline solution with 0.1% Tween (TBST) containing 2.5 g of skim milk powder for 1 h at room temperature with shaking. The membrane was then incubated with the appropriate primary antibody at 4 °C overnight, followed by incubation with horseradish peroxidase (HRP)-conjugated secondary antibody. A 1:1 mix of Super-Signal lumino/enhancer and hydrogen peroxide was added to the membranes to visualize the enhanced chemiluminescence (Thermo Fisher Scientific Inc., USA) as per the manufacturer's instruction. For LMP1, the primary antibody used was mouse anti-LMP1 (clone CS1-4; 1:1000 dilution; Abcam) along with anti-mouse IgG HRP-linked antibody (1:3000 dilution; cell signaling). For LMP2A, the primary antibody used was rat anti-LMP2A (clone 15F9; 1:1000 dilution; Abcam), along with anti-rat IgG HRP-linked antibody (1:3000 dilution; Santa Cruz). For EBNA3A, the primary antibody used was sheep anti-EBNA3A (1:1000 dilution; Abcam) along with rabbit polyclonal anti-sheep lgG HRP-linked antibody (1:5000 dilution; Abcam).

38.4 Results and Discussion

38.4.1 LMP2A and LMP1 Were Optimally Detected Following Heating and Lysis at 70 °C for 45 min and 15 min, Respectively

We first proceeded to optimize the immunoblotting procedure for LMP targets by testing a variety of temperatures and heating durations on the EBV-positive AMAI LCL. Levels of EBNA3A in AMAI cells remained the same throughout treatment with different temperatures and duration under the assumption that the decrease at 45 min and 60 min at 70 °C is insignificant. However, levels of LMP2A peaked when the AMAI cells were heated for 45 min at 70 °C; levels of LMP1 peaked when the AMAI cells were heated for 15 min at 70 °C. (Fig. 38.1).

Heating samples at different times and temperatures affect the extent of protein denaturation. This is because heat ensures that proteins such as LMP1, LMP2A, and EBNA3A are completely denatured into their linear forms from their 3-dimensional structures. Since the samples were mixed with a loading buffer which contains beta-mercaptoethanol (5%), these reducing agents break disulphide bonds in the protein, disrupting intra- and inter-molecular bonding [17]. This also allows the linearized protein to have a uniformly negative charge that is proportional to their molecular weight. However, in the case of LMP2A and LMP1, heating it at high temperatures causes aggregation as they are large molecules with hydrophobic domains [18]. Thus, duration and temperature at which cells are lysed were varied to ensure that the detection of the same amount of proteins is at its optimum.

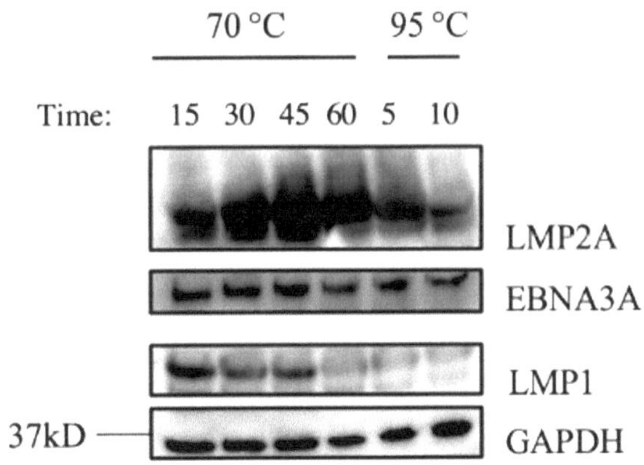

Fig. 38.1 Optimization of Western blotting for LMP1, LMP2A, EBNA3A on EBV-positive AMAI LCL with GAPDH as positive control

38.4.2 Decitabine Treatment Results in Enhanced LMP Expression with LMP2A being Increased to a Greater Extent than LMP1

The viral latent membrane proteins, LMP1 and LMP2A, are mimics of host cellular signaling proteins and can therefore usurp the cell signaling pathway responsible for B-cell activation and survival. Both of them are transcribed from the same promoter LMPp and are coexpressed in latency II and III in NKTL [19]. Hence, it is interesting to observe from our result that the demethylation effect of decitabine on LMPp was seemingly more significant for LMP2A than LMP1 despite them being transcribed from the same promoter LMPp (Fig. 38.2).

One possible reason that LMP2A was detected at higher levels than LMP1 could be the stabilizing effect of LMP2A on LMP1. Recent studies suggest that LMP2A could be stabilizing the LMP1 protein by influencing its degradation and turnover either by sequestering the ligases which target LMP1 or by enhancing the degradation of factors which negatively regulate the degradation of LMP1. This is evident as the half-life of LMP1 almost doubles in the presence of LMP2A [20].

Another possible reason that LMP2A was detected at higher levels than LMP1 could be related to efficiency of antibodies. Since immunoblotting of LMP1 and

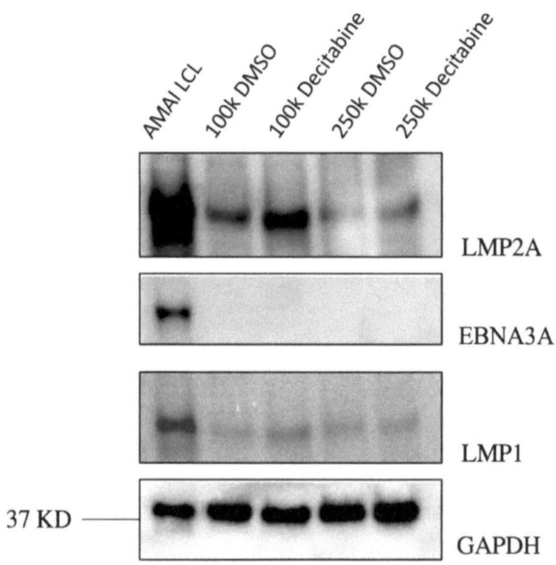

Fig. 38.2 Western blot analysis on effects of decitabine on LMP2A, EBNA3A, and LMP1 expressions on different cell densities of NK-YS cells. EBV-positive AMAI LCL acts as a reference to show that EBNA3A, LMP1, and LMP2A are indeed expressed by the EBV-positive NK-YS cells of latency III when LMPp and Cp are successfully activated. (The AMAI LCL lane is not meant to show the changes in viral protein expression after the drug treatment since EBV-positive AMAI LCL cells are of a different latency program from that of NK-YS cells.)

LMP2A requires different antibodies, where LMP1 required mouse anti-LMP1 antibody and LMP2A required rat anti-LMP2A antibody, there might be a difference in the efficiency of detection by antibodies. To clarify this, construction of standard curves with recombinant LMPs may facilitate more accurate quantitation and importantly cross-comparison of LMP1 and LMP2A levels within the same sample.

38.4.3 Decitabine Treatment Upregulates LMP1 and LMP2A Expression but not EBNA3A

While LMP1 and LMP2A expressions were upregulated by decitabine, the level of EBNA3A expression remained undetectable (Fig. 38.2). The transcription of EBNA3A is controlled by Cp of the EBV genome while the transcription of the LMPs is controlled by the LMPp of the EBV genome. Hence, we postulate that decitabine treatment has a more pronounced demethylating effect on LMPp than Cp.

Cp activation is not prominent in NK-YS cells. The previous studies showed that treating Burkitt lymphoma (BL) cells with azacitidine, which is a hypomethylating agent and chemical analog for cytidine similar to decitabine, leads to demethylation of the viral genome and subsequent activation of EBNA gene transcription from Cp and Wp [21]. The difference in hypomethylation effect may be due to the difference in the mechanisms of the two hypomethylating agents. Upon entry into the cells, azacitidine and decitabine both exist in the form of their respective phosphates. Decitabine triphosphate, the phosphate of decitabine, is incorporated into DNA only, while the majority of azacitidine is incorporated into RNA and only 10–20% of it is converted to 5-aza-2′-deoxycytidine triphosphate can be used by DNA. After the triphosphates are incorporated into DNA, adducts can be formed between DNA and DNMT-1 which enables the hypomethylation effect of the two drugs [22]. Hence, we would expect the efficiency of decitabine to be higher than that of azacitidine of the same concentration given that the majority of azacitidine is incorporated into RNA instead of DNA. Surprisingly, the concentration of azacitidine used in the previous study was 4 μmol per ml [23] which was lower than 5 μmol per ml in our study. Hence, the less efficient azacitidine of a lower concentration successfully demethylated Cp in BL cell lines, whereas the more efficient and the more concentrated decitabine treatment failed to do so. We postulate that this difference may be due to the different genome structures of the different cell types.

Another interesting finding is that LMPp activation occurs to a larger extent than Cp activation. (Fig. 38.2) The length of Cp is around 1400 bp [24] while the length of LMPp is 400–600 bp [25]. The more compact LMPp is possibly more sensitive to decitabine and activates more easily even though the two promoters have similar numbers of CpG sites [26, 27]. Another possibility is that the context sequences in the two promoters may also require different sets of transcription factors, and the ones that bind at Cp may be more restrictive than the ones that bind at LMPp.

These explanations may account for the more pronounced demethylation effect of decitabine on LMPp than Cp.

38.4.4 Seeding Cells at Lower Densities Led to Higher Levels of LMP2A and LMP1

We observed a higher level of LMP2A and LMP1 on NK-YS cells seeded at a density of 100 k/ml compared to that of a density of 250 k/ml (Fig. 38.2).

One possible explanation could be about cell proliferation with respect to different cell density and the mechanism of decitabine.

Cell proliferation is important to this experiment as decitabine can only exert its effect during the S phase of the cell division cycle. This is where decitabine inhibits methylation of the promoter by binding covalently to DNMT1's active site, as well as the nitrogen atom from the modified thymine of the newly formed DNA during DNA replication [28]. This results in a reduction of methylation in DNA of the daughter cells, while the pattern of DNA methylation remains unchanged after the DNA replication.

Since relatively denser cells undergo cell death due to lactic acid accumulation, proliferation of cells at higher cell density is limited, thus explaining why decitabine is thought to have a reduced net effect on the newly formed cells, resulting in less proteins such as LMP1 and LMP2A being expressed. This is evidenced by our results, where the band intensity for LMP1 and LMP2A of 100 k/ml cells is significantly greater than that of 250 k/ml cells.

38.5 Conclusion

In the present study, decitabine significantly increased latency gene expression in EBV-positive NK-YS cells, in particular, LMP1 and LMP2A. We further found that decitabine increased LMP2A expression to a greater extent than that of LMP1.

The previous studies have focused on the cancer arising from EBV-positive B cells, and the mechanism of EBV latency in B cells, the germinal center model, has been well studied and constantly reviewed and refurbished. However, despite NKTL being an aggressive form of cancer common in parts of Asia and South America, the mechanism of EBV latency in NK/T cells is still largely unknown. As a result, NKTL still has poor prognosis today and has yet to have targeted therapy for the patients. Therefore, insights into decitabine as a latency perturbing agent in Epstein-Barr virus (EBV) positive natural killer/T cell lymphoma (NKTL) are critical in paving the way for more targeted clinical therapeutics and better disease prognosis.

As the previous studies have almost exclusively focused on LMP1, this research paper provides insights regarding the optimization of temperature and time of

LMP2A through analysis of the immunoblot. This is of significance as LMP2A is a critical protein that promotes the malignant transformation, especially by affecting the apoptosis of cells and the cell cycle, thereby playing an important role in EBV-associated malignancies. Being able to upregulate LMP2A with decitabine, a clinically approved drug, may provide an additional target for activated T cells in the development of immunotherapy and LMP2A vaccines [29].

In order to ensure that the upregulation of LMP1 and LMP2A protein expression was due to decitabine's on-target effects in inhibiting DNMTs, and not due to unaccounted off-target effects of decitabine, future work will be focused on determining if decitabine indeed reduces methylation of the promoters, or even demethylates it. To do so, restriction enzymes will be used to digest the total cellular DNA into smaller fragments. Fragments will be immunoprecipitated with 5-methylcytosine-specific antibodies and tested for fragment abundance via qualitative PCR (qPCR) with LMP-specific primers. We expect that decitabine will reduce LMPp methylation, resulting in weaker qPCR signals in treated samples. Such analysis will be coupled with bisulfite sequencing, which would allow us to ascertain methylation sites. These, along with our findings, will help verify the mechanism of decitabine-mediated upregulation of viral protein expression in NKTL.

Acknowledgements We thank all staff members and mentors from Singapore Immunology Network (SIgN), A*STAR, Singapore, for their mentorship and guidance.

Author Contributions This paper was drafted and revised by Dr. Wang Liang Wei, C. J. E, L. J. Y and L. M. M. All authors have read and agreed to the published version of the manuscript.

References

1. Jin, X., Xu, Y., Zhang, J., Li, G., Huang, D., Yang, Y., & He, H. (2017, June 17). *Aggressive natural killer cell leukemia or extranodal NK/T cell lymphoma? a case with nasal involvement.* Retrieved January 20, 2021, from https://www.ncbi.nlm.nih.gov/pmc/articles/PMC5474018/
2. White, M., Pagano, J., & Khalili, K. (2014, July). *Viruses and human cancers: A long road of discovery of molecular paradigms.* Retrieved January 20, 2021, from https://www.ncbi.nlm.nih.gov/pmc/articles/PMC4135891/
3. Henle, W., Hummeler, K., & Henle, G. (1966, July). *Antibody coating and agglutination of virus particles separated from the EB3 line of Burkitt lymphoma cells.* Retrieved January 20, 2021, from https://www.ncbi.nlm.nih.gov/pmc/articles/PMC276225/
4. Kempkes, B., & Robertson, E. S. (2018). *Epstein-barr virus latency: Current and future perspectives.* Retrieved January 22, 2021, from https://www.ncbi.nlm.nih.gov/pmc/articles/PMC5868753/
5. Price, A. M., Luftig, M. A. (2015). To be or not IIb: A multi-step process for Epstein-Barr virus latency establishment and consequences for B cell tumorigenesis. *PLOS Pathogens*. Accessed July 16, 2021 from https://journals.plos.org/plospathogens/article?id=10.1371%2Fjournal.ppat.1004656.
6. Shannon-Lowe, C., & Rickinson, A. (2019, July 18). *The global landscape of EBV-associated tumors.* Retrieved January 20, 2021, from https://www.frontiersin.org/articles/https://doi.org/10.3389/fonc.2019.00713/full#h4

7. Somasundaram, N., Lim, J., Ong, C., & Lim, S. (2019). Pathogenesis and biomarkers of natural killer T cell lymphoma (NKTL). *Journal of Hematology & Oncology, 12*(1). https://doi.org/10.1186/s13045-019-0717-6
8. Thorley-Lawson, D. A. (2001). Epstein-Barr virus: Exploiting the immune system. *Nature reviews. Immunology, 1*(1), 75–82. https://doi.org/10.1038/35095584
9. Johannsen, E., Luftig, M., Chase, M.R., Weicksel, S., Cahir-McFarland, E., Illanes, D., Sarracino, D., Kieff, E. (2004, November 16). *Proteins of purified Epstein-Barr virus.* Retrieved January 27, 2021, from https://www.pnas.org/content/101/46/16286
10. Patel, K., Dickson, J., Din, S., Macleod, K., Jodrell, D., & Ramsahoye, B. (2010, July). *Targeting of 5-aza-2'-deoxycytidine residues by chromatin-associated DNMT1 induces proteasomal degradation of the free enzyme.* Retrieved January 21, 2021, from https://www.ncbi.nlm.nih.gov/pmc/articles/PMC2910061/
11. Cancian, L., Bosshard, R., Lucchesi, W., Karstegl, C. E., & Farrell, P. J. (2011). C-terminal region of EBNA-2 determines the superior transforming ability of type 1 Epstein-Barr virus by enhanced gene regulation of LMP-1 and CXCR7. *PLoS Pathogens.* Accessed July 16, 2021 from https://www.ncbi.nlm.nih.gov/pmc/articles/PMC3145799/.
12. Tao, Q., & Robertson, K. D. (2003). *Stealth technology: How Epstein–Barr virus utilizes DNA methylation to cloak itself from immune detection.* Retrieved January 22, 2021, from https://www.sciencedirect.com/science/article/abs/pii/S1521661603001980?via%3Dihub
13. Duymich, C. E., Charlet, J., Yang, X., Jones, P. A., Liang, G. (2021). *DNMT3B isoforms without catalytic activity stimulate gene body methylation as accessory proteins in somatic cells.* https://www.ncbi.nlm.nih.gov/pmc/articles/PMC4853477/. Published 2016. Accessed July 16, 2021.
14. Li, E., & Zhang, Y. (2014, May). *DNA methylation in mammals.* Retrieved January 22, 2021, from https://www.ncbi.nlm.nih.gov/pmc/articles/PMC3996472/
15. Malik, P., & Cashen, A. F. (2014, February). *Decitabine in the treatment of acute myeloid leukemia in elderly patients.* Retrieved January 22, 2021, from https://www.ncbi.nlm.nih.gov/pmc/articles/PMC3917925/
16. Sorrentino, V. G., Thota, S., Gonzalez, E. A., Rameshwar, P., Chang, V. T., Etchegaray, J.-P. (2021). Hypomethylating chemotherapeutic agents as therapy for myelodysplastic syndromes and prevention of acute myeloid Leukemia. *MDPI.* Accessed July 16, 2021 from https://www.mdpi.com/1424-8247/14/7/641/htm.
17. Raynal, N. J., & Issa, J. J. (2016). *DNA methyltransferase inhibitors.* Retrieved January 22, 2021, from https://www.sciencedirect.com/science/article/pii/B9780128022085000072
18. Tomaszewski-Flick, M. J., Rowe, D. T. (2007). Minimal protein domain requirements for the intracellular localization and self-aggregation of Epstein-Barr Virus Latent Membrane Protein 2. *Virus Genes.* Retrieved July 15, 2021 from https://link.springer.com/article/https://doi.org/10.1007/s11262-007-0118-8
19. Price, A. M., & Luftig, M. A. (2014). Dynamic Epstein-Barr virus gene expression on the path to B-cell transformation. *Advances in Virus Research, 88*, 279–313. https://doi.org/10.1016/B978-0-12-800098-4.00006-4
20. Dawson, C. W., George, J. H., Blake, S. M., Longnecker, R., & Young, L. S. (2001). The Epstein-Barr virus encoded latent membrane protein 2A augments signaling from latent membrane protein 1. *Virology, 289*(2), 192–207. https://doi.org/10.1006/viro.2001.1142
21. Altiok, E., Minarovits, J., Hu, L. F., Contreras-Brodin, B., Klein, G., & Ernberg, I. (1992). Host-cell-phenotype-dependent control of the BCR2/BWR1 promoter complex regulates the expression of Epstein-Barr virus nuclear antigens 2–6. *Proceedings of the National Academy of Sciences of the United States of America, 89*(3), 905–909. https://doi.org/10.1073/pnas.89.3.905
22. Derissen, E. J., Beijnen, J. H., & Schellens, J. H. (2013). Concise drug review: Azacitidine and decitabine. *The Oncologist, 18*(5), 619–624. https://doi.org/10.1634/theoncologist.2012-0465
23. Masucci, M. G., Contreras-Salazar, B., Ragnar, E., Falk, K., Minarovits, J., Ernberg, I., & Klein, G. (1989). 5-Azacytidine up regulates the expression of Epstein-Barr virus nuclear antigen 2 (EBNA-2) through EBNA-6 and latent membrane protein in the Burkitt's lymphoma line rael. *Journal of Virology, 63*(7), 3135–3141. https://doi.org/10.1128/JVI.63.7.3135-3141.1989

24. Evans, T. J., Farrell, P. J., & Swaminathan, S. (1996). Molecular genetic analysis of Epstein-Barr virus Cp promoter function. *Journal of Virology, 70*(3), 1695–1705. https://doi.org/10.1128/JVI.70.3.1695-1705.1996
25. Demetriades, C., & Mosialos, G. (2009). The LMP1 promoter can be transactivated directly by NF-kappaB. *Journal of Virology, 83*(10), 5269–5277. https://doi.org/10.1128/JVI.00097-09
26. Falk, K. I., Szekely, L., Aleman, A., & Ernberg, I. (1998). Specific methylation patterns in two control regions of Epstein-Barr virus latency: The LMP-1-coding upstream regulatory region and an origin of DNA replication (oriP). *Journal of Virology, 72*(4), 2969–2974. https://doi.org/10.1128/JVI.72.4.2969-2974.1998
27. Paulson, E. J., & Speck, S. H. (1999). Differential methylation of Epstein-Barr virus latency promoters facilitates viral persistence in healthy seropositive individuals. *Journal of Virology, 73*(12), 9959–9968. https://doi.org/10.1128/JVI.73.12.9959-9968.1999
28. Raynal, N., & Issa, J. (2016). DNA methyltransferase inhibitors. In *Drug discovery in cancer epigenetics* (pp. 169–190). https://doi.org/10.1016/b978-0-12-802208-5.00007-2
29. Zhu, S., Xue, X., Liu, J., Lu, L., Zhao, P., Wang, J., Li, W., & Zhang, L. (2010, July 5). *Expression of HPV6b L1/EBV LMP2 multiepitope and immunogenicity in mice.* Retrieved January 21, 2021, from https://academic.oup.com/abbs/article/42/8/515/824

Chapter 39
AI Panorama Art of Proteins with Image Blending Deep Learning

Adele Lim Yi Dawn and Chen Yu Zong

Abstract Interdisciplinarily weaving science into art, this project aims to highlight both the esthetic beauty of protein structures, as well as the essential and extensive roles they play in organisms, hence promoting a greater appreciation of science, making it more accessible to the larger community. The creative and unique merging of protein structures with panoramic backgrounds can arouse interest in the general public, accessibly promoting science through art. Visualization and rendering of protein structures three-dimensionally were done using software, like BIOVIA Discovery Studios. Then, Adobe Photoshop CS5.1 was used to create and edit a mask, destination and source image for input into a Gaussian–Poisson generative adversarial network (GP-GAN), which has been trained to carry out high-resolution image blending. The use of GP-GAN image blending helps to produce realistic images, with fewer bleedings and unpleasant artifacts.

39.1 Introduction

Background

Proteins are the most abundant biological macromolecules, occurring in all cells and all parts of cells. Essential to life, proteins perform diverse functions enabling living species to explore, adapt, and conquer the vast land, ocean, and outer space. These functions can only be made possible by the great variety of specific 3D protein conformation. Through creative thinking and image blending software, the intrinsic relationships between protein structures and life's exploration capabilities can be presented through novel and beautiful panorama art—showcasing the integration of protein structures into the natural wonders of the land, ocean, and outer space.

A. Lim Yi Dawn (✉)
Hwa Chong Institution, Singapore, Singapore
e-mail: alyd0915@gmail.com

C. Y. Zong
Department of Computational Sciences, National University of Singapore, Singapore, Singapore

H. Guo et al. (eds.), *IRC-SET 2021*, https://doi.org/10.1007/978-981-16-9869-9_39

Panorama Art

Panorama art is characterized by the continuous narrative scene or landscape, revealing a wide, all-encompassing view of a particular subject. Since the mid-nineteenth century, it has been popularized as a way to represent landscapes, topographic views, or historical events [1]. Enabling viewers to have an elongated field of view of the environment depicted from a particular height, angle, or distance, panorama art provides viewers a carefully controlled deep sense of immersion.

Image Blending

High-resolution image blending aims to create a smooth transition and seamlessly blend an object from a source image onto a target image with lightly mask adjustment [2]. This can be achieved by reducing the color or illumination differences between foreground and background, as well as improving spatial and color consistencies between source and target images. Hence, not only generating natural images but also capturing high-frequency image details like textures and edges, without the presence of unpleasant artifacts [3].

GP-GAN: Gaussian–Poisson Generative Adversarial Network

Generative adversarial networks (GANs) aim to form a zero-sum game between the discriminator and generator. The generator aims to produce more realistic images from random and irrelevant noises, while the discriminator aims to identify generated images from real ones [3]. GP-GAN seeks to minimize loss function to generate realistic and well blended images. This is done by having a color constraint and gradient constraint. The color constraint helps to make the generated blended image more realistic and natural, while the gradient constraint helps to crisply showcase details like textures and edges [3].

Supervised training

The algorithm learns on a labeled dataset, which contains input paired with correct outputs. These will be used by the algorithm to evaluate its accuracy on training data. When data are newly input, the supervised learning model aims to predict the correct label [4].

Unsupervised training

Unlabeled labeled data will be presented to the unsupervised learning model. The algorithm tries to make sense of the data by extracting features and patterns on its own [5].

Key Objectives

This research project aims to create novel and beautiful panoramic artworks incorporating protein structures, to raise awareness about the beauty of science.

39.2 Materials and Methods

39.2.1 Materials and Software Used

BIOVIA Discovery Studio:

A modeling and simulation software that allows for the clear visualization of various protein structures, which will be incorporated into the final panoramic art. This facilitates the modification of protein—in terms of color, size, display style (ball and stick, line, flat, solid ribbons, etc.), lighting among many more features.

Adobe Photoshop CS5.5:

An imaging and graphing software that allows for the synthesis of layers and masks, overlays, as well as trimming of protein, which are essential in rendering the source, destination and mask images, needed for input into the GP-GAN.

Google Colab:

A notebook which allows people to write and execute arbitrary Python code through a browser and is highly suitable for machine learning, given that programs can be run using an inbuilt GPU.

Protein Database Bank (PDB) [6]:

PDB is a database for the three-dimensional structural data of large biological molecules, such as proteins. All protein structures used have been identified from this database bank, according to their PDB ID.

39.2.2 General Methodology

1. Conduct molecular visualization using BIOVIA Discovery Studio, to view and study the proteins' molecular 3D structure, conformation, and interactions.
2. Select a panoramic image to include in the final product.
3. After conceptualizing the envisioned panorama art with protein, modify and customize the protein structure's display style using the various software accordingly.
4. Using Adobe Photoshop CS5.5, edit the source (desired protein image), destination (panoramic image), and mask images.
5. Download and clone the source code of GP-GAN [3] into Google Colab. Ensure that the runtime type is set to "GPU."
6. Import mask, source, and destination images into the code and run it. This will generate realistic and novel artworks of panoramas incorporating structures of proteins, through image blending.

39.2.3 Procedure for Creating Mask Image Within Adobe Photoshop

1. Place the images of panoramic backgrounds and protein structures you want to combine into the same document.
2. Trim the protein structure out from its image.
3. Align and superimpose the trimmed protein structure against the panorama background at the desired position.
4. Duplicate the trimmed protein structure layer.
5. Create a new layer beneath the previous.
6. Fill the layer with black color.
7. Merge the layers in steps 5 and 6 together.
8. Using the magic wand and quick selection tools, select the trimmed protein structure.
9. Delete the protein structure from the selection.

Resultantly, there will be a mask image where the entire layer is black with exception of the trimmed protein structure.

39.3 Results and Discussion

39.3.1 Fossils of a Giant Protein on a Seaside Hill

Inspired by the report: Molecules are fossils [7], this image portrays fossils of a giant protein on a seaside hill. The image is of a natural monument—the elephant rock in Hongdo, South Korea, which is named this way simply because it resembles an elephant. The protein incorporated into the panorama is the crystal structure of Asian elephant (elephas maximus) cyano-metmyoglobin (PDB 1EMY) at 1.78-A resolution. Unlike usual myoglobins, this crystal structure has glutamine instead of usual distal site histidine [8]. Myoglobin (Mb) is a small, mostly α-helical, globular hemeprotein, which is abundant in the heart and present in skeletal muscles [9]. Commonly said to act as an oxygen storage, Mb is found to facilitate oxygen unloading from hemoglobin and oxygen diffusion to the mitochondrial surface. It also brings about oxidative phosphorylation [10] (Fig. 39.1 and 39.2).

39.3.2 A Universe of Giant Protein-Shaped Galaxy Cluster

From ancient times, people associate galaxy clusters to various animals as seen through horoscopes, zodiac signs among others. In this creative universe, a protein-like galaxy cluster may well exist in some remote corner of the vast universe. *A*

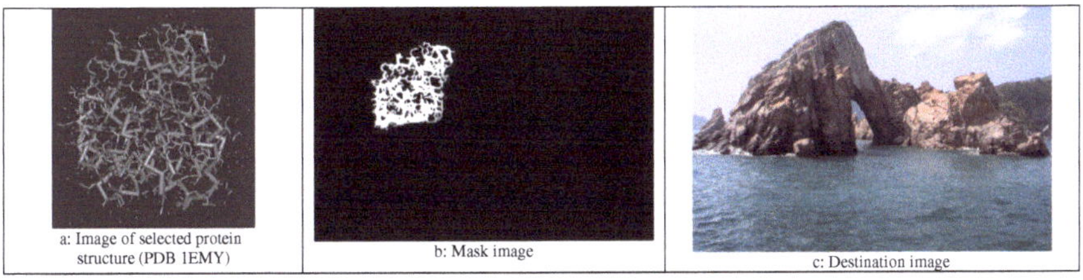

Fig. 39.1 **a** Image of selected protein structure (PDB 1EMY), **b** Mask image, **c** Destination image

universe of giant protein-shaped galaxy clusters showcases proteins blended as constellations in the night sky of a forest with an aurora visible. In this panorama image, two different protein structures relating to photosystems and photosynthesis have been incorporated: the crystal structure of photosystem II (PDB 1FE1) and the excited state dynamics in photosystem II (PDB 1ILX). Photosystem II is where several hundred chlorophyll molecules and accessory pigments—carotene and xanthophyll—harvest light energy that is transferred to the reaction center [11] Water is oxidized, providing electrons to the photosynthesis process and generating oxygen into the atmosphere. This happens at photosystem II, one of the sites where light-dependent reactions occur (Figs. 39.3 and 39.4).

By showcasing these protein structures as constellations illuminating the beautiful night sky, it serves to highlight the important role they play in the photosynthesis process.

39.3.3 Protein Fish Roams Through the Ocean

Lastly, *protein fish roams through the ocean* showcases the underwater scene of zebrafishes. Zebrafishes are commonly used as animal models for research, to study the function of genes, human diseases and to develop new drugs. This can be attributed to their extremely fast breeding rate, with up to 200 embryos every week, coupled with the fact that their embryos and larvae are small, transparent and rapidly develop externally. Furthermore, 70% of human genes is found in them [12] (Figs. 39.5 and 39.6).

Hence, to inspire people to be more appreciative of these model organisms in advancing biomedical research, a "zebrafish" made of zebrafish-related proteins was created. The idea comes directly from the remarkable resemblance of the protein structure to fish.

Proteins used include:

1. Zebrafish alphaA crystalline (PDB 3N3E), which protects vertebrate eye lens proteins from detrimental protein aggregation [13]
2. Zebrafish interferon II (PDB 3PIW), which are signaling proteins made and released by cells in response to the presence of virus [14].

a: Generated blended image (supervised training)

b: Generated blended image (unsupervised training)

Fig. 39.2 **a** Generated blended image (supervised training), **b** Generated blended image (unsupervised training)

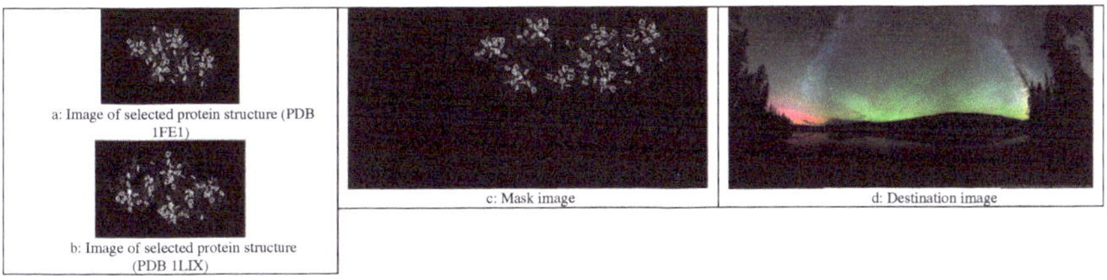

Fig. 39.3 **a** Image of selected protein structure (PDB 1FE1), **b** Image of selected protein structure (PDB 1LIX), **c** Mask image, **d** Destination image

Fig. 39.4 **a** Generated blended image (supervised training), **b** Generated blended image (unsupervised training)

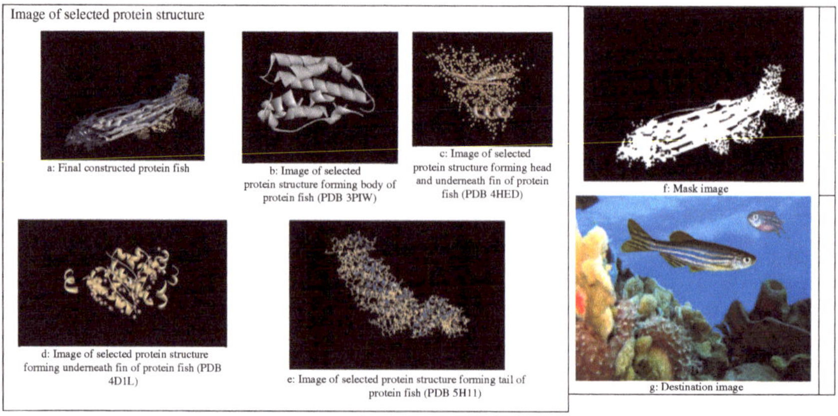

Fig. 39.5 **a** Final constructed protein fish, **b** Image of selected protein structure forming body of protein fish (PDB 3PIW), **c** Image of selected protein structure forming head and underneath fin of protein fish (PDB 4HED), **d** Image of selected protein structure forming underneath fin of protein fish (PDB 4D1L), **e** Image of selected protein structure forming tail of protein fish (PDB 5H11), **f** Mask image, **g** Destination image

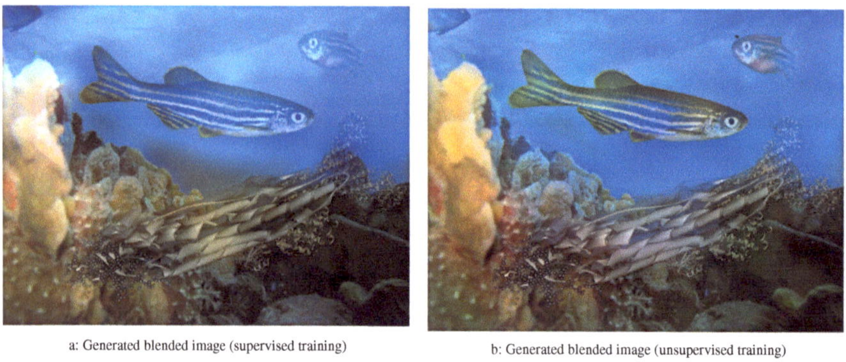

Fig. 39.6 **a** Generated blended image (supervised training), **b** Generated blended image (unsupervised training)

3. Zebrafish chemokine CXL1 (PDB 4HED), which are signaling proteins essential for cell migration and inducing chemotaxis [15].
4. Crystal structure of zebrafish sec10 (PDB 5H11), which is part of the exocyst complex. The exocyst complex mediates vesicle trafficking, including the secretion of biologically active molecules to the extracellular environment and transport of lipids and membrane proteins to the plasma membrane in eukaryotic cells.

39.4 Conclusion

Creatively, there are endless possibilities in terms of the artworks that one could possibly create using image blending with GP-GAN. GP-GAN has allowed for the seamless blending of protein structures into panoramic backgrounds, proving its ability to create natural and realistic blended images with little to no visual artifacts. The various art products presented in this report highlight the diverse range and function of proteins, showing their value in enabling organisms to function properly and normally.

Acknowledgements I would like to thank my mentor, Professor Chen Yu Zong, for both providing me with this valuable opportunity to engage in research under him, and the guidance and support he has given me throughout this project. I would also like to thank my teacher-in-charge, Dr Erkan Polatdemir, who has tirelessly helped with administrative duties and provided me with support. Lastly, I would also like to thank my friends and family for their encouragement.

References

1. Selin, S. (2019). *Panoramas: 19th-century virtual reality*, March 28, 2019. shannonselin.com/2016/11/panoramas-19th-century/
2. Zhang, L., Wen, T., & Shi, J. (2020). Deep image blending. In: *2020 IEEE winter conference on applications of computer vision (WACV)*. https://doi.org/10.1109/wacv45572.2020.9093632
3. Wu, H., Zheng, S., Zhang, J., & Huang, K. (2019). GP-GAN: towards realistic high-resolution image blending. *ACMMM*. https://doi.org/10.1145/3343031.3350944
4. Wilson, A. (2019). A brief introduction to supervised learning. *Medium*, Towards Data Science, Oct 1, 2019. medium.com/m/global-identity?redirectUrl=https%3A%2F%2Ftowardsdatascience.com%2Fa-brief-introduction-to-supervised-learning-54a3e3932590
5. Mishra, S. (2017). Unsupervised learning and data clustering. *Medium*, Towards Data Science, May 21, 2017. towardsdatascience.com/unsupervised-learning-and-data-clustering-eeecb78b422a
6. Berman, H. M., Westbrook, J., Feng, Z., Gilliland, G., Bhat, T. N., Weissig, H., Shindyalov, I. N., Bourne, P. E. (2000). The protein data bank. *Nucleic Acids Research*, *28*, 235–242
7. Goforth, S. *Molecules are fossils, too*. NSF, www.nsf.gov/discoveries/disc_summ.jsp?cntn_id=106951&org=NSF
8. Bisig, D. A., et al. (1995). Crystal structure of asian elephant (elephas maximus) cyano-metmyoglobin at 1.78-Å resolution: Phe29(B10) accounts for its unusual ligand binding properties. *Journal of Biological Chemistry*, Edited by Kay Diederichs, *270*(35), 20754–20762
9. Millikan, G. A. (1939). *Physiology Review, 19*, 503.
10. Wittenberg, J.B., Wittenberg, B. A. (1990). *Annual Review Biophysics Chemistry, 19*, 217–241.
11. Zouni, W., et al. (2001). CrystaStructure of photosystem II from synechococcus elongatus at 3.8 A resolution. *Nature, 409*, 739–743.
12. Tavares, B., & Lopes, S. S. (2013). The importance of zebrafish in biomedical research. *Acta Medica Portuguesa, 26*(5).
13. Peschek, J., et al. (2009). The eye lens chaperone α-Crystallin forms defined globular assemblies. *PNAS*.

14. Hamming, O. J., et al. (2011). Crystal structure of zebrafish interferons I and II reveals conservation of type I interferon structure in vertebrates. *National Center for Biotechnology Information*.https://doi.org/10.1128/JVI.00521-11
15. Bussmann, J., & Raz, E. (2015). Chemokine-guided cell migration and motility in Zebrafish development. *The Embo Journal*. https://doi.org/10.15252/embj.201490105

Chapter 40
A Study in Factors Affecting the Production and Properties of Biocellulose Produced by *Acetobacter Xylinum*

Benrard Kwee Tze Wei, Yihan Li, and Yiting Quak

Abstract Cellulose extracted from plants employ environmentally unfriendly methods which involve the use of corrosive chemicals. Bacterial cellulose, however, is purer, more cost-effective and environmentally friendly. This study aims to synthesise biocellulose with the bacteria *Acetobacter xylinum*, from carbon sources like glucose and fructose, as well as sugarcane bagasse and spent coffee bean extract which are common food wastes rich in carbohydrates. Varying ethanol concentration and presence of metal ion supplements in the culture medium were investigated. The tensile strength and drug release properties of the biocellulose synthesised were evaluated. Biocellulose synthesised from glucose-fructose mixture gave the highest yield, followed by sugarcane bagasse which gave a higher yield than spent coffee bean. Metal ion supplements generally increased the tensile strength of the cellulose. Notably, Mg^{2+} ions increased biocellulose yield and also resulted in the biocellulose having a tensile strength 2.4 times that of the control setup. Ethanol concentration of 0.8% (v/v) was found to be optimal for biocellulose production. Sugarcane biocellulose was the strongest in tensile strength, showing great promise should it be further processed for paper products. Lastly, the rate of ibuprofen release from the synthesised biocellulose was found to be generally steady over 3 h, with the most promising averaging around 2 h for 90% of the drug to be released, demonstrating great potential to be used to conduct further studies on drug release like skin permeation. The controlled release ensures a steady intake of the drug into the bloodstream without much risk of cell death.

Keywords Bacterial cellulose · Biocellulose · Drug release · Acetobacter xylinum

B. K. T. Wei (✉) · Y. Li · Y. Quak
Hwa Chong Institution, 661 Bukit Timah Road, Singapore 269734, Singapore
e-mail: 161679x@student.hci.edu.sg

H. Guo et al. (eds.), *IRC-SET 2021*, https://doi.org/10.1007/978-981-16-9869-9_40

40.1 Introduction

Cellulose is a biological polymer commonly found in plants. It consists of a few hundred to a thousand (1–4) D-glucose monomers. Due to its high tensile strength, it has many industrial uses and is a crucial ingredient in several industries. It can be used to produce paper, clothing, and even food items. Traditionally, cellulose is extracted from plants. To obtain pure cellulose, plant-derived cellulose is purified using various corrosive chemicals like sodium hydroxide. These processes generate a large amount of liquid waste with high biological oxygen demand (BOD) and chemical oxygen demand (COD) that causes severe environmental pollution [1]. Nowadays, a more cost-effective and environmentally friendly way of production is by using bacteria to culture biocellulose. Several types of bacteria, including *Acetobacter Xylinum* (*A. xylinum*), have the ability to polymerise short chain carbohydrates into cellulose pellicles composed of 97% water. *A. xylinum* can also synthesise biocellulose from other carbon sources, such as coconut water and konjac powder. Additional growth factors such as adding metal ions and organic compounds such as caffeine to boost the production of biocellulose have also been investigated. Compared to plant cellulose, biocellulose has higher tensile strength of Young's modulus of 138 GPa and tensile strength of at least 2 GPa, which are almost equal to those of aramid fibre [2] owing to its larger microfibril diameter (over 50 nm), and more extensive cross-links between fibrils [3]. When arranged in the form of a sheet, biocellulose has high crystallinity and is composed of ultrafine fibres. Moreover, unlike plant cellulose, it has no lignin impurities. Its higher purity allows it to better fulfil demand in industrial processes [4] and healthcare purposes. Its thin, porous structure has earmarked it as a potential drug delivery gel. For example, biocellulose, when applied to wounds, can provide controlled release of medicinal drugs into the bloodstream over a period of time whilst protecting the wound from external infection [5]. This is due to its highly hydrophilic nature, excellent biocompatibility and ability to control drug loading and release kinetics [6].

Hypotheses

Biocellulose synthesised from a mixture of glucose and fructose produced the highest yield, followed by sugarcane bagasse which gave a higher yield than spent coffee bean; the addition of metal ion supplements and a higher ethanol concentration resulted in a higher biocellulose yield; the tensile strength of the biocellulose produced was strongest with sugarcane bagasse biocellulose. The release of the drug, ibuprofen, from the synthesised biocellulose into a pH 7.4 buffer solution was steady. Within the duration of 3 h, almost 100% of the absorbed drug was released.

40.2 Materials and Methods

40.2.1 Preparation of Food Waste Extract

For sugarcane bagasse and coffee bean extracts, 250 g of food waste was boiled with 500 cm^3 of deionised water for 2 h. The resultant extract was filtered. The filtrate was topped up with deionised water to 1 dm^3. All food waste was obtained from Newton Hawker Centre in Singapore.

40.2.2 Preparation of Growth Media

The medium made with mixing 18 cm^3 0.44% (w/v) glucose, 4% (w/v) fructose; and 20 cm^3 1% (w/v) yeast extract, 0.54% (w/v) $Na_2HPO_4 . 12H_2O$, 0.23% (w/v) citric acid, 1.2% (v/v) ethanol was prepared. This is the control setup. For media containing the food waste extracts, 18 cm^3 of the prepared extract was added instead of the fructose-glucose solution, replacing it as a carbon source. All chemicals were procured from Sigma-Aldrich.

40.2.3 Inoculation of Acetobacter xylinum *and Fermentation of Biocellulose*

The medium was autoclaved under 15 lbs pressure, 121 °C for 1 h for sterilisation. 2 cm^3 of *A. xylinum* was then inoculated into each medium and statically incubated at 26 °C for 7 days.

40.2.4 Varying the Metal Ions

The fructose-glucose medium in Sect. 2.2 was used in all experiments in this section. To the 40 cm^3 medium, 1 cm^3 of 0.0184 mol dm^{-3} $MgCl_2$ was added. $CaCl_2$ and KCl were used replacing $MgCl_2$ to vary the metal ion. The control setup had no metal ions. All chemicals were procured from Sigma-Aldrich.

40.2.5 Varying Ethanol Concentration

The fructose-glucose medium in Sect. 2.2 was used in all experiments in this section. The concentration of ethanol was varied at concentrations of 0, 0.4, 0.8, and 1.2% (v/v).

40.2.6 Harvesting of Biocellulose

The biocellulose films obtained from 2.2, 2.3, 2.4, and 2.5 were washed with deionised water for 2–3 times and boiled with 0.5 mol dm^{-3} K_2CO_3 at 90 °C for 15 min to remove any remaining bacteria. The films were washed with deionised water until pH neutral. All experiments were conducted in 5 replicates.

40.2.7 Characterisation of Biocellulose

Fourier-Transform Infra-red (FTIR) spectra of biocellulose were obtained using the KBr disc technique with *Bruker Alpha II FTIR Spectrometer*. Pellets were made from biocellulose with KBr. Transmissions were measured at the wave number range of 4000–600 cm^{-1}.

40.2.8 Testing of Tensile Strength of Biocellulose

From Sect. 2.5, a washed film was dried at 80 °C until constant mass. Regular biocellulose strips of 10 mm width and 30 mm length were cut out with a circular indentation of radius 2.5 mm. The strips were placed in a tensile strength tester. The tensile force is the force which the biocellulose broke where the circular indentation was cut out. Experiments were conducted in 5 replicates.

40.2.9 Testing of Drug Release Properties via Dissolution Assays

The biocellulose films obtained from 2.3, 2.4, and 2.5 were dried on filter paper to remove 50–60% of water content. 4 cm^3 of 0.01 mol dm^{-3} phosphate buffer solution (pH 7.4) containing 2% ibuprofen-sodium salt (w/v) and 1% glycerol (v/v) was added to each film, which was shaken at 80 rpm for 48 h to ensure complete absorption of the drug. The film was then shaken at 100 rpm in a large conical flask containing 500 cm^3 of buffer solution. For 3 h, 1 cm^3 of the solution was withdrawn at 15-min intervals and topped up to 10 cm^3 with the buffer solution. This solution was then tested for the concentration of ibuprofen-sodium salt present. 1 cm^3 of the buffer was added back to the jug to maintain constant volume. Experiments were conducted in 3 replicates. The concentration of ibuprofen-sodium salt was measured using UV–Vis spectrophotometry (Shimazu UV-1800) at a wavelength of 222 nm, with a linear calibration curve of $y = 0.0343x$ ($R^2 = 0.9997$), where x ($0 \leq x \leq 20$) is the concentration in ppm.

40.3 Results and Discussion

40.3.1 Characterisation of Biocellulose

Figure 40.1a shows the FTIR graph obtained for the biocellulose synthesised from the fructose-glucose mixture. The broad peak at 3383 cm^{-1} is due to O–H stretching, the peak at 2920 cm^{-1} is due to C-H stretching, the peak at 1651 cm^{-1} can be attributed to O–H bending of adsorbed water, 1427 cm^{-1} is due to the CH_2 scissoring motion in cellulose, 1372 cm^{-1} is due to C-H bending, 1164 cm^{-1} corresponds to the C–C ring stretching, and 1059 cm^{-1} is due to C–O–C ring stretching [7].

The FTIR spectra from biocellulose synthesised from the different food waste are similar to that of commercial cellulose in Fig. 40.1b, showing that the synthesis of biocellulose was successful.

40.3.2 Yield of Biocellulose

From Fig. 40.2a, the biocellulose synthesised from the fructose-glucose solution (control setup) is higher in yield than using a carbon source from sugarcane and spent coffee beans. Preliminary results showed that the rate of formation of biocellulose using spent coffee beans as a carbon source was too slow to form a film of biocellulose in 7 days. As such, the duration of incubation for biocellulose involving a varying carbon source was extended to 12 days. The relatively high biocellulose yield from sugarcane bagasse may be due to the presence of the storage parenchyma tissue in the stem, as they store high levels of sucrose. The enzyme neutral invertase in the tissue [8] hydrolyses sucrose into glucose and fructose for subsequent polymerisation into biocellulose [9]. Spent coffee beans contain mono- and oligosaccharides capable of being polymerised into bacterial cellulose [10]. However, these carbohydrates are rapidly converted to Maillard and pyrolysis products via thermal degradation during roasting [11]. The low monosaccharide content hence limited the production of biocellulose [12], resulting in the lowest yield.

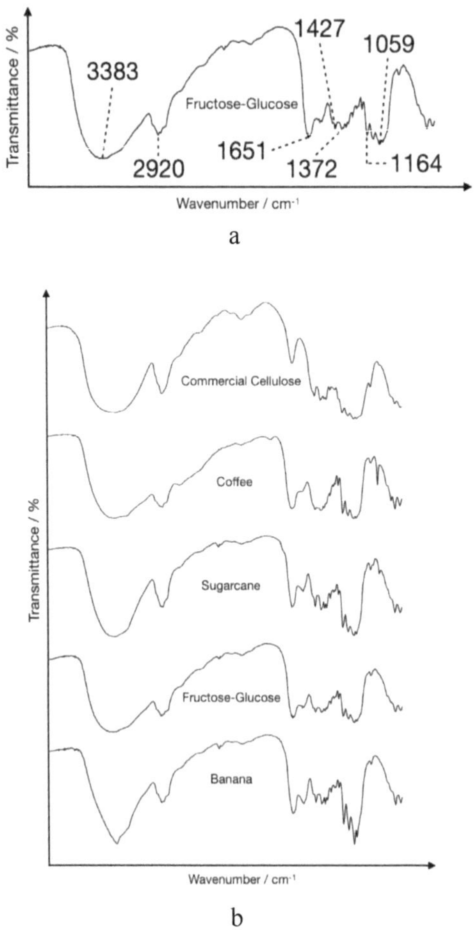

Fig. 40.1 **a** Labelled peaks of FTIR spectra for fructose-glucose cellulose, **b** FTIR Spectra for Cellulose produced under different conditions

From Fig. 40.2b, amongst different metal ion supplements added, it was found that Mg^{2+} increased the yield of biocellulose, whilst Ca^{2+} and K^+ decreased the yield. A specific guanyl cyclase is activated by Mg^{2+}, producing more cyclic di-GMP, an second messenger that activates cellulose synthase, stimulating the production of biocellulose. However, it is unsure currently how Ca^{2+} decreased the yield of biocellulose. Lastly, the potassium ion supplement decreased the yield of biocellulose, as a K^+ rich medium is proven to reduce the motility of *A. Xylinum* [13], which directly decreases cellulose production rate as *A. Xylinum* moves due to the inverse force of extrusion of the biocellulose fibre [14].

Finally, from Fig. 40.2c, biocellulose synthesised with ethanol concentration of 0.8% (v/v) was the highest, followed by ethanol concentration of 1.2% (v/v) (the

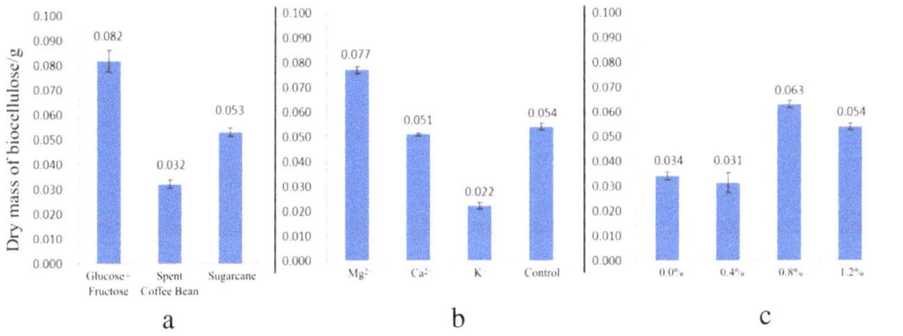

Fig. 40.2 Effect of different variables on yield of biocellulose. **a** Various carbon sources, **b** Various metal ions, **c** Various ethanol concentration

control). Ethanol is slowly oxidised to ethanoic acid during fermentation by atmospheric oxygen, causing the pH to decrease from the optimal pH of 4.5 [15] for biocellulose production and *A. Xylinum* growth. A higher concentration at 1.2% (v/v) causes the pH to drop faster, hence explaining the decrease in biocellulose yield. When the ethanol concentration is below 0.80%, the yield drops too. This is because ethanol acts as an additional energy source for *A. Xylinum* [16] and prevents mutations of defunct cells unable to produce biocellulose, supplementing biocellulose production [17].

40.3.3 Tensile Strength of Biocellulose

From Fig. 40.3a, it was observed that biocellulose produced from sugarcane extract achieved the highest tensile strength. This can be attributed to the cellulose nanofibrils present in the sugarcane extract, which is interwoven with bacterial cellulose produced by *A. Xylinum* to form a heavily cross-linked three-dimensional structure

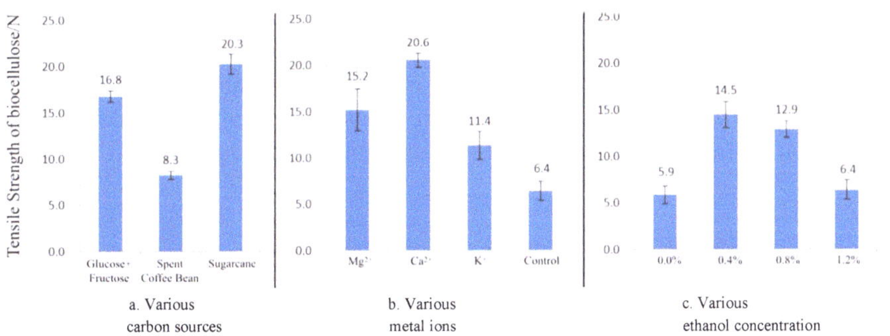

Fig. 40.3 Effect of different variables on tensile strength of biocellulose

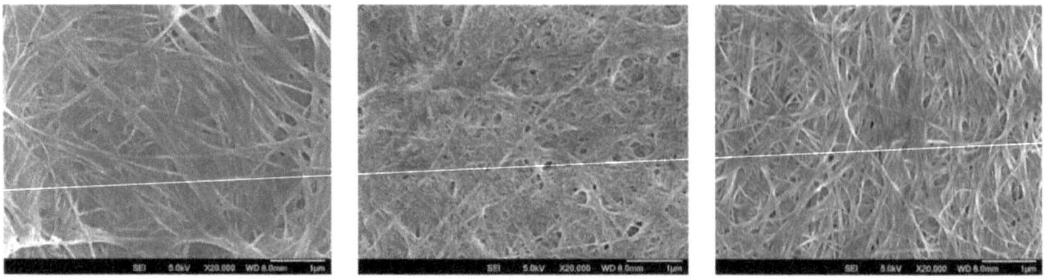

a b c

Fig. 40.4 SEM results for biocellulose produced from different carbon sources. **a** Glucose and fructose carbon source, **b** Coffee bean extract, **c** Sugarcane bagasse

[18], as seen in Fig. 40.4c. On the other hand, biocellulose synthesised from coffee bean extract exhibited the weakest tensile strength. Caffeine in the coffee bean extract acts as an inhibitor of the enzyme diguanyl cyclic phosphodiesterase, prolonging the half-life of cyclic di-GMP [19], resulting in a larger concentration of active cellulose synthase. However, limited by the amount of carbon source in the coffee bean extract, fibrils produced form weaker hydrogen bonds with each other, and are less compact (Fig. 40.4b) with lower tensile strength, compared to the control setup (Fig. 40.4a).

From Fig. 40.3b, it was observed that in general, the addition of metal ion supplements improved the strength of microbial cellulose. Amongst them, Ca^{2+} proved to best improve the tensile strength of the cellulose, owing to its stimulation of cellulose synthase in *A. Xylinum*. Ca^{2+} inhibits a membrane enzyme responsible for the hydrolysis of the cellulose synthase activator [20], whilst Mg^{2+} is able to activate guanyl cyclase, producing more cyclic di-GMP, stimulating the production of biocellulose [19]. As a result, with a sufficient amount of carbon source present as substrate, more complete polymerisation occurs, producing a stronger fibrous structure. Additionally, K^+ ions inhibit the action of bacterial glycogen phosphorylase [21]. Hence, more glucose monomers are available for polymerisation into bacterial cellulose. Longer and more completely polymerised cellulose fibrils are formed, increasing the tensile strength of biocellulose produced in K^+ rich medium.

From Fig. 40.3c, tensile strength of biocellulose synthesised decreases as ethanol concentration increases, with the exception of 0.0% (v/v). The hydroxyl group in ethanol disrupts hydrogen cross-links between cellulose fibres, reducing its tensile strength [22]. However, in the absence of ethanol, the tensile strength decreases significantly. This is as ethanol acts as an important stimulant for biocellulose production [23] and an additional energy source for *A. Xylinum* [16].

40.3.4 Drug Release Property of Biocellulose

Figure 40.5a shows a higher initial rate of ibuprofen release by biocellulose produced from sugarcane bagasse. Additionally, less than 90% of ibuprofen was released after

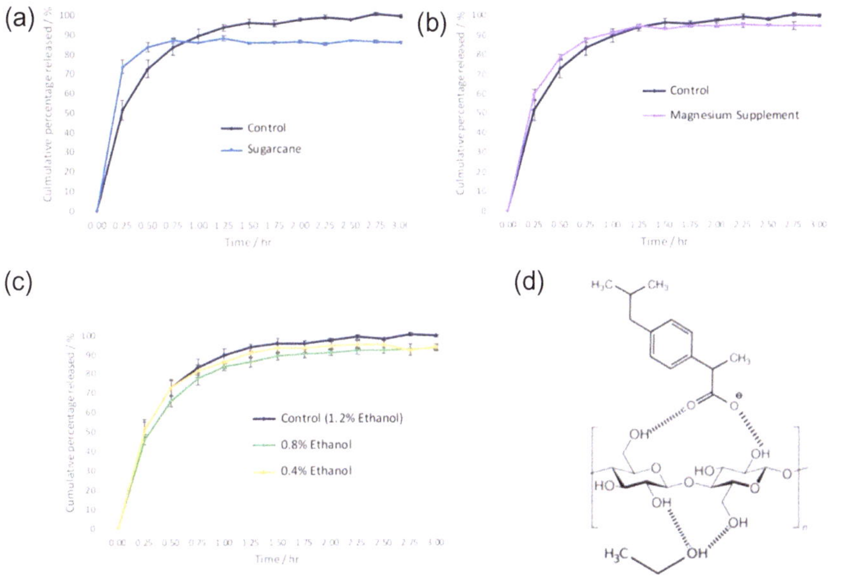

Fig. 40.5 **a**–**c** Graphs of percentage of ibuprofen released/% against time elapsed/hours. **a** Comparing between carbon sources, **b** Comparing between the presence of magnesium ions, **c** Comparing between various ethanol concentration, **d** Hydrogen bonds between the carboxylate group of ibuprofen-sodium salt, ethanol and biocellulose

3 h. This could be attributed to the reduced porosity of the cellulose due to the interweaving of plant-based cellulose nanofibrils in the bagasse and the bacterial cellulose [18], as seen in the compact cellulose macrofibrils in Fig. 40.4c, resulting in incomplete absorption and fast release of ibuprofen.

Figure 40.5b shows a higher initial rate of ibuprofen release with biocellulose produced with Mg^{2+}, but around 95% of ibuprofen was released after 3 h. Inside the fibre matrix, charged Mg^{2+} ions disrupt hydrogen bonds between cellulose fibrils by forming ion–dipole interactions with the fibres. This increases porosity of the bacterial cellulose, allowing higher rates of ibuprofen release. However, Mg^{2+} can form ionic bonds with the ibuprofen anion, causing efficient retention and thus incomplete release of ibuprofen.

Figure 40.5c showed that near 100% of the ibuprofen in the control setup using 1.2% ethanol was released by 2.75 h. Comparing the various concentrations of ethanol used to produce biocellulose, Fig. 40.5c shows that biocellulose produced with 0.8% ethanol released ibuprofen the slowest. When the concentration of ethanol decreases from 0.8%, less ethanol acts as an additional energy source for *A. Xylinum* [16], leading to a significantly lower yield (Fig. 40.2). Thus, less hydroxyl groups are available to bind to ibuprofen, resulting in faster release. At high concentrations beyond 0.8% ethanol, ethanol molecules bind to a large amount of hydroxyl groups on the produced cellulose. This reduces the number of hydroxyl groups available to bind to ibuprofen, resulting in faster release.

The biocellulose synthesised with 0.8% ethanol shows a remarkably steady release of ibuprofen, where it took 1 h 45 min for 90% of ibuprofen to be released. This is advantageous as cells receive controlled concentrations of drug over time, preventing sluggish wound healing or drug overdose. The gentle diffusion gradient hence speeds up recovery whilst preventing cell death.

40.4 Conclusion

Cellulose was successfully synthesised from glucose and fructose, sugarcane bagasse and spent coffee beans. The addition of Ca^{2+} and Mg^{2+} into the media increased the yield of biocellulose, whilst the addition of K^+ decreased the yield. The optimal concentration of ethanol for the culture medium was 0.8% (v/v). The tensile strength of sugarcane bagasse biocellulose was higher than that of the control, whilst the tensile strength of the spent coffee bean biocellulose was significantly lower than that of the control. Addition of Ca^{2+}, Mg^{2+}, and K^+ also increased the tensile strength of biocellulose. Meanwhile, the biocellulose produced using 0.4% (v/v) ethanol solution gave the highest tensile strength amongst biocellulose synthesised using various ethanol concentrations.

The kinetics of the release of ibuprofen was successfully demonstrated through a dissolution assay. The biocellulose produced using the solution containing 0.8% (v/v) ethanol demonstrated a gradual, promising rate of ibuprofen release. Addition of Mg^{2+} increased the rate of release of biocellulose initially, but limited its complete release over 3 h. Lastly, sugarcane bagasse biocellulose had a much faster release of ibuprofen, but not all was released within 3 h. The biocellulose produced using the solution containing 0.8% (v/v) ethanol has high potential to be used for further drug release mechanism studies of ibuprofen into the bloodstream.

Acknowledgements We would like to thank our mentor, Dr. Ella Sze, and the Science Research Centre lab staff, both from Hwa Chong Institution, for their guidance and support.

References

1. Pokalwar, U. S., Mishra, M. K., & Manwar, A. V. (2010). Production of cellulose by Gluconacetobacter sp. *Recent Research in Science and Technology, 2*(7), 14–19.
2. Mohammedi, Z. (2017). Structure, properties and medical advances for biocellulose applications: A review. *American Journal of Polymer Science and Technology, 3*(5), 89. https://doi.org/10.11648/j.ajpst.20170305.12
3. Kennedy, J., & Knill, C. (2002). Handbook of hydrocolloids. *Carbohydrate Polymers, 49*(4), 516. https://doi.org/10.1016/s0144-8617(01)00360-5
4. Mohammad, S.M., Rahman, N.A., Khalil, M.S., Abdullah, S.R., & Malaysia, S. (2014). *An overview of biocellulose production using Acetobacter xylinum culture.*

5. Czaja, W., Krystynowicz, A., Bielecki, S., & Brown, R., Jr. (2006). Microbial cellulose—the natural power to heal wounds. *Biomaterials, 27*(2), 145–151. https://doi.org/10.1016/j.biomaterials.2005.07.035
6. Plackett, D., Letchford, K., Jackson, J., & Burt, H. (2014). A review of nanocellulose as a novel vehicle for drug delivery. *Nordic Pulp & Paper Research Journal, 29*(1), 105–118. https://doi.org/10.3183/npprj-2014-29-01-p105-118
7. Kumar, A., Singh Negi, Y., Choudhary, V., & Kant Bhardwaj, N. (2020). Characterization of cellulose nanocrystals produced by acid-hydrolysis from sugarcane bagasse as agro-waste. *Journal of Materials Physics and Chemistry, 2*(1), 1–8. https://doi.org/10.12691/jmpc-2-1-1
8. Hatch, M., & Glasziou, K. (1963). Sugar accumulation cycle in sugar cane. II. Relationship of Invertase activity to sugar content & growth rate in storage tissue of plants grown in controlled environments. *Plant Physiology, 38*(3), 344–348. https://doi.org/10.1104/pp.38.3.344
9. Moore, P. (1995). Temporal and spatial regulation of sucrose accumulation in the sugarcane stem. *Functional Plant Biology, 22*(4), 661. https://doi.org/10.1071/pp9950661
10. Kim, Y.-J., Choi, J., Lee, G., & Lee, K.-G. (2020). Analysis of furan and monosaccharides in various coffee beans. *Journal of Food Science and Technology*. https://doi.org/10.1007/s13197-020-04600-5
11. Oosterveld, A., Voragen, A., & Schols, H. (2003). Effect of roasting on the carbohydrate composition of Coffea arabica beans. *Carbohydrate Polymers, 54*(2), 183–192. https://doi.org/10.1016/s0144-8617(03)00164-4
12. Puri, V. (1984). Effect of crystallinity and degree of polymerization of cellulose on enzymatic saccharification. *Biotechnology and Bioengineering, 26*(10), 1219–1222. https://doi.org/10.1002/bit.260261010
13. Brown, I., Galperin, M., Glagolev, A., & Skulachev, V. (1983). Utilization of energy stored in the form of Na^+ and K^+ Ion gradients by bacterial cells. *European Journal of Biochemistry, 134*(2), 345–349. https://doi.org/10.1111/j.1432-1033.1983.tb07573.x
14. Tomita, Y., & Kondo, T. (2009). Influential factors to enhance the moving rate of Acetobacter xylinum due to its nanofiber secretion on oriented templates. *Carbohydrate Polymers, 77*(4), 754–759. https://doi.org/10.1016/j.carbpol.2009.02.022
15. Kasim, N., & Rahman, N. (2016). Design and production control of biocellulose from Acetobacter xylinum. *Indian Journal of Science and Technology, 9*(21), 1–10. https://doi.org/10.17485/ijst/2016/v9i21/95241
16. Aswini, K., Gopal, N., & Uthandi, S. (2020). Optimized culture conditions for bacterial cellulose production by Acetobacter senegalensis MA1. *BMC Biotechnology, 20*(1). https://doi.org/10.1186/s12896-020-00639-6
17. Chawla, P., Bajaj, I., & Survase, S. (2009). Microbial cellulose: Fermentative production and applications. *Food Technology And Biotechnology, 47*(2), 107–124.
18. Erbas Kiziltas, E., Kiziltas, A., Blumentritt, M., & Gardner, D. (2015). Biosynthesis of bacterial cellulose in the presence of different nanoparticles to create novel hybrid materials. *Carbohydrate Polymers, 129*, 148–155. https://doi.org/10.1016/j.carbpol.2015.04.039
19. Fontana, J., Franco, V., De Souza, S., Lyra, I., & De Souza, A. (1991). Nature of plant stimulators in the production of Acetobacter xylinum ("tea fungus") biofilm used in skin therapy. *Applied Biochemistry and Biotechnology, 28–29*(1), 341–351. https://doi.org/10.1007/bf02922613
20. Hong, F., & Qiu, K. (2008). An alternative carbon source from konjac powder for enhancing production of bacterial cellulose in static cultures by a model strain Acetobacter aceti subsp. xylinus ATCC 23770. *Carbohydrate Polymers, 72*(3), 545–549. https://doi.org/10.1016/j.carbpol.2007.09.015
21. Hue, L., Bontemps, F., & Hers, H. G. (1975). The effect of glucose and of potassium ions on the interconversion of the two forms of glycogen phosphorylase and of glycogen synthetase in isolated rat liver preparations. *Biochemical Journal, 152*(1), 105. https://doi.org/10.1042/bj1520105

22. Ray, D., & Sain, S. (2016). In situ processing of cellulose nanocomposites. *Composites Part A: Applied Science and Manufacturing, 83*, 19–37. https://doi.org/10.1016/j.compositesa.2015.09.007
23. Lu, Z., Zhang, Y., Chi, Y., Xu, N., Yao, W., & Sun, B. (2011). Effects of alcohols on bacterial cellulose production by Acetobacter xylinum 186. *World Journal Of Microbiology And Biotechnology, 27*(10), 2281–2285. https://doi.org/10.1007/s11274-011-0692-8

Chapter 41
Synthesis of Magnetically Responsive Photocatalytic CuBTC Metal–Organic Frameworks (MOFs) for Degradation of Organic Textile Pollutants

Ng Yan Bin Lucas, Peng Zikang, and Eun Chin Sze Gerald

Abstract CuBTC (copper benzene-1,3,5-tricarboxylate) is a metal–organic framework (MOF) which has the potential to purify organic pollutants present in textile effluents via photocatalysis. In this study, non-magnetic and magnetic CuBTC-MOF were synthesized via precipitation, and removal of organic pollutants including azo dyes methyl orange and direct red 80 and organic pollutant acetone via CuBTC-MOF was investigated. CuBTC-MOF displays enhanced photocatalytic activity under visible light, with photodegradation of organic pollutants further catalysed by the addition of hydrogen peroxide. Magnetization of CuBTC doped with mercapto-acetic acid (MAA) as a capping agent reduced agglomeration and further enhanced dye degradation efficacy. The band gap of magnetic CuBTC and non-magnetic CuBTC-MOF synthesized were 1.98 eV and 2.05 eV respectively, which are both lower than that of conventional catalysts such as titanium dioxide (TiO_2) and zinc oxide (ZnO). Hence, both magnetic and non-magnetic CuBTC-MOF are able to harness visible light and outperformed both TiO_2 and ZnO, which can only function under UV light. In addition, magnetic CuBTC-MOF removed close to 100% of all organic pollutants in the presence of hydrogen peroxide, significantly higher than removal of TiO_2 and ZnO. Reusability tests conducted also highlight that magnetic CuBTC was still able to remove more than 90% of organic pollutants across 3 cycles of regeneration, with high retrievability of 96% due to its magnetization. Hence, magnetic CuBTC-MOF is a promising alternative to conventional photocatalysts in degrading organic pollutants.

Keywords Photocatalyst · MOF · CuBTC · Magnetic · Methyl orange · Acetone

---

N. Y. B. Lucas (✉) · P. Zikang · E. C. S. Gerald
Hwa Chong Institution (High School), Singapore, Singapore
e-mail: lucasnyb@gmail.com

H. Guo et al. (eds.), *IRC-SET 2021*, https://doi.org/10.1007/978-981-16-9869-9_41

41.1 Introduction

The textile industry is worth around US$1 trillion worldwide and makes up 7% of the total world exports [4]. The main damages caused by the textile industry to the environment result mainly from the discharge of untreated textile effluents containing azo dyes like methyl orange and direct red and organic pollutants like acetone into the water bodies [2]. Azo dyes prevent the penetration of light through water [8], which leads to a reduction in the rate of photosynthesis, affecting the entire aquatic biota [10]. The azo dyes are also toxic, mutagenic and carcinogenic [1] and are very difficult to degrade as it consists of one or more azo bonds (–N=N–) in association with one or more aromatic systems [17]. Acetone, another prominent example of the organic pollutants found in textile effluents, is often used as a solvent in the textile industry. Acetone is highly volatile at room temperature and is also toxic to wildlife and contributes to air pollution at high concentrations [7]. For removal of azo dyes and acetone from industrial wastewater, activated carbon is the preferred adsorbent because of its efficiency, capacity, and scalability for commercial usage [11]. However, adsorption using activated carbon does not degrade the dye and acetone and adsorptive properties of the activated carbon will gradually reduce after continuous dye and acetone removal [19].

In recent years, photodegradation has emerged as an efficient and green method to degrade dyes from wastewater. Photocatalysts can be employed for the photomineralization of azo dyes and organic pollutants. One of the most common photocatalyst used is titanium dioxide (TiO_2) nanoparticles. TiO_2, under ultraviolet (UV) light, can catalyse the photocatalytic decomposition of organic pollutants by light-induced electron/hole pairs and highly oxidizing hydroxyl anions [15]. However, the large band gap energy of TiO_2 (around 3.2 eV) limits its absorption of solar radiation to the UV light range [12], reducing its effectiveness. Therefore, there is a need to develop a quick and efficient method for the synthesis of an inexpensive photocatalyst that works under visible light irradiation.

CuBTC-MOF is a type of metal–organic framework (MOF) which has photocatalytic properties. MOFs are crystalline materials with ultrahigh porosity and enormous internal surface areas. CuBTC (Copper(II) benzene-1,3,5-tricarboxylate) has a moderate band gap and is potentially able to harness visible light for photodegradation. Additionally, it is possible to magnetize CuBTC to enhance its reusability. As with the case with other MOFs, CuBTC also has been extensively studied for gaseous adsorption [20]. To date, there have been limited studies on the use of CuBTC to degrade dyes, especially azo dyes. Studies on acetone degradation have yet to be done. Hence, this study aims to synthesize and magnetize CuBTC and to investigate its effectiveness in degrading azo dyes (methyl orange, direct red 80) and acetone. The effectiveness of the MOF in the degradation of these pollutants was then compared with conventional photocatalysts (TiO_2 and ZnO).

Hypothesis

The presence of visible light irradiation, magnetite and H_2O_2 will improve the photodegradation of azo dyes and acetone by CuBTC-MOF. Magnetic CuBTC-MOF will outperform conventional photocatalysts.

41.2 Materials and Methods

41.2.1 Materials

Copper(II) nitrate was purchased from GCE Chemicals. Trimesic acid (H_3BTC), magnetite, mercapto-acetic acid (MAA), methyl orange and direct red 80 were procured from Sigma Aldrich.

41.2.2 Synthesis of Copper (II) Benzene 1,3,5-Tricarboxylate (CuBTC) MOF

1.82 g of $Cu(NO_3)_2$ and 0.875 g of H3BTC were dissolved in 100 ml of deionized water in separate beakers. The solution of H3BTC was then added to the dissolved $Cu(NO_3)_2$, with a constant flow rate of 1 ml min^{-1}, and then topped up to 250 ml with absolute ethanol and stirred for 8 h. The reaction leading to the formation of CuBTC is shown in Eq. 41.1.

$$3Cu(NO_3)_2 + 2H_3BTC + 3H_2O \rightarrow [\mathbf{Cu_3BTC_2}].3H_2O + 6HNO_3 \quad (41.1)$$

The suspension was then centrifuged, and the residue was washed with absolute ethanol, before leaving it to dry in the oven at 70 °C until constant mass.

41.2.3 Magnetization of CuBTC-MOF

0.50 g of magnetite (Fe_3O_4) was added to 1.0 ml of mercapto-acetic acid (MAA) and dispersed in 9.0 ml of ethanol. The suspension was shaken for 24 h, then centrifuged. 0.5 g of MAA-functionalized magnetite was added to the synthesized CuBTC-MOF and dispersed in 100 ml of absolute ethanol before being sonicated for 2 h. The resultant magnetic CuBTC-MOF was washed and dried in the oven at 70 °C until constant mass.

41.2.4 Determination of Effectiveness of CuBTC-MOF in Degrading Dyes and Acetone

0.050 g of magnetic and non-magnetic CuBTC-MOF and 1.5 ml of 6% hydrogen peroxide were added to 20 ml of 50 ppm dye or acetone and stirred for 24 h in the dark or under visible light irradiation. The concentration of methyl orange, direct red 80 and acetone remaining in the supernatant was measured using a UV–Vis Spectrophotometer at 464 nm, 507 nm and 267 nm respectively. A control with only H_2O_2 and dye or acetone was included in the set up. Experiments were repeated with conventional photocatalysts, TiO_2 and ZnO for comparison. The following formula was used to calculate the percentage of pollutant removed.

41.2.5 Reusability Study on CuBTC-MOF

$$P = \frac{C_i - C_f}{C_i} \times 100\%$$

where

$P = \%$ of pollutant removed

$C_i =$ initial [pollutant]

$C_f =$ final [pollutant].

After first cycle of degradation, a magnet was used to retrieve the magnetic CuBTC-MOF from the suspension while the non-magnetic CuBTC-MOF was retrieved via centrifugation. Both catalysts were desorbed with ethanol and dried before subjecting them to another batch of organic pollutant with the same concentration and volume The percentage of organic pollutants removed were analysed over 3 cycles.

41.3 Results and Discussion

41.3.1 Characterization of CuBTC-MOF by X-Ray Diffraction (XRD) and Scanning Electron Microscope (SEM)

Figure 41.1 reveals four theta peaks at 6.56°, 9.18°, 11.95° and 13.81°, which are characteristic of CuBTC-MOF [16], indicating the successful synthesis of CuBTC-MOF.

Non-magnetic CuBTC-MOF (Fig. 41.2a) shows evidence of agglomeration on the crystal surface and the observation agrees with that synthesized by [6]. On the other

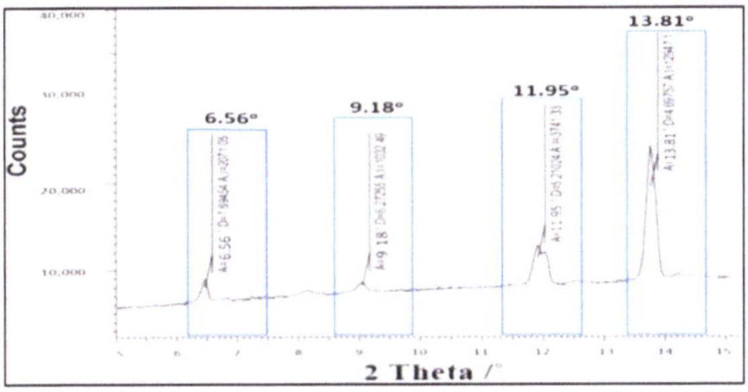

Fig. 41.1 XRD pattern of non-magnetic CuBTC

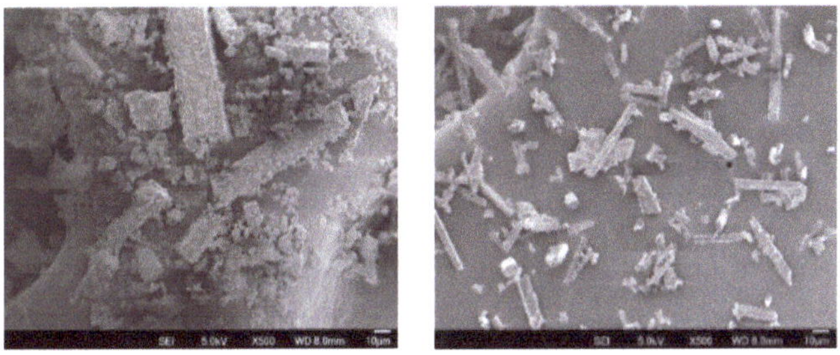

a: SEM image of non-magnetic CuBTC-MOF at 500x magnification

b: SEM image of magnetic CuBTC-MOF at 500x magnification

Fig. 41.2 **a** SEM image of non-magnetic CuBTC-MOF at 500× magnification. **b** SEM image of magnetic CuBTC-MOF at 500× magnification

hand, SEM image of magnetic CuBTC-MOF reveals distinct and regular shaped crystals (Fig. 41.2b).

41.3.2 Effect of Magnetization of CuBTC-MOF on Percentage Removal of Dyes

Figure 41.3 shows that the presence of magnetite significantly enhances degradation of both dyes (p-value of Mann–Whitney U Test = $0.012 < 0.05$). This is likely due to iron in magnetite binding to the dangling carboxylic acid groups in CuBTC-MOF with surface defects, allowing for more complete crystal growth and increasing the MOF's pore size and overall surface area, as supported by the SEM images (Fig. 41.2).

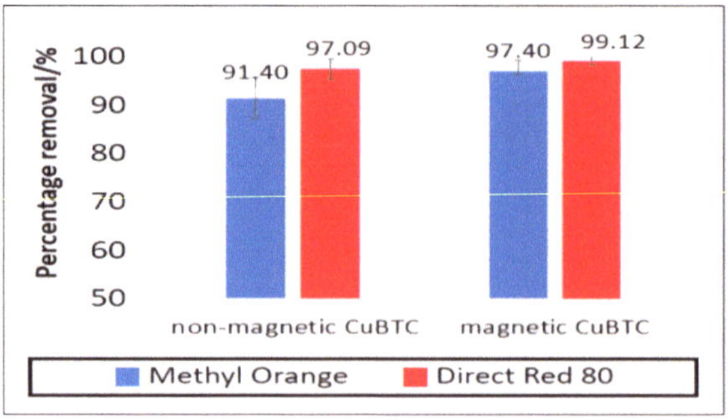

Fig. 41.3 Percentage removal of azo dyes by magnetic and non-magnetic CuBTC-MOF

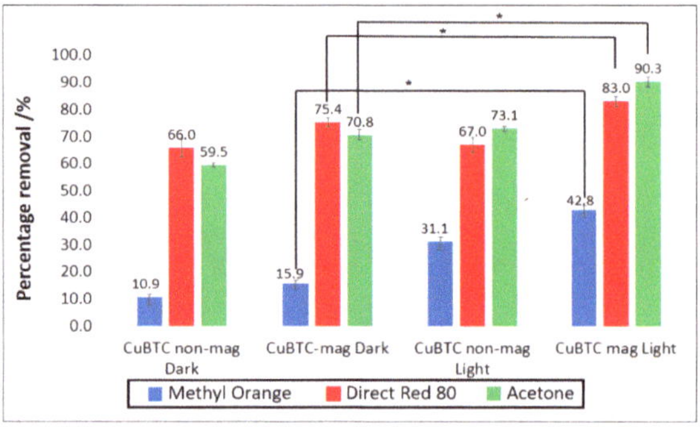

Fig. 41.4 Comparison of the effects of the presence of visible light on the presence removal of dyes. * denotes statistical significance based on Mann Whitney U Test a significance level of 0.05

41.3.3 Effect of Presence of Visible Light on Degradation of Dyes and Acetone

To confirm the photocatalytic capability of the CuBTC-MOF, degradation experiments were conducted both in the dark and under visible light irradiation. Removal of organic pollutants in the dark suggests that both magnetic and non-magnetic CuBTC-MOF can remove dyes and acetone, likely by adsorption, but to a lesser extent. Under visible light irradiation, both magnetic and non-magnetic CuBTC-MOF had a significantly higher percentage removal of azo dyes and acetone as compared to in the dark (Fig. 41.4). This highlights that photodegradation occurs on top of adsorption under visible light irradiation.

41.3.4 Proposed Mechanisms for Photocatalytic Degradation of Organic Pollutants

The mass spectra of degradation residues in dark and light conditions are presented in Fig. 41.5a, b respectively. The intensity of methyl orange at peak 304 is lower in light than in dark condition, suggesting that it had been photodegraded. New peaks that

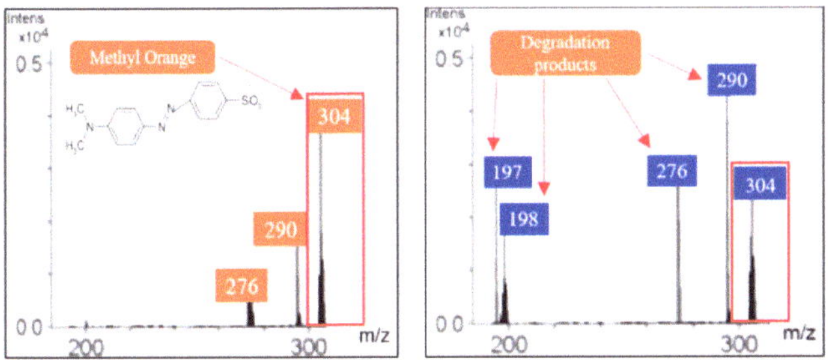

a: Mass spectrum of degradation residue in dark condition

b: Mass spectrum of degradation residue under visible light irradiation

Fig. 41.5 **a** Mass spectrum of degradation residue in dark condition. **b** Mass spectrum of degradation residue under visible light radiation

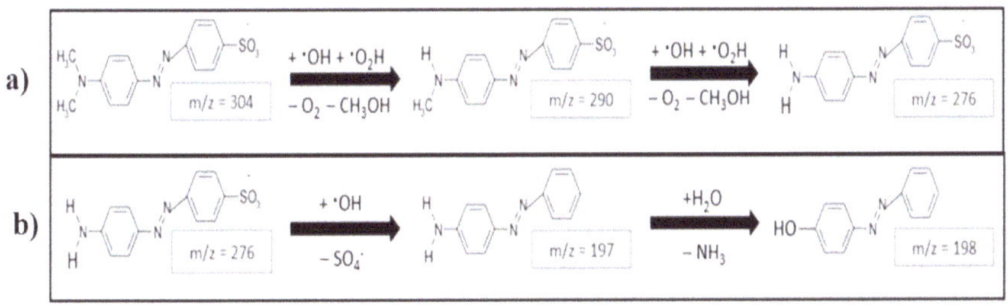

Fig. 41.6 Proposed degradation of pathways of methyl orange: **a** Demethylation, **b** Desulfonation followed by Deamination

correspond to degradation products in light condition at m/z of 197 and 198 were also observed, implying that photodegradation has taken place in light condition. Similar degradation products by Ag/ZnO were reported by [3].

The mechanism of photodegradation of methyl orange was proposed as shown in Fig. 41.6.

When exposed to light, negative-electron (e^-) and positive-hole (h^+) pairs are created in CuBTC-MOF. The positive-hole breaks apart water molecules, producing hydroxyl radicals; the negative-electrons react with water and oxygen to form super-oxide anion, which further produces hydroxyl radicals in the presence of H_2O_2. The hydroxyl radicals attack one of the methyl groups on methyl orange via N-demethylation (Fig. 41.6)] to form Product 1 ($m/z = 290$). This process is then repeated to form Product 2 ($m/z = 276$). Desulfonation followed by deamination occurs to obtain the final degraded product ($m/z = 198$) (Fig. 41.6b), which is less toxic than methyl orange [9, 13].

41.3.5 *Kinetic Studies of Photodegradation of Methyl Orange by Magnetic CuBTC-MOF*

In addition to the mass spectra analysis, kinetic studies were conducted on the magnetic CuBTC-MOF under different conditions for 360 min. From Fig. 41.7, the initial rate of methyl orange removal by magnetic CuBTC-MOF is higher under visible light irradiation than in the dark, suggesting that photodegradation has indeed occurred. The results are in agreement with the mass spectra analysis. Presence of H_2O_2 also enhanced the photodegradation rate of methyl orange. This is likely because the addition of H_2O_2 increases the production of hydroxyl radicals by reacting with the electrons produced and with the superoxide anions to produce more hydroxyl radicals [18], hence enhancing the rate and percentage of dye removed (Fig. 41.8).

$$H_2O_2 + e^- \rightarrow \cdot OH + OH^- \quad (41.2)$$

$$H_2O_2 + O_2^-\cdot \rightarrow \cdot OH + OH^- + O_2 \quad (41.3)$$

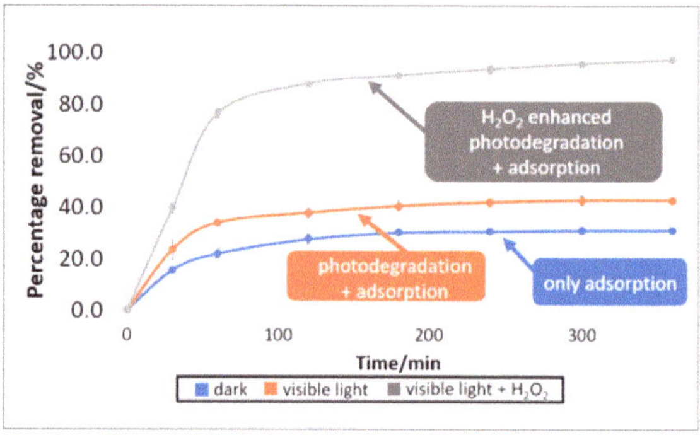

Fig. 41.7 Kinetics studied of magnetic CuBTC-MOF on methyl orange

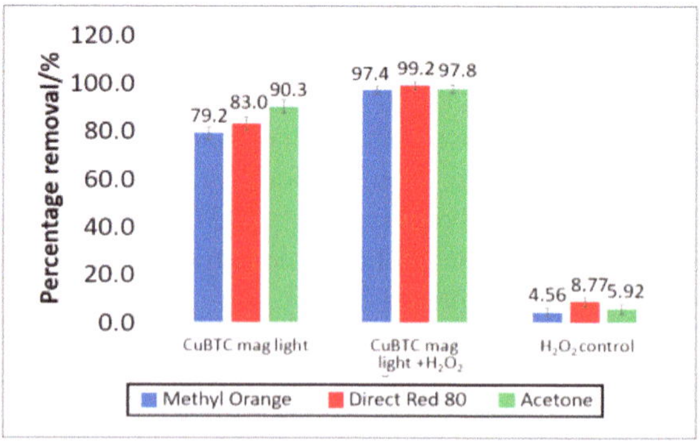

Fig. 41.8 Effect of H_2O_2 on photodegradation of dyes and acetone

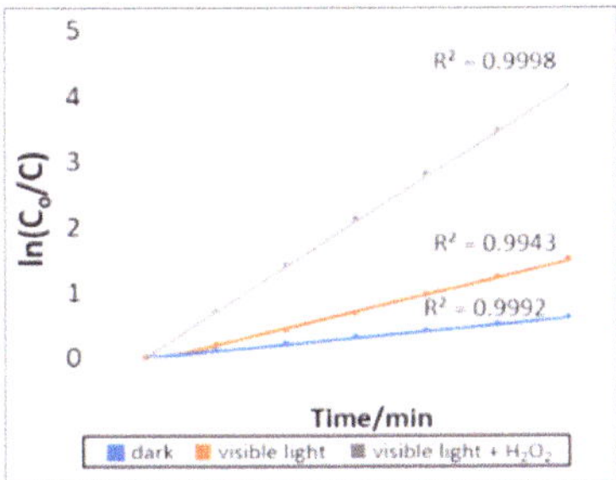

Condition	First order rate constant, k_a
Dark	0.1012
Light	0.2554
Light + H_2O_2	0.6886

Fig. 41.9 Kinetic analysis of photocatalytic degradation of methyl orange

Furthermore, the Langmuir–Hinshelwood model (Eq. 41.4) can be used to describe the kinetics of the photocatalytic degradation of aqueous methyl orange on a heterogeneous catalytic system [14].

$$\ln\left(\frac{C_o}{C}\right) = k_r K_{\text{ad}} t = k_a t \tag{41.4}$$

C_o = initial concentration in ppm;
C = final concentration in ppm;
k_r = rate constant;
K_{ad} = adsorption equilibrium constant;
k_a = pseudo-first-order rate constant;
t = time.

When adsorption is weak and/or methyl orange concentration is low, the Langmuir–Hinshelwood model simplifies to a pseudo-first-order kinetic model, where photocatalytic degradation would supersede, indicated by a linear relationship.

The linear relationship in Fig. 41.9 confirms that the photocatalytic degradation of methyl orange obeys pseudo-first-order kinetics according to the Langmuir–Hinshelwood model. Figure 41.9 reveals that magnetic CuBTC-MOF under visible light irradiation with H_2O_2 has the highest first-order rate constant (k_a) of 0.6886, which signifies greatest photocatalytic activity in degrading organics compared to dark and visible light conditions without H_2O_2.

41.3.6 Comparison of Photocatalytic Degradation Performance of CuBTC with Conventional Photocatalysts

It is also worth noting that both magnetic CuBTC-MOF and non-magnetic CuBTC-MOF showed higher percentages of degradation of all 3 organic pollutants studied

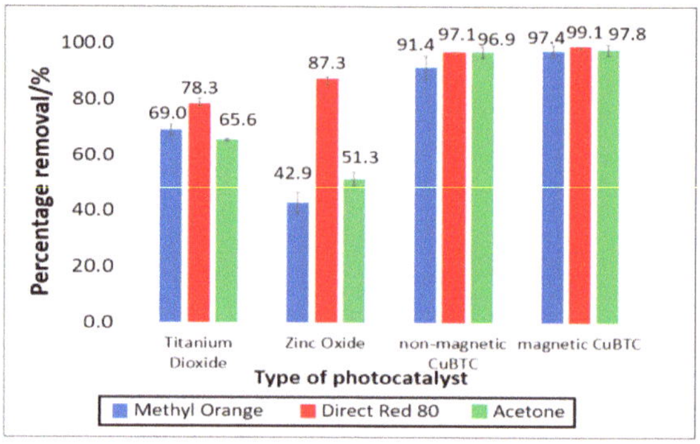

Fig. 41.10 Comparing the performance of CuBTC-MOF with conventional catalysts, TiO_2 and ZnO

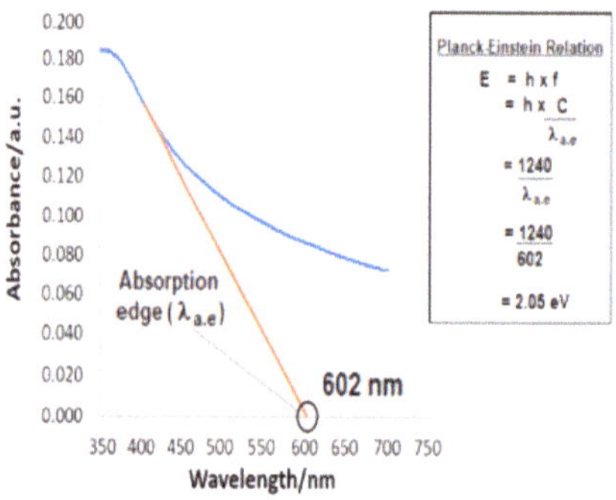

a: Calculation of band gap (E_g) for non-magnetic CuBTC MOF

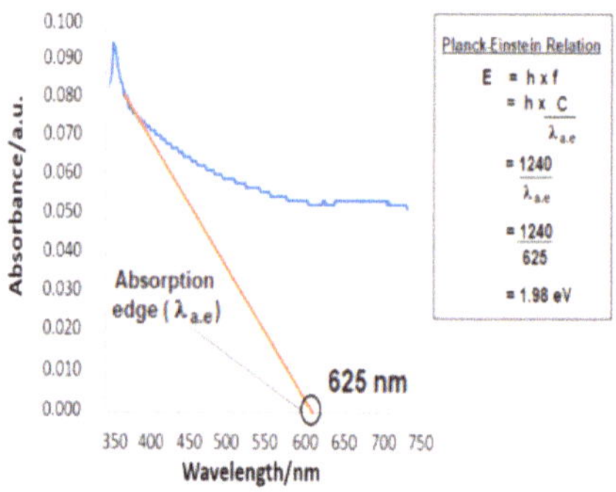

b: Calculation of band gap (E_g) for magnetic CuBTC-MOF

Fig. 41.11 **a** Calculation of band gap (E_g) for non-magnetic CuBTC-MOF. **b** Calculation of band gap (E_g) for magnetic CuBTC-MOF

than both zinc oxide and titanium dioxide, both of which are conventional photocatalysts, under visible light irradiation (Fig. 41.10). Both types of CuBTC-MOF attained a percentage removal of more than 90% for each type of pollutant which clearly outperforms the conventional photocatalysts in photodegradation.

The band gap of semiconductor photocatalysts is a crucial factor in determining their photocatalytic efficiency. Magnetic CuBTC-MOF synthesized in this study was determined to have a band gap of 1.98 eV (Fig. 41.11b), which is lower than the non-magnetic CuBTC-MOF with a band gap of 2.05 eV (Fig. 41.11a). Both band gaps are lower than zinc oxide and titanium dioxide with high band gaps of 3.2 eV [5].

41.3.7 Reusability Studies of Magnetic CuBTC-MOF

Figure 41.12a shows that after 3 cycles of photodegradation, magnetic CuBTC-

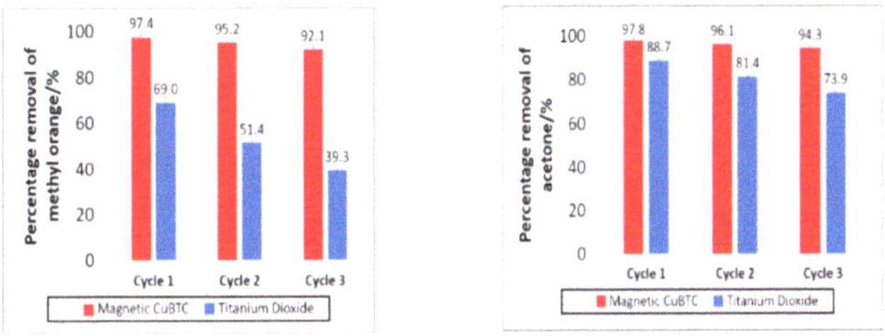

a: Regeneration of magnetic CuBTC-MOF and TiO_2 for degradation of methyl orange

b: Regeneration of magnetic CuBTC-MOF and TiO_2 for degradation of acetone

Fig. 41.12 **a** Regeneration of magnetic CuBTC-MOF and TiO_2 for degradation of methyl orange. **b** Regeneration of magnetic CuBTC-MOF and TiO_2 for degradation of acetone

MOF is still able to effectively degrade methyl orange, with the 5.4% decrease in the percentage of dye removed at third cycle being less than the 43.0% decrease for TiO_2. Similarly, Fig. 41.13b highlights that the decrease in percentage of acetone removed by magnetic CuBTC-MOF is still less than that of TiO_2. This suggests that magnetic CuBTC-MOF has greater reusability in removing methyl orange and acetone than TiO_2. Another advantage of magnetic CuBTC-MOF over TiO_2 is that as much as 96% of magnetic CuBTC-MOF could be retrieved after each cycle of photodegradation via a magnet, while only about 57% of TiO_2 could be retrieved via filtration. Hence, magnetic CuBTC-MOF outperforms commercial TiO_2 in terms of reusability and sustainability and is much easier to retrieve after photodegradation than commercial TiO_2.

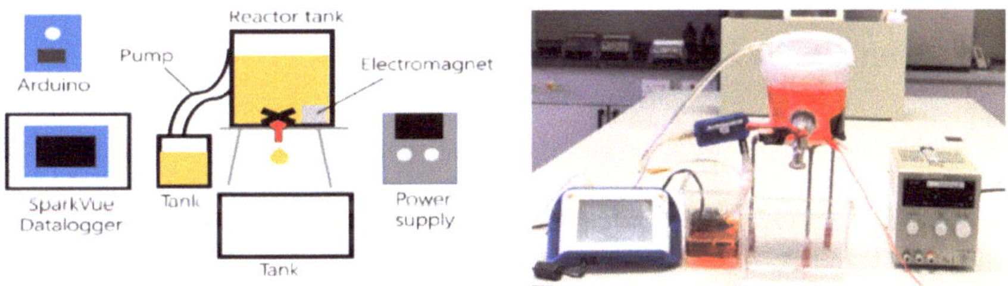

a: Blueprint of CuBTC-MOF in large scale water treatment plants

b: Actual constructed prototype

Fig. 41.13 **a** Blueprint of CuBTC-MOF in large-scale water treatment plants. **b** Actual constructed prototype

41.3.8 Construction of Prototype for Photocatalytic Treatment of Wastewater

Figure 41.13a, b shows our prototype, which consists of two rectangular tanks to store the untreated wastewater and the purified water. A pump would draw wastewater from the tank which would enter the central treatment cell where photocatalytic degradation occurs. A 25 mm hole was cut in the cylinder-shaped container and a motor was fitted through it, connected to a DC supply. A SPARKvue datalogger was then connected to an optic light sensor (PS-3213) that sends a signal to an Arduino control board which controls the draining of purified water into the tank. Finally, an electromagnet was installed to attract the magnetic MOF during the draining process.

The wastewater treatment process kicks off when the main supply is turned on. The pump is then activated which pumps wastewater into the cylinder-shaped container. Visible light easily penetrates through the transparent container to reach the solution mixture. The spinning motor helps to mix magnetic CuBTC-MOF with the wastewater to allow photocatalytic degradation to occur effectively. After 2 h, when the pollutant has been completely degraded, the optic light sensor detects the significantly higher light intensity and sends a signal to the integrated Arduino control board. The tap valve then opens to drain the purified water into the collection tank. At the same time, the motor stops spinning, and the electromagnet is automatically turned on to prevent magnetic CuBTC-MOF from being drained out. For subsequent cycles, the process repeats itself as the wastewater is automatically replenished. This renders the process facile and fully automated.

41.4 Conclusion

In conclusion, CuBTC-MOF synthesized is able to photodegrade azo dyes and acetone under visible light irradiation. Magnetization of CuBTC-MOF with MAA-functionalized Fe_3O_4 improved the degradation of dyes and acetone. Both magnetic and non-magnetic CuBTC-MOF, with a lower band gap than ZnO and TiO_2, was able to degrade dyes and acetone more effectively than TiO_2 and ZnO under visible light irradiation. Furthermore, magnetic CuBTC-MOF can be reused for at least 3 cycles of photodegradation without a significant drop in effectiveness, potentially saving cost and making its use in water treatment even more convenient.

A model on how photocatalytic CuBTC-MOF treatment system that can be used in industrial wastewater treatment plants has been proposed and constructed. The prototype has been fully automated to continuously channel wastewater into the central treatment cell for photodegradation to occur. The solution is then drained via a tap valve into another tank which is automatically drained away for industrial use. At the same time, the electromagnet helps to attract the MOF, preventing it from being drained out together with the purified water. On an industrial scale, such a prototype could be potentially used for a large-scale treatment process for textile wastewater, allowing for a facile treatment process.

In future, the ability of CuBTC-MOF in degrading other organic pollutants such as atrazine (a herbicide) and malathion (an insecticide) will be investigated. The concentrations of organic pollutants studied can also be scaled up to simulate industrial conditions.

Acknowledgements We would like to thank Mrs Sow-Peh Yoke Keow for her guidance and support.

References

1. Aquino, J. M., Rocha-Filho, R. C., Ruotolo, L. A., Bocchi, N., & Biaggio, S. R. (2014). Electrochemical degradation of a real textile wastewater using β-PbO2 and DSA® anodes. *Chemical Engineering Journal, 251*, 138–145.
2. Bhatia, S. C. (2017). *Pollution control in textile industry*. Woodhead Publishing India Ptv.
3. Chen, T., Zheng, Y., Lin, J. M., & Chen, G. (2008). Study on the photocatalytic degradation of methyl orange in water using Ag/ZnO as catalyst by liquid chromatography electrospray ionization ion-trap mass spectrometry. *Journal of the American Society for Mass Spectrometry, 19*(7), 997–1003.
4. Desore, A., & Narula, S. A. (2017). An overview on corporate response towards sustainability issues in textile industry. *Environmental Developmental and Sustainability, 20*(4), 1439–1459.
5. Dodd, A., Mckinley, A., Tsuzuki, T., & Saunders, M. (2009). Tailoring the photocatalytic activity of nanoparticulate zinc oxide by transition metal oxide doping. *Materials Chemistry and Physics, 114*(1), 382–386.
6. Fang, Z. L., Bueken, B., De Vos, D., & Fischer, R. A. (2015). Defect engineered metal-organic-frameworks. *Angewandte Chemie International Edition, 54*, 7234–7254.
7. Gadipelly, C. R., Perez-Gonzalez, A., Yadav, G. D., & Ortiz, I. (2014). Pharmaceutical industry wastewater: review of the technologies for water treatment and reuse. *Pharmaceutical Industry Wastewater: Review of the Technologies for Water Treatment and Reuse, 53*(29), 11571–11952.
8. Hassan, M. M., & Carr, C. M. (2018). A critical review on recent advancements of the removal of reactive dyes from dyehouse effluent by ion-exchange adsorbents. *Chemosphere, 209*, 201–219.
9. Holcomber, G. W., Phipps, G. L., Knuth, M. L., & Falber, T. (1984). The acute toxicity of selected substituted phenols, benzenes and benzoic acid esters to fathead minnows. *Environmental Pollution (Series A), 35*(4), 367–381.
10. Imran, M., Crowley, D. E., Khalid, A., Hussain, S., Mumtaz, M. W., & Arshad, M. (2014). Microbial biotechnology for decolorization of textile wastewaters. *Reviews in Environmental Science and Bio/Technology, 14*, 73–92.
11. Mattson, J. S., & Mark, H. B. (1971). *Activated carbon*. Marcel Dekker.
12. Moma J., & Baloyi J. (2018). *Modified titanium dioxide for photocatalytic applications*. IntechOpen.
13. Prival, M. J., Bell, S. J., Mitchell, V. D., Peiperl, M. D., & Vaughan, V. L. (1984). Mutagenicity of benzidine and benzidine-congener dyes and selected monoazo dyes in a modified Salmonella assay. *Mutation Research, 136*(1), 33–47.
14. Ragai, J., & Yacoub, N. (2013). Synergistic effect between TiO_2, zirconium and carbon in the photodegradation of methyl orange and methylene blue. *Adsorption Science and Technology, 31*(2–3)
15. Saquib, M., & Muneer, M. (2003). TiO_2-mediated photocatalytic degradation of a triphenylmethane dye (gentian violet), in aqueous suspension. *Dyes and Pigments, 56*, 37–49.

16. Sulistiono, D. O., Santoso, E., & Ediati, R. (2019). Enhanced adsorption of methylene blue and congo red from aqueous solutions by MCM-41/HKUST-1 composites. *Asian Journal of Chemistry, 31*(8), 1675–1682.
17. Thi, D. N., Ngoc, H. P., Manh, H. D., & Kim, T. N. (2011). Magnetic Fe_2MO_4 (M:Fe, Mn) activated carbons: Fabrication, characterization and heterogeneous Fenton oxidation of methyl orange. *Journal of Hazardous Materials, 185*(2–3), 653–661.
18. Tseng, D. H., Juang, L. C. & Huang, H. H. (2012). Effect of oxygen and hydrogen peroxide on the photocatalytic degradation of monochlorobenzene in TiO_2 aqueous suspension. *International Journal of Photoenergy, 2012.*
19. Wawrzkiewicz, M. (2012). Anion exchange resins as effective sorbents for acidic dye removal from aqueous solutions and wastewaters. *Solvent Extraction and Ion Exchange, 30*(5), 507–523.
20. Zhao, P., Fang, H., Mukhopadhyay, S., et al. (2019). Structural dynamics of a metal–organic framework induced by CO_2 migration in its non-uniform porous structure. *Nature Communications, 10*, 999.

Chapter 42
Object Motion Detection

Can Hu and Zilin Tao

Abstract This project aims to provide a robust and informative algorithm to detect object motion in the presence of camera motion. Our algorithm detects moving objects in a frame, generates bounding boxes around them and interprets them accurately and logically. Consequently, motion detection can be more readily applied in fields like autonomous vehicles and robots where computational resources are scarce. To detect motion from a moving camera, we identify the deviation of a region's optical flow from that of the surrounding objects. Next, we determine moving objects by analyzing pixel motions using two-dimensional histogram and image morphology. We use histogram of orientated gradient feature descriptor and intersection over union to identify moving objects, increase the robustness of our identification approach and establish connections of identical objects between frames. Even in instances where the detection fails, temporal consistency is maintained through the identification functionality. Nevertheless, false positive results do occur sometimes as our assumption—the pixel intensity of the same object is consistent between consecutive frames—may not always be satisfied in real-life situations, such as when shadows appear.

Keywords Computer vision · Motion detection · Optical flow · OpenCV

42.1 Introduction

Detection of object motion is important to both humans and vehicles for situation awareness and collision-free navigation in scenes. Such defection is especially important in real-life applications, such as autonomous vehicles, mobile robots, and unmanned aerial vehicles, all of which require motion detection from a moving camera. However, motion detection from a moving camera is challenging as the camera motion and object motion are mixed.

C. Hu (✉) · Z. Tao
Hwa Chong Institution, Singapore, Singapore
e-mail: 162027y@student.hci.edu.sg

H. Guo et al. (eds.), *IRC-SET 2021*, https://doi.org/10.1007/978-981-16-9869-9_42

Various research has been done to detect motion from a moving camera, among which two approaches are commonly adopted [1]. The first approach measures the deviation of the target object's radial optical flow field from that of the surrounding objects, which indicates whether the direction of motion and that of the speed of an object is consistent with its background motion patterns. The second approach utilizes adaptive background modeling by background subtraction. While research suggests that it works well with no abrupt changes in the background, high computational complexity is involved with the complex background modeling. Hence, our motion detection algorithm is largely based on the first approach.

Nevertheless, our algorithm differs from previous research by incorporating an identification process that can refine the detection result generated from the computation of optical flow. In particular, we employ the histogram of orientated gradient (HOG) feature descriptor [2] together with the intersection over union (IoU) evaluation metric [3], which make the detection algorithm more robust and informative. Nevertheless, since there are numerous possible camera motions, this project will focus on camera motion which is parallel to the surface of the stationary scene only.

The rest of this paper is organized as follows: Sect. 42.2 provides the algorithm of the motion detention process and the identification process, Sect. 42.3 presents the results of our algorithm, Sect. 42.4 discusses the limitations, and Sect. 42.5 concludes.

42.2 Algorithm

An overview of our algorithm is shown in Fig. 42.1. The algorithm can be divided

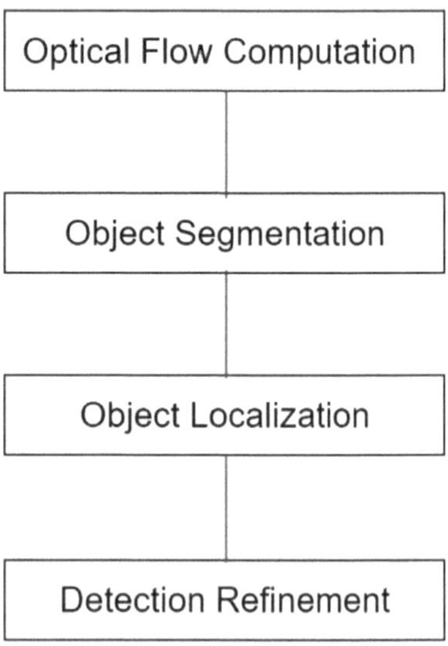

Fig. 42.1 Overview of the algorithm

into four main steps. Firstly, we calculate the frame-to-frame optical flow vectors. Secondly, we perform object segmentation to determine whether the pixels in the frame belong to object motion or camera motion. Thirdly, we locate each detected moving object with a bounding box. Lastly, we refine the detection result by re-establishing connections between frames to achieve temporal consistency.

42.2.1 Optical Flow Computation

Optical flow presents an apparent change of a moving object's location or deformation between frames [4–6]. Motion estimation via optical flow yields a two-dimensional vector field that represents velocities and directions of each pixel in a time sequence of two consequent images. The estimation is done under the assumption that pixels belonging to the same object between consecutive frames share the same intensity (i.e., pixels belonging to the same object have the same optical flow). For the actual calculation of the optical flow, the Lucas and Kanade method is used and implemented using the OpenCV library.

42.2.2 Object Segmentation

To achieve object segmentation in videos filmed from a moving camera, we need to differentiate between object motion and scene motion (i.e., camera motion). To calculate scene motion, we assume that the appeared area of the stationary scene is greater than that of the objects in motion. This assumption is likely to hold true since the optical flow of the stationary scene is uniform when camera motion is parallel to the surface of the stationary scene (i.e., when the direction of camera motion falls within the scope of this project). Therefore, scene motion can be obtained by finding the dominant motion in a video. For implementation, a two-dimensional histogram [7] is then used to find the optical flow vector of the most occurrence. Subsequently, we classify pixels whose motions are different from that of the camera motion as those belonging to the moving objects.

After object segmentation, we implement morphological operations [8, 9] to extract from images significant features that are useful in the representation and description of the shapes. To fill holes in the regions while keeping the initial region sizes, we implement erosion followed by dilation. In this way, separate parts of a moving object detected can be joined together and identified as one object, producing more accurate and logical output.

42.2.3 Object Localization

To locate the moving objects detected, we draw a contour around each object and generate a bounding box that best fits each contour. With the localization, we are also able to determine the number of moving objects in the scene.

42.2.4 Detection Refinement

Hitherto, the detection is computed independently for each frame. To apply the detection across time, improve the detection accuracy and to obtain a better interpretation of the objects in motion, we implement a simple identification of detected objects in our algorithm—where objects are identified in terms of identical objects, missing objects, reappeared (missing) objects, and new objects.

To that end, we use histogram of orientated gradient (HOG) feature descriptor [10] to extract the distribution of direction gradients of the detected objects as the appearance features for them. These features can facilitate the identification of objects, as a unique object that appears in individual frames is likely to share the same appearance feature. We express these features in terms of vectors with a dimension of (3780, 0) and calculate the Euclidean distance between them to determine their similarity. The resultant distance is compared with a threshold value to conclude the identity of the two objects. The threshold value is determined via several experimental trials.

Moreover, we use intersection over union (IoU), which measures the ratio of the area of overlap to the area of union between two object detection results, to enhance the identification obtained from HOG. In general, objects in neighboring frames will approximately be in the same location as the camera frame rate is high. Therefore, detected objects with a high IoU value are likely the same moving object between frames.

For the implementation of identification, two objects with a low Euclidean distance and a high IoU value are identified as identical objects. Otherwise, our algorithm would search the missing object list for a match: if there is a match, the missing object is considered to reappear in the current frame; if not, a new object is deemed to have appeared. When an identical object is not found in the next frame, the object is added to the missing list.

42.3 Result

The result of the computation of optical flow is shown in Fig. 42.2. Optical flow vectors in different directions are displayed in different colors for easy visualization. When the camera is stationary, only moving objects are colored. This means that only pixels belonging to moving objects possess observable optical flow vectors.

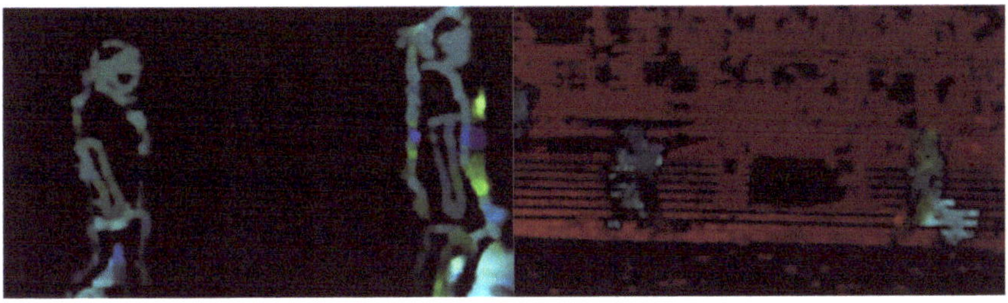

Fig. 42.2 Optical flow for a stationary camera (left) and a moving camera (right)

On the other hand, when the camera is moving, the majority of pixels are colored and deemed as moving in computer vision. Nevertheless, it can be clearly seen that the pixels belonging to the moving objects have different colors compared to those belonging to the stationary scene.

The object segmentation is shown in Fig. 42.3 where the region in white represents the moving objects detected and the region in black represents the stationary scene. We applied Image morphology to fill the holes with the outline of objects identified (Fig. 42.3).

We first tested our algorithm in the presence of obstruction. The detection of moving objects is successful as shown in Fig. 42.4: the bounding box on each moving object identified moves in tandem with the object motion throughout the video. Moreover, the algorithm was also able to identify the same person by labeling the same number in different frames, such as "Item 2." Moreover, when a person (is

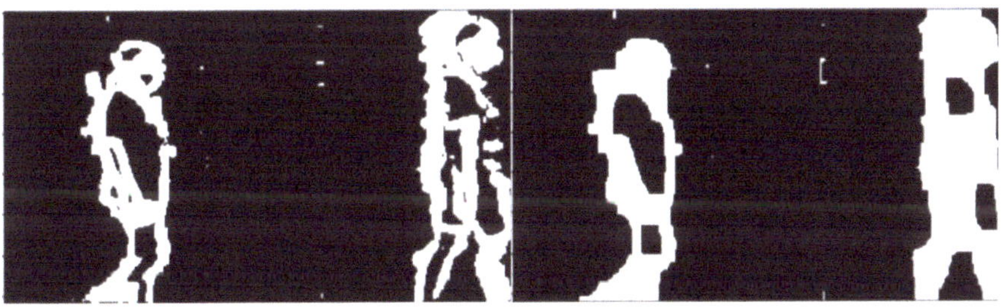

Fig. 42.3 Result of segmentation (left) and image morphology (right)

Fig. 42.4 Three output frames in a chronological order from a video with obstruction

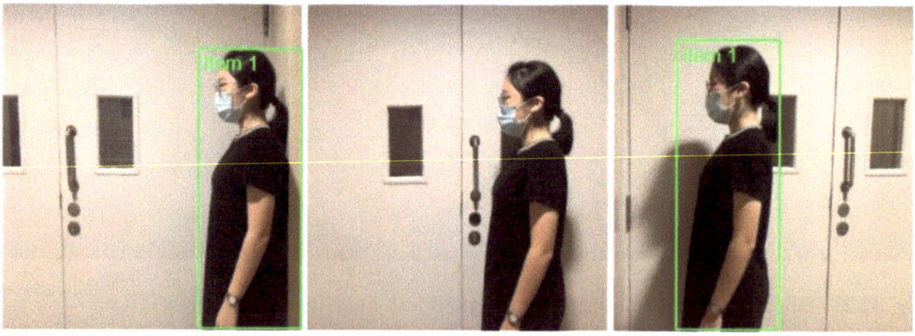

Fig. 42.5 Three consecutive output frames with failed detection in the second frame and successful detection in the first and third frame

blocked but) reappears after a few frames, the person would not be identified as a new person, but recognized as one of missing people appeared previously and thus inherited the original labeling.

We also tested our algorithm in cases where the motion detection via the computation of optical flow fails. While detection did fail in certain frames when the object motion is not apparent enough to generate detectable optical flow factors, the identification process in our algorithm was able to build a connection between consecutive frames with successful detections through the labeling of bounding boxes, hence mitigating the effect of false positive cases (Fig. 42.5).

42.4 Discussion

As shown in the results of the computation of optical flow above, optical flow can be used to detect object motion and differentiate object motion from scene motion effectively. However, the performance of such detection varies in different situations: motion is best detected on the outlines due to sudden changes in pixel intensities, whereas failure to detect observable motion often occurs in the middle of objects, away from the outlines, due to the aperture problem. Aperture problems arise when the motion of an object cannot be determined unambiguously if it is viewed from a small aperture such that the boundaries are not visible. For example, while the motions in the first and second situation in Fig. 42.6 are visible, the motion of the same object is invisible in the third situation as the boundaries can no longer be seen in the limited aperture. Hence, one of the two limitations of our algorithm is the inability to detect objects whose motions are invisible to computer vision due to the aperture problem.

The identification process equips our algorithm with the functionality to paint a more complete picture of the moving objects detected. This is especially useful in situations where objects (is obstructed and) become invisible in the video and when detection fails due to a failure in detecting distinct thus observable object

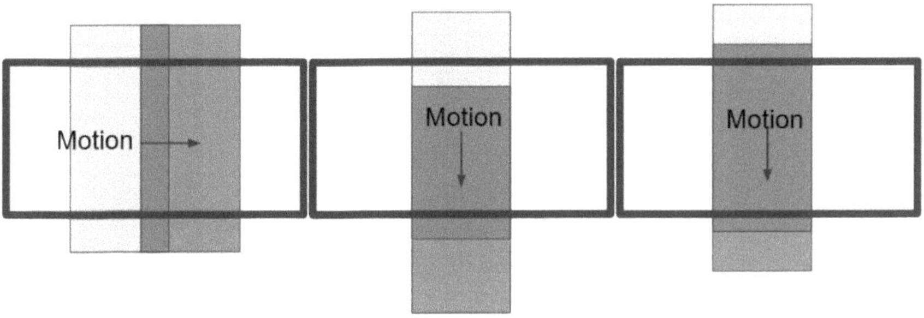

Fig. 42.6 Motions in a small aperture: visible (left and middle) and invisible (right)

motion from the camera motion. By determining the identity of objects detected, connections can be re-established between frames with successful detection results, thus mitigating the effects of imperfect detection.

The effectiveness of the identification process in our algorithm can be illustrated in the success of differentiating two people with similar features and the correct identification of the "new" person in the video (Fig. 42.4). Our algorithm is hence more robust and has achieved better temporal consistency. The findings of this algorithm can be applied in real life in various industries. For example, in the application of object detection for security purposes, the detection of a second person and the detection of the one person appearing again in surveillance cameras would lead to drastically different interpretations. The same principle is also true for the detection of walking pedestrians, driving cars, and other moving obstacles for manless vehicles and robots, where the number of distinct obstacles is closely related to the safety level.

Despite the aforementioned strengths of our algorithms, false positive cases do exist when the assumption in using optical flow does not hold true (i.e., the intensity of pixels representing the same object is consistent between consecutive frames). In Fig. 42.7, the stationary ground is falsely identified as moving objects due to the presence of shadow. Although videos which satisfy this assumption can be selected to test the validity of the research hypothesis, such assumptions may not always be satisfied in real-life scenarios, hence undermining the applicability of our algorithm.

42.5 Conclusion

Our project has presented an algorithm for object motion detection from videos taken by a moving camera. The detection of the moving objects is achieved through optical flow computation, object segmentation, and object localization. The detection result is further refined through the identification process via a feature descriptor and an evaluation metric. In general, our algorithm has achieved robust and informative motion detection.

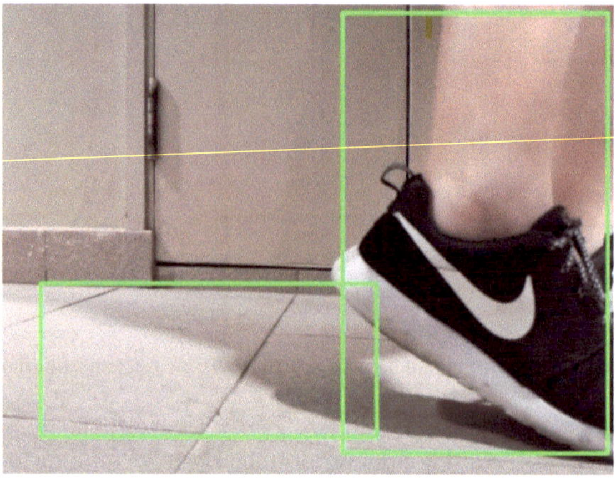

Fig. 42.7 False positive case due to the presence of shadow

Acknowledgements We thank Dr. Bai Jiamin and Dr Erkan Polatdemir for their mentorship, comments and support.

References

1. Chen, T., & Lu, S. (2017). Object-level motion detection from moving cameras. *IEEE Transactions on Circuits and Systems for Video Technology, 27*(11), 2333–2343.
2. Chopra, E. (2019). Using histogram of oriented gradients (HOG) for object detection. OpenGenus IQ: Learn Computer Science. OpenGenus IQ: Learn Computer Science, August 5, 2019. https://iq.opengenus.org/object-detection-with-histogram-of-oriented-gradients-hog/
3. Sheremet, O. (2020). Intersection over Union (IoU) Calculation for Evaluating an Image Segmentation Model. Medium. Towards Data Science, September 7, 2020. https://towardsdatascience.com/intersection-over-union-iou-calculation-for-evaluating-an-image-segmentation-model-8b22e2e84686
4. Optical Flow. (2020). OpenCV. Accessed June 24, 2020. https://docs.opencv.org/3.4/d4/dee/tutorial_optical_flow.html
5. Optical Flow. (2020). OpenCV. Accessed June 24, 2020. https://opencv-pythontutroals.readthedocs.io/en/latest/py_tutorials/py_video/py_lucas_kanade/py_lucas_kanade.html
6. Pal, S. (2020). OpenCV Functions: OpenCV For Computer Vision. Analytics Vidhya, June 24, 2020. https://www.analyticsvidhya.com/blog/2019/03/opencv-functions-computer-vision-python/
7. Holtz, Y. (2017). #83 Basic 2D Histograms with Matplotlib. The Python Graph Gallery, August 24, 2017. https://python-graph-gallery.com/83-basic-2d-histograms-with-matplotlib/
8. Morphological Image Processing. Accessed June 28, 2020. https://www.cs.auckland.ac.nz/courses/compsci773s1c/lectures/ImageProcessing-html/topic4.htm
9. Morphological Transformations. OpenCV. Accessed July 2, 2020. https://docs.opencv.org/trunk/d9/d61/tutorial_py_morphological_ops.html
10. Mallick, S. (2018). Histogram of Oriented Gradients: Learn OpenCV." Learn OpenCV|OpenCV, PyTorch, Keras, Tensorflow examples and tutorials, February 5, 2018. https://www.learnopencv.com/histogram-of-oriented-gradients/

Chapter 43
Spectral Analysis of Public Transport Networks

See Min Lim, Kai Ting Ho, and Yuanfei Li

Abstract In this project, the effects of changes to public transport networks (PTNs) on the centrality of stations and the overall connectivity of the network were analysed using spectral graph theory. We began by modelling Singapore's PTN as a graph. To determine the effects of alterations to the PTN on centrality and connectivity, we analysed the relative eigenvalues and eigenvectors of the adjacency, Laplacian and random walk matrices of the modified graphs. Additionally, to determine the optimal network topology, random transport networks (TNs) with varying graph topologies were generated based on actual TNs around the world and analysed. It was found that stations in the northeast region had relatively low centrality and that stations on the Downtown Line and Circle Line were generally, relatively, underutilised. Overall connectivity increased with the addition of lines, and the addition of a straight line that cuts across Singapore from east to west improved connectivity most significantly, as compared to adding lines of other shapes. In the randomly generated TNs, the star (or hub-and-spoke) topology generally displayed the highest connectivity, as compared to the grid and wheel topologies. Our project shows that spectral graph theory allows for efficient measurement of critical properties of TNs such as centrality and connectivity, which is useful for city planners in planning and optimising PTNs.

Keywords Spectral graph theory · Transport networks · Random walks · Graph Laplacian · Topology

43.1 Introduction

Transportation is a system of conveying people or goods from one place to another [1]. In the physical world, this has utmost significance and relevance as it is the basis of virtually all social, economic and political activities [2]. A robust public

S. M. Lim (✉) · K. T. Ho · Y. Li
Hwa Chong Institution (College Section), Singapore, Singapore
e-mail: lsm.limseemin@gmail.com

K. T. Ho
e-mail: kaitinghh@gmail.com

H. Guo et al. (eds.), *IRC-SET 2021*, https://doi.org/10.1007/978-981-16-9869-9_43

transport system, including buses and trains, becomes especially important in the context of providing mobility for people who do not have access to private transport [3]. In today's era where we are looking towards moving to a greener lifestyle, greater adoption of public transport is also encouraged all around the world, due to the role of public transport in conserving energy and reducing emissions [4].

Network route design is the first step in building efficient and well-connected public transport networks, followed by timetable development and vehicle and crew scheduling [5].

In network route design, city planners analyse the network structure of a public transport network to identify bottlenecks or assess the connectivity of the city as a whole. This also helps them predict the consequences of temporary events such as disruptions, or long-term events such as adding a new train line. With this information, city planners can then decide the best course of action that can be taken to improve the public transport network.

In this project, we investigate public transport networks using spectral graph theory. We compute the eigen-centrality, algebraic connectivity and convergence rate of graphs derived from public transport networks, and we use these to gain insights on how to plan such networks.

43.2 Spectral Graph Theory

Spectral graph theory is the study of a graph's properties by looking at eigenvectors and eigenvalues of matrices associated with the graph. A weighted graph, G, can be represented as an $n \times n$ **adjacency matrix**, A, where n is the number of nodes in G. The entries of A are$a_{ij} = b$, where b is the sum of weights on all the edges between nodes i and$j$. The **degree matrix**, D, is an $n \times n$ diagonal matrix with $d(1), d(2), ..., d(n)$ along the diagonal, where $d(i) = \sum_{j=1}^{n} a_{ij}$. From the matrices A and D, we can generate the **random walk matrix**, $W = AD^{-1}$, which is a matrix with entries $w_{ij} = \frac{a_{ij}}{d(j)}$ representing the probability of ending up at node i from node j after a movement away from node j with equal probability among the neighbours of node$j$. We can also generate the **Laplacian matrix**, $L = D - A$, which is used in computing algebraic connectivity.

We focus on certain eigenvalues and eigenvectors of the matrices A, W and L. An **eigenvector** of a matrix, M, is a non-zero vector x which satisfies the equation $Mx = cx$ for some value c, which is called an **eigenvalue** of M. The eigenvector centrality or **eigen-centrality** of node $\boldsymbol{i}$, denoted $\boldsymbol{c(i)}$ is the ith entry of the strictly positive normalised eigenvector of A corresponding to the largest eigenvalue. Such an eigenvector that is strictly positive will always exist for an adjacency matrix A according to the Perron-Frobenius Theorem [6]. The eigen-centrality is a positive number, and the larger the value, the more central the node is [7]. The **convergence rate** of a random walk on the graph is $\rho = 1 - \omega_2$, where ω_2 is the second largest eigenvalue of W (the largest eigenvalue of W is always 1). The larger the convergence rate, the faster a random walk on G will converge to equilibrium [8], i.e. the probability vector, p, of the nodes converges to the equilibrium distribution of the random walk. The **algebraic connectivity** of a graph is given by $\alpha = \lambda_2$, where λ_2 is the second smallest eigenvalue of L(the smallest eigenvalue of L is always 0). The larger the algebraic connectivity, the more connected a graph is; in fact, adding more edges to G will always increase the algebraic connectivity [9].

43.3 Graphs from Transport Networks

The LTA DataMall API provides information on all of Singapore's bus, MRT and LRT routes, stations and passenger volume.

Using the LTA data, we generated different graphs representing Singapore's public transport network. We started with the bus network as a base to prevent having unconnected nodes in our graphs and subsequently added edges to represent MRT/LRT lines. For the bus network, bus stops on opposite sides of the road and "staggered" bus stops, designed to reduce congestion in busy areas such as Orchard and which are right next to each other, were considered to be the same node in the graph, resulting in a graph with 2840 nodes. We assigned edges corresponding to bus routes and MRT routes a weight of 1 and 35, respectively. We attained this ratio of weights by multiplying our estimates of the relative capacity, frequency and speed of MRT trains versus buses. We generated two types of graphs based on the public transport route data: the **adjacent graph** and the **transfer graph**. In the adjacent graph, two nodes are connected by an edge if they are one stop after the other on some MRT/LRT route. In the transfer graph, two nodes are connected as long as they lie on the same MRT/LRT route. Figure 43.1 illustrates the differences between these two graphs.

The purpose of the transfer graph is to capture the transfers that a passenger has to make when travelling from one node to another node in the graph. The bus network graphs in both graphs were generated with the adjacent graph method since its purpose is to serve as a base, and we will not be investigating the effects of any modifications to bus routes.

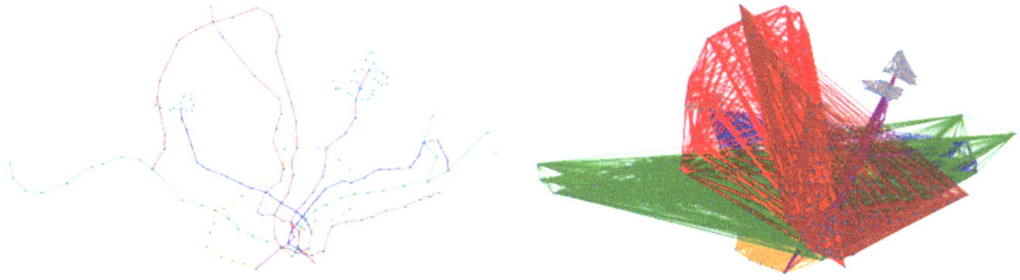

(a) Adjacent graph *(b) Transfer graph*

Fig. 43.1 The adjacent and transfer graphs of Singapore's MRT/LRT networks

43.4 Effect of MRT Lines on Centrality

We started by investigating the centrality of MRT stations, as well as how adding and removing MRT lines affected the centrality. We added each MRT line to the base bus network graphs and computed the eigen-centrality of nodes after adding each MRT line. Figure 43.2 shows the eigen-centrality of nodes on the adjacent and transfer graph of the current transport network. In Fig. 43.2a, the nodes in the city centre are the most central, while in Fig. 43.2b, nodes with high centrality are more spread out and correspond to MRT stations. The transfer graph shows that the East–West Line (EWL) and Downtown Line (DTL) dominate in centrality, meaning that they require the least transfers to get to other nodes. Stations in the northeast region had very low

(a) Adjacent graph

(b) Transfer graph

Fig. 43.2 Geographical plot of bus stops and MRT/LRT stations in Singapore, with sizes and colours of nodes scaled by eigen-centrality. Darker and larger nodes are more central

centrality compared to other regions, suggesting that enhancing connectivity to the region will improve convenience significantly.

Figure 43.3 illustrates how the centralities of nodes changed with the addition of each MRT line. The North–South Line (NSL) and EWL greatly increase the centrality of City Hall and Raffles Place, which then decreases as more lines are added. This suggests that the NSL and the EWL are vital in connecting City Hall and Raffles Place to the rest of Singapore.

To test the effect of adding parallel lines, we compared the centrality of stations on the NSL before and after the adding the TEL. We hypothesised that the centrality of stations which are on both lines would increase, while those only on the original line would decrease. A pronounced increase in the centrality of stations that lie on both the TEL and NSL was observed, which aligns with our hypothesis. To test the effect of adding cycle or circular lines, we compared the current MRT network with the Circle Line (CCL) closed up and the MRT network with CCL removed.

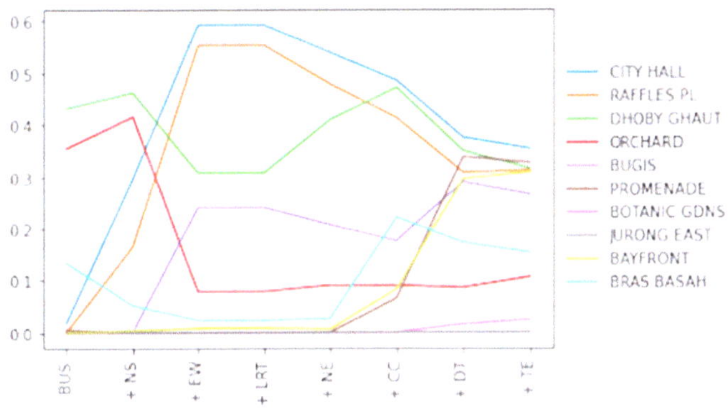

(a) Adjacent graph

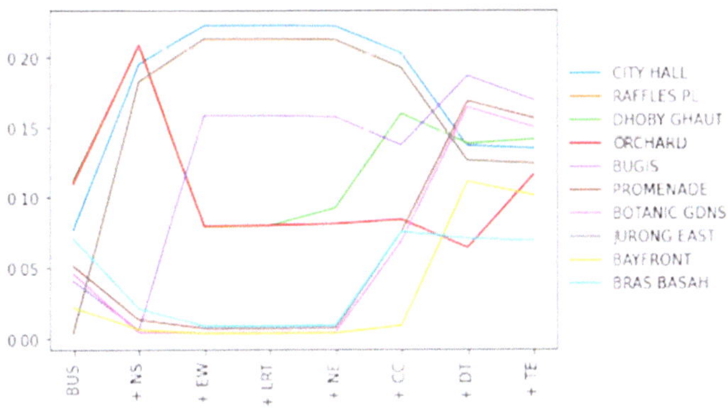

(b) Transfer graph

Fig. 43.3 Eigen-centrality of some nodes as MRT lines are added. (In **(b)**, the lines for City Hall and Raffles Place overlap)

The centrality of some stations "surrounded" by the circle such as Little India and City Hall decreased when the CCL was added or closed up. However, we found that for some stations such as Bugis, the centrality increases in the adjacent graph but decreases in the transfer graph when the CCL was added or closed up. We speculate that this is because Bugis is close to the CCL and thus more connected to the CCL, while stations like Little India and City Hall are further within the circle.

We also wished to identify stations prone to congestion or underutilisation. To investigate the relationship between passenger volume and centrality, we plotted the combined tap-in/tap-out volume at each MRT node against its eigen-centrality, as shown in Fig. 43.4. We hypothesised an increasing straight-line trend, since a more central and well-connected station is likely to have higher passenger volume. Points above the line would indicate that the node is congested while points below the line indicate that the node is underutilised. This conjecture links the centrality-passenger volume ratio to utilisation based on the idea that a more central station, with better opportunities to transit to other routes and more routes to reach the station, can handle a higher passenger volume. However, Fig. 43.4 does not show an obvious trend, though the transfer graph does have a slight increasing trend. In the transfer graph, nodes on the same MRT line tend to have similar centrality values. The passenger volumes at NSL and EWL were quite spread out, while the passenger volumes on the CCL or DTL were low compared to their centrality, suggesting that stations on these lines are underutilised.

43.5 Effect of MRT Lines on Connectivity

Next, we wish to investigate the effect of adding or removing MRT lines on the overall connectivity of Singapore's transport network. We quantified connectivity in two ways: the convergence rate ρ and the algebraic connectivity α, both computed using the transfer graph. We hypothesised that removing existing MRT lines will decrease the overall connectivity of the public transport network. We experimentally verified this by removing each individual MRT line and seeing its effect on connectivity; results are shown in Table 43.1.

Using algebraic connectivity, all results agree with the hypothesis, which is consistent with the theorem that adding edges to a graph increases the algebraic connectivity [9]. Using convergence rate, the removal of all lines except LRT and the TEL causes a decrease in connectivity, which is generally consistent with our hypothesis. The results also show that the EWL contributes the most to the overall connectivity, suggesting that a breakdown of the EWL is likely to have the most detrimental effect compared to other lines. We speculated that it is because the EWL has many stations (33, only less than that of the DT line, which has 35) and connects distant east and west stations together.

To investigate whether shape was part of the reason why the removal of the EWL has the most impact on the overall connectivity, we added two fictional straight and circular lines and hypothesised that adding a circular line will increase the overall

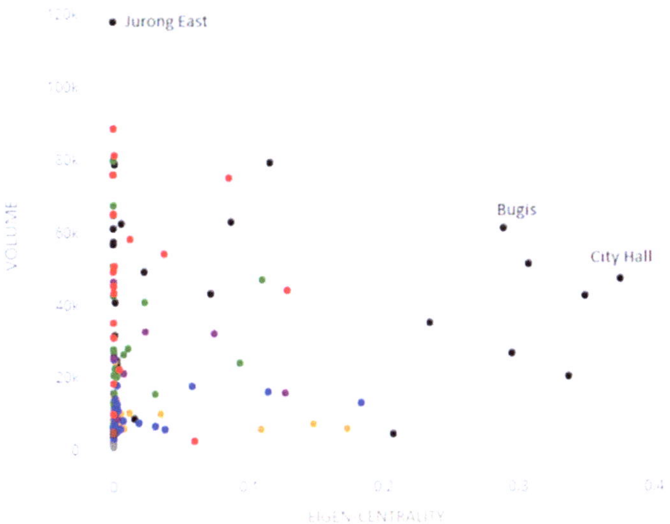

(a) Adjacent graph

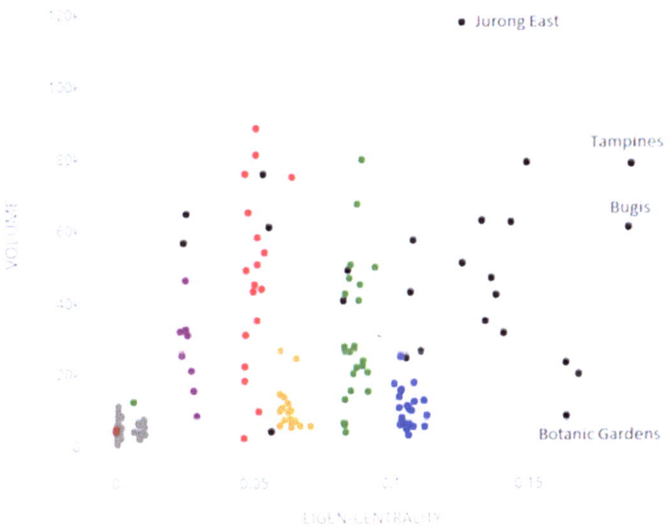

(b) Transfer graph

Fig. 43.4 Passenger volume against eigen-centrality on the current MRT network. Nodes are coloured by the MRT line they are on; interchanges are black, while LRT stations are grey

Table 43.1. Effect of removing each of the MRT lines on overall connectivity

Removed line	Algebraic connectivity	Convergence rate
None (MRT network now)	0.99952311	0.0641863711
BP Bukit Panjang LRT	0.99952296	0.0641864270
SK Sengkang LRT	0.99952288	0.0641949089
NE North East Line	0.99952211	0.0641821534
PG Punggol LRT	0.99952171	0.0641957182
TE Thomson-East Coast Line	0.99952067	0.0641935214
CC Circle Line	0.99951915	0.0641664579
DT Downtown Line	0.99951739	0.0641479296
NS North-South Line	0.99951518	0.0641324507
EW East-West Line	0.99750404	0.0365902257
All (only bus network)	0.98167060	0.0366587967

Darker green colour represents lower connectivity after removing the line, which means that the line contributes *more* to connectivity; yellow colour represents the reverse

connectivity more than adding a straight line that cuts across the network. We generated two hypothetical MRT lines using existing bus stops as MRT stations. The straight MRT line ***(CR line)*** that cuts across the network was generated partly using the LTA plan for Cross Island Line. The circular MRT line ***(YF line)*** was fictional and selected so that it does not overlap with existing lines. To minimise the effect of factors other than the shape of the MRT lines, we kept the number of stations roughly the same, at 21 and 20, respectively. Figure 43.5 shows the MRT network with CR line (in magenta) and YF line (in cyan), while Table 43.2 displays the connectivity results. The straight CR line significantly increases connectivity while the circular YF line only increases it by a small amount.

The reason behind this dramatic difference is likely due to a lack of MRT network in the West Jurong residential area and an abundance of MRT lines in the central area. Hence, our result shows that the construction of the Cross Island Line would, in theory, significantly boost the connectivity of Singapore's public transport network.

Fig. 43.5 Singapore's MRT/LRT network (grey) along with the CR (magenta) and YF (cyan) lines

Table 43.2 Effect of adding CR/YF MRT lines on the overall connectivity

Added line	Algebraic connectivity	Convergence rate
None (MRT network now)	0.999523105	0.06418637114
CR (straight)	1.630404901	0.06427608787
YF (circular)	0.999524452	0.06419152805
CR, YF	1.630429018	0.06427666457

Darker colour represents lower connectivity

43.6 Effect of Network Topology on Connectivity

Finally, we investigated how overall network topology affects the connectivity of transport networks. We generated six random cities with transport networks of 50 nodes, 7 lines and 60 edges each. The nodes were taken and modified from existing public transport networks in various cities [10]. For each of these cities, we connected the nodes using the following network topologies—**grid**, hub-and-spoke ("**star**") and concentric circles ("**wheel**"). Figure 43.6 shows an illustration of these topologies.

These network topologies are displayed in public transport networks in real cities: Adelaide's Train and Tram network (Fig. 43.7) is an example of a star topology; Tokyo's Metro network (Fig. 43.8) is a combination of the grid and star topology; Berlin's Transit system (Fig. 43.9) is a combination of the wheel and star topology.

(a) Grid

(b) Star

(c) Wheel

Fig. 43.6 Example of different topologies on the same randomly generated city

The convergence rate and algebraic connectivity of each configuration are shown in Table 43.3. For the adjacent graph, the star and wheel graphs tended to have the highest convergence rate and algebraic connectivity. For the transfer graph, the star topology had the highest convergence rate, followed by the grid and wheel, while the grid and the star networks had the highest algebraic connectivity. From our results, the star network by and large had a higher degree of connectivity than the other two topologies. This is likely because the structure of the star networks included 'hubs' which connected 3 lines or more. These hubs ensured that a commuter would be able to transfer between lines to get to different locations more easily. This is in contrast to the grid and wheel networks which might not have as many opportunities for line transfers, when the number of nodes and edges were kept constant, and hence would have lower degrees of connectivity. These results suggest that in theory, a star network topology is the most ideal network topology in terms of connectivity, compared to the other common topologies such as grid and wheel.

Fig. 43.7 Adelaide Metro Train and Tram [11]

43.7 Conclusion

Spectral graph theory allows for efficient measurement of critical properties of transport networks such as centrality and connectivity, the analysis of which can lead to an optimal transportation system. The adjacent graph tended to overemphasise the impact of each edge, while the transfer graph overemphasised lines connecting Singapore lengthwise. Locally, it can be observed that newer lines reduce the centrality of City Hall and Raffles Place while lines connecting east and west influence connectivity the most. It can be observed that the star topology appears to be the best for random cities.

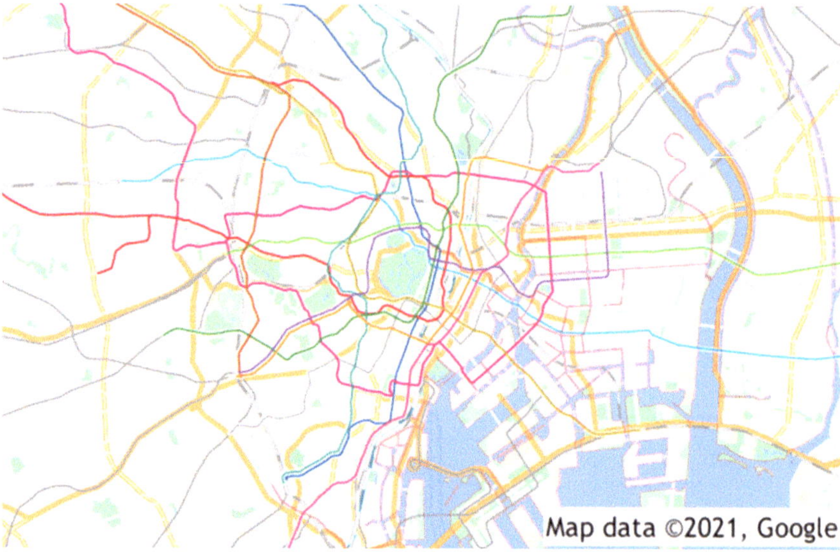

Fig. 43.8 Tokyo Metro [12]

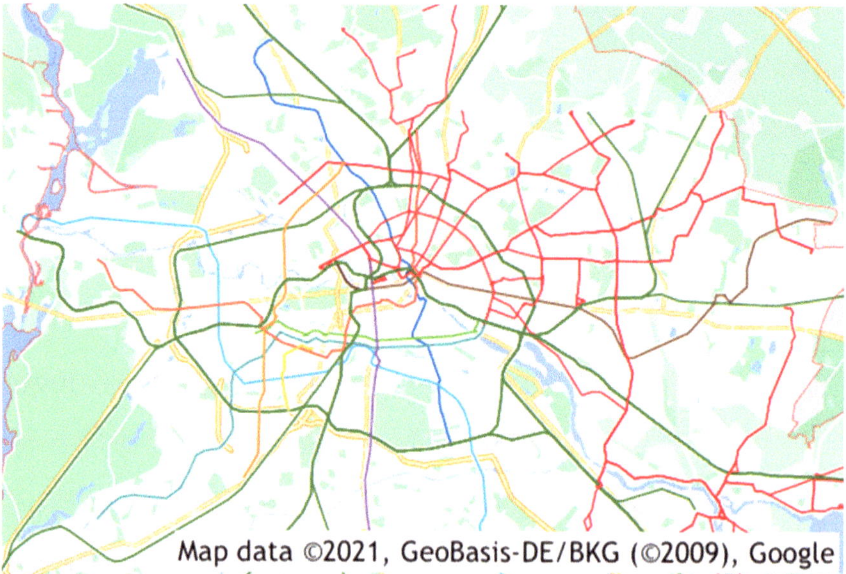

Fig. 43.9 Berlin transit system [13]

Table 43.3 Connectivity of different topologies

	Network topology		
City	Grid	Star	Wheel
(a) Convergence rate of adjacent graph			
1	0.034128	0.046172	0.050791
2	0.028799	0.042919	0.056498
3	0.023258	0.055251	0.047955
4	0.034391	0.045770	0.036295
5	0.024806	0.033140	0.018922
6	0.024051	0.025500	0.054148
(b) Convergence rate of transfer graph			
1	0.132572	0.144666	0.073330
2	0.101115	0.165486	0.065653
3	0.116859	0.157947	0.068176
4	0.178928	0.205928	0.056845
5	0.053445	0.168920	0.019680
6	0.095898	0.121986	0.072681
(c) Algebraic connectivity of adjacent graph			
1	0.067554	0.079345	0.107851
2	0.059313	0.075628	0.122034
3	0.046182	0.097935	0.100916
4	0.070556	0.078050	0.074980
5	0.046215	0.057329	0.040076
6	0.048374	0.045627	0.116892
(d) Algebraic connectivity of transfer graph			
1	1.082551	1.341144	0.442669
2	1.126870	1.300561	0.546980
3	1.405127	1.384322	0.522752
4	1.836886	1.798760	0.701273
5	0.428298	1.788097	0.285505
6	1.093711	1.238285	0.666752

Darker cells indicate higher connectivity within each row

In future work, the combined effect of distance and transfers can be investigated through a transfer graph where each MRT/LRT route edge is weighted based on distance. In addition, passenger volume can be more accurately deduced through data on origin–destination passenger volume and Dijkstra's shortest-path algorithm.

Acknowledgements We thank our research mentor Dr. Wong Liang Ze for mentoring us in understanding spectral graph theory and for close supervision and guidance of this project, our teacher mentor Dr. Wee Juan Dee for providing us with administrative help and the many opportunities throughout this research journey, and our school Hwa Chong Institution (College Section) for providing us with this opportunity to embark on this research project. This work contains information from LTA DataMall API for Bus Routes and Passenger Volume by Train Stations accessed on 19 September 2020 and 19 November 2020 respectively from LTA DataMall which is made available under the terms of the Singapore Open Data Licence version 1.0. https://datamall.lta.gov.sg/content/datamall/en/SingaporeOpenDataLicence.html.

References

1. Collins, W. (2004). *Transportation definition and meaning: Collins english dictionary*. HarperCollins.
2. Rodrigue, J. P. (2016). *The geography of transport systems*. Taylor & Francis.
3. Banister, D. (2011). Cities, mobility and climate change. *Journal of Transport Geography, 19*(6), 1538–1546.
4. Shapiro, R. J., Hassett, K. A., & Arnold, F. S. (2016). Conserving energy and preserving the environment: The role of public transportation.
5. Ceder, A. (2016). *Public transit planning and operation: Modeling, practice and behavior*. CRC Press.

6. Tudisco, F., & Higham, D. J. (2021). Node and edge eigenvector centrality for hypergraphs. arXiv preprint arXiv:2101.06215
7. Straffin, P. D. (1980). Linear algebra in geography: Eigenvectors of networks. *Mathematics Magazine, 53*(5), 269–276.
8. Lovász, L. (1993). Random walks on graphs: A survey. *Combinatorics, Paul erdos is eighty, 2*(1), 1–46.
9. Fiedler, M. (1973). Algebraic connectivity of graphs. *Czechoslovak mathematical journal, 23*(2), 298–305.
10. Kujala, R., Weckström, C., Darst, R. K., Mladenović, M. N., & Saramäki, J. (2018). A collection of public transport network data sets for 25 cities. *Scientific Data, 5*, 180089. https://doi.org/10.1038/sdata.2018.89
11. Google. (n. d.). [Google Maps Adelaide, SA, Australia with Transit layer]. Retrieved October 7, 2021, from https://www.google.com/maps/@-34.9998809,138.3302859,10z/data=!5m1!1e2
12. Google. (n. d.). [Google Maps Tokyo, Japan with Transit Layer]. Retrieved October 8, 2021, from https://www.google.com/maps/@35.6873306,139.7183754,12z/data=!5m1!1e2
13. Google. (n. d.). [Google Maps Berlin, Germany with Transit Layer]. Retrieved October 7, 2021, from https://www.google.com/maps/@52.5069296,13.1438639,10z/data=!5m1!1e2

Chapter 44
Antimicrobial, Antioxidant, Toxicity and Phytochemical Screening of *Dieffenbachia Camilla* and the Synthesis of a Novel and Green Topical Delivery Method for It

Mahendran Ravindran, Ethan Herng Rwei Lim, and Ming Jie Kuan

Abstract Chronic wounds are prone to bacterial infections. These have gained multidrug resistance, creating an urgent need for novel, medically-applicable compounds. We investigated the antimicrobial, antioxidant, toxicity and phytochemical composition of *Dieffenbachia camilla* (DC). Toxic compounds such as calcium oxalate were removed via hexane phase separation to prepare DC extracts for relevant tests. Such DC extracts proved to be non-toxic to nauplii in brine shrimp lethality assays. Well diffusion and time point assays ascertained DC's antibacterial properties, whilst MIC assays showed only a low concentration of DC extract was needed to inhibit bacterial growth, likely due to its abundance of phytochemicals which mainly consisted of flavonoids such as Epigallocatechin 3-O-gallate. These flavonoids were identified by ultraviolet spectroscopic analysis, and are known to bind to porin proteins on bacterial cell membranes, inhibiting the transport of hydrophilic molecules, hence arresting cellular respiration, inducing cell death. DC also demonstrated antioxidative and free radical scavenging activity in DPPH assays as potent antioxidants such as quercetin were present. DC increased membrane permeability when tested via UV-spectrofluorimetric analysis using propidium iodide solution, disrupting homeostasis in bacterial cells, causing cell death. DC was proven fungicidal in an antifungal assay, likely due to the activation of antifungal compounds via hexane. DC inhibited biofilm formation and showed quorum quenching potential due to the presence of the flavonoid, chrysin that impeded cell-signalling molecules. Chrysin binds to the bacterial LasR protein, significantly reducing its ability to encode Quorum Sensing-regulated promoters, thus achieving quorum quenching potential. The DC extracts after hexane phase separation were synthesised into a zein fibroin mesh via electrospinning, and with DC's antimicrobial, non-toxic and membrane permeability properties, this zein fibroin mesh has the potential to be an inexpensive and effective topical delivery agent.

Keywords Dieffenbachia camilla · Antimicrobial · Topical delivery agent

---

R. Mahendran (✉) · E. H. R. Lim · M. J. Kuan
Hwa Chong Institution, 661 Bukit Timah Rd, Singapore 269734, Singapore
e-mail: 181745h@student.hci.edu.sg

H. Guo et al. (eds.), *IRC-SET 2021*, https://doi.org/10.1007/978-981-16-9869-9_44

44.1 Introduction

Chronic wounds are prone to multidrug-resistant bacterial infections [5], creating an urgent need for novel, medically-applicable antimicrobials [7].

Dieffenbachia camilla (DC) is a common plant distributed around Southeast Asia and thought to contain many many bioactives that might be antimicrobial. Unfortunately, DC's potential application has been overshadowed by potentially harmful effects of the calcium oxalate [3] it contains, and is thus not often explored. DC is filled with phytochemicals, used for protection and may confer benefits during wound healing [32]. Various methods are also being investigated for the topical delivery of these compounds. Zein, for example, is a naturally occurring Maize protein that can be electrospun to produce fine meshes of less than 0.2 nm [15]. It shows great potential for sustained diffusion of drugs, making it a good candidate to produce a biodegradable topical delivery agent [2].

44.2 Objectives and Hypothesis

This study aims to investigate the antimicrobial, antioxidant, toxicity properties and phytochemical composition of DC and a "green" topical delivery method for these compounds. This study postulates that active compounds in DC show potent antimicrobial and antioxidant activity, and can be integrated into an environmentally sustainable nanomesh for sustained topical application.

44.3 Methodology

44.3.1 Extraction and Preparation of Samples

Leaves of DC were blended with either deionised water or 1% dimethyl sulfoxide (DMSO) solution and were then centrifuged at 8000 rpm for 10 min at 15 °C. The supernatants were decanted, filtered with a 0.2 μm syringe driven filter, aliquoted, and stored at 0 °C for future use. 50% (m/v) extracts were also dialysed against sterile water, whilst some were also acidified and neutralised with 2 M of hydrochloric acid and 2 M sodium hydroxide, respectively.

44.3.2 Hexane Phase Separation

Hexane and DC extracts were added to a separating funnel in a 1:1 ratio and left overnight to remove toxic compounds. The hexane fractions were physically separated from the remaining DC extracts. The hexane was removed using a rotary evaporator at 40 °C. Compounds previously removed by hexane were separately redissolved in water and DMSO.

44.3.3 Well Diffusion Assay

Escherichia coli and *Staphylococcus epidermidis* were separately cultured in nutrient broth at 37 °C overnight. The bacteria were spread evenly onto sterile agar plates and wells were punched. 100 μL of either DC extracts, Ampicillin or sterile water was introduced into each well, then incubated overnight. Any resultant zones of inhibition generated were documented [23].

44.3.4 Time Point Assay

E. coli and *S. epidermidis*, were cultured overnight in nutrient broth, quantified and its numbers were adjusted to a starting concentration of 10^6 CFU/ml. The bacterial suspension was then mixed with either DC extracts, 10% bleach, or sterile water. Samples of the mixtures were taken at 10-min intervals, serially diluted, and plated on agar plates. These were incubated at 37 °C overnight. The resultant colonies at various time points were counted.

44.3.5 2,2diphenyl-1-Picrylhydrazyl (DPPH) Radical Scavenging Activity Assay

1.9 mL of methanol was mixed with 0.1 mL of DC extract and 1 mL of DPPH solution. Positive control of 1.9 mL methanol mixed with 0.1 mL of ascorbic acid and 1 mL DPPH solution, and negative control of 2 mL methanol and 1 mL of DPPH solution was also prepared. The tubes were placed in the dark for 10 min before spectrophotometric measurements were taken at 517 nm wavelength of light. By calculating the DPPH left, the antioxidant effects of the extracts were determined.

44.3.6 Minimum Inhibitory Concentration Assay

Bacteria were cultured overnight in nutrient broth using a shaking incubator. The resultant bacterial suspension was mixed with decreasing concentrations of DC extract or sterile water as the negative control and incubated overnight. Spectrophotometric measurement of each sample was taken at 517 nm wavelength of light, and the minimum inhibitory concentration was calculated.

44.3.7 Antifungal Assay

Aspergillus niger (Carolina) was cultured on potato dextrose agar (PDA) at room temperature. 5 mL of DC extract was added to 95 mL of molten PDA. After solidification, a thin layer of *A. niger* agar was cut out and placed in the middle of the Petri dishes containing PDA with or without DC extract. The resultant diameter of fungal surface growth was monitored over a week.

44.3.8 Brine Shrimp Lethality Assay (BSL)

100 μL of freshly-hatched brine shrimp mixture was mixed with 50 μL of extract or 1% DMSO. The brine shrimps were examined under a light microscope immediately after the addition of extract, and again after 24 h. The average number of shrimps that survived was determined.

44.3.9 Membrane Permeability Assay

Bacteria were cultured and incubated overnight. The bacterial mixture was centrifuged to obtain a cell pellet, rinsed with Phosphate-Buffered Saline whilst the supernatant was aspirated. 100 μL of extract, sterile water or ethanol was added to bacterial culture, 30 μM of propidium iodide and 700 μL of sterile saline. The resultant mixture was incubated in the dark for 15 min, then excited under 533 nm of fluorescent light. Emissions ranging from 555 to 655 nm were measured.

44.3.10 Biofilm Microtiter Assay

S. epidermidis and *E. coli* were cultured overnight at 37 °C, then diluted in a 1:100 ratio and incubated at 37 °C with 100 μL of extracts. The wells were washed with a

Table 44.1 Phytochemical analysis tests

Secondary metabolite	Test
Flavonoids	Alkaline Reagent Test [30]
	Shinoda Test [12]
Phenols	Ferric Chloride Test [24]
Quinones	H_2SO_4 Test [26]
Steroids	Liebermann Burchard Test [4]
Saponins	Froth Test [10]
Cardiac Glycosides	Keller-Kiliani Test [14]
	Salkowski's Test [1]
Essential Oils	NaOH - HCl Test [20]

PBS buffer and stained with 125 μL of 0.1% crystal violet. Finally, the cell-bound crystal violet was dissolved in 33% acetic acid. Biofilm was monitored at 570 nm using a microplate reader.

44.3.11 *Phytochemical Analysis*

Phytochemical analysis for 8 phytochemicals and secondary metabolites using 8 different tests as shown in Table 44.1 was performed.

44.3.12 *Flash Column Chromatography*

Cotton wool, sand and a homogenous hexane-silica gel slurry was added to a glass column. 3 mL of extracts was added and allowed to settle into the silica column. The solvent's polarity was increased by mixing hexane with ethyl acetate in a 5:1, 2:1, 1:1 ratio. Different fractions were collected and analysed using Ultraviolet spectroscopy.

44.3.13 *Synthesis of DC/Fibroin via Electrospinning*

0.50 g of zein and 0.375 g of extracts containing the lower molecular weight fraction and hexane, previously concentrated using a centrivap, was magnetically stirred with 15.0 mL of absolute ethanol. The solutions were drawn into a 20 mL syringe. A flattened 23-gauge needle was screwed onto the syringe and secured into a syringe pump. The live wire of the high-voltage DC power supply was attached to the needle near its tip whilst the ground wire was attached to the collector. After the syringe pump was turned on at 4 ml/h and the DC power supply was turned on at 10 kV,

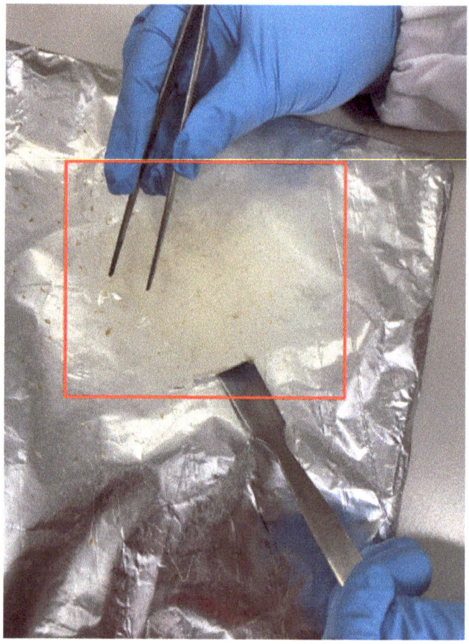

Fig. 44.1 Electrospun mesh of zein fibroin

a charged jet of zein solution was sprayed from the needle tip and zein nanofibers were deposited onto the collector plate as shown in Fig. 44.1.

44.4 Results and Discussion

44.4.1 Hexane Phase Separation Removes Toxic Compounds

Non-polar hexane dissolves calcium oxalate, arguably responsible for the topical irritation caused by DC, at a higher efficacy than polar compounds such as water [11]. Reference [33] also suggests that a proteolytic enzyme is another toxic compound present in DC. Hexane alters the three-dimensional conformation of the enzyme's active site, denaturing it [19]. From the brine shrimp lethality assay, the brine shrimp in the H_2O and DMSO extracts containing the lower molecular weight fraction have a low mortality rate of 8% and 4%, respectively, as shown in Table 44.2 indicating that the toxin-free (TF) extracts are safe for topical applications following hexane phase separation.

Table 44.2 Mortality rate of brine shrimp after 24 h

	50% H_2O	50% DMSO	TF H_2O	TF DMSO
Dead	9	10	2	1
Survivor after 24 h	16	15	23	24
Original	25	25	25	25
Mortality (%)	36	40	8	4

44.4.2 TF Extracts Induce Antibacterial and Bactericidal Potency via Phytochemicals

Well diffusion and quantitative time point assays were conducted to elucidate the antibacterial properties of DC. The TF extracts had an average of 60 mm zones of inhibition and were also able to almost fully eliminate *S. epidermidis* and *E. coli* within 10 min of addition to the mixtures, shown in Table 44.3 and Fig. 44.2. The highly potent antibacterial activity of the TF extracts is likely because of the abundant phytochemicals present in the extracts, notably polyphenols, found to be present using the Alkaline Reagent Test as shown in Table 44.4. Potent antibacterial molecules are commonly less than 40 kDa in size and are not soluble in organic solvents like hexane [27], which could result in the loss of these compounds during

Table 44.3 Well diffusion of DC TF extracts

Bacteria	Diameter of zone of inhibition (mm)
E. coli (ATCC 25922)	47
E. coli (K-12)	52
S. epidermidis	72

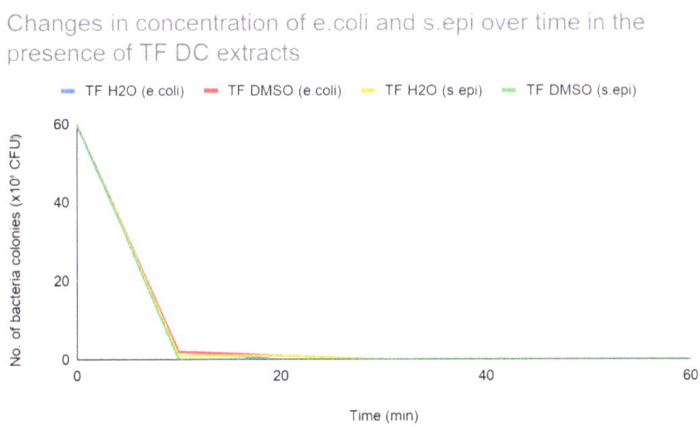

Fig. 44.2 Time point assay of TF DC extracts against *Staphylococcus epidermidis* and *Escherichia coli*

Table 44.4 Phytochemical composition of different DC extracts

Secondary metabolite	50% H_2O	50% DMSO	TF H_2O	TF DMSO
Flavonoids	+	+	+	+
Phenols	+	+	+	+
Quinones	+	+	–	–
Steroids	+	+	-	-
Saponins	+	+	–	–
Tannins	–	–	–	–
Cardiac Glycosides	–	–	–	–
Essential oils	+	+	+	+

Key: Positive (+) Absent (−)

the hexane phase separation and thus decreased antibacterial activity of the separated fraction containing hexane and higher molecular weight fraction. Ultraviolet spectroscopic (UVS) analysis showed a peak at 275 nm, suggesting the flavonoids contained may be Epigallocatechin 3-O-gallate (EGCg). EGCg reacts with porin protein on the outer membrane of *E. coli* and *S. epidermidis*. Catechin moieties in EGCg enter the porin pore and bind to arginine residues present in the inner surface of the porin via hydrogen bonding [21]. This inhibits the transportation of hydrophilic molecules such as glucose, which arrests cellular respiration.

44.4.3 Compounds of DC Recognise Conserved Elements of Lipopolysaccharides

E. coli K-12 presents the most conserved form of lipopolysaccharides (LPS) chains [16] and this binds optimally with the antibacterial compounds of the extracts. The TF extracts showed a much higher average zone of inhibition on *E. coli* K-12 than *E. coli* ATCC 25922, as shown in Table 44.4. In previous studies, the additional sugar moieties of LPS block and hinder this binding, reducing the antibacterial efficacy of the extract. This suggests that compounds in DC recognise pathogen associated molecular patterns (PAMPS) and bind to LPS favourably, resulting in antibacterial activities.

44.4.4 Compounds of DC Are Potent Antioxidative Agents

DPPH radical scavenging activity assay was used to elucidate the antioxidative effect of DC. From Table 44.5, the TF extracts showed an average radical scavenging activity of 85%, whilst the 50% crude extracts showed an average radical scavenging

Table 44.5 DPPH radical scavenging activity of extracts

Extract	DPPH radical scavenging activity (%)
50% H_2O	86.2
50% DMSO	78.1
TF H_2O	88.0
TF DMSO	82.0
Ascorbic acid (positive control)	72.9

activity of 82%, higher than that of ascorbic acid, a known stable antioxidant, at 73%. This suggests that the antioxidative molecules in DC are soluble in organic solvents and within 40 kDa molecular weight cut off. These antioxidative molecules can prevent free radicals and reactive oxygen species (ROS) from attacking the peptide bonds in the protein backbone of epithelial cells. UVS analysis suggests the presence of quercetin, a potent antioxidant compound. In vivo, quercetin can induce glutathione disulfide (GSH) synthesis, which acts as a hydrogen donor in transforming free radicals into hydrogen peroxide (H_2O_2) using superoxide dismutase (SOD). The resultant H_2O_2 decomposes into H_2O via an enzyme-catalysed reaction, preventing the cascade of lipid peroxidation [18], allowing for efficient wound healing.

44.4.5 Compounds of DC Are Fungicidal

DC was tested against *Aspergillus niger* for its fungicidal potential. 50% crude plant extracts showed minimal inhibition against the fungi, showing a resultant growth of 7 cm. However, the TF phase showed enhanced antifungal properties, showing a resultant growth of about 1 cm. This could be caused by the hexane activating and releasing potent fungicidal molecules originally bound in a complex. Such a mechanism would be similar to the oligomerisation of CRP within the innate immune system and its subsequent decomplexation following the detection of pathogen associated molecular patterns (PAMPs) [17], accounting for antimicrobial activities shown by TF extracts (Table 44.6).

Table 44.6 Resultant diameter of fungal growth

Extract	Resultant diameter of fungal growth (cm)			
	Day 1	Day 3	Day 5	Day 7
TF H_2O	0.20	0.41	0.62	0.84
TF DMSO	0.28	0.57	0.87	1.87
Sterile water	2.00	4.10	6.25	8.55

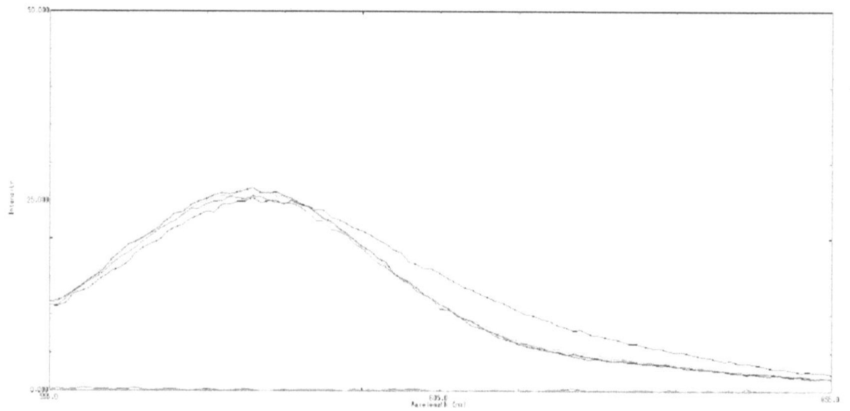

Fig. 44.3 Spectrofluorimetric analysis of DC extracts, sterile water and ethanol. Key: DC extracts—red, blue and green graphs. Sterile water (negative control)—pink graph. 70% ethanol (positive control)—black graph

44.4.6 Compounds of DC Increase Membrane Permeability Inducing Cell Death

The membrane permeability assay was conducted to further elucidate DC's antibacterial mechanisms. TF and crude plant extracts increase membrane permeability, with similar fluorescence as ethanol, a known membrane disruptor as shown in Fig. 44.3. UVS analysis presented a peak at 402 nm, suggesting the presence of Kappa-carrageenan (κ-CGN). κ-CGN is commonly found in haploid plants, such as DC [9]. It can cause morphological changes of the cell, induce cytoplasmic swelling and the spilling of intracellular content, leading to subcellular events which cause the release of hydrolytic enzymes into the cellular content, and membrane permeabilization. This disrupts homeostasis in the bacteria allowing antibacterial molecules in the extracts to penetrate the bacterial cell envelope via general diffusion porins [8], inducing cell death.

44.4.7 Compounds of DC Have Quorum Quenching Potential

In vitro biofilm, microtiter plate assay was conducted to check the quorum quenching (QQ) potential of DC in inhibiting the secretion of exopolysaccharides and dampening virulence factors [22]. DC extracts showed excellent quorum quenching potential as shown in Table 44.7. When placed under UVS analysis, it presented a peak at 279.50 nm, suggesting the presence of chrysin, a flavonoid. This result is consistent with the preliminary phytochemical analysis (Table 44.1) which suggests the presence of flavonoids. Chrysin present in DC inhibits LasR protein, preventing

Table 44.7 Mean absorption of extracts

	Mean absorption
Ampicillin	0.0556
Distilled Water	0.3414
TF H_2O	0.0372
TF DMSO	0.0392

Table 44.8 Well diffusion of electrospun nanomesh

	Diameter of zone of inhibition (mm)		
	E. coli (K-12)	*E. coli* (ATCC 25922)	*S. epidermidis*
TF extract mesh	37	36	41
Ampicillin	20	17	19
Sterile water	0	0	0

LasR from binding to the Quorum Sensing-regulated promoters [6]. Thus the inhibition of LasR would inhibit biofilm formation, impede cell-signalling molecules and virulence factors, causing quorum quenching activities, making the bacteria more susceptible to the ameliorating powers of DC.

44.4.8 TF Extract Zein Electrospun Mesh Was Able to Induce Antibacterial Activities

TF extract zein electrospun mesh was able to induce a zone of inhibition greater than ampicillin (Table 44.8). This suggests that the resultant scaffold can incorporate and subsequently effectively release antibacterial substances of DC.

44.5 Conclusion and Limitations

44.5.1 Conclusion

Five main conclusions could be drawn from this study. Firstly, DC can exhibit potent antibacterial activity. EGCg found in DC can bind to porin protein via hydrogen bonding causing an arrest in cellular respiration. Also, κ-CGN found in DC was able to disrupt bacterial membranes, and cause subcellular events inducing necrosis. Furthermore, molecules found in DC may bind to rough LPS chains of bacterial cell walls. This multifaceted approach causes multiple microscopic events inducing

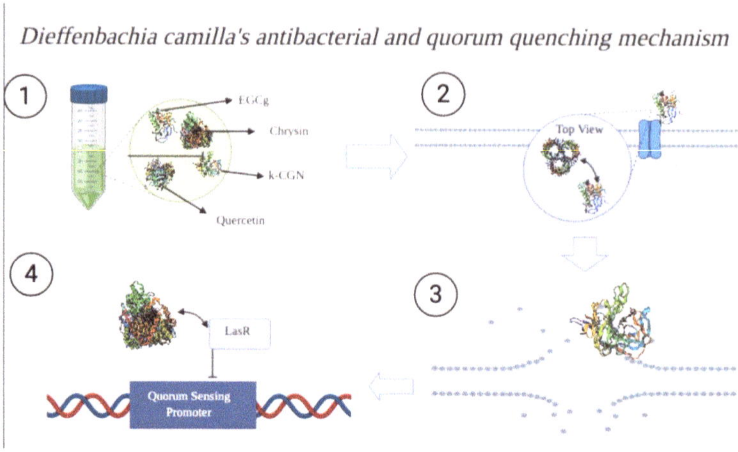

Fig. 44.4 Antibacterial and quorum quenching mechanism of DC

bactericidal and antibacterial effects. Secondly, DC has quorum quenching potential. Chrysin present in DC binds to the LasR protein, preventing it from binding to Quorum Sensing-Regulated promoters, thus producing Quorum Quenching potential, providing chronic wounds with a pathogen-free space to recover. Thirdly, compounds of DC have potent antioxidant properties due to the presence of quercetin and can ameliorate free radicals and ROS present in the wound, thereby promoting wound healing. Fourthly, compounds of DC are fungicidal, it can recognise PAMPs on fungal cell walls, ameliorating it. EGCg in extracts causes multiple subcellular events leading to cell death. Lastly, hexane phase separation can remove toxic deposits present in the extracts, and the resultant TF extracts were successfully integrated into the electrospun nanomesh. The mechanisms of DC's antibacterial activity are illustrated in Fig. 44.4.

44.5.2 Limitations

The use of the rotary evaporator might lead to the denaturation of proteinaceous compounds in DC. This study hopes to extend its research into the compounds postulated from UVS via GC–MS and further its research on electrospinning meshes effect on animal or human models via clinical trials.

Acknowledgements We would like to thank our mentor Dr. Tan Seok Hwee Sandra for her invaluable guidance during the whole course of this research project, as well as the school for providing us with the resources and the opportunity to embark on this research project.

References

1. Ajiboye, B. O., Ibukun, E. O., Edobor, G., Ojo, A. O., & Onikanni, S. A. (2013). Qualitative and quantitative analysis of phytochemicals in Senecio biafrae leaf. *International Journal of Pharmaceutical Science Invention, 1*(5), 428–432.
2. Ameer, J. M., PR AK, & Kasoju, N. (2019). Strategies to tune electrospun scaffold porosity for effective cell response in tissue engineering. *Journal of Functional Biomaterials, 10*(3), 30.
3. Arditti, J., & Rodriguez, E. (1982). Dieffenbachia: Uses, abuses and toxic constituents: A review. *Journal of Ethnopharmacology, 5*(3), 293–302.
4. Bargah, R. K. (2015). Preliminary test of phytochemical screening of crude ethanolic and aqueous extract of Moringa pterygosperma Gaertn. *Journal of Pharmacognosy and Phytochemistry, 4*(1), 07–09.
5. Bowler, P. G., Welsby, S., Towers, V., Booth, R., Hogarth, A., Rowlands, V., Joseph, A., & Jones, S. A. (2012). Multidrug-resistant organisms, wounds and topical antimicrobial protection. *International Wound Journal, 9*(4), 387–396.
6. Chong, Y. M., How, K. Y., Yin, W. F., & Chan, K. G. (2018). The effects of Chinese herbal medicines on the Quorum sensing-regulated virulence in *Pseudomonas aeruginosa* PAO1. *Molecules (Basel, Switzerland), 23*(4), 972.
7. D'Andrea, M., Fraziano, M., Thaller, M., & Rossolini, G. (2019). The urgent need for novel antimicrobial agents and strategies to fight antibiotic resistance. *Antibiotics, 8*(4), 254. https://doi.org/10.3390/antibiotics8040254
8. Delcour, A. H. (2009). Outer membrane permeability and antibiotic resistance. *Biochimica Et Biophysica Acta (BBA)—Proteins and Proteomics, 1794*(5), 808–816. https://doi.org/10.1016/j.bbapap.2008.11.005
9. Dhooghe, E., Laere, K., Eeckhaut, T., Leus, L., & Huylenbroeck, J. (2011). Mitotic chromosome doubling of plant tissues in vitro. *Plant Cell Tissue and Organ Culture (PCTOC), 104*, 359–373.
10. Djaafar, Z., & Ridha, O. M. (2014). Phytochemical study of selected medicinal plant, Solanum nigrum, the Algerian Dessert. *International Letters of Chemistry, Physics and Astronomy, 1*, 25–30.
11. Fasihzadeh, S., Sadeghimanesh, A., Wibowo, J.P., Lorigooini, Z., Hosseinabadi, T & Torki, A. (2018). Solubility of calcium oxalate and calcium phosphate crystallization in the presence of crude extract and fractions from Kelussia odoratissima Mozaff. *Pharmacognosy Research, 10*(4).
12. Harborne, A. J. (1998). *Phytochemical methods: A guide to modern techniques of plant analysis* (3rd ed.). Springer.
13. IOSR, *Journal of Applied Chemistry, 7*(1), 10–13.
14. Jaradat, N., Hussen, F., & Ali, A. A. (2015). Preliminary phytochemical screening, quantitative estimation of total flavonoids, total phenols and antioxidant activity of Ephedra alata Decne. *Journal of Materials and Environmental Science, 6*(6), 1771–1778.
15. Kayaci, F., & Uyar, T. (2012). Electrospun zein nanofibers incorporating cyclodextrins. *Carbohydrate Polymers, 90*(1), 558–568. https://doi.org/10.1016/j.carbpol.2012.05.078
16. Klena, J., Zhang, P., Schwartz, O., Hull, S., & Chen, T. (2005). The core lipopolysaccharide of Escherichia coli is a ligand for the dendritic-cell-specific intercellular adhesion molecule nonintegrin CD209 receptor. *Journal of bacteriology, 187*(5), 1710–1715.
17. Kresl, J., Potempa, L., & Anderson, B. (1998). Conversion of native oligomeric to a modified monomeric form of human C-reactive protein. *The International Journal of Biochemistry & Cell Biology, 30*(12), 1415–1426.
18. Kumar, P., Sharma, S., Khanna, M., & Raj, H. (2003). Effect of Quercetin on lipid peroxidation and changes in lung morphology in experimental influenza virus infection. *International Journal of Experimental Pathology, 84*(3), 127–134.
19. Meng, Y., Yuan, Y., Zhu, Y., Guo, Y., Li, M., Wang, Z., Pu, X., & Jiang, L. (2013). Effects of organic solvents and substrate binding on trypsin in acetonitrile and hexane media. *Journal of Molecular Modeling, 19*(9), 3749–3766. https://doi.org/10.1007/s00894-013-1900-2

20. Mir, S. A., Mishra, A. K., Reshi, Z. A., & Sharma, M. P. (2013). Preliminary phytochemical screening of some pteridophytes from District.
21. Nakayama, M., Shimatani, K., Ozawa, T., Shigemune, N., Tsugukuni, T., Tomiyama, D., Kurahachi, M., Nonaka, A., & Miyamoto, T. (2013). A study of the antibacterial mechanism of catechins: Isolation and identification of Escherichia coli cell surface proteins that interact with epigallocatechin gallate. *Food Control, 33*(2), 433–439. https://doi.org/10.1016/j.foodcont.2013.03.016
22. Paluch, E., Rewak-Soroczyńska, J., Jędrusik, I., Mazurkiewicz, E., & Jermakow, K. (2020). Prevention of biofilm formation by quorum quenching. *Applied Microbiology and Biotechnology, 104*, 1871–1881.
23. Purkayastha, S., & Dahiya, P. (2012). Phytochemical screening and antimicrobial activity of some medicinal plants against multi-drug resistant bacteria from clinical isolates. *Indian Journal of Pharmaceutical Sciences, 74*(5), 443. https://doi.org/10.4103/0250-474x.108420
24. Raaman, N. (2006). *Phytochemical techniques*. New India Publishing.
25. Rahman, M. M., Islam, M. B., Biswas, M., & Alam, A. H. M. K. (2015). In vitro antioxidant and free radical scavenging activity of different parts of *Tabebuia pallida* growing in Bangladesh. *BMC Research Notes, 8*, 621. https://doi.org/10.1186/s13104-015-1618-6
26. Rajesh, H., Rao, S. N., Rani, M. N., Shetty, P. K., Rejeesh, E. P., & Chandrashekar, R. (2013). Phytochemical analysis of methanolic extract of Curcuma longa Linn rhizome. *International Journal of pharma and Bio Sciences, 2*(2), 39–45.
27. Sahariah, P., Cibor, D., Zielińska, D., Hjálmarsdóttir, M., Stawski, D., & Másson, M. (2019). The effect of molecular weight on the antibacterial activity of N, N, N-trimethyl chitosan (TMC). *International Journal of Molecular Sciences, 20*(7), 1743.
28. Sahgal, G., Ramanathan, S., Sasidharan, S., Mordi, M. N., Ismail, S., & Mansor, S. M. (2010). Brine shrimp lethality and acute oral toxicity studies on Swietenia mahagoni (Linn.) Jacq. Seed methanolic extract. *Pharmacognosy Research, 2*(4), 215–220. https://doi.org/10.4103/0974-8490.69107
29. Sarah, Q. S., Anny, F. C., & Misbahuddin, M. (2017). Brine shrimp lethality assay. *Bangladesh Journal of Pharmacology, 12*(2), 5. https://doi.org/10.3329/bjp.v12i2.32796
30. Sheel, R., Nisha, K., & Kumar, J. (2014). Preliminary phytochemical screening of methanolic extract of Clerodendron infortunatum.
31. Shopian (J & K). *International Journal of Pharmacy and Pharmaceutical Sciences, 5*(4), 632–637.
32. Thangapazham, R., Sharad, S., & Maheshwari, R. (2016). Phytochemicals in wound healing. *Advances in Wound Care, 5*(5), 230.
33. Walter, W., & Khanna, P. (1972). Chemistry of the Aroids I. Dieffenbachia seguine, amoena and picta. *Economic Botany, 26*(4), 364–372.

Chapter 45
Epigenetic Modification of *fruitless* in the Protocerebrum Influences Male *Drosophila* Courtship Behaviour

You Chen Roy Quah and Daiqin Li

Abstract Epigenetic modifications are known to influence phenotypes such as morphology, behaviour and learning in insects. However, the role of the epigenome in courtship behaviour in *Drosophila melanogaster* remains unexplored. Epigenetic control such as acetylation and deacetylation of histones play an important role in the expression profile of *fruitless*, a gene implicated in the male courtship response. To investigate how epigenetic modifications of *fruitless* would affect courtship behaviour, we modified the epigenome of *D. melanogaster* using CRISPR-dCAS9. Here we show that *fruitless* followed an all-or-none response to elicit male courtship behaviour by creating transgenic lines of *D. melanogaster*. We assessed transcription states of *fruitless* in the transgenic lines using RT-qPCR, and behavioural changes using the RING assay and the Courtship Index. Our results demonstrated that males overexpressing *fruitless* displayed more sexually aggressive behaviour, whilst males underexpressing *fruitless* became behaviourally sterile. These findings show that epigenetics is implicated in the genetic control of courtship behaviour in *D. melanogaster*. Together, these results will contribute to the understanding of neurological memory as well as behavioural epigenetics in *D. melanogaster*.

Keywords Animal sexual behaviour · Behavioural epigenetics · Courtship · Drosophila melanogaster · Molecular ethology

45.1 Introduction

The courtship behaviour is a behavioural prezygotic barrier in *Drosophila melanogaster*. *D. melanogaster* courtship behaviour involves chemical, visual, auditory and tactile stimuli from males when courting females [1], and primarily includes

Y. C. R. Quah (✉)
Hwa Chong Institution, Singapore, Singapore
e-mail: Qycroy@gmail.com

D. Li
Department of Biological Sciences, National University of Singapore, Singapore, Singapore
e-mail: daiqin.li@nus.edu.sg

H. Guo et al. (eds.), *IRC-SET 2021*, https://doi.org/10.1007/978-981-16-9869-9_45

four main responses occurring in the order listed, orientation, tapping, singing and licking [1]. This leads to copulation if the female is receptive to the male's courtship attempts. The mushroom body found in the protocerebrum of *D. melanogaster* regulates these behaviours and their conditioning [2], as well as interpretation of olfactory cues, learning and memory [3]. Male-specific P1 neurons are a neuron cluster associated with the mushroom body, found in the protocerebrum that regulates information processing during mate recognition and tapping. These P1 neurons are implicated as key regulators for this courtship behaviour [4, 5]. Plasticity in courtship behaviour can be caused by extrinsic factors such as competition [6] and rejection from previous courtship attempts [7]. This allows *D. melanogaster* to exhibit considerable plasticity in courtship duration and vigour of the attempts. Two genes, *fruitless* and *doublesex*, involved in sexual differentiation in *D. melanogaster* [8] can also regulate courtship behaviour in males, resulting in courtship behavioural plasticity. These two genes are sex-based alternatively spliced genes, which code for transcription factors [9, 10], primarily expressed in neurons [10]. Whilst *doublesex* is an important sex determination gene in predominately female *D. melanogaster,* the present study exclusively focuses on *fruitless* which is expressed predominantly in male *D. melanogaster*. Male-specific splice variants of *fruitless* (fru^M) are expressed in male *D. melanogaster*. *fruitless* is involved in many aspects of courtship behaviour, such as mate recognition, tapping and the activation of wing muscles during singing [11, 12]. fru^M is primarily involved in the regulation of innate courtship behaviour [13]. However, it has been suggested that *fruitless* also mediates experience-dependent learning in courtship behaviour via chromatin regulation in neurons [14]. Males that have *fruitless* knocked down results in the expression of female behaviour in males [15]. When fru^M levels increase in P2B and pIP10 neuron clusters, which are controlled by P1 neurons, males start initiating courtship and singing even in the absence of females [10]. This suggests that *fruitless* plays a role in the elicitation of male courtship behaviour in protocerebral neurons. Hence, there is strong reason to believe that fru^M expression in P1 neurons regulates male courtship behaviour. fru^M expression may be controlled via epigenetic modifications in the regulation of courtship behaviour.

Epigenetic modifications are heritable changes in phenotypes not due to a change in DNA sequence [16]. Epigenetic modifications are known to lead to behavioural plasticity in insects. For example, epigenetic modifications in *D. melanogaster* play a role in foraging behaviour [17] as well as in memory and learning [18]. *Fruitless* encodes a BTB (broad-complex, tramtrack and bric-à-brac), zinc-finger motif transcription factor, strongly suggesting that it has activity related to gene expression. Although the role of *fruitless* in the courtship behaviour of *D. melanogaster* has been studied, how the epigenetic regulation of *fruitless* affects *D. melanogaster* courtship behaviour remains unstudied. *D. melanogaster* exhibits considerable epigenetic plasticity, as well as an extremely complex epigenome [19]. Hence, this study aims to contribute to the growing body of knowledge in order to understand the *D. melanogaster* epigenome. It is hypothesised that *fruitless* is involved in altering transcription responses in neurons expressing fru^M. Pull-down assays have shown that fru^M is precipitated together with chromatin regulatory proteins such as Histone

Deacetylase 1 (HDAC1) and Heterochromatin 1α (HP1α) [20], which suggest that fru^M may be involved in transcriptional repression of certain genes. In this study, we investigated how epigenetic modifications of fru^M would affect male D. melanogaster courtship behaviour.

We hypothesised that fru^M would promote motor neuron activity, resulting in shortened courtship latency. We also hypothesised that histone acetylation in *fruitless* in protocerebral neurons would result in a stronger manifestation of male courtship behaviour, whereas histone deacetylation would result in weaker manifestation of male courtship behaviour.

45.2 Materials and Methods

45.2.1 Drosophila Melanogaster Cultures

A culture of homozygous Wild Type (WT) *D. melanogaster* was obtained from the NUS Insectary, whilst the *w*[*]; *P*{*w*[+mW.hs] = GawB}fru[NP0021]/TM6, *P*{*w*[−] = UAS-lacZ.UW23-1}UW23-1 line was obtained from the Kyoto Fly Stock Centre (DGRC). This line has the GAL4 gene from *Saccharomyces cerevisiae* fused under the promoter of *fruitless*. All flies were raised on the same food medium in the laboratory under a controlled environmental (25 °C, 60–80% relative humidity, and 12:12 h photoperiod). Upon eclosion, the adults were then isolated and raised in vials to prevent sexual learning in males [21], which may influence patterns of epigenetic modifications.

45.2.2 CRISPR-dCAS9 Constructs

CRISPR-dCAS9 was used to modify the epigenome of the DGRC line as the system can be used to target specific regions of the genome using a complementary guide RNA. dCAS9 is a catalytically inactive form of CAS9 due to two-point mutations and has been used in epigenetic modifications due to its inactive ability to modify genomic DNA [22]. p300, a histone acetyltransferase was fused to the C terminus of dCAS9. Histone acetyltransferases acetylate histones, loosening chromatin structure, increasing expression levels of a gene. In this case, we induced the epigenetic activation of *fruitless* using the dCAS9-p300. pcDNA-dCas9-p300 Core was a gift from Charles Gersbach (Addgene plasmid #61357; http://n2t.net/addgene:61357; RRID: Addgene_61357) [23]. Human Histone Deacetylase 3 (hHDAC3) was fused to the C terminus of dCAS9. HDACs deacetylate histones, condensing chromatin structure, decreasing expression levels of a gene. We induced the epigenetic silencing of *fruitless* using the dCAS9-hHDAC3. Similarly, this plasmid was also obtained from Addgene. dCas9-hHDAC3 was a gift from Zhaolan Zhou (Addgene plasmid

98591; http://n2t.net/addgene:98591; RRID: Addgene_98591) [24]. Each plasmid was ligated to a pACU2 empty backbone for P-element mediated transgenesis in *D. melanogaster*. pACU2 was a gift from Yuh-Nung Jan (Addgene plasmid # 31223; http://n2t.net/addgene:31223; RRID: Addgene_31223) [25]. The bacteria stabs were cultured using the protocol provided by Addgene to obtain sufficient amounts of plasmid for P-element mediated transgenesis. We subcloned the dCAS9 constructs into the pACU2 expression vector under the hsp70-5xUAS promoter. The Hsp70-5xUAS promoter ensures *fruitless* dependent expression of the dCAS9 constructs as GAL4 was expressed simultaneously with *fruitless* in the DGRC line. This ensures neuron-specific epigenetic modification of *fruitless*. As no guide RNA is synthesised in vitro in this line, dCAS9 was unable to target the *fruitless* gene until sgRNA (single guide RNA) was injected into the cranial tissue.

45.2.3 CRISPR-dCAS9 Construct Expression

We transformed the DGRC line via P-element mediated transgenes is using the subcloned plasmids to obtain three lines, one line expressing dCAS9-p300, the other expressing dCas9-hHDAC3 and another line transformed with an empty pACU2 vector. Afterwards, we maintained all three lines of *D. melanogaster* cultures for the use in experiments. We raised flies of the same line in their larval stages collectively but isolated the adult flies upon eclosion to prevent sexual learning [21].

45.2.4 Single Guide RNA (sgRNA) Design

The P1 promoter of *fruitless* was chosen as the sgRNA design candidate as the dCAS9 fusion proteins are directed to the local chromatin structure for epigenetic modification. Furthermore, the P1 promoter is involved in the temporal and spatial control of *fruitless* expression [26], hence specifically targeting *fruitless* in a context dependent manner. The most effective target for dCAS9 fusion proteins was located ~100 nucleotides upstream and downstream for CRISPRa(p300) and CRISPRi(hHDAC3), respectively, from the promoter element and was taken into consideration for the selection of sgRNA design candidates. The sequence was retrieved from FlyBase (FlyBase ID: FBgn0004652) and was sent to Crisprdirect (https://crispr.dbcls.jp/) for sgRNA design [27].

45.2.5 Rapid Iterative Negative Geotaxis (RING) Assay

We performed the RING assay to assess the functionality of motor neurons of all lines (WT, dCAS9-p300 and dCAS9-hHDAC3) of *D. melanogaster*. We randomly collected 10 male *D. melanogaster* using CO_2 anesthesia and placed them into 10 ml Falcon tubes. This was repeated until a rack containing 5 Falcon tubes was filled. We shook the rack vigorously to ensure that all the flies fell to the bottom, initiating the negative geotaxis response. We then gave the flies 3 s to climb the wall of the Falcon tube and recorded the height each fly climbed. We repeated this assay with the other lines.

45.2.6 Epigenetic Modification

We raised the flies under the same conditions mentioned previously. We injected 4 nmol of sgRNA into the cranial tissue of the dCAS9-p300 and the dCAS9-hHDAC3 *D. melanogaster* line flies 30 min before courtship. The sgRNA was synthesised by following a protocol [28]. After the males recovered, the courtship assay was performed.

45.2.7 Epigenetic Analysis

To assess the chromatin state of fruitless, we performed Reverse Transcription-quantitative Polymerase Chain Reaction (RT-qPCR). If the *fruitless* chromatin structure was open, RNA transcripts would be produced, which would be extracted and converted to cDNA, and amplified. However, if the *fruitless* chromatin was condensed, fewer RNA transcripts would be produced, and qPCR would yield a lower CT value. RT-qPCR was performed on WT, dCAS9-p300 and dCAS9-hHDAC3 lines. Following the courtship trial, we dissected the protocerebrum of male *D. melanogaster* flies and extracted RNA using the ReliaPrep™ miRNA Cell and Tissue Miniprep System kit from Promega. Three samples of each line were taken, each with three replicates consisting of protocerebrum extracted from 20 flies each.

45.2.8 Reverse Transcription-Quantitative Polymerase Chain Reaction (RT-qPCR)

Primers generated by Primer3Plus were used. After inputting the *fruitless* gene into the tool, the primer pair that displayed the least self-complementarity and the least self 3' complementarity was chosen. The corresponding primers chosen were AGCGAGAAAGAGAGCGTGAG as the forward primer and CTGAGAGTC-CTGCGGGTAGT as the reverse primer. The reference gene *ACT42A* was chosen as a baseline as *ACT42A* shares similar expression profiles with *fruitless*. A set of primers for *ACT42A* was obtained from the DRSC FlyPrimerBank. The corresponding primers are ATGGTAGGAATGGGACAAAAGGA and CTCAGTAAG-CAAGACGGGGTG as the forward and reverse primers, respectively. All primers supplied were provided by Integrated DNA Technologies (Singapore). We first performed PCR to verify that the primers could amplify the sequences of interest (See Appendix). RT-qPCR was then performed using Bio-Rad (CFX96 Touch Deep Well Real-Time PCR system). The expression level of *fruitless* was then quantified by comparing the percentage increase of the ratio of *fruitless* and *ACT42A*. To compare a fold change in expression levels, we used the $\Delta\Delta$Ct analysis[4].

45.3 Results and Discussion

45.3.1 RING Assay

The WT *D. melanogaster* line, DGRC Line, Empty vector line, dCas9-hHDAC3 line, and dCas9-p300 line flies climbed an average of 4.09 cm, 4.10 cm, 4.12, 4.09 cm, and 4.08 cm in 3 s (Fig. 45.1), respectively. There was no significant difference in the height climbed in RING assays amongst three fruit fly lines (One-way ANOVA: F4, 245 = 1.00, $p = 0.416$; Fig. 45.1). This indicates that alterations in the expression of *fruitless* does not impact *D. melanogaster* motor neuron activity, thus inconsistent with our earlier hypothesis on motor neuron function. This also shows that the insertion of the transgene and the vector has no significant effect on motor neuron activity. The insertion of the dCAS9 constructs also did not have a significant effect on motor

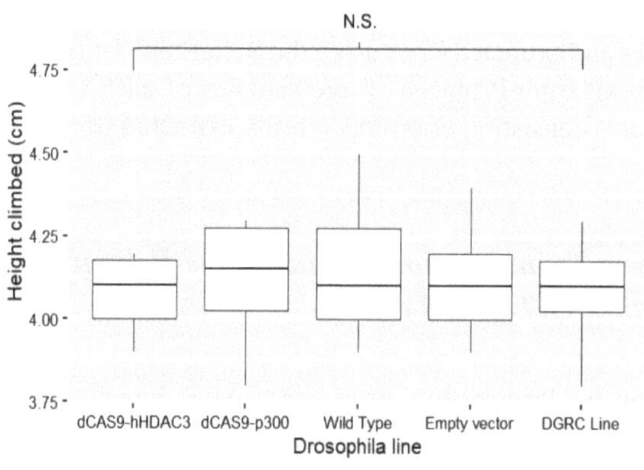

Fig. 45.1 Boxplots of the RING Assay results showing the height (cm) climbed in 3 s of *D. melanogaster* lines. Each line had five replicates, each of which had 10 flies. Boxplots show the median (central line), first and third quartiles (box), and the minimum and maximum values (whiskers). Square bracket is inclusive of all groups. Legend: N.S. = No significant difference, * $= p < 0.05$, ** $= p < 0.01$, *** $= p < 0.001$

neuron activity. Hence, we conclude that histone acetylation or deacetylation has no effect on motor neuron function. The non-significant scores on the RING assay suggest that expression levels of *fruitless* does not affect motor neuron function or activity. This is corroborated by the fact that P1 neurons are sensory neurons [29], and their specific role as chemosensory receptors shows that P1 neurons modulate motor neuron activity, but do not take part in locomotion directly. *Fruitless* may not act unless in the presence of the appropriate stimuli, thus acting as a conditional promoter of neuron activity. This indicates that P1 neurons do not directly partake in the neuronal circuitry regulating movement and behaviour, and instead, they play a regulatory role. However, there remain other factors to be explored, such as the assessment of locomotory function in the presence of certain chemical cues, which may allow us to predict other neuronal circuits.

45.3.2 Courtship Index

WT *D. melanogaster* flies displayed a mean courtship latency of 0.69, which was used as the baseline value for comparing the courtship latency of the other lines of *D. melanogaster*. The DGRC and empty vector line displayed similar mean courtship latency as the WT line of 0.68 and 0.62, respectively. The dCas9-p300 line displayed a mean courtship latency of 0.77 whilst the dCas9-hHDAC3 line displayed a mean courtship latency of 0.19. The flies in the dCas9-hHDAC3 line did not attempt to copulate with the females. Accordingly, instead of using the time taken until copulation, the CI was calculated using the time spent in courtship behaviour over a period of 10 *min*. There was a significant difference in CI amongst five lines of *D. melanogaster*. (One-way ANOVA: $F_{5,40} = 154$, $p < 0.0001$; Fig. 45.2). Since a significant difference exists, we used Tukey's HSD test to determine significant differences between any two groups. Post hoc comparisons using Tukey's HSD test revealed a significant difference in CI between the WT and the dCAS9-p300 line ($p = 0.01711$) as well as between the WT and dCAS9-hHDAC3 line ($p < 0.0001$). No significant difference existed between the WT, DGRC and empty vector lines. This suggests that the insertion of the transgene and the vector has no significant effect on P1 neuron activity.

An overexpression of fru^M could trigger the aggression pathway to act, even when this pathway may be turned off in the courtship context. This may be attributed to the activity of fru^M, as demonstrated in another study, showing that fru^M plays a role in the elicitation of sex-specific aggression [30]. This increases the competitiveness of the male *D. melanogaster*, hence investing more time and energy for a courtship attempt. This is evident as induced competitiveness in male *D. melanogaster* in the dCAS9-p300 line led to a higher CI, and generally greater success for copulation. It was noted that the dCas9-p300 line behaved more aggressively in courtship, spending a larger proportion of time singing, with more vigorous wing beating. The male was also observed occasionally attempting forced copulation. However, this increased success rate for courtship may come at a cost. Virgin females were used in the

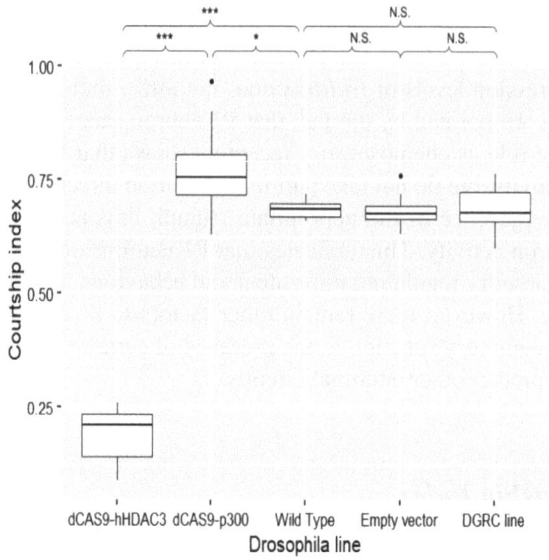

Fig. 45.2 Courtship index (CI) of three *D. melanogaster* lines (dCas9-hHDAC3, dCas9-p300 and WT). Boxplots show the median (central line), first and third quartiles (box), and the minimum and maximum values (whiskers). Legend: N.S. = No significant difference, * = $p < 0.05$, ** = $p < 0.01$, *** = $p < 0.001$

trial, when in reality, male *D. melanogaster* could encounter mated females that are unreceptive [31], wasting energy and time spent that could be spent courting other females. On the other hand, the silencing of *fruitless* and hence *fru*^{M} expression led to behavioural sterility in male *D. melanogaster*. A weak expression of *fruitless* prevents P1 neurons from activating the courtship pathway, leading to no manifestation of courtship behaviour in the dCAS9-hHDAC3 line. This behaviour is similar to *fruitless* knockouts in male flies [32]. There was a drastic decrease in courtship latency as the male *D. melanogaster* did not attempt to court the female or were markedly failed in their attempts to. This indicates that the silencing of expression of *fruitless* leads to male behavioural sterility.

45.3.3 RT-qPCR

Using the ΔΔCt analysis, the fold change of *fruitless* was calculated and tabulated (See Appendix). The difference in fold change amongst the WT *D. melanogaster* line and the two dCAS9 lines was significant (One-Way ANOVA: $F_{2, 24} = 28.2$, $p < 0.0001$; Fig. 45.3). The mean fold change in the dCAS9-p300 line was 2.96, indicating that the chromatin state of *fruitless* was mostly closed, and acetylation of the local chromatin structure opened up *fruitless* for expression. The difference

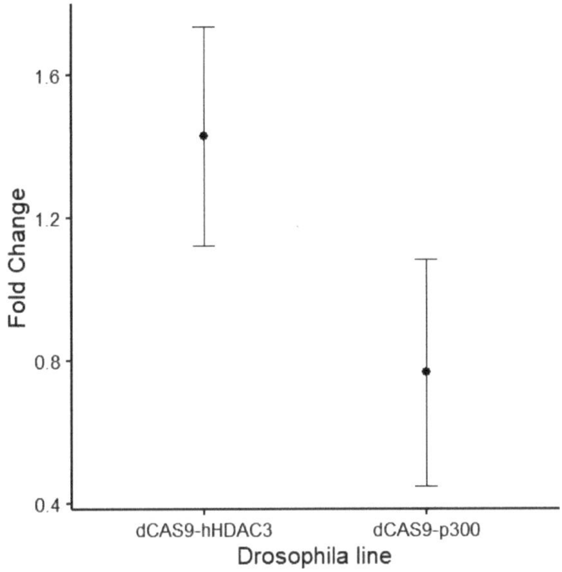

Fig. 45.3 RT-qPCR quantification results. The black dot shows the mean, whilst the line depicts the standard error

in fold change between the dCAS9-p300 and WT line was found to be statistically significant ($t_8 = 8.04$, $p < 0.0001$). The result shows that acetylation of the histones in the local chromatin area increased access for transcription machinery, increasing transcription of *fru*^{M} and hence the expression of *fruitless* increased, allowing the courtship behaviour to manifest itself more aggressively. Interestingly, a small difference in the expression of *fruitless* led to behavioural sterility in the dCAS9-hHDAC3 line. This suggests that *fruitless* may already be at a basal expression rate and cannot be silenced much further. The difference in fold change in the dCAS9-hHDAC3 line was not statistically significant ($t_8 = -0.71$, $p = 0.249$). We conclude that there is a threshold level for the expression of fruitless for the male courtship behaviour to manifest itself. Without meeting the threshold level, there may be insufficient transcription and repression of other related genes involved in the neuronal circuitry. This thus creates an all-or-none stimulus for *fruitless* to activate the P1 neurons and to elicit the courtship response. From this, we infer that *fruitless* triggers conditional firing of the P1 neurons, thus eliciting courtship behaviour only when the P1 neurons are fully committed to firing.

45.4 Conclusion

Our results show that epigenetic modifications of *fruitless* do play an important role in the regulation of the courtship response in male *D. melanogaster*, thus supporting our second hypothesis where we hypothesised that histone acetylation in *fruitless* in protocerebral neurons would result in a stronger manifestation of male courtship behaviour. The acetylation of the surrounding chromatin region of *fruitless* resulted in increased expression of *fruitless*, and hence more aggressive behaviour, suggesting that the courtship conditioning memory is encoded via the acetylation of the histones. However, if a male *D. melanogaster* becomes too aggressive in its pursuit, there is a diminishing return of time and energy. Hence, we propose that the courtship response is attenuated by the deacetylation or potentially methylation of the surrounding histones to get an optimal success to energy invested ratio, which then further fine-tunes the expression of *fruitless*. This study also provides further evidence of *fruitless* controlling courtship behaviour in male *D. melanogaster*. Not only is *fruitless* involved in the sexual dimorphism of neuronal circuitry in *D. melanogaster* [32], we show that *fruitless* is implicated in the regulation of courtship behaviour as well, controlling the vigour of courtship via expression level. We also found that *fruitless* follows an all-or-none stimulus, where it has to reach a critical threshold of expression in the P1 neuron cluster as well as its associated neurons for the male courtship behaviour to manifest itself, the intensity of which is then determined by the expression level of *fruitless*. However, this critical threshold has yet to be determined, which could be the subject of future research. Post-transcriptional control of *fruitless* mRNA transcripts could also be investigated and how it impacts the expression profile of *fruitless*. These results put epigenetics in perspective in the *D. melanogaster* connectome, as the conditional firing of neurons mapped to specific behaviours demonstrates the complexities of *D. melanogaster* neuronal circuitry. We believe that these findings will be able to further the study of the genetic control of courtship behaviour in *D. melanogaster*, as well as provide further insights into the behavioural epigenetics of *D. melanogaster*. Clearly, both nature and nurture play a key role in the courtship vigour of *D. melanogaster*.

Whilst this study provided insights into the epigenetic control of *fruitless*, the individual effects on *fruitless* targets remained unstudied. Moreover, this study only looks at P1 neuron system in the protocerebrum. Many other neurons also express *fruitless* and may play crucial roles in the neuronal circuitry of *D. melanogaster*. Further investigations should explore other neurons expressing *fruitless* to provide a better picture of the *D. melanogaster* connectome. A comparison of epigenetic modifications of *fruitless* before and after courtship, as well as epigenetic inheritance patterns of *fruitless* in *D. melanogaster*, should be conducted to ascertain whether epigenetic memory could be a function of reproductive success.

Acknowledgements We would like to thank Rachel Seah, Bernetta Kwek, Tirtha Banerjee, Suriya Narayanan Murugesan, Jocelyn Wee, Tan Lu Wee, Dr. Antonia Monteiro and other members of the Butterfly/Spider Lab for guiding and overseeing us throughout our experiments. This study was partially supported by the Ministry of Education AcRF grant (R-154-000-B72-114).

Appendix

Note

1. P1 promoter sequence: 5'TGCTGCTGTCGCCGTGCTGCGC-CTGCTGCTGCTGCGTTTCGTTGGTGCTGCGACAGCA-GAGTCGTTGCAACAATCGCAGCTTGCACATCCAGCAGCAGTCAGTTG GACGCGGCGGCTCCTTCAGCAAGGACGCACAGCTCACACAATCC-CTTCGAAGGAAATCAGCAGCCGACATACCGAACGACCTGCCACAA CATTCCA3'
2. For CRISPRa, the target sequence was CAGCAGCAGTCAGTTGGACGCGG (+). There were no 20 mer or 12 mer off-target sites. The corresponding sgRNA was: 5′GAAATTAATACGACTCACTATAGGCAGCAG CAGTCAGTTGGACGGTTTTAGAGCTAGAAATAGC3′
3. For CRISPRi, the target sequence was CCGACATACCGAACGACCT-GCCA (−). There were no 20 mer or 12 mer off-target sites. The most corresponding sgRNA was: 5′GAAATTAATACGACTCACTATAGGT GGCAGGTCGTTCGGTATGTGTTTTAGAGCTAGAAATAGC3′.
4. Gene expression quantification formula (ΔΔCT method): The fold change in gene expression is given by the formula $2^{(\Delta(\Delta CTE - \Delta CTC))}$ where ΔCTE and ΔCTC are defined as the difference between the gene of interest and the reference gene (Δ(GOI-RG)) of the experiment and control respectively. In this case, the gene of interest was *fruitless*, whilst the reference gene was *ACT42A*.

Supplementary Fig. 45.4.

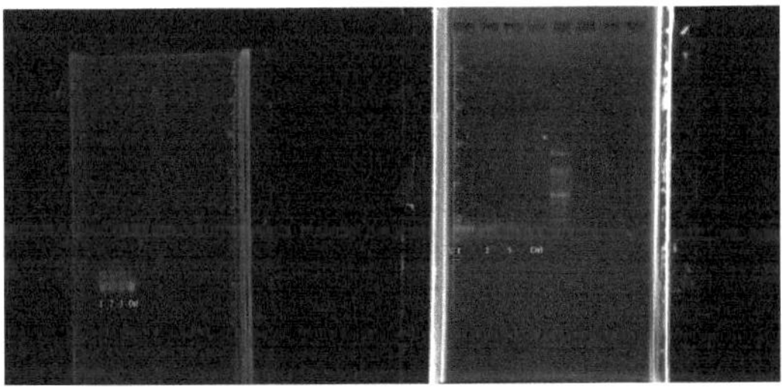

Fig. 45.4 *ACT42A* (left) and *fruitless* (right) PCR amplification in WT as proof of concept. As the gene sequences were conserved in the dCAS9-p300 and dCAS9-hHDAC3 lines, the RT-qPCR amplification results remains valid. The top bands show amplification of the target sequence, whilst the bottom bands indicate primer dimers

Table 45.1 Results of RT-qPCR expression quantification

	C_q		SD		ΔCt	ΔΔCt	Fold change	SD
Line (mean)	ACT42A	Fruitless	ACT42A	Fruitless				
Wild type	24.28	25.50	0.957	0.788	1.22	–	–	–
dCAS9-p300	29.47	29.50	0.950	0.586	−0.35	−1.57	2.96	0.178
dCAS9-hHDAC3	29.41	31.05	0.395	0.167	1.52	0.30	0.81	0.300

Legend:
C_q: The cycle number where the sample's reaction curve intersects the threshold line
ΔCt: The difference in C_q between the Gene of Interest (*fruitless*) and the Reference Gene (*ACT42A*)
$\Delta\Delta Ct$: The difference in ΔCt between the Gene of Interest and the Reference Gene
Fold Change: The expression ratio between the Gene of Interest and the Reference Gene

Legend:

Samples 1, 2 and 3 are mRNA extracts from WT protocerebrum converted to cDNA, which was used as the template for PCR. The control (ctrl) used was PCR performed without any template DNA.

Supplementary Table 45.1.

References

1. Spieth, H. T. (1974). Courtship behavior in Drosophila. *Annual Review of Entomology, 19*, 385–405. https://doi.org/10.1146/annurev.en.19.010174.002125
2. Zars, T. (2000). Behavioral functions of the insect mushroom bodies. *Current Opinion in Neurobiology, 10*(6), 790–795. https://doi.org/10.1016/s0959-4388(00)00147-1
3. Guven-Ozkan, T., & Davis, R. L. (2014). Functional neuroanatomy of Drosophila olfactory memory formation. *Learning & Memory, 21*(10), 519–526. https://doi.org/10.1101/lm.034363.114
4. Kohatsu, S., Koganezawa, M., & Yamamoto, D. (2011). Female contact activates male-specific interneurons that trigger stereotypic courtship behavior in Drosophila. *Neuron, 69*, 498–508. https://doi.org/10.1016/j.neuron.2010.12.017
5. Kimura, K.-I., Hachiya, T., Koganezawa, M., Tazawa, T., & Yamamoto, D. (2008). fruitless and doublesex coordinate to generate male-specific neurons that can initiate courtship. *Neuron, 59*, 759–769. https://doi.org/10.1016/j.neuron.2008.06.007
6. Kim, W. J., Jan, L. Y., & Jan, Y. N. (2012). Contribution of visual and circadian neural circuits to memory for prolonged mating induced by rivals. *Nature Neuroscience, 15*, 876–883. https://doi.org/10.1038/nn.3104
7. Siegel, R. W., & Hall, J. C. (1979). Conditioned responses in courtship behavior of normal and mutant Drosophila. *PNAS, 76*, 3430–3434. https://doi.org/10.1073/pnas.76.7.3430
8. Rideout, E. J., Billeter, J. C., & Goodwin, S. F. (2007). The sex-determination genes fruitless and doublesex specify a neural substrate required for courtship song. *Current Biology, 17*, 1473–1478. https://doi.org/10.1016/j.cub.2007.07.047
9. Mackay, T. F. C., Heinsohn, S. L., Lyman, R. F., Moehring, A. J., Morgan, T. J., & Rollmann, S. M. (2005). Genetics and genomics of Drosophila mating behavior. *PNAS, 102*(Supplement 1), 6622–6629. https://doi.org/10.1073/pnas.0501986102

10. Pavlou, H. J., & Goodwin, S. F. (2013). Courtship behavior in Drosophila melanogaster: Towards a 'courtship connectome.' *Current Opinion in Neurobiology, 23*(1), 76–83. https://doi.org/10.1016/j.conb.2012.09.002
11. von Philipsborn, A. C., Liu, T., Yu, J. Y., Masser, C., Bidaye, S. S., & Dickson, B. J. (2011). Neuronal control of Drosophila courtship song. *Neuron, 69*, 509–522. https://doi.org/10.1016/j.celrep.2013.09.039
12. Pan, Y., Robinett, C. C. & Baker, B. S. (2011). Turning males on: activation of male courtship behavior in Drosophila melanogaster. *PLoS One, 6*, e21144.https://doi.org/10.1371/journal.pone.0021144
13. Pan, Y., & Baker, B. S. (2014). Genetic identification and separation of innate and experience-dependent courtship behaviors in Drosophila. *Cell, 156*(1–2), 236–248. https://doi.org/10.1016/j.cell.2013.11.041
14. Yamamoto, D., & Kohatsu, S. (2017). What does the fruitless gene tell us about nature vs. nurture in the sex life of Drosophila? *Fly, 11*(2), 139–147. https://doi.org/10.1080/19336934.2016.1263778
15. Villella, A., Gailey, D. A., Berwald, B., Ohshima, S., Barnes, P. T., & Hall, J. C. (1997). Extended reproductive roles of the fruitless gene in Drosophila melanogaster revealed by behavioral analysis of new FRU mutants. *Genetics, 147*, 1107–1130.
16. Dupont, C., Armant, D. R., & Brenner, C. A. (2009). Epigenetics: Definition, mechanisms and clinical perspective. *Seminars in Reproductive Medicine, 27*(5), 351–357. https://doi.org/10.1055/s-0029-1237423
17. Anreiter, I., Kramer, J. M., & Sokolowski, M. B. (2017). Epigenetic mechanisms modulate differences in Drosophila foraging behavior. *PNAS, 114*(47), 12518–12523. https://doi.org/10.1073/pnas.1710770114
18. Xu, S., Wilf, R., Menon, T., Panikker, P., Sarthi, J., & Elefant, F. (2014). Epigenetic control of learning and memory in Drosophila by Tip60 HAT action. *Genetics, 198*(4), 1571–1586. https://doi.org/10.1534/genetics.114.171660
19. Riddle, N. C., Minoda, A., Kharchenko, P. V., Alekseyenko, A. A., Schwartz, Y. B., Tolstorukov, M. Y., Gorchakov, A. A., Jaffe, J. D., Kennedy, C., Linder-Basso, D., Peach, S. E., Shanower, G., Zheng, H., Kuroda, M. I., Pirrotta, V., Park, P. J., Elgin, S. C., & Karpen, G. H. (2011). Plasticity in patterns of histone modifications and chromosomal proteins in Drosophila heterochromatin. *Genome Research, 21*(2), 147–163. https://doi.org/10.1101/gr.110098.110
20. Ito, H., Sato, K., Koganezawa, M., Ote, M., Matsumoto, K., Hama, C., & Yamamoto, D. (2012). fruitless recruits two antagonistic chromatin factors to establish single-neuron sexual dimorphism. *Cell, 149*(6), 1327–1338. https://doi.org/10.1016/j.cell.2012.04.025
21. Saleem, S., Ruggles, P. H., Abbott, W. K., & Carney, G. E. (2014). Sexual experience enhances Drosophila melanogaster male mating behavior and success. *PLoS One, 9*(5), e96639.https://doi.org/10.1371/journal.pone.0096639
22. Brocken, D. J. W, Tark-Dame, M., & Dame, R. T. (2018). dCas9: A versatile tool for epigenome editing. *Current Issues in Molecular Biology, 26*, 15–32. https://doi.org/10.21775/cimb.026.015
23. Hilton, I. B., D'Ippolito, A. M., Vockley, C. M., Thakore, P. I., Crawford, G. E., Reddy, T. E., & Gersbach, C. A. (2015). Epigenome editing by a CRISPR-Cas9-based acetyltransferase activates genes from promoters and enhancers. *Nature Biotechnology, 33*(5), 510–517. https://doi.org/10.1038/nbt.3199
24. Kwon, D. Y., Zhao, Y.-T., Lamonica, J. M., & Zhou, Z. (2017). Locus-specific histone deacetylation using a synthetic CRISPR-Cas9-based HDAC. *Nature Communications, 8*, 15315. https://doi.org/10.1038/ncomms15315
25. Han, C., Jan, L. Y., & Jan, Y.-N. (2011). Enhancer-driven membrane markers for analysis of nonautonomous mechanisms reveal neuron-glia interactions in Drosophila. *Proceedings of the National Academy of Sciences, 108*(23), 9673–9678. https://doi.org/10.1073/pnas.1106386108
26. Billeter, J. C., & Goodwin, S. F. (2004). Characterization of Drosophila fruitless-GAL4 transgenes reveals expression in male-specific fruitless neurons and innervation of male reproductive

structures. *The Journal of Comparative Neurology, 475*(2), 270–287. https://doi.org/10.1002/cne.20177

27. Naito, Y., Hino, K., Bono, H., & Ui-Tei, K. (2015). CRISPRdirect: Software for designing CRISPR/Cas guide RNA with reduced off-target sites. *Bioinformatics, 31*, 1120–1123. https://doi.org/10.1093/bioinformatics/btu743
28. Banerjee, T., & Monteiro, A. (2018). CRISPR-Cas9 mediated genome editing in *Bicyclus anynana* butterflies. *Methods and Protocols, 1*(2), 16. https://doi.org/10.3390/mps1020016
29. Kallman, B.R., Kim, H., & Scott, K. (2015). Excitation and inhibition onto central courtship neurons biases *Drosophila* mate choice. eLife. 4, e11188. https://doi.org/10.7554/eLife.11188
30. Wohl, M., Ishii, K., & Asahina, K. (2020). Layered roles of fruitless isoforms in specification and function of male aggression-promoting neurons in *Drosophila*. eLife. 9, e52702 https://doi.org/10.7554/elife.52702
31. Aranha, M. M., & Vasconcelos, M. L. (2018). Deciphering Drosophila female innate behaviors. *Current Opinion in Neurobiology, 52*, 139–148. https://doi.org/10.1016/j.conb.2018.06.005
32. Neville, M. C., Nojima, T., Ashley, E., Parker, D. J., Walker, J., Southall, T., Van de Sande, B., Marques, A. C., Fischer, B., Brand, A. H., Russell, S., Ritchie, M. G., Aerts, S., & Goodwin, S. F. (2014). Male-specific fruitless isoforms target neurodevelopmental genes to specify a sexually dimorphic nervous system. *Current Biology, 24*(3), 229–241. https://doi.org/10.1016/j.cub.2013.11.035

Chapter 46
An Edge-Based Dustbin for Smart Compacting and Segregating

Kamya Johar, Aryan Rajput, Ramneek Kalra, and Nithyavathy

Abstract Due to the increase of waste from numerous sources including humans, kitchen, industrial, sanitary, etc. There's a warning alert across the planet for management and monitoring of same either it is on domestic or public space. Due to this alarming situation, authors are proposing solution to the problems namely, improper segregation, disposal, transportation and collection of waste. The citizens many times are not willing to dispose-off the waste properly and segregation is not done on the consumer level which in turn poses a threat at the later levels. In this proposed study, authors are solving the problem of waste segregation at an initial level using Solar Energy and Edge Computing. The application used in the proposed system rewards the users with a credit which can be redeemed after the end of every month after the inputting of waste into proposed dustbin model. This enhances the proper segregation and use of public dustbins for cleanliness and waste management.

Keywords Edge computing · Solar energy · Waste management · Industrial waste

46.1 Introduction

As per report by Indian Ministry of New and renewable energy, nation has been focusing on using renewable energy in every sector since past few years. There is 23.39% share of renewable energy of 369.98 GW in India [1]. India has the largest E commerce market. With so many people shopping online, there are large number of plastic packages produced and thrown every day.

K. Johar · A. Rajput
Department of Electrical and Electronics Engineering, HMR Insitute of Technology & Management, Delhi, India
e-mail: kamya.johar@ieee.org

R. Kalra (✉)
Sonipat, Haryana, India
e-mail: ramneek@ieee.org

Nithyavathy
Department of Mechatronics Engineering, Kongu Engineering College, Erode, India

H. Guo et al. (eds.), *IRC-SET 2021*, https://doi.org/10.1007/978-981-16-9869-9_46

In this research study, authors present a concrete idea to increase the willingness of the consumer to put the waste in the dustbin. Waste generation and collection is one of the biggest environmental challenge faced by India and other countries across the planet. In a metropolitan city like Delhi (Capital of India), waste is collected from door to door or the New Delhi Municipal Corporation (NDMC) waste management van. Waste is being collected but disposed-off in a landfill so it does not make much of a difference on how it is collected. Some cities like Indore, Madhya Pradesh have managed to become one of the cleanest cities in India through its efficient waste management policies. There is no need for bio-mining if the waste is segregated at an initial level. Bio-mining means screening the garbage so that the green and the plastic waste separates out. Despite being 3 operational wastes to power plants in Delhi, Bhalswa landfill in Delhi is 40 m high garbage landfill of untreated waste [2]. Though it contains rubber tires and synthetic clothes to a major level but if waste is segregated at initial level only, it would not have created such impact on the soil and the Yamuna river. The growth of Indian plastic industry is the fastest in the world. It has surged from 14 to 311 MT in 2014 [3]. We happen to see scrap shops in every nook and corner of our streets. They take plastic from us at INR 2 per Kg which includes the shampoo bottles, plastic boxes but nobody credits you for the e-commerce plastic bags we just throw away down the streets. Animals tend to eat them which choke them. The proposed system will reduce the pollution spread due to the garbage on the roads [4]. Segregation of waste is only possible at an initial level by the consumer.

46.2 Literature Survey

The information about the dustbins developed by some companies is elaborated in Table 46.1. along with the year range and their countries.

Every dustbin uses real time data to monitor the levels of the dustbin. The oldest smart dustbin was developed in year 2004 in Colorado in USA [5]. Most of the dustbins are using compactors and the fill sensors to make more space in the dustbin. Each idea is similar but still contrasting at its core. In [6] the technology used was CLEAN software which was used to monitor the levels of bin. The waste management system has become more extensive from year 2010–2015 as there was more need of waste management in these years. In 2011, the system proposed by Intelligent Routing [7] uses a solar powered compactor. After 2011 solar energy is being used in every dustbin. The system proposed by [8] uses IOT to connect the dustbin with other devices. In 2015 the dustbin developed by Big Belly Solar Compactors [9] in The United States segregates different kinds of materials from the recyclables. This model is better than that of the other models because if the materials are recycled on that moment only, they are of no use to any company. The recyclables are contaminated so they cannot be recycled without processing. In 2015 [10] debuted in Australia with a compactor and a recycler. In 2017, Bin-e uses artificial intelligence to segregate 4 different types of materials. In 2018, a system was developed in Russia which consists

Table 46.1 Numerous case studies of country based smart dustbin models

Country	Year range	Company	Case study
USA (Colarado)	2004	Big belly	Uses a software "CLEAN" to monitor real time garbage in the dustbin when a pedestrian disposes off the garbage. The capacity of the bin is around 567L
Canada	2011	ECUBE Labs	This product of E CUBE laboratories known as CLEAN CUBE is a solar powered waste compactor. Analyzation of waste can be easily done by a platform known as Clean City Networks (CCN). It is available in two capacities,one is 120 l and the other one is 240 l. CLEAN CUBE uses a fill level sensor in the dustbin to identify and monitor the levels of waste
Pennsylvania	2015	Clean robotics (Trashbot)	Automatically separates the recyclable materials from the waste using AI. It is a segregator and not a recycler. Trashbot offers cloud connectivity which lets the company know when the bin is getting filled. It consists of a monitor for corporate communications, education and advertising
Denmark	2015	Nord sense	Nord sense runs on IoT. It shows the real time data on why the waste is being generated and how can it be prevented. Nordsense uses a fill sensor to sense the waste
Australia	2015	Big belly	Big belly in Australia is a Wi-Fi enabled solar compactor and recycler. The compactor reduces collection and waste overflows. It uses a software "CLEAN" to analyze the level of the dustbin

(continued)

Table 46.1 (continued)

Country	Year range	Company	Case study
Poland	2017	Bin-e	Bin-e uses AI and big data to recognize the waste. After which the waste is segregated in four different bins labeled as metal, paper, plastic and glass. The capacity of the dustbin is 303L. Waste is then sorted and compressed
Russia	2018	Binology (smart city separation system-3)	Binology consists of 2 -10 different containers for segregation of waste. It runs on IoT and uses a software called Smart city management to recognize the amount of the waste segregated in each container. The waste is then compacted. All the satellite containers are non-compacting. The only compacting container is the container used for landfill waste
Finland	2019	Enevo	Enevo optimizes waste collection by using the real time data monitored by their revolutionary sensors. The sensors autonomously ensures container fill levels and records waste for recycling disposal and collections. Enevo's advanced analytical software anticipates the waste behavior
India	2020	Antariksh (Air Bin)	Air bins are installed near hospitals, public schools and quarantine zones. They were specially manufactured to reduce the risks of COVID-19 in highly affected areas. Air Bin is powered by technologies like LoRa and Narrow Band – IoT. It can be used for four months on a single charge

of 2–10 different containers for waste segregation and use IoT to determine the level of waste. In 2019, Enevo [11] uses a software to analyze the waste behavior. In 2020, India developed Air bin [12] particularly to prevent the spread of COVID-19.

Further, Rohit et al. [13] have proposed a system with two dustbins A and B. The Bin B does not open until bin A is full. At the moment when dustbin gets full,

GSM system informs the municipal corporation. The automated model of the dustbin gives more time to the municipal corporation to empty the waste from the dustbin, thereby saving the fuel costs. The system presented by Raj et al. [14] depicts the use of different sensors to segregate three different types of waste: biodegradable, Non-biodegradable and metal waste. The processing of data is done by the cloud.

These all previous proposed models work on the waste management and monitoring aspects. With respect to, communication and data processing for the execution of compacting, Chaudhari et al. [15] represents a system where the dustbin uses esp8266 module with two ultrasonic sensors at the top and the bottom of the lid. Singhvi et al. [16] depicts a system where the dustbin level is indicated by the ultrasonic sensor and interfaced with GSM system. The proposed system uses a website to indicate real time data with the help of IoT. When the dustbin exceeds a threshold value, the GSM system sends a message to the corporation. As from all previous research studies, all have focused over implementation of Internet or bandwidth consumption for data transmission and processing. This makes our proposed model as a separate one with the introduction to Edge-Based smart compacting and segregation techniques.

46.3 Methodology

46.3.1 Physical Architecture

The physical view of the proposed dustbin as depicted by Fig. 46.1 consists of two large alloy boxes or vessels. One vessel is a compactor and other is a segregator. The dustbin gets its energy from the solar panel installed on the top of the segregator and the compactor. When the sun sets, the battery installed is capable of powering the dustbin. The dustbin does not use any handle or pedal to open. It uses sensors which sense the presence of the waste from a distance and opens the flap of the dustbin. The compactor is particularly used for biodegradable waste and the segregator contains slots basically for plastic. One slot in segregator is for plastic bottles and the other slot is for e-commerce plastic bags.

46.3.2 Proposed System

As depicted from Fig. 46.2, the solar panel gives power to the dustbin. The solar panel is connected to the buck converter which steps down the DC voltage as well as improves the dynamic performance. The buck converter in turn is connected to the microcontroller which controls all the sensors. The battery is connected directly to the solar panel so that the power is stored directly from the solar panel to the battery.

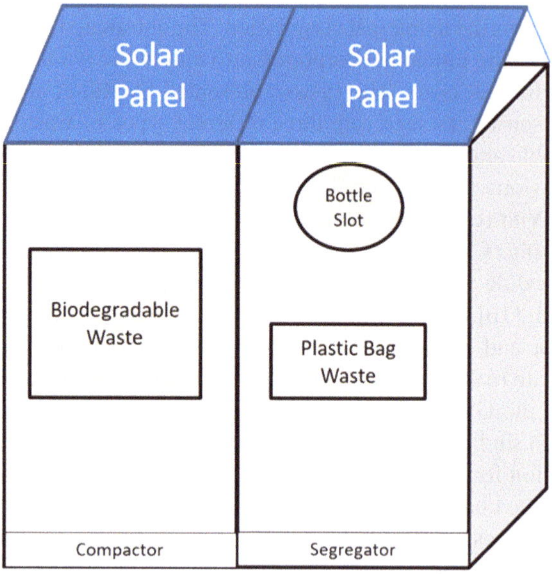

Fig. 46.1 Physical view of proposed model

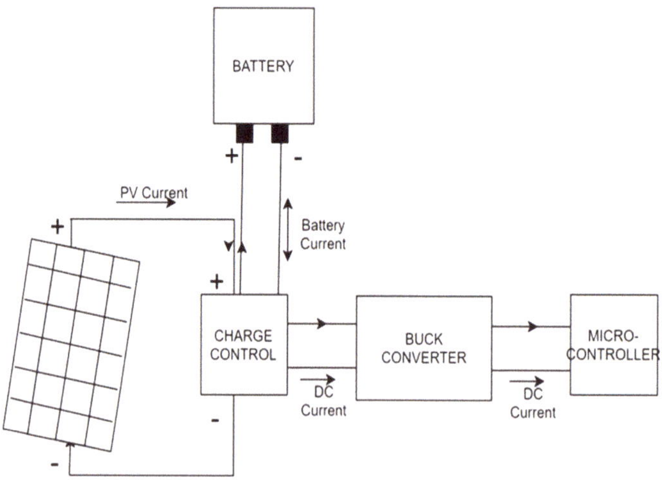

Fig. 46.2 Electrical circuit of the proposed model

46.3.3 *Use Case Study for Proposed System*

The user scans the unique QR code provided to him in the app. If the waste is wet as depicted from Fig. 46.3, the moisture sensor detects the waste and slot opens up. The slot remains open for 10 s in which the user can easily put the waste. The weight

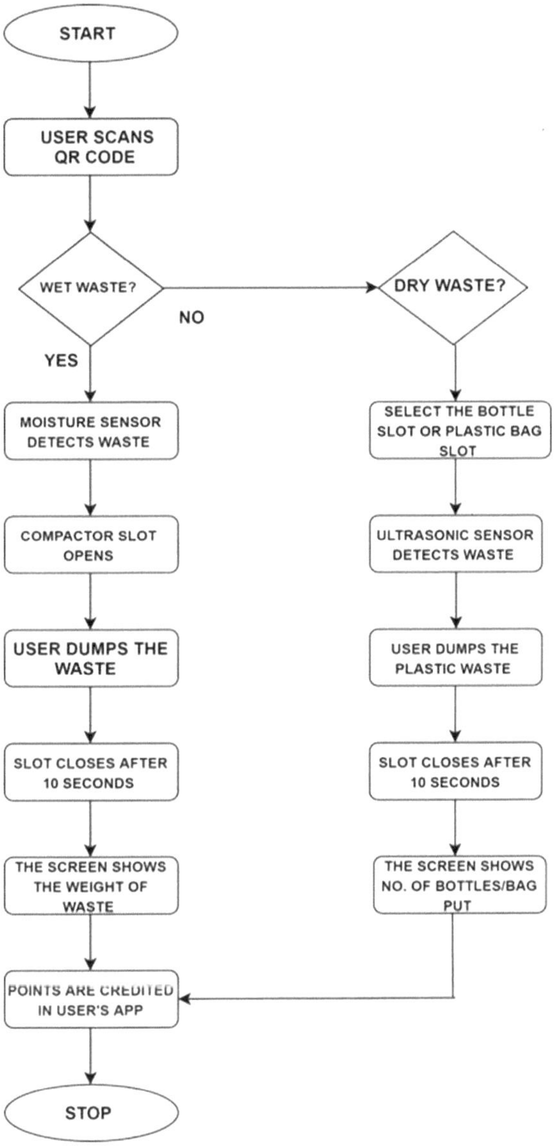

Fig. 46.3 Use case study diagram of proposed model

sensor measures the weight and credits the points in the app. If the waste is dry the user needs to enter the slot, he wants to put the waste in. Once he chooses the slot and the number of bottles or bags to be put in, the ultrasonic sensor detects the waste and the points are credited in the app.

46.3.4 Flow Illustration When "Dustbin is Full"

When dustbin is full as depicted from Fig. 46.4, the infrared sensor gives a signal to the compactor to compress the waste. The compactor moves down and compacts the waste. In turn the space is formed to put more waste. When the compactor cannot compact more waste, the microcontroller sends the signal to the government office to empty the bin. When the segregator column is full; the infrared sensor informs the municipal community to pick up the waste.

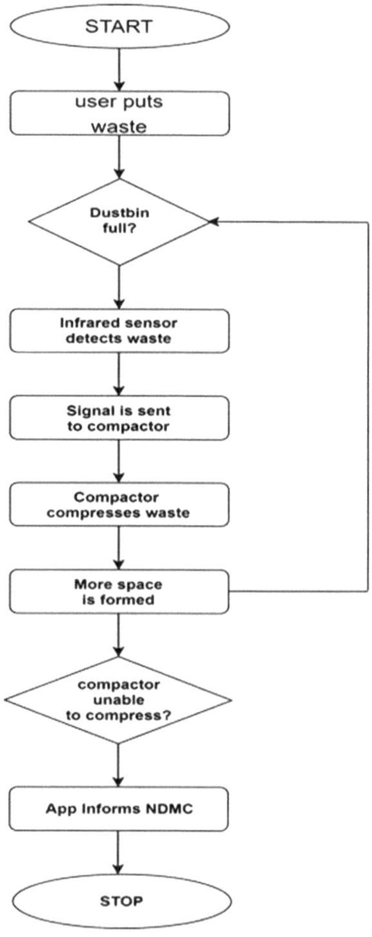

Fig. 46.4 Flow diagram when the "dustbin is full"

46.3.5 Components Used

The hardware implementation is shown below in Fig. 46.5. The proposed model of dustbin uses a microprocessor, various sensors namely, moisture, ultrasonic, infrared and load cell, motors and an LCD display for its execution.

46.3.5.1 Raspberry Pi 3 (Model B ARM V8)

In the proposed system the microcontroller used is Raspberry Pi 3 as it has 40 General Purpose Input/Output (GPI/O) pins. It has built in Wi-Fi and Bluetooth modules. The microprocessor can easily be powered with 5.1 V. It has 1 GB RAM.

Fig. 46.5 Hardware View of the proposed model

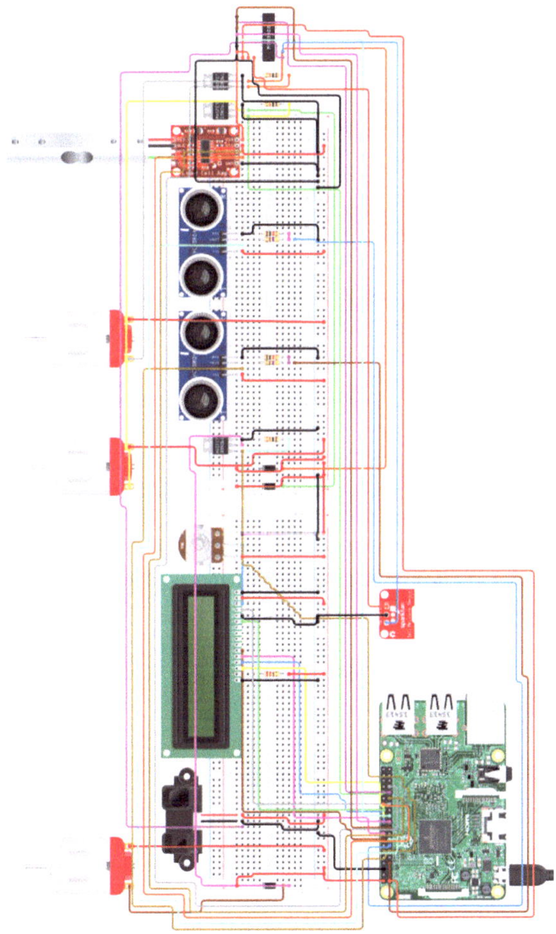

This is being used for the data collection and processing for edge-based communication.

46.3.5.2 Ultrasonic Sensor (HCSR04)

In the proposed system, the ultrasonic sensors are used in the non-biodegradable slot. It detects the waste and opens the slot. The plastic bottles and the e-commerce bags can easily be put in the dustbin without any pedal or the usage of handles. The contactless design of the proposed system promotes safety.

46.3.5.3 Moisture Sensor

The moisture sensor basically detects the presence of moisture in the soil. In the proposed system the moisture sensor detects the presence of moisture in the wet waste. The slot of the biodegradable dustbin opens when the wet waste is detected. The moisture sensor shows an error if the waste is not biodegradable.

46.3.5.4 Infrared Sensor (Infrared Proximity Sensor Long Range (Sharp GP2YA02YK0F)

In the proposed system, the infrared sensor detects the level of the waste. When the level of waste in the dustbin is full, the infrared sensor gives a signal to the compactor and the compactor compresses the waste to make more space in the dustbin.

46.3.5.5 Weight Sensor (HX711 Load Cell)

The weight sensor converts the force applied into resistance values which can be easily measured. The electrical signals change with the weight applied. The weight sensor measures the weight of the garbage inserted and rewards points as per the weight.

46.3.5.6 Two 16 x 2 LCD Displays

One LCD display is fitted on the compactor and the other LCD display on the segregator. It shows the weight of waste put in the compactor and the number of bottles or bags inserted into the segregator. When the process is completed, the screen displays that the points are successfully added to one's application of proposed system.

46.3.6 Network Architecture

As mentioned in Section E, the proposed model contains the data processing and collection for the communication with the application layer regarding the conditions/states of the dustbin as per the waste collected.

This section explains the network architecture of the proposed model using CISCO Packet Tracer simulator.

As depicted in Fig. 46.6, simulation of Hardware components is being illustrated by connecting required sensors and LCDs as mentioned in Section E, furthermore, Raspberry Pi 3 is wirelessly connected with User Application, i.e., Smartphone for data sharing and UI.

As depicted in Fig. 46.7, Solar powered battery helps all the connected motors of Compactor and Segregator slots as mentioned in Section E.

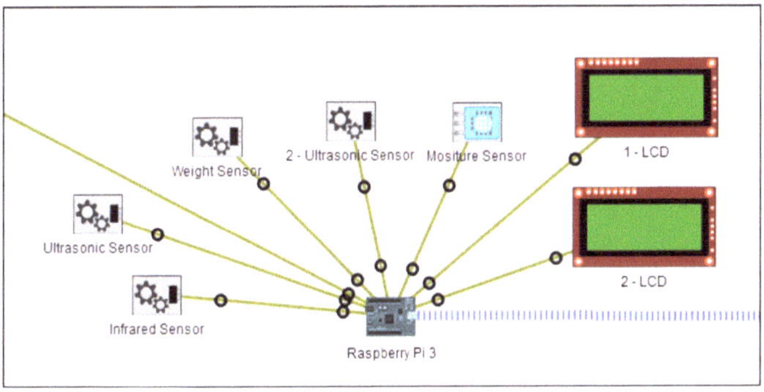

Fig. 46.6 Physical view diagram for wired communication

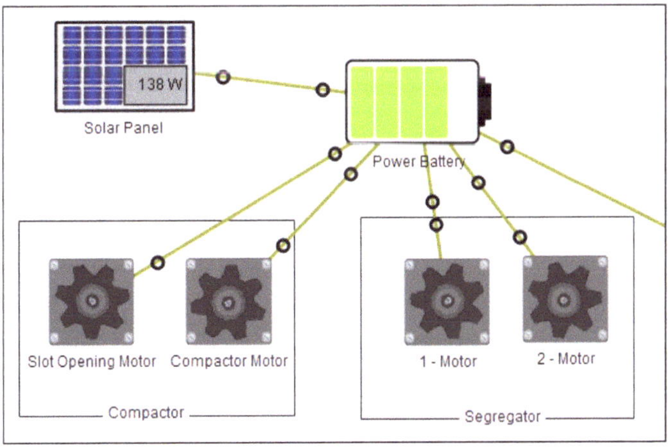

Fig. 46.7 Solar powered simulation for compactor and segregator

The communication part among the User Application and Raspberry Pi 3 is managed by Edge Computing where real time data processing and collection is done over the so-called edge node. The data processed and analyzed is being sent without consuming bandwidth to the connected user application/smartphone.

46.4 Conclusions

Through the proposed model, the wet waste is easily compacted which can be used as a material for pre processed manure for agricultural purpose. The plastic waste is segregated at the consumer level only which can be easily given to the companies for recycling. The users are rewarded with the points which are credited into their account and can be redeemed into your bank account at the end of the month.

The proposed model uses open-source technology which is inexpensive. As well it requires low energy consumption as it runs on renewable energy. The proposed model can be installed in workplaces, universities and streets.

46.5 Future Scope

With the consideration of proposed study, hardware security for the system and Anti-Theft System can be adopted and installed.

References

1. Initiatives and Achievements. Accessed March 27, 2021 [Online] Available: http://www.mnre.gov.in
2. Department of Environment—Waste Management. [Online]. Available: http://web.delhi.gov.in/wps/wcm/connect/environment/Environment/Home/Environmental+Issues/Waste+Management. Accessed March 27, 2021
3. The problem of plastics. Accessed March 27, 2021 [Online] Available: http://www.cseindia.org/the-problem-of-plastics-9712
4. Sinha, T, Mugesh Kumar, K., & Saisharan, P. (2015). Smart dustbin. *International Journal of Industrial Electronics and Electrical Engineering, 3*(5). ISSN: 2347–6982.
5. Accessed March 25, 2021 [Online] Available: https://bigbelly.com/
6. How BIN-E WORKS. Accessed March 25, 2021 [Online] Available: https://www.bine.world/howitworks/
7. Intelligent Routing. Accessed March 26, 2021 [Online] Available: https://nordsense.com/
8. Accessed March 26, 2021 [Online] Available: https://cleanrobotics.com/trashbot/
9. Big Belly Compactors. Accessed July, 2022 [online] Available: https://bigbelly.com/solution/city/
10. Big Belly Solar Compactors. Accessed March 26, 2021 [Online] Available: https://www.solarbins.com.au/

11. How our tech suite reduced collections by 61%for a cleaner greener campus at top university. Accessed March 26, 2021 [Online] Available: https://www.enevo.com/
12. Accessed March 25, 2021 [Online] Available: https://airb.in/
13. Rohit, G. S., Chandra, M. B., Saha, S., & Das, D. (2018). Smart dual dustbin model for waste management in smart cities. In *2018 3rd International Conference for Convergence in Technology (I2CT)*. https://doi.org/10.1109/i2ct.2018.8529600
14. Raj, J. R., Rajula, B. I. P., Tamilbharathi, R., & Srinivasulu, S. (2020). AN IoT based waste segreggator for recycling biodegradable and non-biodegradable waste. In *2020 6th International Conference on Advanced Computing and Communication Systems (ICACCS)*. https://doi.org/10.1109/icaccs48705.2020.9074251
15. Chaudhari, M. S., Patil, B., & Raut, V. (2019). IoT based waste collection management system for smart cities: An overview. In *2019 3rd International Conference on Computing Methodologies and Communication (ICCMC)*. https://doi.org/10.1109/iccmc.2019.8819776
16. Singhvi, R. K., Lohar, R. L., Kumar, A., Sharma, R., Sharma, L. D., & Saraswat, R. K. (2019). IoT BasedSmart waste management system: India prospective. In *2019 4th International Conference on Internet of Things: Smart Innovation and Usages (IoT-SIU)*. https://doi.org/10.1109/iot-siu.2019.8777698

Chapter 47
Self-supervised Learning with Deep Neural Networks for Computer Vision

Tan Huan Xi Gregory, Neo Souw Chuan, and Shen Bingquan

Abstract Self-supervised learning has gained popularity in recent years due to a need to avoid the expensive costs of large-scale data annotation. This field has had huge developments in recent months, with the state-of-the-art self-supervised learning methods achieving results that surpass their supervised counterparts on the ImageNet dataset. In this study, we investigate and implement 2 forms of self-supervised learning: Momentum Contrast (MoCo) and autoencoders, which are contrastive and generative methods, respectively. Through several experiments, we analyze the quality of the latent representations that are learnt by each method and assess whether they allow for increased performance when labels are scarce. We also propose a dual head network for self-supervised learning, combining elements from both the above methods and study the effectiveness of such a method.

47.1 Introduction

Deep neural networks, specifically convolutional neural networks, have seen great success in Computer Vision, with networks such as ResNet being able to classify images with great accuracy. The current convention for training deep neural networks is supervised learning, where the network is trained using a large amount of labeled data, split into train, test and validation sets. However, a bottleneck is arising in this process. Annotating and labeling image data remains a time consuming and expensive process that is prone to human error. Furthermore, supervised networks suffer from generalization errors and are prone to adversarial attacks. As such, self-supervised learning is an increasingly popular field in machine learning, which is the process of utilizing the large amounts of unlabeled data available to extract effective representations from images. In recent months, notable successes in self-supervised

T. H. X. Gregory (✉) · N. S. Chuan
Hwa Chong Institution, Singapore, Singapore
e-mail: 171432j@student.hci.edu.sg

S. Bingquan
DSO National Laboratories, Singapore, Singapore

H. Guo et al. (eds.), *IRC-SET 2021*, https://doi.org/10.1007/978-981-16-9869-9_47

representation learning with images include MoCo [1], SIMCLR [2], BYOL [3] and SimSiam [4]. The latent representations learnt through self-supervised pre-training can then be transferred into downstream tasks, such as Object Detection, Semantic Segmentation and Image Classification. For example, a model trained using self-supervised learning can allow for fewer labels to be used to achieve high classification accuracy. This can be especially helpful in fields where labeled data is incredibly limited, such as medical imaging or robotics. Our aim is to investigate and explore various methods for self-supervised learning and propose our own method.

In this paper, we study 2 different methods of self-supervised learning: MoCo and autoencoders. These 2 training methods vary greatly, as MoCo utilizes contrastive learning approach and is a more recent, state-of-the-art method, whereas autoencoders are a more traditional form of self-supervised learning, which uses a generative approach. We also propose a dual head network, which adds an additional decoder head to the query encoder of the MoCo model so as to produce a weighted loss that combines the contrastive loss of MoCo with the reconstruction loss from the autoencoders, to see if combining these two contrastive and generative would result in a more robust model. We use ResNet18 as our backbone for the models and train them on the CIFAR-10 dataset.

We then evaluate their performance on downstream image classification tasks to analyze how strong the learnt representations are. We also study how this classification accuracy scales with the amount of labels, so as to find out if self-supervised learning is effective at reducing the amount of required labels for high classification accuracy.

47.2 Related Work

A large body of research in Computer Vision is dedicated to training convolutional neural networks without the use of labeled data. There have been many proposed methods for self-supervised learning, which can be broadly split into two categories: Pretext Tasks and Contrastive Learning.

47.2.1 Pretext Tasks

One of the most popular methods for self-supervised learning for Computer Vision involve automatically creating "pseudo-labels" based on data alone and training the deep neural networks on auxiliary tasks using these labels. Many pretext tasks have been used for self-supervised learning. One example is autoencoders, which involves the pretext task of reconstructing an image based on a latent representation, treating the original image as the pseudo-label. Other works include colorization of grayscale images [5] and predicting the rotation of an image [6]. Even though these methods have proven effective at learning image representations, the representations are far from their supervised counterparts and are unable to achieve state-of-art performance.

47.2.2 *Contrastive Learning*

Recent developments in self-supervised learning mostly involve contrastive self-supervised learning, focusing on the task of Instance Discrimination for training. In short, Instance Discrimination task treats each image and its augmented version as its own class. A contrastive loss, such as InfoNCE [7], is used to maximize the similarity between the different augmentations of the same image, while at the same time minimize the similarity between the augmented images and entirely different images. Very aggressive image augmentation has proven to be integral to the success of the instance discrimination task. Many works have taken varying approaches to this task. CPC [7] and CPC v2 [3] propose using patches within an image as instances. Wu et al. [8] uses a memory bank to store a large amount of negative instances, which MoCo [1] expands upon by maintaining a queue of negative examples. SIMCLR [2] and BYOL [3] draw negative instances from the same mini-batch as the positive sample, requiring large batch sizes for successful representations to be learnt. SimSiam [4] uses a Siamese architecture, proposing that doing so eliminates the need for the negative examples used in SIMCLR [2] and the momentum encoder used in BYOL [3] and MoCo [1]. Currently, contrastive self-supervised learning methods have state-of-the-art performance, with some methods even surpassing supervised methods on the ImageNet dataset.

47.3 Methods

47.3.1 *Dataset and Model*

In this study, we used the CIFAR-10 dataset, which contains 50 000 32×32 images consisting of 10 classes of different objects and animals. The dataset was chosen due to its popularity and relatively small size in comparison to other commonly used benchmark datasets such as ImageNet, which contains over 1 million images. This smaller dataset size made it feasible for us to train our models, given that we lacked access to powerful hardware for training. We also chose ResNet18 [9] as the main encoder network to be trained using the self-supervised training method. We chose this model as opposed to the more commonly used ResNet50 as its shallower depth was sufficient for a relatively small dataset like CIFAR-10. The model architecture was slightly changed to better suit the CIFAR-10 dataset, with the first MaxPooling layer being removed.

47.3.2 Experiment 1: Momentum Contrast

One of the state-of-the-art methods for the self-supervised learning that we have chosen to study in detail is Momentum Contrast (MoCo) [1], a method that uses contrastive learning. MoCo defines self-supervised contrastive learning as a dictionary lookup problem, where encoders are trained to encode similar images into embeddings that are similar to each other and dissimilar to the embeddings of different images. MoCo trains a visual representation encoder by matching an encoded query q to a dictionary of encoded keys using a contrastive loss [7]. The dictionary is built as a queue, which is constantly updated by enqueuing the current batch and dequeuing the oldest batch. The use of a dictionary decouples the number of keys, K, from the batch size, allowing for a large number of keys without a large batch size, which is memory-intensive. The keys are encoded by a slowly progressing momentum encoder, which derives its weights from moving average of the weights of the query encoder for consistency. Denoting the θ_k as the parameters of the momentum encoder f_k and θ_q as the parameters of the query encoder f_q, θ_k is updated by:

$$\theta_k \leftarrow m\theta_k + (1 - m)\theta_q$$

where m is the momentum hyperparameter. It was found that a relatively large momentum worked much better suggesting that a slowly progressing key encoder is integral to the use of a dictionary (Figs. 47.1 and 47.2).

In the training procedure, the MoCo model randomly performs a series of strong augmentations (Color Jitter, Gaussian Blur, Random Resized Crop, Random Horizontal Flip, Random Grayscale) on the same image twice to produce two different views of the same image, x^q and x^k, which are then fed into the query

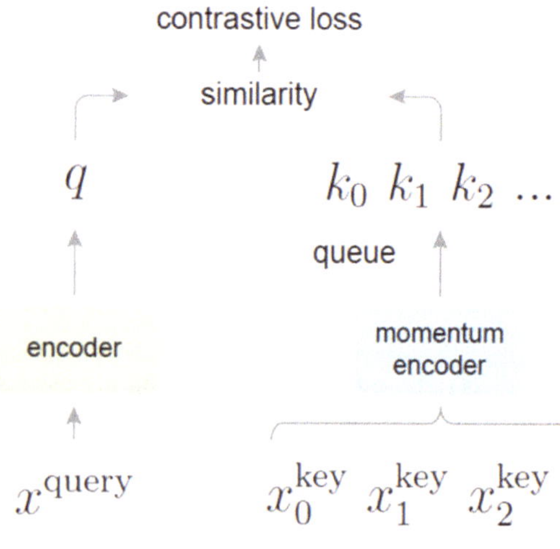

Fig. 47.1 MoCo maintains a dictionary of encoded keys which serve as negative samples for the contrastive loss function

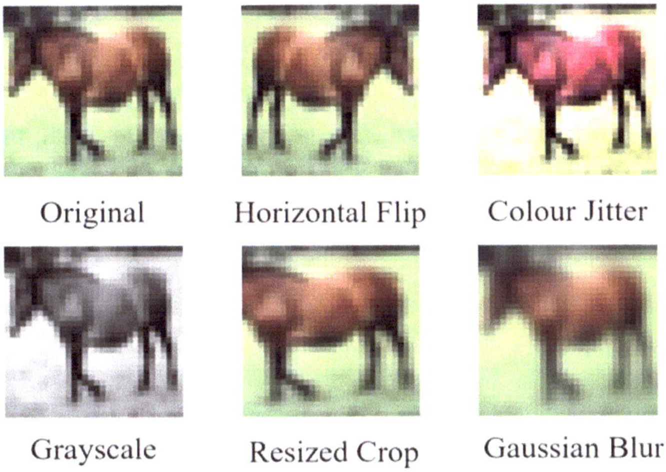

Fig. 47.2 Augmentations used on CIFAR-10 images for MoCo

encoder f_q and momentum encoder $f_{k,}$ respectively, to produce encoded query q and its positive key k^+.

A contrastive loss function known as InfoNCE [7] is then applied. The function is defined as follows:

$$L_q = -\log\frac{\exp(q \cdot k^+/\tau)}{\Sigma_{i=0}^{K}\exp(q \cdot k/\tau)}$$

where τ is a temperature hyperparameter. The sum is over one positive sample and K negative samples from the dictionary. This loss can be seen as the log loss of a $(K+1)$-way softmax-based classifier that tries to classify q as k^+. Minimizing this loss function helps the query encoder to learn to discriminate between the positive and negative samples, thus allowing it to learn strong image representations in a self-supervised manner. After training, the query encoder is kept as a feature extractor for downstream tasks. We trained our MoCo model using ResNet18 [9] as the base encoder, with $K = 4096$. The linear head was replaced with a projection MLP head which contained a hidden layer with 512 neurons and a ReLU activation, a feature from [2] that improved performance. A batch size of 256 was used and the model was trained for 800 epochs using SGD with momentum as the optimizer. The learning rate was initially set at 0.03 and was divided by 10 at epoch 120 and epoch 160.

47.4 Results and Discussion

We evaluated the quality of the learnt representations by freezing the ResNet18 encoder and training a single fully-connected layer on top of it to classify the images. Our linear classification protocol is as follows: the linear layer on top of the ResNet18 was trained for 100 epochs, using SGD optimizer with momentum 0.9. The learning rate was set at 0.3 and divided by 10 at the 60th and 80th epoch. No image augmentations were applied. As a comparison, a fully supervised ResNet18 was also trained using basic augmentations (Random Resized Crop, Horizontal Flip). We also trained a linear layer on top of a randomly initialized ResNet18. The optimizer, learning rate and learning rate schedule were the same for all tests (Table 47.1).

The above results prove that MoCo is indeed a highly effective method for self-supervised learning as it achieves a high linear classification accuracy, one that is rather close to the accuracy of a fully supervised network. This is consistent with the successful results that other contrastive self-supervised learning methods such as SIMCLR and BYOL have achieved on the ImageNet dataset (Fig. 47.3).

The t-SNE plot of the output of from the final Average Pooling Layer of the model shows clear clustering of the classes in the CIFAR-10 dataset, with the ship, truck, airplane and automobile classes being the most linearly separable. This reaffirms that a good representation has been learnt, allowing the encoder to properly discriminate between images of each class.

47.4.1 Experiment 2: Autoencoder

Autoencoders are one of the earliest forms of self-supervised learning where the image is compressed into a small latent space through an encoder before being decompressed back into the image with a decoder. The loss is then taken by taking the mean squared error of the reconstructed image and the original image. For self-supervised learning, the more common approach is to use a denoising autoencoder rather than a regular autoencoder, as shown in [6]. The input images are corrupted with either noise or occlusion and the autoencoder is expected to restore the image to its original form. The architecture of the denoising encoder is identical to that of a basic autoencoder, with the only difference being that the input image is a corrupted one.

Table 47.1 Results of linear classifier for MoCo

Training method	Train accuracy/%	Test accuracy/%
MoCo	86.59	86.62
Fully supervised	91.61	93.08
Randomly initialized	36.21 ± 0.47	34.14 ± 0.95

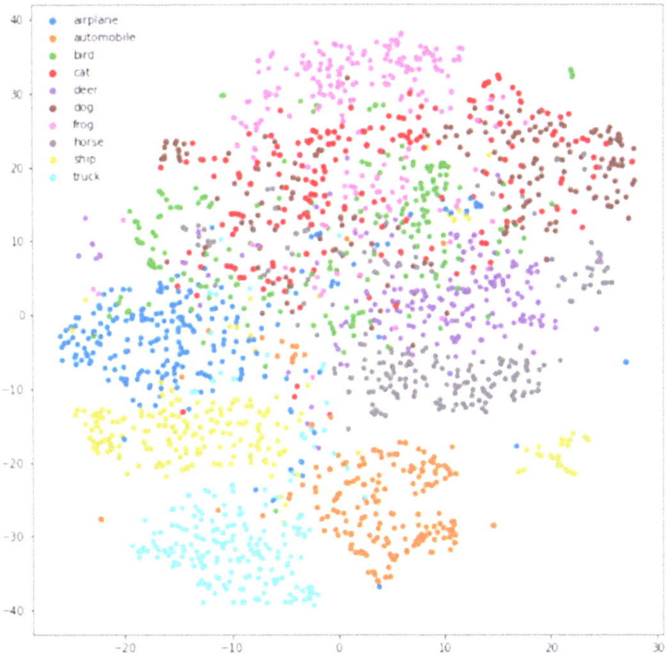

Fig. 47.3 t-SNE visualization of representations learnt by MoCo

We implemented our autoencoders using ResNet18 [9] as the base encoder to allow for a consistent comparison with MoCo. For the decoder portion of the autoencoder, we simply reversed the layers of the regular ResNet18, with the addition of up sampling layers, which used interpolation to resize the compressed latent representation back to the size of the original image. For a more in-depth study, we tested 3 variations of autoencoders: a basic autoencoder, a denoising autoencoder and a denoising autoencoder with strong augmentations. The first denoising autoencoder applied pixel-level Gaussian noise to then images (with a standard deviation of 0.15). We only applied gaussian noise as opposed to other works which cut out portions of the image, as the low resolution of CIFAR-10 would render such image corruption ineffective. The second denoising autoencoder applied the same strong augmentation that was used for our MoCo model (shown in Fig. 47.2). Each autoencoder model was trained for 200 epochs with Adam optimizer with an initial learning rate of 0.001, which was divided by 10 at the 80 and 160 epochs.

47.5 Results and Discussion

We evaluated the effectiveness of autoencoders for self-supervised representation learning by applying the same linear classification method that we used for MoCo.

After training, the decoder portion of the autoencoder was discarded and the encoder was frozen. A linear layer was then trained on top of these frozen features, using the same hyperparameters as the linear classification for MoCo (Table 47.2).

From the results above, autoencoders prove to be a less effective method for self-supervised learning, with classification accuracies significantly lower than MoCo. However, the linear classification accuracy of all 3 autoencoder methods do still show a large improvement in comparison to the randomly initialized network, indicating that image representations have indeed been learnt by the encoder. The results also indicate that the quality of the learnt representations tend to increase with the amount of augmentation added to the image, even if the increased augmentations harm performance on the pretext task, as shown below. The increase from the basic autoencoder to the denoising autoencoder is relatively small, whereas the increase from the denoising autoencoder to the denoising autoencoder with strong augmentation is quite large. This indicates the importance of strong image augmentation in self-supervised learning and shows that it is incredibly helpful not only for contrastive methods, but also for generative ones (Figs. 47.4, 47.5 and 47.6).

Even though the performance of the autoencoder with the strongest augmentation on the reconstruction task was the poorest, the learnt representations proved to be the strongest, showing that the performance on the pretext task and the performance on downstream tasks can be completely unrelated.

Table 47.2 Results of linear classifier for autoencoders

Training method	Train accuracy/%	Test accuracy/%
Regular autoencoder	54.79	54.83
Denoising autoencoder	58.50	58.62
Autoencoder with augmentation	67.71	67.73
Fully supervised	91.61	93.08
Randomly initialized	36.21 ± 0.47	34.14 ± 0.95

Fig. 47.4 Original images (top) and reconstructed images (bottom) for basic autoencoder

Fig. 47.5 Original images (top), noisy images (middle) and reconstructed images for denoising autoencoder

Fig. 47.6 Original images (top), augmented images (middle) and reconstructed images for autoencoder with augmentation

47.5.1 Experiment 3: Dual Head Network

After studying and implementing the above methods of self-supervised learning, we wanted to explore the possibility of combining multiple methods (pretext task and contrastive) to see if it would result in better performance. To study this, we created a dual head network, which feeds the encoded query from the query encoder of the MoCo network into a decoder to produce an MSE loss. This loss is then combined with the contrastive loss of the original MoCo network via the following equation:

$$L = \alpha L_{\text{contrastive}} + (1 - \alpha) L_{\text{MSE}}$$

where α is the hyperparameter adjusting the weightage of each individual loss function (Fig. 47.7).

Due to the inclusion of the regressive MSE loss in the loss function, the learning rate had to be lowered from 0.03 to 0.003 to prevent the loss from increasing exponentially. We trained the model with SGD optimizer with momentum 0.9 for 800 epochs, dividing the learning rate by 10 at the 160th, 240th and 320th epoch.

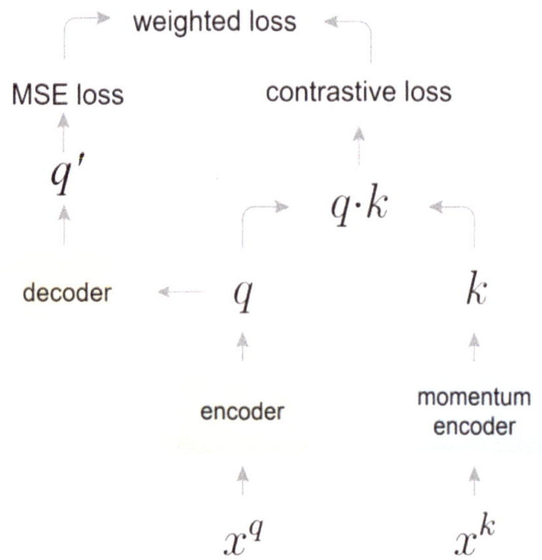

Fig. 47.7 Diagram explaining the structure of the dual head network

47.6 Results and Discussion

The results above show that as alpha values increase and the influence of the contrastive loss increases, the accuracy increases, then stagnates after alpha = 0.7. This is contradictory to what we expected as we hoped that the addition of the autoencoder would increase the accuracy since new latent features are detected allowing for better classification. However, the results show otherwise, and we hypothesize that this is due to the addition of the two loss functions, which results in the gradient descent being in two not necessarily similar directions. Concatenating the outputs of the alpha = 1.0 and alpha = 0.0 and classifying the combined features yielded an accuracy of 73.6%, which was very close to the accuracy of the alpha = 1.0 network, suggesting that the features learnt by the autoencoder and MoCo sections ended up learning were similar. The 10 × lower learning rate used could also be a factor contributing to the dual head network being less effective at learning image representations than a pure MoCo. The lower learning rate had to be used to prevent the regressive MSE Loss from exploding, but may have caused the model to get stuck in local minima and become unable to effectively minimize the loss function. Possible improvements to this dual head network include having separate weights for the two loss functions, to give the MSE Loss a smaller weightage in order to prevent it from exploding even with a greater learning rate. Further tuning of the learning rate and momentum hyperparameters and use of other optimizers such ADAM and AdaGrad could also help to avoid local minima (Fig. 47.8).

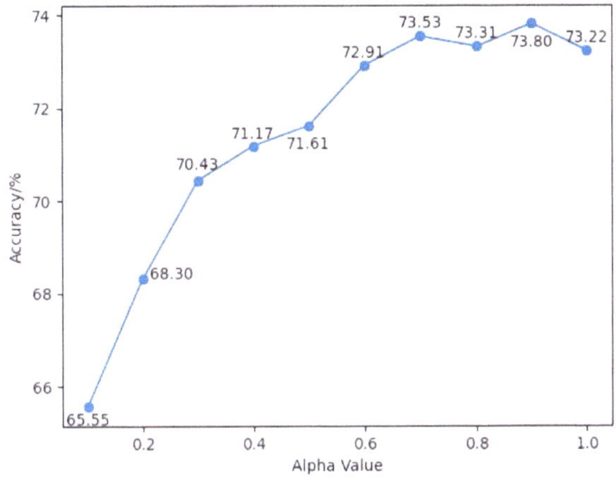

Fig. 47.8 Graph of test accuracy against alpha value for dual head network

47.6.1 *Experiment 4: Varying Amounts of Labels*

After implementing various methods of self-supervised learning, we wanted to analyze the effectiveness of the trained encoders on one of the key goals of self-supervised representation learning: reducing the reliance on large amounts of labeled data. We ran tests on 4 different label benchmarks: 1000, 2000, 4000 and 10,000 labels. We ensured that an equal number of labeled images from each of the 10 classes. We chose to run this test with our MoCo network, the autoencoder with augmentation, as well as the dual head network with an alpha value of 0.5, which meant an equal representation of both loss functions. We once again applied our evaluation protocol that was used for the linear classification with all the labels, training a single fully-connected layer on top of the frozen encoder. Since the labeled images were chosen at random, we ran the test 5 times, each with a different random seed. This was benchmarked against a supervised ResNet18, which was trained with the basic augmentations of Horizontal Flip and Random Resized Crop.

47.7 Results and Discussion

As the above results show, the accuracy of the self-supervised methods is much greater when the amount of available data is sparse. With 1000 labels, the accuracy of the encoder trained with MoCo is almost double that of the supervised method. Even the autoencoder achieves greater accuracy at this label benchmark. This clearly shows that self-supervised learning can lead to far superior results with limited data. The results also prove that the accuracy of the supervised method greatly increases with more labeled data, whereas the accuracy of the self-supervised methods only increase slightly. It is important to note that the main reason for this trend is that only the linear head of the self-supervised encoders was fine-tuned to the labels, with

Table 47.3 Results for linear classification using varying amounts of labeled data

Model	10 k labels	4 k labels	2 k labels	1 k labels
MoCo	81.80 ± 0.11	80.64 ± 0.21	79.78 ± 0.28	78.87 ± 0.19
Autoencoder with augmentation	62.64 ± 0.46	60.73 ± 0.46	58.42 ± 0.54	56.22 ± 0.80
Fully supervised	83.10 ± 0.11	74.12 ± 0.59	62.96 ± 1.32	49.45 ± 0.75
Dual head (alpha = 0.5)	67.00 ± 0.21	64.56 ± 0.17	62.42 ± 0.26	59.57 ± 0.50

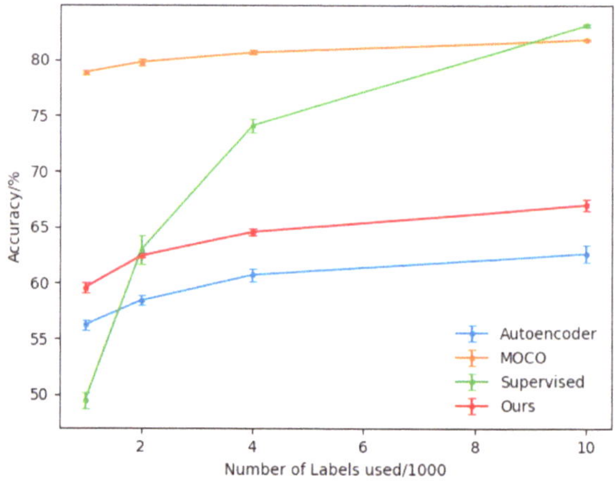

Fig. 47.9 Graph of classification accuracy against amount of labeled images, including all accuracy with all 50 k labels

the rest of the network remaining frozen. This is also the reason for the accuracy with all the labels being greater with the supervised method. We surmise that the classification accuracy would be even greater if the entire model was unfrozen and allowed to be fine-tuned on the labels, rather than only the final layer, as such results have been shown with other self-supervised training methods on the ImageNet dataset [7]. Using the pre-trained self-supervised weights as an initialization and training the entire model would likely surpass the supervised baseline by a significant margin at all label benchmarks (Table 47.3 and Fig. 47.9).

47.8 Conclusion

In this paper, we implemented MoCo and 3 variations of autoencoders for self-supervised representation learning. We verify that contrastive self-supervised learning is highly effective for learning deep image representations, with performance on image classification coming close to the supervised benchmark. Our study also shows that contrastive self-supervised learning is also effective with lower resolution images (32×32 images in CIFAR-10, compared to 224×224 images in the more commonly used ImageNet) and with shallower models (ResNet18 rather than the standard ResNet50). We also proposed a method for integrating contrastive methods with generative methods, through adding a secondary head. Although Dual Head networks do not provide a promising result, we believe that it is due to the combination of Autoencoder and MoCo, which results in similar latent representations thus preventing it from being able to achieve higher accuracy. Due to similar representations being learnt despite the networks using two different methods for self-supervised learning, we also believe that there is a need for models to focus on smaller features as well.

47.9 Future Work

Even though our dual head network using a decoder head did not help performance, we believe that more research can be done in combining generative and contrastive methods for self-supervised learning, as well as training the encoder to recognize alternative representations as well. The decoder could be replaced with other pretext tasks, such as a self-supervised CutMix [10] which criteria is how accurate the encoder can recognize the replaced section of the image allowing it to learn a diversity of representations.

References

1. He, K., Fan, H., Wu, Y., Xie, S., & Girshick, R. (2020, March 23). Momentum contrast for unsupervised visual representation learning. Retrieved December 21, 2020, from https://arxiv.org/abs/1911.05722
2. Chen, T., Kornblith, S., Norouzi, M., & Hinton, G. (2020, March 30). A simple framework for contrastive learning of visual representations. Retrieved June 30, 2020, from https://arxiv.org/abs/2002.05709
3. Grill, J., Strub, F., Altché, F., Tallec, C., Richemond, P., Buchatskaya, E., . . . Valko, M. (2020, June 13). Bootstrap Your Own Latent: A New Approach to Self-Supervised Learning. Retrieved June 30, 2020, from https://arxiv.org/abs/2006.07733
4. Chen, X., & He, K. (2020, November 20). Exploring simple Siamese representation learning. Retrieved December 26, 2020, from https://arxiv.org/abs/2011.10566
5. Zhang, R., Isola, P., & Efros, A. (2016, October 05). Colorful Image colorization. Retrieved December 24, 2020, from https://arxiv.org/abs/1603.08511
6. Gidaris, S., Singh, P., & Komodakis, N. (2018, March 21). Unsupervised Representation Learning by Predicting Image Rotations. Retrieved December 24, 2020, from https://arxiv.org/abs/1803.07728
7. Hénaff, O., Srinivas, A., De Fauw, J., Razavi, A., Doersch, C., Eslami, S., & Oord, A. (2019, December 06). Data-efficient image recognition with contrastive predictive coding. Retrieved June 30, 2020, from https://arxiv.org/abs/1905.09272
8. Wu, Z., Xiong, Y., Yu, S., & Lin, D. (2018, May 05). Unsupervised feature learning via non-parametric instance-level discrimination. Retrieved December 26, 2020, from https://arxiv.org/abs/1805.01978v1

9. He, K., Zhang, X., Ren, S., & Sun, J. (2015, December 10). Deep Residual Learning for Image Recognition. Retrieved December 22, 2020, from https://arxiv.org/abs/1512.03385
10. Yun, S., Han, D., Oh, S., Chun, S., Choe, J., & Yoo, Y. (2019, August 7) CutMix: Regularization strategy to train strong classifiers with localizable features. [online] Retrieved December 28, 2020, from https://arxiv.org/pdf/2003.05991.pdf

Chapter 48
IoT-Enabled Smart Window with Environmental Sensors

Jia Cheng Raymand Tey, Jiok Duan Jadie Tan, and Rusli

Abstract This project aims to design an IoT-enabled smart window that operates with the aid of electrochromic material and environmental sensors to improve energy efficiency and optimize comfort for the building's occupants. A control algorithm is developed to regulate the temperature and light intensity within their respective ranges set by the users, ensuring comfort for users while reducing the reliance on electrical appliances, saving energy for the building. The smart system is IoT-enabled, and it allows users to read and set the desired room temperature and light intensity, as well as directly control the electrical appliances over the Internet using the Blynk application.

Keywords Engineering · Internet of Things · Embedded Systems

48.1 Introduction

Glass windows are an integral part of homes and offices as they provide open views and natural lighting for the living environment. However, they also allow a large part of infrared radiation to pass through and heats up the air in the room, as the overall heat transfer coefficient of a window is approximately five times greater compared to buildings that are enclosed with other structures such as walls and doors [1]. The consequence is that the indoor temperature with glass windows can rise rapidly, reaching as high as 35 °C, making it unbearable for the occupants. This issue currently requires greater attention as more people are now working from home and spending most of their day time at home due to the COVID-19 pandemic, where working from home is the new normal.

J. C. R. Tey (✉) · J. D. J. Tan
Hwa Chong Institution, Singapore, Singapore
e-mail: 162336a@student.hci.edu.sg

Rusli
Nanyang Technological University School of Electrical and Electronic Engineering, Singapore, Singapore

H. Guo et al. (eds.), *IRC-SET 2021*, https://doi.org/10.1007/978-981-16-9869-9_48

The above heating problem can be alleviated by blocking external heat using curtains or blinds, which however will also shut off natural lighting. It is also possible to apply a permanent glazing to the window, the drawback of which is that it may not result in an optimum compromise between heat blocking and indoor lighting, as it is dependent on the outdoor weather and lighting conditions that change continuously throughout a day. Alternatively, the occupants can address the window heating problem by turning on fans and air conditioners for cooling, which however will result in high electricity demand. Indeed, a lot of energy is primarily used for temperature control in a building [2, 3], which is environmentally unfriendly and can contribute to climate change.

To address the above issue, we propose the use of smart windows in this project, which are glass windows coated with electrochromic material having tunable transparency, to automatically control and regulate the indoor temperature and lighting condition. The aim is to maintain the indoor environmental condition at a comfortable level and concurrently minimize the use of electrical appliances such as fans and air conditioners for cooling and light bulbs for lighting so as to reduce energy consumption. An app will also be developed that allows interactive control of the smart film system over smartphones using the Internet of Things (IoT) so that the indoor condition can be regulated remotely, which is important in the development of future smart homes. The proposed smart window can be applied to both homes and offices.

Electrochromic (EC) material was first commercialized in 1987 [4]. It is a thin film material whose transparency can be controlled by the application of an electric voltage across the film. It is commonly applied on glass panels to regulate their transparency. One common application is to tune the transparency of glass panels in offices and conference rooms for privacy consideration. Hitherto, the material has been mostly used to enhance privacy and security in the architectural and automotive markets [4, 5] but not yet widely developed for residential application [6].

The use of EC material with tunable transparency for window glazing is effective in ensuring the comfort for the building's occupants and improving energy efficiency for several reasons. It is capable of reducing the heat transfer from solar energy to the building [7]. EC glazing is also effective in protecting the user from discomfort glares by blocking it from the daylight source [8]. EC windows ensure a proper balance between energy saving and visual comfort [9]. Besides, EC devices allow for interactive control by users, according to their needs. While EC material is utilized to block out sunlight, it usually only offers the function of being fully transparent or translucent. Even if there is a separate function to tune the transparency, users are required to do it manually. Therefore, our project to implement EC smart windows coupled with environmental sensors such as temperature and light sensors in an IoT system, to automatically regulate the indoor environmental condition, can maximize users' comfort and enhance the energy efficiency of homes and buildings.

48.2 Hypothesis

An IoT-enabled smart window can be designed and implemented to remotely tune the transparency of the window through the use of smart film, according to the feedback from individual environmental sensors, such as temperature and light sensors. As part of a smart home initiative, the usage of smart film to automatically regulate indoor temperature and vary light intensity will ensure the most comfortable environment for the occupants, while minimizing energy consumption from appliances.

48.3 Methodology

48.3.1 Materials and Devices

A4-sized electrochromic films and a dimmer control (transformer) are used to tune the transparency of the films, the specifications of which are given in Appendix 1. By varying an AC voltage that is applied across the films, their transparency can be tuned over a wide range. For reference, translucent refers to the state of the EC film with the lowest possible transmissivity, as covered in Appendix 1.

The microprocessor used in this project is the Arduino Uno Wi-Fi Rev 2, the specifications of which are given in Appendix 2. It has sufficient digital pins to accommodate many sensors deployed and the use of LED for signaling functions as well. Furthermore, it is optimized for IoT projects and is able to connect to Blynk, an IoT platform that we used for app development to allow manual user control to override the system where necessary.

Light and temperature sensors are used to collect data about the environmental conditions. They are the BH1750 light sensor module and DS18B20 temperature sensor, respectively, and their specifications are provided in Appendix 3. The BH1750 light sensor is the preferred choice as it is a digital light sensor offering a wide range of light intensity detection with high resolution. The waterproofed DS18B20 temperature sensor is the preferred choice as multiple DS18B20s can coexist on the same one-wire bus since each DS18B20 has a unique serial code. This will allow deployment of several sensors to better measure the room temperature. In our project, three light sensors are used indoors, spread out across the room, so that the average light intensity can be deduced. Three temperature sensors are also placed indoors, with one temperature sensor placed on the window to account for the approximate outdoor temperature outside the glass window. The remaining two temperature sensors are spread across the room so that the average temperature can be deduced. UV sensors were initially considered as part of the system to regulate the UV level in the room by tuning the transparency of the films. However, it was ultimately not used as our experiments have shown that the EC film itself is capable of blocking the UV rays and reducing it to a safe UV index level. Figure 48.1 shows the sensors and microprocessors chosen for the prototype setup.

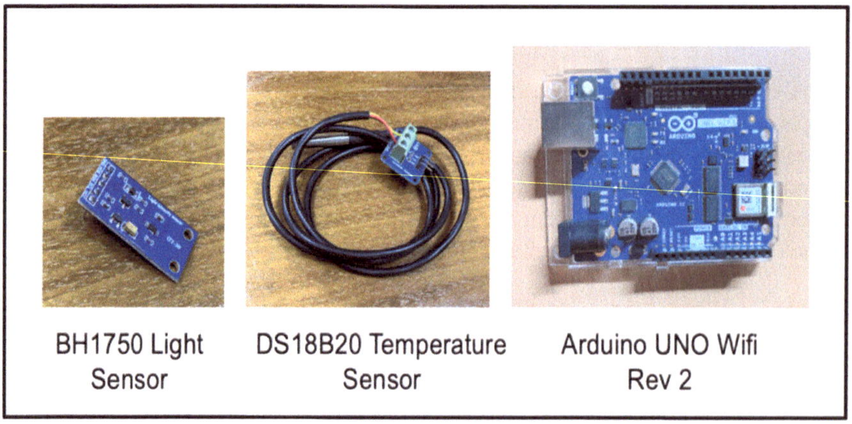

Fig. 48.1 Microprocessor and environmental sensors used in the project

48.3.2 Control Algorithm

Temperature condition is considered to be the main priority in our control of the environment as compared to lighting conditions, as temperature regulators like air conditioners and fans consume much more power than LED light. The temperature variables that are involved are T_{in}, T_{out}, T_{set}, where T_{in} and T_{out} are recorded readings of the indoor and outdoor temperatures, respectively, as measured by the temperature sensors. An average reading for indoor temperature from two sensors will be used for T_{in}.

T_{set} is the temperature set point in the room as determined by the user. Similarly, the light variables involved are L_{avg} and L_{set}; the former is the average light intensity in the room measured by the three light sensors, while the latter is the user predetermined set point for light intensity.

Figure 48.2 shows a detailed flowchart of the smart window system's control algorithm programmed in the microprocessor. For $T_{\text{in}} = T_{\text{set}}$ to be considered true, it should fall in a range of $\pm 10\%$ of the set point. Similar to the case of temperature control, the light intensity set point is deemed to have been reached if L_{avg} is within $\pm 10\%$ of L_{set}.

After reading the temperature values from temperature sensors, the algorithm will consider different scenarios and take appropriate actions. For instance when $T_{\text{in}} \leq T_{\text{set}}$, it proceeds to save energy by reducing the reliance on electrical appliances, such as the fan and the air conditioner if they have been turned on.

Subsequently, it proceeds with increasing the transparency of the smart film to introduce natural lighting if $T_{\text{in}} = T_{\text{set}}$. This is done all while continuously monitoring the indoor temperature to ensure it is maintained within the optimal range.

Another case is when $T_{\text{in}} > T_{\text{set}}$ where necessary actions will be taken to reduce the indoor temperature, with priority given to controlling the film rather than increasing fan speed or cooling from the air conditioner. The algorithm will determine if T_{out}

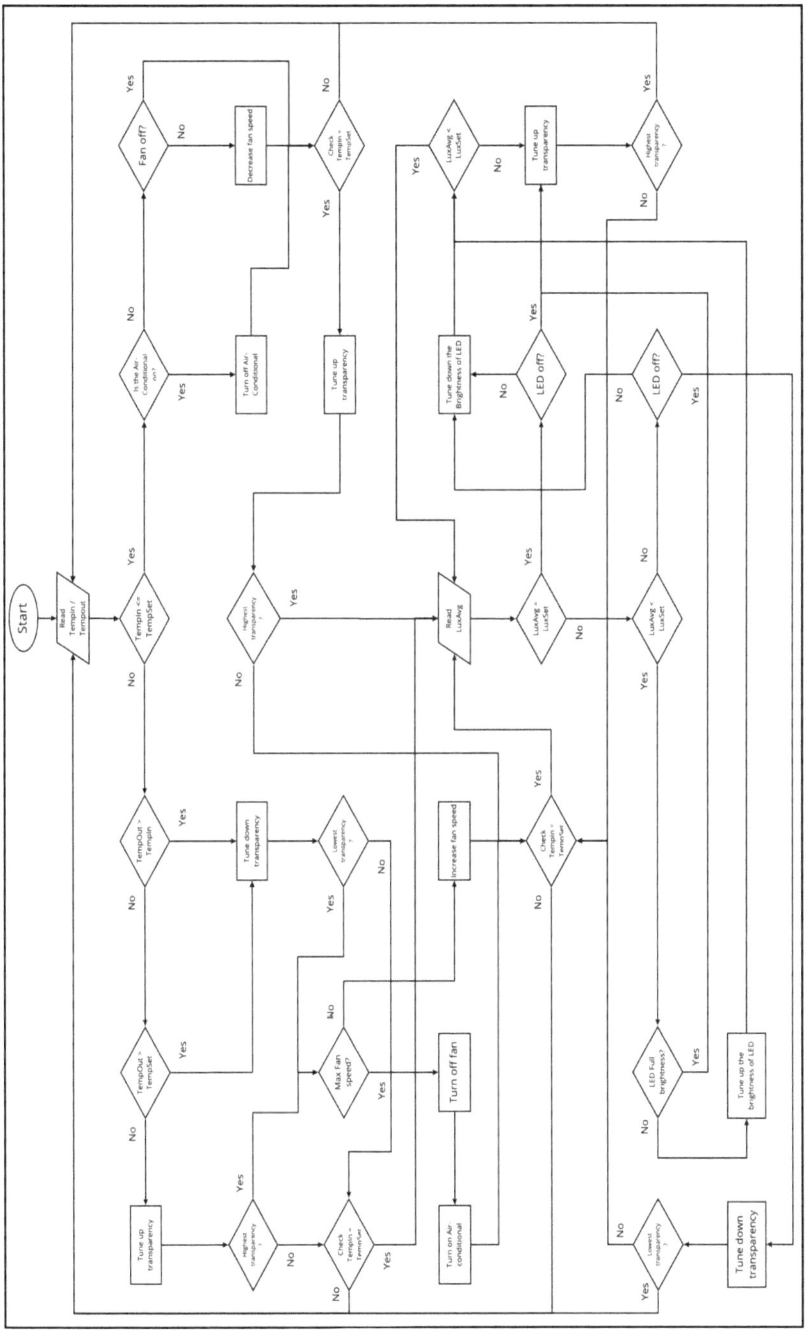

Fig. 48.2 Algorithm flowchart

< T_{in} and $T_{out} < T_{set}$, before increasing the transparency of the film for heat transfer from indoors to outdoors. Subsequently, fan and air conditioner control may be implemented if necessary. Otherwise, if $T_{out} > T_{in}$ or $T_{out} > T_{set}$, the film is made less transparent to reduce heat transfer. The fan is then set to spin at a higher speed or the air conditioner will be switched on if needed.

Once the indoor temperature is regulated to a reading within the optimal temperature range, the second environmental condition to take into account is light intensity. Likewise, the algorithm will identify if actions need to be taken to increase, decrease, or maintain the indoor light intensity. Possible actions to take would be tuning the transparency of the film to allow more or less light in, and further regulating the brightness in the room by controlling LED lights indoors if needed, while ensuring that the temperature control is not disrupted so as to minimize the use of appliances.

In every case, prioritizing temperature control of the film is deemed as the first condition to be checked on, before the algorithm considers the adjustments and controls of other appliances like air conditioners, fans, and LED lights. This ties in with our aim of improving the efficiency of the smart window system, saving energy where possible. After the end of every case and its subsequent processes, the algorithm loops back to continue reading temperature and light values to keep both T_{in} and L_{avg} values within their respective optimal ranges.

48.4 Results and Discussion

48.4.1 Prototype

Figure 48.3 is an overall display of our prototype, which simulates the surroundings of an enclosed room with a glass window, sensors, fan, and LED light strips.

Figure 48.4 presents our specific setup for the placement of the two types of environmental sensors, with one temperature sensor and one light sensor mounted on the inner surface of the window, and other sensors are generally spread out across the room to collect the average reading.

Figures 48.5 and 48.6 show the positions of the appliances. A fan is used for cooling the room and LED strips for lighting up the room. Both are controlled by the microprocessor based on the control algorithm. In the prototype, fan speed and light intensity are varied by controlling the speed of the motor and percentage brightness of LED strips (refer to Appendix 4). Due to the complexity, air conditioners were not included in the prototype demonstration. In practice, if the room temperature is still high despite the fan operating at maximum speed, air conditioners will be activated if available.

The above covers the description and uses of appliances we have used in the prototype. We have tested the prototype and demonstrated that the proposed control algorithm is able to automatically regulate the temperature and the lighting conditions

Fig. 48.3 Overall prototype setup

Fig. 48.4 Prototype setup with temperature and light sensors

in the room based on the user-determined set points, with the primary control on achieving the desired temperature through controlling the smart film transparency.

48.4.2 IoT Applications

Figure 48.7 shows the Blynk app's interface, which is created for the user to interact with the smart film system. The app allows the user to enter the set points for the temperature and light intensity in the room and provides a real-time display of the measured temperature and light intensity. The user is also able to override the smart system by directly controlling the fan speed and the light intensity of the

Fig. 48.5 Prototype setup with fan

Fig. 48.6 LED placement

LEDs. Blynk is controlled via the user's Wi-Fi network, enabling a more convenient implementation of IoT control of the smart film system in home and office environments.

48.5 Conclusions

The IoT-enabled smart window system implemented seeks to extend the applications of EC material to residential areas and office buildings. Overall, the algorithms implemented were able to monitor and maintain the indoor lighting and temperature

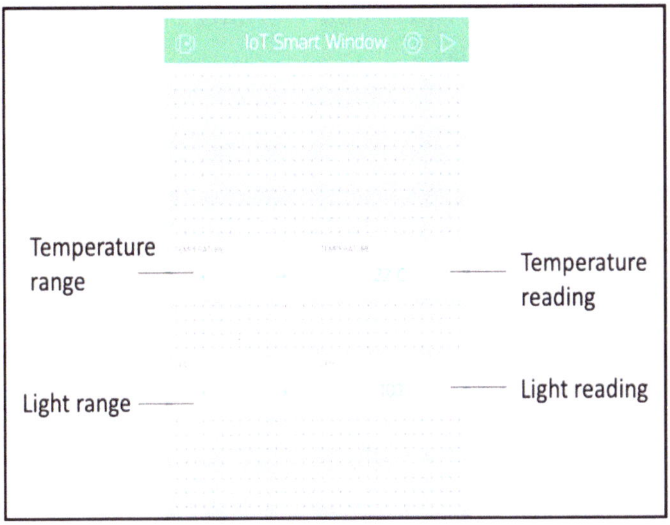

Fig. 48.7 Adjust range and display readings on Blynk app

conditions according to the set points determined by the user while minimizing the use of electrical appliances. Additionally, an app was successfully developed using Blynk that allows interactive control of the smart film system using smartphones. The user is able to directly monitor and set the desired temperature and light intensity according to their preferences.

48.6 Further Applications

Implementation of the smart window system can be carried out in greenhouses in other countries that have four seasons. Rather than using shade clothes or manual covering to reduce lighting, it is more convenient to use smart windows to automatically regulate adequate sunlight to enter the greenhouses.

Acknowledgements We would like to express our gratitude to our project supervisor, Professor Rusli, for his patient guidance, support, and encouragement throughout our research journey. We would also like to thank Genson Chua for tutoring and assisting us in our prototype design and overall setup. Lastly, we would like to extend our gratitude to our NRP teacher coordinator from Hwa Chong Institution, Mr. Low Kay Siang, for taking care of administrative and project supervision over the course of our project.

Appendix 1: Electronic Specs and Working Principles of Electrochromic Films

See Table 48.1.

Without a voltage applied, the liquid crystal molecules of the EC film (micro-droplets) are disordered. This prevents light from penetrating the film, rendering it translucent. When a voltage is applied to the smart film, the liquid crystal molecules are forced into alignment, making it transparent.

Figure 48.8 shows the EC film under the conditions of fully transparent, semi-translucent, and fully translucent

Table 48.1 Electronic specifications of EC films

Optical properties	Parallel light transmission	On	More than 75%
		Off	Less than 5%
	Visible light transmission	On	More than 80%
		Off	More than 60%
	Visual angle	On	150°
	UV blocking	On/Off	More than 99%
Electrical properties	Operating voltage	On	60 V A.C
	Frequency	On	50–60 Hz
	Current	On	2 mA/sq mt
	Power consumption	On	10 W/sq mt
	Switching time	On to Off	Tr.2 ms
		Off to On	Td.10 ms
Specification	Thickness	0.37 mm	
	Max size	1500 mm × 3000 mm	
	Glass color:	Milky white, light gray, dark gray	
Other index	Working temperature	– 10~60 °C	
	Preserve temperature	– 20~75 °C	
	Lifetime	less than 50,000 h	

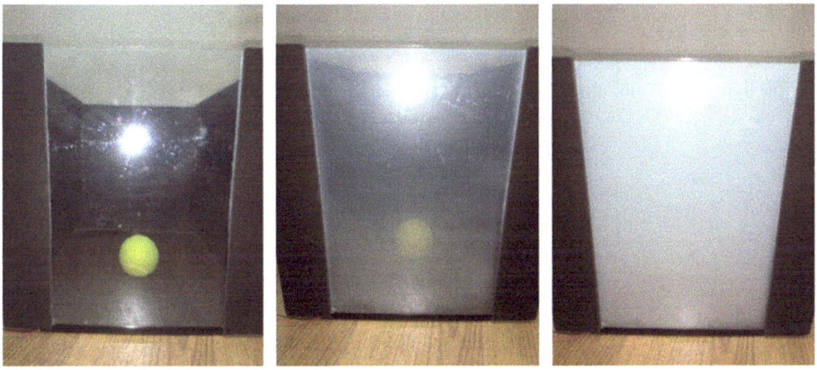

Fig. 48.8 Different transparency states of EC film

Appendix 2: Electronic Specs of Arduino Uno Wi-Fi Rev2

See Table 48.2.

Table 48.2 Electronic specifications of Arduino Wi-Fi Rev 2

Microcontroller	ATmega4809
Operating voltage	5 V
Input voltage (recommended)	7–12 V
Digital I/O pins	14—5 provide PWM output
PWM digital I/O pins	5
Analog input pins	6
DC current per I/O pin	20 mA
DC current for 3.3 V pin	50 mA
Flash memory	48 KB (ATmega4809)
SRAM	6144 bytes (ATmega4809)
EEPROM	256 bytes (ATmega4809)
Clock speed	16 MHz
Radio module	u-blox NINA-W102
Secure element	ATECC608A
Inertial measurement unit	LSM6DS3TR
LED_BUILTIN	25
Length	68.6 mm
Width	53.4 mm
Weight	25 g

Table 48.3 Electronic specifications and features of environmental sensors

Sensors	Features
BH1750 light sensor module	• Power supply: 2.4–3.6 V (typically 3.0 V) • Less current consumption: 0.12 mA • Measuring Range: 1–65535 lx • Communication: I2C bus • Accuracy: ± 20% • Built in A/D converter for converting analog illuminance in the digital data • Very small effect of IR radiation • Highly responsive near to the human eye
DS18B20 temperature sensor	• Unique 1-Wire® interface requires only one port pin for communication • Measures temperatures from −55 to +125 °C (−67 to +257°F) • ±0.5 °C Accuracy from −10 to +85 °C • Programmable resolution from 9 to 12 bits • Each device has a unique 64-bit serial code, stored in on-board ROM • Available in 8-pin SO (150 mils), 8-pin μSOP, and 3-pin TO-92 packages

Appendix 3: Features of Environmental Sensors

See Table 48.3.

Appendix 4 Code for Control of Appliances via Sensor Feedback

Figures 48.9 and 48.10 are the sample control codes for various appliances (fan and LED light strips) via feedback from different sensors.

Fig. 48.9 Sample control of LED via multiple light sensors

```
#include <Arduino.h>
#include <Wire.h>
#include <BH1750.h>
#include <Adafruit_NeoPixel.h>

BH1750 lightMeter;
const int dinPin = 4;
const int numOfLeds = 15;
Adafruit_NeoPixel pixels = Adafruit_NeoPixel(numOfLeds, dinPin, NEO_GRB + NEO_KHZ800);

int red = 255;
int green = 255;
int blue = 255;

float lux;

void TCA9548A(uint8_t bus){
  Wire.beginTransmission(0x70);
  Wire.write(1 << bus);
  Wire.endTransmission();
}

void setup(){
  Wire.begin();
  Serial.begin(9600);
  TCA9548A(0);
  TCA9548A(1);
  TCA9548A(2);
  if (!lightMeter.begin()){
    Serial.println("Could not find a valid light sensor, check wiring!");
    while(1){}
  }
  pixels.begin();
  pixels.setBrightness(100);
}

void loop(){
  TCA9548A(0);
  float lux1 = lightMeter.readLightLevel(true);
  Serial.print("Light sensor 1:");
  Serial.print(lux1);
  Serial.print(" lx");
  delay(1000);

  TCA9548A(1);
  float lux2 = lightMeter.readLightLevel(true);
  Serial.print("Light sensor 2:");
  Serial.print(lux2);
  Serial.print(" lx");
  delay(1000);

  TCA9548A(2);
  float lux3 = lightMeter.readLightLevel(true);
  Serial.print("Light sensor 3:");
  Serial.print(lux3);
  Serial.print(" lx");
  delay(1000);

  float lux = (lux1 + lux2 + lux3)/3;

  if (lux < 500){
      for(int i=0;i<numOfLeds;i++){
      pixels.setPixelColor(i, pixels.Color(red,green,blue));


      pixels.show(); // This sends the updated pixel color to the hardware.
      //delay(1000);   // Delay for a period of time to change the next led
      pixels.setBrightness(80); // Value from 0 to 100%
      }
  }

   if (lux > 1000){
      for(int i=0;i<numOfLeds;i++){
      pixels.setPixelColor(i, pixels.Color(red,green,blue));
      pixels.show(); // This sends the updated pixel color to the hardware.
      //delay(1000);   // Delay for a period of time to change the next led
      pixels.setBrightness(20); // Value from 0 to 100%
      }
   }
}
```

Fig. 48.10 Sample control of fan via multiple temperature sensors

```
#include <OneWire.h>
#include <DallasTemperature.h>

// Data wire is plugged into port 2 on the Arduino
#define ONE_WIRE_BUS 2

// Setup a oneWire instance to communicate with any OneWire devices
OneWire oneWire(ONE_WIRE_BUS);

// Pass our oneWire reference to Dallas Temperature.
DallasTemperature sensors(&oneWire);

// Addresses of 3 DS18B20s
//uint8_t sensor1[8] = { 0x28, 0x71, 0xEB, 0x33, 0x3C, 0x19, 0x01, 0xBE };
uint8_t sensor2[8] = { 0x28, 0xAA, 0x61, 0xBC, 0x3F, 0x14, 0x01, 0xDA };
uint8_t sensor3[8] = { 0x28, 0x39, 0x4A, 0x8C, 0x3C, 0x19, 0x01, 0x9F };


// initialise motor pins and temperature
int motorPin = 3;
//float temp1;
float temp2;
float temp3;
float finalTemp = 0;


void setup(void)
{
  sensors.begin(); //start up temp sensor library
  pinMode(motorPin, OUTPUT);
  Serial.begin(9600);
  pinMode(motorPin, OUTPUT);
}


void loop(void)
{
  Serial.print("Requesting temperatures...");
  sensors.requestTemperatures();  // send command to get temperatures
  Serial.println("DONE");

  //temp1 = sensors.getTempC(sensor1)
  //Serial.print("Sensor 1(*C): ");
  //Serial.print(temp1);

  temp2 = sensors.getTempC(sensor2);
  Serial.print("Sensor 2(*C): ");
  Serial.print(temp2);

  temp3 = sensors.getTempC(sensor3);
  Serial.print("Sensor 3(*C): ");
  Serial.print(temp3);

  // average
  finalTemp = (temp2 + temp3) / 2;
  Serial.print("Final Temp(*C): ");
  Serial.print(finalTemp);

  // algorithm to on/off the motor
  if (finalTemp < 26) {
    analogWrite(motorPin, 100);
  }

  if (finalTemp > 26){
    analogWrite(motorPin, 255);
  }
  delay(500);

}
```

References

1. Lee, J.-W., Jung, H.-J., Park, J.-Y., Lee, J., & Yoon, Y. (2013). Optimization of building window system in Asian regions by analyzing solar heat gain and daylighting elements. *Renewable Energy, 50*, 522–531.
2. Omer, A. M. (2009). Energy use and environmental impacts: A general review. *Journal of Renewable and Sustainable Energy, 1*, 053101.
3. Granqvist, C.-G., Green, S., Niklasson, G. A., Mlyuka, N. R., Von Kraemer, S., & Georen, P. (2010). Advances in chromogenic materials and devices. *Thin Solid Films, 518*(11), 3046–3053.
4. Hakemi, H. (2017). Polymer-dispersed liquid crystal technology 'industrial evolution and current market situation'. Taylor & Francis.
5. Lampert, C. M. (1998). Smart switchable glazing for solar energy and daylight control. *Solar Energy Materials and Solar Cells, 52*(3–4), 207–221.
6. Santamouris, M. (2016). Cooling the buildings–past, present and future. *Energy and Buildings, 128*, 617–638.
7. DeForest, N., Shehabi, A., Selkowitz, S., & Milliron, D. J. (2017). A comparative energy analysis of three electrochromic glazing technologies in commercial and residential buildings. *Applied Energy, 192*, 95–109.
8. Detsi, M., Manolitsis, A., Atsonios, I., Mandilaras, I., & Founti, M. (2020). Energy savings in an office building with high WWR using glazing systems combining thermochromic and electrochromic layers. *Energies, 13*(11), 3020.
9. Cannavale, A., Ayr, U., Fiorito, F., & Martellotta, F. (2020). Smart electrochromic windows to enhance building energy efficiency and visual comfort. *Energies, 13*(6), 1449.

Chapter 49
Synthesis of Magnetic Chitosan Hydrogel from Crab Shells as an Environmentally Friendly Adsorbent for Water Purification

Yu Zhenning, Fu Wenbo, and Ethan John Lim

Abstract Water pollution is a serious problem in the world. Crab shells are abundant bio-wastes which can be harnessed to produce useful chitosan. In this project, chitosan was synthesized from chitin extracted from crab shells. The degree of deacetylation and molecular weight of chitosan synthesized were determined by Fourier transformation infrared spectroscopy (FTIR) and mass spectrophotometry, respectively. A novel method of magnetic chitosan hydrogel (MCH) synthesis was devised based on literature concepts, allowing MCH to be synthesized one-pot and using a greener procedure of chitosan synthesis. MCH was as effective as commercial activated carbon (AC) and chitosan in removing direct red and acid blue dye and outperformed both adsorbents in removing methyl orange. The maximum adsorption capacity of MCH on direct red and acid blue derived from the Langmuir isotherm were higher than that of both unconverted chitosan and commercial AC and also several other adsorbents reported in literature. The unique magnetic property of both MCH renders it reusable while sustaining more than 90% removal of acid blue even after five cycles, which outperforms unconverted chitosan by a great margin. MCH is thus a promising and environmentally friendly adsorbent which is able to remove dyes rapidly from water.

Keywords Chitosan hydrogel · Magnetite · Water purification · Dyes · Isotherm studies

49.1 Introduction

Industries are the major sources of pollutants to the environment, and various levels of pollutants are being discharged into the environment either directly or indirectly [1]. Many industries such as the textile industry utilise dyes to colour their products. However, 50% of the dyes used are lost and discharged in wastewater due to the low levels of dye–fibre fixation [2]. The removal of dyes from wastewater discharge has

Z. Yu (✉) · W. Fu · E. J. Lim
Hwa Chong Institution (College Section), Singapore, Singapore
e-mail: zhenningyu1@gmail.com

H. Guo et al. (eds.), *IRC-SET 2021*, https://doi.org/10.1007/978-981-16-9869-9_49

become increasingly important as even a small quantity of dye in water can be toxic and highly visible [3]. Discharge of these dyes into water sources affects the people who use these sources for living purposes such as washing, bathing and drinking [4]. Dyes impart toxicity to aquatic life and are found to be mutagenic, carcinogenic and may cause severe damage to organs [5]. Dyes can also deleteriously affect aquatic plants because they inhibit the penetration of sunlight into the receiving water bodies [6]. Due to the environmental challenges that dyes pose and the imposition of government legislation requiring textile wastewater to be treated, there is significant demand for an effective solution to remove such dyes from wastewater.

To tackle this issue, hyperfiltration, ultrafiltration and nanofiltration were investigated to recover the dyes and chemicals from industrial wastewater [7]. Although these methods are efficient, they are expensive and commercially undesirable [8]. Adsorption has been found to be superior to other techniques for water reuse in terms of manufacturing cost, speed, ease of operation and insensitivity to toxic pollutants [9]. However, many of the current adsorbents, such as activated carbon, are limited by the high costs involved in both the precursors and the production process [10]. Thus, it is imperative for production of efficacious and cost-effective adsorbents for dye removal to be researched.

Chitosan is a type of natural polysaccharide derived from the partial deacetylation of chitin, and it is the second-most prevalent polymer in nature after the lignocellulose group [11]. Chitosan can be derived from crustacean shells such as crab shells, an undesirable by-product of the food industry [12]. The high chitin content of crab shells [13] renders it a suitable precursor for the extraction of chitin and the subsequent production of chitosan from it. The food industry generates 6 million to 8 million tonnes of crustacean shells every year [14]. A large proportion of the wastes are either burnt or jettisoned in landfills, which perpetuates environmental degradation.

The chemical structure of chitosan monosaccharide is very similar to that of cellulose, the only difference being the amino group at the C-2 position instead of a hydroxyl group. Chitosan displays a high adsorption capacity for many types of dyes, but powdered chitosan has limitations which prevents practical applications of it. It is soluble in acids and has poor mechanical strength. In addition, the separation and recovery of chitosan powder is still problematic for researchers [15].

In this study, a rapid and efficacious method for the dye removal using magnetic chitosan hydrogel, synthesized with cheap and environmentally friendly chitosan from crab shell was proposed. The use of crab shells could potentially reduce the cost of the hydrogel and at the same time alleviate the impacts that these waste crab shells pose to the environment. Magnetizing the hydrogel renders the separation of the hydrogel from treated wastewater convenient and effective.

49.2 Objectives and Hypothesis

This study aims to extract chitin from crab shells and convert it to chitosan. The chitosan produced in this study will be converted to magnetic chitosan hydrogel (MCH) via NaOH precipitation. The effectiveness of MCH in removing anionic dyes (acid blue, direct red and methyl orange) was compared with commercial activated carbon and the unconverted chitosan synthesized. The reusability of MCH over several cycles was also evaluated and compared.

This study hypothesizes that magnetic chitosan hydrogel (MCH) will be successfully synthesized from chitin extracted from crab shell. The effectiveness of MCH in removing anionic dyes will be comparable to commercial activated carbon and better than unconverted chitosan. MCH will also be able to sustain their effectiveness in dye removal for at least five cycles.

49.3 Materials and Methods

49.3.1 Materials

Crab shells of species *Scylla serrata* were obtained from a market. Sodium hydroxide, trisodium citrate, iron(II) sulphate and iron(III) chloride were purchased from GCE Chemicals. Direct red, acid blue and methyl orange dyes were procured from Sigma Aldrich.

49.3.2 Extraction of Chitin and Synthesis of Chitosan from Crab Shell

There are three main components in this process: demineralisation, deproteinization and deacetylation. Demineralisation of the crab shells was carried out with 3% (v/v) HC*l* at room temperature with a solid to solvent ratio of 1:5 (w/v) for 8 h. The residue was washed until neutral pH. Deproteinization of the crab shells was carried out with 4% (w/v) NaOH at room temperature with a solid to solvent ratio of 1:5 (w/v) for 20 h. The residue was washed until neutral pH. Purified chitin was then dried and ground to small particles to expedite deacetylation. Deacetylation of chitin was carried out using 10 M NaOH at 65 °C with a solid to solvent ratio of 1:10 (w/v) for 20 h. Acetic acid was added to dissolve the chitosan, after which the mixture was centrifuged, and NaOH was added until pH 12. The mixture was then centrifuged and the obtained chitosan was washed, dried and ground. It was then washed until pH neutral and dried.

49.3.3 Determination of Degree of Deacetylation (%DDA) of Chitosan Synthesized by Fourier Transform Infrared Spectroscopy (FTIR)

During the process of deacetylation, the acetyl group present in chitin is removed, converting the amide functional group (−CO–NH) into an amine (-NH_2). Therefore, by taking the ratio of the absorbance reading of the amide group (changed during deacetylation) at 1655 cm^{-1} to the reading of the hydroxyl group (unchanged during deacetylation) at 3450 cm^{-1}, the percentage of deacetylation (DDA) of the chitosan synthesized can be determined by the following formula:

$$\mathrm{DDA}(\%) = [1 - (A_{1655}/A_{3450})/1.33] \times 100$$

The ratio of 1.33 represents the ratio of A_{1655}/A_{3450} for fully N-acetylated chitosan.

49.3.4 Synthesis of Magnetic Chitosan Hydrogel (MCH)

The chitosan synthesized in this study was converted to MCH, and its effectiveness in removing dye was compared with unconverted chitosan. Commercial chitosan was prepared using 80% (w/v) sodium hydroxide [16] which is extremely corrosive and environmentally unfriendly. It is, hence, necessary for this study to explore a more environmentally friendly method to produce chitosan. A method reported in literature [17] utilized commercial chitosan to produce a hydrogel by precipitation with ethanol. However, due to a lower concentration of sodium hydroxide being used in this study, the chitosan produced is of lower molecular weight and degree of deacetylation. Hence, an alternative method of producing MCH needs to be devised where NaOH, instead of ethanol, is used to precipitate chitosan of low molecular weight. Coincidentally, the precipitation of magnetite is also induced by NaOH, thus magnetizing the hydrogel homogenously. This is in contrast to the separate addition of magnetite nanoparticles into the solution in the method reported in literature [17], which both utilises more corrosive solutions during the synthesis of the nanoparticles, and may also result in a non-homogenous dispersion of the nanoparticles. 3.87 g of iron(II) sulphate, 7.97 g of iron(III) chloride and 1.0 g of chitosan synthesized in this study were dissolved in 50 ml of 0.5 M $FeCl_3$ solution. The mixture was heated to 80 °C for 30 min. The pH of the solution was adjusted to 10 using 1 M NaOH, after which 6.34 g of trisodium citrate was added and the solution was then stirred for 12 h. The resulting MCH was obtained by centrifugation and crosslinking was carried out by submerging the MCH in 20 ml of 5% (v/v) glutaraldehyde solution for 24 h. The chemical reaction for the co-precipitation of magnetite is shown below:

$$Fe^{2+}(aq) + 2Fe^{3+}(aq) + 8OH^{-}(aq) \rightarrow Fe_3O_4(s) + 4H_2O \quad (49.1)$$

49.3.5 Batch Adsorption Studies

Adsorption tests were carried out by adding 0.1 g of chitosan, MCH or commercial activated carbon (AC) to 20 ml of 50 mg/L dye solution into a conical flask, and the mixture was agitated by the orbital shaker for 24 h. The adsorbent was then separated either by using a magnet for MCH or by centrifuging for chitosan and AC, and the supernatant was analysed. Acid blue, direct red and methyl orange solutions were analysed using a UV–VIS Spectrophotometer (Shimadzu UV 1800) at 565 nm, 526 nm and 462 nm, respectively. The percentage of dye adsorbed was determined.

49.3.6 Isotherm Studies

Adsorption was carried out with 50–400 mg/L of acid blue and direct red dye and the equilibrium concentration data was fitted into the linearized forms of the Langmuir and Freundlich isotherms (Appendix, Pg. 11–14). The adsorption capacity (q_m) was determined by the following formula:

$$q_m = \frac{(C_i - C_f)(V)}{M}$$

where

C_i initial concentration in mg/l,
C_f final concentration in mg/l,
V volume of solution in dm^3,
M mass of MCH in g.

49.3.7 Regeneration of MCH/Chitosan and Reusability Tests

Adsorption and regeneration of acid blue was repeated for 5 cycles. After each adsorption cycle, MCH was magnetically separated and washed with 1 M NaOH, followed by deionized water until the colour of the dye was removed. To regenerate the hydrogel, 1 M HCl was added. Chitosan was desorbed with ethanol, washed and dried before being reused for the next cycle.

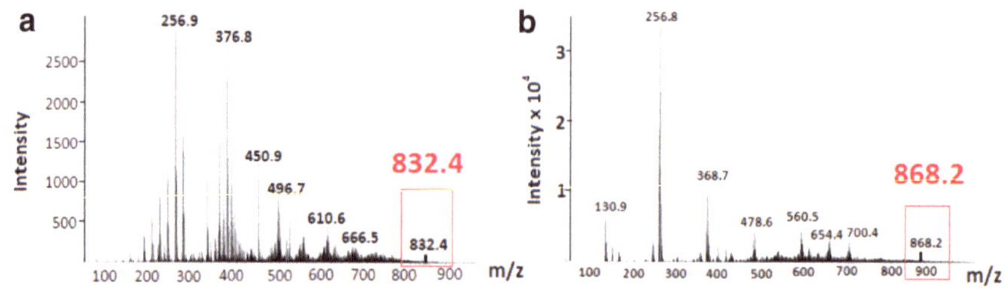

Fig. 49.1 **a** Mass spectrum (positive polarity) of chitosan synthesized (left), **b** Mass spectrum (negative polarity) of chitosan synthesized (right)

49.4 Results and Discussion

49.4.1 Characterization of Chitosan Synthesized

49.4.1.1 By FTIR

Percentage of deacetylation of the chitosan synthesized was determined from its FTIR spectrum (Appendix 1, Pg 11) to be 41.3%, which was lower than that of the commercial chitosan, which was 80%, indicating that the binding of the chitosan to the central iron(III) ion may not be as strong as that in the commercial chitosan. This may result in lower adsorption capacity, as less sites are available for dyes to be adsorbed.

49.4.1.2 By Mass Spectrophotometry

Analysis by mass spectra reveals that the chitosan synthesized has a molecular weight of 800–900 Daltons, which is much lower than that of the commercially produced chitosan, which ranges from 30,000 to 150,000 Daltons (Fig. 49.1). Thus, the chitosan synthesized would not be able to be precipitated by ethanol, due to the low molecular weight and would have to be precipitated by NaOH as described in Sect. 49.3.4.

49.4.2 Characterization of MCH

49.4.2.1 By X-Ray Diffraction (XRD)

XRD pattern of MCH (Fig. 49.2) reveals sharp peaks at 30.36°, 35.79°, 43.39°, 53.70°, 57.44 and 62.88°, which are similar to that reported by Loh et al. [18]. This suggests that magnetite has been successfully incorporated into MCH.

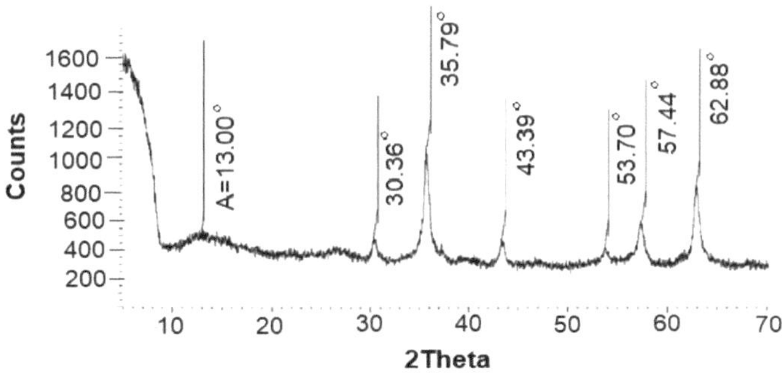

Fig. 49.2 XRD pattern of MCH-E

Fig. 49.3 SEM image of MCH (×5000 zoom)

49.4.2.2 By Scanning Electron Microscopy (SEM)

From the SEM images (Figs. 49.3 and 49.4), it is observed that a layer of magnetite nanoparticles of 10–50 nm is coated onto the surface of the hydrogel, which allows it to be magnetic.

49.4.3 Batch Adsorption Studies Results

MCH showed excellent percentage removal of more than 95% for all three dyes, while commercial activated carbon (AC) and chitosan pales in comparison in removing methyl orange (Fig. 49.5). This significant difference between the performance of commercial activated carbon (AC), chitosan and MCH (p-value of Kruskal–Wallis

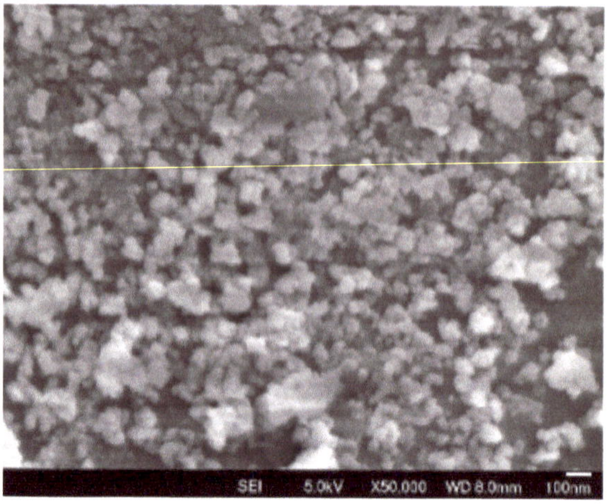

Fig. 49.4 SEM image of MCH (×50,000 zoom)

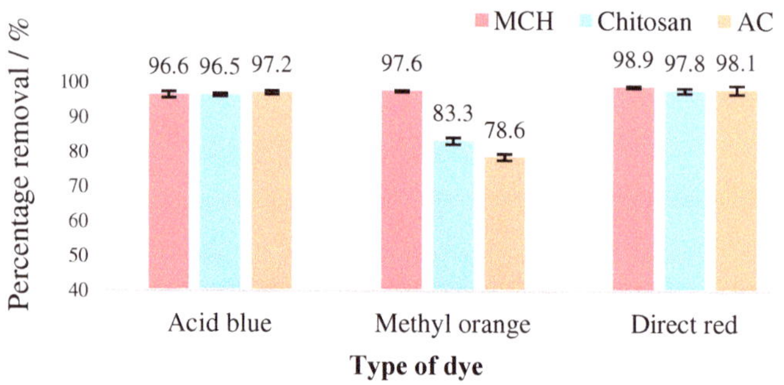

Fig. 49.5 Adsorption of dyes by MCH, chitosan and AC

test = 0.00193 < 0.05) in removing methyl orange is due to the difference in the mechanism of adsorption of these adsorbents.

Commercial AC removes dyes mainly through pi–pi interactions between the aromatic rings of the dyes and the aromatic rings in the AC (Fig. 49.6) [19]. Acid blue and direct red are easily removed by AC due to the large number of aromatic rings present in the organic dye, with 6 and 8 rings, respectively. On the other hand, methyl orange only has two aromatic rings, explaining its low percentage removal by AC.

MCH and chitosan have a different adsorption mechanism. Acid blue, methyl orange and direct red are anionic dyes containing -SO_3Na groups, which dissociate in water to form $-SO_3^-$. Chitosan adsorbs these dyes by electrostatic forces of attraction using its amine group (Fig. 49.7), which is protonated in water, causing it to have a positive charge. Although better than AC, chitosan does not outperform MCH, as MCH adsorbs these dyes by allowing the dyes to act as ligands, binding

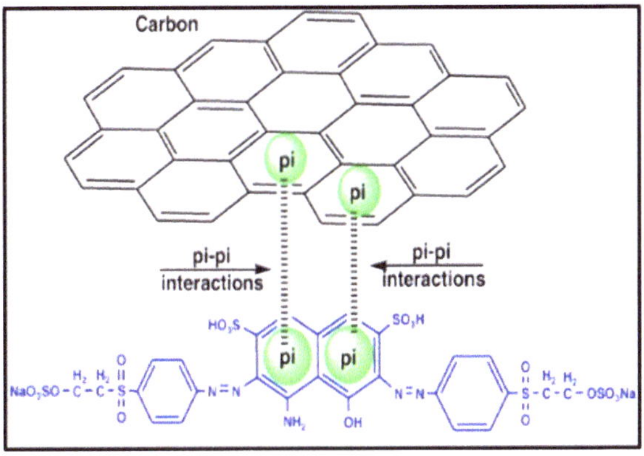

Fig. 49.6 Interaction of AC with dyes

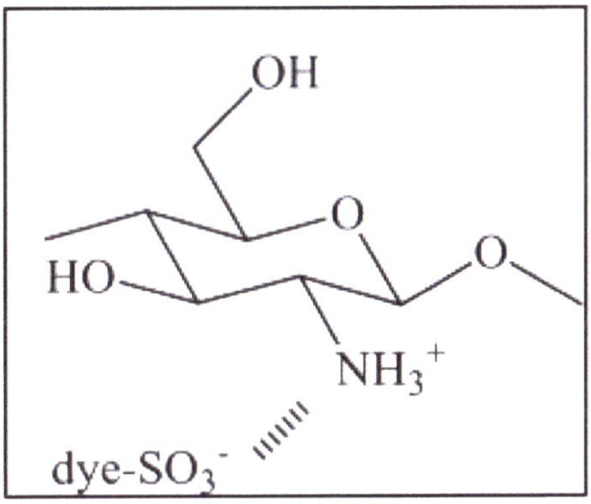

Fig. 49.7 Interaction of chitosan with dyes

to the central Fe^{3+} ion in the MCH complex (Fig. 49.8), hence removing them from the solution.

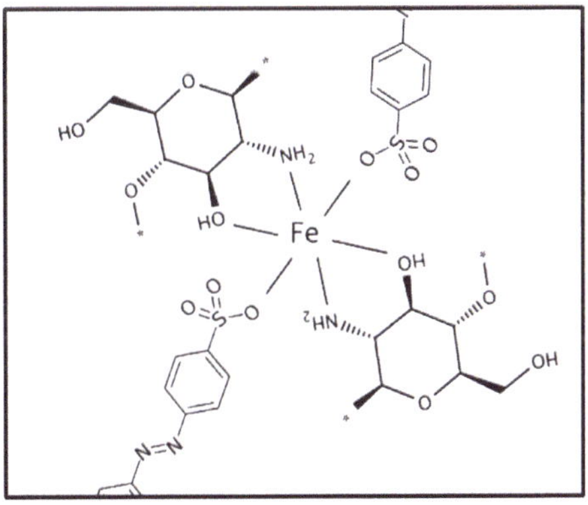

Fig. 49.8 Interaction of MCH with dyes

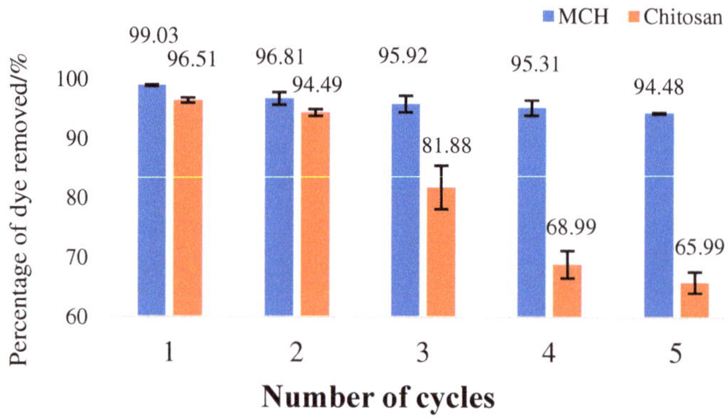

Fig. 49.9 Reusability of MCHs and chitosan in adsorbing acid blue

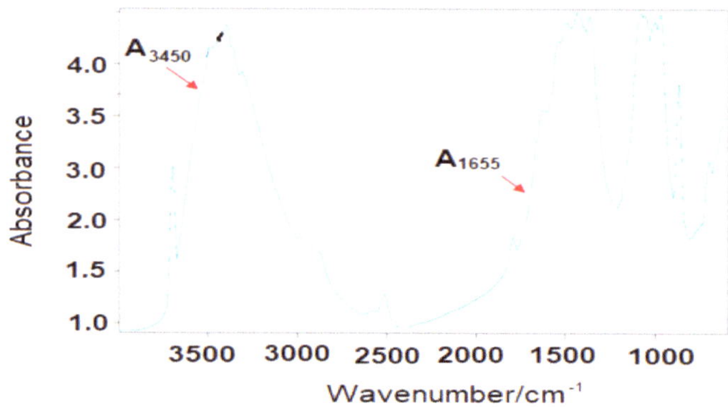

Fig. 49.10 FTIR spectrum of chitosan synthesized

49.4.4 Isotherm Studies

Maximum adsorption capacity (q_m) of both MCH and chitosan derived from Langmuir isotherm studies (Appendix 2, pg. 11–14) on acid blue and direct red were compared with other studies; two of which were targeted at removing acid blue dye, while the other two were targeted at removing direct red dye (Table 49.1).

MCH outperforms unconverted chitosan in its maximum adsorption capacity (q_m) for both acid blue and direct red, proving that the polymerisation and magnetisation has definitely improved the effectiveness of the chitosan. MCH also outperforms commercial activated carbon (AC) and the adsorbents synthesized by several other researchers in removing both dyes.

Table 49.1 Comparison of maximum adsorption capacity (q_m) of MCH with other adsorbents

Type of adsorbent	Acid blue q_m/mg/g	Direct red q_m/mg/g	References
MCH	113.6	125.0	This study
Chitosan	94.3	55.2	This study
Commercial AC	89.3	101.0	This study
Modified bentonite	45.0	–	[20]
Activated red mud	83.3	–	[21]
Activated rice husks	–	74.6	[22]
Cationized sawdust	–	68.5	[23]

49.4.5 Reusability Tests

Results of the reusability tests on acid blue reveal that both MCH is able to sustain a percentage removal of more than 90% for at least 5 cycles, demonstrating its ability to be reused multiple times and still be effective in removing dyes from water (Fig. 49.9). This is possible due to the MCH being magnetic, resulting it being easily retrieved after adsorption via the use of a magnet. The retrieved MCH can then be easily desorbed and regenerated for subsequent use. Unpolymerized chitosan, on the other hand, was unable to sustain its initial removal as the dye was unable to be effectively desorbed after each cycle. Retrieval of the adsorbent by filtration is also inefficient, with only 59% of it being retrieved after the fifth cycle, which is contrasted to the 92% of MCH.

49.5 Conclusions and Future Work

Magnetic chitosan hydrogel (MCH) was successfully synthesized from crab shell chitosan, with magnetite nanoparticles coating the surface of the hydrogel. The effectiveness of MCH in removing acid blue and direct red was comparable to commercial activated carbon and unconverted chitosan, while the removal of methyl orange by MCH surpasses both that of commercial activated carbon and the unconverted chitosan. The successful synthesis of MCH using the one-pot procedure described in this paper not only made use of a more environmentally friendly method of production, but it also requires less sodium hydroxide and other environmentally damaging chemicals, potentially reducing cost of synthesis and rendering the use of it is more eco-friendly as sodium hydroxide is a corrosive chemical. Isotherm studies reveal that the maximum adsorption capacity of MCH is much higher than unconverted chitosan and also outperforms several adsorbents synthesized by other researchers.

MCH is capable of being reused and regenerated, consistently removing more than 90% of acid blue across five cycles of adsorption and regeneration, with a high retrieval rate superior to the unconverted chitosan.

In future, isotherm and kinetic studies can be extended to methyl orange. Adsorption of other anionic species such as phosphate by MCH can be explored. It would also be interesting to investigate the effect of pH on adsorption of dyes by MCH. Mass of magnetite used in the synthesis may affect adsorption capacity and optimum mass of magnetite which results in greatest adsorption of MCH could be determined. Finally, it is important to determine if there is any significant leaching of iron from magnetite present in MCH as iron is a secondary pollutant.

Appendix 1: FTIR of Chitosan Synthesized

The percentage of deacetylation (DDA) of the chitosan synthesized was determined by the following formula:

$$\mathrm{DDA}(\%) = [1 - (A_{1655}/A_{3450})/1.33] \times 100$$

Using FTIR software, A_{1655} and A_{3450} were determined to be 2.88 and 3.69 respectively (Fig. 49.10). DDA(%) was hence calculated to be 41.3%.

Appendix 2: Langmuir and Freundlich Isotherm

The equilibrium concentration data obtained from initial concentration tests on acid blue and direct red were fitted into Langmuir isotherm and Freundlich isotherm. The Langmuir isotherm assumes that the adsorbed material (such as acid blue) is adsorbed over a uniform adsorbent surface at a constant temperature.

The linear form of Langmuir isotherm equation is given by:

$$\frac{C_e}{q_e} = \frac{1}{bq_m} + \frac{C_e}{q_m}$$

where C_e is the equilibrium concentration of dye (mg/L), q_e is the equilibrium capacity of the sorbents (mg/g), b is the Langmuir constant that indicates the sorption intensity, and q_m is the maximum sorption capacity (mg/g).

The Freundlich isotherm assumes that the adsorption occurs on a heterogeneous surface.

The linear form of Freundlich equation is given by:

$$\log(q_e) = \log(K_F) + \frac{1}{n}\log(C_e)$$

Table 49.2 R^2 value comparison of Langmuir and Freundlich isotherms for acid blue and direct red

	Acid blue		Direct red	
Adsorbent	R^2 value for Langmuir isotherm	R^2 value for Freundlich isotherm	R^2 value for Langmuir isotherm	R^2 value for Freundlich isotherm
Chitosan	0.9980	0.9111	0.9987	0.9693
MCH	0.9940	0.9684	0.9927	0.9743
Commercial AC	0.9996	0.8173	0.9985	0.9547

where C_e is the equilibrium concentration of dye (mg/L), q_e is the equilibrium capacity of the sorbents (mg/g), K_F, a constant, is related to sorption capacity and n corresponds to sorption intensity.

If the equilibrium concentration data fits the Langmuir isotherm, adsorption can be inferred to be monolayer. Maximum adsorption capacity of the MCH and chitosan can be derived from the inverse of the gradient of Langmuir linear equations.

In contrast, if the equilibrium concentration data fits the Freundlich isotherm, adsorption can be inferred to occur on a heterogeneous surface and adsorption is multilayer. A comparison between the R^2 values of the graph can determine which model the data set fits better (Table 49.2).

All data sets had a better fit for the Langmuir isotherm, implying that both MCH and chitosan have a homogenous surface with monolayer adsorption.

References

1. Ado, A., Tukur, A. I., Ladan, M., Gumel, S. M., Muhammad, A. A., Habibu, S., & Koki, I. B. (2015). A review on industrial effluents as major sources of water pollution in Nigeria. *Chemistry Journal, 1*(5), 159–164.
2. Mohan, N., Balasubramanian, N., & Basha, C. A. (2007). Electrochemical oxidation of textile wastewater and its reuse. *Journal of hazardous materials, 147*(1–2), 644–651.
3. Malik, R., Ramteke, D. S., & Wate, S. R. (2007). Adsorption of malachite green on groundnut shell waste based powdered activated carbon. *Waste Management, 27*(9), 1129–1138.
4. Sharma, M. K., & Sobti, R. C. (2000). Rec effect of certain textile dyes in Bacillus subtilis. *Mutation Research/Genetic Toxicology and Environmental Mutagenesis, 465*(1–2), 27–38.
5. Kadirvelu, K., Kavipriya, M., Karthika, C., Radhika, M., Vennilamani, N., & Pattabhi, S. (2003). Utilization of various agricultural wastes for activated carbon preparation and application for the removal of dyes and metal ions from aqueous solutions. *Bioresource Technology, 87*(1), 129–132.
6. Shen, D., Fan, J., Zhou, W., Gao, B., Yue, Q., & Kang, Q. (2009). Adsorption kinetics and isotherm of anionic dyes onto organo-bentonite from single and multisolute systems. *Journal of Hazardous Materials, 172*(1), 99–107.
7. Avlonitis, S. A., Poulios, I., Sotiriou, D., Pappas, M., & Moutesidis, K. (2008). Simulated cotton dye effluents treatment and reuse by nanofiltration. *Desalination, 221*(1–3), 259–267.
8. Sulak, M. T., Demirbas, E., & Kobya, M. (2007). Removal of Astrazon Yellow 7GL from aqueous solutions by adsorption onto wheat bran. *Bioresource technology, 98*(13), 2590–2598.

9. Crini, G., & Badot, P. M. (2008). Application of chitosan, a natural aminopolysaccharide, for dye removal from aqueous solutions by adsorption processes using batch studies: A review of recent literature. *Progress in polymer science, 33*(4), 399–447.
10. Srivastava, V. C., Mall, I. D., & Mishra, I. M. (2007). Adsorption thermodynamics and isosteric heat of adsorption of toxic metal ions onto bagasse fly ash (BFA) and rice husk ash (RHA). *Chemical Engineering Journal, 132*(1–3), 267–278.
11. Ngan, W. S. W., Teong, L. C., & Hanafiah, M. A. K. M. (2011). Adsorption of dyes and heavy metal ions by chitosan composites: A review. *Carbohydrate Polymers, 83*(4), 1446–1456.
12. Ngan, W. S. W., & Isa, I. M. (1998). Comparison study of copper ion adsorption on chitosan, Dowex A-1, and Zerolit 225. *Journal of Applied Polymer Science, 67*(6), 1067–1070.
13. Toliba, A. O., Rabie, M. A., & El-Araby, G. M. (2014). Extending the shelf-life of cold-stored strawberry by chitosan and carnauba coatings. *Zagazig Journal of Agricultural Research, 41*(5), 1067–1076.
14. Drahl, C. (2019). How seafood shells could help solve the plastic waste problem. *Science News*. Retrieved from https://www.sciencenews.org/article/seafood-shells-chitin-plastic-food-waste
15. Zhu, H. Y., Jiang, R., Fu, Y. Q., Jiang, J. H., Xiao, L., & Zeng, G. M. (2011). Preparation, characterisation and dye adsorption properties of γ-Fe_2O_3/SiO_2/chitosan composite. *Applied Surface Science, 258*(4), 1337–1344.
16. Takarina, N. D., Indah, A. B., Nasrul, A. A., Nurmarina, A., Saefumillah, A., Fanani, A. A., & Loka, K. D. P. (2017). Optimisation of deacetylation process for chitosan production from Red Snapper (Lutjanus sp.) scale waste. *Journal of Physics: Conference Series, 812*(1), 012110.
17. Shen, C., Shen, Y., Wen, Y., Wang, H., & Liu, W. (2011). Fast and highly efficient removal of dyes under alkaline conditions using magnetic chitosan-Fe (III) hydrogel. *Water Research, 45*(16), 5200–5210.
18. Loh, K., Lee, Y., Musa, A., Salmah, A., & Zamri, I. (2008). Use of Fe_3O_4 Nanoparticles for enhancement of biosensor response to the herbicide 2,4-dichlorophenoxyacetic acid. *Sensors, 8*(9), 5775–5791. https://doi.org/10.3390/s8095775
19. Lupul, I., Yperman, J., Carleer, R., & Gryglewicz, G. (2015). Adsorption of atrazine on hemp stem-based activated carbons with different surface chemistry. *Adsorption, 21*(6–7), 489–498.
20. Jeeva, M., & Wan Zuhairi, W. Y. (2018, April). Adsorption of Acid Blue 25 dye by bentonite and surfactant modified bentonite. In *AIP Conference Proceedings* (Vol. 1940, No. 1, p. 020030). AIP Publishing LLC.
21. Shirzad-Siboni, M., Jafari, S. J., Giahi, O., Kim, I., Lee, S. M., & Yang, J. K. (2014). Removal of acid blue 113 and reactive black 5 dye from aqueous solutions by activated red mud. *Journal of Industrial and Engineering Chemistry, 20*(4), 1432–1437.
22. Safa, Y., Bhatti, H. N., Bhatti, I. A., & Asgher, M. (2011). Removal of direct Red-31 and direct Orange-26 by low cost rice husk: Influence of immobilisation and pretreatments. *The Canadian Journal of Chemical Engineering, 89*(6), 1554–1565.
23. Hebeish, A., Ramadan, M. A., Abdel-Halim, E., & Abo-Okeil, A. (2011). An effective adsorbent based on sawdust for removal of direct dye from aqueous solutions. *Clean Technologies and Environmental Policy, 13*(5), 713–718.

Chapter 50
Electrochemical Treatment of an Effluent with Azo Dyes and Heavy Metals Using a Graphene-Coated Electrode

Xinyan Liao and Jiecong Tan

Abstract Textile wastewaters have exacerbated water scarcity, and its pollutants, including azo dyes and heavy metals, have detrimental effects on human health. Fortunately, electrochemical processes are gaining traction for their exceptional ability to simultaneously mineralize dyes and recover metals from wastewaters, with electro-Fenton (EF) and electrodeposition being particularly promising for combined removal of organic and inorganic compounds. In this context, this study aims to optimize these processes and maximize the treatment efficiency of synthetic textile wastewaters. Each setup contained 300 mL of synthetic wastewater at pH 3, a graphene-coated carbon brush cathode and a boron-doped diamond (BDD) anode. To determine the optimal potential for H_2O_2 generation, the O_2 reduction process was conducted at 4, 5 and 6 V. Thereafter, the optimal potential for Fe, Zn and Ni metal recovery was identified among 5, 6 and 7 V. Next, the effect of initial dye concentration on degradation and mineralization efficiencies was investigated with 25, 50 and 100 ppm of Reactive Red-120 (RR120) dye. Moreover, the effectiveness of BDD was compared to that of the more affordable dimensionally stable anode (DSA). Finally, high-performance liquid chromatography analysis of the degradation by-products and reusability tests were conducted too. These studies revealed that electrochemical processes were remarkably effective even with numerous heavy metals present, demonstrating their capability to treat effluents with mixtures of organic and inorganic pollutants. To maximize the treatment efficiency, the optimal potential should be 5 V for degradation and 7 V for metal recovery, with 100 ppm of RR120. Notably, the use of cost-effective DSA and highly reusable graphene cathode can lower the treatment cost tremendously. Furthermore, electrochemical processes cause negligible sludge generation. Evidently, these low-cost and eco-friendly processes have immense potential in resolving the perennial water pollution crisis.

Keywords Electro-Fenton · Dye pollution · Anodic oxidation · Electrodeposition · Heavy metal · Azo dyes

X. Liao (✉) · J. Tan
Hwa Chong Institution, Singapore, Singapore
e-mail: liaoxinyan03@gmail.com

H. Guo et al. (eds.), *IRC-SET 2021*, https://doi.org/10.1007/978-981-16-9869-9_50

50.1 Introduction

Due to the industrial boom, water bodies are being increasingly polluted by industrial effluents; in particular, textile effluents are the most prevalent. These effluents contain a myriad of high-risk pollutants like synthetic dyes and heavy metal ions which cause dire health consequences [1]. Azo dyes, such as Reactive Red-120 (RR120), are most widely used in textile and paper products manufacturing [2]. Furthermore, past studies have shown that RR120 is often broken down into carcinogenic metabolites in mammals and accelerates tumor formation [3, 4]. Ni^{2+}, Cu^{2+} and Fe^{2+} are most commonly found in textile effluents for their pigmentation and tanning properties [5, 6]. Unfortunately, they lead to a range of debilitating illnesses like neurodegenerative diseases (Ni^{2+}), hemolysis (Cu^{2+}) and severe liver damage (Fe^{2+}) [7–9].

Conventional methods for the removal of azo dyes include adsorption, membrane filtration and biological treatment [10]. Notably, adsorption using activated carbon is one of the most effective methods, generating >90% removal in previous studies on RR120 [11, 12]. However, adsorption does not degrade the recalcitrant dye molecule which still poses environmental risks, and the disposal or regeneration process is also costly [13]. In addition, heavy metal ions are most commonly removed by chemical precipitation and ion exchange [14]. However, such treatment methods merely convert the dissolved heavy metals into sludge [15], thereby incurring further costs for solid waste disposal. Moreover, little research has been conducted on treatments capable of simultaneously removing mixtures of heavy metals and azo dyes, as it would typically be found in textile wastewaters.

In this context, electrochemical advanced oxidation processes (EAOPs), such as electro-Fenton (EF) and anodic oxidation (AO), alone or paired, have attracted great interest in recent years due to their ability to simultaneously mineralize organic dyes. Meanwhile, electrodeposition, a simple electrochemical process which consists of using the negative potential of the cathode to obtain the metallic form of the ion, can be applied to recover heavy metals (Eq. 50.1) [16].

EF involves the in situ generation of H_2O_2 via the two-electron cathodic reduction of O_2 (Eq. 50.2), where H_2O_2 is thereafter utilized to produce highly reactive hydroxyl radicals (·OH) in the presence of a Fe^{2+} catalyst (Eq. 50.3) [17]. These ·OH radicals are very effective against organic dyes, while the Fe^{2+} ions can eventually be removed by electrodeposition. Furthermore, the coupling of EF with AO using a non-active anode such as boron-doped diamond (BDD) can improve the mineralization efficiency, due to the generation of additional ·OH radicals at the electrode's surface [18].

$$M^{x+} + Xe^- \rightarrow M^0 \tag{50.1}$$

$$O_2 + 2e^- + 2H^+ \rightarrow H_2O_2 \tag{50.2}$$

$$H_2O_2 + Fe^{2+} \rightarrow Fe^{3+} + {}^-OH + \cdot OH \tag{50.3}$$

$$Fe^{3+} + e^- \rightarrow Fe^{2+} \tag{50.4}$$

The EF process minimizes the need for additional catalysts to be added [19], by promoting the regeneration of Fe^{2+} ions via cathodic reduction of Fe^{3+} (Eq. 50.4). Moreover, Fe^{2+} itself can be subsequently recovered as Fe metal via electrodeposition to be recycled into different products. This not only reduces the need for sludge management but also recovers the Fe^{2+} ions in their most valuable state for other industrial processes.

Currently, EF in the presence of heavy metal ions other than Fe^{2+} has not been studied. Therefore, this study aims to optimize EF for the treatment of textile wastewater by simultaneously degrading organic compounds and recovering heavy metals. Various experimental parameters such as the cell potential, concentration of RR120 and anode material were carefully investigated. Repeated electrolysis cycles were conducted to assess the reusability of the electrodes, while identification of the degradation by-products by high-performance liquid chromatography (HPLC) is useful to understand the mechanisms and identify possible post-treatment options.

50.2 Hypotheses

- The presence of heavy metal ions other than Fe^{2+} in textile wastewater will promote Fenton-like reactions.
- Effective degradation of RR120 in a mixture of heavy metal ions will be rapidly achieved by EF.
- A combination of different applied potentials will be an effective way to remove both organic compounds and heavy metals from wastewater.
- Boron-doped diamond (BDD) will result in greater efficiency of the EF process compared to dimensionally stable anodes (DSA), at the expense of higher material cost.

50.3 Materials and Methods

A. *Materials*

Carbon fibers coated with graphene and shaped in the form of a brush were used as the cathode [20], while the anode consisted of either BDD or DSA. Reactive Red-120 (RR120) was used as a model pollutant. Sodium chloride, iron (II) sulfate, copper (II) sulfate, nickel (II) sulfate, sulfuric acid, sodium hydroxide, titanium oxysulfate and potassium sulfate were purchased commercially and used as it is.

B. *Assembly of Electrochemical Cell*

Each cell contained 300 mL of electrolyte, and the electrodes were connected to a power supply. The pH was adjusted to 3 using 98% H_2SO_4, and air bubbling was supplied 40 min before the start of and throughout each experiment (refer to Appendix 1).

C. *Optimization of Applied Potential for Maximum Generation of H_2O_2*

The optimization of applied potential was carried out using 17.42 g of K_2SO_4 (50 mM) as electrolyte. Three different potentials were supplied: 4, 5 and 6 V. The electrochemical generation of H_2O_2 was evaluated using a colorimetric method, by taking 4.5 mL sample aliquots at different time intervals from 0 to 60 min, and mixing them with 0.5 mL of titanium oxysulfate and 5 mL of water. A Hach DR6000 UV–Vis spectrophotometer was used to measure the absorbance of the samples, with the wavelength set at 408 nm.

D. *Degradation and Mineralization of Reactive Red-120*

The dye solutions were prepared using 0.877 g of NaCl (50 mM) as the electrolyte and varying mass of RR120 amounting to 25, 50 and 100 ppm. The presence of heavy metals was simulated with 0.00747 g of $FeSO_4 \cdot 7H_2O$, 0.00590 g of $CuSO_4 \cdot 5H_2O$ and 0.00672 g of $NiSO_4 \cdot 6H_2O$ (5 ppm of Fe, Cu and Ni). Electrolysis was conducted at 5 V. Degradation efficiency was assessed by extracting 2.5 mL samples at 15-min intervals up to 60 min, with 2.5 mL of water added to these samples, and analyzing them using a Hach DR6000 UV–Vis Spectrophotometer, with wavelength set at 512 nm. Mineralization efficiency was assessed by extracting 5 mL samples at 1 h intervals up to 6 h, with 15 mL of water added to these samples, and analyzing the total organic carbon (TOC) concentration using a Shimadzu TOC Analyzer.

E. *Electrodeposition of Metals*

Electrodeposition optimization was evaluated using 0.8766 g of NaCl (50 mM) as the electrolyte, and three different applied potentials: 5, 6 and 7 V. The initial metal concentration was 25 ppm for all the metals, by adding 0.0374 g of $FeSO_4 \cdot 7H_2O$, 0.0295g of $CuSO_4 \cdot 5H_2O$ and 0.0336 g of $NiSO_4 \cdot 6H_2O$ to 300 mL of DI water. To analyze the remaining metal concentration in the solution, 2.5 mL of each solution was extracted every 5 min until 1 h, and 2.5 mL of water and 0.25 mL of dilute HNO_3 were added to each of these samples, which were then tested using inductively coupled plasma optical emission spectrometry (ICP-OES).

F. *Comparison of BDD and DSA Anodes and Identification of Degradation By-products*

Anode material was evaluated using 100-ppm dye solutions. 5 mL aliquots of each solution were extracted at 0, 1, 2, 3, 4, 5 and 6 h, and 15 mL of water was added to these samples for TOC analysis. 0.5 mL of each solution was extracted at the same timings, and 0.5 mL of water was added to these samples for high-performance liquid chromatography (HPLC) analysis.

G. *Reusability of Graphene-coated Carbon Brush*

Several electrolysis cycles were carried out using the same electrodes with the same experimental conditions to evaluate the reusability of the electrodes. A 100-ppm dye solution was used for this experiment. 5 mL and 20 mL aliquots were taken at 45 min and 180 min for analysis of remaining dye concentration and TOC analysis, respectively.

50.4 Results and Discussion

A. *Optimization of Applied Potential for H_2O_2 Generation*

In the beginning of the study, the ability of the electrolytic cell to produce H_2O_2 at different potentials via the 2-e$^-$ reduction of O_2 (Eq. 50.2) was investigated. Optimal H_2O_2 production was observed at 5 $_V$, with the highest H_2O_2 concentration of almost 40 ppm (Fig. 50.1). However, as the applied potential increased beyond the optimal, as observed in the case of 6 V, H_2O_2 concentration was observed to decrease. This is due to higher applied potentials accelerating waste reactions (Eqs. 50.5 and 50.6) that hinder the availability of O_2 and electrons for H_2O_2 generation [21]. Also, the concentration of H_2O_2 reached a plateau after approximately 40 min because the rate of generation at the cathode (Eq. 50.2) and degradation at the anode (Eq. 50.7) became equal [22].

$$O_2 + 4e^- + 4H^+ \rightarrow 2H_2O \tag{50.5}$$

$$2H^+ + 2e^- \rightarrow H_2 \tag{50.6}$$

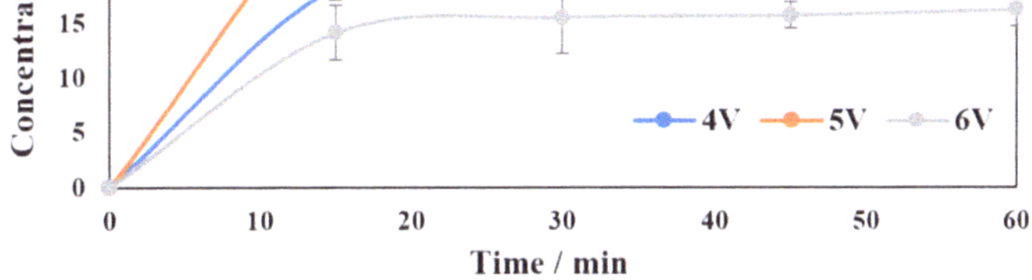

Fig. 50.1 Production of H_2O_2 at different potentials. Time = 1 h; volume 300 mL; pH = 3; cathode = carbon brush; anode = BDD

$$2H_2O_2 \rightarrow 2H_2O + O_2 \quad (50.7)$$

B. *Effect of Initial Dye Concentration on Reactive Red-120 Degradation and Mineralization*

The initial dye concentration is another crucial factor that determines the efficiency of its degradation and mineralization. The concentrations of dye and heavy metals utilized in this experiment are similar to those commonly found in textile dye wastewaters [23]. Evidently, the EF process is highly efficient as all three concentrations of dye achieved >99% degradation within 30 min (Fig. 50.2). Also, the degradation rate constant for 25 ppm was the highest, at 0.095 min^{-1} (refer to Appendix 2). The quicker degradation for 25 and 50 ppm can be attributed to the destruction of a larger proportion of organic load with a similar quantity of ·OH [24]. As depicted in Fig. 50.3, the 100-ppm solution achieved the highest percentage of mineralization because the greater concentration of degradation by-products increased the frequency of effective collisions with the ·OH radicals. The rate of mineralization then decreased with time as dye molecules formed short-chain carboxylic acids, such as oxalic acid, which are more resistant to ·OH radicals [25].

As shown in Table 50.1, compared to other treatment methods in previous studies, the EF process is much more effective in both degrading and mineralizing RR120. This showcases the immense potential of EF for the cost-effective and eco-friendly treatment of textile wastewater.

C. *Effect of Applied Potential on Metal Electrodeposition*

Electrochemical processes are unique as they simultaneously remove heavy metals and degrade organic dyes in wastewaters. Hence, this study also investigated the effect of different applied potentials (5, 6 and 7 V) to achieve the highest efficiency

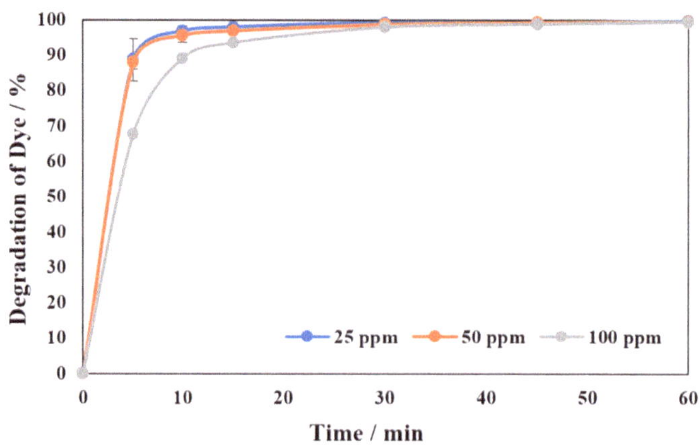

Fig. 50.2 Electrochemical degradation of reactive red 120 at different concentrations. Time = 1 h; volume = 300 mL; potential = 5 V; $[Fe^{2+}]$ = 5 ppm; pH = 3; cathode = carbon brush; anode = BDD

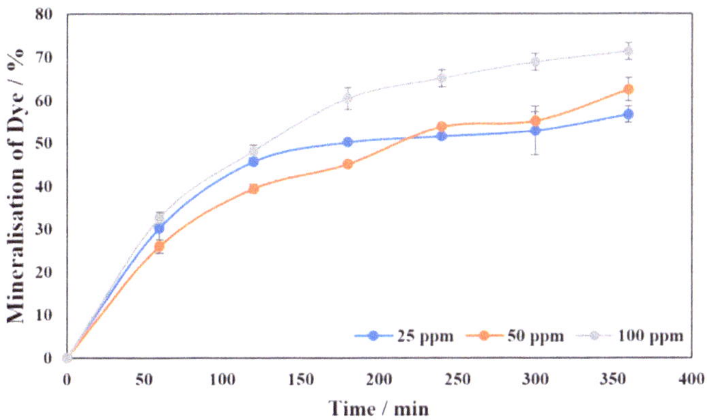

Fig. 50.3 Electrochemical mineralization of reactive red 120 at different concentrations. Time = 6 h; volume = 300 mL; potential = 5 V; $[Fe^{2+}]$ = 5 ppm; pH = 3; cathode = carbon brush; anode = BDD

Table 50.1 Comparison of maximum degradation and mineralization percentages of RR120 between various types of treatment

Type of treatment	Maximum RR120 degradation (%)	Maximum RR120 mineralization (%)	References
Electro-Fenton	>99%; 0.5 h required	75%; 6 h required	This study
Biological (*Bacillus lentus*)	98%; 12 h required	17.5%; 12 h required	[26]
Ozonation	96%, 1 h required	15.3%; 48 h required	[27]

of electrodeposition in solutions containing just metal ions. From Fig. 50.4, the trend is that a higher applied potential corresponds to higher removal of heavy metals, by increasing the rates of reduction reactions portrayed in Table 50.2.

Moreover, it can be seen from Fig. 50.5 that removal efficiency follows the trend of $Cu^{2+} > Fe^{2+} > Ni^{2+}$. Cu^{2+} achieved the highest percentage removal at >99% because it has the most positive standard electrode potential (E°) (Table 50.2). However, although E° of Ni^{2+} is more positive than Fe^{2+}, iron experienced higher percentage removal. This likely occurs because most of the iron could be in the form of Fe^{3+}, due to the constant oxidation of Fe^{2+} to produce Fe^{3+} during the EF process (Eq. 50.3). Hence, as the E° of Fe^{3+} is more positive than that of Ni^{2+}, reduction of Fe^{3+} is more favorable; thus, the removal of Fe is higher overall.

D. *Effect of Anode Material on Efficiency of EF Process*

As shown in Fig. 50.6, BDD was slightly more effective than DSA for both mineralization and degradation of RR120. This occurs because BDD is less active compared to DSA, generating more ·OH radicals at the surface of the anode, thus minimizing the occurrence of waste reactions at lower potentials (3–4 V) [28]. As such, the pairing of EF and AO maximizes the production of ·OH radicals since they are produced

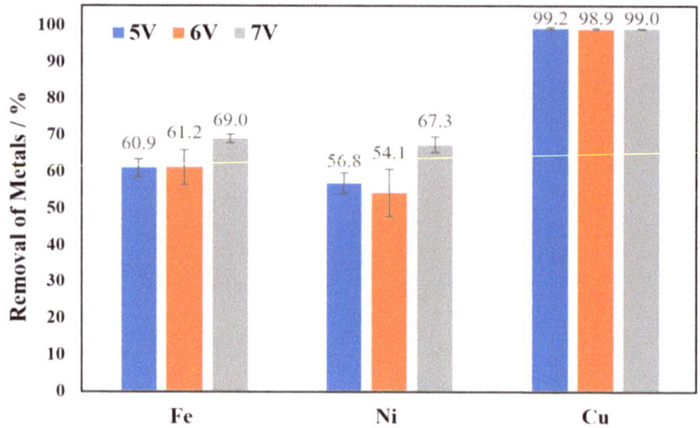

Fig. 50.4 Maximum removal of various heavy metals (25 ppm) at different potentials. Time = 1 h; volume = 300 mL; pH = 3; cathode = carbon brush; anode = BDD

Table 50.2 Standard electrode potentials of various electrodeposition reactions

Reaction	$Ni^{2+} + 2e^- \rightarrow Ni$ (s)	$Fe^{2+} + 2e^- \rightarrow Fe$ (s)	$Fe^{3+} + 3e^- \rightarrow Fe$ (s)	$Cu^{2+} + 2e^- \rightarrow Cu$ (s)
E°/V	−0.25	−0.44	−0.037	+0.15

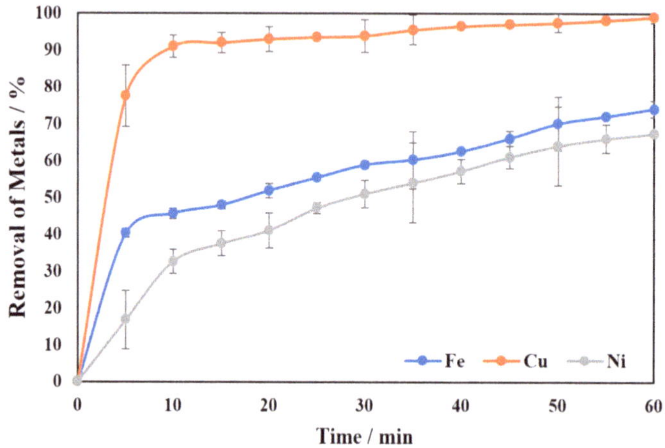

Fig. 50.5 Electrodeposition of various heavy metals (25 ppm) at 7 V. Time = 1 h; volume = 300 mL; potential = 7 V; pH = 3; cathode = carbon brush; anode = BDD

at both the anode surface and in the bulk solution, thus increasing the efficiency of degradation and mineralization.

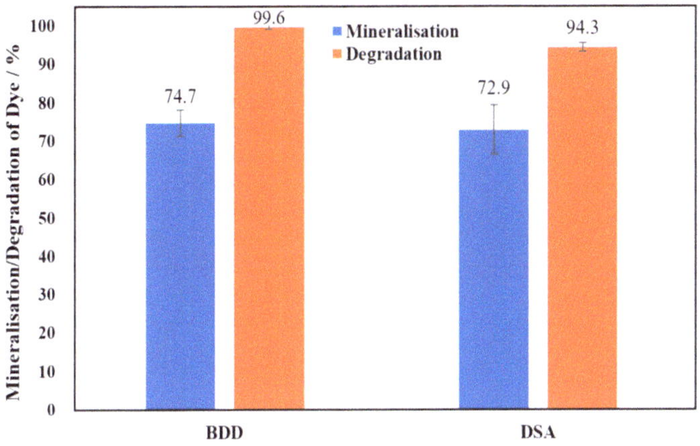

Fig. 50.6 Maximum mineralization and degradation by different anodes. Time = 3 h; volume = 300 mL; potential = 5 V; pH = 3; cathode = carbon brush; anode = BDD or DSA

E. *Reusability Studies*

As observed from Fig. 50.7, the percentage degradation and mineralization of RR120 remained relatively constant even after multiple cycles. This certifies the ability of the graphene-coated carbon brush to be repeatedly reused as a cathode in the EF process, greatly decreasing the fixed capital costs.

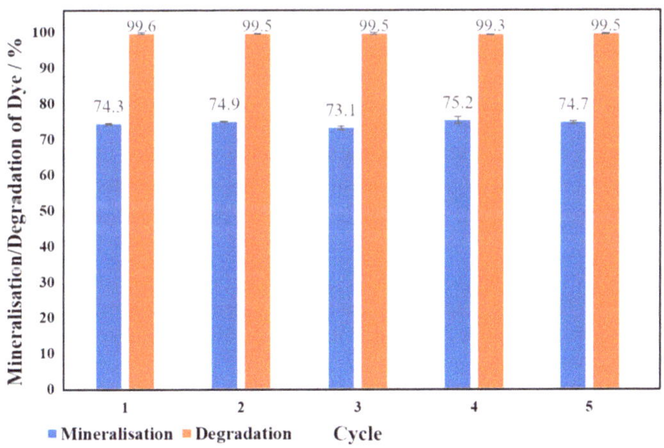

Fig. 50.7 Dye mineralization and degradation by different cycles. Time = 3 h; volume = 300 mL; potential = 5 V; pH = 3; cathode = carbon brush; anode = BDD

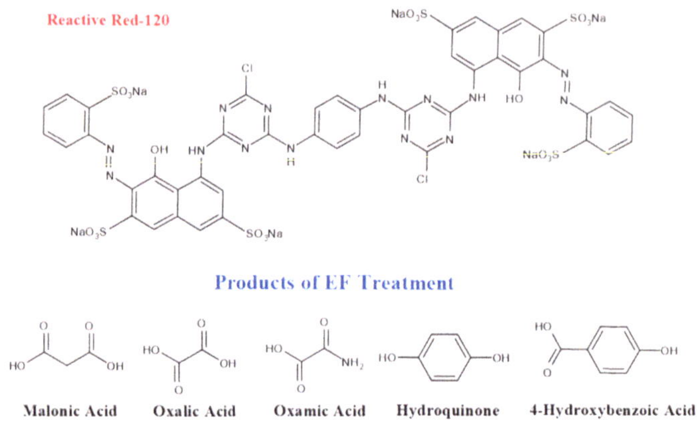

Fig. 50.8 Degradation products identified through HPLC

F. *Identification of Degradation Intermediates*

Based on the HPLC analysis, five organic compounds (Fig. 50.8) were identified as degradation by-products, some of which, like carboxylic acids, are notably difficult to degrade by EF. However, biological treatment is highly effective in the treatment of such organic compounds [29, 30]. Therefore, EF could be used as a pre-treatment to effectively break down RR120 into smaller and more biodegradable compounds.

50.5 Conclusion

This study places EF as a highly effective solution for the simultaneous removal of organics and heavy metals in textile effluents. Future work will include the comprehensive investigation of the degradation pathway of RR120 and its degradation intermediates, and testing with real wastewater to verify the effects of the matrix in real applications.

Acknowledgements The authors appreciate the support given by Dr. Orlando García-Rodríguez and Dr. Olivier Lefebvre from the National University of Singapore, Department of Civil and Environmental Engineering.

Appendix 1: Electrochemical Cell Setup

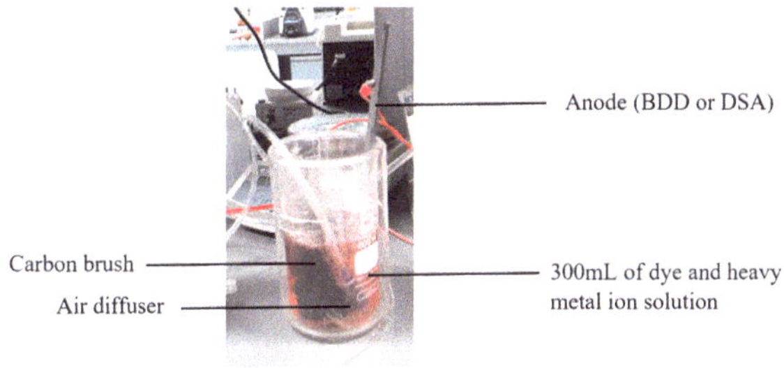

Appendix 2: Degradation Rate Constant for Different Dye Concentrations

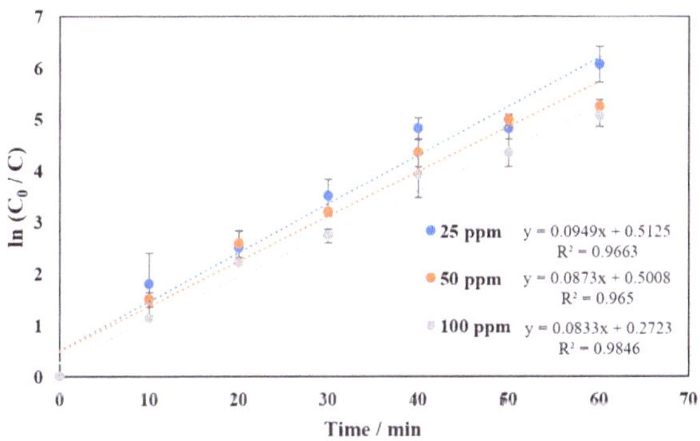

C_0 refers to the initial concentration of dye, and C refers to the concentration at each time. As shown in the graph above, the degradation rate constant, which is equal to the gradient of the line, is the greatest for 25 ppm, with a value of 0.0949 min^{-1}. This is followed by 0.0873 min^{-1} for 50 ppm and 0.0833 min^{-1} for 100 ppm.

References

1. Pal, P. (2017). Chapter 6—Industry-specific water treatment: Case studies. *Industrial Water Treatment Process Technology*, 243–511. https://www.sciencedirect.com/science/article/pii/B9780128103913000060
2. Ghaly, A. E., Ananthashankar, R., Alhattab, M., & Ramakrishnan, V. V. (2013). Production, characterization and treatment of textile effluents: A critical review. *The Journal of Chemical Engineering & Process Technology, 05*, 1e19. https://doi.org/10.4172/2157-7048.1000182
3. Platzek, T., Lang, C., Grohmann, G., Gi, U.-S., & Baltes, W. (1999). Formation of a carcinogenic aromatic amine from an azo dye by human skin bacteria in vitro. *Human & Experimental Toxicology, 18*, 552e559. https://doi.org/10.1191/096032799678845061
4. Chung, K. (2016). Azo dyes and human health: A review. *Journal of Environmental Science and Health, Part C, 34*(4), 233–261. https://doi.org/10.1080/10590501.2016.1236602
5. Bhardwaj, V., Kumar, P., & Singhai, G. (2014). Toxicity of heavy metals pollutants in textile mills. *International Journal of Scientific & Engineering Research, 5*(7).
6. Wilson, H., Carr, C., & Hacke, M. (2012). Production and validation of model iron-tannate dyed textiles for use as historic textile substitutes in stabilisation treatment studies. *ChemCentJournal, 6*. 10.1186%2F1752-153X-6-44
7. Das, K. K., & Das, S. N. (2008). Nickel, its adverse health effects and oxidative stress. *The Indian Journal of Medical Research, 128*, 412–425.
8. Bozynski, C. C. (2009) *et al.* Copper Toxicosis with hemolysis and hemoglobinuric nephrosis in three adult Boer goats. *Journal of Veterinary Diagnostic Investigation*. 10.1177%2F104063870902100319
9. Kowedly, K. V. (2016). Iron overload in patients with chronic liver disease. *Gastroenterologia y Hepatologia, 12*, 695–698. PMID: 28035198.
10. Ahmad, A., Mohd-Setapar, S. H., Chuong, C. S., Khatoon, A., Wani, W. A., Kumar, R., & Rafatullah, M. (2015). Recent advances in new generation dye removal technologies: novel search for approaches to reprocess wastewater. *RSC Advances, 5*(39), 30801–30818. https://pubs.rsc.org/is/content/articlepdf/2015/ra/c4ra16959j
11. Senthilkumaar, S., et al. (2006). Adsorption of dissolved Reactive red dye from aqueous phase onto activated carbon prepared from agricultural waste. *Bioresource Technology, 97*(14), 1618–1625. https://doi.org/10.1016/j.biortech.2005.08.001
12. Al-Degs, Y. S., et al. (2008). Effect of solution pH, ionic strength, and temperature on adsorption behavior of reactive dyes on activated carbon. *Dyes and Pigments, 77*(1), 16–23. https://doi.org/10.1016/j.dyepig.2007.03.001
13. Visa, M., Bogatu, C., & Duta, A. (2010). Simultaneous adsorption of dyes and heavy metals from multicomponent solutions using fly ash. *Applied Surface Science, 256*(17), 5486–5491. https://www.sciencedirect.com/science/article/abs/pii/S0169433209018674
14. Fu, F., & Wang, Q. (2011). Removal of heavy metal ions from wastewaters: A review. *Journal of Environmental Management, 92*(3), 407–418. https://www.sciencedirect.com/science/article/pii/S0301479710004147
15. Maarof, H. I., Daud, W. M. A. W., & Aroua, M. K. (2017). Recent trends in removal and recovery of heavy metals from wastewater by electrochemical technologies. *Reviews in Chemical Engineering, 33*(4), 359–386. https://www.degruyter.com/view/journals/revce/33/4/article-p359.xml?language=en
16. Anglada, A., Urtiaga, A., & Ortiz, I. (2009). Contributions of electrochemical oxidation to waste-water treatment: Fundamentals and review of applications. *Journal of Chemical Technology & Biotechnology, 84*(12), 1747–1755. https://doi.org/10.1002/jctb.2214
17. Zhou, M., Oturan, M. A., & Sires, I. (Eds.). (2018). Electro-fenton process: new trends and scale-up. In *Handbook of environmental chemistry* (Vol. 61, p. 430). Springer Nature. https://www.springer.com/gp/book/9789811064050
18. Chu, Y. Y., Qian, Y., Wang, W. J., & Deng, X. L. (2012). A dual-cathode electro-Fenton oxidation coupled with anodic oxidation system used for 4-nitrophenol degradation. *Journal of Hazardous Materials, 199–200*, 179–185. https://doi.org/10.1016/j.jhazmat.2011.10.079

19. Moreira, F. C., Boaventura, R. A., Brillas, E., & Vilar, V. J. (2017). Electrochemical advanced oxidation processes: A review on their application to synthetic and real wastewaters. *Applied Catalysis B: Environmental, 202*, 217–261. https://doi.org/10.1016/j.apcatb.2016.08.037
20. Mousset, E., Wang, Z., Hammaker, J., & Lefebvre, O., Electrocatalytic phenol degradation by a novel nanostructured carbon fiber brush cathode coated with graphene ink. *Electrochimica Acta, 258*, 607–617.
21. Deng, F., Qiu, S., Olvera-Vargas, H., Zhu, Y., Gao, W., Yang, J., & Ma, F. (2019). Electrocatalytic sulfathiazole degradation by a novel nickel-foam cathode coated with nitrogen-dopedporouscarbon. *Electrochimica Acta,* 297, 21–30. https://doi.org/10.1016/j.electacta.2018.11.180
22. Garcia-Rodriguez, O., et al. (2018). Mineralization of electronic wastewater by electro-Fenton with an enhanced graphene-based gas diffusion cathode. *Electrochimica Acta, 276*, 12–20.
23. Yaseen, D. A., & Scholz, M. (2019). Textile dye wastewater characteristics and constituents of synthetic effluents: A critical review. *International Journal of Environmental Science and Technology, 16*, 1193–1226. https://doi.org/10.1007/s13762-018-2130-z
24. Sirés, I., Brillas, E., Oturan, M., Rodrigo, M., & Panizza, M. (2014). Electrochemical advanced oxidation processes: Today and tomorrow. A review. *Environmental Science and Pollution Research, 21*, 8336–8367.
25. Oturan, M. A., Pimentel, M., Oturan, N., & Sirés, I. (2008). Reaction sequence for the mineralization of the short-chain carboxylic acids usually formed upon cleavage of aromatics during electrochemical Fenton treatment. *Electrochimica Acta, 54*(2), 173–182.
26. Oturkar, C. C., et al. (2011). Mechanistic investigation of decolorization and degradation of Reactive Red 120 by Bacillus lentus BI377. *Bioresource Technology, 102*(2), 758–764. https://doi.org/10.1016/j.biortech.2010.08.094
27. Kurosumi, A., Kaneko, E., & Nakamura, Y. (2008). Degradation of reactive dyes by ozonation and oxalic acid-assimilating bacteria isolated from soil. *Biodegradation, 19*, 489–494. https://doi.org/10.1007/s10532-007-9153-3
28. Garcia-Segura, S., & Brillas, E. (2011). Mineralization of the recalcitrant oxalic and oxamic acids by electrochemical advanced oxidation processes using a boron-doped diamond anode. *Water Research, 45*, 2975–2984. https://doi.org/10.1016/j.watres.2011.03.017
29. McGuinness, M., & Dowling, D. (2009). Plant-associated bacterial degradation of toxic organic compounds in soil. *International Journal of Environmental Research and Public Health., 6*(8), 2226–2247. https://doi.org/10.3390/ijerph6082226
30. Seo, J.-S., Keum, Y.-S., & Li, Q. X. (2009). Bacterial degradation of aromatic compounds. *International Journal of Environmental Research and Public Health, 6*(1), 278–309. https://doi.org/10.3390/ijerph6010278

Chapter 51
Detection and Removal of Obstructions in Still Images Utilising Machine Learning

Donggeon Lim, Marvyn Chia, and Clarissa Koh

51.1 Introduction

Photos play a vital role in any individual life by capturing memorable moments. However, non-optimal imaging conditions often present themselves as obstructions in otherwise perfect photos. Such visual obstructions—such as fences—are often impossible to avoid and difficult to edit out.

Hence, we propose a computational approach where users can remove obstructions in their photographs automatically and easily. This is broken down into three steps: creation of training data, detection and removal of the obstruction, and reconstruction (inpainting) of the image.

Existing inpainting solutions are generally unsatisfactory results due to the limitations of simple algorithms [1]. Motion-based decomposition is also inapplicable on still images and thus not very flexible [2]. Our solution is novel in the idea that by harnessing machine learning, we are able to create a sophisticated algorithm to produce results far more realistic and reliable than existing solutions, while managing to make it completely automated.

We hypothesise that our solutions will be more realistic and seamless than existing solutions.

51.2 Existing Solutions

PatchMatch is a non-evolutionary randomised correspondence algorithm [1]. It was developed back in 2017 and uses nearest neighbour approximations to fill in gaps in the image. However, such algorithms that are manually coded will adapt poorly

D. Lim (✉) · M. Chia · C. Koh
Singapore, Singapore
e-mail: topgeon@gmail.com

H. Guo et al. (eds.), *IRC-SET 2021*, https://doi.org/10.1007/978-981-16-9869-9_51

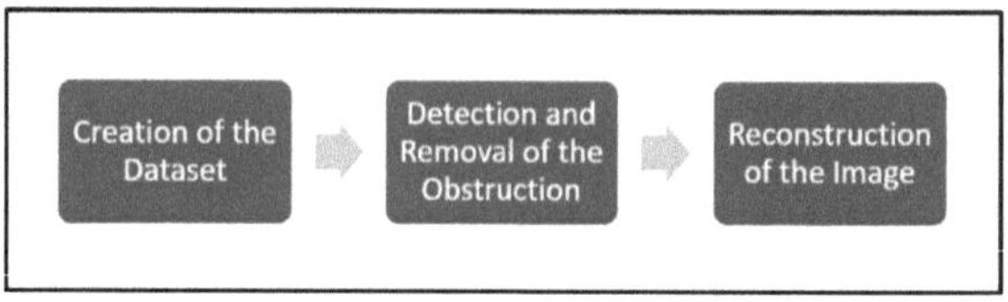

Fig. 51.1 An overview of our methodology

with input that it was not created for, making PatchMatch a solution to a niche group of applications. It additionally requires user's input to detect the area to be removed and reconstructed, although this was done by design.

A video-based solution was developed in 2015 [2]. It requires moving video footage as input and works by capturing the differences in movement between the subject and the obstruction caused by parallax. It then separates and outputs solely the subject. This solution promises clean results that are widely applicable, but its limitations on input means it cannot be utilised to remove obstructions in photographs that were taken in the past. Additionally, the high number of frames to process translates to high demand of processing power, as well as increased storage burden.

51.3 Methodology

See Fig. 51.1.

51.4 Creation of Data set

For the detection and removal of the obstruction, a data set to train an algorithm was required. However, this is specific to an application, and thus, there were no readily available sets of 'masked' images. As such, a program to create a custom data set was created using Python, which was chosen due to its expansive library ecosystem. Python Imaging Library (PIL) and numPy among others allowed the creation of the required data set. A mask image of the obstruction was overlain on common images. The image would then undergo data augmentation, where the mask image was resized, reorientated, and recoloured under a random parameter [3]. The process was then automated to repeat itself so that a large data set can be created relatively quickly (Fig. 51.2).

In most situations, a variety of data is usually better, as it avoids overfitting. Overfitting is the process of making spurious correlations that occur in training data but does not apply to real life. Increasing the size of our data set reduces these spurious correlations and improves the performance of the program.

Fig. 51.2 Non-comprehensive demonstration of implemented data augmentation

51.5 Detection and Removal of Obstruction

The proposed obstruction detection is based on the U-Net architecture [4]. A popular method is to utilise how chain fences often have a consistent colour and fence regions can be estimated using the colour pattern recognised [5]. However, this is often inaccurate and is insufficient for more complex cases [5]. Hence, extracting both global features and local features was required. Global features describe the image as a whole, while local features describe the images in parts. To achieve this, a novel descriptor that extracts the colour-texture features via two information types was used [6]. This is based on the combination of global features using wavelet transform and a modified version of local ternary patterns, whereas for the local features, a bag of words model was used.

Task-specific edge detection using convolutional neural networks and a discriminatively trained domain transform was also carried out [7]. Edge detection is an image processing technique for finding the boundaries of objects within images. It notably decreases the amount of data while maintaining crucial structural properties in an image [6]. The edge detection utilised consists of three stages as shown in Fig. 51.3.

In the case of detected chain fences, many models define fence joints which ultimately are crossed wires in an '*X*' shape as positive data, while negative data is defined as non-joints. This approach is insufficient for distorted fences or fences that do not contain chain nodes [8]. A regression-based fence detection where cropped parts of fences constitute a data set was implemented (Fig. 51.4).

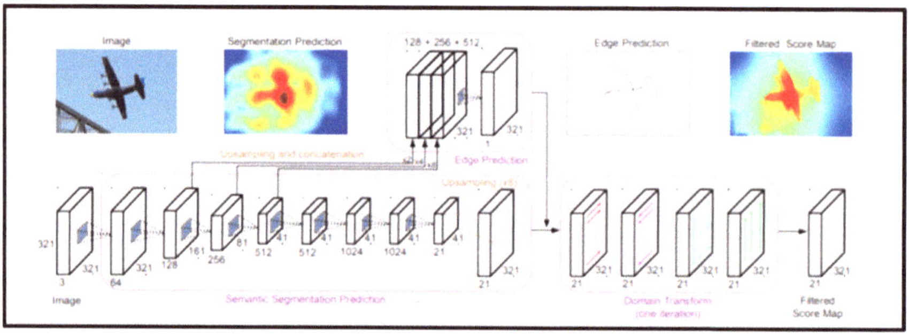

Fig. 51.3 DeepLab for semantic segmentation prediction, EdgeNet for edge prediction, and domain transform to accurately align segmentation scores with object boundaries

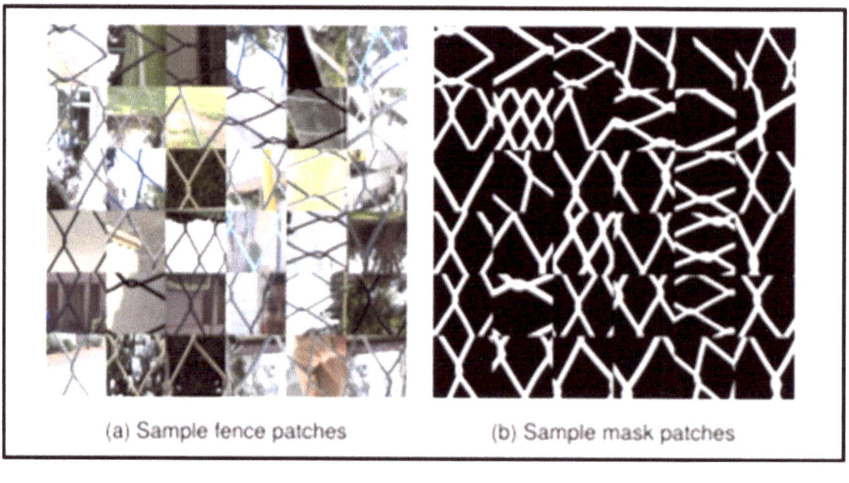

Fig. 51.4 Sample images of positive and negative data of the fence

After being able to segment the area the fence is, we created a simple algorithm to change the RGB channel of the pixel segmented to (255, 255, 255), or absolute white.

Other past methods require the use of video sequences to pre-train their model. This is a major limitation as the video sequence is not available online and cannot be computationally generated, thus requiring prohibitive amounts of time to obtain [2, 9]. Additionally, those methods demand extremely high processing power. Furthermore, alternative methods are heavily dependent on the colour of the fence and are unable to distinguish fences and background that have similar colours [10].

The aforementioned tools complement each other in successfully detecting obstructions in images.

51.6 Reconstructing the Image

Existing solutions use a standard algorithm conditioned to insert values in the masked holes, either by using nearest neighbours or the like [1]. This often leads to unsatisfactory results like blurriness due to the limitations of such algorithms. Motion-based decomposition and reconstruction are also inapplicable on still images. By harnessing machine learning to optimise the algorithm, as well as employing a sophisticated model that relies on partial convolution, we are able to produce results far more reliable and seamless than existing solutions, while managing to make it completely automated [11].

A novel solution had to be created not only in the interest of optimisation, but also application. Many existing solutions merely assume a regular shaped and relatively small hole in the centre of the image, and thus can hardly accommodate obstructions likely to be found in everyday photographs—such as fences.

Procedure: The model was based on a partial convolution model, along with optimisations and implementation of a double layer—a comparison between the original and predicted product in order to train the algorithm [11]. Training was done with TensorFlow, and ImageNet was used as the data set [12, 13]. While training is a GPU-heavy process that required a cloud-based rendering farm to achieve, the end product is a simple algorithm that can be run both on local machines and hosted on the Web without high demand for processing power.

51.7 Results

In this scenario, our model managed to produce an obstruction-free image very similar to ground zero. This is in comparison to NVIDIA's Inpainting, which is widely considered to be the most sophisticated current solution out there. Something to note is that NVIDIA's Inpainting requires manual masking and input. However, it is worth considering that the input images in these tests use fabricated data set we created, and results may be misrepresentative of our model's capabilities in real scenarios (Figs. 51.5 and 51.6, 51.7).

In a test scenario that purely tests reconstruction, our model managed to perform slightly better when it came to filling in the void of the lamppost. It also avoided strange visual artefacts in unrelated areas. As for removing the tree from the image,

Fig. 51.5 Results of our algorithm in comparison to NVIDIA's Inpainting tool

Fig. 51.6 Manually masked images with intent to purely test reconstruction potential

Fig. 51.7 Testing 3D perspective warp with automatic detection enabled in our model

while it does seem like our model produced marginally better results than Inpainting, both leaves much to be desired.

A real-life scenario involving low contrast between fence and image, as well as 3D perspective shift of the fence. Both produced very convincing images, but the structure on the left has been detected and removed by our model. This overcorrection is intentional in order to create a cleaner image.[1]

51.8 Limitations, Reflection, and Conclusion

Some limitations of our approaches include limited data sets used for training, limited mask variety, and lack of post-processing.

The ImageNet data set is generally very capable and suitable for a variety of applications, but lacks enough samples on faces for example. This means that the reconstruction AI will have increased difficulty when trying to inpaint portrait images—or any other that are not remotely close to the images found one the ImageNet network.

Mask detection was also trained with detection of fence-like objects in the foreground as the main focus, and thus could fail to work in the case of lens flares or lampposts for example.

[1] Overcorrection is favoured over undercorrection if contrast between obstruction and background is high.

The lack of post-processing (such as anti-aliasing) presents itself as a double edged sword, as while it reduces computing power demands, post-processing is generally useful in removing minor imperfections in image reconstruction [11].

There is great potential in our program, and it could be amplified with further optimisations and iterations. For example, the entire system architecture could be edited to fit within a generative adversarial network (GAN), which allows the program to look at its output and readjust the algorithms accordingly, infinitely improving upon itself [14].[2]

Despite the limitations and areas for expansions considered, the final product was considered an overall success. More importantly, we believe that the results shown will speak for themselves.

References

1. Barnes, C., Shechtman, E., Finkelstein, A., & Goldman, D. B. (2009). PatchMatch. *ACM Transactions on Graphics, 28*(3), 1–11. https://doi.org/10.1145/1531326.1531330
2. Xue, T., Rubinstein, M., Liu, C., & Freeman, W. T. (2015). A computational approach for obstruction-free photography. *ACM Transactions on Graphics, 34*(4), 1–11. https://doi.org/10.1145/2766940
3. Li, S. Y. (2020, April 24). *Automating data augmentation: Practice, theory and new direction.* The Stanford AI Lab Blog.
4. Ronneberger, O., Fischer, P., & Brox, T. (2015). U-Net: Convolutional networks for biomedical image segmentation. In *Lecture notes in computer science* (pp. 234–241). Springer International Publishing. https://doi.org/10.1007/978-3-319-24574-4_28
5. Varalakshmamma, M., & Venkateswarlu, T. (2019). Detection and restoration of image from multi-color fence occlusions. *Pattern Recognition and Image Analysis, 29*(3), 546–558. https://doi.org/10.1134/s1054661819030209
6. Chen, L.-C., Barron, J. T., Papandreou, G., Murphy, K., & Yuille, A. L. (2016, June). semantic image segmentation with task-specific edge detection using CNNs and a discriminatively trained domain transform. In *2016 IEEE conference on computer vision and pattern recognition (CVPR).* https://doi.org/10.1109/cvpr.2016.492
7. International Journal of Computer Science and Information Technology (n.d.). Academy and Industry Research Collaboration Center (AIRCC). https://doi.org/10.5121/ijcsit
8. Matsui, T., & Ikehara, M. (2020). Single-image fence removal using deep convolutional neural network. *IEEE Access, 8*, 38846–38854. https://doi.org/10.1109/access.2019.2960087
9. Khasare, V. S., Sahay, R. R., & Kankanhalli, M. S. (2013, September). Seeing through the fence: Image de-fencing using a video sequence. In *2013 20th IEEE international conference on image processing (ICIP).* https://doi.org/10.1109/icip.2013.6738278
10. Farid, M. S., Mahmood, A., & Grangetto, M. (2016). Image de-fencing framework with hybrid inpainting algorithm. *Signal, Image and Video Processing, 10*(7), 1193–1201. https://doi.org/10.1007/s11760-016-0876-7
11. Liu, G., Reda, F. A., Shih, K. J., Ting-Chun, W., Tao, A., & Catanzaro, B. (2018). Image Inpainting for irregular holes using partial convolutions.
12. https://www.tensorflow.org/
13. http://www.image-net.org/
14. Pearton, S. J., Zolper, J. C., Shul, R. J., & Ren, F. (1999). GaN: Processing, defects, and devices. *Journal of Applied Physics, 86*(1), 1–78. https://doi.org/10.1063/1.371145

[2] In practice, this is highly unlikely due to diminishing returns.

Chapter 52 Investigating the Application of the Separation Capabilities of a Zirconium MOF b-oriented MFI Zeolite Membrane

Kew Cheng Joon and Qi Xuan Tan

Abstract Many people from third world countries still lack access to clean and safe drinking water, leading to a rise in water borne diseases like cholera. Zirconium zeolite membranes show promising capabilities in terms of heavy metal ion and salt separation. This project investigated the water purification capabilities of the Zirconium MOF b-oriented MFI Zeolite Membrane in comparison with the commercial—alumina membrane with the aim of improving the purification capabilities of these membranes. The following tests were carried out to investigate the separation capabilities of the membranes in terms of NaCl, copper (II) ions and zinc ions, respectively. Firstly, the membrane was tested in a 35 ppt NaCl solution and 100 ppm copper and zinc solutions, respectively. In the NaCl separation tests, the zirconium membrane removed 43.1% of NaCl from water, whereas the—alumina membrane only managed to remove 18.7% of NaCl; for copper (II) ions, the percentage removal for the—alumina membrane and zirconium membrane were 14.3% and 63.3%, respectively; for zinc ions, the percentage removal were 36.9% and 37.3%, respectively. To optimize the percentage removal results, we tested the zirconium membranes in 17.5 ppt salt solution, which showed a higher percentage removal. The zirconium membrane showed promising results compared to the—alumina membrane in most aspects and showed great potential in water purification.

Keywords Water · Zirconium membrane · Waste management · Ceramic · Wastewater · Zeolite

52.1 Introduction

Today, 785 million people—1 in 9 people—lack access to safe drinking water [1]. This has led to the spread of waterborne diseases where there are more than 3.4

K. C. Joon · Q. X. Tan (✉)
Hwa Chong Institution, Bukit Timah, Singapore
e-mail: 171538B@student.hci.edu.sg

K. C. Joon
e-mail: 171420N@student.hci.edu.sg

H. Guo et al. (eds.), *IRC-SET 2021*, https://doi.org/10.1007/978-981-16-9869-9_52

million deaths from waterborne diseases every year, making it the leading cause of disease and death around the world [2]. Desalination, a water purification technique, is a reliable method for obtaining clean water from seawater, thus increasing the availability of clean, safe drinking water. Desalination removes the salt and minerals from seawater, making it no longer saline and suitable for consumption [3]. There has been an exponential growth of desalination over the years as it has been shown to be a cheap and effective method of obtaining safe and clean drinking water [4]. Desalination requires reverse osmosis and a nanoporous membrane is needed [5]. More effective and efficient membranes are hence important.

Additionally, due to their high chemical stability and their peculiar ability to reject ions, zeolite membranes with smaller sized pores demonstrate promising separation performance that make them potential membrane candidates in water desalination [6].

The zirconium zeolite membrane leverages on the Adsorption-Diffusion Mechanism, where the zeolite membrane would be able to filter out contaminants from the water by leaving the contaminants outside the tube and only allowing pure water to diffuse into the tube.

52.2 Objectives and Hypothesis

This paper aims to determine whether b-oriented MFI zeolite membranes were effective and efficient in the separation of NaCl from water and the adsorption of heavy metal ions such as zinc and copper (II) ions, as well as to compare its separation capabilities with the a-alumina support structure ceramic membrane.

It was hypothesized that a b-oriented MFI zeolite membrane can be synthesized on a porous support. It was also predicted that the membrane would be able to adsorb heavy metal ions and separate NaCl from seawater and that the effectiveness of the zeolite membrane will be greater than that of the alpha alumina support structure ceramic membrane.

52.3 Materials and Methods

52.3.1 Materials

Zirconyl chloride octahydrate, sodium chloride, tetramethylammonium hydroxide, hydrated copper (II) sulfate, polyvinyl alcohol and hydrated zinc sulfate were purchased from Sigma Aldrich.

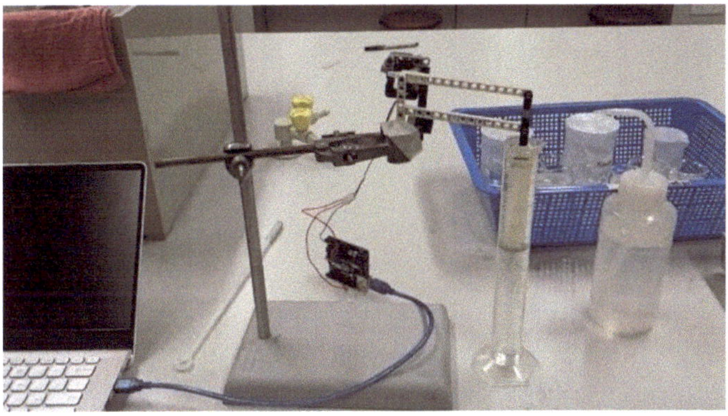

Fig. 52.1 Picture showing the dip coating set-up made

52.3.2 Preparation of Zirconia Membrane

1.00 g of zirconyl chloride octahydrate was added to 48.7 cm^3 deionized water and sent to ultrasound treatment for 5 min to obtain zirconyl chloride octahydrate in aqueous form. 0.557 cm^3 of aqueous tetramethylammonium hydroxide was added dropwise to the solution and was left for 15 min under ultrasound treatment. The sols were then obtained. The sols were then sintered in an oven overnight at 600 degrees Celsius. Immediately after, 10% mass concentration of polyvinyl alcohol was added to adjust the viscosity of the sols. The colloidal sols were then transferred into a 100 cm^3 measuring cylinder and the a-alumina support structure ceramic membrane were then dip coated. The membrane was then removed after placing them for 30 s in the sol. The wet membranes were then placed in air conditioning for 12 h before drying in an oven at 110 °C for 12 h.

52.3.3 Preparation of Dip Coating Set-Up

As there was no cheap commercial dip coating machine and as the laboratories did not have such a device, a custom device was programmed and created. An Arduino chip was used to control a servo motor mechanism that raised and lowered the a-alumina membrane to carry out dip coating of the membrane (Fig. 52.1).

52.3.4 Preparation of Test Solutions

3.5% aqueous sodium chloride solution was prepared by dissolving 35.0 g of sodium chloride NaCl in 1.00 dm^3 of deionized water in a 1dm^3 volumetric flask. 100 ppm

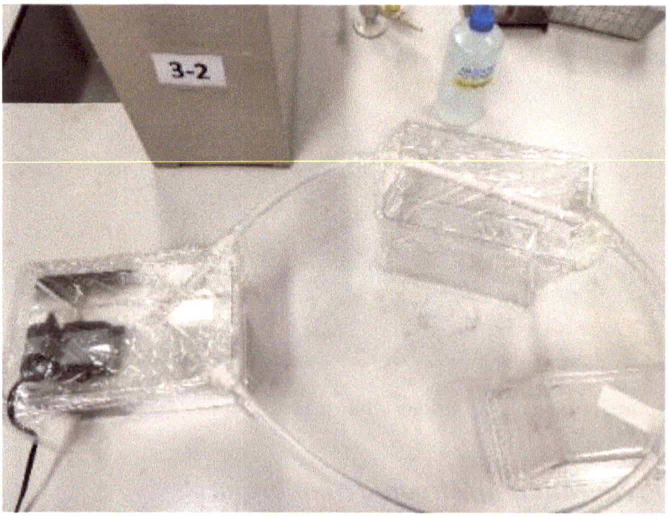

Fig. 52.2 Picture showing top view of the filtration set-up

(~ 0.04%) of copper (II) ions was prepared from copper (II) sulfate by dissolving 0.393 g of copper (II) sulfate in 1dm^3 of deionized water in a 1.00 dm^3 volumetric flask. 100 ppm (~ 0.04%) of zinc ions from zinc sulfate was prepared by dissolving 0.439 g of zinc sulfate in 1 dm^3 of deionized water in a 1.00 dm^3 volumetric flask.

52.3.5 *Preparation of Filtration Set-Up*

A set-up that was able to incorporate a tubular ceramic membrane and mostly airtight was designed. The set-up consists of a water pump, 2 mini fish tanks, plastic tubes and the membrane itself. (Figs. 52.2 and 52.3).

52.4 Results and Discussion

52.4.1 *NaCl Separation Capabilities*

These were the results of our desalination tests conducted on our test and control membranes. Using the salinity meter, a probe was inserted into the solutions to measure the concentration of NaCl in the test solutions. The average concentration changes of NaCl from the starting solution and the filtrate, when the alumina ceramic membrane was used, was − 6.74 parts per thousand and the average percentage removal of NaCl was 18.7%. The alpha alumina ceramic membrane was not very

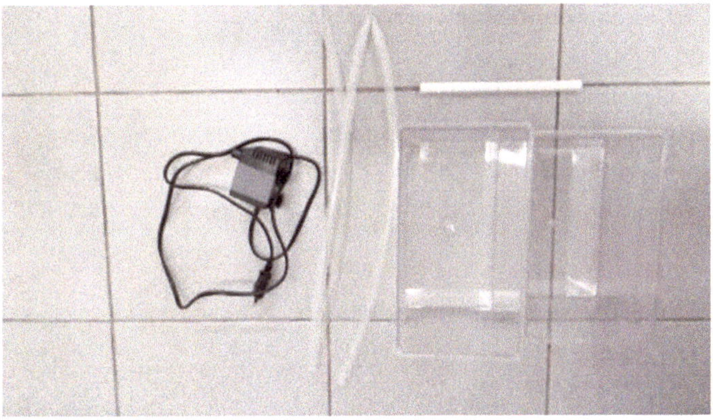

Fig. 52.3 Picture showing materials used for the filtration set-up

effective in desalination. The average concentration changes of NaCl from the starting solution and the filtrate, when the zirconium membrane was used, was − 15.00 parts per thousand and the average percentage removal of NaCl was 43.1%. Thus, it was observed that the zirconium zeolite membrane was significantly more effective in desalination. The zirconium membrane showed a clear reduction in the concentration of NaCl after filtration. There was an average percentage concentration removal of 43.1, 24.4% more than the alpha alumina ceramic membrane. A Mann–Whitney U Test was conducted and it was found that the results were significant as P value was smaller than 0.05 (Chart 52.1; Tables 52.1, 52.2).

The bar chart shows that the zirconium zeolite membrane was more than twice as effective as the alpha alumina support membrane in filtering off NaCl from deionized water.

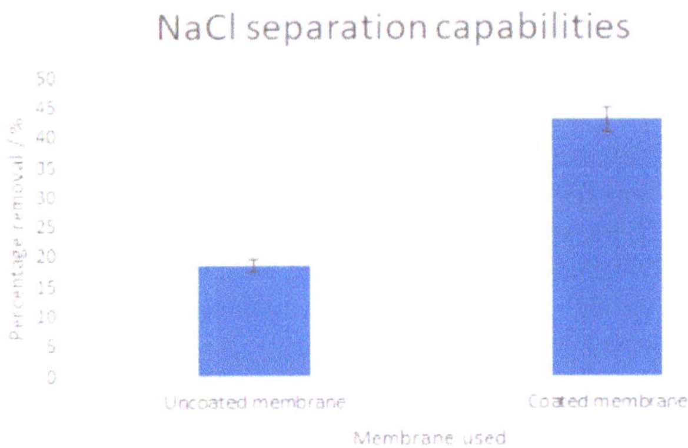

Chart 52.1 Comparison between uncoated and coated for zirconium separation

Table 52.1 NaCl separation of alumina membrane

Replicates	1	2	3	4	5
[NaCl] in starting solution/ppt	35.0				
[NaCl] in filtrate/ppt	28.5	28.3	27.8	28.5	28.3
Change in concentration/ppt	– 6.54	– 6.74	– 7.24	– 6.50	– 6.67
Percentage removal/%	18.7	19.3	20.7	18.6	19.1

Table 52.2 NaCl separation of zirconium membrane

Replicates	1	2	3	4	5
[NaCl] in starting solution/ppt	35.0				
[NaCl] in filtrate/ppt	19.5	20.1	19.4	20.4	20.5
Change in concentration/ppt	– 15.5	– 14.9	– 15.6	– 14.6	– 14.5
Percentage removal/%	44.2	42.5	44.6	41.7	41.4

52.4.2 *Heavy Metal Ions Adsorption*

52.4.2.1 Copper (II) Ions Separation Capabilities

These were the results of our copper (II) ion separation tests conducted on our test and control membranes. The average concentration changes of copper (II) ions from the starting solution and the filtrate, when the alumina ceramic membrane was used, was – 14.3 mg per dm^3 and the average percentage removal of copper (II) ions was 14.3%. The average concentration changes of copper (II) ions from the starting solution and the filtrate, when the zirconium membrane was used, was – 63.3 mg per dm^3 and the average percentage removal of copper (II) ions was 63.3%. Thus, it was observed that the zirconium membrane was significantly more effective in copper (II) ion separation. The zirconium membrane showed a clear reduction in the concentration of copper (II) ions after filtration. There was an average percentage concentration removal of 63.3, 49.0% more than the alpha alumina ceramic membrane. A Mann–Whitney U Test was conducted and it was found that the results were significant as P value was smaller than 0.05. Thus, the zirconium zeolite membrane showed great potential in copper (II) ion separation (Chart 52.2; Tables 52.3, 52.4).

This bar chart shows that the zirconium zeolite membrane was more than three times as effective as the alpha alumina support membrane in filtering off copper (II) ions from deionized water.

52.4.2.2 Zinc Ions Separation Capabilities

These were the results of our zinc ion separation tests conducted on our test and control membranes. The average concentration changes of zinc ions from the starting

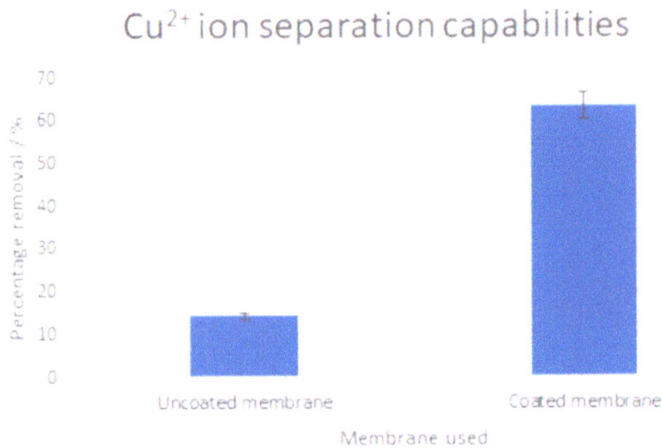

Chart 52.2 Comparison between uncoated and coated for copper (II) ion separation

Table 52.3 Copper(II) ion separation of alumina membrane

Replicates	1	2	3	4	5
$[Cu^{2+}]$ in starting solution/ppm	100				
$[Cu^{2+}]$ in filtrate/ppm	77.1	88.2	87.3	88.6	87.5
Change in concentration/ppm	– 22.9	– 11.8	– 12.7	– 11.4	– 12.5
Percentage removal/%	22.9	11.8	12.7	11.4	12.5

Table 52.4 Copper(II) ion separation of zirconium membrane

Replicates	1	2	3	4	5
$[Cu^{2+}]$ in starting solution/ppm	100				
$[Cu^{2+}]$ in filtrate/ppm	34.2	35.3	38.4	38.0	37.5
Change in concentration/ppm	– 65.8	– 64.7	– 61.6	– 62.0	– 62.5
Percentage removal/%	65.8	64.7	61.6	62.0	62.5

solution and the filtrate, when the alumina ceramic membrane was used, was – 36.9 mg per dm^3 and the average percentage removal of zinc ions was 36.9%. Thus, it was observed that the alpha alumina ceramic membrane was not very effective in zinc ion separation. The average concentration changes of zinc ions from the starting solution and the filtrate, when the zirconium membrane was used, was – 37.3 mg per dm^3 and the average percentage removal of zinc ions was 37.3%. Thus, it was observed that the zirconium zeolite membrane was rather ineffective in the separation of zinc ions as the percentage removal was way lesser compared to copper (II) ions and NaCl. The zirconium zeolite membrane showed a disappointing reduction in the concentration of zinc ions when compared to the control membrane after filtration. There was an average percentage concentration removal of 37.3%, which was only 0.4% more than the alpha alumina ceramic membrane (Tables 52.5 and 52.6).

Table 52.5 Zinc ion separation of alumina membrane

Replicates	1	2	3
$[Zn^{2+}]$ in starting solution/ppm	100		
$[Zn^{2+}]$ in filtrate/ppm	64.9	62.7	61.9
Change in concentration/ppm	– 35.1	– 37.3	– 38.1
Percentage removal/%	35.1	37.3	38.1

Table 52.6 Zinc ion separation of zirconium membrane

Replicates	1	2	3
$[Zn^{2+}]$ in starting solution/ppm	100		
$[Zn^{2+}]$ in filtrate/ppm	62.2	63.8	62.0
Change in concentration/ppm	– 37.8	– 36.2	– 38.0
Percentage removal/%	37.8	36.2	38.0

The results obtained for the separation capabilities of the zirconium zeolite, however, were deemed inconclusive. There might be a multitude of reasons why the zirconium membrane had failed. There might be some errors that occurred when carrying out the testing for its separation capabilities. Due to these systematic errors as a result of only three replicates conducted, the separation capabilities of zinc ions were inconclusive unless further testing was conducted using multiple new membranes and newly made set-ups (Chart 52.3).

This bar chart, with error bars, shows the results of zinc ion separation for the uncoated and coated membrane. The efficiency of the coated membrane was very close to that of the uncoated membrane, showing how the zirconium membrane is ineffective in the separation of zinc ions.

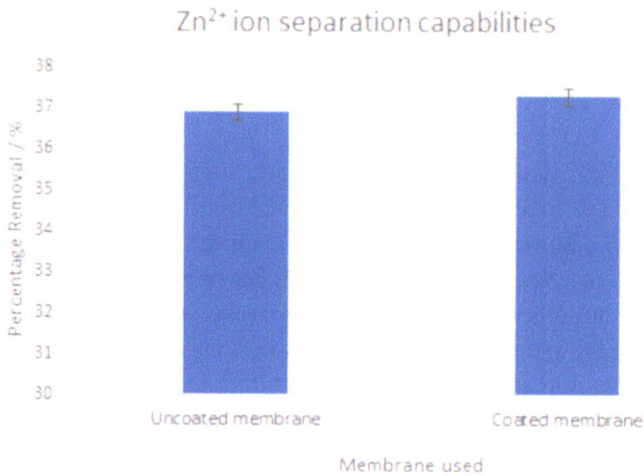

Chart 52.3 Comparison between uncoated and coated for zinc ion separation

52.4.3 Potential Experimental Errors

Although the coated membrane did way better than the commercial uncoated membrane in terms of NaCl and copper (II) ion separation, the percentage removal was only tested to be around 40% and was not ideal. Moreover, the separation capabilities of the zinc ions were rather disappointing too. There were 2 main problems that could have been attributed to this.

Firstly, the set-up might have microleaks at the joints, where the membrane was connected to the pipes. Although the connections between the pipes and the membrane were seemingly airtight, there might be unidentified microleaks which allowed some solution to flow through unfiltered, resulting in the low percentage removal. Secondly, the membrane might not be evenly coated. Even though we used a dip coating machine to dip coat the membrane, some of the sols from the top of the membrane might have dripped off while drying and this might result in uneven coating, allowing some of the pollutants to pass through the unevenly coated part of the membrane.

To minimize the occurrence of these systematic errors, we decided to carry out optimisation and one way to do this was to reduce the concentration of our starting solution. In our optimisation process, we reduced the concentration from 100 to 50 ppm. We also used completely new set-ups and membranes to attain the best results possible.

52.4.4 Optimisation of Separation Capabilities

From the results shown above, the average percentage removal of the zirconium membrane for NaCl is that desirable, only achieving a 43.1% percentage removal of NaCl. We decided to set our starting solution of NaCl to be at 17.5 ppt instead to ensure less of these systematic errors were in place. These were the results of the desalination tests conducted on our test membranes. From the table, the average concentration changes of NaCl from the starting solution and the filtrate, when the zirconium membrane was used, was − 11.3 ppt and the average percentage removal of NaCl was 64.4%. The zirconium membrane showed a clear reduction in the concentration of NaCl after filtration (Chart 52.4; Table 52.7).

This bar chart shows the difference in percentage removal of NaCl when different concentrations of NaCl starting solutions were used. There is a distinct increase in percentage removal of NaCl when 17.5 ppt NaCl solution was used as compared to when 35.0 ppt NaCl solution was used. The zirconium membrane is much more effective at NaCl separation when lower concentrations of NaCl solution is used.

Chart 52.4 Comparison between separation of NaCl before and after optimisation

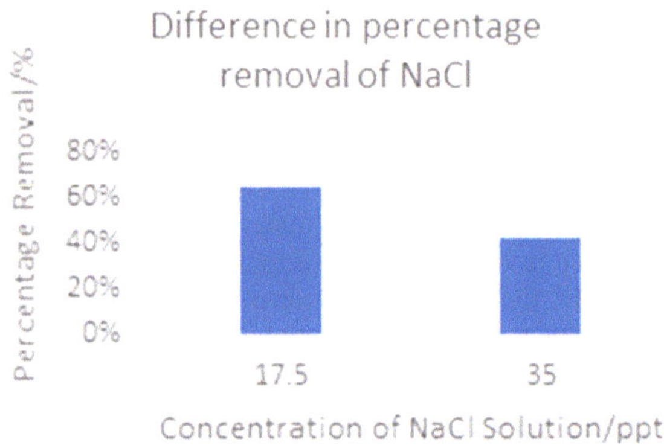

Table 52.7 Separation capabilities of coated membrane after optimisation

Replicates	1	2	3
[NaCl] in starting solution/ppt	17.5		
[NaCl] in filtrate/ppt	6.45	6.01	6.32
Change in concentration/ppt	11.1	11.5	11.2
Percentage removal/%	63.4	65.7	64.0

52.4.5 *Overall Comparison*

The graph shows the overall comparison of the coated and uncoated membranes in terms of the separation of NaCl, copper (II) and zinc ions. The coated membrane had been seen to be consistently more effective in separating harmful copper (II) ions and NaCl from water as compared to the uncoated membrane (Chart 52.5).

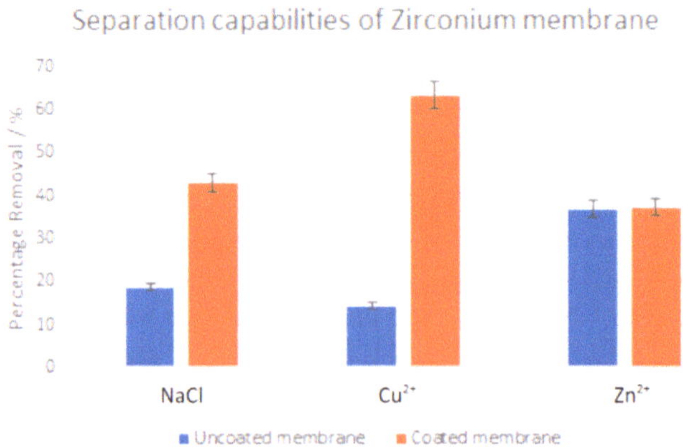

Chart 52.5 Comparison between separation capabilities of uncoated and coated membranes

52.5 Conclusions and Future Work

It could be concluded that the zirconium membrane was effective in separating NaCl and copper (II) ions from water, but was rather inconclusive in the separation of zinc ions. Overall, since the percentage removal of the zirconium zeolite membrane for NaCl, copper (II) ions and zinc ions varied only around 50 ppm, starting the solution at 100 ppm might have been too ambitious. The starting solution should be at 50 ppm instead. The number of coated membranes in each filtration set-up can attain a greater percentage removal. Set-ups can also be equipped with a higher pressure water pump to increase the rate of cross-flow filtration. Characterization of the zirconium zeolite membrane under a SEM can be conducted to measure the pore sizes of the membrane and to inspect the adsorption capabilities of the membrane. More heavy metals ions, such as lead (II) ions, can also be tested using the Zirconium Zeolite membrane to ensure that the zirconium membrane is a universal membrane that is effective in absorbing most harmful pollutants from water. Instead of only using the α alumina ceramic membrane as the support membrane, more ceramic membranes can be tested, such as titania ceramic membranes or silicon carbide ceramic membranes. This way, it could ensure that the zirconium coat will be effective on various support structures and not solely on the aluminum support structure.

References

1. World Health Organization and UNICEF Joint Monitoring Program (2017). Progress on drinking water and sanitation. Retrieved from https://water.org/our-impact/water-crisis/global-water-crisis/
2. Hawthorne, J. (2018). Critical facts about waterborne diseases in the united states and abroad. Retrieved from https://businessconnectworld.com/2018/02/15/critical-facts-waterborne-diseases-us/
3. United States Geological Survey (n.d.). Desalination. https://www.usgs.gov/special-topic/water-science-school/science/desalination?qt-science_center_objects=0#qt-science_center_objects
4. Robbins, J. (2019, June 11). As water scarcity increases, desalination plants are on the rise. *Yale E360*. https://e360.yale.edu/features/as-water-scarcity-increases-desalination-plants-are-on-the-rise
5. Castelletto, S., & Boretti, A. (2021, February 18). Advantages, limitations, and future suggestions in studying graphene-based desalination membranes. *RSC Advances*. https://pubs.rsc.org/en/content/articlehtml/2021/ra/d1ra00278c
6. Kazemimoghadam, M., & Mohammadi, T. (2007). Synthesis of MFI zeolite membranes for water desalination. *Desalination, 206*, 547–553. https://doi.org/10.1016/j.desal.2006.04.063

Additional References

7. Carbon Capture and Storage Association (2011). *Storage—The Carbon Capture and Storage Association (CCSA)* [online]. Available at http://www.ccsassociation.org/what-is-ccs/storage/. Accessed 17 December 2019.
8. Global CCS Institute (2018). *Understanding CCS—Global CCS Institute* [online] Available at https://www.globalccsinstitute.com/why-ccs/what-is-ccs/. Accessed 17 December 2019.
9. Jamali, S. H., Vlugt, T. J., & Lin, L. C. (2017). Atomistic understanding of zeolite nanosheets for water desalination. *Journal of Physical Chemistry, 121*, 11273–11280. https://doi.org/10.1021/acs.jpcc.7b00214
10. Khulbe, K. C., & Matsuura, T. (2018). Removal of heavy metals and pollutants by membrane adsorption techniques. *Applied Water Science, 8*, 19. https://doi.org/10.1007/s13201-018-0661-6
11. Levin, K. (2019). *New global CO_2 emissions numbers are in. They're not good* [online]. World Resources Institute. Available at https://www.wri.org/blog/2018/12/new-global-co2-emissions-numbers-are-they-re-not-good. Accessed 17 December 2019.
12. Mirjalili, F., Hasmaliza, M., & Abdullah, L. C. (2010). Size-controlled synthesis of nano α-alumina particles through the sol-gel method. *Ceramics International, 36*, 1253–1257. https://doi.org/10.1016/j.ceramint.2010.01.009
13. Mulligan, J., Ellison, G., & Levin, K. (2018). *6 ways to remove carbon pollution from the sky* [online]. World Resources Institute. Available at https://www.wri.org/blog/2018/09/6-ways-remove-carbon-pollution-sky. Accessed 17 December 2019.
14. Schmidt, G., & Schmunk, R. (2019). *Data GISS: GISS surface temperature analysis (GISTEMP v4)* [online]. Data.giss.nasa.gov. Available at https://data.giss.nasa.gov/gistemp/. Accessed 17 December 2019.
15. Shaftel, H., Jackson, R., & Callery, S. (2019). *Arctic sea ice minimum|NASA global climate change* [online]. Climate Change: Vital Signs of the Planet. Available at https://climate.nasa.gov/vital-signs/arctic-sea-ice/. Accessed 17 December 2019.
16. Sima, F., Ristoscu, C., Duta, L., Gallet, O., Anselme, K., & Mihailescu, I. N. (2016). Laser thin films deposition and characterization for biomedical applications. *Laser Surface Modification of Biomaterials*, 77–125.https://doi.org/10.1016/B978-0-08-100883-6.00003-4
17. University of California, Berkley (2009). Flexible MOFs in polymer membrane support. Retrieved from http://www.cchem.berkeley.edu/molsim/teaching/spring2013/CCS/Group8/membrane.html
18. Voiland, A. (2009). NASA—2009: Second warmest year on record; end of warmest decade. Retrieved from https://www.nasa.gov/topics/earth/features/temp-analysis-2009.html
19. Wang, D. K., Bauer, R. A., Yan, K., Mergos, I. A., Yang, Z., Zhou, Y., & Verweij, H. (2019). High selectivity gas separation by interfacial diffusion membranes. *Advanced Materials Interfaces, 6*, 1970008. https://doi.org/10.1002/admi.201801273
20. Yan, Q., Qiu, M., Chen, X., & Fan, Y. (2019). Ultrasound assisted synthesis of size-controlled aqueous colloids for the fabrication of nanoporous zirconia membrane. *Frontiers in Chemistry, 7*, 337. https://doi.org/10.3389/fchem.2019.00337

Chapter 53
Investigating the Effects of Herbs on Reducing Symptoms Associated with Schizophrenia in the Model Organism *Caenorhabditis elegans*

Kertin Siaw and Leow Ee-K

53.1 Background and Purpose of Research Area

Schizophrenia is a chronic mental disorder characterised by an inability to connect with reality. Whilst the genetic cause of schizophrenia is unknown, a major hypothesis investigates the dysregulation of the neurotransmitter glutamate in patients with schizophrenia The glutamate hypothesis states that negative and cognitive symptoms of schizophrenia, such as social withdrawal and memory defects, are associated with increased glutamate levels in the brain, which causes neurodegeneration through excitotoxicity [9, 11].

One model organism that neuroscience has been increasingly using is the nematode *Caenorhabditis elegans*. The chemical properties of *C. elegans* neurones are similar to those of mammalian neurones with synapses utilising chemical neurotransmitters such as acetylcholine, glutamate and dopamine. Despite the simplicity of its nervous system, *C. elegans* shows well-defined behaviours, such as the ability to detect and escape noxious substances, display social or solitary feeding, and locate optimal oxygen concentration and temperature [1].

Caenorhabditis elegans isolates exhibit either social or solitary feeding on bacteria. Solitary feeders such as the wild type *C. elegans* N2 strain disperse across a bacterial lawn, whilst social feeders tend to aggregate together [4] or accumulate on the border of a bacterial lawn. Social feeding is induced by nociceptive neurones that detect adverse or stressful conditions, hence social feeding is probably due to the sensation of noxious chemicals by certain sensory neurones [5]. According to [3] social feeding is a suitable protophenotype for social withdrawal, a component of schizophrenia. *C. elegans* CB1372 *daf-7(e1372)* has lower daf-7 activity. Since daf-7 negatively regulates daf-3 activity, *C. elegans* CB1372 *daf-7(e1372)* has higher daf-3 activity, leading to an overexpression of tph-1 in ADF neurons (sensory neurone).

K. Siaw · L. Ee-K (✉)
Hwa Chong Institution, Bukit Timah, Singapore
e-mail: 171164b@student.hci.edu.sg

H. Guo et al. (eds.), *IRC-SET 2021*, https://doi.org/10.1007/978-981-16-9869-9_53

This further leads to higher serotonin levels which act as a stress signal for *C. elegans*. Hence, *C. elegans* CB1372 *daf-7(e1372)* shows increased agitation, physical discomfort and is more hyperoxia (excess oxygen) avoidant and aggregate more to reduce exposure to ambient oxygen. Excess oxygen or hyperoxia is toxic for *C. elegans* as it promotes the production of toxic oxygen species [11]. Thus, aggregation in *C. elegans* is not mediated via bonding or positive motivation, but by avoidance and safety-in-numbers considerations.

Moreover, likewise in *C. elegans*, stress-induced serotonergic overdrive leads to symptoms of schizophrenia, like psychosis and cognitive impairment in humans [7]. Antipsychotic drugs like clozapine used to treat schizophrenia in humans also inhibit serotonin receptors due to their high affinity to serotonin receptors. Moreover, these same drugs inhibit social feeding in *C. elegans* with the *daf-7(e1372)* mutation [6]. Furthermore, according to [10], the hypofunction of NMDA receptors in humans causes excessive glutamate release and hyper-glutamatergic functions, leading to disease progression from prodromal symptoms to psychosis in humans. *C. elegans* CB1372 *daf-7(e1372)* with the *daf-7* mutation have increased GLR-1 transcription, leading to increased glutamatergic signalling as well. *C. elegans* CB1372 *daf-7(e1372)* have a disrupted mechanosensation and a disrupted escape response to mechanical stimuli to the nose and, as a result, rapidly alternate between forward and backward movement. Hence, symptoms of schizophrenia manifest as more random locomotion and a lower average moving velocity in *C. elegans* CB1372 *daf-7(e1372)*. Essentially, through the social feeding and locomotion protophenotypes, *C. elegans* CB1372 *daf-7(e1372)* serves as an excellent model organism to replicate schizophrenia in humans.

Rhodiola rosea and *Curcuma longa* are two natural compounds which have been shown to help with depression and anxiety in the past and have also been found to decrease glutamate-induced excitotoxicity [2, 14].

Hence, the objective of this study is to investigate the extent in which the extracts of *C. longa* and *R. rosea* are able to reduce social feeding and normalise the locomotion patterns in *C. elegans* CB1372 *daf-7(e1372)* by determining the number of *C. elegans* aggregates on the plates and the increase in its average velocity of locomotion.

53.2 Hypotheses of the Research

The hypotheses are that both *Curcuma longa* and *R. rosea* are able to reduce social feeding and normalise the locomotion patterns of *C. elegans* CB1372 *daf-7(e1372)* by reducing the number of aggregates observed and increasing its average velocity.

53.3 Methods

53.3.1 Materials

Rhodiola rosea herb extract was procured from Swedish Herbal Institute. Aquacumin (water-soluble *Curcuma longa* herb extract) was procured from Ava Plant. Luria–Bertani (LB) agar broth powder was procured from Becton, Dickinson and Company. Bacto-peptone powder was procured from Oxoid Company.

53.3.2 Preparation of Herb Extracts

Aquacumin powder was dissolved in water at 5 mg/ml. *Rhodiola rosea* powder was dissolved in water at 5 mg/ml. Filtration was conducted to remove undissolved residue.

53.3.3 Preparation of Nematode Growth Medium (NGM)

0.9 g NaCl, 7.5 g agar, 0.75 g bacto-peptone were added with 291.6 ml water. After autoclaving, 0.3 ml of cholesterol (5 mg/ml), 0.3 ml of $MgSO_4$ solution (1 M), 0.3 ml $CaCl_2$ solution (1 M), 7.5 ml of KH_2PO_4 solution pH 6.0 (1 M) were added.

53.3.4 Growth of Bacterial Culture

Escherichia coli OP50 was inoculated and grown in 10 ml of LB broth overnight at 30 °C in a shaking incubator. The absorbance of the bacterial culture at 600 nm was standardised at 0.8 units.

53.3.5 Growth of E. coli OP50 on NGM Plates

100 μl of extract solution (test) or sterile water (control) was spread on the NGM plates and left to dry. 50 μl of *E. coli* OP50 was added in the centre of the NGM plates. The plates were incubated for 1 day at 32 °C.

Table 53.1 Set-ups prepared

	Rhodiola rosea treatment	*Curcuma longa* treatment	No treatment (sterile water)
Healthy organisms-*C. elegans N2*	Control (*group 1*)	Control (*group 2*)	Control (*group 3*)
Mutated organisms-*C. elegans CB1372daf-7 (e1372)*	Experimental (*group 4*)	Experimental (*group 5*)	Control (*group 6*)

53.3.6 Growth of C. elegans on Prepared NGM Plates

A block of agar containing *C. elegans* N2 or *C. elegans* CB1372 *daf-7(e1372)* was cut and placed in the centre of the NGM plates. The plates were incubated at 20 °C for 2 days. The number of aggregates of *C. elegans* was then determined. 6 set-ups were prepared, as shown in Table 53.1.

The purpose of the N2 controls (set-up 1 and 2) is to compare with the wild type N2 *C. elegans* to test for side effects of using the herb extracts. Experimental set-ups 4 and 5 will be used to test our hypothesis. Control set-up 6 will be used to compare with control set-up 3 to show the presence of our dependent variables that we will be testing.

53.3.7 Data Analysis

For the social feeding assay, the number of aggregates of *C. elegans* were counted using a microscope. The standard error was determined and plotted as error bars. Celleste Image analysis was used to study the locomotion patterns of *C. elegans*. The Kruskal–Wallis statistical test was conducted to determine if there were significant differences between the mean values of the set-ups.

53.4 Results

53.4.1 Social Feeding Assay

Figure 53.1 shows the mean number of *C. elegans* aggregates observed in the NGM plates across all 6 set-ups. It can be observed that after treatment with *C. longa* and *R. rosea*, the social feeding behaviour of *C. elegans* CB1372 *daf-7(e1372)* was significantly altered and the worms were more dispersed. A p-value of $0.011 < 0.05$ suggests that there is a significant difference between the mean values of the test set-ups and control set-ups for *C. elegans* CB1372 *daf-7(e1372)*. Whereas, a p-value

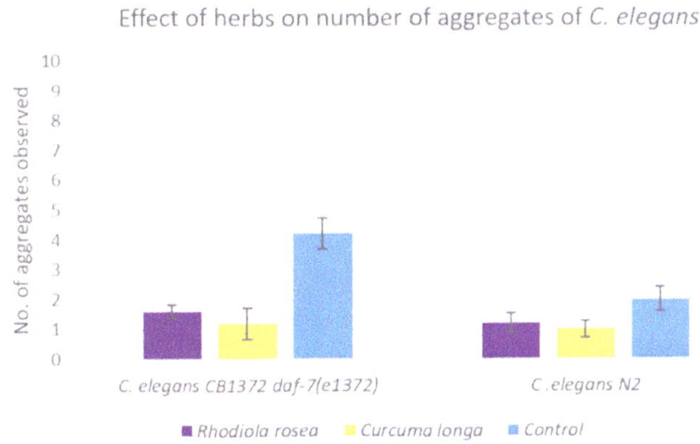

Fig. 53.1 Graph of number of *C. elegans* aggregates observed in respective NGM plates

of 0.242 > 0.05 suggests that there is no significant difference between the mean values of the control and test set-ups for wild type *C. elegans* N2 (Fig. 53.2).

53.4.2 Locomotion Assay

As observed from Fig. 53.3, both *C. longa* and *R. rosea* are able to increase the average velocity of *C. elegans* CB1372 *daf-7(e1372)*. The Kruskal–Wallis statistical test indicated a *p*-value of 0.016 suggests that there is indeed a significant difference between the mean velocity of set-ups for mutated *C. elegans*. Additionally, the *p*-value of 0.365 for the mean velocity of set-ups with wild type *C. elegans* N2 suggests that the herbs do not have a significant side effect on locomotion.

53.5 Conclusion and Discussion

In conclusion, both *Curcuma longa* and *R. rosea* are able to effectively reduce social feeding in *C. elegans* CB1372 *daf-7(e1372)* to levels seen in the control set-up of *C. elegans* N2. Both herbs are also effective in increasing the average velocity of *C. elegans* CB1372 *daf-7(e1372)*.

Curcumin, an active ingredient of *Curcuma longa*, has shown to have anti-inflammatory and antioxidant effects. Research done by [12] also suggests that curcumin reduces inflammatory responses and activates neuroprotective genes in *C. elegans*. By its chemical structure, curcumin may act as a natural free radical (ROS) scavenger. Acting through the Neurotrophic factor κB (NF- κB), curcumin can decrease the release of different interleukins (inflammatory compounds). Curcumin could act as a stress response mimetic that induces protein homeostasis network.

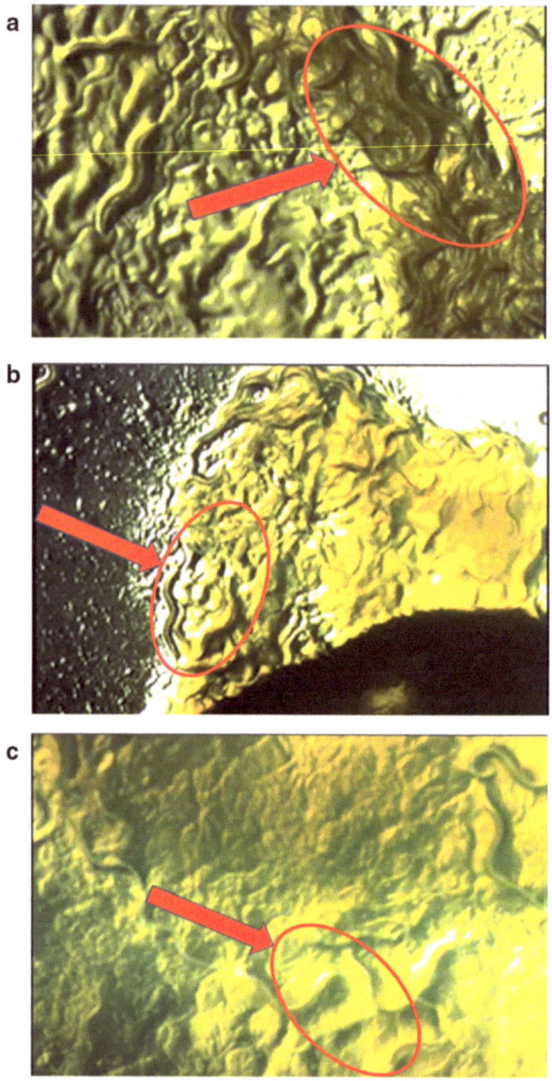

Fig. 53.2 **a** Photograph illustrating social feeding behaviour in the control set-up of *C. elegans* CB1372 *daf-7(e1372)*. **b** Photograph illustrating solitary feeding behaviour in the experimental set-up of *C. elegans* CB1372 *daf-7(e1372)* with *C. longa* treatment. **c** Photograph illustrating diminished social feeding behaviour in the experimental set-up of *C. elegans* CB1372 *daf-7(e1372)* with *R. rosea* treatment

This induction requires the transcription factor SKN-1 in *C. elegans*. Additionally, curcumin could act as a DR mimetic to activate this transcription factor through the AMPK pathway. Hence, *Curcuma longa* treatment through reducing inflammation (which has been previously linked to high glutamate levels and depression in humans by [8] and offering neuroprotection may be able to decrease the ill-effects of increased glutamate levels as a result of the aforementioned *daf-7* mutation. This reduces the social feeding behaviour observed in *C. elegans* CB1372. Research by Weigant et al. [13] suggests that *R. rosea* induces translocation of the DAF-16 transcription factor

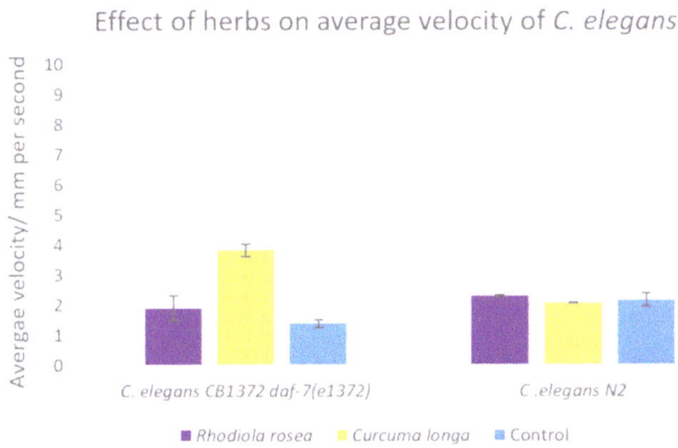

Fig. 53.3 Graph illustrating the average velocity of *C. elegans* N2 and *C. elegans* CB1372 *daf-7(1372)* across all 6 experimental groups

from the cytoplasm into the nucleus in *C. elegans*, leading to a reprogramming of transcriptional activities favouring the synthesis of proteins involved in stress resistance and longevity. The increased stress resistance of *C. elegans* CB1372 after undergoing *R. rosea* treatment means that it is now less aversive to the threat of hyperoxia and, hence, has reduced social feeding.

Research done by Satapathy and Salim [12] suggests that curcumin treatment is able to restore dopamine levels in *C. elegans* previously exposed to neurotoxins, hence, curcumin offers neuroprotective efficacy to *C. elegans*. Increased neuroprotection accrued by curcumin may be able to reduce the ill-effects of increased glutamate levels, leading to more normalised locomotion patterns in *C. elegans* CB1372. This manifests as a higher average velocity for *C. elegans* CB1372 after *C. longa* treatment as compared to the control set-up.

Research by Weigant. et al. [13] suggests that *R. rosea* increases stress resistance and longevity of *C. elegans*. Increased stress resistance may mean that *C. elegans* CB1372 is less susceptible to the ill-effects of increased glutamate levels and have normalised locomotion patterns. *C. elegans* CB1372 treated with *R. rosea* had a higher average velocity as compared to the control set-up. However, it remains unknown as to why *C. elegans* CB1372 treated with *C. longa* had a significantly higher average velocity than the control set-up with *C. elegans* N2.

In the future, the optimum dosage and concentration of *C. longa* and *R. rosea* to reduce symptoms associated with schizophrenia in *C. elegans* can be determined. This can be done by altering dosage and concentration of herbs, and subsequently determining the exact dosage and concentration which reduces social feeding and increases average velocity most effectively.

References

1. Burne, T., Scott, E., Swinderen, B. V., Hilliard, M., Reinhard, J., Claudianos, C., Eyles, D., & McGrath, J. (2011). Big ideas for small brains: what can psychiatry learn from worms, flies, bees and fish? *Molecular Psychiatry, 16*, 7–16. https://doi.org/10.1038/mp.2010.35
2. Chen, K., An, Y., Tie, L., Pan, Y., & Li, X. (2015). Curcumin protects neurons from glutamate-induced excitotoxicity by membrane anchored AKAP79-PKA interaction network. *Evidence-Based Complementary and Alternative Medicine: ECAM, 2015*, 706207. https://doi.org/10.1155/2015/706207
3. Chang, A. J., Chronis, N., Karow, D. S., Marletta, M. A., & Bargmann, C. I. (2006). A distributed chemosensory circuit for oxygen preference in *C. elegans. PLoS Biology, 4*(9), e274. https://doi.org/10.1371/journal.pbio.0040274
4. de Bono, M., & Bargmann, C. I. (1998). Natural variation in a neuropeptide Y receptor homolog modifies social behaviour and food response in *C. elegans. Cell, 94*, 679–689. https://doi.org/10.1016/S0092-8674(00)81609-8
5. de Bono, M., Tobin, D.M., Davis, M.W., Avery, L. & Bargmann C. I. (2002). Social feeding in *Caenorhabditis elegans* is induced by neurons that detect aversive stimuli. *Nature, 419*, 899–903. https://doi.org/10.1038/nature01169
6. Dwyer, D. S. (2017). Crossing the worm-brain barrier by using *Caenorhabditis elegans* to explore fundamentals of human psychiatric illness. *Molecular Neuropsychiatry, 3*(3), 170–179. https://doi.org/10.1159/000485423
7. Eggers, A. E. (2013). A serotonin hypothesis of schizophrenia. *Medical Hypotheses, 80*(6), 791–794. https://doi.org/10.1016/j.mehy.2013.03.0
8. Emory Health Sciences (2016). Inflammation markers could guide depression treatments: Inflammation connected to elevated glutamate in brain. *ScienceDaily.* Retrieved August 10, 2020 from www.sciencedaily.com/releases/2016/01/160112091414.htm
9. Howes, O., McCutcheon, R., & Stone, J. (2017). Glutamate and dopamine in schizophrenia: An update for the 21st century. *Journal of Psychopharmacology, 29*(2), 97–115. https://doi.org/10.1177/0269881114563634
10. Mei, Y.-Y., Wu, D. C., & Zhou, N. (2018).Astrocytic regulation of glutamate transmission in schizophrenia. *Frontiers in Psychiatry, 9*.https://doi.org/10.3389/fpsyt.2018.00544
11. Monroy, A., Lithgow, G. J., & Alavez, S. (2013). Curcumin and neurodegenerative diseases. *BioFactors (Oxford, England), 39*(1), 122–132. https://doi.org/10.1002/biof.1063
12. Satapathy, P., & Salim, C. (2016). Attenuation of dopaminergic neuronal dysfunction in *Caenorhabditis elegans* by hydrophilic form of curcumin. *Neurochemistry & Neuropharmacology, 2*(1). https://doi.org/10.4172/2469-9780.100011
13. Wiegant, F. A., Surinova, S., Ytsma, E., Langelaar-Makkinje, M., Wikman, G., & Post, J. A. (2008). Plant adaptogens increase lifespan and stress resistance in C. elegans. *Biogerontology, 10*(1), 27–42. https://doi.org/10.1007/s10522-008-9151-9
14. Zheng, Y., Brockie, P. J., Mellem, J. E., Madsen, D. M., & Maricq, A. V. (1999). Neuronal control of locomotion in *C. elegans* is modified by a dominant mutation in the GLR-1 ionotropic glutamate receptor. *Neuron, 24*(2), 347–361. https://doi.org/10.1016/s0896-6273(00)80849-1

Chapter 54
Investigating the Effect of Sugarcane Extract and Lotus Root Extract on the Growth of *Lactobacillus* Spp. and Its Effect on Microbial Antagonism

Lok Ting Yuan, Wong Kai Ray, and Yap Jia Le

Abstract The beneficial properties of *Lactobacillus* spp. have been widely researched all over the world. It is also a known fact that prebiotics like xylo-oligosaccharide (XOS), galacto-oligosaccharide (GOS), and fructo-oligosaccharide (FOS) are can effectively proliferate the growth, and hence further boost the beneficial properties of probiotics. This project aims to recycle and harvest the prebiotic contents present in sugarcane bagasse and lotus roots, in order to mitigate copious food wastage. Sugarcane bagasse is an abundant agricultural waste and has high XOS content. Lotus root is a very widely produced plant in both Southeast Asia and Australia, and has high polyphenolic content, which also serves as prebiotics. The prebiotic extracts were prepared from powdered sugarcane bagasse and lotus roots separately. These extracts were mixed with *Lactobacillus gasseri* (*L. gasseri*), *Lactobacillus plantarum* (*L. plantarum*), and *Lactobacillus rhamnosus* (*L. rhamnosus*) in different set-ups and grown in de Man, Rogosa and Sharpe (MRS) agar. All three *Lactobacillus* spp. had higher mean absorbance and mean colony counts in the experimental set-ups than in the control set-ups where no extracts were added. This indicates that the addition of sugarcane bagasse and lotus root extracts were able to effectively proliferate the growth of *L. gasseri*, *L. plantarum*, and *L. rhamnos*us. Hence, they are viable prebiotics. This study was extended with the usage of the proliferated *L. gasseri* and *L. plantarum* to inhibit the growth of *Escherichia coli* (*E. coli*). They inhibited the growth of *E. coli* to a greater extent than without extracts, as the colony counts of *E. coli* were lower in the former, indicating a higher antimicrobial activity.

Keywords Lactobacillus · Sugarcane bagasse extract, lotus root extract · Escherichia coli · Proliferation of Lactobacillus · Inhibition of Escherichia coli

---

L. T. Yuan (✉) · W. K. Ray · Y. J. Le
High School Hwa Chong Institution, Singapore, Singapore
e-mail: 181739H@student.hci.edu.sg

W. K. Ray
e-mail: 181869U@student.hci.edu.sg

Y. J. Le
e-mail: 181886b@student.hci.edu.sg

H. Guo et al. (eds.), *IRC-SET 2021*, https://doi.org/10.1007/978-981-16-9869-9_54

54.1 Introduction

According to Singapore's National Environment Agency's website [1], in 2019, out of the 744,000 tons of food waste generated, only a mere 18% (136,000 tons) gets recycled. Food still containing beneficial content is usually thrown away and wasted. Therefore, recycling sugarcane bagasse and lotus roots, and making use of the high prebiotic contents available in them to proliferate the growth of *Lactobacillus* spp. to inhibit the growth of *Escherichia coli* (*E. coli*) can be a feasible solution for the waste problem.

Lactobacillus is a genus of bacteria that is defined as Gram-positive, facultative anaerobic bacteria that mainly converts sugars into lactic acid [2]. La Fata et al. [3] reported that *Lactobacilli* can be used as probiotics to modify the gut microbiota, which is prominent in the treatment of diseases [4]. Hence, it is of interest to find methods to promote the growth of *Lactobacillus*.

Escherichia coli is a rod-shaped, Gram-negative bacteria. Many pathogenic strains of *E. coli* that can be deadly to humans [5]. For example, the Shiga-toxin producing *E. coli* (STEC), causing foodborne diseases. According to the World Health Organization, [6], the STEC outbreak in Europe in 2011 took 50 lives in Germany alone. Hence making the study to inhibit the growth of *E. coli* valuable.

Sugarcane bagasse (Saccharum officinarum) has high xylan content at 21.46%. Xylan hydrolysate samples were found to have increased effect on growth of all species of *Lactobacillus* as compared to standard galacto and fructo-oligosaccharide (GOS & FOS) [7]. Hence, it is of interest to study the effects of sugarcane extract on the growth of *Lactobacillus* spp.

Lotus (Nelumbo nucifera Gaertn.) is grown in large scales across Southeast Asia and Australia for commercial purposes [8]. Studies show that lotus root contains high levels of polyphenolic compounds, and they can be used as prebiotics to enhance the growth of beneficial gut microbes such as *Bifidobacterium* and *Lactobacillus* spp. [9]. Hence, it is of interest to study the effects of lotus root extract on the growth of *Lactobacillus* spp.

In addition, according to Rao and Sreenivasa [10], *Lactobacillus* strains can inhibit the growth of and attachment of pathogens, such as *E. coli*, onto the intestinal surfaces and boost the immune system of the host. They do so by producing lactic acid and enzymes, which have been proven to possess antimicrobial properties against bacteria like *Enterococcus faecalis* and *Enterococcus durans* [11].

By recycling sugarcane bagasse and lotus roots to produce prebiotic extracts, this helps to reduce the amount of food waste generated. Since the production of these prebiotic extracts is inexpensive, food production companies can add these prebiotic extracts into their food at a lower cost. This increases the nutritional benefit for the consumers at a cheap price.

54.2 Objective

The objective of this study is to investigate if sugarcane extract and lotus root extract are viable prebiotics that proliferate growth of *Lactobacillus gasseri* (*L. gasseri*), *Lactobacillus plantarum* (*L. plantarum*), and *Lactobacillus rhamnosus* (*L. rhamnosus*). To also investigate the microbial antagonism properties of the proliferated *L. gasseri* and *L. plantarum* on *E. coli*.

54.3 Hypotheses

The hypothesis of this study is that sugarcane and lotus root extracts are viable prebiotics that will proliferate the growth of *L. gasseri*, *L. plantarum*, and *L. rhamnosus*, and a higher growth rate of *L. gasseri* and *L. plantarum* will result in stronger antibacterial activity against *Escherichia coli*.

54.4 Materials and Methods

54.4.1 Materials

Lactobacillus gasseri, *Lactobacillus plantarum*, *Lactobacillus rhamnosus*, and *Escherichia coli* were obtained from American Type Culture Collection (ATCC). Sugarcane bagasse (Saccharum officinarum) was obtained from Bukit Timah Market and Food Center. Lotus root (Nelumbo nucifera Gaertn.) was purchased from NTUC FairPrice Co. (Singapore). de Man, Rogosa and Sharpe (MRS) broth, Luria Broth (LB), MRS agar, and LB agar from BD (Becton, Dickinson and Company) were used.

54.4.2 Methods

(1) *Preparation of Prebiotic Extracts*

Sugarcane bagasse and lotus root were washed and cut into small pieces. Then, they were dried in a hot air oven at 70 °C for 24 h. Dried samples were blended to powder form and added separately to deionized water in an extraction ratio of 1 g of powder to 25 ml of water. Subsequently, they were submerged in a water bath at 85 °C for 30 min, then the supernatant, after centrifugation at 8000 rpm for 10 min, was collected and autoclaved at 10 psi for 10 min. Extracts are stored in the refrigerator until needed.

(2) *Growth of pre-cultures of Lactobacillus*

Lactobacillus gasseri ATCC 19,992, *L. plantarum* ATCC 8014, and *L. rhamnosus* ATCC 7469 were inoculated separately into 10 ml of MRS broth and grown for 24 h at 30 °C in a shaking incubator. The absorbance of each culture at 600 nm was standardized at 0.8 the next day.

(3) *Growth of pre-cultures of E. coli*

Escherichia coli was inoculated into two 10 ml portions of LB broth and grown for 24 h at 30 °C in a shaking incubator. The absorbance of each culture at 600 nm was standardized at 0.8 the following day.

(4) *Testing effects of prebiotic extracts on growth of Lactobacillus spp.*

In the experimental set-ups, equal volumes of double strength MRS broth were mixed separately with sugarcane and lotus root extracts. 40 μl of *Lactobacillus* culture supernatants were added to 9.96 *ml* of the mixture. Five replicates of this mixture were prepared for each type of prebiotic extract and *Lactobacillus* pre-culture added. In the control set-ups, 40 μl of *Lactobacillus* pre-culture was added to 9.96 *ml* of MRS broth. Five replicates of this mixture were prepared for each type of *Lactobacillus* pre-culture. All replicates were placed in a shaking incubator at 30 °C for 24 h. 6 ml of culture is separated from each of the 10 ml cultures, they are to be used in the microbial antagonism tests. With the remaining 4 ml of culture, serial tenfold dilution was carried out with normal saline to appropriate dilution factor. 0.1 ml of the diluted culture was spread on MRS agar plates, then incubated at 30 °C for 24 h. Next, the number of colony forming units on each MRS agar plate was determined.

(5) *Microbial antagonism test*

In the experimental set-up, 0.5 ml of *E. coli* culture was added to 5 *ml* of filter-sterilized culture supernatant (FSCS) of the 15 *Lactobacillus* cultures and 4.5 ml of LB broth. In the MRS control set-ups, 0.5 ml of *E. coli* culture was added to 5 ml of MRS broth and 4.5 ml of LB broth. In the extract control set-ups, 0.5 ml of *E. coli* culture was added to 2.5 ml of double strength MRS broth, 2.5 ml of either sugarcane extract or lotus root extract, and 4.5 ml of LB broth. Five replicates of each set-up were prepared. There were 6 different set-ups with 30 total mixtures. All replicates were placed in a shaking incubator at 30 °C for 24 h. Serial tenfold dilution was carried out with normal saline to appropriate dilution factor. 0.1 ml of the diluted culture was spread on LB agar plates and incubated overnight at 30 °C for 24 h. The colony counts were recorded.

54.5 Results and Discussion

54.5.1 *Lactobacillus gasseri*

From Fig. 54.1, the mean colony count of *L. gasseri* in the control set-ups were also lower than that in the sugarcane and lotus set-ups (31.6 vs. 105.6 and 559.0 [$\times 10^8$], respectively). The results signify that *L. gasseri* has had higher growth rates in the experimental set-ups than in the control. Since the dilution factor for the control and experimental set-ups are not the same ($\times 10^{-6}$ vs. $\times 10^{-7}$, respectively), a visual comparison in Fig. 54.2 is not accurate. A $\times 10^{-7}$ dilution factor for the experimental set-ups (instead of $\times 10^{-6}$ that was used for control set-ups) was used to be able to accurately determine the number of colonies on the agar plates. However, the data between the experimental set-ups and control set-ups were compared at the same dilution factor by multiplying the data of the different experiments by the correct factor.

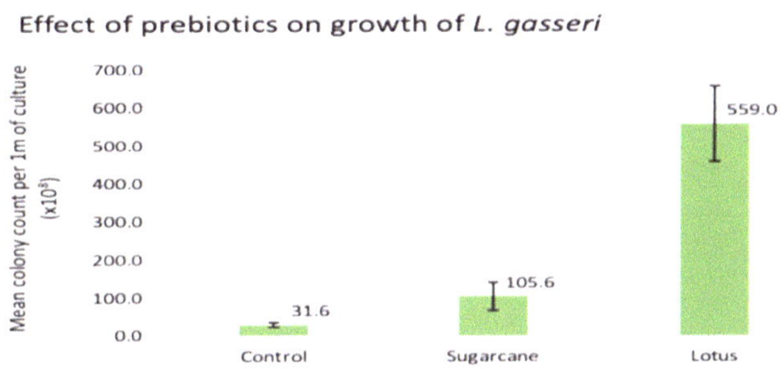

Fig. 54.1 Mean colony count of *L. gasseri* per 1 ml of culture in the control and experimental set-ups (1×10^8)

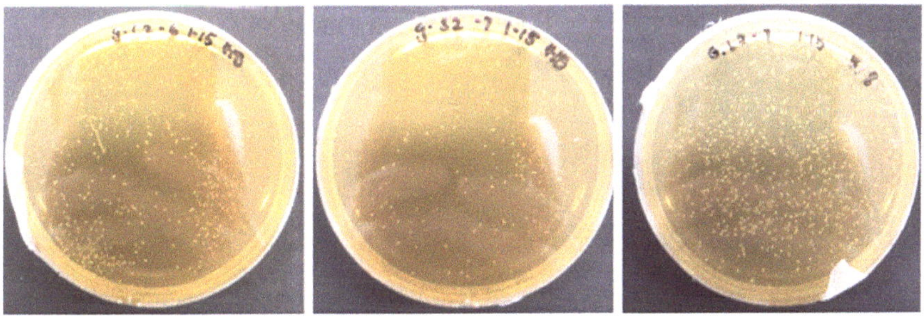

Fig. 54.2 *L. gasseri* colony count in control, sugarcane and lotus root set-ups (left to right) replicate #2. *P*-values are 0.016 and 0.025, respectively

54.5.2 *Lactobacillus plantarum*

From Fig. 54.3, the mean colony count of *L. plantarum* in the control set-ups were also lower than that in the sugarcane and lotus set-ups (12.9 vs. 101.8 and 42.3 [$\times 10^8$], respectively). The results signify that *L. plantarum* has had higher growth rates in the experimental set-ups than in the control. Since the dilution factor for the control and experimental set-ups are not the same ($\times 10^{-6}$ vs. $\times 10^{-7}$, respectively), a visual comparison in Fig. 54.4 is not accurate. A $\times 10^{-7}$ dilution factor for the experimental set-ups (instead of $\times 10^{-6}$ that was used for control set-ups) was used to be able to accurately determine the number of colonies on the agar plates. However, the data between the experimental set-ups and control set-ups were compared at the same dilution factor by multiplying the data of the different experiments by the correct factor.

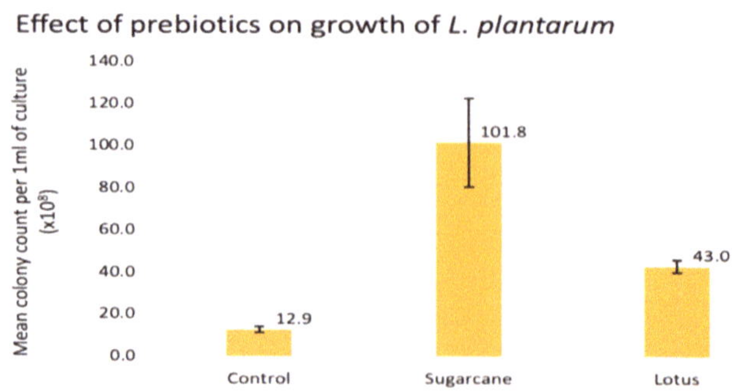

Fig. 54.3 Mean colony count of *L. plantarum* per 1 ml of culture in the control and experimental set-ups (1×10^8)

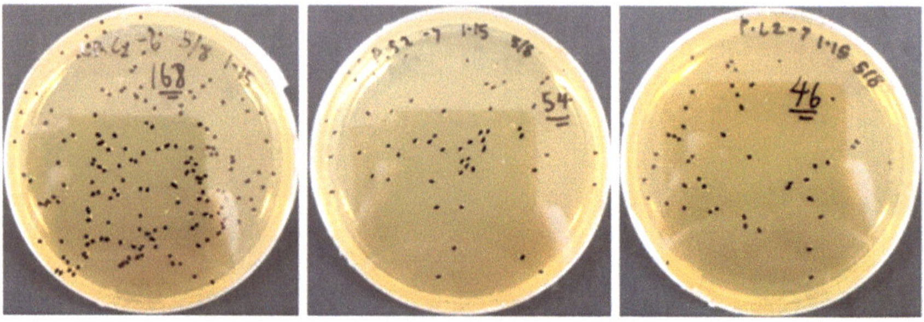

Fig. 54.4 *L. plantarum* colony count in control, sugarcane and lotus root set-ups (left to right) replicate #2. *P*-values are 0.012 and 0.036, respectively

54.5.3 *Lactobacillus rhamnosus*

From Fig. 54.5, the mean colony count of *L. rhamnosus* in the control set-ups were also lower than that in the sugarcane and lotus set-ups (16.2 vs. 172.0 and 70.8 [$\times 10^8$], respectively). The results signify that *L. rhamnosus* has had higher growth rates in the experimental set-ups than in the control. Since the dilution factor for the control and experimental set-ups are not the same ($\times 10^{-6}$ vs. $\times 10^{-7}$, respectively), a visual comparison in Fig. 54.6 is not accurate. A $\times 10^{-7}$ dilution factor for the experimental set-ups (instead of $\times 10^{-6}$ that was used for control set-ups) was used to be able to accurately determine the number of colonies on the agar plates. However, the data between the experimental set-ups and control set-ups were compared at the same dilution factor by multiplying the data of the different experiments by the correct factor.

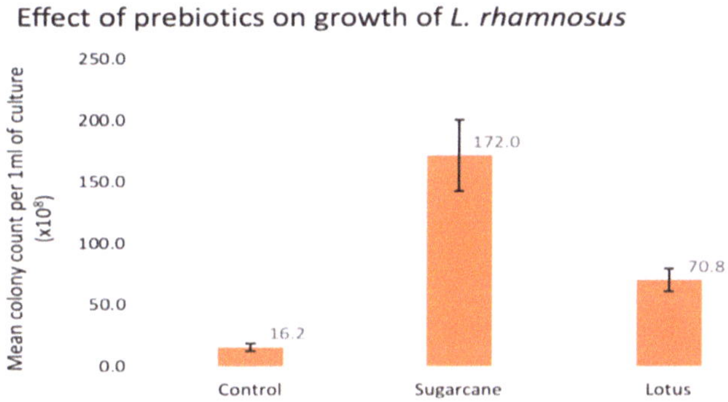

Fig. 54.5 Mean colony count of *L. rhamnosus* per 1 ml of culture in the control and experimental set-ups (1×10^8)

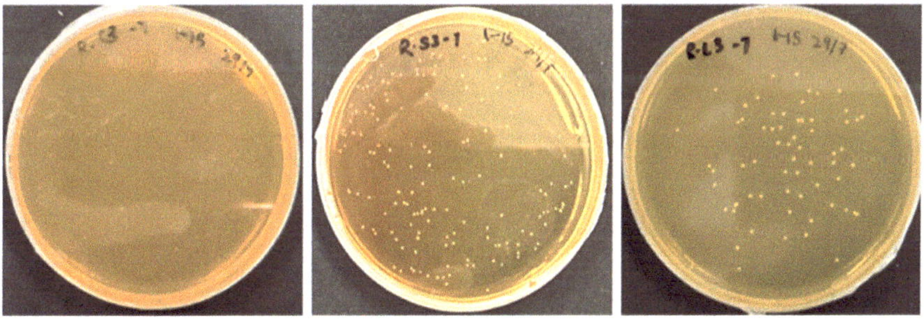

Fig. 54.6 *L. rhamnosus* colony count in control, sugarcane and lotus root set-ups (left to right) replicate #3. *P*-values are 0.012 and 0.016, respectively

54.5.4 Effects of Proliferated L. gasseri and L. plantarum on Growth of E. coli

From Figs. 54.7 and 54.10, two conclusions can be made. (I): The mean colony count of *E. coli* in the control set-ups were higher than that in the experimental set-ups. This signifies that the extract-proliferated *L. gasseri* and *L. plantarum* had greater antimicrobial effects against *E. coli* compared to the control set-ups. (II): The mean colony count of *E. coli* in the set-ups with FSCS of *L. gasseri* and *L. plantarum* were lower than the set-ups without, regardless of the presence of extracts, proving that the inhibition was due to the proliferated *L. plantarum* and *L. gasseri*, not the extracts themselves. Since the dilution factor for the control and experimental set-ups are not the same ($\times\ 10^{-6}$ vs. $\times\ 10^{-5}$, respectively), a visual comparison in Figs. 54.8, 54.9, 54.11, and 54.12 is not accurate. A different dilution factor for experimental set-ups was needed to be used to accurately determine the colony counts of *E. coli* on the agar plates. However, the data between control and experimental set-ups were multiplied to the same factor for comparison.

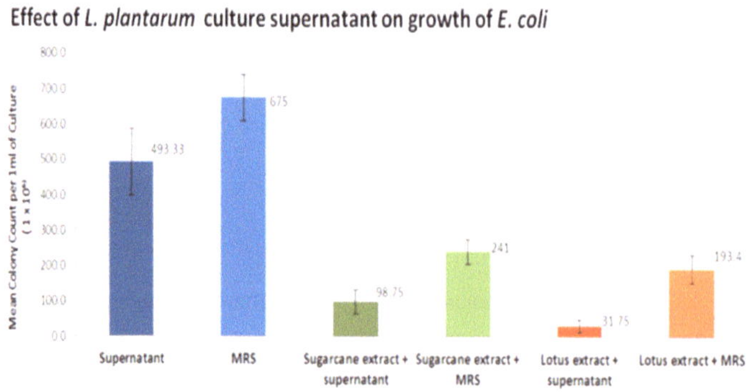

Fig. 54.7 Mean colony count of *E. coli* in the control and experimental set-ups (with/without FSCS). P-value between the set-ups is 0.008

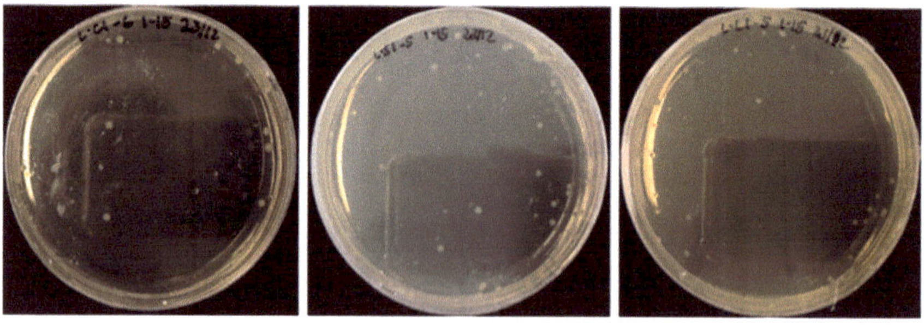

Fig. 54.8 *E. coli* colony count in control, sugarcane and lotus root set-ups with supernatant (left to right) replicate #1

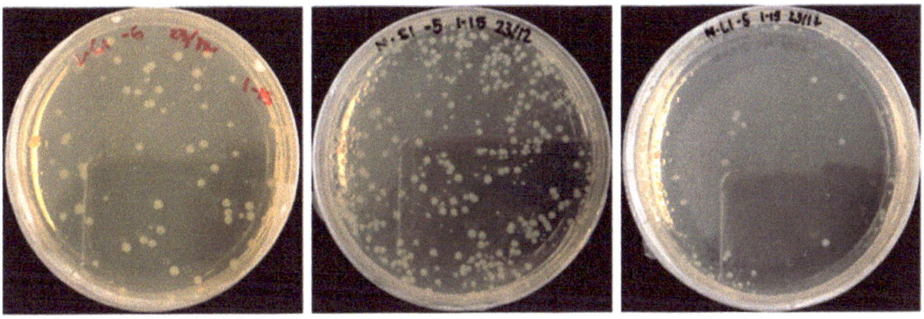

Fig. 54.9 *E. coli* colony count in control, sugarcane and lotus root set-ups with MRS (left to right) replicate #1

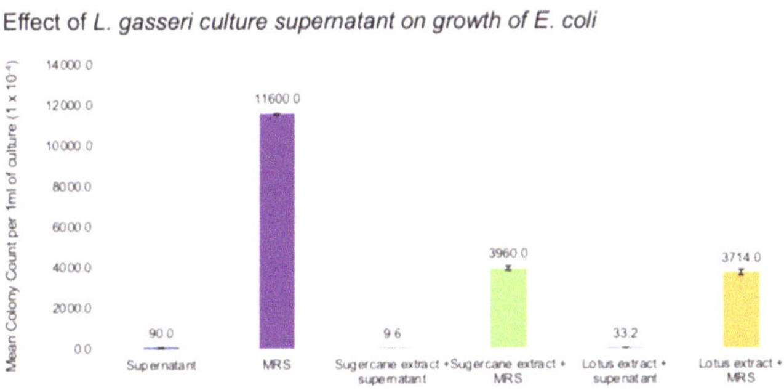

Fig. 54.10 Mean colony count of *E. coli* in the control and experimental set-ups (with/without FSCS). *P*-value between the set-ups is 0.000

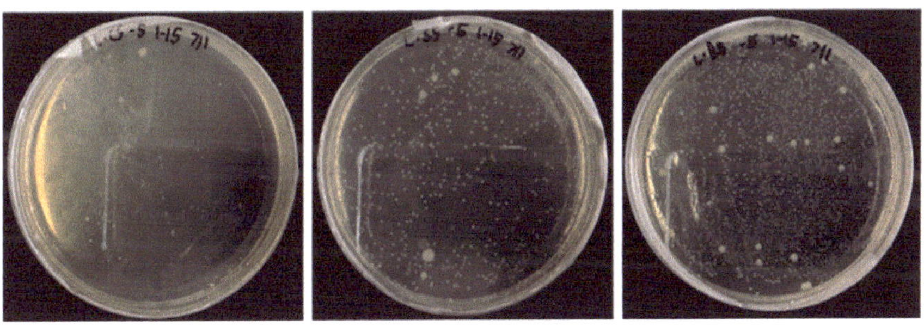

Fig. 54.11 *E. coli* colony count in control, sugarcane and lotus root set-ups with supernatant (left to right) replicate #5

54.5.5 Data Analysis

The Mann–Whitney U test was carried out and there were significant differences in the data, all *p*-values were lower than 0.05.

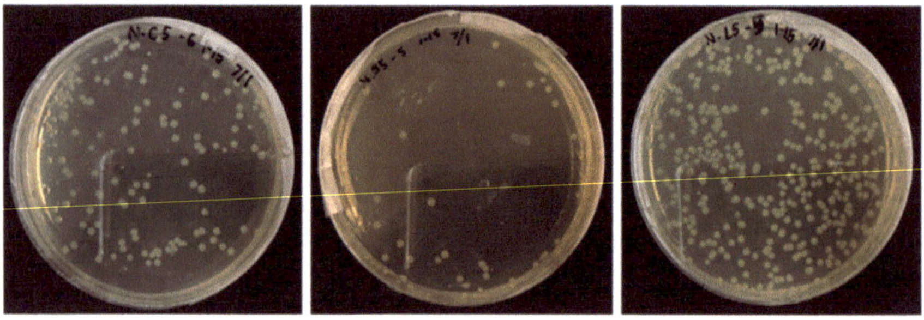

Fig. 54.12 *E.coli* colony count in control, sugarcane and lotus root set-ups with MRS (left to right) replicate #5

In other studies, xylo-oligosaccharide (XOS) extracted from sugarcane bagasse proliferated growth of *Lactobacillus brevis*, *Lactobacillus acidophilus* and *Lactobacillus viridescens* [7] and lotus root extract was reported to be able to proliferate the growth of *Lactobacillus casei subsp. rhamnosus* in fermented milk [12]. This project however, made used powdered sugarcane bagasse and was carried out in an MRS medium. Similarly, a positive result was obtained in this study with sugarcane and lotus root extract on *L. gasseri*, *L. plantarum,* and *L. rhamnosus.* The proliferated *L. gasseri* and *L. plantarum* were also able to increase the production of lactic acid in the medium, hence the higher concentration of lactic acid was able to more effectively inhibit the growth of *E. coli.* These results agree with the research done by Chen et al. [13], where several *Lactobacillus* strains displayed strong antimicrobial activities against carbapenem-resistant *E. coli* in well diffusion assays.

54.6 Conclusion

From all data collected, it can be concluded that sugarcane and lotus root extracts are able to proliferate the growth of *L. gasseri*, *L. plantarum,* and *L. rhamnosus. L. gasseri* and *L. plantarum* proliferated by sugarcane extract and lotus root extract is also able to better inhibit the growth of *E. coli*, compared to non-proliferated *L. gasseri* and *L. plantarum* itself. The waste products used in this study can be recycled, and used for production of prebiotic extracts, decreasing overall food wastage in Singapore. Since it is economically feasible to prepare the prebiotic extracts, food production companies can add them into probiotic nutritional supplements at a low cost, increasing nutritional benefits of food for consumers. Sugarcane extract and lotus root extract can be tested on other *Lactobacillus* species to determine if they are able to proliferate the growth of the other *Lactobacillus* species as well, they can then be tested on other common pathogens.

References

1. National Environmental Agency (2020). *Food waste management*. 3R Programmes and Resources. https://www.nea.gov.sg/our-services/waste-management/3r-programmes-and-resources/food-waste-management.
2. Makarova, K., Slesarev, A., Wolf, Y., Sorokin, A., Mirkin, B., Koonin, E., Pavlov, A., Pavlova, N., Karamychev, V., Polouchine, N., Shakhova, V., Grigoriev, I., Lou, Y., Rohksar, D., Lucas, S., Huang, K., Goodstein, D. M., Hawkins, T., Plengvidhya, V., & Mills, D. (2006). Comparative genomics of the lactic acid bacteria. *Proceedings of the National Academy of Sciences, 103*(42), 15611–15616. https://doi.org/10.1073/pnas.0607117103
3. La Fata, G., Weber, P., & Mohajeri, M. H. (2017). Probiotics and the GUT immune system: Indirect regulation. *Probiotics and Antimicrobial Proteins, 10*(1), 11–21. https://doi.org/10.1007/s12602-017-9322-6
4. Azad, M. A. K., Sarker, M., Li, T., & Yin, J. (2018). Probiotic species in the modulation of gut microbiota: An overview. *BioMed Research International, 2018*, 1–8. https://doi.org/10.1155/2018/9478630
5. Kaper, J., Nataro, J., & Mobley, H. (2004, February). Pathogenic *Escherichia coli. Nature Reviews Microbiology, 2*, 123–140. Retrieved from https://doi.org/10.1038/nrmicro818
6. World Health Organization (2011). Outbreaks of *E. coli* O104:H4 infection: Update 30. Retrieved December 31, 2020, from https://www.euro.who.int/en/health-topics/disease-prevention/food-safety/news/news/2011/07/outbreaks-of-e.-coli-o104h4-infection-update-30
7. Kaur, R., Uppal, S. K., & Sharma, P. (2018). Production of xylooligosaccharides from sugarcane bagasse and evaluation of their prebiotic potency in vitro. *Waste and Biomass Valorization, 10*(9), 2627–2635. https://doi.org/10.1007/s12649-018-0266-1
8. Wang, H.-X., Yi, Y., Sun, J., Lamikanra, O., & Min, T. (2018). Fingerprint profiling of polysaccharides from different parts of lotus root varieties. *RSC Advances, 8*(30), 16574–16584. https://doi.org/10.1039/c8ra01104d
9. Thilakarathna, W. W., Langille, M. G., & Rupasinghe, H. V. (2018). Polyphenol-based prebiotics and synbiotics: Potential for cancer chemoprevention. *Current Opinion in Food Science, 20*, 51–57. https://doi.org/10.1016/j.cofs.2018.02.011
10. Rao, P. K., & Sreenivasa, M. Y. (2016). Probiotic *Lactobacillus* strains and their antimicrobial peptides to counteract biofilm-associated infections—A promising biological approach. *SM Journal of Bioinformatics and Proteomics*, *1*(2), 1–7. https://smjournals.com/bioinformatics-proteomics/fulltext/smjbp-v1-1009.php
11. Arias-Moliz, M., Baca, P., Ordonez-Becerra, S., Gonzalez-Rodriguez, M., & Ferrer-Luque, C. (2012). Eradication of *enterococci* biofilms by lactic acid alone and combined with chlorhexidine and cetrimide. *Medicinal Oral Patologia Oral y Cirugia Bucal, 6*, 902–906. https://doi.org/10.4317/medoral.18133
12. Muadsri, P. (2017). The influence of lotus root extract on *Lactobacillus casei subsp. rhamnosus* growth in fermented milk. *Science and Technology Nakhon Sawan Rajabhat University Journal*, 9(9), 20–25. Retrieved from https://ph02.tci-thaijo.org/index.php/JSTNSRU/article/view/70388
13. Chen, C.-C., Lai, C.-C., Huang, H.-L., Huang, W.-Y., Toh, H.-S., Weng, T.-C., Chuang, Y.-C., Lu, Y.-C., & Tang, H.-J. (2019). Antimicrobial activity of lactobacillus species against carbapenem-resistant enterobacteriaceae. *Frontiers in Microbiology, 10*. https://doi.org/10.3389/fmicb.2019.00789

MIX
Papier aus verantwortungsvollen Quellen
Paper from responsible sources
FSC® C105338

If you have any concerns about our products,
you can contact us on
ProductSafety@springernature.com

In case Publisher is established outside the EU,
the EU authorized representative is:
Springer Nature Customer Service Center GmbH
Europaplatz 3, 69115 Heidelberg, Germany

Printed by Libri Plureos GmbH
in Hamburg, Germany